Chemistry of the Planets

Saint Jacut de la Mer,
Brittany, France
14–16 June 2010

FARADAY DISCUSSIONS
Volume 147, 2010

RSC Publishing

The Faraday Division of the Royal Society of Chemistry, previously the Faraday Society, founded in 1903 to promote the study of sciences lying between Chemistry, Physics and Biology.

EDITORIAL STAFF

Editor
Philip Earis

Deputy editor
Jane Hordern

Senior publishing editor
Nicola Nugent

Development editor
Amaya Camara-Campos

Publishing editors
Michael Spencelayh
Paul Cooper

Publishing assistant
Kate Bandoo

Publisher
Niamh O'Connor

Faraday Discussions (Print ISSN 1359-6640, Electronic ISSN 1364-5498) is published 4 times a year by the Royal Society of Chemistry, Thomas Graham House, Science Park, Milton Road, Cambridge, UK CB4 0WF. Volume 147 ISBN-13: 978 1 84973 0570

2010 annual subscription price: print+electronic £622, US $1,160; electronic only £560, US $1,045. Customers in Canada will be subject to a surcharge to cover GST. Customers in the EU subscribing to the electronic version only will be charged VAT. All orders, with cheques made payable to the Royal Society of Chemistry, should be sent to RSC Distribution Services, c/o Portland Customer Services, Commerce Way, Colchester, Essex, UK CO2 8HP.
Tel +44 (0) 1206 226050;
E-mail sales@rscdistribution.org

If you take an institutional subscription to any RSC journal you are entitled to free, site-wide web access to that journal. You can arrange access via Internet Protocol (IP) address at www.rsc.org/ip. Customers should make payments by cheque in sterling payable on a UK clearing bank or in US dollars payable on a US clearing bank. Periodicals postage is paid at Rahway, NJ and at additional mailing offices. Airfreight and mailing in the USA by Mercury Airfreight International Ltd., 365 Blair Road, Avenel, NJ 07001, USA.

US Postmaster: send address changes to *Faraday Discussions*, c/o Mercury Airfreight International Ltd., 365 Blair Road, Avenel, NJ 07001. All despatches outside the UK by Consolidated Airfreight.

PRINTED IN THE UK

Faraday Discussions documents a long-established series of *Faraday Discussion* meetings which provide a unique international forum for the exchange of views and newly acquired results in developing areas of physical chemistry, biophysical chemistry and chemical physics.

ORGANISING COMMITTEE, Volume 147

Chair
Ian Sims (University of Rennes 1, France)

Bruno Bézard (Observatoire de Paris-Site de Meudon, France)
André Canosa (University of Rennes 1, France)
Helen Fraser (University of Strathclyde, UK)
John Plane (University of Leeds, UK)
Jonathan Tennyson (University College London, UK)

FARADAY STANDING COMMITTEE ON CONFERENCES

Chair
D E Heard (Leeds, UK)

W A Brown (UCL, UK)
I Hamley (Reading, UK)
J Hirst (Nottingham, UK)
A Mount (Edinburgh, UK)

© The Royal Society of Chemistry 2010. Apart from fair dealing for the purposes of research or private study, or criticism or review, as permitted under the Copyright, Designs and Patents Act 1988 and Related Rights Regulations 2003, this publication may only be reproduced, stored or transmitted, in any form or by any means, with the prior permission in writing of the Publishers or in the case of reprographic reproduction in accordance with the terms of licences issued by the Copyright Licensing Agency in the UK. US copyright law applicable to users in the USA. The Royal Society of Chemistry takes reasonable care in the preparation of this publication but does not accept liability for the consequences of any errors or omissions.

Royal Society of Chemistry: Registered Charity No. 207890.

⊚The paper used in this publication meets the requirements of ANSI/NISO Z39.48-1992 (Permanence of Paper).

Chemistry of the Planets

Faraday Discussions
www.rsc.org/faraday_d

A General Discussion on Chemistry of the Planets was held at Saint Jacut de la Mer, France on 14th, 15th and 16th June 2010.

RSC Publishing is a not-for-profit publisher and a division of the Royal Society of Chemistry. Any surplus made is used to support charitable activities aimed at advancing the chemical sciences. Full details are available from www.rsc.org

CONTENTS

ISSN 1359-6640; ISBN 978-1-84973-057-0

Cover
See Kaiser *et al.*, *Faraday Discuss.*, 2010, **147**, 429–478. Unraveling the chemical processing of the atmospheres of planets and their moons by a combination of crossed beams, kinetic studies and theory.

Image reproduced by permission of Professor Ralf Kaiser, from *Faraday Discuss.*, 2010, **147**, 429.

INTRODUCTORY LECTURE

9 **The significance of trace constituents in the solar system**
Sushil K. Atreya

PAPERS AND DISCUSSIONS

31 **Formation of NH_3 and CH_2NH in Titan's upper atmosphere**
Roger V. Yelle, V. Vuitton, P. Lavvas, S. J. Klippenstein, M. A. Smith, S. M. Hörst, J. Cui

51 **Mapping Titan's HCN in the far infra-red: implications for photochemistry**
N. A. Teanby, P. G. J. Irwin, R. de Kok, C. A. Nixon

65 **Upper limits for undetected trace species in the stratosphere of Titan**
Conor A. Nixon, Richard K. Achterberg, Nicholas A. Teanby, Patrick G. J. Irwin, Jean-Marie Flaud, Isabelle Kleiner, Alix Dehayem-Kamadjeu, Linda R. Brown, Robert L. Sams, Bruno Bézard, Athena Coustenis, Todd M. Ansty, Andrei Mamoutkine, Sandrine Vinatier, Gordon L. Bjoraker, Donald E. Jennings, Paul N. Romani, F. Michael Flasar

83   General Discussion

103 **On the abundance of non-cometary HCN on Jupiter**
     Julianne I. Moses, Channon Visscher, Thomas C. Keane, Aubrey Sperier

137 **Photochemical modeling of Titan atmosphere at the "10 percent uncertainty horizon"**
     Zhe Peng, Michel Dobrijevic, Eric Hébrard, Nathalie Carrasco, Pascal Pernot

155 **Experimental measurements of low temperature rate coefficients for neutral–neutral reactions of interest for atmospheric chemistry of Titan, Pluto and Triton: Reactions of the CN radical**
     Sébastien B. Morales, Sébastien D. Le Picard, André Canosa, Ian R. Sims

173 **An experimental and theoretical investigation of the competition between chemical reaction and relaxation for the reactions of $^1CH_2$ with acetylene and ethene: implications for the chemistry of the giant planets**
     Kelly L. Gannon, Mark A. Blitz, Chi-Hsiu Liang, Michael J. Pilling, Paul W. Seakins, David R. Glowacki, Jeremy N. Harvey

189 **Formation of nitriles and imines in the atmosphere of Titan: combined crossed-beam and theoretical studies on the reaction dynamics of excited nitrogen atoms $N(^2D)$ with ethane**
     Nadia Balucani, Francesca Leonori, Raffaele Petrucci, Massimiliano Stazi, Dimitris Skouteris, Marzio Rosi, Piergiorgio Casavecchia

217 **Structural and spectroscopic characterization of mixed planetary ices**
     Nuria Plattner, Myung Won Lee, Markus Meuwly

231 **Isomer specific spectroscopy of $C_{10}H_n$, $n = 8–12$: Exploring pathways to naphthalene in Titan's atmosphere**
     Joshua A. Sebree, Vadim V. Kislov, Alexander M. Mebel, Timothy S. Zwier

251 General Discussion

283 **H_3^+ cooling in planetary atmospheres**
     Steve Miller, Tom Stallard, Henrik Melin, Jonathan Tennyson

293 **Negative ions at Titan and Enceladus: recent results**
     Andrew J. Coates, Anne Wellbrock, Gethyn R. Lewis, Geraint H. Jones, David T. Young, Frank J. Crary, J. Hunter Waite, Robert E. Johnson, Thomas W. Hill, Edward C. Sittler Jr

307 **Chemical origins of the Mars ultraviolet dayglow**
     David L. Huestis, Tom G. Slanger, Brian D. Sharpee, Jane L. Fox

323 **Laboratory chemistry relevant to understanding and modeling the ionosphere of Titan**
     Nigel G. Adams, L. Dalila Mathews, David Osborne, Jr

337 **Fast ion–molecule reactions in planetary atmospheres: a semiempirical capture approach**
     Alexandre Faure, Véronique Vuitton, Roland Thissen, Laurent Wiesenfeld, Odile Dutuit

349 **Meteoric ion layers in the Martian atmosphere**
     Charlotte L. Whalley, John M. C. Plane

369 **Exploring extrasolar worlds: from gas giants to terrestrial habitable planets**
     Giovanna Tinetti, Caitlin A. Griffith, Mark R. Swain, Pieter Deroo, Jean Philippe Beaulieu, Gautam Vasisht, David Kipping, Ingo Waldmann, Jonathan Tennyson, Robert J. Barber, Jeroen Bouwman, Nicole Allard, Linda R. Brown

379 General Discussion

405 **Titan and habitable planets around M-dwarfs**
     Jonathan I. Lunine

419 **The fate of aerosols on the surface of Titan**
S. I. Ramírez, P. Coll, A. Buch, C. Brassé, O. Poch, F. Raulin

429 **Untangling the chemical evolution of Titan's atmosphere and surface–from homogeneous to heterogeneous chemistry**
Ralf I. Kaiser, Pavlo Maksyutenko, Courtney Ennis, Fangtong Zhang, Xibin Gu, Sergey P. Krishtal, Alexander M. Mebel, Oleg Kostko, Musahid Ahmed

479 **Mechanisms of formation of nitrogen-containing polycyclic aromatic compounds in low-temperature environments of planetary atmospheres: A theoretical study**
Alexander Landera, Alexander M. Mebel

495 **Very high resolution mass spectrometry of HCN polymers and tholins**
Véronique Vuitton, Jean-Yves Bonnet, Maeliss Frisari, Roland Thissen, Eric Quirico, Odile Dutuit, Bernard Schmitt, Léna Le Roy, Nicolas Fray, Hervé Cottin, Ella Sciamma-O'Brien, Nathalie Carrasco, Cyril Szopa

509 **Volatile inventories in clathrate hydrates formed in the primordial nebula**
Olivier Mousis, Jonathan I. Lunine, Sylvain Picaud, Daniel Cordier

527 **General Discussion**

CONCLUDING REMARKS

553 **Closing remarks**
Darrell F. Strobel

ADDITIONAL INFORMATION

561 **Poster titles**
565 **List of participants**
569 **Index of contributors**

The significance of trace constituents in the solar system

Sushil K. Atreya

Received 30th June 2010, Accepted 2nd July 2010
DOI: 10.1039/c005460g

Trace or minor constituents are key to the origin, maintenance, and the eventual fate of atmospheres of solar system objects. In this Introductory Paper, I illustrate this point by discussing certain cross cutting themes, including the chemistry of the formation and stability of a nitrogen atmosphere on Titan and the Earth, the chemical and biochemical origin of methane on the terrestrial planets and Titan, production and role of photochemical haze and aerosols, especially on Titan, and the significance of electro-photochemistry for habitability of Mars.

1. Introduction

Nitrogen is ubiquitous on terrestrial planets, comprising nearly three quarters by volume of the Earth's atmosphere and three percent each of the atmospheres of Mars and Venus today. However, the detection of a massive atmosphere of nitrogen with a surface pressure of 1500 millibar on a comparatively small and very cold moon Titan of Saturn by Voyager in 1980 was surprising. Even-smaller objects, Neptune's moon Triton and Pluto, are known to have a nitrogen atmosphere, but with a smaller and seasonally variable pressure ranging from less than 10 microbar to as large as 50 microbar. Plumes above the tiger stripes of Saturn's moon Enceladus might also contain traces of nitrogen. Nitrogen on planets and satellites could either have a primordial origin, or it could be secondary, *i.e.* a dissociation product of another nitrogen-bearing primordial molecule such as ammonia.

Nitrogen on Triton and Pluto is believed to be primordial, *i.e.* it was delivered as N_2 by the planetesimals that formed these objects. Triton and Pluto formed in the extreme cold regions of the solar nebula where the temperatures were below 40 K, allowing nitrogen to be trapped directly. Cold trapping of nitrogen in the Titan forming planetesimals was suggested also,[48] but is not supported by the Huygens Gas Chromatograph Mass Spectrometer (GCMS) measurement of primordial argon (^{36}Ar). Direct capture of N_2 would be associated with ^{36}Ar also since the trapping temperature of argon is similar to that of nitrogen. This would result in the solar ^{36}Ar/N_2 = 0.11, whereas the GCMS measured this ratio to be 2.1×10^{-7}.[46] This clearly rules out direct capture of N_2 as the origin of Titan's nitrogen. It also implies that the planetesimals that formed Saturn's satellites, including Titan and Enceladus were warm, certainly too warm to trap nitrogen directly. This is even more of the case for the much warmer terrestrial planets. Thus the nitrogen on these bodies must be secondary.

The amount of nitrogen in the plumes of Enceladus is controversial, perhaps below 0.1% based on the Cassini Ion and Neutral Mass Spectrometer (INMS) data on the plume composition (approximately 90% H_2O, 5% CO_2, 4% CO, 0.8% NH_3, *etc.*[64]) and perhaps variable. The presence of volatiles above Enceladus is

---

Department of Atmospheric, Oceanic, and Space Sciences Space Research Building, University of Michigan, 2455 Hayward Street, Ann Arbor, MI, 48109-2143, USA

tied to active plume sites, since the "atmosphere" of Enceladus is not permanent due to the moon's low gravity and hence low escape velocity, such that its gases would disappear if the plumes became inactive.[3] Any amount of nitrogen in the plumes requires an explanation. Matson et al.[41] have proposed a mechanism for producing nitrogen from ammonia by endogenic process in the interior of Enceladus. The heating of the interior is assumed to result from tidal and radiogenic processes. For temperatures between 575 K and 850 K, they find rapid conversion of NH_3 to N_2:

$$2NH_3 \Leftrightarrow N_2 + 3H_2$$

with a 70% efficiency at 800 K and 35 MPa, and, for temperatures below 575 K the process works best with metal or clay mineral catalysis:

$$2NH_3(aq) \Leftrightarrow N_2(aq) + 3H_2(aq)$$

where the reverse reaction is inhibited due reactant dispersal.

The above endogenic process is a straightforward thermal decomposition mechanism, except for the chemistry of metal or clay mineral catalysis, which is somewhat poorly constrained at high temperatures. The endogenic process could have had a role in producing some nitrogen on the terrestrial planets and Titan also, but photochemical conversion of ammonia was inescapable and far more effective, as will be discussed later. Ammonia is expected to have been a minor constituent of the primordial atmospheres on these bodies.

Another minor constituent of the atmosphere of Earth is methane, but it carries a profound significance as it is largely of biological origin. A small amount of methane is also reported to be present in the atmosphere of Mars, where it could be either geological or biological in origin. Methane and other organics are absent from Venus, which is not surprising considering the planet's sulfur and halogen dominated thermochemistry and photochemistry. On Titan, methane has a particularly special role as the haze forming from the chemistry between methane and nitrogen in the ionosphere and the neutral atmosphere below result in up to 100 K warming of the stratosphere, and the collision-induced opacity of CH_4–N_2, H_2–N_2 and N_2–N_2 result in a 20 K increase in temperature at the troposphere. Without such warming Titan's atmosphere would cool to the point that nitrogen would condense out as liquid, causing the atmosphere to collapse to very low pressure levels. Thus the very existence of an atmosphere on Titan is tied critically to trace constituents, which in turn result from the methane that comprises 5.65% (ref. 46, value supersedes ref. 47) of Titan's atmosphere. The chemistry of the origin and fate of methane will also be reviewed in this paper.

2. The chemistry of the formation of nitrogen on terrestrial planets and Titan

Nitrogen is of a secondary origin on the terrestrial planets, Mars, Venus and Earth, as well as on Titan, i.e. it was produced originally from another nitrogen bearing compound, most likely ammonia. Thermal decomposition of ammonia in the past magmatic processes contributed to the ammonia of the terrestrial planets. Such a process was probably active in the interior of Titan also. However, the temperatures of primordial Mars, Earth, Venus and Titan were large enough to outgas large quantities of ammonia from their interiors to the atmosphere. This would naturally lead to the dissociation of ammonia by solar ultraviolet radiation. The solar UV flux in the dissociating region below 300 nm was as much as 10^4 times greater than today in the Sun's pre-main sequence phase,[72] which resulted in large photolysis of NH_3 and a relatively fast production of N_2 as a consequence. The chemistry of the formation of nitrogen from ammonia for Titan is described below, and a similar process would have been at work on the terrestrial planets.

Atreya et al.[5] developed a photochemical model that showed production of large quantities of nitrogen from ammonia on primordial Titan. Voyager discovered nitrogen on Titan a couple of years later in 1980. Speculations of a nitrogen atmosphere resulting from the photolysis of ammonia existed as early as the early 1970s,[34] however. Considering Titan's density of 1.8 g cm^{-3}, it is believed that Titan is made of roughly 40% ice by mass, and the rest rock. Ammonia was trapped initially in ice of the Titan forming planetesimals. During the accretionary heating phase the volatiles were released to the atmosphere. Large quantities of water vapor, methane and ammonia, presumably in the solar O/C/N ratio, are believed to have been present in Titan's primordial atmosphere.[36] (In a relative sense, though, ammonia was a minor constituent of Titan's primordial atmosphere.) The photolysis of NH_3 takes place at wavelengths below 300 nm, and peaks at around 195 nm. The photochemical scheme of ammonia on primordial Titan is shown in Fig. 1, after Atreya et al.[5,6] The model also contained methane, which played a central role in the radiative transfer hence the thermal structure of Titan's primordial atmosphere.[1] However, methane does not interfere with the chemical production of nitrogen from ammonia, as the photolysis of the two gases is both spectrally (CH_4 largely at 121.6 nm, and NH_3 largely around 200 nm) and spatially separated. As shown in Fig. 1, the photolysis of ammonia produces amidogen radicals (NH_2). About a third of the ammonia so destroyed is recycled right back by the reaction of these radicals with hydrogen atoms also produced in the NH_3 photolysis. The self-recombination of the remaining NH_2 radicals results in hydrazine (N_2H_4) molecules. At temperatures above 150 K, sufficient quantities of N_2H_4 remain in the vapor phase, allowing its photolysis to proceed and form an intermediate radical, hydrazyl (N_2H_3). The self-recombination reaction of these radicals leads to the production of N_2. Below 150 K, little NH_3 is in the vapor form, and even the small amount of hydrazine formed from it condenses, preventing production of N_2. At temperatures greater than 250 K, a large quantity of water vapor coexists with the ammonia vapor. The photolysis of H_2O vapor results in the highly reactive hydroxyl molecules (OH) and H atoms. The latter react with NH_2 and recycle ammonia. Although some production of N_2 from NH (from OH + NH_2) and any small amount of N_2H_4 that did form is still likely but the yield turns out to be extremely low. The ideal temperature range for producing N_2 from NH_3 on primordial Titan is thus 150–250 K,[6] slightly revised from the earlier range of 150–200 K[5] due to the inclusion of a radiatively controlled structure in the atmospheric temperature rather than an isothermal temperature. The temperature range also has an impact on the time scale for the formation of the currently observed pressure of nitrogen on Titan.

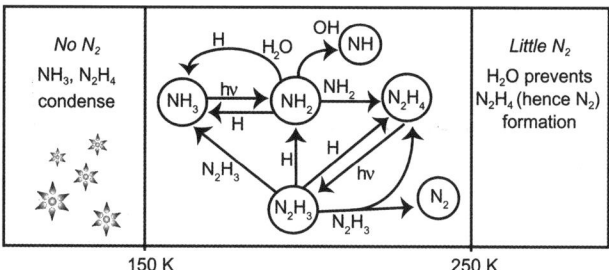

Fig. 1 Photochemical production of N_2 from NH_3 in Titan's primordial atmosphere (updated from ref. 5. Below 150 K, little NH_3 is in the vapor phase, and whatever small intermediate product hydrazine (N_2H_4) forms from it condenses, thus preventing N_2 formation. Above 250 K, water vapor has a sufficiently large vapor pressure, so that OH from its photolysis reacts with NH_2 drastically decreasing the subsequent yield of N_2, while some recycling of NH_3 occurs by the reaction of NH_2 with H from H_2O.

The present nitrogen on Titan may be a relic of its original abundance because of escape from the atmosphere[33,37] that would result in the fractionation of nitrogen isotopes. The Huygens GCMS measures $^{14}N/^{15}N = 167.7$ in Titan's N_2.[46] Assuming the Earth's $^{14}N/^{15}N = 272$ as the reference (Fig. 2), this implies that Titan should have started out with approximately 5–10 bars of N_2 to account for the current 1.5 bars present today.[6] The time scale for producing 10 bars of nitrogen photochemically from ammonia is calculated to be approximately 30 Myr.[1] Evolutionary models predict that Titan's surface temperature was in the 150–250 K range (the ideal temperature range, as explained above) for over 100 Myr,[35] which is more than that required for producing 10 bars of nitrogen thus allowing sufficient wiggle room to accommodate uncertainties in various model calculations.

The above time scale is based on the assumption that Titan's initial $^{15}N/^{14}N$ ratio was the same as the Earth's, since, like the Earth, its nitrogen is secondary that was produced from ammonia photochemistry. As shown in the right panel of Fig. 2, the $^{15}N/^{14}N$ ratio is the same on Venus, Earth and solid Mars (as reflected in ALH84001 meteorite). The higher $^{15}N/^{14}N$ value in the atmosphere of Mars (ATM) indicates escape of nitrogen, which favors the lighter isotope. Likewise, escape would render the $^{15}N/^{14}N$ value on Titan to be greater than terrestrial ($^{15}N/^{14}N = 3.7 \times 10^{-3}$ or

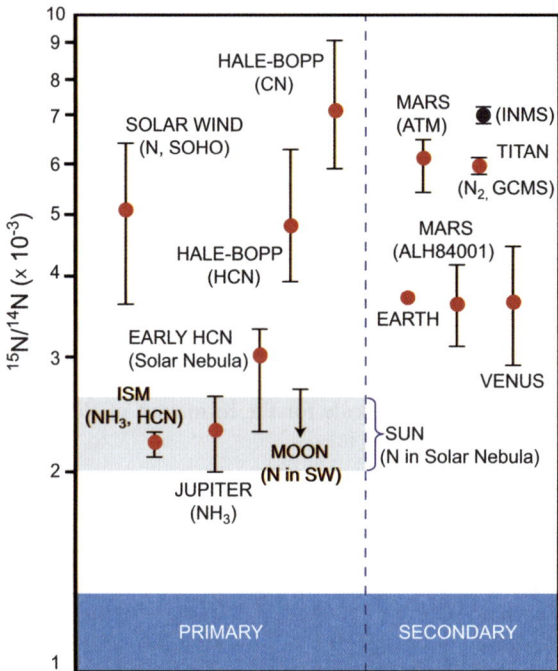

Fig. 2 Comparison of Titan's nitrogen isotope abundances from GCMS and INMS with those on the terrestrial planets and primary sources including the interstellar medium (ISM), Sun (SW: solar wind, SOHO: Solar and Heliospheric Observatory), Jupiter (which represents the protosolar value) and comet Hale-Bopp. For Titan, the GCMS value is more appropriate as the INMS result represents a model dependent extrapolation from the ionosphere at ~1000 km to the troposphere below 45 km. The $^{15}N/^{14}N$ ratio from HCN in comet Hale-Bopp is based on reanalysis of the HCN data of Jewitt et al.[31] and Ziurys et al.[74] by Bockelee-Morvan et al.,[11] according to which the previous $^{15}N/^{14}N$ ratio was revised up by about 35%. Bockelee-Morvan et al.[11] determine the nitrogen isotopes in another comet 17P/Holmes and find them to be quite different from Hale-Bopp (see text). For Mars, two values are shown, in the atmosphere (ATM) and Allan Hills meteorite (ALH84001), the latter presumably represents the solid Mars value. Prepared in consultation with Paul R. Mahaffy.

$^{14}N/^{15}N = 272$) provided that the initial nitrogen isotope ratio on Titan was same as that in the Earth's nitrogen. According to one model, however, hydrodynamic and sputtering loss may not have resulted in the fractionation in Titan's nitrogen isotopes so that the current $^{15}N/^{14}N$ value is indicative of the initial value.[40] This idea needs to be investigated further as early loss processes especially during the T-Tauri phase are quite plausible.

Although there is consensus that Titan's nitrogen was produced from ammonia, the original source of ammonia is less clear. The Titan forming planetesimals could have originated in Saturn's subnebula or as comets in the outer solar nebula. If comets supplied the ammonia to Titan, the $^{15}N/^{14}N$ ratio on Titan should be the same as on comets, in principle, provided that no escape of nitrogen occurred over geologic time. The $^{15}N/^{14}N$ ratio from HCN and CN of comet Hale-Bopp (one of only two comets where the value has been determined from both molecules) shown in Fig. 2 does not match that on Titan as measured by the GCMS. (The INMS value shown in Fig.2, $^{15}N/^{14}N = 7.0 \times 10^{-3}$ ($^{14}N/^{15}N = 143$), is about 15% greater than the GCMS value of $^{15}N/^{14}N = 6.0 \times 10^{-3}$ ($^{14}N/^{15}N = 167.7$). It could be due to the fact that the INMS value represents a model-dependent extrapolation from the ionosphere around 1000 km to the troposphere, unlike the *in situ* value in the troposphere measured by the GCMS. Thus it is more appropriate to use the GCMS value.) On another comet, 17P/Holmes, the nitrogen isotope ratio from HCN is about 30% different from that in Hale-Bopp ($^{15}N/^{14}N = 7.2 \times 10^{-3}$ or $^{14}N/^{15}N = 139$, compared to $^{15}N/^{14}N = 4.9 \times 10^{-3}$ or $^{14}N/^{15}N = 205$ in Hale-Bopp) and about 20% different from CN ($^{15}N/^{14}N = 6.1 \times 10^{-3}$ or $^{14}N/^{15}N = 165$, compared to $^{15}N/^{14}N = 7.15 \times 10^{-3}$ or $^{14}N/^{15}N = 140$ in Hale-Bopp, or a dozen other comets in which the value is available from CN) although the error bars around the central values are large.[11] The above example illustrates that not all comets are alike. Moreover, any comparison with cometary nitrogen isotopes must be carried out in the comet's NH_3, not its photochemical products (with CH_4), HCN and CN, as parent–daughter fractionation in isotopes can and does occur. However, no measurement of nitrogen isotopes in the ammonia of comets is yet available. Even when it does become available, utmost caution needs to be exercised in comparing that value to Titan's, as the value in one or even a handful of comets may not reflect the diversity of cometary nitrogen (or other) isotopes.

The dissociation of ammonia (or a nitrogen bearing organic molecule) by impacts associated with or without shock induced chemistry during Titan's accretion or the late heavy bombardment has also been suggested as a possible origin of Titan's nitrogen. Although an intriguing idea, it does not satisfactorily explain a number of relevant observations such as (1) H_2 and D/H: the cometary impact would result in excessive molecular hydrogen that cannot be removed from the atmosphere without introducing spurious fractionation in the hydrogen isotopes (D/H), not seen in the Huygens GCMS data, as discussed by Atreya *et al.*,[6] (2) conservation of CHONPS elements, *i.e.* absence or enrichment of species: the impacts would exhume many more trace or minor constituents besides NH_3, such as xenon and krypton (if sequestered in surface or subsurface), sulfur and phosphorus compounds, amongst others, and would create or enhance other constituents such as oxygen bearing species including CO, but the Cassini–Huygens data do not show evidence of any such effects of cometary impact either in the atmosphere or the surface of Titan, and (3) presence of species: certain species present on Titan such as primordial argon (^{36}Ar) are absent from comets, which would be puzzling if comets were responsible for supplying Titan's volatiles either in the process of forming Titan or during the late heavy bombardment. (Except for a single report of ^{36}Ar in comet Hale-Bopp,[75] which could not be confirmed independently, ^{36}Ar has not been detected in any other comet.[76]) Considering the above difficulties, the origin of Titan's nitrogen (and methane) atmosphere from comets remains as an interesting hypothesis that needs testing with appropriate measurements in future missions to Titan.

In summary, the nitrogen of Earth, Mars, Venus and Titan is secondary. Its origin from ammonia, a minor constituent of their primordial atmospheres, to its subsequent chemical formation is controlled by trace constituents $-N_2H_4$, OH and H in the case of Titan, and OH, H, Cl and perhaps sulfur and oxygen species in the case of the terrestrial planets.

3. The chemistry of the formation of methane on Earth, Mars and Titan

3.1 Biotic and abiotic methane on Earth

Methane is considered as the most significant potential biomarker in the solar system, except where it is formed by inorganic chemical processes such as in the interiors of the giant planets or the primordial solar or planetary nebulae. On Earth, 90–95% of the 1775 ppbv CH_4 present in the atmosphere is biogenic in origin, due either to current or past biology. A distribution of current sources of methane on Earth is shown in Fig. 3. Only oceans and lakes (3%) can be considered as essentially abiotic sources, whereas a portion of the methane from gas hydrates (1%) and the geologic sources (~8%, with about 1% each from mud volcanoes and other geothermal sources and 6% from seepage) could also be abiotic. Fully a quarter of all methane comes from swamps and wetlands. Grass eating ungulates belch out about 15% of the total terrestrial methane as a metabolic byproduct of the bacteria breaking down the organic matter in the guts of such animals. Leakage of natural gas, which is about 85% methane and the result of past or fossilized biology, and coal mines together make a similar contribution. Rice paddies, biomass burning and waste and landfill are responsible for about a tenth each. Termites make up the rest (4%). Thus, it is only natural to wonder whether biology is responsible when methane is detected on another planet or satellite. Such is the case with Mars where small amounts of methane are reported, and Titan where the gas comprises about 6% of the atmosphere.

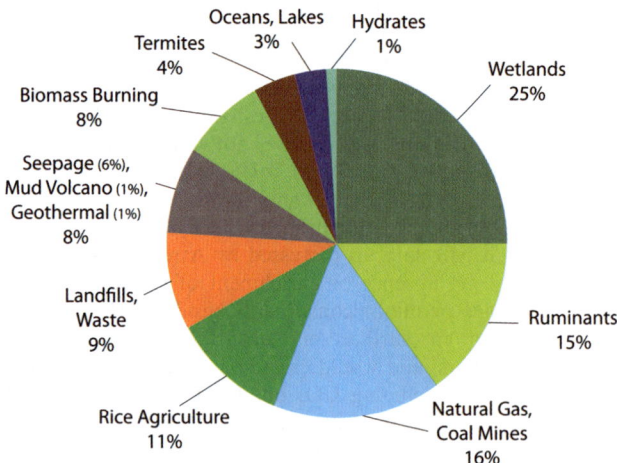

Fig. 3 Methane sources on Earth. Values used in this chart were compiled largely from the IPCC Report[20] and some recent updates (supersedes earlier values (ref. 53)).

3.2 Chemistry of geologic and biogenic origin of methane on Mars

Ground-based[32,44] and Mars Express orbital observations[26] indicate trace amounts of methane in the atmosphere of Mars, with a global average abundance of 10–15 ppbv. Zahnle et al.[73] have questioned the validity of detection of methane and its seasonal variability reported from the high resolution ground-based observations,[44] because of possible contamination of the blue-shifted Martian $^{12}CH_4$ spectral signature by the twenty-times stronger telluric $^{13}CH_4$ isotope signature. The criticism does not apply to the observations done from Mars orbit by the Planetary Fourier Spectrometer on Mars Express.[26] On the other hand, the latter data suffer from low resolving power issues including non- or ambiguous identification of several spectral features. Nevertheless, the above observations seem consistent with a small abundance of methane on Mars, amounting to a global average of 10–15 ppbv. However, claims of rapid variability and seasonal changes from either data set are far less convincing considering the marginal quality of all available data. One should recognize, however, that measurement of such trace quantities of methane in the Martian atmosphere is highly challenging. Dedicated, high precision techniques such as the tunable laser spectrometer with enrichment at front end by mass spectrometer on the 2011 Mars Science Laboratory rover[38,66] are required to unambiguously determine the presence, amount, isotope abundance, and variability of methane on Mars. The origin of even trace amounts of methane indicated by available observations needs to be understood. The associated chemical processes could provide some clues. Following is a brief discussion of the mechanisms for the production and fate of possible methane on Mars.

An extraplanetary origin such as that from comets, meteorites or interplanetary dust particles has been ruled out for the Martian methane as it would contribute less than 1% of the global value.[7] Volcanoes do not help either, as they put out 100–1000 times greater SO_2 at Earth, but SO_2 is actually less than one hundredths of the methane amount in the Martian atmosphere, besides the fact that the volcanoes have been extinct for over 100 Myr on Mars. Only geologic or biogenic origin is likely.[7] The production from either mechanism could be taking place now in the interior of Mars, or it could have occurred on the warmer and wetter Mars in the past. In either case, liquid water is essential for both biotic and geologic production of methane. The MARSIS radar on the Mars Express spacecraft and the SHARAD radar on Mars Reconnaissance Orbiter so far have found no trace of liquid water down to a few kilometres depth, but the water table could actually lie much deeper. Models of the Mars geotherms indicate a range of somewhere between 2 km and 25 km for the depth of aquifers,[50] deeper than the current radar data. Thus, current production of methane on Mars cannot be ruled out. Whether produced in the past or now, methane can be stored as clathrate-hydrate ($CH_4 \cdot 6H_2O$ molar proportions) in the subsurface ice and released from time to time when the clathrates are destabilized by any number of means including magmatic (large or small), thermal or mechanical stresses and impacts.

The geologic process is based on a set of hydrogeochemical reactions, involving first the reaction between water and rock (serpentinization) to liberate H_2, followed by the reduction of primordial carbon in the form of CO, CO_2 or carbon grains to CH_4 by metal-catalyzed Fisher–Tropsch type reactions.[7,50] The principal chemical pathways are listed below.

1. Serpentinization: hydration of ultramafic silicates (Mg, Fe-rich) such as olivine and pyroxene results first in the formation of serpentine (Mg, Fe)$_3$Si$_2$O$_5$(OH)$_4$, and molecular hydrogen ($H_{2(aq)}$),

$$(Mg, Fe)_2SiO_4 + H_2O \rightarrow Mg_3Si_2O_5(OH)_4 + Mg(OH)_2 + Fe_3O_4 + H_{2(aq)}$$

i.e., olivine/pyroxene + water → serpentine + brucite + magnetite + hydrogen, where the key hydrogeochemical reactions are:

$$6Fe_2SiO_4 + 7H_2O \rightarrow H_{2(aq)} + 3Fe_3Si_2O_5(OH)_4 + Fe_3O_4$$

$$2Mg_2SiO_4 + 3H_2O \rightarrow Mg_3Si_2O_5(OH)_4 + Mg(OH)_2$$

followed by:

$$2Fe_3Si_2O_5(OH)_4 + 6Mg(OH)_2 \rightarrow 2H_{2(aq)} + 2Mg_3Si_2O_5(OH)_4 + 2Fe_3O_4 + 4H_2O$$

2. Methane by Fischer–Tropsch process: The $H_{2(aq)}$ liberated in above serpentinization process reacts with CO_2, CO or C of the crustal rocks/pores to produce methane and possibly higher order hydrocarbons in decreasing abundances, *i.e.*

$$CO_{2(aq)} + [2 + (m/2n)]H_{2(aq)} \rightarrow (1/n)\ C_nH_m + 2H_2O$$

$$CO_{2(aq)} + 4H_{2(aq)} \rightarrow CH_4 + 2H_2O$$

$$C + 2H_{2(aq)} \rightarrow CH_4$$

The above Fischer–Tropsch processes require metal catalysis by either Mg, Fr, Cr or their oxides, all of which are components of the ultramafic silicate rocks. Production of large quantities of methane is known to take place by above mechanism in the Black Smoker hydrothermal vents (300–500 °C, 300–400 bar) in terrestrial oceans as well as Lost City hydrothermal vents about 15 km from the spreading center where the temperatures are much milder, on the order of 30–120 °C (the upper range reflects the temperature of the underlying peridotite rock in contact with ocean water).

The above geochemical mechanism for producing methane can work in the interior of Mars if liquid water is present now or in the past when it almost certainly was. Both low and high temperature serpentinization processes are likely in either case, considering the low-high range of temperatures in the water table, according to the Martian geotherm models.[50] Finally, the recent detection of serpentine[21] on Mars lends some credibility to the above hydrogeochemical process for the formation of methane now or in the past. A biogenic source cannot be ruled out either.

Anaerobic, subterranean and oceanic chemolithotrophic microbial ecosystems on Earth are known to produce methane as a product of metabolism.[15,57,51] Coincidently, the biochemical pathways involved[67] are similar to the above geochemical mechanism, and both require liquid water as the reactant, medium or solvent:

$$4CO + 2H_2O \rightarrow CH_4 + 3CO_2$$

$$4H_2 + CO_2 \rightarrow CH_4 + 2H_2O$$

At Mars, the atmosphere is a large source of CO (~700 ppmv) and H_2 (40–50 ppmv). The diffusion of these gases through the regolith to the subsurface aquifers would provide the nutrients to any microbial colonies that might exist in the subpermafrost aquifer environment of Mars. An abundant supply of H_2 may also be available locally from serpentinization as discussed above. Fermentation is another process on Earth where various groups of microorganisms convert organic matter through a series of consecutive metabolic steps into acetic acid (CH_3COOH) and then CH_4 and CO_2. It is perhaps less likely at Mars, even if biology existed. As in the case of hydrogeochemical processes, the biotic production of methane on Mars need not be current. If microorganisms did exist on Mars in the past during its warmer and wetter phase, or if they exist in the aquifers now, the methane resulting from their metabolic activity can be stored in the permafrost or other ice for later release.

Irrespective of the production mechanism, once in the atmosphere the removal of methane occurs by photolysis largely above 50 km and by oxidation below, by OH from water vapor and $O(^1D)$ from ozone. The combined loss processes determine a timescale of 300–600 years against photochemical loss of methane in the Martian atmosphere.[70] A much more rapid loss is possible *via* heterogeneous processes involving a reactive surface, as will be discussed later (section 4.3). The photochemistry of methane in the atmosphere of Mars results mainly in the formation of ethane (C_2H_6), formaldehyde (CH_2O) and methanol (CH_3OH). However, their yield is very low with most products at pptv level or less, even if CH_4 were present at hundreds of ppmv levels.[70] Thus, for all practical purposes, methane is of little significance for the main chemistry of the Martian atmosphere, which is controlled by the CO_2–H_2O chemical pathways as will be discussed later (section 4.3).

3.3 Hydrogeochemical, primordial, and biochemical origin of methane on Titan

Potential sources of methane on Titan include: (i) direct delivery of CH_4 to Titan by the Titan forming planetesimals; (ii) formation of CH_4 on Titan either from primordial (non-methane) carbon; or (iii) biological. The abiotic production of methane in Titan's interior can be considered to be similar to the hydrogeochemical process discussed above for the Martian methane. During the accretionary heating phase of Titan the liquid water ocean in the interior is expected to have extended all the way to the core, where the rocks are. During that period, a good likelihood existed for both the high temperature serpentinization initially and the low temperature serpentinization later as the interior cooled. This was then followed by a Fischer–Tropsch process to produce sufficient quantities of methane.[8] The original carbon delivered to Titan in this scenario was in the form of primordial CO, CO_2 or carbon grains. As Titan cooled, the water ocean was gradually separated from the rocky core by a high-pressure ice barrier. Initially this ice, and subsequently the ice at shallower depths above the ocean, allowed the sequestration of methane as clathrate hydrate in Titan's interior. Methane was released to the atmosphere, perhaps episodically,[60] by cryovolcanism, impacts or other clathrate destabilizing processes, as also suggested above for Mars.

The non detection of heavy noble gases, Xe and Kr, by Huygens GCMS seems to provide support to the above hypothesis of production of methane on Titan. Had methane arrived at Titan directly as CH_4, the same planetesimals would have delivered much greater quantities of Xe and Kr than the GCMS upper limit of 10 ppbv. However, suggestions have been made that Xe and Kr might be sequestered as clathrates in the ice below Titan's purported ocean several hundred kilometres beneath the surface. It is surprising that no outgassing of these volatiles occurred in Titan's 4.5 Gyr geologic history, considering that even ^{40}Ar produced in Titan's rocks some 2000 km below the surface gets out to the atmosphere where it has been detected. Sequestration of Xe and Kr in the aerosols in the stratosphere is another idea, but it is not evident why degassing does not occur once the aerosols descend to the surface where the temperature is twenty degrees warmer. Moreover, impacts of micrometeorites and charged particles such as the galactic cosmic rays (GCR) are also likely to release them. Perhaps there is a way out of this dilemma, but none satisfactory has been found yet.

Assuming that the absence of Xe and Kr in Titan's atmosphere has some as yet unknown explanation, another model has been proposed for the origin of Titan's methane. According to this model, Titan forming planetesimals originated from the outer solar nebula, not Saturn's subnebula, and they delivered methane directly to Titan.[43] In their best case scenario, these authors[43] find that if methane was produced by the abovementioned serpentinization process, to explain the current D/H ratio in Titan's CH_4 (1. 32 × 10^{-4}) the D/H ratio in Titan's initial H_2O ice would be only 0.7 times the D/H found from H_2O on Enceladus, and worse yet if photochemistry enriched primordial D/H in Titan's methane. The model is based on several assumptions, notably: (a) D/H in Titan's ice must be same as in Enceladus' ice, which is far from certain as the volatile history of these two objects is very different; (b) Enceladus and the comets have the same D/H in their H_2O, which is based on only three comets (Halley, Hyakutake, Hale-Bopp) all belonging to the same family of Oort cloud comets; (c) fractionation in D/H did not occur in the chemistry of serpentinization or the Fischer–Tropsch process or post CH_4 formation, which still needs to be demonstrated. Finally, since ^{36}Ar is not present in comets, the cometary origin of Titan's volatiles hypothesis also implies that no measurable amount of primordial argon (^{36}Ar) can be present on Titan, contrary to the Huygens GCMS detection of 200 ppb of ^{36}Ar in Titan's atmosphere (see also end of Sec. 2). On the other hand, the direct detection of CO_2 in the surface of Titan[46] may be an indicator of the formation of methane on Titan. Nevertheless, direct delivery of methane by planetesimals from the outer solar nebula is an attractive hypothesis that deserves testing by determining the D/H ratio in the surface ice of Titan in future missions, as well as the D/H ratio in the ice of multiple and diverse comets. Much theoretical and laboratory work under Titan-specific conditions is also needed to fully comprehend the various chemical fractionation scenarios.

It has also been suggested that Titan's methane might be produced by methanogens.[42,55] Life as we know it requires a source of energy, nutrients, and water – a solvent or medium for the biochemical reactions and the transport of nutrients. McKay and Smith[42] proposed new types of microorganism that use Titan's liquid methane, instead of water, which is frozen solid at 94 K, as the medium, and utilize H_2 and C_2H_2 from Titan's atmosphere as nutrients. Methane is released in turn. They also calculated an energy release of 334 kJ mol^{-1} in the above biochemical process:

$$C_2H_2 + 3H_2 \rightarrow 2CH_4 \; (+334 \text{ kJ mol}^{-1})$$

Considering that this energy is sufficient for survival of methanogenic bacteria on Earth, they suggested the possibility of widespread methanogenic life in Titan's methane liquid.

Even if the above scenario works, it would not be responsible for the origin of methane on Titan, because in order for the bacteria to generate methane, they need methane in the first place as the nutrients, $C_2H_2 + 3H_2$, are photochemical products of methane.[8] Titan's methane must have another origin. Whether or not methanogens are present on Titan is another matter. If they did exist, they would severely deplete the nutrients, H_2 and C_2H_2. The Huygens GCMS determines a uniform mole fraction of 0.001 for H_2,[46] in good agreement with the Cassini Composite Infrared Radiometer Spectrometer value[18] and the Voyager value,[17] as well as various photochemical models according to which H_2 arises from CH_4 and is expected to be uniformly mixed because of its one million year lifetime in Titan's atmosphere. Furthermore, the Huygens GCMS detected acetylene (amongst other molecules) evaporating from Titan's surface in the measurements made for over an hour after the probe landed. Thus, no perceptible depletion in the nutrients, C_2H_2 and H_2, is recorded, which argues against widespread methanogenic life in Titan's surface, at least in the Huygens landing site, as was suspected previously from preliminary analysis of the Cassini-Huygens data.[6] It should be mentioned, however, that the INMS reported an H_2 mole fraction in the lower thermosphere[65] that is four times greater than every other measurement or models. This puzzling result warrants an independent analysis of the INMS H_2 data, especially considering its potential implication to Titan's biology.

4. Chemistry in the atmosphere and ionosphere, formation of aerosols, surface processes

In the atmospheres of planets, numerous trace gases result from the chemistry of and between major and minor constituents. Here I will limit the discussion to the two constituents nitrogen and methane, whose chemistry is intertwined at Titan but is rather separate at Mars. As mentioned previously, the chemistry of methane in the atmosphere of Mars can produce higher order hydrocarbons, *etc.* but in insignificant quantities and with relatively short lifetimes. In contrast to Mars, a rich chemistry is triggered by the solar ultraviolet flux and the charged particles in Titan's neutral atmosphere below approximately 800 km and the ionosphere above.

4.1 The neutral atmosphere of Titan – hydrocarbon and nitrile chemistry and hazes

The chemistry of Titan's neutral atmosphere has been modeled by several groups (see ref. 59 for a review). Here I will highlight the salient points. Fig. 4 illustrates the basic chemical processes at work in the stratosphere and the upper atmosphere of Titan. On the left side are shown the pathways for an isolated CH_4 chemistry, which are in fact very similar to those in the atmosphere of Jupiter. Methane undergoes photolysis largely at the Lyman-alpha wavelength of 121.6 nm due to the large solar flux at this wavelength. Nearly 50% of the products are in the form of methyl radicals (CH_3), while the rest are in the form of $^1CH_2$, $^3CH_2$ and CH. The quantum yields of these latter radicals are not well constrained, especially at the non Lyman-alpha wavelengths. CH_3 is the precursor to all complex organics at Titan and the giant planets. Photolysis in the upper atmosphere accounts for nearly a third of the CH_4 destruction rate on Titan (4.8 × 10^9 cm^{-2} s^{-1}, or 8.7 × 10^9 cm^{-2} s^{-1} when referenced to surface.[68] The other two-thirds is due to catalytic destruction by C_2H_2 in the stratosphere below. The escape from the atmosphere results in an additional 15–20% loss of methane from Titan's atmosphere. Thus, the total irreversible loss of methane takes place at a rate of approximately 1.3 × 10^{10} cm^{-2} s^{-1}, referenced to the surface.[68] At this rate methane would be irreversibly converted to other forms or lost from the atmosphere in approximately 10–30 Myr.[68] The continued presence of methane in the atmosphere, which is critical for maintaining the very atmosphere on Titan, thus requires its replenishment from time to time.[60] It is estimated that the interior of Titan may hold tens of billions of years worth of supply of methane in the form of clathrate hydrates, while only about 5% of the available supply has been consumed by photochemistry in Titan's geologic history.[6] Even accounting for uncertainties in the storage estimate, it is safe to assume there is little likelihood of an energy crisis on Titan to power Titan's chemistry through the life of the solar system. In Titan's present atmosphere, the two above-mentioned main dissociating chemical pathways of methane are elaborated below. They convert methane to ethane at a rate of about 40%, whereas ethane condenses out of the atmosphere, first as solid ice particles in the stratosphere and then turning to liquid droplets in the middle troposphere.

$2(CH_4 + h\nu \rightarrow CH_3 + H)$ and $2(C_2H_2 + h\nu \rightarrow C_2H + H)$

$CH_3 + CH_3 + M \rightarrow C_2H_6 + M$ $2(C_2H + CH_4 \rightarrow C_2H_2 + CH_3)$

$CH_3 + CH_3 + M \rightarrow C_2H_6 + M$

Net : $2CH_4 \rightarrow C_2H_6 + 2H$ Net : $2CH_4 \rightarrow C_2H_6 + 2H$

[M is the concentration of the background atmosphere]

As shown in Fig. 4, further chemistry results in the formation of stable heavier hydrocarbons, including C_2H_2, C_2H_4, C_3H_8, C_4H_{10}, polyacetylenes or polyynes ($C_{2n+2}H_2$, n = 1,2,3...), *etc.* Laboratory and theoretical work by Gu *et al.*[28] shows formation of triacetylene (C_6H_2) from a reaction between C_2H and C_4H_2, *i.e.* by the usual pathway for polyyne formation. Polyacetylenes higher than diacetylene have not been detected yet, perhaps because of their low vapor phase concentrations. Polymerization of the polyynes is expected to result in haze layers at higher elevations in the upper atmosphere. Benzene (C_6H_6) results on self-recombination of the propargyl radicals (C_3H_3). The continued H-abstraction C_2H_2-addition reaction sequence subsequent to the formation of benzene as well as the continued

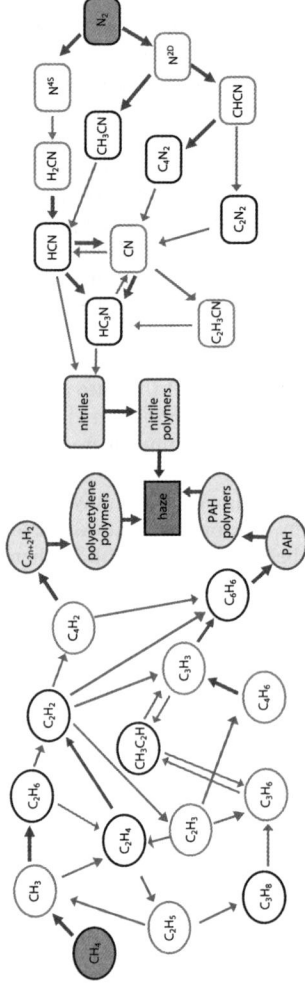

Fig. 4 Basic schematic of coupled methane–ammonia chemistry in the neutral atmosphere of Titan below ~800 km, showing the production of stable hydrocarbons and nitriles, and subsequently the photochemical haze.[8]

C_2H_2-addition to close the second ring are expected to form polycyclic aromatic hydrocarbons (PAH) that subsequently polymerize to form haze at lower elevations in the stratosphere, according to photochemical models (e.g. ref. 69). The presence of nitrogen as Titan's main atmospheric component couples the chemistry of methane with that of nitrogen, as illustrated on the right side of Fig. 4. Reactive nitrogen atoms (N^{2D} and N^{4S}) form from the solar UV and charged particle dissociation of N_2. Their reaction with methane and its products results in the most abundant nitrile, HCN, as well as smaller abundances of CH_3CN, HC_3N, C_2N_2, C_4N_2, etc. The polymerization of HCN and other nitriles results in additional haze layers. The total hydrocarbon and nitrile polymer haze formed in the neutral atmosphere photochemistry accounts for about a 10% rate of destruction of methane, while the condensation of non-ethane organics as aerosols accounts for 30%.[68] The combined polymer–aerosol haze is key to maintaining a stable atmosphere of nitrogen on Titan, as discussed earlier.

4.2 Titan's ionospheric processes – cations, anions, hazes and aerosols

In addition to the hazes formed in the neutral atmosphere, the ionosphere may be a significant source, perhaps even the major source, of haze in Titan's atmosphere. The fast ion–molecule reactions in the ionosphere can ultimately result in heavy hydrocarbon and nitrile positive ions. In fact, the Cassini INMS detects such ions in the instrument's 100 Da mass range, with a one-to-one correspondence between the neutrals below and the resulting ions above, such as CH_4 and CH_5^+, C_6H_6 and $C_6H_5^+/C_6H_7^+$, etc.[63] A significant finding of the INMS was the first ring molecule, benzene, at a mole fraction that is up to 1000 times greater than in the neutral atmosphere below. Even after considering the lower densities of the ionospheric region, it is obvious that the resulting haze from PAHs and other heavy organics would compete in importance with similar hazes produced in the neutral atmosphere.[4] Ionospheric reactions also have the potential of introducing nitrogen into the PAHs, thus creating PANHs. Vuitton et al.[61] have developed a comprehensive model to interpret the above INMS data. The kinetics of reactions beyond simple molecules is poorly constrained, however. Nevertheless it is apparent the heavy molecules formed from cation–molecule reactions would make a significant contribution to Titan's haze. Even larger molecules and subsequent aerosols and soot could result from the anion chemistry, as discussed below.

One of the most intriguing findings about the ionosphere came from observations made with the Electron Spectrometer (ELS) of the Cassini Plasma Spectrometer (CAPS), which provided evidence for negative ions ranging to 10 000 Da at ~1000 km in Titan's ionosphere.[13,14,63] Distinct anion peaks were identified at 22 ± 4 Da, 44 ± 8 Da, and a less certain identification at 82 ± 14 Da (ELS was not designed for detecting ions, hence the low mass resolution). Model calculations (e.g. ref. 62) identified, these peaks, respectively, as CN^-, C_3N^-/C_4H^- and C_5N^-. The concentration of the negative ions was determined to be about 200 cm^{-3} [13,14] or about 10% of the positive ion concentration. The presence of massive negative ions has important implications for Titan's haze.

While negative ions have been known to be prevalent in the relatively dense D-region of the Earth's ionosphere below 80 km, and indeed small densities of the negative ions CH_3^- and H^- were predicted below 200 km in Titan's atmosphere from the impact of galactic cosmic rays,[12] the existence of massive organic anions at nanobar levels was unexpected. In the Earth's D-region, 3-body electron attachment to atoms and molecules results in the formation of O_2^-, and halogens Cl^- and Br^-, all with large electron affinity. Subsequent reactions with neutral atoms and molecules result in other anions. For example, O_2^- reactions form O^-, O_3^-, O_4^-, CO_3^-, NO_3^-, etc. The loss of negative ions occurs primarily by: (i) photodetachment ($X^- + h\nu \rightarrow X + e^-$), such as H^- in stellar atmospheres;[10] (ii) associative detachment ($X^- + Y \rightarrow XY + e^-$); or (iii) cation–anion recombination. The latter (iii) can be more important at Titan, where the solar flux is only 1% of that at the Earth. In Titan's ionosphere, dissociative electron attachment of HCN seems to be the trigger for more complex anions, *i.e.*

$$HCN + e \rightarrow CN^- + H$$

followed by charge transfer to neutral nitriles and hydrocarbons, such as

$$CN^- + HC_3N \rightarrow C_3N^- + HCN$$

and the loss of the resulting anions mainly by associative detachment.

The simple negative ions so formed and identified from the CAPS/ELS data (CN^-, C_3N^-/C_4H^- and C_5N^-) could be precursors to more complex negative ions and subsequently high mass neutrals in the ionosphere.

The very large mass of negative ions is still puzzling. It is likely that cluster ions form in Titan's thermosphere, as they do in the Earth's atmosphere. Once formed, the cluster ions would eventually lead to macromolecules, nanometre size aerosols and soot in Titan's thermosphere. The chemistry of cluster ion formation in the Earth's atmosphere has been studied extensively since the 1960s, observationally, theoretically and experimentally. Water molecules participate in first forming mainly positive cluster ions, such as $O_2^+ \cdot H_2O$ below 80 km that go on to produce H_3O^+ and $H_5O_2^+$ in subsequent reactions, while $NO^+ \cdot (H_2O)_n$ clusters dominate above 80 km.[45,30] Negative cluster ions such as $O_2^- \cdot (H_2O)_2$ are seen in the D-region.[56] High electric fields can produce $OH^- \cdot (H_2O)_n$ negative cluster ions.[58] Several molecules at ionospheric heights of Titan including NH_3 and H_2O as well as aerosols can serve as ideal candidates for cluster ion formation, both positive and negative. Negative ions detected in the plume of Enceladus during Cassini's E3 encounter plausibly correspond to water clusters.[14] As mentioned earlier, ion-ion recombination is expected to be an important process in Titan's ionosphere, and it could result in large molecular clusters. Nigel Adams suggests (see the Discussion section) that since ion–ion recombination is an order of magnitude slower and less energetic than electron–ion recombination, there is more time for clusters to build up rapidly on both positive and negative ions. Much of the relevant laboratory chemical kinetics data on the formation of ion clusters and just ordinary ions at Enceladus and Titan is presently lacking other than that for the very simplest ones. However, progress has been made recently. For example, the Selected Ion Flow Tube measurements of a series of ring molecules, benzene, toluene and pyridine, with CH_3^+ and $C_3H_3^{+2}$ represent an important step in the right direction.

4.3 Electrochemistry, negative ion pairs, oxidants, and organics at Mars

At Mars, negative ion chemistry could have profound implications for habitability of the planet. Storm electric fields on Mars create negative ion pairs that might lead to excessive oxidant production. An oxidized surface was invoked to explain the data from Life Sciences Experiment on the 1976 Mars Viking landers, but the amount of surface oxidizer required was too large. The background of this problem is that the Viking GCMS found no trace of organics on the surface of Mars – which was surprising even if Mars has no indigenous organics, since 9000 tons of micrometeoritic material enters Mars each year (300 g s^{-1}), a quarter of which survives to the surface,[25] and a few percent of that is organic material – while the other Life Sciences Experiments discovered a highly reactive surface. When nutrients were added to the soil, the Labeled Release (LR) experiment found that gas was given off and the Gas Exchange (GEX) found that both O_2 and CO_2 were released from the soil. Microbes were suggested, but the rapidity of reactions implied an unrealistic metabolism. The seemingly contradictory findings of the Viking GCMS: absence of organics, and the LR and GEX experiments: possible presence of microorganisms, hence organics, could be reconciled if a powerful oxidizer were present in the Martian surface. Oyama et al.[49] proposed hydrogen peroxide (H_2O_2) as that oxidizer. A quarter century later, hydrogen peroxide was eventually detected in the atmosphere.[16,22] However, the observed abundance of 20 ppbv (and highly variable seasonally) was too low by a factor of 1000 (ref. 39) to destroy the surface organics (see ref. 7 for a detailed discussion). The answer could lie in the ion chemistry driven by triboelectric processes in convective dust storms and dust devils that are ubiquitous on Mars.

Taking a cue from their measurements of electric fields in terrestrial dust devils, Delory et al.[19] and Farrell et al.[23] developed a collisional plasma model for the Martian dust devils and storms. They find that negatively charged dust, not sand (ref. 24), particles are critical for generating large electrostatic fields of up to 25 kV m^{-1} that extend to high altitudes. In the ensuing collisional plasma, three new product sets result from two key molecules of the Mars atmosphere, CO_2 and H_2O. Electron impact of CO_2 results in CO_2^+ ions (>14 eV). More importantly to the oxidant question, dissociative attachment by electrons produces CO/O$^-$ from CO_2 (4.4 eV) and OH/H$^-$ from H_2O (6.5, 8.6, 11.8 eV). These ion pairs are crucial to the electrochemical production of H_2O_2, as discussed below.

Atreya et al.[9] developed an electrochemical model, based on the ion pair production rates of Delory et al.,[19] and found an H_2O_2 production rate that exceeded its photochemical production rate by up to a factor of 10^4 for the highest electrostatic fields of 25 kV m^{-1}. The chemical scheme is illustrated in Fig. 5. The photolysis of CO_2 below 320 nm produces CO and O. The recombination of O atoms produces O_2 molecules. The reaction between O and O_2 yields O_3 molecules as in the Earth's atmosphere. The photolysis of H_2O below 200 nm produces OH and H. The reaction between the H atoms (from H_2O) and O_2 molecules (from CO_2) results in peroxy radicals (HO_2). The self-recombination of two HO_2 radicals produces H_2O_2 in the atmosphere of Mars. In electrochemistry, triboelectric processes break down H_2O to produce additional OH, and CO_2 to produce additional CO. Since the process takes place near the surface, a maximum amount of water vapor is available to produce the OH, unlike photochemistry, which occurs well above the water

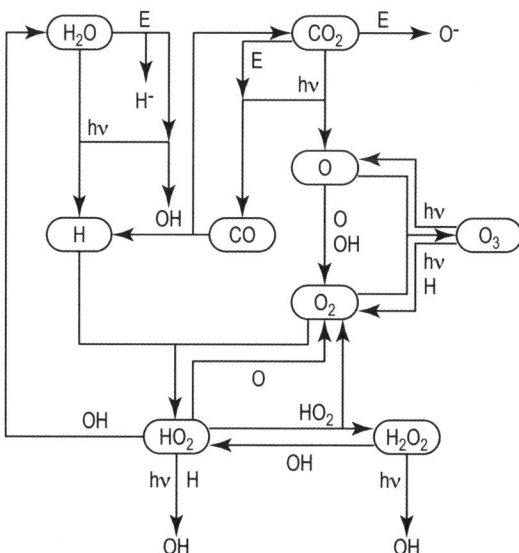

Fig. 5 Electro-photochemical scheme for production of hydrogen peroxide in the atmosphere of Mars. CO/O$^-$, OH/H$^-$ ion pairs result from action of triboelectric fields generated in the Martian dust devils and storms.

condensation level on Mars. The addition OH so produced combine with the additional CO to produce additional H atoms. The rest of the pathways are the same as in photochemistry, with the end result being excess H_2O_2 production. The H^- and O^- ions produced in the above electrochemical process are neutralized by photodetachment and added to the volume affected. The H atoms produced here are also part of the above excess production of H_2O_2 by electrochemistry. In the steady state there is little change in the CO–CO_2 balance, as the excess CO is simply recycled into CO_2 that previously underwent triboelectric destruction. This would occur from a CO + OH reaction both in the affected volume and subsequently in the surface/subsurface upon the breakdown of H_2O_2.

The electrochemical production of peroxide can be so large that most of the peroxide would rapidly diffuse into the regolith where it could be sequestered for hundreds to millions of years depending on the depth of penetration, which can be several meters.[9] Laboratory experiments show that peroxide destroys organics quite efficiently.[54,27] In a generic sense, peroxide could also destroy simple organic molecules, such as methane. Observations indicate localized destruction and rapid temporal and seasonal changes of methane on Mars, implying a lifetime on the order of a less than a year to a few years instead of the photochemical lifetime of a few hundred years. As mentioned earlier, one needs to be cautious not to over interpret the marginal data available now, but if the above characteristics of methane distribution are confirmed by precise measurements in the future, the answer might lie in its destruction also by surface oxidants as the gas diffuses into the regolith from the atmosphere. It is important to note the boundaries and location of the Martian aeolian processes that are proposed for the excess oxidant production electrochemically also shift over time, which would result in a non uniform distribution of oxidants in the surface.

Direct oxidation of CH_4 by hydrogen peroxide is kinetically inhibited in the atmosphere of Mars,[70] and the encounter of CH_4 with H_2O_2 in the soil would not change that. However, mineral reactions in the soil are likely to change H_2O_2 into much more potent oxidants,[52,71] such as superoxides ($O2^-$), peroxy (HO_2), or hydroxyl (OH) radicals. It is these latter oxidants, produced from the mineral processing of H_2O_2 in the regolith, that are likely to destroy simpler organics such as CH_4, as well as more complex ones. The long lifetime of H_2O_2 in the regolith is a major advantage for the mineral reactions to proceed to completion to above oxidants. Although peroxide was proposed to explain the absence of organics in the surface of Mars, it has not yet been detected in the surface because of the lack of appropriate instruments on spacecraft missions to Mars. On the other hand, another type of oxidant, perchlorate was detected in the north polar region of Mars with the Wet Chemistry Laboratory (WCL) on Phoenix Mars Lander.[29] No perchlorate data are available at other sites. WCL found 0.4–0.6 wt% of perchlorate salts ($Mg(ClO_4)_2$, $NaClO_4$, and smaller amounts of $Ca(ClO_4)_2$ and $KClO_4$), with the maximum comparable to the largest deposits on Earth such as those in the perchlorate flats of the Atacama Desert in Chile. Unlike hydrogen peroxide, perchlorates are, paradoxically, highly stable, a characteristic that makes them suitable as an oxidizer in rocket fuel where controlled combustion, not explosion, is essential. This is most likely the reason why direct destruction of organics by perchlorates is not seen.[27] On the other hand, there is the possibility that they too could also change to more reactive forms by mineral processing in the Martian surface. Mineralogical and chemical processing of potential oxidants at Mars is an area in need of laboratory work.

5. Summary and concluding remarks

Trace constituent chemistry plays a critical role in planetary atmospheres, from the very formation of the atmospheres to their evolution over geologic time to their current structure, composition and stability. A minor component of the primordial atmosphere, ammonia, together with its trace products was responsible for the atmosphere of nitrogen we have on Titan and the Earth today and for the nitrogen component of Venus an Mars atmospheres. Although biogenic methane dominates on Earth, either hydrogeochemical or biochemical or both processes could be at work on Mars. Trace gases produced in the photochemistry of the atmosphere as well as the primordial ones are believed to participate in the relevant processes to produce methane geologically or biochemically in the interior of Mars. Methane is not generated by photochemistry in the atmosphere of Mars. The origin of methane at Titan is not biogenic, and the relevant data sets all taken with the same instrument rule out widespread methanogenic life, at least in the Huygens landing site. Hydrogeochemical, primordial and endogenic processes could each contribute to the methane on Titan, but critical data are presently lacking to determine their relative contributions. Chemistry triggered by charged particles plays a crucial role in numerous planetary phenomena. On Mars, triboelectric processes in convective dust activity create negative ion pairs that are predicted to produce strong oxidants by electrochemistry, which would subsequently destroy the organics. Negative ion chemistry in the aeolian processes could have implications for other (non oxidant) chemistry also. Negative and positive ion chemistry in Titan's upper atmosphere is a major, perhaps even dominant, contributor to the moon's upper atmospheric haze layer/s as well as the total atmospheric aerosol. Although very massive (negative) ions to 10 000 Da are detected in the ionosphere, only simple hydrocarbons and nitriles have been found in the neutral atmosphere below. This is actually not completely surprising considering that the more complex organics are expected to be below the detection limit of current instruments on spacecraft and the ground. Moreover, with the exception of nitrogen, they would all condense in Titan's lower atmosphere and the tropopause (70 K, 100 mb, 45 km), resulting in a drastic reduction in their vapor phase concentrations. On the other hand, aerosols and haze produced in Titan's ionosphere and the neutral atmosphere fall on to the

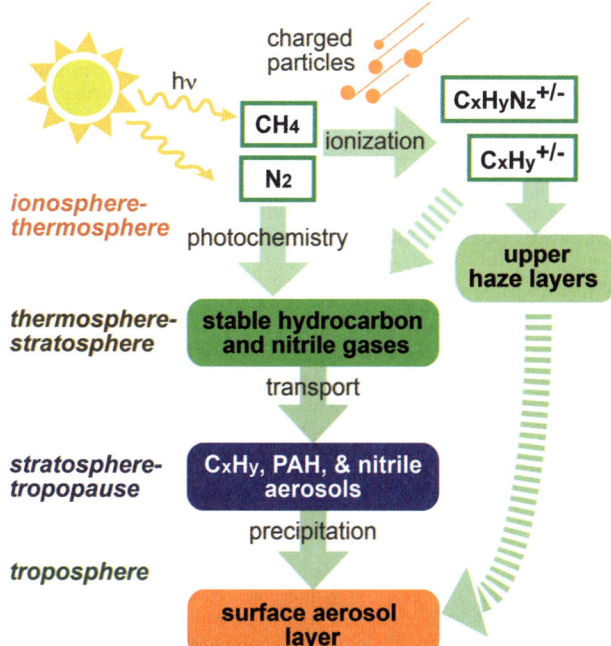

Fig. 6 Cartoon of the photochemical processes in the neutral and ionospheric regions of Titan, showing that all stable hydrocarbons, nitriles, aerosols and hazes either precipitate or sediment out of the atmosphere on to the surface of Titan.

surface, as illustrated in Fig. 6. In fact, the mass spectrum of the material evaporated from the surface after Huygens landed is extremely rich. Since the GCMS was designed as an atmospheric sampler, its mass range was limited to 140 Da and the instrument was not optimized for analyzing the surface material. Thus identification of the large mass peaks in the surface spectrum, many of which undoubtedly represent fractionation of molecules much heavier than 140 AMU, is presently not possible. Nevertheless the spectrum clearly demonstrates the presence of complex organic molecules in the surface, perhaps even prebiotic molecules, which began their journey from above but built up in far greater concentration at the surface over time. The secrets of Titan's atmosphere and the ionosphere lie in its surface![4] Processing of the surface material by galactic cosmic rays (GCR) and impact gardening is a possibility that needs to be investigated carefully in the laboratory and by modeling. The GCRs could be important also in shaping the nitrile chemistry in the lower stratosphere where the GCRs reach their unit optical depth (*e.g.* ref. 69). The oxygen chemistry in Titan's atmosphere is puzzling. While various photochemical models predict detectable abundance of several molecules including aldehydes, it is surprising that other than the simple oxygen bearing molecules, H_2O in the thermosphere and CO and CO_2 in the lower atmosphere and the surface, none other have yet been identified. Clearly, deficiencies exist in either modeling or the relevant laboratory and theoretical work. Great strides are being made recently in all areas of planetary chemistry, however, as is evident from papers in this *Faraday* volume. Studies that mimic as closely as possible the relevant environmental conditions of the atmosphere and ionosphere including temperature, pressure, bath gas, thermodynamic phase and conductivity, and, where applicable, the geochemical and mineralogical characteristics of the surface and the interior are most relevant and valuable to the interpretation of planetary and satellites observations.

References

1 E. Y. Adams, 2006. *Titan's Thermal Structure and the Formation of a Nitrogen Atmosphere.* Thesis, The University of Michigan, Ann Arbor, USA.
2 N. Adams, L. D. Mathews and D. Osborne, Laboratory chemistry relevant to understanding and modeling the ionosphere of titan, *Faraday Discuss.*, 2010, **147**, DOI: 10.1039/c003233f (this publication).
3 S. K. Atreya, 2010. *Atmospheric moons Galileo would have loved*, in *Galileo's Medicean Moons – Their Impact on 400 Years of Discovery*, ed. C. Barbieri, S. Chakrabarti, M. Coradini & M. Lazzarin, Cambridge University Press, in press.
4 S. K. Atreya, Titan's organic factory, *Science*, 2007, **316**, 843–845.
5 S. K. Atreya, T. M. Donahue and W. R. Kuhn, Evolution of a nitrogen atmosphere on Titan, *Science*, 1978, **201**, 611–613.
6 S. K. Atreya, R. D. Lorenz, & J. H. Waite, 2009. Volatile Origin and Cycles: Nitrogen and Methane, in: *Titan from Cassini-Huygens*, ed. R. H. Brown, J.-P. Lebreton, J. H. Waite, pp 177–199, Springer-Verlag, Dordrecht.
7 S. K. Atreya, P. R. Mahaffy and A. S. Wong, Methane and related trace species on Mars: Origin, loss, implications for life and habitability, *Planet. Space Sci.*, 2007, **55**, 358–369.
8 S. K. Atreya, E. Y. Adams, H. B. Niemann, J. E. Demick-Montelara, T. C. Owen, M. Fulchignoni, F. Ferri and E. H. Wilson, Titan's methane cycle, *Planet. Space Sci.*, 2006, **54**, 1177–1187.
9 S. K. Atreya, A. S. Wong, N. O. Renno, W. M. Farrell, G. T. Delory, D. D. Sentman, S. A. Cummer, J. R. Marshall, S. C. R. Rafkin and D. Catling, Oxidant enhancement in Martian dust devils and storms: Implications for life, and habitability, *Astrobiology*, 2006, **6**(3), 439–450.
10 D. R. Bates, Other men's flowers—classical treatment of collisions; Massey's adiabatic criterion and ionization in flames; ion–molecule reactions; bound–free and free–free transitions of electrons in ambient atomic hydrogen, *Phys. Rep.*, 1978, **35**, 305–372.
11 D. Bockelée-Morvan, N. Biver, E. Jehin, A. L. Cochran, H. Wiesemeyer, J. Manfroid, D. Hutsemékers, C. Arpigny, J. Boissier, W. Cochran, P. Colom, J. Crovisier, N. Milutinovic, N. Moreno, J. X. Prochaska, I. Ramirez, R. Schulz and J.-M. Zucconi, Large excess of heavy nitrogen in both hydrogen cyanide and cyanogen from comet 17P/Holmes, *Astrophys. J.*, 2008, **679**, L49–L52.

12 L. A. Capone, R. C. Whitten, J. Dubach, S. S. Prasad and W. T. Huntress, The lower ionosphere of Titan, *Icarus*, 1976, **28**, 367–378.
13 A. J. Coates, F. J. Crary, G. R. Lewis, D. T. Young, J. H. Waite and E. C. Sittler, Discovery of heavy negative ions in Titan's ionosphere. Geophys, *Geophys. Res. Lett.*, 2007, **34**, L22103.
14 A. J. Coates, A. Wellbrock, G. R. Lewis, G. H. Jones, D. T. Young, F. J. Crary, J. H. Waite, R. E. Johnson, T. W. Hill and E. C. Sittler Jr, Negative ions at Titan and Enceladus: recent results, *Faraday Discuss.*, 2010, **147**, DOI: 10.1039/c004700g (this publication).
15 F. H. Chapelle, K. O'Neill, P. M. Bradley, B. A. Methe, S. A. Ciufo, L. L. Knobel and D. R. Lovley, A hydrogen-based subsurface microbial community dominated by methanogens, *Nature*, 2002, **415**, 312–315.
16 R. T. Clancy, B. J. Sandor and G. H. Moriarty-Schieven, A measurement of the 362 GHz absorption line of Mars atmospheric H_2O_2, *Icarus*, 2004, **168**, 116–121.
17 R. Courtin, D. Gautier and C. P. McKay, Titan's thermal emission spectrum: Reanalysis of the Voyager infrared measurements, *Icarus*, 1995, **114**, 144–162.
18 R. Courtin, C. K. Sim, S. J. Kim and D. Gautier, The tropospheric abundance of Titan from the Cassini CIRS investigation, *Bull. Am. Astron. Soc.*, 2007, **39**, 529.
19 G. T. Delory, W. M. Farrell, D. D. Sentman, N. O. Renno, S. K. Atreya, A. S. Wong, S. A. Cummer, J. R. Marshall, S. C. R. Rafkin and D. Catling, Oxidant enhancement in Martian dust devils and storms: Storm electric fields and electron dissociative attachment (Special issue on Space Physics, Mars, and Life), *Astrobiology*, 2006, **6**(3), 451–462.
20 K. L. Denman *et al.*, Couplings between changes in the climate system and biogeochemistry, in *Climate Change 2007: The Physical Science Basis. Contribution of Working Group I to the Fourth Assessment Report of the Intergovernmental Panel on Climate Change (IPCC)*, ed. S. Solomon, *et al.*, 2007, Cambridge University Press, Cambridge, UK and New York, USA.
21 B. L. Ehlmann, J. F. Mustard and S. L. Murchie, 2010. Geologic setting of serpentine-bearing rocks on Mars, *Geophys. Res. Lett.*, 2010, **37**, L06201, DOI: 10.1029/2010GL042596.
22 Th. Encrenaz, B. Bézard, T. Greathouse, M. Richter, J. Lacy, S. K. Atreya, A. S. Wong, S. Lebonnois, F. Lefevre and F. Forget, Hydrogen peroxide on Mars: spatial distribution and seasonal variations, *Icarus*, 2004, **170**, 424–429.
23 W. M. Farrell, G. T. Delory and S. K. Atreya, Martian dust storms as a possible sink of atmospheric methane, Geophys, *Geophys. Res. Lett.*, 2006, **33**, L21203, DOI: 10.1029/2006GL027210.
24 W. M. Farrell, personal communication, 2009.
25 G. Flynn, The delivery of organic matter from asteroids and comets to the early surface of Mars, *Earth, Moon, Planets*, 1996, **72**, 469–474.
26 V. Formisano, S. K. Atreya, T. Encrenaz, N. Ignatiev and M. Giuranna, Detection of methane in the atmosphere of Mars, *Science*, 2004, **306**, 1758–1761.
27 R. V. Gough, J. J. Turley, G. R. Ferrell, K. E. Cordova, S. E. Wood, D. O. DeHaan, C. P. McKay, O. B. Toon and M. A. Tolbert, Can rapid loss, high variability of Martian methane be explained by surface H_2O_2?, *Planet. Space Sci*, 2010, in press.
28 X. Gu, Y. S. Kim, R. I. Kaiser, A. M. Mebel, M. C. Liang and Y. L. Yung, Chemical dynamics of triacetylene formation and implications to the synthesis of polyynes in Titan's atmosphere, *Proc. Natl. Acad. Sci. U. S. A.*, 2009, **106**, 16078.
29 M. H. Hecht, S. P. Kounaves, R. C. Quinn, S. J. West, S. M. M. Young, D. W. Ming, D. C. Catling, B. C. Clark, W. V. Boynton, J. Hoffman, L. P. DeFlores, K. Gospodinova, J. Kapit and P. H. Smith, Detection of perchlorate and the soluble chemistry of Martian soil at the Phoenix lander site, *Science*, 2009, **325**, 64–67.
30 B. G. Hunt, Cluster ions and nitric oxide in the D-region, *J. Atmos. Terr. Phys.*, 1971, **33**, 929–942.
31 D. C. Jewitt, H. E. Matthews, T. Owen and R. Meier, Measurements of $^{12}C/^{13}C$, $^{14}N/^{15}N$ and $^{32}S/^{34}S$ ratios in comet Hale-Bopp (C/1995 O1), *Science*, 1997, **278**, 90–93.
32 V. A. Krasnopolsky, J. P. Maillard and T. C. Owen, Detection of methane in the Martian atmosphere: Evidence for life?, *Icarus*, 2004, **172**, 537–547.
33 H. Lammer, H. Stumptner, G. J. Molina-Cuberos and S. J. Bauer, Owen, T.: Nitrogen isotope fractionation and its consequence for Titan's atmosphericevolution, *Planet. Space Sci.*, 2000, **48**, 529–543.
34 J. S. Lewis, Satellites of the outer planets: Their physical and chemical nature, *Icarus*, 1971, **15**, 174–185.
35 J. I. Lunine and D. J. Stevenson, Thermodynamics of clathrate hydrate at low and high pressures with application to the outer solar system, *Astrophys. J. Suppl.*, 1985, **58**, 493–531.
36 J. I. Lunine and D. J. Stevenson, Clathrate and ammonia hydrates at high pressure – application to the origin of methane on Titan, *Icarus*, 1987, **70**, 61–77.

37 J. I. Lunine, Y. L. Yung and R. D. Lorenz, On the volatile inventory of Titan from isotopic abundances in nitrogen and methane, *Planet. Space Sci.*, 1999, **47**, 1291–1303.
38 P. R. Mahaffy, 2009. *Sample Analysis at Mars: Developing Analytical Tools to Search for a Habitable Environment on the Red Planet, Geochem.* News 141, October issue.
39 R. L. Mancinelli, Peroxides and the survivability of microorganisms on the surface of Mars, *Adv. Space Res.*, 1989, **9**, 191–195.
40 K. E. Mandt, J. H. Waite, W. Lewis, B. A. Magee, J. M. Bell, J. Lunine, O. Mousis and D. Cordier, Isotopic evolution of the major constituents of Titan's atmosphere based on Cassini data, *Planet. Space Sci.*, 2009, **57**, 1917–1930.
41 D. L. Matson, J. C. Castillo, J. I. Lunine and T. V. Johnson, Enceladus' plume: Compositional evidence for a hot interior, *Icarus*, 2007, **187**, 569–573.
42 C. P. McKay and H. D. Smith, Possibilities for methanogenic life in liquid methane on the surface of Titan, *Icarus*, 2005, **178**, 274–276.
43 O. Mousis, J. I. Lunine, M. Pasek, D. Cordier, J. H. Waite, K. E. Mandt, W. S. Lewis and M.-J. Nguyen, A primordial origin for the atmospheric methane of Saturn's moon Titan, *Icarus*, 2009, **204**, 749–751.
44 M. J. Mumma, G. L. Villanueva, R. E. Novak, T. Hewagama, B. P. Bonev, M. A. DiSanti, B. P. Bonev, N. Dello Russo, A. M. Mandell and M. D. Smith, Strong release of methane on Mars in northern summer 2003, *Science*, 2009, **323**, 1041–1044.
45 R. S. Narcisi and A. D. Bailey, *J. Geophys. Res.*, 1965, **70**, 3687.
46 H. B. Niemann, S. K. Atreya, J. E. Demick, J. Gautier-Haberman, D. N. Harpold, W. T. Kasprzak, J. I. Lunine, T. C. Owen and F. Raulin, The composition of Titan's lower atmosphere and simple surface volatiles as measured by the Cassini-Huygens probe gas chromatograph mass spectrometer experiment, *J. Geophys. Res.*, 2010, in press.
47 H. B. Niemann and the GCMS Team, The abundances of constituents of Titan's atmosphere from the GCMS instrument on the Huygens probe, *Nature*, 2005, **438**, 779–784.
48 T. C. Owen, The composition and origin of Titan's atmosphere, *Planet. Space Sci.*, 1982, **30**, 833–838.
49 V. I. Oyama, B. J. Berdahl and G. C. Carle, Preliminary findings of Viking gas-exchange experiment and a model for Martian surface-chemistry, *Nature*, 1977, **265**, 110–114.
50 C. Oze and M. Sharma, Have olivine, will gas: Serpentinization and the abiogenic production of methane on Mars, *Geophys. Res. Lett.*, 2005, **32**, L10203.
51 R. J. Parkes, B. A. Cragg, S. J. Bale, J. M. Getliff, K. Goodman, P. A. Rochelle, J. C. Fry, A. J. Weightman and S. M. Harvey, Deep bacterial biosphere in Pacific Ocean sediments, *Nature*, 1994, **371**, 410–413.
52 R. C. Plumb, R. Tantayanon, M. Libby and W. W. Xu, Chemical-model for Viking biology experiments – Implications for the composition of the Martian regolith, *Nature*, 1989, **338**(6217), 633–635.
53 E. A. Paul, F. E. Clark, 1996. *Soil Microbiology and Biochemistry*, 2nd edn, Academic Press.
54 R. C. Quinn and A. P. Zent, Peroxide-modified titanium dioxide: A chemical analog of putative Martian soil oxidants, *Origins Life Evol. Biosphere*, 1999, **29**(1), 59–72.
55 D. Schulze-Makuch and D. H. Grinspoon, Biologically enhanced energy and carbon cycling on Titan?, *Astrobiology*, 2005, **5**, 560–567.
56 Smirov, Cluster ions in gases, *Usp. Fiz. Nauk*, 1977, **121**, 231–258.
57 T. O. Stevens and J. P. McKinley, Lithoautotrophic microbial ecosystems in deep basalt aquifers, *Science*, 1995, **270**, 450–454.
58 A. Stintz and J. A. Panitz, Negative cluster ion formation from water ice in high electric fields, *Int. J. Mass Spectrom. Ion Processes*, 1994, **133**, 59–64.
59 D. F. Strobel, S. K. Atreya, B. Bézard, F. Ferri, F. M. Flasar, M. Fulchignoni, E. Lellouch, I. Muller-Wodarg. Atmospheric composition and structure, in *Titan From Cassini-Huygens*, 2009, ed. R. H. Brown, J.-P. Lebreton and J. H. Waite, pp. 235–257, Springer-Verlag, Dordrecht.
60 G. Tobie, J. I. Lunine and C. Sotin, Episodic outgassing as the origin of atmospheric methane on Titan, *Nature*, 2006, **440**, 61–64.
61 V. Vuitton, R. V. Yelle and M. J. McEwan, Ion chemistry and N-containing molecules in Titan's upper atmosphere, *Planet. Space Sci.*, 2008, **191**, 722–742.
62 V. Vuitton, P. Lavvas, R. V. Yelle, M. Galand, A. Wellbrock, G. R. Lewis, A. J. Coates and J.-E. Wahlaund, Negative ion chemistry in Titan's upper atmosphere, *Planet. Space Sci.*, 2009, **57**, 1558–1572.
63 J. H. Waite, D. T. Young, T. E. Cravens, A. J. Coates, F. J. Crary, B. Magee and J. Westlake, The process of tholin formation in Titan's upper atmosphere, *Science*, 2007, **316**, 870–872.
64 J. H. Waite, W. S. Lewis, B. A. Margee, J. I. Lunine, W. B. McKinnon, C. R. Glein, O. Mousis, D. T. Young, T. Brockwell, J. Westlake, M.-J. Nguyen, B. D. Teolis,

H. B. Niemann, R. L. McNutt, M. Perry and W.-H. Ip, Liquid water on Enceladus from observations of ammonia and ^{40}Ar in the plume, *Nature*, 2009, **460**, 487.
65 J. H. Waite Jr, *et al.*, Ion Neutral Mass Spectrometer (INMS) results from the first flyby of Titan, *Science*, 2005, **308**, 982–986.
66 C. R. Webster and P. R. Mahaffy, Determining the local abundance of Martian methane and its $^{13}C/^{12}C$ and D/H isotopic ratios for comparison with related gas and soil analysis on the 2011 Mars Science Laboratory (MSL) Mission, *Planet. Space Sci*, 2010, in press.
67 B. P. Weiss, Y. L. Yung and K. H. Nealson, Atmospheric energy for subsurface life on Mars, *Proc. Natl. Acad. Sci. U. S. A.*, 2000, **97**, 1395–1399.
68 E. H. Wilson and S. K. Atreya, Titan's carbon budget and the case of the missing ethane, *J. Phys. Chem. A*, 2009, **113**, 11221–11226.
69 E. H. Wilson, S. K. Atreya and J. Geophys. Res., *J. Geophys. Res.*, 2004, **109**, E06002, DOI: 10.1029/2003JE002181.
70 A. S. Wong, S. K. Atreya and Th Encrenaz, Chemical markers of possible hot spots on Mars, *J. Geophys. Res.*, 2003, **108**(e4), 5026, DOI: 10.1029/2002JE002003. Correction and updated reaction in *J. Geophys. Res.*, 2005, **110**, E10002, 10.1029/2005JE, 002509.
71 A. S. Yen, S. S. Kim, M. H. Hecht, M. S. Frant and B. Murray, Evidence that The reactivity of the Martian soil is due to superoxide ions, *Science*, 2000, **289**(5486), 1909–1912.
72 K. J. Zahnle and J. C. G. Walker, The evolution of solar luminosity, *Rev. Geophys.*, 1982, **20**, 280–292.
73 K. Zahnle, R. Freedman and D. C. Caitling, Is there methane on Mars?, 41st Lunar and Planetary Science Conference, LPI Contrib. No. 1533, 2010, p. 2456.
74 L. M. Ziurys, C. Savage, M. A. Brewster, A. J. Apponi, T. C. Pesch and S. Wyckoff, Cyanide chemistry in comet Hale-Bopp (C/1995 O1), *Astrophys. J.*, 1999, **527**, L67–L71.
75 S. A. Stern, D. C. Slater, M. C. Festou, J. W. Parker, G. R. Gladstone, M. F. A'Hearn and E. Wilkinson, The discovery of argon in comet C/1995 O1 (Hale-Bopp), *Astrophys. J.*, 2000, **544**, L169–L172.
76 H. A. Weaver, P. D. Feldman, M. R. Combi, V. Krasnopolsky, C. M. Lisse and D. E. Shemansky, A search for argon and O VI in three comets using the Far Ultraviolet Spectroscopic Explorer, *Astrophys. J.*, 2002, **576**, L95–L98.

Formation of NH_3 and CH_2NH in Titan's upper atmosphere

Roger V. Yelle,[a] V. Vuitton,[b] P. Lavvas,[a] S. J. Klippenstein,[c] M. A. Smith,[a,d] S. M. Hörst[a] and J. Cui[e]

Received 30th March 2010, Accepted 20th April 2010
DOI: 10.1039/c004787m

The large abundance of NH_3 in Titan's upper atmosphere is a consequence of coupled ion and neutral chemistry. The density of NH_3 is inferred from the measured abundance of NH_4^+. NH_3 is produced primarily through reaction of NH_2 with H_2CN, a process neglected in previous models. NH_2 is produced by several reactions including electron recombination of $CH_2NH_2^+$. The density of $CH_2NH_2^+$ is closely linked to the density of CH_2NH through proton exchange reactions and recombination. CH_2NH is produced by reaction of $N(^2D)$ and NH with ambient hydrocarbons. Thus, production of NH_3 is the result of a chain of reactions involving non-nitrile functional groups and the large density of NH_3 implies large densities for these associated molecules. This suggests that amine and imine functional groups may be incorporated as well in other, more complex organic molecules.

1. Introduction

Measurements of the composition of Titan's ionosphere provide a sensitive probe of the composition of the neutral atmosphere. Analysis of the ion mass spectrum reveals the presence of numerous nitrogen-bearing molecules.[1-3] In addition to nitriles, the chemistry of which has been well studied with photochemical models, the ionospheric measurements indicate substantial densities of $CH_2NH_2^+$ and NH_4^+, which in turn imply the presence of substantial quantities of CH_2NH and NH_3 in the upper atmosphere.[1-3] The chemistry of these species is important because the nitrogen functional groups (imines, amines, *etc.*) may be incorporated into larger organic molecules of biological interest, such as amino acids or nucleic acid bases.[4] We therefore present here an investigation into the photochemistry of non-nitrile nitrogenous species, constrained by Cassini observations of Titan's upper atmosphere.

The distributions of NH_4^+ and NH_3 and $CH_2NH_2^+$ and CH_2NH are closely related. One of the main chemical processes in Titan's ionosphere is proton exchange, where charge flows to the species with the largest proton affinity.[2,3,5] Thus, NH_4^+ and $CH_2NH_2^+$ are created by reaction of NH_3 and CH_2NH with other protonated molecules, while recombination of NH_4^+ and $CH_2NH_2^+$ produce NH_3 and CH_2NH. The densities of the neutral and protonated species are tightly connected by this chemistry and the observed ion densities along with a model for the chemistry predict that NH_3 and CH_2NH are present in Titan's upper atmosphere

[a] *Department of Planetary Sciences, University of Arizona, Tucson, AZ, 85721, USA*
[b] *Laboratoire de Planétogie de Grenoble, Université J. Fourier, Grenoble, France*
[c] *Chemical Sciences and Engineering Division, Argonne National Laboratory, Argonne, IL, 60439, USA*
[d] *Department of Chemistry and Biochemistry, University of Arizona, Tucson, AZ, 85721, USA*
[e] *Department of Physics, Imperial College, Prince Consort Road, London, SW7 2BW, U.K.*

with a mole fraction of several ppm at an altitude of 1100 km, near the ionospheric peak.[1-3]

The presence of several ppm of NH_3 in Titan's upper atmosphere was not predicted by photochemical models.[6-11] The NH_3 mole fraction in the stratosphere must be much smaller than in the upper atmosphere, because at several ppm spectral emission features would be apparent, but have not been seen; thus, the mole fraction of NH_3 must increase with altitude and there is a flux of NH_3 from the upper to lower atmosphere. This indicates that NH_3 is formed in the upper atmosphere. The situation is similar to that of benzene on Titan, which has a mole fraction of several ppm near 1000 km and is synthesized by chemistry in the ionosphere.[12] Here, we show that NH_3 is synthesized by a combination of neutral and ion chemistry in the upper atmosphere. Our investigation also predicts significant levels of N-bearing radicals in Titan's upper atmosphere.

2. Observations

Measurements of the ion densities in Titan's upper atmosphere have been described extensively by Cui *et al.*[13] and Cui *et al.*[14] and we use essentially the same data set here. The ion densities depend on the spacecraft potential and we use the procedure outlined in Cui *et al.*[13] to correct for this effect. The observations were recorded during 40 flybys of the Cassini spacecraft through Titan's upper atmosphere. These data are collected along the spacecraft track over which altitude, latitude, longitude, solar zenith angle, *etc.* all vary considerably. Cui *et al.*[14] averaged and interpolated these data set to produce mean altitude profiles of constituent densities for several ranges of solar zenith angle. Binning by solar zenith angle is motivated by the fact that ion and electron densities are observed to be well correlated with solar input.[13,15] Fig. 1 shows the mean NH_4^+ and $CH_2NH_2^+$ densities for the dayside and nightside. As pointed out in Cui *et al.*,[13] NH_4^+ displays little diurnal variation while $CH_2NH_2^+$ actually has a slightly larger density on the nightside than the dayside. These characteristics are related to the fact that both species are terminal ions, lost primarily through electron recombination, and characterized by relatively long time constants.

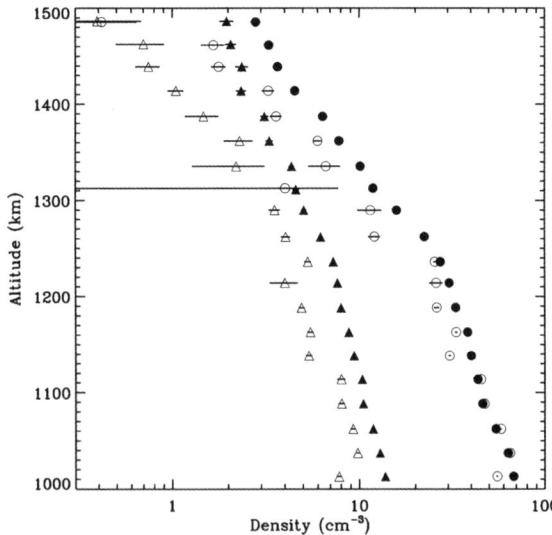

Fig. 1 Circles and triangles represent the $CH_2NH_2^+$ and NH_4^+ densities, respectively. Filled symbols represent average dayside values and open symbols average nightside values. The error bars include only uncertainties due to counting statistics.

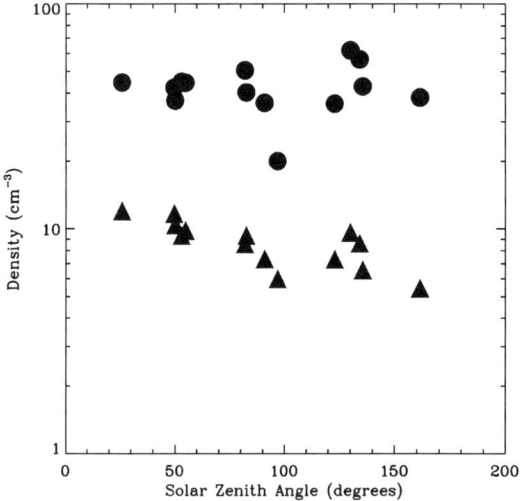

Fig. 2 Variation of $CH_2NH_2^+$ (circles) and NH_4^+ (triangles) with solar zenith angle.

As the ionosphere moves to larger solar zenith angles because of the rotation of the satellite and strong winds, the short-lived ions transfer their charge to longer-lived ions through ion–neutral reactions. Chemical production of terminal ions therefore continues on the nightside, explaining how some ions can be more abundant at night than during the day.[13]

Fig. 2 shows the densities of NH_4^+ or $CH_2NH_2^+$ for each of the passes used in this analysis. The pass-to-pass variations are fairly small, although $CH_2NH_2^+$ does exhibit some outliers. The NH_4^+ data shows a small but clear trend of decreasing density with increasing solar zenith angle. The $CH_2NH_2^+$ shows no clear correlation with solar zenith angle. The lack of strong variability in the data implies that they can be adequately interpreted with a 1D model.

The identification of the signals at $m/z = 18$ and 30 as NH_4^+ and $CH_2NH_2^+$ is discussed by Vuitton *et al.*,[2] Vuitton *et al.*,[3] and Cravens *et al.*[1] For $m/z = 18$, the only alternative to NH_4^+ is H_2O^+; however, the main loss for H_2O^+ is reaction with neutrals to produce H_3O^+, while H_3O^+ recombines with electrons at a slower rate. Thus, the lack of a strong signal at $m/z = 19$, implies a negligible contribution at $m/z = 18$ from H_2O^+. For $m/z = 30$, the options are $CH_2NH_2^+$, NO^+, and $C_2H_6^+$. The latter species is a radical ion and therefore highly reactive and chemical models imply that it should have a small density.[3] NO^+ is stable, but should also have a low density, essentially because the O density in Titan's atmosphere is low.[3]

3. Chemistry

Fig. 3 illustrates the chemical pathways leading to production of NH_3. To keep the diagram simple and readable we show only major chemical reactions. There are two main routes to production of NH_3. The lower path relies exclusively on addition of H to NH_x^+ through reactions with CH_4 and H_2 and has been suggested previously by Atreya.[16] In fact, there is a very tight connection between NH_3 and NH_4^+ because the proton exchange reaction and recombination both proceed rapidly; however, this does not represent a change in the NH_3 abundance, but only a change in its form (protonated or not). The rate for the ion chemistry channel is not limited by production or recombination of NH_4^+, but production of NH_2^+ through reactions of N^+ with H_2 and NH^+ with CH_4. The former reaction proceeds rapidly; however, most of the NH^+ formed by reaction of N^+ with H_2 reacts with N_2 to form N_2H^+,

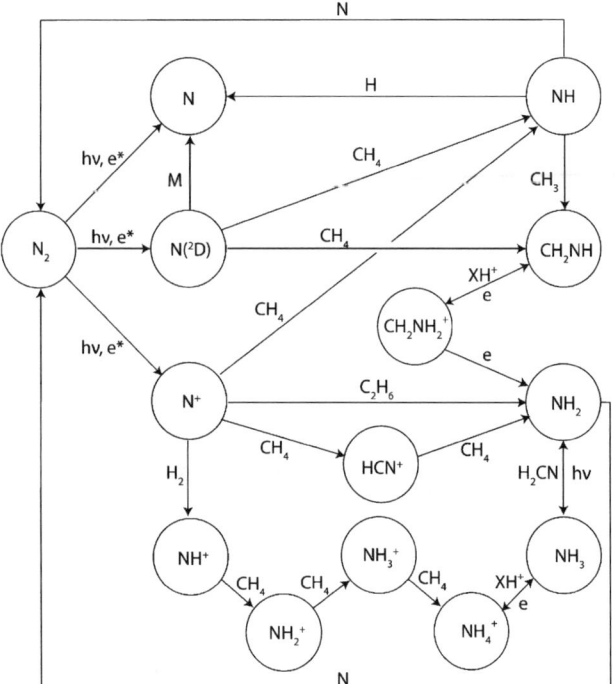

Fig. 3 Chemical pathways for production of NH_3 and CH_2NH.

which then reacts with CH_4 and HCN to produce CH_5^+ and $HCNH^+$. This effectively short-circuits production of NH_3 through this sequence. Calculations described in sections 5 and 6 show that this channel is a minor source of NH_3 on Titan.

The other pathway shown in Fig. 3 relies on conversion of NH_2 to NH_3. The NH_2 radical does not react with any of the stable molecules in the upper atmosphere (N_2, CH_4, C_2H_2, C_2H_4, *etc.*). The primary chemical loss for NH_2 must be reaction with other radicals. H, CH_3 and N are the most abundant radicals in Titan's upper atmosphere; however, NH_2 does not undergo two-body reactions with H. Three-body reactions do occur, but happen at too high a pressure to affect the ionosphere. NH_2 does react with N, leading to production of N_2, which, along with NH + N → N_2 + H, is a main channel for loss of active nitrogen on Titan. NH_2 does undergo a three-body reaction with CH_3, producing CH_3NH_2, but the two-body reaction has not been measured. The H_2CN radical also has a fairly large abundance in the upper atmosphere. It is produced by reaction of N and CH_3, two of the main products from photodissociation of N_2 and CH_4. H_2CN is also the precursor of HCN, the most abundant nitrile in Titan's atmosphere. The difficulty with this proposition is that the reaction rate for $NH_2 + H_2CN \rightarrow NH_3 + HCN$ has not been measured. Nevertheless the reaction is exothermic and, as a radical–radical reaction, should proceed rapidly. In section 4 we present calculations of the rate coefficient based on transition state theory that show that it is quite rapid. This implies that $NH_2 + H_2CN \rightarrow NH_3 + HCN$ is indeed the dominant pathway for production of NH_3 in Titan's upper atmosphere.

With this approach, to produce NH_3, we first need NH_2. The amino radical is also produced by two reactions, but both involve ionospheric chemistry, recombination ($CH_2NH_2^+ + e \rightarrow CH_2 + NH_2$) and ion–neutral reaction ($N^+ + C_2H_4 \rightarrow NH_2 + C_2H_2^+$ or $N^+ + C_2H_6 \rightarrow NH_2 + C_2H_4^+$). N^+ for the latter channels is produced directly from dissociative ionization of N_2 by solar photons and suprathermal electrons. $CH_2NH_2^+$ can be produced by proton exchange reactions with any of the protonated species in Titan's ionosphere. There are many candidates because the ionosphere is composed predominantly of such ions,[2,3] the most abundant of which is $HCNH^+$. CH_2NH has a proton affinity of 853.7 kJ mol^{-1}, which is larger than that for most of the nitrile species (including HCN) and all of the hydrocarbon species in Titan's atmosphere;[3] therefore all of these species will react with CH_2NH to produce $CH_2NH_2^+$. These considerations indicate that there is a direct connection among the abundances of $CH_2NH_2^+$, CH_2NH, NH_4^+, and NH_3.

Methanimine on Titan is produced by two reactions: $NH + CH_3 \rightarrow CH_2NH + H$ and $N(^2D) + CH_4 \rightarrow CH_2NH + H$. The metastable $N(^2D)$ atoms, which are produced by photo or electron impact dissociation of N_2, plays an essential role in the nitrogen chemistry on Titan.[17] NH is produced from $N(^2D)$ through reaction with CH_4 and from reaction of N^+ with CH_4. CH_2NH may also be recycled through electron recombination of $CH_2NH_2^+$, though the products of this reaction have never been measured. This has a small effect on the chemistry because CH_2NH simply cycles between neutral and protonated forms until NH_2 is produced.

There are two ways that production of NH_2 might not follow production of CH_2NH. One possibility is if recombination of $CH_2NH_2^+$ produces HCN. This is energetically possible, but remains to be verified by theory or experiment. The other possibility is that CH_2NH is photo-dissociated into HCN + 2H. In fact, according to Nguyen et al.[18] this is the dominant channel for dissociation. We consider both these possibilities in our numerical model, described below. Neither alter the conclusion that NH_3 is produced primarily from NH_2. Photolysis of NH_3 (R4) also produces NH_2, but this is important primarily at lower altitudes.

4. Calculation of the rate coefficient for $NH_2 + H_2CN \rightarrow NH_3 + HCN$

The mechanism for the reaction of NH_2 with H_2CN was explored at the QCISD(T)/CBS//B3LYP/6-311++G(d,p) level, and is illustrated in Fig. 4. In these calculations, the rovibrational properties of the stationary points were mapped out with B3LYP (Becke-3 Lee–Yang–Parr) density functional theory employing the 6-311++G(d,p) basis set. Complete basis set (CBS) RQCISD(T) (spin-restricted quadratic configuration interaction with perturbative inclusion of triplets) energy estimates are then

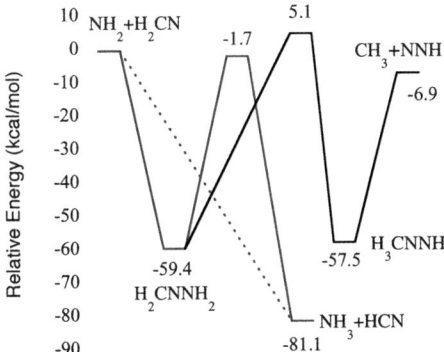

Fig. 4 Schematic plot of the potential energy surface for the reaction of NH_2 with H_2CN. The numbers denote QCISD(T)/CBS//B3LYP/6-311++G(d,p) zero-point corrected energies (in kcal mol^{-1}) relative to reactants. The blue dotted line represents a direct abstraction pathway, the red solid line denotes an addition–elimination pathway to form NH_3 + HCN, and the black solid line denotes a pathway that leads to the formation of CH_3 + NNH. Other pathways, to produce H_2 + H_2CNN for example, involve high energy saddle points and so are not shown.

obtained from basis set extrapolation of calculations with Dunning's correlation-consistent polarized-valence triple-zeta (cc-pVTZ) and quadruple-zeta (cc-pVQZ) basis sets;[19] Kendall et al.[20]).

The plot in Fig. 4 indicates that, at least at low temperature, the $NH_2 + H_2CN$ reaction will involve two primary pathways. One pathway involves the simple addition to form an H_2CNNH_2 adduct, which may then proceed on to NH_3 + HCN via a tight transition state at −1.7 kcal mol^{-1}. Alternatively, further collisions may simply stabilize the initial adduct. A second pathway involves direct abstraction to immediately form NH_3 + HCN. A third pathway, with a saddle point at 5.1 kcal mol^{-1} for isomerization of H_2CNNH_2 to $HCNNH_3$, is sufficiently high in energy that it will make little contribution under the conditions in Titan's atmosphere.

The simple doublet radical nature of each of the reactants suggests that the addition reaction will be barrierless. Meanwhile, the fact that the transition state for the isomerization from H_2CNNH_2 to NH_3 + HCN lies below the reactants suggests that the overall addition–elimination reaction should be quite rapid at low temperatures and low pressures. At higher temperatures it may be somewhat slower due to the low entropy for the isomerization transition state. The highly exothermic nature of the direct abstraction suggests that the abstraction channel is also likely to be barrierless and to occur with a rate coefficient approaching the collision limit.

Multi-reference second order perturbation theory (CASPT2) calculations indicate that both the simple addition and direct abstraction channels are indeed barrierless, as illustrated in Fig. 5. These CASPT2/CBS calculations employ a 4-electron 4-orbital (4e,4o) active space consisting of the NH_2 and H_2CN radical orbitals in addition to the H_2CN π, π^* orbitals. The plots are for the interaction between NH_2 and H_2CN as a function of either the NH (for abstraction) or NN (for addition) separation, with the two radicals in fixed orientations (appropriate for either the abstraction or the addition channels) and with their fixed asymptotic structures. Allowing for relaxation of the orientations and the internal structures of the reacting moieties would simply yield modestly more attractive interaction potentials for these two channels. Clearly, the addition and direct abstraction pathways are indeed barrierless.

Here we implement the direct variable-reaction coordinate (VRC) transition state theory (TST) approach[21–24] in predicting the kinetics for the addition and abstraction channels. The VRC-TST approach was designed to accurately treat the effect of

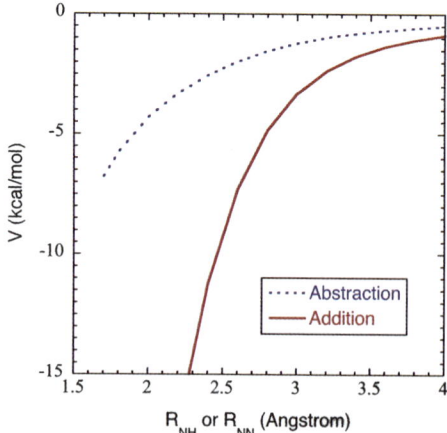

Fig. 5 Plot of the CASPT2(4e,4o)/CBS interaction potentials for the abstraction (blue dashed line) and addition (red solid) channels in the $NH_2 + H_2CN$ reaction.

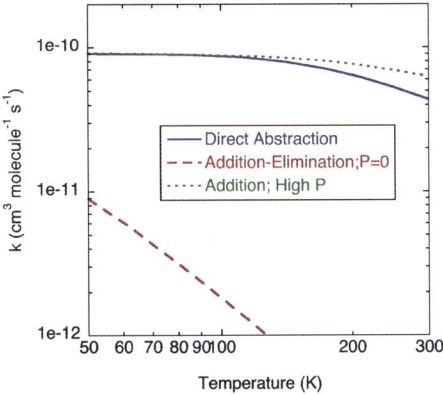

Fig. 6 Plot of the temperature dependence of the rate coefficients for direct abstraction (blue solid line), for addition–elimination (red dashed line), and for addition in the high pressure limit (green dotted line) in the NH$_2$ + H$_2$CN reaction.

anharmonicities and mode couplings for such barrierless reactions, and has been shown to yield accurate kinetic predictions for various radical–radical reactions.[25,26] Here we employ direct CASPT2(4e,4o) calculations of the orientation dependence of the interaction energies. These calculations were done for both the cc-pVDZ and aug-cc-pVDZ basis sets. The final estimates for the interaction energies are obtained by adding one-dimensional CASPT2 complete basis set and geometry relaxation corrections. The kinetic predictions for the corrected CASPT2/cc-pvdz and CASPT2/aug-cc-pvdz samplings differed by only a few percent. The results reported here employ the average of these two results and incorporate a dynamical correction factor of 0.85, which is based on dynamical evaluation of the transition state recrossing for the related CH$_3$ + CH$_3$ recombination reaction.[26]

For the addition process, it is also important to consider the branching between stabilization, elimination, and back dissociation from the initially formed H$_2$CNNH$_2$ adduct. Sample master equation simulations suggest that stabilization of the complex is insignificant for the temperature and pressures of relevance to Titan's upper atmosphere. In this case, the addition–elimination rate constant is equal to its collisionless limit value, essentially independent of pressure.

The CASPT2 calculations were done using the formalism of Celani and Werner[27] as implemented in the MOLPRO08 electronic structure software package. The QCISD(T) calculations also use the MOLPRO08 package while the B3LYP calculations were done with the GAUSSIAN98 software package.[28–31]

The temperature dependent rate coefficients for the direct abstraction, high pressure addition, and addition–elimination reactions are plotted in Fig. 6 Interestingly, the direct abstraction and high pressure addition rate coefficients are roughly equivalent. However, the addition–elimination rate coefficient is greatly reduced from the high pressure addition rate coefficient even at a temperature of 50 K. Apparently, the tight transition state for the isomerization from H$_2$CNNH$_2$ to NH$_3$ + HCN is a significant bottleneck down to rather low temperature. The modified Arrhenius expression, $5.42 \times 10^{-11}(T/300)^{-1.06} \exp(-60.8/T)$ cm^3 s^{-1}, with T in K, reproduces the present predictions for the total rate coefficient for formation of NH$_3$ + HCN in the low pressure limit over the 40 to 400 K temperature range.

5. Photochemical model

The model used in this investigation is adapted from several elements used in previous investigations. The chemistry of N-bearing molecules is closely coupled to the ion chemistry in Titan's upper atmosphere. We therefore model the ion and neutral chemistry in a coupled, self-consistent manner. This is an improvement over the approach taken in Vuitton et al.,[12] where the ionosphere was treated as a source for neutral molecules, but the influence of the neutral composition on the ionosphere was not included self-consistently. The ionospheric part of the model is based on the reaction list described by Vuitton et al.[3] Some aspects of the neutral photochemistry as well as the treatment of eddy and molecular diffusion are discussed in Hörst et al.[32] The neutral N chemistry is adapted largely from Lavvas et al.,[33] with important additions discussed below. Our calculations extend up to 1500 km but we emphasize the region near the ionospheric peak at 1100 km and therefore neglect ion diffusion and assume local chemical equilibrium. This assumption is accurate near the ionospheric peak but breaks down near $\sim$1250 km.[14] This should not have any effect on our conclusions. Diffusion is included for all neutral species. Diffusion coefficients for most species come from Mason and Marrero.[34] For NH_3–N_2 diffusion we use the coefficients from Massman.[35] No data were found for CH_2NH–N_2 diffusion coefficients, so we assume the value is equal to that for C_2H_4–N_2, scaled by the square-root of the reduced mass. The eddy diffusion profile is taken from Yelle et al.[36]

Characteristics of the background neutral atmosphere used in our calculations are shown in Fig. 7 and 8. The neutral densities and temperatures are based on Cassini INMS data[13,37] and the electron temperature from the Langmuir probe channel of the Cassini RPWS experiment.[15] Hydrocarbon abundances are based on our photochemical calculations[12] and are in good agreement with observational constraints.[64] The model used here is appropriate for northern mid-latitudes.

Table 1 presents the important reactions for this investigation. When available, reaction rate data are taken from the literature, but in several cases no measurements are available and rates coefficients are estimated. The sensitivities of our results to these assumptions are discussed in section 6. In this work we are interested primarily in the upper atmosphere, where two-body reactions dominate. Our nitrogen reaction

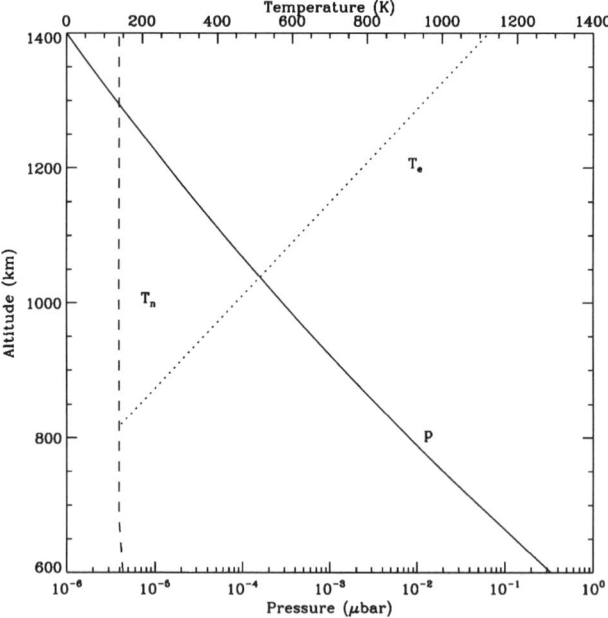

Fig. 7 The altitude variation of pressure, neutral temperature, and electron temperature used in the photochemical calculations.

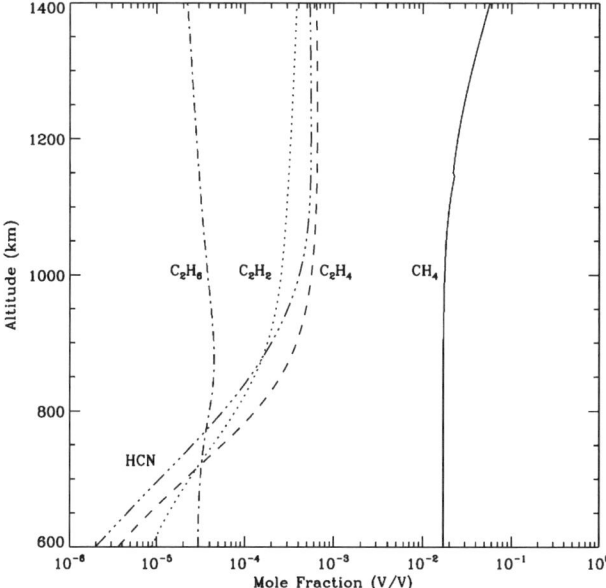

Fig. 8 Mole fractions of stable neutral species used in the photochemical calculations.

list therefore emphasizes two-body processes and likely neglects some three-body reactions that could be important in the stratosphere.

The model is one dimensional and uses globally averaged photolysis rates. This is justified by the fact that the observed diurnal variations of NH_4^+ and $CH_2NH_2^+$ are quite small (Fig. 2); moreover, latitudinal and diurnal variations in the background atmosphere near 1100 km are small.[37] Also, we are interested here primarily in identifying the chemical pathways for production of NH_3 and CH_2NH, rather than precise modeling of their density profiles. A 1D model is adequate for this purpose. The complexity and computational expense of a 3D model, along with the associated uncertainty in the circulation patterns, suggest that the 3D models be deferred until the chemistry is well established and the dynamics of the upper atmosphere better understood.

Dissociation and ionization of N_2 and CH_4 are modeled in detail, including both photon and electron induced processes. Neutral photodissociation of N_2 is calculated making use of newly determined high-resolution cross sections.[62,63] The supra-thermal electron distribution is calculated based on a local energy deposition approximation that has been validated through comparison with a supra-thermal electron transport code. The reader is referred to [17] for an in depth discussion of the photolysis of CH_4 and N_2.

Several of the rate coefficients in Table 1 had to be estimated, because of a lack of laboratory measurements or theoretical calculations. In section 4, we describe our calculations of the rate coefficient for $NH_2 + H_2CN$. The most important reaction for which we are missing required data is electron recombination of $CH_2NH_2^+$ (R19–R21). Based on analogy with electron recombination of other complex hydrocarbon ions, we expect the rate coefficient to be large and adopt in our baseline model a value of 2.1×10^{-6} cm^3 s^{-1} for the net rate with equal probabilities for three branches: $CH_2NH + H$, $CH_2 + NH_2$, and $HCN + H + H_2$. This value is chosen because it provides the best match between predicted and observed densities of $CH_2NH_2^+$; however, we also consider in the next section the sensitivity of our model to the value assumed for the rate coefficient and branching ratios. Also uncertain is the photolysis rate for CH_2NH.[33] The products have been estimated theoretically,[18]

Table 1 Selected reactions

	Reaction	Rate coefficient[a]	Reference
Photolysis			
R1	$N_2 + h\nu \rightarrow N^+ + N + e$	4.9×10^{-10}	38, 39
R2	$N_2 + h\nu \rightarrow N(^2D) + N$	2.2×10^{-9}	17
R3	$CH_2NH + h\nu \rightarrow HCN + H + H$	1.1×10^{-8}	18, 40
R4	$NH_3 + h\nu \rightarrow NH_2 + H$	8.2×10^{-7}	41, 42
Ion reactions			
R5	$N^+ + CH_4 \rightarrow CH_3^+ + NH$	5.75×10^{-9}	43
R6	$\rightarrow CH_4^+ + N$	5.75×10^{-10}	43
R7	$\rightarrow HCN^+ + H_2 + H$	1.15×10^{-10}	43
R8	$N^+ + H_2 \rightarrow NH^+ + H$	5.00×10^{-10}	43
R9	$N^+ + C_2H_4 \rightarrow C_2H_3^+ + NH$	3.25×10^{-10}	44, 43
R10	$\rightarrow C_2H_4^+ + NH_2$	1.30×10^{-10}	44, 45
R11	$N^+ + C_2H_6 \rightarrow C_2H_3^+ + NH_3$	2.50×10^{-10}	44, 45
R12	$HCN^+ + CH_4 \rightarrow C_2H_3^+ + NH_2$	1.27×10^{-10}	43
R13	$HCNH^+ + CH_2NH \rightarrow CH_2NH_2^+ + HCN$	2.7×10^{-9}	46
R14	$C_2H_5^+ + CH_2NH \rightarrow CH_2NH_2^+ + C_2H_4$	2.7×10^{-9}	3
R15	$CH_5^+ + CH_2NH \rightarrow CH_2NH_2^+ + CH_4$	3.0×10^{-9}	3
R16	$HCNH^+ + NH_3 \rightarrow NH_4^+ + HCN$	2.30×10^{-9}	43
R17	$C_2H_5^+ + NH_3 \rightarrow NH_4^+ + C_2H_4$	2.00×10^{-9}	43
R18	$HC_3NH^+ + NH_3 \rightarrow NH_4^+ + HC_3N$	2.09×10^{-9}	47
R19	$CH_2NH_2^+ + e \rightarrow CH_2NH + H$	$0.5 - 1.4 \times 10^{-6} (300/T_e)^{0.7}$	Estimated (see text)
R20	$\rightarrow CH_2 + NH_2$	$0.5 - 1.4 \times 10^{-6} (300/T_e)^{0.7}$	Estimated (see text)
R21	$\rightarrow HCN + H + H_2$	$0.5 - 1.4 \times 10^{-6} (300/T_e)^{0.7}$	Estimated (see text)
R22	$NH_4^+ + e \rightarrow NH_3 + H$	$8.02 \times 10^{-7} (300/T_e)^{0.605} e^{-510/T_e}$	48
R23	$\rightarrow NH_2 + H + H$	$1.23 \times 10^{-7} (300/T_e)^{0.605} e^{-510/T_e}$	48
Neutral reactions			
R24	$H + CH_2NH \rightarrow H_2CN + H_2$	4.0×10^{-14}	49
R25	$N + CH_3 \rightarrow H_2CN + H$	$4.3 \times 10^{-10} e^{-420/T}$	50

Table 1 (Contd.)

	Reaction	Rate coefficient[a]	Reference
R26	$N + H_2CN \rightarrow HCN + NH$	$1.0 \times 10^{-10} e^{-200/T}$	51
R27	$N(^2D) \rightarrow N + h\nu$	2.3×10^{-5}	52
R28	$N(^2D) + N_2 \rightarrow N + N_2$	1.7×10^{-14}	53
R29	$N(^2D) + CH_4 \rightarrow CH_2NH + H$	$3.84 \times 10^{-11} e^{-750/T}$	53
R30	$\rightarrow NH + CH_3$	$9.6 \times 10^{-12} e^{-750/T}$	53
R31	$N(^2D) + H_2 \rightarrow N + H_2$	2.28×10^{-12}	54
R32	$\rightarrow NH + H$	$4.2 \times 10^{-11} e^{-880/T}$	53
R33	$N(^2D) + C_2H_2 \rightarrow HC_2N + H$	$1.6 \times 10^{-10} e^{-270/T}$	55, 53
R34	$N(^2D) + C_2H_4 \rightarrow CH_3CN + H$	4.4×10^{-11}	53
R35	$N(^2D) + C_2H_6 \rightarrow NH + C_2H_5$	3.8×10^{-12}	53
R36	$N(^2D) + HCN \rightarrow N_2 + CH$	$1.6 \times 10^{-10} e^{-270/T}$	Estimated, based on R35
R37	$NH + C_2H_2 \rightarrow HC_2N + H_2$	$2.01 \times 10^{-9} T^{-1.07}$	57
R38	$NH + C_2H_4 \rightarrow CH_3CN + H_2$	$2.3 \times 10^{-12} (T/300)^{-1.09}$	57
R39	$NH + C_2H_6 \rightarrow C_2H_5N + H_2$	6.8×10^{-12}	57
R40	$NH + C_4H_2 \rightarrow C_4HN + H_2$	$8.24 \times 10^{-9} T^{-1.23}$	57
R41	$NH + H \rightarrow N + H_2$	$3.12 \times 10^{-16} T^{1.55} e^{-103/T}$	56
R42	$NH + CH_3 \rightarrow CH_2NH + H$	$3.12 \times 10^{-16} T^{1.55} e^{-103/T}$	Estimated, based on R43
R43	$NH + N \rightarrow N_2 + H$	2.49×10^{-11}	58
R44	$NH + NH \rightarrow NH_2 + N$	$9.9 \times 10^{-22} T^{2.89} e^{1021/T}$	59
R45	$NH_2 + N \rightarrow N_2 + H + H$	1.2×10^{-10}	60
R46	$NH_2 + NH_2 \rightarrow N_2H_4$	$8.97 \times 10^{-20} T^{-3.9}$	61
R47	$NH_2 + H2CN \rightarrow NH_3 + HCN$	$5.42 \times 10^{-11} (T/300)^{-1.06} e^{-60.8/T}$	This work

[a] Photolysis rates are diurnally averaged optically thin values at 9.5 AU. Units are s^{-1}. Other rate coefficients have units of $cm^3 \, s^{-1}$.

but the cross section has only been measured over a small wavelength range.[40] Using this information we estimate a rate at 1 AU of 1.0×10^{-6} s^{-1}, but also consider models with other values.

6. Model results

Calculated densities for a selection of ion and neutral species in the baseline model are shown in Fig. 9, along with the measured densities of NH_4^+ and $CH_2NH_2^+$. Agreement is adequate over most of the altitude range and sufficient to conclude that the primary production and loss mechanisms for NH_3 and CH_2NH have been properly identified. Radical species, N, H_2CN, NH and N(^{2}D) dominate near the ionospheric peak at ~1100 km. At lower altitudes these give way to the more stable species, NH_3 and CH_2NH.

Reactions rates shown in Fig. 10a–d, can be used to follow the chemical cycles. NH is produced primarily by ion chemistry through R5: $N^+ + CH_4 \rightarrow CH_3^+ + NH$ and lost through reaction with C_2H_2 and C_2H_4 (R37, R38) which produce nitrile species and through reaction with CH_3, which produces CH_2NH (R42). The column-integrated rate of NH production through R5 is 2.6×10^7 cm^2 s^{-1} and roughly 40% of the NH so produced results in CH_2NH production through R42. There is no direct route of any significance from NH to NH_2, but we discuss below how production of CH_2NH can lead to NH_2. Loss due to diffusion is not significant for NH and the density is close to photochemical equilibrium.

CH_2NH plays a dual role in the chemistry as both an intermediary for NH_2 production and as an important, stable product itself. As shown in Fig. 10b, CH_2NH is produced by reaction of NH with CH_3 (R42) and by electron recombination of $CH_2NH_2^+$ (R19). However, $CH_2NH_2^+$ is produced primarily by proton transfer reactions of several species with CH_2NH (R13–15), so production of CH_2NH through R19 is a recycling of CH_2NH rather than production of new molecules. Electron recombination of $CH_2NH_2^+$ may also produce CH_2 and NH_2 (R20), which eventually leads to formation of NH_3. At lower altitudes, CH_2NH is lost by reaction with H (R24), which produces H_2CN. The H_2CN is eventually converted into HCN,

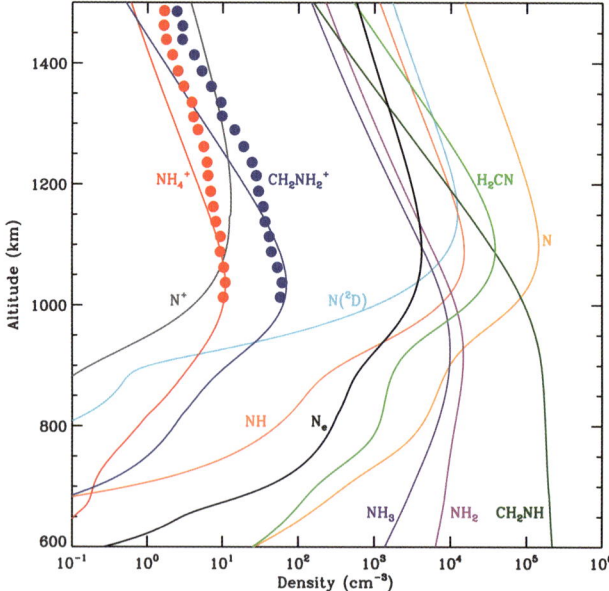

Fig. 9 Densities of significant nitrogen-bearing species calculated in the baseline model. The data points represent INMS measurements, solid lines the model calculations.

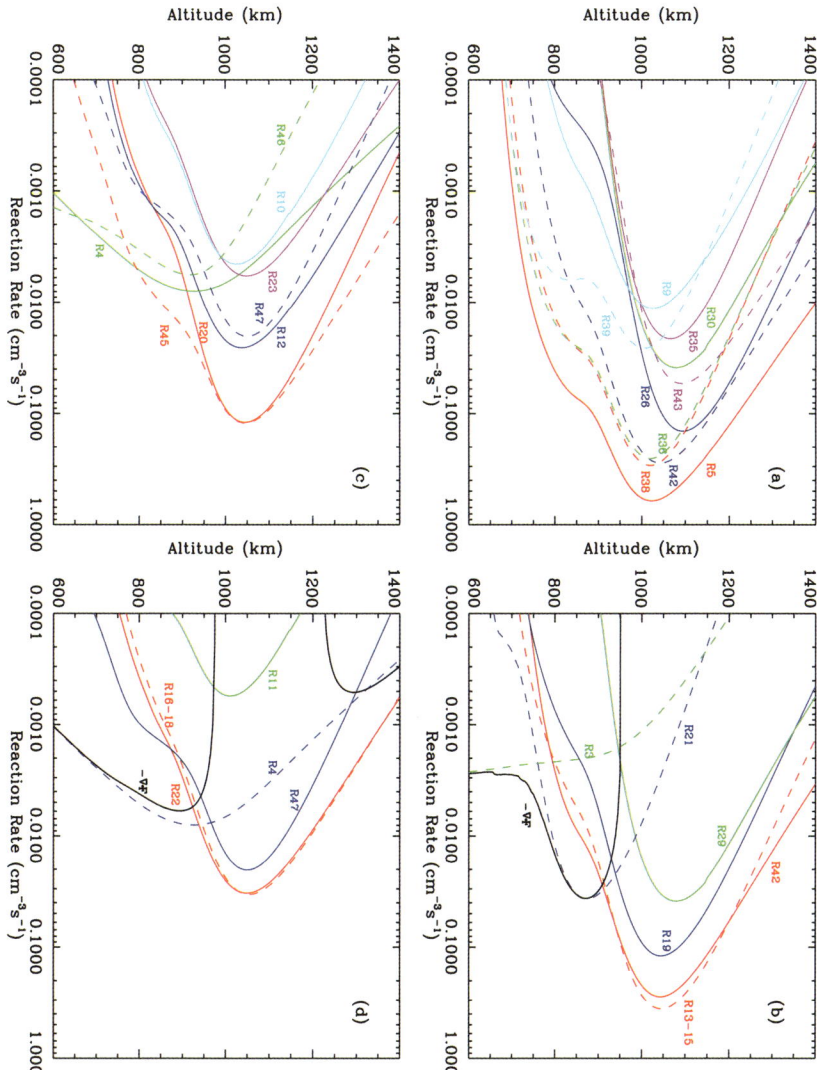

Fig. 10 Primary production and loss rates in our model. Solid and dashed curves represent production and loss, respectively. $-\nabla \cdot F$ represents local production due to diffusion. Labels for other curves refer to Table 1. Panels a–d show the dominant production and loss processes for a: NH, b: CH$_2$NH, c: NH$_2$, and d: NH$_3$.

and also aids in the production of NH$_3$. The column-integrated rate for R20 is 4.2×10^6 cm^2 s^{-1}, about 40% of the value for R42.

NH$_2$ is produced by reaction of HCN$^+$ with CH$_4$ (R12), in addition to R20. The former reaction contributes about 25% to the net rate and the latter 75%. The dominant loss process is reaction with N (R45), that produces N$_2$, destroying active nitrogen. The second most important loss process is reaction with H$_2$CN, which produces NH$_3$ (R47). In total, 81% of the NH$_2$ produced goes back to N$_2$ and 11% is converted to NH$_3$. The remaining 8% is converted to N$_2$H$_4$ by reaction with itself (R46). At lower altitudes, NH$_2$ is produced by photolysis of NH$_3$ (R4). The higher densities at lower altitude favor three-body recombination and the NH$_2$ produced from R4 is converted into N$_2$H$_4$.

Table 2 Model runs

Model	k_{47}/cm^3 s^{-1}	k_{19}/cm^3 s^{-1}	k_{20}/cm^3 s^{-1}	k_{21}/cm^3 s^{-1}	J_3/s^{-1}
A	$5.42 \times 10^{-11}(T/300)^{-1.06}e^{-60.8/T}$	$7.0 \times 10^{-7}(300/Te)^{0.7}$	$7.0 \times 10^{-7}(300/Te)^{0.7}$	$7.0 \times 10^{-7}(300/Te)^{0.7}$	10^{-6}
B	$3.61 \times 10^{-11}(T/300)^{-1.06}e^{-60.8/T}$	$7.0 \times 10^{-7}(300/Te)^{0.7}$	$7.0 \times 10^{-7}(300/Te)^{0.7}$	$7.0 \times 10^{-7}(300/Te)^{0.7}$	10^{-6}
C	$8.13 \times 10^{-11}(T/300)^{-1.06}e^{-60.8/T}$	$7.0 \times 10^{-7}(300/Te)^{0.7}$	$7.0 \times 10^{-7}(300/Te)^{0.7}$	$7.0 \times 10^{-7}(300/Te)^{0.7}$	10^{-6}
D	$5.42 \times 10^{-11}(T/300)^{-1.06}e^{-60.8/T}$	$5.0 \times 10^{-7}(300/Te)^{0.7}$	$5.0 \times 10^{-7}(300/Te)^{0.7}$	$5.0 \times 10^{-7}(300/Te)^{0.7}$	10^{-6}
E	$5.42 \times 10^{-11}(T/300)^{-1.06}e^{-60.8/T}$	$1.4 \times 10^{-6}(300/Te)^{0.7}$	$1.4 \times 10^{-6}(300/Te)^{0.7}$	$1.4 \times 10^{-6}(300/Te)^{0.7}$	10^{-6}
F	$5.42 \times 10^{-11}(T/300)^{-1.06}e^{-60.8/T}$	$7.0 \times 10^{-7}(300/Te)^{0.7}$	$7.0 \times 10^{-7}(300/Te)^{0.7}$	0	10^{-6}
G	$5.42 \times 10^{-11}(T/300)^{-1.06}e^{-60.8/T}$	$7.0 \times 10^{-7}(300/Te)^{0.7}$	$7.0 \times 10^{-7}(300/Te)^{0.7}$	$7.0 \times 10^{-7}(300/Te)^{0.7}$	10^{-7}

As shown in Fig. 10d, there is a precise balance between production of NH_3 by electron recombination of NH_4^+ and loss due to reaction with protonated ions, especially $HCNH^+$, $C_2H_5^+$, and CH_5^+ (R16–18). The balance reflects the fact that these reactions are not destroying NH_3 but only changing its form from the neutral to the protonated ion and back. True production of NH_3 occurs primarily through R47; thus, NH_3 follows directly from NH_2. Production through this channel is 25 times larger than from NH_3^+ considered in previous models.[16] Photolysis of NH_3 also produces NH_2, but this process is unimportant near 1100 km, although it becomes the dominant loss process at lower altitudes. The most likely fate of NH_2 produced from photolysis below ~800 km, is recombination to N_2H_4, which along with NH_3 and CH_2NH diffuse downward to the stratosphere.

As mentioned previously, several of the rate coefficients involved in these chemical cycles are uncertain and we therefore consider how the results of the numerical model will change for reasonable variations of these parameters. The parameters for these runs are summarized in Table 2 and the results are shown in Fig. 11. The key reaction for production of NH_3 is R47. Our baseline model uses the calculated rate coefficient described in section 4, which corresponds to a value of 7.4 × 10^{-11} cm^3 s^{-1} at 150 K (the approximate temperature of Titan's upper atmosphere). The accuracy of the calculated rate coefficient is expected to be 30%, but to be conservative, we also consider in models B and C the consequences of values 50% smaller and larger than our predicted value for the rate coefficient. The results, shown in Fig. 11b reveal that the calculated $CH_2NH_2^+$ density at 1100 km is 30% smaller in model B and 42% larger in model C. The calculated $CH_2NH_2^+$ density does not change significantly for these variations in k_{47}.

The effect of the $CH_2NH_2^+$ recombination rate coefficient is shown in Fig. 11b. Models D and E shows that scaling the net rate coefficient downward by 30% raises the predicted density at 1100 km by 50%, whereas scaling the rate coefficient upward by a factor of 2 lowers the predicted density at 1100 km by 40%. One might suspect that the model would be most sensitive to the branching ratio for production of HCN (R21), because this channel creates a nitrile, thereby removing the N atom from the imine/amine chemistry. However, the density of $CH_2NH_2^+$ in model F does not differ significantly from Model A. We note that none of these variations in the $CH_2NH_2^+$ recombination rate coefficients has a significant effect on the NH_4^+ densities in the models. We also considered uncertainties in the CH_2NH photolysis rate in model G (not shown in Fig. 11). Decreasing this rate by a factor

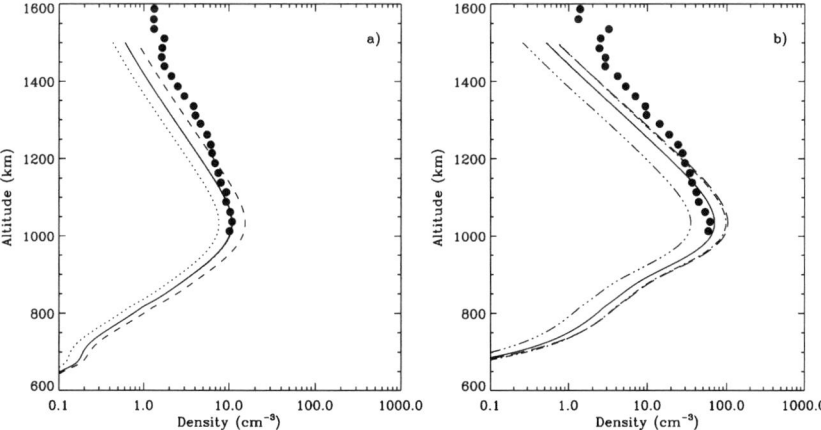

Fig. 11 (a) Calculated NH_4^+ densities for different assumptions about rate coefficients. The data points represent the observations. The solid line represents model A, dotted B, dashed C, dashed-dot D, and dashed-triple dot E. (b) The same as (a), but for $CH_2NH_2^+$.

of 10 caused the CH_2NH density to increase by 20%, which is less than the uncertainty in the data or models. These sensitivity tests support our conclusion that reaction R47 is the primary channel for production of NH_3.

7. Discussion and implications

The distribution of CH_2NH and NH_3 in Titan's upper atmosphere can be understood as the consequence of coupled ion and neutral chemistry. Nitrogen photolysis in Titan's upper atmosphere leads to production of $N(^2D)$ which reacts with CH_4 to either produce CH_2NH directly or produces NH, which reacts with CH_3 to produce CH_2NH. CH_2NH has a large proton affinity, enabling proton-transfer reactions with many species and leading to rapid production of $CH_2NH_2^+$. This ion dissociatively recombines, producing NH_2. Reaction of HCN^+ with C_2H_6 also produces NH_2. Using transition state theory we calculate a rate coefficient for NH_2 + $H_2CN \rightarrow NH_3 + HCN$ of $5.43 \times 10^{-11}(T/300)^{-1.06} \exp(-60.8/T)$. With this coefficient, our photochemical calculations predict densities of NH_4^+ and $CH_2NH_2^+$ in accord with observations.

Much of the nitrogen chemistry used here is based on Lavvas et al.[33] and Lavvas et al.[10] The important improvements include the detailed treatment of ion chemistry and reaction R47. The lack of ion reactions in Lavvas et al.[10] led to an overestimate of the CH_2NH density in those calculations, because of the absence of loss of CH_2NH through proton transfer followed by dissociative recombination. Lavvas et al.[10] speculated that CH_2NH may also be lost through radical–radical reactions (for example $CH_2NH + H_2CN$, leading to polymeric molecular growth). This may well be occurring in Titan's atmosphere but measurements or theoretical calculations of the rate coefficients for these processes are required to quantitatively investigate this possibility. The Lavvas et al.[10] models also under-predicted the density of NH_3. This is also remedied by ion chemistry through production of the NH_2 molecule from $CH_2NH_2^+$ followed by conversion of NH_2 to NH_3. Thus, the ionospheric chemistry results in the conversion of CH_2NH to NH_3, simultaneously solving both problems encountered with the earlier models.

Photochemical models by Krasnopolsky[11] also predicted CH_2NH mole fractions in fairly good agreement with the observations. CH_2NH is produced by our R29 and loss through $CH_2NH + H \rightarrow CH_3 + NH$ in the Krasnopolsky[11] models; however, this latter reaction is endothermic and unlikely to occur at a significant rate in Titan's atmosphere. Loss through proton-transfer followed by dissociative recombination seems more likely, but is not a dominant process in the Krasnopolsky[11] models because a relatively low value (compared to our value) for the electron recombination rate is assumed. The Krasnopolsky[11] model also under predicts the density of NH_3. The NH_2 densities calculated by Krasnopolsky[11] are consistent with those presented here so the lower NH_3 densities are clearly due to the absence of R47 in those models.

Vuitton et al.[12] showed that C_6H_6 in Titan's upper atmosphere was synthesized by a chain of ion–neutral reactions, culminating in dissociative recombination of $C_6H_7^+$, producing C_6H_6. Here, we show that ion chemistry plays a critical role in the chemistry of NH_3 by helping to produce NH_2 from CH_2NH. The Vuitton et al.[12] paper and the investigation described herein serve to emphasize the importance of ion chemistry for the composition of the neutral atmosphere. The existence of high energy photons and electrons in the upper atmosphere results in the opening of chemical pathways that are not possible in Titan's stratosphere, where chemistry is instigated by longer wavelength, less energetic solar radiation. This is clearly seen in our models for C_6H_6, NH_3 and CH_2NH chemistry, but is likely to extend beyond these examples and deserves further, careful investigation.

One of the main results of a study such as this is the identification of laboratory measurements required to improve the photochemical models. The most important deficiencies in laboratory data for the chemistry discussed here are the rate coefficient and products for electron recombination of $CH_2NH_2^+$. In addition, the rate coefficient for $NH_2 + H_2CN \rightarrow NH_3 + HCN$ and the absorption cross section and dissociation products for CH_2NH photodissociation need to be measured.

This research has been supported by the NASA's Planetary Atmospheres Program through grants NNX09AB58G and NNH09AK24I, NASA's exobiology program through grant NNX08AO13G, NASA's Cassini Data Analysis Program through grant NNX08AX62H and NASA Astrobiology Initiative through JPL subcontract 1372177 to the University of Arizona. Computational resources for the kinetics predictions were provided by by the U. S. Department of Energy, Office of Basic Energy Sciences, Division of Chemical Sciences, Geosciences and Biosciences under Contract No. DE-AC02-06CH11357.

References

1 T. E. Cravens, I. P. Robertson, J. Clark, J.-E. Wahlund, J. H. Waite, S. A. Ledvina, H. B. Niemann, R. V. Yelle, W. T. Kasprzak, J. G. Luhmann, R. L. McNutt, W.-H. Ip, V. De La Haye, I. Müller-Wodarg, D. T. Young and A. J. Coates, Titan's ionosphere: Model comparisons with Cassini Ta data, *Geophys. Res. Lett.*, 2005, **32**, L12108–12111.
2 V. Vuitton, R. V. Yelle and V. G. Anicich, The nitrogen chemistry of Titan's upper atmosphere revealed, *Astrophys. J.*, 2006, **647**, L175–L178.
3 V. Vuitton, R. V. Yelle and M. J. McEwan, Ion chemistry and N-containing molecules in Titan's upper atmosphere, *Icarus*, 2007, **191**, 722–742.
4 S. M. Hörst, R. V. Yelle, A. Bauch, N. Carrasco, G. Cernogora, O. Dutuit, E. Quirico, E. Sciamma-O B'rien, M. A. Smith, A. Somogyi, C. Szopa, R. Thissen and V. Vuitton, Formation of prebiotic molecules in a Titan simulation experiment, in preparation.
5 J. L. Fox and R. V. Yelle, Hydrocarbon ions in the ionosphere of Titan, *Geophys. Res. Lett.*, 1997, **24**, 2179–2182.
6 Y. L. Yung, M. Allen and J. P. Pinto, Photochemistry of the atmosphere of Titan – comparison between model and observations, *Astrophys. J. Suppl.*, 1984, **55**, 465–506.
7 D. Toublanc, J. P. Parisot, J. Brillet, D. Gautier, F. Raulin and C. P. McKay, Photochemical modeling of Titan's atmosphere, *Icarus*, 1995, **113**, 2–26.
8 L. M. Lara, E. Lellouch, J. J. López-Moreno and R. Rodrigo, Vertical distribution of Titan's atmospheric neutral constituents, *J. Geophys. Res.*, 1996, **101**, 23261–23283.
9 E. H. Wilson and S. K. Atreya, Current state of modeling the photochemistry of Titan's mutually dependent atmosphere and ionosphere, *J. Geophys. Res.*, 2004, **109**, E06002.
10 P. P. Lavvas, A. Coustenis and I. M. Vardavas, Coupling photochemistry with haze formation in Titan's atmosphere, Part II: Results and validation with Cassini–Huygens data, *Planet. Space Sci.*, 2008, **56**, 67–99.
11 V. A. Krasnopolsky, A photochemical model of Titan's atmosphere and ionosphere, *Icarus*, 2009, **201**, 226–256.
12 V. Vuitton, R. V. Yelle and J. Cui, Formation and distribution of benzene on Titan, *J. Geophys. Res.*, 2008, **113**, E05007.
13 J. Cui, M. Galand, R. V. Yelle, V. Vuitton, J. Wahlund, P. P. Lavvas, I. C. F. Müller-Wodarg, T. E. Cravens, W. T. Kasprzak and J. H. Waite, Diurnal variations of Titan's ionosphere, *J. Geophys. Res.*, 2009, **114**, A06310.
14 J. Cui, M. Galand, R. V. Yelle, J. Wahlund, K. Ångren, J. H. Waite and M. K. Dougherty, Ion transport in Titan's upper atmosphere, *J. Geophys. Res.*, 2010, in press.
15 K. Ågren, J. Wahlund, P. Garnier, R. Modolo, J. Cui, M. Galand and I. Müller-Wodarg, On the ionospheric structure of Titan, *Planet. Space Sci.*, 2009, **57**, 1821–1827.
16 S. K. Atreya, *Atmospheres and Ionospheres of the Outer Planets and their Satellites*, 1986, Springer-Verlag.
17 P. Lavvas, M. Galand, R. V. Yelle, A. N. Heays, B. R. Lewis, G. R. Lewis and A. J. Coates, 2010, Energy deposition and primary chemical products in Titan's upper atmosphere, in preparation.
18 M. T. Nguyen, D. Sengupta and T. K. Ha, Another look at the decomposition of methyl azide and methanimine: how is HCN formed?, *J. Phys. Chem.*, 1996, **100**, 6499–6503.
19 T. H. Dunning, Jr., Gaussian basis sets for use in correlated molecular calculations. I. The atoms boron through neon and hydrogen, *J. Chem. Phys.*, 1989, **90**, 1007–1023.
20 R. A. Kendall, T. H. Dunning, Jr. and R. J. Harrison, Electron affinities of the first-row atoms revisited. Systematic basis sets and wave functions, *J. Chem. Phys.*, 1992, **96**, 6796–6806.

21 S. J. Klippenstein, Variational optimizations in the Rice–Ramsperger–Kassel–Marcus theory calculations for unimolecular dissociations with no reverse barrier, *J. Chem. Phys.*, 1992, **96**, 367–371.
22 Y. Georgievskii and S. J. Klippenstein, Variable reaction coordinate transition state theory: Analytic results and application to the C_2H_3 + H → C_2H_4 reaction, *J. Chem. Phys.*, 2003, **118**, 5442–5455.
23 S. J. Klippenstein, A. L. L. East and W. D. Allen, A high level ab initio map and direct statistical treatment of the fragmentation of singlet ketene, *J. Chem. Phys.*, 1996, **105**, 118–140.
24 S. J. Klippenstein and L. B. Harding, A theoretical study of the kinetics of C_2H_3 + H, *Phys. Chem. Chem. Phys.*, 1999, **1**, 989–997.
25 L. B. Harding, Y. Georgievskii and S. J. Klippenstein, Predictive theory for hydrogen atom–hydrocarbon radical association kinetics, *J. Phys. Chem. A*, 2005, **109**, 4646–4656.
26 S. J. Klippenstein, Y. Georgievskii and L. B. Harding, Predictive theory for the combination kinetics of two alkyl radicals, *Phys. Chem. Chem. Phys.*, 2006, **8**, 1133.
27 P. Celani and H. Werner, Multireference perturbation theory for large restricted and selected active space reference wave functions, *J. Chem. Phys.*, 2000, **112**, 5546–5557.
28 R. D. Amos, A. Bernhardsson, A. Berning, P. Celani, D. L. Cooper, M. J. O. Deegan, A. J. Dobbyn, F. Eckert, C. Hampel, G. Hetzer, P. J. Knowles, T. Korona, R. Lindh, A. W. Lloyd, S. J. McNicholas, F. R. Manby, W. Meyer, M. E. Mura, A. Nicklass, P. Palmieri, R. Pitzer, G. Rauhut, M. Schütz, U. Schumann, H. Stoll, A. J. Stone, R. Tarroni, T. Thorsteinsson and H.-J. Werner, *MOLPRO, a package of ab initio programs designed by H.-J. Werner and P. J. Knowles, Version 2009.1*, 2009.
29 H. Werner and P. J. Knowles, A second order multiconfiguration SCF procedure with optimum convergence, *J. Chem. Phys.*, 1985, **82**, 5053–5063.
30 P. J. Knowles and H. Werner, An efficient second-order MC SCF method for long configuration expansions, *Chem. Phys. Lett.*, 1985, **115**, 259–267.
31 M. J. Frisch, G. W. Trucks, H. B. Schlegel, G. E. Scuseria, M. A. Robb, J. R. Cheeseman, V. G. Zakrzewski, J. A. Montgomery, Jr., R. E. Stratmann, J. C. Burant, S. Dapprich, J. M. Millam, A. D. Daniels, K. N. Kudin, M. C. Strain, O. Farkas, J. Tomasi, V. Barone, M. Cossi, R. Cammi, B. Mennucci, C. Pomelli, C. Adamo, S. Clifford, J. Ochterski, G. A. Petersson, P. Y. Ayala, Q. Cui, K. Morokuma, D. K. Malick, A. D. Rabuck, K. Raghavachari, J. B. Foresman, J. Cioslowski, J. V. Ortiz, A. G. Baboul, B. B. Stefanov, G. Liu, A. Liashenko, P. Piskorz, I. Komaromi, R. Gomperts, R. L. Martin, D. J. Fox, T. Keith, M. A. Al-Laham, C. Y. Peng, A. Nanayakkara, C. Gonzalez, M. Challacombe, P. M. W. Gill, B. G. Johnson, W. Chen, M. W. Wong, J. L. Andres, M. Head-Gordon, E. S. Replogle and J. A. Pople, *GAUSSIAN 98*, Gaussian, Inc., Pittsburgh, PA, 1998.
32 S. M. Hörst, V. Vuitton and R. V. Yelle, Origin of oxygen species in Titan's atmosphere, *J. Geophys. Res.*, 2008, **113**, E10006.
33 P. P. Lavvas, A. Coustenis and I. M. Vardavas, Coupling photochemistry with haze formation in Titan's atmosphere, Part I: Model description, *Planet. Space Sci.*, 2008, **56**, 27–66.
34 E. A. Mason, and T. R. Marrero, The diffusion of atoms and molecules, in *Advances in Atomic and Molecular Physics*, 1970, vol. 6, pp. 155–232.
35 W. J. Massman, A review of the molecular diffusivities of H_2O, CO_2, CO, O_3, SO_2, NH_3, N_2O, NO and NO_2 in Air, O_2, and N_2 near STP, *Atmos. Environ.*, 1998, **32**, 1111–1127.
36 R. V. Yelle, J. Cui and I. C. F. Müller-Wodarg, Methane escape from Titan's atmosphere, *J. Geophys. Res. [Planets]*, 2008, **113**, 10003.
37 I. C. F. Müller-Wodarg, R. V. Yelle, J. Cui and J. H. Waite, Horizontal structures and dynamics of Titan's thermosphere, *J. Geophys. Res.*, 2008, **113**, E10005.
38 J. A. R. Samson, T. Masuoka, P. N. Pareek and G. C. Angel, Total and dissociative photoionization cross sections of N_2 from threshold to 107 eV, *J. Chem. Phys.*, 1987, **86**, 6128–6132.
39 W. C. Stolte, Z. X. He, J. N. Cutler, Y. Lu and J. A. R. Samson, Dissociative photoionization cross sections of N_2 and O_2 from 100 to 800 eV, *At. Data Nucl. Data Tables*, 1998, **69**, 171.
40 A. Teslja, B. Nizamov and P. J. Dagdigian, The electronic spectrum of methyleneimine, *J. Phys. Chem.*, 2004, **108**, 4433–4439.
41 F. Z. Chen, D. L. Judge, C. Y. R. Wu and J. Caldwell, Low and room temperature photoabsorption cross sections of NH_3 in the UV region, *Planet. Space Sci.*, 1999, **47**, 261–266.
42 B. M. Cheng, H. C. Lu, H. K. Chen, M. Bahou, Y. P. Lee, A. M. Mebel, L. C. Lee, M. C. Liang and Y. L. Yung, Absorption cross sections of NH_3, NH_2D, NHD_2, and ND_3 in the spectral range 140–220 nm and implications for planetary isotopic fractionation, *Astrophys. J.*, 2006, **647**, 1535–1542.

43 V. G. Anicich, Evaluated bimolecular ion–molecule gas phase kinetics of positive ions for use in modeling planetary atmospheres, cometary comae, and interstellar clouds, *J. Phys. Chem. Ref. Data*, 1993, **22**, 1469–1569.
44 V. G. Anicich and M. J. McEwan, Ion–molecule chemistry in Titan's ionosphere, *Planet. Space Sci.*, 1997, **45**, 897–921.
45 M. J. McEwan, G. B. I. Scott and V. G. Anicich, Ion–molecule reactions relevant to Titan's ionosphere, *Int. J. Mass Spectrom. Ion Processes*, 1998, **172**, 209–219.
46 S. J. Edwards, C. G. Freeman and M. J. McEwan, The ion chemistry of methylenimine and propionitrile and their relevance to Titan. Inter. J., *Int. J. Mass Spectrom.*, 2008, **272**, 86–90.
47 S. Petrie, C. G. Freeman and M. J. McEwan, The ion-molecule chemistry of acrylonitrile – astrochemical implications, *Mon. Not. R. Astron. Soc.*, 1992, **257**, 438–444.
48 J. Öjekull, P. U. Andersson, M. B. Nagard, J. B. C. Pettersson, A. M. Derkatch, A. Neau, S. Rosén, R. Thomas, M. Larsson, F. Österdahl, J. Semaniak, H. Danared, A. Källberg, M. a. Ugglas and N. Markovicć, Dissociative recombination of NH_4^+ and ND_4^+ ions: Storage ring experiments and ab initio molecular dynamics, *J. Chem. Phys.*, 2004, **120**, 7391–7399.
49 S. Dobe, C. Oehlers, F. Temps, H. G. Wagner and H. Ziemer, Observations of an H/D-isotope exchange channel in the reaction D + H_2CO, *Ber. Bunsen-Ges. Phys. Chem.*, 1994, **98**, 754–757.
50 G. Marston, F. L. Nesbitt and L. J. Stief, Branching ratios in the N + CH_3 reaction – Formation of the methylene amidogen (H_2CN) radical, *J. Chem. Phys.*, 1989, **91**, 3483–3491.
51 F. L. Nesbitt, G. Marston and L. J. Stief, Kinetic studies of the reactions of H_2CN and D_2CN radicals with N and H, *J. Phys. Chem.*, 1990, **94**, 4946–4951.
52 H. Okabe, *Photochemistry of Small Molecules*, 1978, John Wiley and Sons Inc., New York.
53 J Herron, Evaluated chemical kinetics data for reactions of $N(^2D)$ $N(^2P)$, and $N_2(A^3\Sigma^+_u)$ in the gas phase, *J. Phys. Chem. Ref. Data*, 1999, **28**, 1453.
54 H. Umemoto, T. Nakae, H. Hashimoto, K. Kongo and M. Kawasaki, Reactions of $N(^2D)$ with methane and deuterated methanes, *J. Chem. Phys.*, 1998, **109**, 5844–5848.
55 N. Balucani, O. Asvany, Y. Osamura, L. C. L. Huang, Y. T. Lee and R. I. Kaiser, Laboratory investigation on the formation of unsaturated nitriles in Titan's atmosphere, *Planet. Space Sci.*, 2000, **48**, 447–462.
56 L. Adam, W. Hack, H. Zhu, Z.-W. Qu and R. Schinke, Experimental and theoretical investigation of the reaction $NH(X^3\Sigma^-)$ + $H(^2S)$ → $N(^4S)$ + $H_2(X^1\Sigma_g^+)$, *J. Chem. Phys.*, 2005, **122**(11), 114301.
57 C. Mullen and M. A. Smith, Low temperature $NH(X^3\Sigma^-)$ radical reactions with NO, saturated, and unsaturated hydrocarbons studied in a pulsed supersonic laval nozzle flow reactor between 53 and 188 K, *J. Phys. Chem.*, 2005, **109**, 1391–1399.
58 W. Hack, H. Wagner and A. Zaspypkin, Elementary reactions of $NH(^1\Delta)$ and $NH(X^3\Sigma)$ with N, O and NO, *Bunsen-Ges. Phys. Chem.*, 1994, **98**, 156–164.
59 Z.-F. Xu, F. D.-C. and F.X.-Y, Ab initio study on the reaction $2NH(X^3\Sigma^-)$ → $NH_2(X^2B_1)$ + $N(^4S)$, *Chem. Phys. Lett.*, 1997, **275**(3–4), 386–391.
60 P. Dransfeld and H. G. Wagner, Investigation of the gas phase reaction N + NH_2 → N_2 + 2H, *Z. Phys. Chem., Neue Folge*, 1987, **153**, 89–97.
61 K. Fagerstrom, J. T. Jodkowski and E. Ratajczak, Kinetics of the self-reaction and the reaction with OH of the amidogen radical, *Chem. Phys. Lett.*, 1995, **236**, 103–110.
62 M. Liang, A. N. Heays, B. R. Lewis, S. T. Gibson and Y. L. Yung, Source of nitrogen isotope anomaly in HCN in the atmosphere of titan, *Astrophys. J.*, 2007, **664**, L115–L118.
63 B. R. Lewis, S. T. Gibson, J. P. Sprengers, W. Ubachs, A. Johansson and C. Wahlström, Lifetime and predissociation yield of $^{14}N_2$ $b^1\Pi_u$ (v = 1) revisited: Effects of rotation, *J. Chem. Phys.*, 2005, **123**(23), 236101.
64 J. Cui, R. V. Yelle, V. Vuitton, J. H. Waite, W. T. Kasprzak, D. A. Gell, H. B. Niemann, I. C. F. Müller-Wodarg, N. Borggren, G. G. Fletcher, E. L. Patrick, E. Raaen and B. A. Magee, Analysis of Titan's neutral upper atmosphere from Cassini ion neutral mass spectrometer measurements, *Icarus*, 2009, **200**, 581–615.
65 S. Petrie, G. Javahery and D. K. Bohme, Gas-phase reactions of benzenoid hydrocarbon ions with hydrogen atoms and molecules: uncommon constraints to reactivity, *J. Am. Chem. Soc.*, 1992, **114**, 9205.

PAPER

Mapping Titan's HCN in the far infra-red: implications for photochemistry

N. A. Teanby,[*a] P. G. J. Irwin,[b] R. de Kok[c] and C. A. Nixon[d]

Received 27th January 2010, Accepted 6th April 2010
DOI: 10.1039/c001690j

Observations of Titan's far infra-red spectra by the Cassini orbiter's Composite InfraRed Spectrometer have been used to determine the latitude distribution of HCN at 1 mbar by fitting the HCN and CO rotational lines in the 18–60 cm^{-1} (160–550 µm) spectral range. Results confirm the north polar HCN enrichment previously observed using mid-IR data and support the conclusion that Titan's nitrile species are significantly more enriched than hydrocarbons species with similar predicted photochemical lifetimes. This suggests Titan's photochemical cycle includes an additional sink for nitrogen bearing species. The abundance of CO was also determined, and had a mean value of 55 ± 6 ppm at 20 mbar. However, it was not possible to reliably determine the CO latitude variation due to unconstrained temperatures in the north polar lower stratosphere.

1 Introduction

Titan, Saturn's largest moon, is unusual because it has a thick nitrogen and methane atmosphere (1.5 bar surface pressure). Interaction with solar UV photons and magnetospheric electrons in Titan's upper atmosphere creates nitrogen and methane radicals, which form the basis of a rich photochemical cycle—producing a vast array of hydrocarbons and nitriles.[1–3] These photochemical processes are often cited as analogues of processes occurring in the primordial terrestrial atmosphere. Atmospheric trace gases produced by Titan's photochemical cycle have a large range of lifetimes from days to thousands of years. As they descend from the upper atmosphere source region (∼1000 km) to the cold (70 K) tropopause (∼40 km) most species are removed by condensation and photodissociation. Under photochemical equilibrium conditions the high source–low sink configuration gives rise to gas profiles with a positive vertical gradient (*i.e.* mixing ratios increase with altitude), which has been confirmed for many species by both high resolution ground based measurements[4–6] and limb sounding by the Cassini spacecraft.[7–9]

Titan's seasonally variable meridional circulation can modify these basic photochemical profiles. For example, descending cold air at the winter pole advects enriched air from the upper atmosphere production zone, which causes enrichment of most species in the stratosphere, so measurements of gas abundances can be used to diagnose atmospheric circulation. Mid-IR Cassini measurements have recently indicated an inverse relationship between photochemical lifetime and polar enrichment.[10] This is because shorter lifetime species have steeper vertical gradients

[a]*Atmospheric, Oceanic & Planetary Physics, Department of Physics, University of Oxford, Clarendon Laboratory, Parks Road, Oxford, OX1 3PU, UK. E-mail: teanby@atm.ox.ac.uk*
[b]*Atmospheric, Oceanic & Planetary Physics, Department of Physics, University of Oxford, Clarendon Laboratory, Parks Road, Oxford, OX1 3PU, UK. E-mail: irwin@atm.ox.ac.uk*
[c]*SRON, Sorbonnelaan 2, 3584 CA Utrecht, Netherlands. E-mail: R.J.de.Kok@sron.nl*
[d]*Department of Astronomy, University of Maryland, College Park, MD, 20742, USA. E-mail: Conor.A.Nixon@nasa.gov*

and are thus affected more by any vertical motion. Therefore, the observed relative polar enrichment can be used to indirectly determine the atmospheric lifetime of each species and constrain photochemical processes.

The latitudinal variation of trace species for Titan's northern winter season has been extensively studied using mid-IR data from Cassini's Composite InfraRed Spectrometer (CIRS).[10–16] All studies show an enrichment of trace compounds at northern latitudes. However, interpretation of the observed enrichment is complicated by a hot stratopause at Titan's northern winter pole,[17] which causes the previously studied mid-IR data to sound very different atmospheric levels at equator and pole—about 3 and 0.1 mbar respectively[15]—it is difficult to sound much below 1 mbar in polar regions with the mid-IR.

The descending motion of polar air mitigates this problem somewhat, as it tends to produce profiles that are more uniform with altitude, which can be used to assert the assumption that mean mixing ratios at 0.1 mbar are approximately comparable to those at 3 mbar near the north pole. However, this is a large assumption and requires testing. The far-IR is not sensitive to the hot stratopause and sounds a relatively constant atmospheric level (2–10 mbar) at all latitudes. This makes it possible to compare more directly the chemistry of polar and equatorial regions.

Recently Teanby et al.[18] used far-IR CIRS data to determine the relative polar enrichment of two hydrocarbons (C_3H_4 and C_4H_2) and a minor nitrile species (C_2N_2). The hydrocarbon results were consistent with results from mid-IR data, but C_2N_2 has no mid-IR feature for comparison. Hence, in this study we investigate hydrogen cyanide (HCN), which is the major nitrile species on Titan and has prominent spectral emission lines in both the mid- and far-IR. The aim is to validate the mid-IR relative polar enrichment ratios, which have important implications for the overall lifetimes of Titan's trace gas inventory.

2 Cassini CIRS observations

We use spectral observation sequences collected by the Cassini spacecraft's Composite Infra-Red Spectrometer (CIRS)[19–21] instrument over a four year period. CIRS is a Fourier transform spectrometer covering the wavelength range 10–1500 cm^{-1} (7–1000 μm) with an apodised full-width half-maximum (FWHM) spectral resolution adjustable between 0.5 and 15 cm^{-1}. The full spectral range is covered by three separate focal planes (FP): FP1, far-IR 10–600 cm^{-1}; FP3, mid-IR 600–1100 cm^{-1}; and FP4, mid-IR 1100–1500 cm^{-1}. The far-IR FP1 is a single detector with an angular field of view (FOV) of 2.5 mrad FWHM, whereas the mid-IR FP3 and FP4 comprise linear arrays of 10 square pixels, each with a single pixel FOV of 0.27 mrad FWHM. The spectral range of CIRS covers the emission peaks of many IR-active trace species found in Titan's atmosphere, which can be used to accurately determine stratospheric composition and temperature.

Observation sequences used in this paper are specifically designed for far-IR nadir (downward looking) composition measurements, where the far-IR FP1 FOV is positioned over a single latitude and longitude for around 300 min, which results in ∼300 spectra with the same viewing geometry. Spectra were measured at CIRS' highest spectral resolution (0.5 cm^{-1}) to allow discrimination of emission features from different gases. Cassini–Titan distances for these observations were typically 250 000 km, which gave a projected field of view which covered 20 degrees of latitude on Titan. Observations were typically positioned away from the disc centre with emission angles of around 45° to give increased path length through the atmosphere, which increased sensitivity to trace species. Observations used in this study are summarised in Table 1.

Constant viewing geometry throughout each observation allowed the individual spectra to be averaged together to increase the signal-to-noise ratio. Depending on the orientation of the Cassini orbiter the mid-IR FP3/4 focal planes sometimes sampled the same latitude as FP1. Where this was the case we averaged the FP3/4

Table 1 List of observations used in this study. Columns are: Dur, observation dwell time; Rev, Cassini orbit number; θ and ϕ, latitude and longitude in tilted frame; e, emission angle; N, number of spectra in average; and FOV, great circle FOV diameter. Note that observations taken between orbits 43 and 93 suffered from electrical interference in the spectral region around the HCN/CO rotational lines and so were not used

Obs.	Date	Time	Dur./min	Rev.	θ (°N)	ϕ (°W)	e (°)	N	FOV (°)
F1	31/Mar/2005	07:35:16	240	005	−39.3	15.2	42.8	241	21.1
F2	01/Apr/2005	04:05:16	240	005	45.3	−150.3	52.2	240	22.6
F3	16/Apr/2005	07:16:46	205	006	52.1	15.6	51.9	178	23.5
F4	12/Dec/2004	23:38:13	240	00B	15.4	118.6	46.9	223	21.4
F5	21/Aug/2005	22:05:37	198	013	29.2	−27.6	52.2	191	20.6
F6	22/Aug/2005	16:23:37	270	013	−51.6	−140.8	45.6	247	20.5
F7	28/Oct/2005	13:15:25	180	017	19.1	32.5	50.1	119	23.2
F8	26/Dec/2005	07:49:30	130	019	0.2	60.0	50.6	107	21.9
F9	14/Jan/2006	23:41:27	120	020	19.1	−172.0	46.2	104	22.3
F10	15/Jan/2006	19:41:27	380	020	10.7	−27.4	24.4	268	17.7
F11	27/Feb/2006	16:55:19	280	021	−29.0	172.2	49.5	213	23.6
F12	26/Feb/2006	09:55:19	750	021	26.2	2.7	28.1	216	26.0
F13	18/Mar/2006	10:05:57	420	022	−0.1	−171.8	42.2	401	21.0
F14	19/Mar/2006	12:25:57	101	022	24.0	−42.2	47.5	83	28.0
F15	01/May/2006	07:28:15	247	023	−33.8	−150.4	36.3	213	23.9
F16	19/May/2006	20:48:12	390	024	−14.6	125.8	26.8	350	21.2
F17	02/Jul/2006	18:20:47	330	025	23.8	−121.7	40.0	298	22.7
F18	01/Jul/2006	19:50:47	210	025	38.2	19.2	44.6	185	23.3
F19	01/Jul/2006	18:15:47	95	025	12.2	19.0	25.1	94	25.2
F20	21/Jul/2006	00:36:26	889	026	11.7	161.3	22.4	517	25.7
F21	07/Sep/2006	06:16:51	360	028	29.0	144.6	21.5	350	18.7
F22	23/Sep/2006	04:58:49	330	029	10.1	96.7	42.2	312	21.7
F23	10/Oct/2006	03:30:07	351	030	−58.3	−57.2	40.8	339	24.9
F24	09/Oct/2006	03:30:07	300	030	33.7	115.4	22.6	285	20.2
F25	11/Dec/2006	21:11:31	330	035	61.9	129.0	33.4	305	21.6
F26	12/Dec/2006	22:09:31	180	035	−72.1	−53.7	40.2	156	23.1
F27	28/Dec/2006	18:35:22	150	036	−87.2	−42.0	47.2	136	19.9
F28	27/Dec/2006	20:05:22	330	036	77.1	151.7	43.1	319	21.9
F29	12/Jan/2007	19:38:31	180	037	75.2	−163.7	46.1	157	23.2
F30	13/Jan/2007	17:38:31	120	037	−71.6	−141.3	48.6	103	20.9
F31	29/Jan/2007	16:15:55	300	038	−41.2	−102.7	40.9	249	21.6
F32	09/Mar/2007	12:49:00	180	040	−49.8	78.2	28.0	158	21.6
F33	10/Mar/2007	10:49:00	120	040	87.4	−110.4	45.8	106	21.4
F34	10/Mar/2007	14:49:00	134	040	72.8	−83.4	36.8	142	28.1
F35	26/Mar/2007	09:23:27	120	041	61.4	155.5	52.4	102	23.1
F36	26/Mar/2007	13:23:27	495	041	72.2	−95.9	41.8	129	31.7
F37	10/Apr/2007	07:58:00	120	042	−60.0	31.1	26.7	102	26.0
F38	11/Apr/2007	07:58:00	300	042	70.1	−174.8	48.5	272	25.1
F39	26/Apr/2007	06:46:58	286	043	−49.2	29.4	23.3	263	22.6
F40	27/Apr/2007	06:32:58	120	043	72.3	−139.8	53.1	104	23.9
F41	20/Nov/2008	01:56:28	180	093	43.4	−106.7	13.7	160	20.4
F42	18/Nov/2008	19:19:33	457	093	−5.2	45.2	30.9	191	32.5
F43	05/Dec/2008	01:25:45	240	095	−14.9	67.8	23.1	213	20.2
F44	20/Dec/2008	23:59:52	240	097	−12.3	106.5	41.0	231	22.3
F45	21/Dec/2008	21:59:52	240	098	24.6	−75.2	21.9	134	19.7
F46	26/Mar/2009	16:43:36	180	106	−61.0	146.2	41.1	163	21.0
F47	03/Apr/2009	12:47:47	240	107	41.2	82.3	19.0	263	18.7
F48	27/Mar/2009	14:43:36	180	107	35.4	−143.3	50.1	164	26.7
F49	06/May/2009	07:54:16	300	110	−69.7	−166.3	48.3	282	24.0
F50	22/May/2009	06:26:41	180	111	−28.5	1.8	45.0	165	21.8

Table 1 (Contd.)

Obs.	Date	Time	Dur./min	Rev.	θ (°N)	φ (°W)	e (°)	N	FOV (°)
F51	06/Jun/2009	06 : 07 : 49	232	112	49.2	60.4	37.5	218	22.0
F52	07/Jun/2009	05 : 00 : 01	300	112	−59.3	−105.4	46.8	274	23.9
F53	09/Jul/2009	02 : 04 : 03	180	114	−68.2	−20.4	43.4	163	20.5
F54	09/Jul/2009	07 : 04 : 03	540	114	−52.6	−14.6	28.3	108	35.0
F55	24/Jul/2009	02 : 34 : 03	180	115	49.1	101.6	34.9	146	21.7

spectra in a 10° latitude bin centred on the central latitude of the FP1 data. These mid-IR data provide a useful comparison to the far-IR results. Prior to averaging, the latitudes and longitudes of each measured spectra were corrected for line-of-sight effects[15] to be appropriate for the 100 km (FP1) and 150 km (FP3/4) atmospheric levels, which are the approximate altitudes sounded by the data.

3 Composition determination method

HCN abundance was determined using the far-IR rotational lines in the 18–60 cm^{-1} spectral range. The continuum in this region of Titan's spectrum is affected mainly by the tropospheric temperature, nitrogen–nitrogen collision induced absorption (CIA), and to a lesser extent stratospheric haze. The rotational lines of HCN and CO are superimposed on this continuum and overlap significantly. An example of Titan's spectrum in this spectral region is shown in Fig. 1.

The radiative transfer modelling used N_2–N_2 CIA coefficients from Borysow and Frommhold.[22] Spectral linedata for HCN is from HITRAN2004[23] but the linewidths have been multiplied by 1.106 to convert air broadened widths into nitrogen broadened widths after Rinsland et al.[24] Linedata for CO is also from HITRAN2004. The spectral dependence of Titan's stratospheric haze is calculated from Mie theory assuming 0.2 μm radius particles and refractive indices measured by Khare et al.[25] from laboratory tholins.

Measured spectra were fitted using the NEMESIS radiative transfer tool.[26] NEMESIS uses a non-linear iterative constrained retrieval scheme, with gaseous opacity calculated using the correlated-k approximation.[27] NEMESIS starts with an initial guess, or a priori estimate, of the atmospheric temperature and composition and iteratively adjusts this until an optimal fit to the data is obtained. The final retrieved atmospheric state gives the best compromise between fitting the data and remaining close to the initial a priori constraints, which helps prevent unphysical solutions being obtained when retrieving continuous atmospheric profiles.

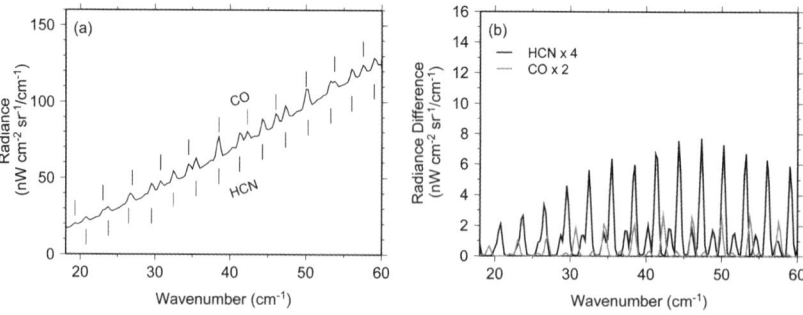

Fig. 1 (a) Example of Titan's far-IR spectrum showing the rotational lines of HCN and CO. (b) Radiance difference caused by increasing the HCN and CO abundances by factors of 4 and 2 respectively.

The *a priori* conditions required for this analysis were: temperature profiles as a function of latitude, a haze abundance profile, and HCN/CO volume mixing ratio (VMR) profiles. The tropospheric temperature is predicted to have a very weak latitude dependence so we assumed the Huygens probe descent profile in the troposphere for pressures greater than 56 mbar.[28] However, stratospheric temperature varies widely with latitude, so for pressures less than 10 mbar we assumed the latitude-dependent temperature structure of Achterberg *et al.*[17] derived from CIRS limb sounding. Between 10 and 56 mbar, temperature is expected to vary with latitude but limited constraints exist. Therefore, we used a linear interpolation in log pressure between these pressures. Typical equatorial and north polar temperature profiles are shown in Fig. 2a. The haze abundance profile is that used by Teanby *et al.*[18] and had a scale height of 65 km consistent with Huygens[29] and CIRS[30] results. HCN is known to have a vertical gradient at equatorial and southern latitudes from Cassini CIRS limb sounding.[7-9] Therefore we use a simplified version of the vertically varying profile derived at 15°S by Teanby *et al.*[7] At high northern latitudes a uniform profile is more appropriate.[8] However, high vertical resolution limb measurements display complex layered structures,[7,31] but such fine layering is not resolvable with nadir data and a uniform VMR profile provides a better approximation than a vertical gradient. Increasing HCN abundance and decreasing stratospheric temperature in the north cause HCN to exceed its saturation vapour pressure (SVP) at lower pressures (higher altitudes). When this occurs the VMR is set to that predicted by the SVP law in Lara *et al.*[32] (HCN SVP in atmospheres is given by $P = \exp(19.63 - 533.7/T)$). Typical HCN VMR profiles are shown in Fig. 2b. CO does not condense in Titan's atmosphere, so we used a uniform profile with an *a priori* VMR of 47 ± 8 ppm from de Kok *et al.*[14]

The strength of HCN and CO spectral emission peaks are determined by a combination of gas abundance and temperature in the line-forming region of the atmosphere. For accurate composition determinations it is essential that the temperature of the line-forming region is well constrained. Therefore, we now consider the information content of our observations at different pressure levels for both composition and temperature.

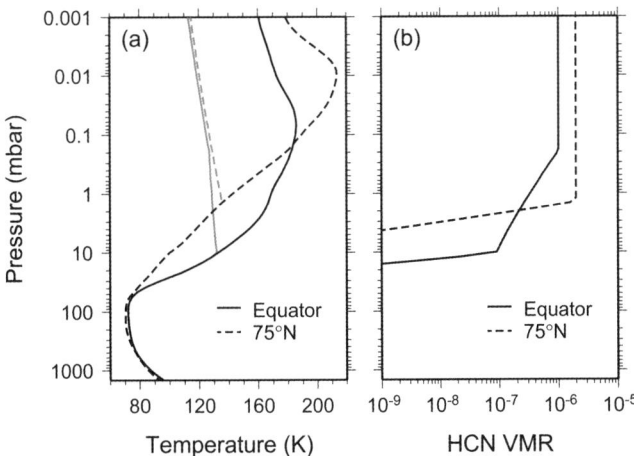

Fig. 2 (a) Typical temperature profiles at the equator and 75°N. Note the hot elevated stratopause (0.01 mbar, 210 K) at 75°N, which is typical of northern latitudes. The grey lines show the temperature at which the HCN volume mixing ratio (VMR) reaches saturation. (b) Typical HCN profiles used to fit the data at equatorial and polar regions. At the equator previous limb studies show a vertical gradient, while in polar regions uniform profiles are more appropriate. Note the saturation level varies between 10 and 2 mbar with latitude.

Fig. 3 shows the contribution functions for temperature and composition for the model atmospheres shown in Fig. 2 for both far- and mid-IR spectral ranges. The temperature of the stratosphere is derived primarily from the ν_4 methane band in FP4 (1240–1360 cm^{-1}), which gives temperature information in the pressure range 10–0.1 mbar at the equator and 1–0.01 mbar near the north pole (due to cold lower stratosphere temperatures and a hot elevated stratopause). The continuum level in the far-IR also contains information about temperature around the tropopause (100 mbar). In between these regions the temperature information is very limited and not well constrained.

HCN condensation in the lower stratosphere means that the HCN line-forming region overlaps with pressure levels where we have good temperature control derived from FP4. This also lies well above the tropopause region that is responsible for the far-IR continuum. The far-IR HCN contribution functions also cover a very narrow pressure range, unlike those for the mid-IR which have much broader peaks and can become double peaked in the north. The far-IR spectral region thus provides stringent constraints on the HCN abundance over a range of pressures from 2–10 mbar.

CO, on the other hand, has its line-forming region around 10–50 mbar, where we have very poor temperature control. Around the equator we can use the Huygens temperature profile for constraint, but toward the north pole, temperature control in this pressure range becomes very poor. Therefore, the far-IR rotational lines of CO only provide moderate constraint on the CO abundance.

Inspection of the CIRS far-IR data also reveals a problem with the continuum in this spectral region. Fig. 4 shows the brightness temperature of the three observations that were closest to the Huygens probe descent latitude (12°S), for which we have excellent temperature information throughout the lower stratosphere and troposphere.[28] The minimum atmospheric temperature recorded by Huygens was 70.3 K at the tropopause. The atmosphere (and surface) is hotter than this at all other pressure levels. Therefore, the minimum possible brightness temperature of the measured spectra should be 70.3 K. However, below 30 cm^{-1} many of the data points lie well below this minimum temperature, by as much as 10–20 K. This indicates that the continuum level below 30 cm^{-1} is affected by baseline calibration issues and it is not possible to fit the continuum using physically sensible temperature and haze profiles. Reasonable perturbations to the low temperature N$_2$–N$_2$ CIA coefficients (by up to ± 50%) also do not improve the fit. Fortunately,

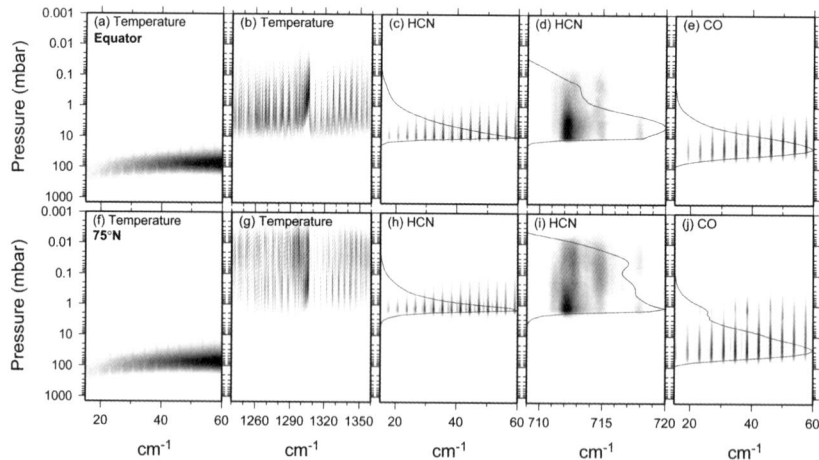

Fig. 3 Contribution functions as a function of pressure and wavenumber using the atmospheric profiles in Fig. 2 for the equator (a–e) and 75°N (f–j). The thin lines in panels (c–e) and (h–j) show the normalised average contribution function obtained by summing contributions over all wavenumbers for each pressure level.

this continuum shift does not affect the HCN emission lines as the HCN line-forming region lies well above the tropopause. However, the CO contribution functions overlap with this region and will be affected.

Considering the limitations of the data at low wavenumbers (< 30 cm^{-1}) we adopted a three stage approach to fitting the data and determining HCN and CO composition from each of the averaged observation sequences:

Stage 1: If coincident FP4 data were available, the 1240–1360 cm^{-1} spectral region was used to update the *a priori* temperature profile using the method described in Teanby *et al.*,[15] which resulted in small temperature changes of less than 1 K. The 18–60 cm^{-1} far-IR region of the far-IR FP1 spectrum was selected and the HCN and CO rotational lines masked out. Tropospheric temperature and haze were then adjusted to give the best fit to the remaining continuum datapoints.

Stage 2: Typically the continuum data points were up to 4 nW cm^{-2} sr^{-1}/cm^{-1} below the fit at low wavenumbers. It was not possible to fit these points using

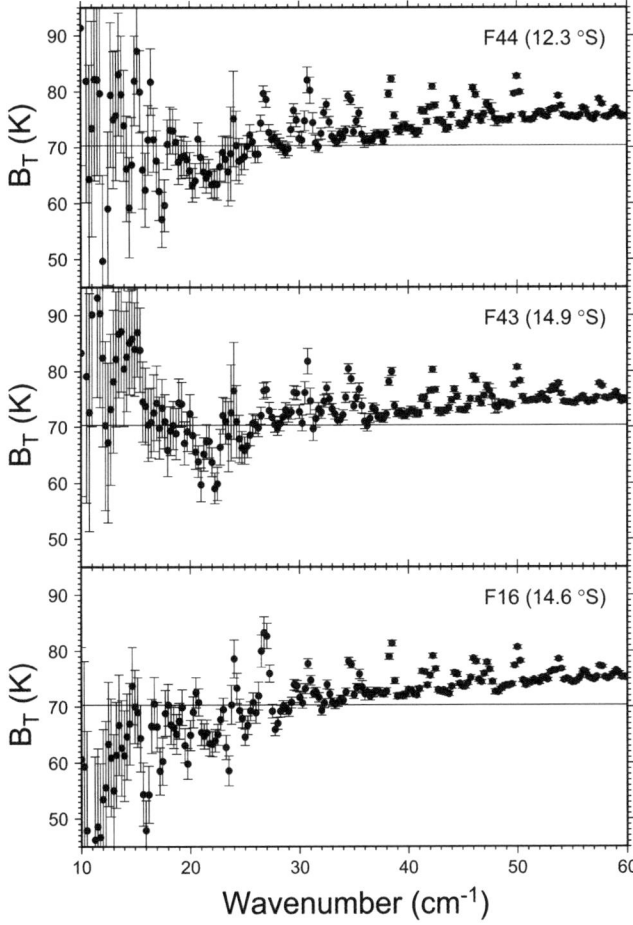

Fig. 4 Equivalent brightness temperature (B_T) of the three observations closest to the latitude of the Huygens probe descent profile (12°S). The horizontal line at 70.3 K corresponds to the coldest temperature measured by Huygens (at the tropopause). Therefore, all brightness temperatures should be at least 70.3 K. However, below 30 cm^{-1} many brightness temperatures are significantly (10–20 K) below this level, which indicates problems with the continuum level in the CIRS data in this wavelength region.

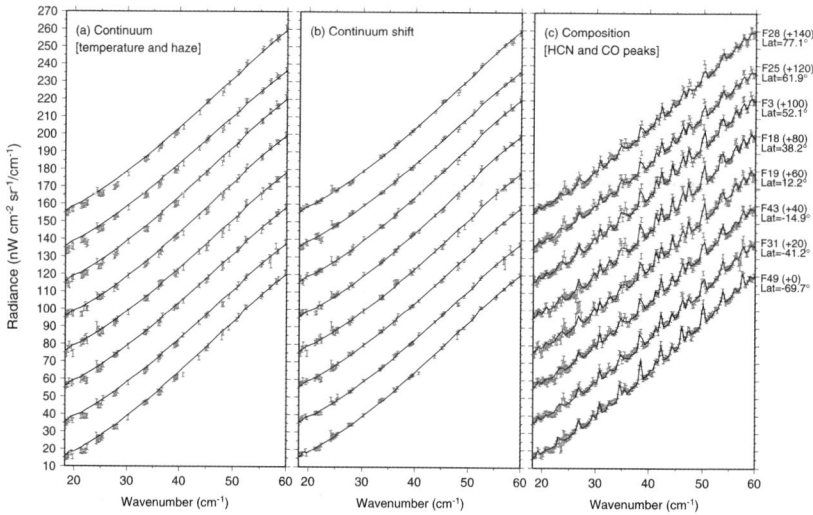

Fig. 5 Example fits to the far-IR CIRS data at eight representative latitudes. The retrieval method has three stages: (a) The HCN/CO rotational lines are masked out and the continuum is fitted by adjusting tropospheric temperature and haze. Note that the data continuum level is up to 4 nW cm^{-2} sr^{-1}/cm^{-1} too low at low wavenumbers, which is most probably due to a calibration effect (Fig. 4). (b) The continuum level is adjusted by adding a smooth fit to the residuals in (a) to the data, which is required to prevent underfitting of the HCN/CO emission peaks at low wavenumbers. (c) The VMR profiles for CO and HCN are scaled to give the best overall fit to the data. For plotting purposes data have been offset vertically by the amount given in brackets.

physically realistic profiles and the discrepancy is assumed to be due to calibration effects. To prevent under fitting of the HCN and CO lines, a cubic spline curve with a knot spacing of 15 cm^{-1} was fitted to the residuals from stage 1 using the method of Teanby[33] and the resulting smooth variation was used to correct the continuum level.

Stage 3: Temperature and haze profiles were fixed to those derived in stage 1 and the HCN and CO profiles were scaled to give the best fit to the corrected measured spectra. This stage was repeated using both uniform and variable HCN profiles. The maximum possible volume mixing ratio of HCN at all pressure levels was not allowed to exceed the saturation vapour pressure. If any coincident mid-IR FP3 data covering the same latitude was available, these data were also fitted using the method in Teanby et al.,[15] for comparison with the far-IR results. Fig. 5 show example fits to the data at each of the three stages.

4 HCN and CO abundance results

Fig. 6 shows the latitude variation of HCN at 1 mbar derived by fitting the far-IR CIRS spectra assuming both variable and uniform with altitude VMR profiles. Also shown is a re-analysis of mid-IR mapping observations from Teanby et al.[10] obtained on Cassini's 36th and 37th orbit for comparison. All latitudes have been projected into the tilted reference frame of Achterberg et al.,[34] which defines the symmetry axis of Titan's stratosphere during the current season and is offset from the solid body rotation axis by 4.1°. This projection is required or scatter in the HCN determinations is increased significantly.[35]

Scaling the variable HCN profile derived from equatorial limb data gives good agreement between mid- and far-IR results for latitudes south of 45°N. However, within the north polar vortex the agreement is poor indicating that such a vertical gradient is not appropriate there.

Scaling a uniform HCN profile gives good agreement between mid- and far-IR results within the north polar vortex, indicating that such a profile is more applicable at high northern latitudes.

Both types of profile result in poor mid/far-IR agreement on the vortex boundary. However, this region is known to exhibit large composition gradients[9] and complex layered structures.[31]

The variation of CO with latitude is shown in Fig. 7. At southern and tropical latitudes a mean mixing ratio of 55 ± 6 ppm gives good agreement with the observations. These measurements are most appropriate for the 20 mbar level, which is the peak of the contribution functions in Fig. 3. The apparent increase in CO in the north is most probably an artifact due to poor temperature control in the lower stratosphere and we do not regard this feature as reliable.

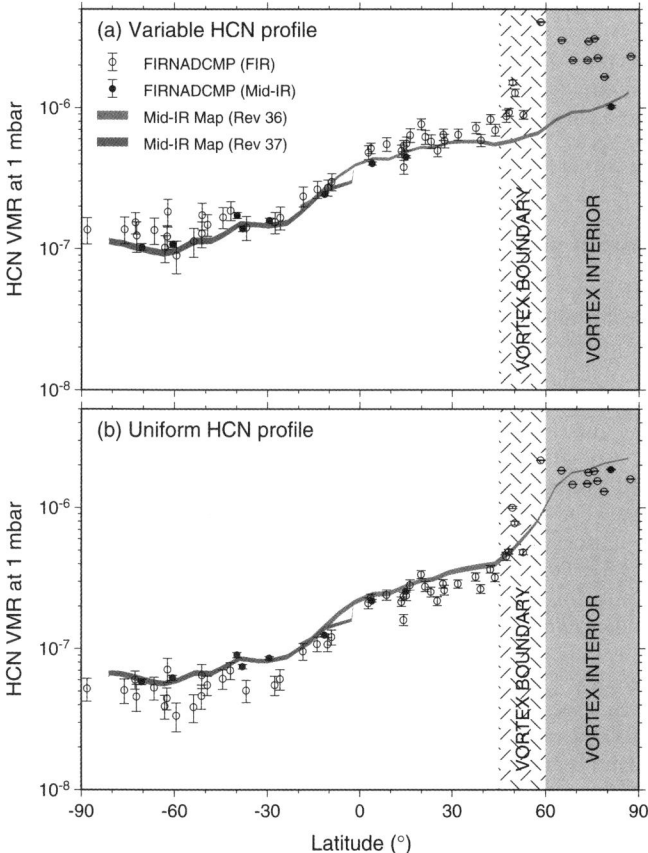

Fig. 6 HCN VMR at 1 mbar obtained assuming a vertical gradient (a) and uniform (b) VMR profiles. Open circles show VMRs derived from the far-IR rotational lines, solid points are from overlapping mid-IR data taken during the same observation (if available), and shaded curves show re-analysed mid-IR data taken from two mapping observations previously analysed by Teanby et al.[10] for comparison. The variable profile is most appropriate for tropical and southern latitudes, which is evident from the good agreement between mid- and far-IR results south of 45°N in (a). Within the north polar vortex a variable profile is not applicable and leads to large disagreements. However, (b) shows that a uniform profile gives very good agreement between the mid- and far-IR determinations within the vortex. Therefore, over the range of altitudes covered by CIRS nadir data the assumed profile should be variable south of 45°N and uniform north of 60°N.

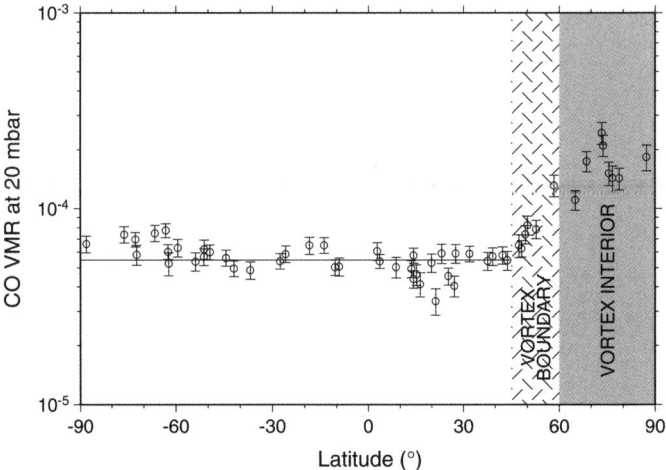

Fig. 7 CO VMR at 20 mbar obtained assuming a uniform VMR profile. An average mixing ratio of 55 ± 6 ppm (horizontal line) fits the equatorial and southern hemisphere data well. North of the vortex boundary the lower stratosphere temperature is not well known so CO VMRs there are not reliable.

5 Discussion and implications for photochemistry

Cassini CIRS far-IR spectra, combined with reasonable assumptions about the HCN vertical profile based on previous limb data studies, can be used to determine the relative polar enrichment of HCN at a constant 1 mbar pressure level. The advantages of using far-IR data compared to higher signal-to-noise and spatial resolution mid-IR measurements is that the far-IR probes a single well defined atmospheric level and there are many more HCN emission features.

The best estimate of HCN abundance at 1 mbar at the equator is 450 ± 45 ppb and is found assuming a HCN profile with a vertical gradient derived from equatorial limb data. The best estimate of HCN at 70°N at 1 mbar is 1600 ± 200 ppb, which assumes a uniform abundance profile. This gives a relative polar enrichment at 1 mbar of 3.6 ± 0.6.

Importantly, this can be compared with and used to validate results from previous mid-IR studies, which sound a more extended pressure range and suffer double peaked contribution functions in polar regions. Fig. 8 compares predicted photochemical lifetimes from Wilson and Atreya[1] to the relative polar enrichment of Titan's trace species determined from previous nadir mid-IR[10,16] and far-IR[18] studies. This type of plot is important when considering the overall lifetimes of trace species in Titan's atmosphere. A greater relative polar enrichment implies a steeper gradient in the stratosphere and mesosphere—assuming the polar enrichment is caused by descending polar air. Steep gradients are characteristic of short lifetime species, as a consequence of competition between the vertical mixing and removal timescales as they are transferred from the high altitude production zone to the lower atmosphere condensation sink. Therefore, large enrichment ratios should correspond to short lifetimes and any deviation from this indicates additional processes are at work.

Interestingly the far-IR HCN polar enrichment ratio found here agrees with those derived from mid-IR data. The enrichments of two hydrocarbon species (C_3H_4 and C_4H_2) determined from far-IR data by Teanby et al.[18] are also in agreement with the mid-IR results. This agreement can be used as an argument for the validity of the mid-IR relative polar enrichment ratios, which cover more trace gas species, despite the difference in north polar mid- and far-IR contribution functions. This suggests

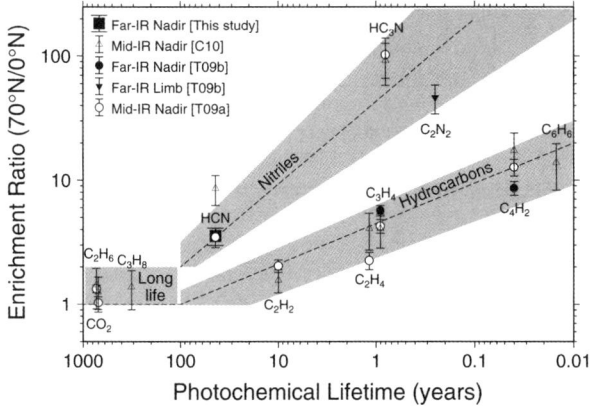

Fig. 8 Summary plot of photochemical lifetime *versus* relative polar enrichment determined from this and other recent composition studies with Cassini CIRS. Photochemical lifetimes are at 300 km altitude and are taken from Wilson and Atreya,[1] Table 7. There is good agreement between far- and mid-IR studies, and also between different authors, which provides validation for the more complete mid-IR data despite the complicating effects of the hot polar stratopause. Generally the shorter lifetime species have larger polar enrichments, with nitriles appearing significantly more enriched than hydrocarbons. Grey regions denote the three distinct enrichment regimes discussed in the text. (Previous studies are: T09a;[10] T09b;[18] and C10 [16]).

that the tendency for descending polar air to create uniform mixing ratio profiles in the north does indeed allow for direct comparison of equatorial and polar abundances. Mid-IR derived relative polar enrichment ratios determined by different authors using different retrieval schemes are also in broad agreement.[10,16]

Thus, armed with our validated plot of photochemical lifetime *versus* enrichment (Fig. 8) we can observe three distinct regimes in lifetime-enrichment space: (1) a very long lifetime region, with virtually no polar enrichment; (2) a trend line for hydrocarbon species; and (3) a separate trend line for nitrile species. Nitriles are significantly more enriched than hydrocarbons with similar predicted photochemical lifetimes. This indicates that current models are perhaps missing some loss process for nitriles, which should cause a steeper than predicted vertical gradient. This is consistent with Vinatier *et al.*,[8] who found that the HCN profile found using limb data was much steeper than could be explained by eddy mixing and photochemical production, and an additional sink term was required. A sink term of similar magnitude was used by Lara *et al.*[36] to produce photochemical HCN profiles consistent with observations. Such a sink could be due to haze formation[37] or lower atmosphere photopolymerisation of nitrogen species.[38] Another possible factor is the suggestion that galactic cosmic rays could be an important source of nitrogen radicals in the lower stratosphere.[39] This is supported by comparisons of the equatorial C_2N_2 abundance with photochemical model predictions.[18] It is clear that much remains to be understood about Titan's nitrogen chemistry.

Fig. 8 can also be used as a predictive tool. By assuming that relative polar enrichments are determined by the trend line defined by the three nitriles measured so far, we can predict the enrichments of other nitrile species whose latitude distribution are yet to be measured by Cassini (Table 2). Verifications of these predictions may be possible in Cassini's extended mission by averaging larger numbers of spectra or *via* dedicated composition observations.

The CO VMR determined from the rotational lines should be reliable at equatorial and southern latitudes, where we have a reasonable idea of the temperature in the lower stratosphere. The determinations from 90°S to 45°N give a weighted average of 55 ± 6 ppm. This is consistent with the value of 47 ± 8 ppm found by de Kok *et al.*[14]

Table 2 Predicted enrichments of species yet to be observed by Cassini based on the nitrile trend line in Fig. 8 given to one significant figure. Lifetime is defined at 300 km and is taken from Table 7 of Wilson and Atreya.[1]

Species	Lifetime/yrs	Predicted enrichment (VMR 70°N/0°N)
CH_3CN	41	3
C_2H_3CN	0.051	300
C_4N_2	0.12	200

The apparent CO increase in the north is most probably an artifact of poor temperature control in the lower stratosphere. This region is very difficult to sound because of very low temperatures and the overlying hot stratopause. Preliminary Cassini radio science observations[40] indicate that the temperatures we have used in the north polar lower stratosphere may in fact be too cold, which would result in the overestimation of CO. It is most likely that the long 25 000 year lifetime[1] of CO results in a well-mixed uniform global and vertical distribution.

Acknowledgements

This research was funded by the UK Science and Technology Facilities Council and the NASA Cassini program. The authors thank the Cassini CIRS instrument team for their continued efforts upon which this research relies.

References

1 E. H. Wilson and S. K. Atreya, *J. Geophys. Res.*, 2004, **109**, E06002.
2 V. A. Krasnopolsky, *Icarus*, 2009, **201**, 226–256.
3 P. P. Lavvas, R. V. Yelle and V. Vuitton, *Icarus*, 2009, **201**, 626–633.
4 T. Hidayat, A. Marten, B. Bézard, D. Gautier, T. Owen, H. Matthews and G. Paubert, *Icarus*, 1997, **126**, 170–182.
5 A. Marten, T. Hidayat, Y. Biraud and R. Moreno, *Icarus*, 2002, **158**, 532–544.
6 M. A. Gurwell, *Astrophys. J.*, 2004, **616**, L7–L10.
7 N. A. Teanby, P. G. J. Irwin, R. de Kok, S. Vinatier, B. Bézard, C. A. Nixon, F. M. Flasar, S. B. Calcutt, N. E. Bowles, L. Fletcher, C. Howett and F. W. Taylor, *Icarus*, 2007, **186**, 364–384.
8 S. Vinatier, B. Bézard, T. Fouchet, N. A. Teanby, R. de Kok, P. G. J. Irwin, B. J. Conrath, C. A. Nixon, P. N. Romani, F. M. Flasar and A. Coustenis, *Icarus*, 2007, **188**, 120–138.
9 N. A. Teanby, R. de Kok, P. G. J. Irwin, S. Osprey, S. Vinatier, P. J. Gierasch, P. L. Read, F. M. Flasar, B. J. Conrath, R. K. Achterberg, B. Bezard, C. A. Nixon and S. B. Calcutt, *J. Geophys. Res.*, 2008, **113**, E12003.
10 N. A. Teanby, P. G. J. Irwin, R. de Kok and C. A. Nixon, *Philos. Trans. R. Soc. London, Ser. A*, 2009, **367**, 697–711.
11 F. M. Flasar, R. K. Achterberg, B. J. Conrath, P. J. Gierasch, V. G. Kunde, C. A. Nixon, G. L. Bjoraker, D. E. Jennings, P. N. Romani, A. A. Simon-Miller, B. Bézard, A. Coustenis, P. G. J. Irwin, N. A. Teanby, J. Brasunas, J. C. Pearl, M. E. Segura, R. C. Carlson, A. Mamoutkine, P. J. Schinder, A. Barucci, R. Courtin, T. Fouchet, D. Gautier, E. Lellouch, A. Marten, R. Prange, S. Vinatier, D. F. Strobel, S. B. Calcutt, P. L. Read, F. W. Taylor, N. Bowles, R. E. Samuelson, G. S. Orton, L. J. Spilker, T. C. Owen, J. R. Spencer, M. R. Showalter, C. Ferrari, M. M. Abbas, F. Raulin, S. Edgington, P. Ade and E. H. Wishnow, *Science*, 2005, **308**, 975–978.
12 N. A. Teanby, P. G. J. Irwin, R. de Kok, C. A. Nixon, A. Coustenis, B. Bézard, S. B. Calcutt, N. E. Bowles, F. M. Flasar, L. Fletcher, C. Howett and F. W. Taylor, *Icarus*, 2006, **181**, 243–255.
13 A. Coustenis, R. K. Achterberg, B. J. Conrath, D. E. Jennings, A. Marten, D. Gautier, C. A. Nixon, F. M. Flasar, N. A. Teanby, B. Bézard, R. E. Samuelson, R. C. Carlson, E. Lellouch, G. L. Bjoraker, P. N. Romani, F. W. Taylor, P. G. J. Irwin, T. Fouchet, A. Hubert, G. S. Orton, V. G. Kunde, S. Vinatier, J. Mondellini, M. M. Abbas and R. Courtin, *Icarus*, 2007, **189**, 35–62.

14 R de Kok, P. G. J. Irwin, N. A. Teanby, E. Lellouch, B. Bezard, S. Vinatier, C. A. Nixon, L. Fletcher, C. Howett, S. B. Calcutt, N. E. Bowles, F. M. Flasar and F. W. Taylor, *Icarus*, 2007, **186**, 354–363.
15 N. A. Teanby, P. G. J. Irwin, R. de Kok, C. A. Nixon, A. Coustenis, E. Royer, S. B. Calcutt, N. E. Bowles, L. Fletcher, C. Howett and F. W. Taylor, *Icarus*, 2008, **193**, 595–611.
16 A. Coustenis, C. Nixon, R. Achterberg, P. Lavvas, S. Vinatier, N. Teanby, G. Bjoraker, R. Carlson, L. Piani, G. Bampasidis, F. Flasar and P. Romani, *Icarus*, 2010, **207**, 461–476.
17 R. K. Achterberg, B. J. Conrath, P. J. Gierasch, F. M. Flasar and C. A. Nixon, *Icarus*, 2008, **194**, 263–277.
18 N. A. Teanby, P. G. J. Irwin, R. de Kok, A. Jolly, B. Bézard, C. A. Nixon and S. B. Calcutt, *Icarus*, 2009, **202**, 620–631.
19 V. Kunde, P. Ade, R. Barney, D. Bergman, J. Bonnal, R. Borelli, D. Boyd, J. Brasunas, G. Brown, S. Calcutt, F. Carroll, R. Courtin, J. Cretolle, J. Crooke, M. Davis, S. Edberg, R. Fettig, M. Flasar, D. Glenar, S. Graham, J. Hagopian, C. Hakun, P. Hayes, L. Herath, L. Horn, D. Jennings, G. Karpati, C. Kellebenz, B. Lakew, J. Lindsay, J. Lohr, J. Lyons, R. Martineau, A. Martino, M. Matsumura, J. McCloskey, T. Melak, G. Michel, A. Morell, C. Mosier, L. Pack, M. Plants, D. Robinson, L. Rodriguez, P. Romani, W. Schaefer, S. Schmidt, C. Trujillo, T. Vellacott, K. Wagner and D. Yun, *Proc. Soc. Photo-Opt. Instrum. Eng.*, 1996, **2803**, 162–177.
20 F. M. Flasar, V. G. Kunde, M. M. Abbas, R. K. Achterberg, P. Ade, A. Barucci, B. Bézard, G. L. Bjoraker, J. C. Brasunas, S. Calcutt, R. Carlson, C. J. C. Esarsky, B. J. Conrath, A. Coradini, R. Courtin, A. Coustenis, S. Edberg, S. Edgington, C. Ferrari, T. Fouchet, D. Gautier, P. J. Gierasch, K. Grossman, P. Irwin, D. E. Jennings, E. Lellouch, A. A. Mamoutkine, A. Marten, J. P. Meyer, C. A. Nixon, G. S. Orton, T. C. Owen, J. C. Pearl, R. Prange, F. Raulin, P. L. Read, P. N. Romani, R. E. Samuelson, M. E. Segura, M. R. Showalter, A. A. Simon-Miller, M. D. Smith, J. R. Spencer, L. J. Spilker and F. W. Taylor, *Space Sci. Rev.*, 2004, **115**, 169–297.
21 C. A. Nixon, N. A. Teanby, S. B. Calcutt, S. Aslam, D. E. Jennings, V. G. Kunde, F. M. Flasar, P. G. J. Irwin, F. W. Taylor, D. A. Glenar and M. D. Smith, *Appl. Opt.*, 2009, **48**, 1912–1925.
22 A. Borysow and L. Frommhold, *Astrophys. J.*, 1986, **311**, 1043–1057.
23 L. S. Rothman, D. Jacquemart, A. Barbe, D. C. Benner, M. Birk, L. R. Brown, M. R. Carleer, C. Chackerian, K. Chance, L. H. Coudert, V. Dana, V. M. Devi, J. M. Flaud, R. R. Gamache, A. Goldman, J. M. Hartmann, K. W. Jucks, A. G. Maki, J. Y. Mandin, S. T. Massie, J. Orphal, A. Perrin, C. P. Rinsland, M. A. H. Smith, J. Tennyson, R. N. Tolchenov, R. A. Toth, J. Vander Auwera, P. Varanasi and G. Wagner, *J. Quant. Spectrosc. Radiat. Transfer*, 2005, **96**, 139–204.
24 C. P. Rinsland, V. M. Devi, M. A. H. Smith, D. C. Benner, S. W. Sharpe and R. L. Sams, *J. Quant. Spectrosc. Radiat. Transfer*, 2003, **82**, 343–362.
25 B. Khare, C. Sagan, E. Arakawa, F. Suits, T. Callcott and M. Williams, *Icarus*, 1984, **60**, 127–137.
26 P. Irwin, N. Teanby, R. de Kok, L. Fletcher, C. Howett, C. Tsang, C. Wilson, S. Calcutt, C. Nixon and P. Parrish, *J. Quant. Spectrosc. Radiat. Transfer*, 2008, **109**, 1136–1150.
27 A. A. Lacis and V. Oinas, *J. Geophys. Res.*, 1991, **96**, 9027–9063.
28 M. Fulchignoni, F. Ferri, F. Angrilli, A. J. Ball, A. Bar-Nun, M. A. Barucci, C. Bettanini, G. Bianchini, W. Borucki, G. Colombatti, M. Coradini, A. Coustenis, S. Debei, P. Falkner, G. Fanti, E. Flamini, V. Gaborit, R. Grard, M. Hamelin, A. M. Harri, B. Hathi, I. Jernej, M. R. Leese, A. Lehto, P. F. L. Stoppato, J. J. Lopez-Moreno, T. Makinen, J. A. M. McDonnell, C. P. McKay, G. Molina-Cuberos, F. M. Neubauer, V. Pirronello, R. Rodrigo, B. Saggin, K. Schwingenschuh, A. Seiff, F. Simoes, H. Svedhem, T. Tokano, M. C. Towner, R. Trautner, P. Withers and J. C. Zarnecki, *Nature*, 2005, **438**, 785–791.
29 M. G. Tomasko, L. Doose, S. Engel, L. E. Dafoe, R. West, M. Lemmon, E. Karkoschka and C. See, *Planet. Space Sci.*, 2008, **56**, 669–707.
30 R de Kok, P. G. J. Irwin, N. A. Teanby, C. A. Nixon, D. E. Jennings, L. Fletcher, C. Howett, S. B. Calcutt, N. E. Bowles, F. M. Flasar and F. W. Taylor, *Icarus*, 2007, **191**, 223–235.
31 N. A. Teanby, P. G. J. Irwin and R. de Kok, *Icarus*, 2009, **204**, 645–657.
32 L. Lara, E. Lellouch, J. López-Moreno and R. Rodrigo, *J. Geophys. Res.*, 1996, **101**, 23261–23283.
33 N. A. Teanby, *Math. Geol.*, 2007, **39**, 419–434.
34 R. K. Achterberg, B. J. Conrath, P. J. Gierasch, F. M. Flasar and C. A. Nixon, *Icarus*, 2008, **197**, 549–555.
35 N. A. Teanby, P. G. J. Irwin and R. de Kok, *Planet. Space Sci.*, 2010, **58**, 792–800.
36 L. M. Lara, E. Lellouch and V. Shematovich, *Astron. Astrophys.*, 1999, **341**, 312–317.

37 C. P. McKay, *Planet. Space Sci.*, 1996, **44**, 741–747.
38 D. W. Clarke and J. P. Ferris, *J. Geophys. Res.*, 1996, **101**, 7575–7584.
39 L. A. Capone, J. Dubach, S. S. Prasad and R. C. Whitten, *Icarus*, 1983, **55**, 73–82.
40 P. J. Schinder, F. M. Flasar, E. A. Marouf, R. G. French and C. A. McGhee, *Bull. Am. Astron. Soc.*, 2007, **39**, 475.

PAPER

Upper limits for undetected trace species in the stratosphere of Titan†

Conor A. Nixon,*[ab] Richard K. Achterberg,[ab] Nicholas A. Teanby,[c] Patrick G. J. Irwin,[c] Jean-Marie Flaud,[d] Isabelle Kleiner,[d] Alix Dehayem-Kamadjeu,[de] Linda R. Brown,[f] Robert L. Sams,[g] Bruno Bézard,[h] Athena Coustenis,[h] Todd M. Ansty,[i] Andrei Mamoutkine,[j] Sandrine Vinatier,[b] Gordon L. Bjoraker,[b] Donald E. Jennings,[b] Paul. N. Romani[b] and F. Michael Flasar[b]

Received 2nd March 2010, Accepted 17th April 2010
DOI: 10.1039/c003771k

In this paper we describe the first quantitative search for several molecules in Titan's stratosphere in Cassini CIRS infrared spectra. These are: ammonia (NH_3), methanol (CH_3OH), formaldehyde (H_2CO), and acetonitrile (CH_3CN), all of which are predicted by photochemical models but only the last of which has been observed, and not in the infrared. We find non-detections in all cases, but derive upper limits on the abundances from low-noise observations at 25°S and 75°N. Comparing these constraints to model predictions, we conclude that CIRS is highly unlikely to see NH_3 or CH_3OH emissions. However, CH_3CN and H_2CO are closer to CIRS detectability, and we suggest ways in which the sensitivity threshold may be lowered towards this goal.

1 Introduction

The atmosphere of Titan, Saturn's largest moon, exhibits the greatest diversity of chemicals found in any planetary atmosphere outside of the Earth's. The first of these – CH_4, H_2, and C_2H_2 – were detected through spectroscopy using ground-based telescopes,[1-3] an arduous task requiring removal of telluric absorption. The first 'golden age' of multiple gas discoveries however came with the Voyager 1 flyby

---

[a]Department of Astronomy, University of Maryland, College Park, MD, 20742, USA. E-mail: conor.a.nixon@nasa.gov; Fax: (+1) 301 286-0212; Tel: (+1) 301 286-6757
[b]Planetary Systems Laboratory, NASA Goddard Space Flight Center, Greenbelt, MD, 20771, USA
[c]Atmospheric, Oceanic and Planetary Physics, University of Oxford, Oxford, OX1 3PU, UK
[d]LISA, CNRS, UMR 7583, Universités Paris 7 et Paris Est, 61 Avenue Général de Gaulle, 94010 Créteil, France
[e]Department of Physics, College of Biological and Physical Sciences, University of Nairobi, P.O. Box 30197, Nairobi, Kenya
[f]Jet Propulsion Laboratory, California Institute of Technology, Pasadena, CA, 91109, USA
[g]Pacific Northwest National Laboratory, P.O. Box 999, Richland, WA, 99352, USA
[h]LESIA, Observatoire de Paris-Meudon, 92195 Meudon Principal Cedex, France
[i]Department of Space Science, Cornell University, Ithaca, NY, 14853, USA
[j]Adnet Systems, Inc., Rockville, MD, 20852, USA

† This paper is dedicated to Virgil G. Kunde as he approaches his 75th birthday, in recognition of his more than forty years of contributions to spectroscopic measurements of planetary atmospheres, including his leadership in the design, fabrication, and operation of the Cassini CIRS spectrometer.

in 1980, when many new species were detected in the stratosphere using the IRIS infrared spectrometer, including HCN, C_2H_4 and C_2H_6;[4] HC_3N, C_2N_2, C_4H_2;[5] C_3H_4 and C_3H_8;[6] and CO_2.[7] N_2 was infered from ultraviolet spectroscopy.[8] Later CO and CH_3CN were added to the roster using ground-based telescopes,[9,10] while H_2O and C_6H_6 were found in the stratosphere using the Infrared Space Observatory (ISO).[11,12]

The second 'golden age' of gas discovery on Titan came with the arrival of the Cassini Saturn orbiter in 2004, but this time the detections did not flow from infrared spectroscopy of the middle atmosphere, or indeed from any remote sensing technique. Rather, it was in the ionosphere where the Cassini mass spectrometer (INMS) showed a veritable bounty of molecular species occupying all niches in the mass spectrum to the detection limit at 100 amu;[13] while the CAPS plasma spectrometer showed heavy negative ions extending to many thousands of amu beyond.[14] Detailed modelling of the INMS cracking patterns based on open-source ion (OSI) spectra[15] has proposed the presence of species unseen in the stratosphere such as ammonia (NH_3), toluene (C_7H_8), tri- and tetra-cetylene (C_6H_2, C_8H_2), methylenimine (CH_2NH), and several nitriles including vinyl cyanide (acrylonitrile, C_2H_3CN) and ethyl cyanide (propionitrile, C_2H_5CN). However, a later evaluation of the closed-source neutral (CSN) mass spectra, which are less contaminated by wall reactions, found confirmation only for ammonia on the above list – beyond the species already known from the infrared.[16]

Most published models[17–20] of Titan's chemistry have focused on neutral chemistry in the upper and middle atmosphere – prior to the surprise Cassini revelations of the complex ions. These invoked radical species to initiate the chemical cycle, derived from the precursor molecules N_2, CH_4 (bulk atmospheric constituents) and H_2O (from external flux) by either solar photolysis or magnetospheric electron impacts. These basic radicals, including CH_3, CH_2, CH, N, H and OH, are the building blocks of the more complex species: $e.g.$ $CH_3 + CH_3 \rightarrow C_2H_6$. The models necessarily include many species not yet observed in Titan's neutral atmosphere: some are likely unstable intermediaries, $e.g.$ C_4H_6, but others are stable and yet remain undetected, such as C_3H_6 and C_2H_3CN.

In this paper we report on a preliminary search for four of these predicted species in Titan's stratosphere by analysis of infrared spectra from the Cassini Composite Infrared Spectrometer (CIRS). These are the gases ammonia (NH_3), formaldehyde (H_2CO), methanol (CH_3OH) and the previously-discovered CH_3CN (acetonitrile), which has so far eluded detection in the infrared portion of the spectrum. An infrared route to measurement would permit efficient global and temporal mapping, which cannot be achieved by either $in\ situ$ mass spectrometry, or ground-based submillimetre observations that do not resolve Titan's disk.

In the next section we introduce the CIRS instrument and describe the observations, followed by the data analysis method and then the results. This is followed by a discussion, and then conclusions, including the needed future work to expand this study.

2 Experimental

2.1 The Cassini composite infrared spectrometer

CIRS is a dual Fourier-Transform Spectrometer (FTS), sensitive to a far-infrared spectral range (10–600 cm^{-1}, 1000–17 μm) using a polarization-splitting (Martin–Puplett) interferometer, and to a mid-infrared range (600–1400 cm^{-1}, 17–7 μm) using a conventional amplitude splitting (Michelson-type) interferometer. Both interferometers share a common 50 cm diameter telescope, fore-optics, scan mechanism and reference laser, and can achieve identical spectral resolutions ranging from 0.5 to 15 cm^{-1}, depending on the commanded scan time (mirror displacement). In the far-infrared, a single bolometer (known as focal plane 1, or FP1) detects the incident radiation, having a relatively large, circular field-of-view with a Gaussian response of Full Width to Half Maximum (FWHM) 2.5 mrad. The mid-infrared focal plane in contrast consists of two linear 1 × 10 HgCdTe pixel arrays sensitive to different sub-ranges: focal plane 3 (FP3) is operational from 600–1100 cm^{-1} (17–9 µm), whilst focal plane 4 (FP4) detects the shortest wavelengths from 1000–1400 cm^{-1} (9–7 µm). The CIRS focal planes are depicted in Fig. 1. In this paper, only spectra from the mid-infrared range are considered (FP3 and FP4). Further information regarding the instrument can be found in the literature.[21,22]

2.2 Observations

For this study, we expect to be limited in our sensitivity to the weak species by the random noise level in the spectrum, which defines the smallest emission that we can detect at the 1-σ level (one standard deviation). A 3-σ signal is normally considered the minimum to claim a detection. As our instrumental noise should in general be uncorrelated (White or Gaussian noise), we expect that by co-adding spectra we can reduce the noise level considerably; *i.e.*, if we define the noise equivalent spectral radiance (NESR) on a single spectrum as $\sigma_1(\nu)$, then the spectral noise on a set of N spectra should be reduced to $\sigma_n(\nu) = \sigma_1(\nu)/\sqrt{N}$. This assumes the absence of non-random effects such as instrumental artefacts, which are in fact a significant limitation at some wavelengths.

Therefore, we chose for our upper limits study to use several recent observations designed for the purpose of maximising the signal-to-noise ratio (SNR) of trace atmospheric species as much as possible, by simultaneously increasing the signal and decreasing the noise as much as possible. The increased signal is attained by pointing the instrument at Titan's limb, *i.e.* along a path that does not intersect the surface, while the decreased noise is attained by co-adding large numbers of spectra.

Limb observations by CIRS have previously been used to detect very weak isotopic variants of major gases, including H^{13}CN, HC^{15}N, ^{13}CH$_3$D, ^{13}CO$_2$, CO^{18}O, H^{13}CCCN and C$_2$HD.[23–27] However, these prior detections used relatively

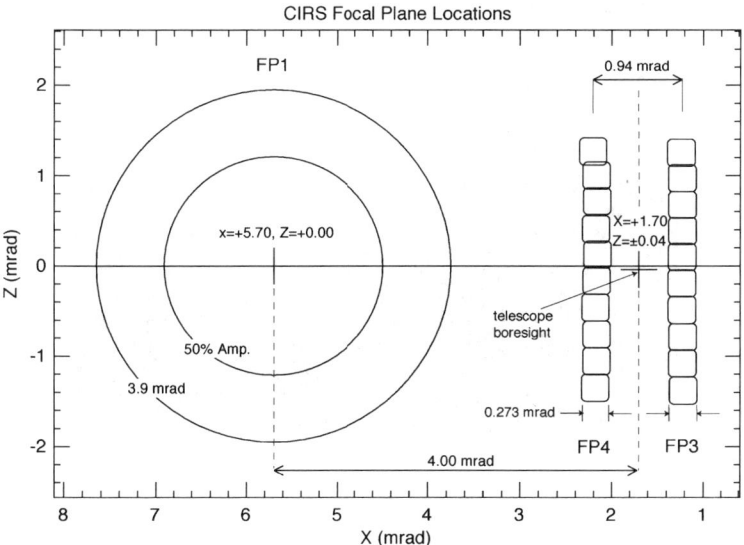

Fig. 1 CIRS instrument focal plane schematic, showing relative sizes and positions of the three detector arrays and projected footprint on sky.

small spectral sets recorded at a single latitude, or were obliged to average across wide ranges of latitude to reduce the noise level, because there are typically few spectra recorded by CIRS at identical altitudes, latitudes and times. This follows from the second major advantage of limb-sounding compared to nadir (surface-intersecting) viewing, which is the ability to make resolved measurements over a wide vertical range. This arises because the opacity for trace species is usually less than unity, and therefore the radiance contribution arises almost entirely from the localized atmospheric region at the tangent altitude of the limb ray. CIRS performs around one dozen standard types of Titan observations, designed for various specific purposes such as measuring vertical aerosol opacity variation in the far-infrared, or global mapping of temperatures in the mid-infrared.

For limb measurements of gas vertical abundance profiles, the usual paradigm is to place the mid-infrared detector arrays perpendicular to the disk edge (*i.e.* radially from Titan's centre) to enable simultaneous measurements over a range of altitudes by virtue of the 10-element detector arrays. These observations (known as MIRLM-BINTs), while excellent for vertical mapping of the more abundant trace gases, are not however ideally suited for new detections, as the spectra from different altitudes must be modelled separately and cannot be co-added to produce a single very low noise spectrum. Moreover, the atmosphere thins with altitude and the temperature changes, therefore for most minor gases the absolute signal peaks in the lower stratosphere (depending on latitude) and declines below and above.

In this work, we analyse observations of a new type (known as MIRLMPAIRs) that were custom-designed for the purpose of enabling new molecular detections – or putting low limits on non-detected species. In these observations (see Fig. 2) the CIRS mid-infrared arrays are placed parallel to the limb, with the advantage that the spectra from all ten detectors in each array fall at nearly the same altitude and within a narrow range of latitudes (depending on distance from Titan) and can therefore be cautiously co-added into a single long-path and low-noise spectrum. The caution arises from the need to consider lateral variations in gases and temperature, which are small at equatorial latitudes but change quickly at mid and high northern latitudes, at the current season. CIRS has only five electronic channels

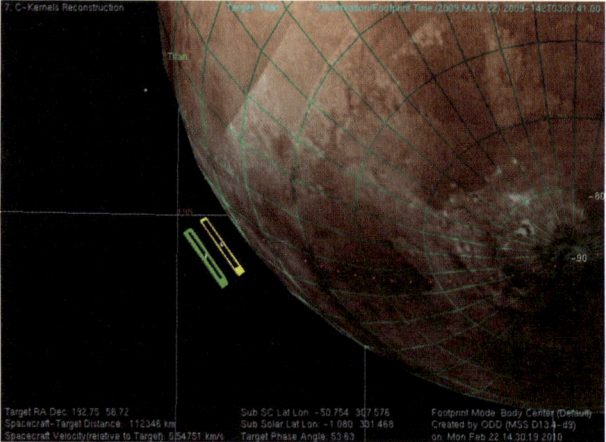

Fig. 2 Pictorial representation of the CIRS Titan mid-infrared limb pair (MIRLMPAIR) observation on the T55 flyby, May 22nd 2009. The FP3 array, fixed at 107 km altitude and oriented horizontally is shown in yellow and FP4 (0.94 mrad above) is in green. During the course of the observation the range from Titan increases from 109 000 to 178 000 km (113 000 depicted), and therefore the FP4 altitude steadily increases from 230 to 290 km. All detectors are used in pair mode, so the arrays are composed of effectively five oblong pixels rather than ten squares.

for each ten detectors, therefore an onboard pairing process is used to co-add adjacent pixels before the spectrum is returned.

Two such MIRLMPAIR sequences were performed, centred on 25°S (flyby T55) and 76°N (flyby T64) latitude, with the lower FP3 arrray positioned near 110 km (8 mbar) and 225 km (0.3 mbar) respectively. A drawback of this type of design is that, compared to the radial mode, the FP3 and FP4 detector arrays no longer cover the same range of altitude. This means the temperature sensing that is usually achieved from the methane ν_4 band at 1304 cm^{-1} on FP4 no longer corresponds to the right altitude to give modelling input for the FP3 spectrum, which is targeted lower. Therefore in order to model the FP3 spectra, additional temperature measurements must be used. For the T55 sequence we derived lower-stratosphere temperatures from a nadir mapping sequence (MIDIRTMAP), while for the T64 sequence we used a vertical limb measurement (MIRLMBINT, Fig. 3) that covered the full range of needed altitudes. See Table 1 for details.

3 Model

3.1 Atmospheric model

Our Titan initial model atmosphere was similar to that used in our previous study of Titan's propane bands,[28] with 100 layers equally spaced in log pressure from 1.45 bar (0 km) to 4×10^{-8} bar (about 700 km), and is shown in Fig. 4. The T–p (temperature-pressure) profile is derived from the Huygens HASI results,[29] with altitudes computed from hydrostatic equilibrium. The initial abundance profiles of major known gases are mostly assumed to be uniformly mixed with altitude in the stratosphere, at values appropriate for Titan's equatorial regions.[30,31] The exception is acetylene, which has a vertical stratospheric gradient at low latitudes.[32] All the gases with the exception of the diatoms, H_2 and N_2, and ethylene (C_2H_4), condense near the tropopause and their abundance drops to low values. Isotopologues of these gases were also included as needed, i.e. $^{13}CH_4$ and CH_3D. Not all known Titan species or isotopologues were included in the model (e.g. CO, HC_3N, C_4H_2), if they did not have bands near the spectral regions of interest for the sensitivity study (see §3.6).

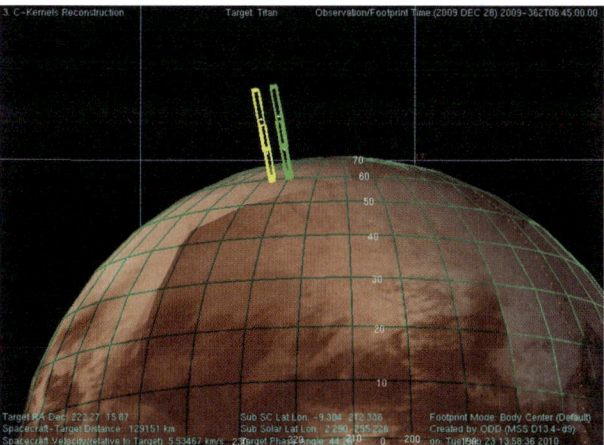

Fig. 3 Pictorial representation of the CIRS Titan mid-infrared limb integration (MIRLM-BINT) observation on the T64 flyby, December 28th 2009. The FP3 array is shown in yellow and FP4 in green (side-by-side). Two vertical positions centred on 350 km, and ~100 km are used sequentially to measure the vertical profile of gas abundances in the atmosphere. The range is from 113 000 to 177 000 km (129 000 km depicted) and the arrays cover a vertical extent of ~400 km (each position).

Table 1 Cassini CIRS observations analysed in this report.

Flyby #	Observation name	Start date and time	Duration	CIRS focal plane	Latitude range (mean)	Altitude range (mean)	Pressure range of sensitivity	Number of spectra
Temperature retrievals								
T55	MIDIRTMAP002	22-MAY-2009 11:26:41	8 h	4	90°S–40°N	120–220 km	5.0–0.5 mbar	6263
T64	MIRLMBINT002	28-DEC-2009 05:16:59	4 h	4	75°N–76°N (75.5°N)	100–500 km (50 km bins)	3.8–0.0014 mbar	86, 126, 131, 141, 57, 67, 78, 61
Upper limits calculations								
T55	MIRLMPAIR002	22-MAY-2009 02:26:41	4 h	3	25 ± 2°S (25.5°S)	97–122 km (107 km)	7.6 mbar	1213
				4	25 ± 2°S (25.5°S)	225–275 km (247 km)	0.27 mbar	941
T64	MIRLMPAIR001	27-DEC-2009 15:16:59	4 h	3	75.8 ± 2.0°N (76.0°N)	204–254 km (224 km)	0.26 mbar	517
				4	75.8 ± 2.0°N (75.7°N)	325–375 km (348 km)	0.018 mbar	491

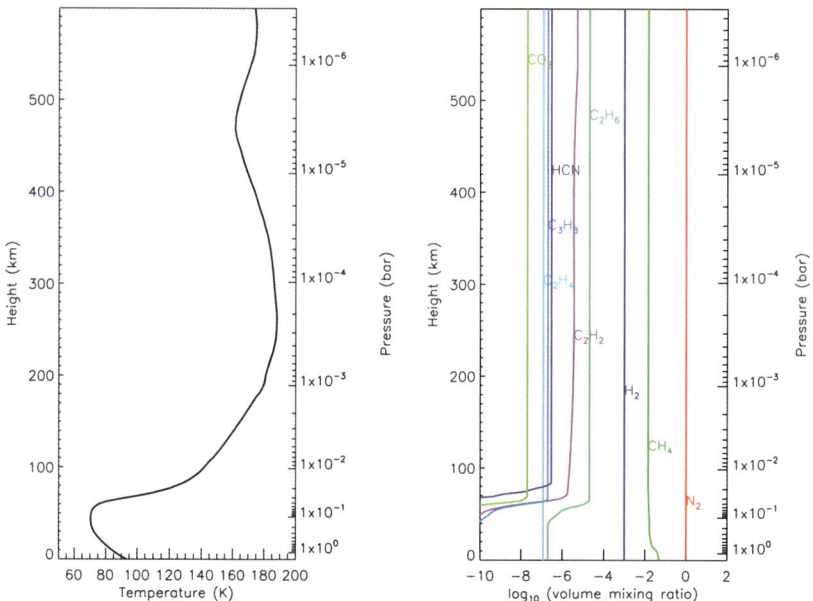

Fig. 4 Titan model atmosphere initial temperature (left) and abundance profiles of major known gases (right).

Profiles for the four test gases were later introduced in a very simple form: uniformly mixed above the tropopause and dropping quickly to an insignificant value at 45 km. The stratospheric abundance was varied as required in the sensitivity study. The final element of the atmospheric model was a haze absorber, uniformly mixed above the tropopause but variable in particle density, and using optical properties derived from a study of laboratory tholins,[33] which allowed a good fitting to the continuum over small intervals.

3.2 Spectral synthesis

The emerging radiance was computed based on the model atmosphere and spectral atlases for the required gases, using the NEMESIS computer code developed at Oxford University.[34] This code has been extensively validated, being previously use to model spectra of planets including Venus, Mars, Jupiter, Saturn, Uranus, and also many studies of Titan CIRS data.

The spectral atlases used for gaseous opacity are as described in a previous paper,[28] except for the addition of the four new gas species as noted here. For NH_3 and CH_3OH we extracted line data from the HITRAN 2008 atlas.[35] For H_2CO we used a new spectral line atlas containing ~5600 transitions from 877 to 1500 cm^{-1}, subsetted from a more comprehensive study of the ν_2, ν_3, ν_4 and ν_6 bands,[36] to include only the spectral range overlapping with CIRS. The ν_6 band (CH_2 wag) centred on 1167 cm^{-1} is the one of interest in this study. Finally, we used a new unpublished list for the ν_7 band (1041 cm^{-1}) of CH_3CN created for CIRS use as described in the next section.

The correlated-k method[37,38] for fast computation of radiances was used, due to the prohibitive slowness of calculating the Voigt lineshape over substantial spectral ranges in real time for use in retrievals. Therefore we computed k-tables for each gas over the necessary range of temperatures and pressures of the Titan atmospheric profile, including convolution with the CIRS highest spectral resolution of 0.48 cm^{-1} and using 50 g-ordinate sampling of the k-distribution.

Finally, to model each recorded CIRS spectrum, the spatial extent of the detector footprints on Titan's limb had to be considered. A means of accurately modelling the MIRLMBINT type observations has been reported, which begins with laboratory-measured spatial responses for each detector[39] and co-adds these at appropriate altitudes to arrive at an 'exact' spatial weighting function.[28] We followed this treatment, sampling the spatial weighting function at 10 km intervals (about 1/5 of a Titan atmospheric scale height in the stratosphere) to enable computation in a reasonable time, while maintaining sufficient accuracy.

However, the MIRLMPAIR observations are somewhat different, as the detectors are placed horizontally on the limb, and because the detector signals are added pair-wise on the spacecraft, so that for example detectors 1 and 2 become a virtual detector '1 + 2', with a spatial footprint about double the size of each individual detector (ignoring a small gap between them). In the MIRLMPAIR observations, the long direction of these detector pairs is also along the array direction (Z). To model these, we simply assumed that they have a boxcar response cross-section in the short direction (X) of 0.273 mrad, *i.e.* vertically. We then convolved this function scaled appropriately for distance, with a sampling of rays emerging from the limb at 10-km vertical increments as before.

3.3 Creation of CH_3CN line atlas

The CH_3CN linelist is based on a very preliminary analysis of the ν_7 fundamental near 10 μm. In 2006, experimental line positions were obtained by peak finding from one high resolution room temperature FTS spectrum (see description of our previous study[40] of the 11 μm region for details). Assignments were extended and modelled up to $J = 56$ to produce a prediction of the spectrum with intensities normalized to the published integrated intensity.[41] The resulting catalogue contained over 5300 transitions between 960 to 1135 cm^{-1}. The accuracies of the calculated positions ranged from 0.001 cm^{-1} for assigned lines, but these deviated sharply (0.1 cm^{-1}) at the highest J and K, because the present model does not take into account all the interactions between the ν_4, ν_7 and $2\nu_8$ bands presently assigned, or with the $3\nu_8$ band, in the 10–14 μm range.[42] The accuracies of individual line intensities were not characterized, and hot band transitions were not included in the database. However, the sum of the linelist intensities (3.99×10^{-19} cm molecule^{-1} at 296 K) fell within 3% of the reported ν_7 band strength.[41] Nitrogen-broadened half widths were set to a constant value of 0.09 cm^{-1} atm^{-1} at 296 K, but these were later shown to be too low, by 30% to 200%.[40]

3.4 Retrieval method

The fitting of the spectral data by the model is achieved by successive iterations using a non-linear least squares optimal estimation method.[43] At each iteration, a cost function is computed, similar to the weighted χ^2 test between the data and model spectrum, but including an additional term to allow for an *a priori* constraint, which smooths the solution and prevents it from gravitating towards an unphysical parameter set. Model parameters (temperature and aerosol opacity, or gas abundance and aerosol opacity) are adjusted after each iteration along calculated downhill gradients to reduce the χ^2 on the next iteration, until an arbitrary convergence has been reached. The retrieval formalism computes final errors on retrieved parameters, based on the initial spectral error and *a priori* error (amount of constraint). A detailed description of the method as implemented in the NEMESIS code is available in the literature.[34]

3.5 Upper limits method

The calculation of the upper limits proceeded in three stages. In the first stage, the ν_4 band of methane from 1225–1325 cm^{-1} was modelled to retrieve temperatures.

NEMESIS was used to model the FP4 limb data (three of the sets described in §2.2), while the lower spectral resolution FP4 nadir data was modelled using a different correlated-k model.[44] The combination of these four data sets was sufficient to give temperature information at the various altitudes required to later model both FP3 and FP4 MIRLMPAIR observations. The temperature profiles were then fixed and the FP3 and FP4 MIRLMPAIR spectra were then fitted for gas and haze abundance in the spectral ranges of interest, at 9–11 μm as described in §4. Finally, the abundances of known gases and isotopes, and haze opacity were also fixed, and the new gases introduced to the profile at a range of presumed abundances. The forward model was calculated directly in a single iteration (no retrieval) to show the change to the spectrum resulting.

We follow the method used previously in a CIRS study of C_2N_2 detection/upper limits.[45] We first define the error weighted χ^2 measure of agreement between model and data as follows:

$$\chi_j^2 = \sum_{i=1}^{M} \frac{(I_{\text{data}}(\nu_i) - I_{\text{model}}(\nu_i, q_j))^2}{\sigma_i^2} \quad (1)$$

where $I_{\text{data}}(\nu_i)$ and $I_{\text{model}}(\nu_i)$ are the data and model spectra respectively at wavenumber ν_i, σ_i is the random noise estimate at this wavenumber, and q_j is the test abundance of the new gas that we have introduced. We also define χ_0^2 as the reference case when $q_j = 0$.

We sample the parameter space by trying model calculations over a wide range of q_j, in each case calculating the change to the χ^2, defined as $\Delta\chi^2 = \chi_0^2 - \chi_j^2$. If $\Delta\chi^2$ decreases to a minimum, then a positive detection of 1-σ, 2-σ, and 3-σ is made when $\Delta\chi^2$ reaches -1, -4 or -9 respectively. Similarly, a rejection is made at the 1-σ, 2-σ, and 3-σ level if $\Delta\chi^2$ increases monotonically to reach a level of $+1$, $+4$ or $+9$ respectively.[46]

3.6 Spectral range selection

When investigating the spectra for the purpose of making gas detections or placing upper limits, several considerations affect the choice of spectral range used: (i) location of the strongest emission bands of each test gas within the CIRS spectral range; (ii) the intrinsic noise level of each of the three CIRS focal planes (FP1 highest, FP4 lowest); (iii) other features that might mask the gas signatures, including weak propane bands that are not currently modelled,[28] non-random electrical interference ('spikes') on the spectrum at known frequencies that are hard to remove, and very strong gas bands that are modellable (*e.g.* C_2H_2 ν_5 and C_2H_6 ν_9) but generally leave some residual unfitted emission above the noise level, due to slight inaccuracies in line widths and other parameters.

A good range for Titan trace gas searches using CIRS spectra is the region between 900–1150 cm^{-1} (11–9 μm), where the spectrum is mostly free of other strong emission bands (the comparatively weak/sparse ν_7 band of C_2H_4 at 949 cm^{-1} is the most prominent, and models well), and the lower noise of FP3 (up to 1100 cm^{-1}) and FP4 (1000 cm^{-1} and beyond) is available. An added bonus is the overlap in spectral range between FP3 and FP4 at 1000–1100 cm^{-1}, which in principle means that measurements could be made simultaneously in this spectral range at two altitudes, for the MIRLMPAIR observations. However, in practice we found that the FP3 data was noisy and suffered from some calibration problems at 1000–1100 cm^{-1}, therefore we exclusively used FP4 data at wavenumbers above 1000 cm^{-1}.

These considerations led us to select the following bands for the upper limit study: ammonia ν_2 centred on 950 cm^{-1}; methanol ν_8 (1033 cm^{-1}); acetonitrile ν_7 (1041 cm^{-1}); and formaldehyde ν_6 (1167 cm^{-1}). This leaves us with a mixture of weak and strong bands. For example, the ν_8 of CH_3OH is very strong, with a peak intensity of 1×10^{-18} cm molecule^{-1},[47] and excellent for our purposes; while CH_3CN ν_7 is much weaker, reaching an intensity of just 8×10^{-20} cm molecule^{-1}.[40] In the conclusions we will address how in future studies this work might be extended to include other bands in the CIRS range, with the potential to improve the upper limits in some cases.

4 Results

We will not here describe the results of fitting Titan's known stratospheric gases, as the retrieval of spatial and temporal variations of temperature and gas abundance from Cassini CIRS data have been extensively published in the literature.[31,32,44,48–51] These studies used a variety of line-by-line and correlated-k forward models, including NEMESIS, and have generally very close agreement of results.

Fig. 5 (25°S) and Fig. 6 (75°N) show our results. In each case, the left-hand column shows in black the residual data spectrum in the four gas ranges after the fitting and subtraction of the emissions of known species (principally C_2H_4 centred on 949 cm^{-1} and CH_3D centred on 1156 cm^{-1}). In the case of CH_3OH and CH_3CN, we avoided the ν_{20} band of propane at 1054 cm^{-1}, for which currently no line atlas is available and hence we cannot model, by restricting our search to the region 1030–1050 cm^{-1}. The coloured lines in each case again show the residual, but after a large amount of the test gas is added to the model calculation, for the purpose of identifying the locations of the strongest features of the test gas. The right-hand column shows the curve of $\Delta\chi^2$ for each species, the change in χ^2 of fit that occurs as the abundance of the test gas is varied, computed across the whole range to improve sensitivity.

In none of the eight cases was any convincing minimum seen (*i.e.* $\Delta\chi^2 = -9$ for a 3-σ detection), although two very slight minima at the 1-σ confidence level occurred: for CH_3OH at 25°S ($\Delta\chi^2 = -1.68$, abundance 3×10^{-9}) and CH_3CN at 76°N ($\Delta\chi^2 = -1.82$, abundance 3×10^{-7}). As the significance is so low we infer non-detections in all cases based on the present sampling, although these indications should encourage further study. Vertical dashed lines on the charts show the abundances corresponding to the 1-σ, 2-σ, and 3-σ upper limits respectively.

Table 2 enumerates the derived upper limits at each level of significance. The most restrictive upper limits resulted for NH_3, despite the higher NESR on FP3, whereas the least constraining limits resulted for CH_3CN due to the much weaker spectral band used. The upper limits were also lower (better) for the 25°S spectra than those at 76°N, due to the lower altitude of this observation, resulting in a significantly higher atmospheric opacity.

5 Discussion

5.1 Ammonia

NH_3 is the likely source of Titan's present N_2-dominated atmosphere, *via* photolysis of out-gassed ammonia shortly after Titan's formation. In the present atmosphere ammonia could be present from either episodic outgassing, that also releases the methane needed to refuel the carbon-photochemistry, or else as an intermediate product of the chemistry that continuously processes N_2 and CH_4 into more complex substances, including nitriles (–C≡N).

The 1-D photochemical model of Wilson and Atreya (hereafter WA04)[20] shows two sources of NH_3 production: one in the lower stratosphere at ~120 km from cosmic-ray ionization of N_2 resulting in a mole fraction of 9.6×10^{-13}, and a second in the upper atmosphere due to electron recombination of ammonium ions. This second source results in a relatively high abundance in the ionosphere (~10^{-8}–10^{-7}), which rapidly decreases to lower altitudes, reaching a minimum below 10^{-14} near 300 km. The ionospheric source was confirmed by the Cassini INMS experiment, which found a volume mixing ratio (VMR) of 7×10^{-6} at 1100 km using

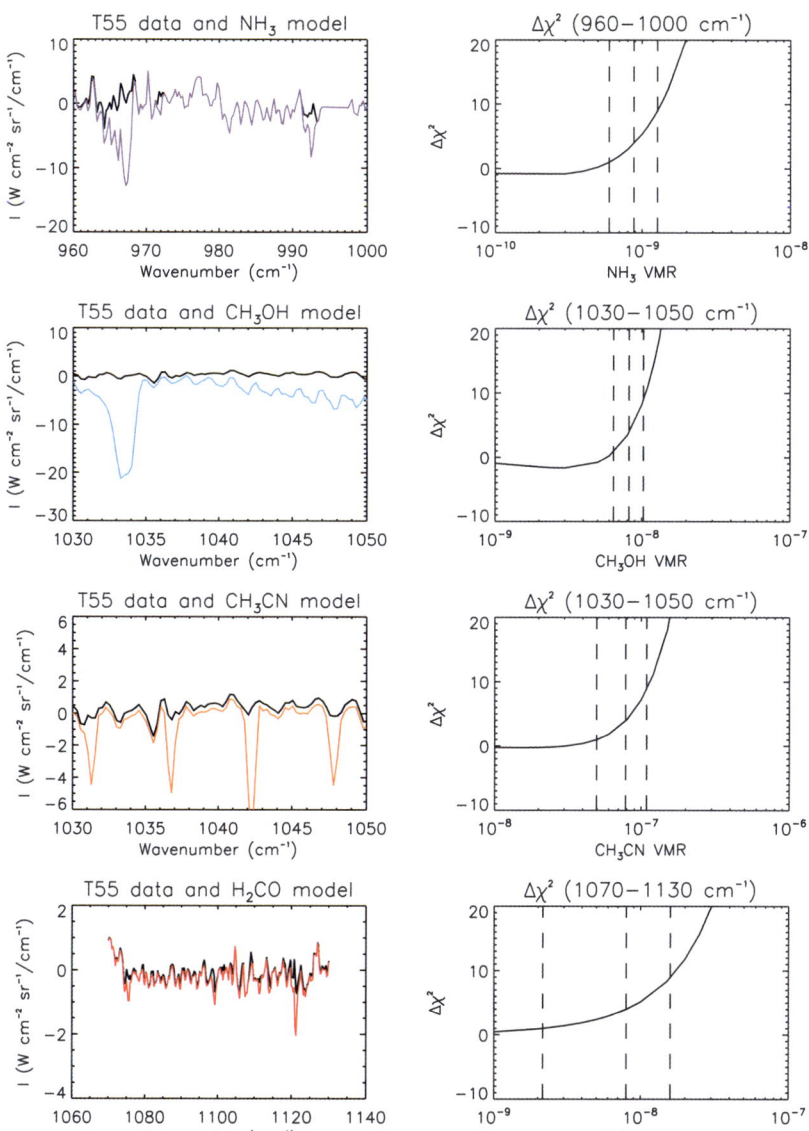

Fig. 5 Upper limits derived from T55 MIRLMPAIR data (25°S). Left column: Titan residual (data-model) spectrum in selected ranges after removal of known gases (black), and the residual found after addition of an arbitrary amount of trial gas to the model, showing location of spectral features (colored lines). Right column: the change $\Delta\chi^2$ in the goodness of fit measure χ^2 plotted over a wide range of trial gas abundances (solid line). Vertical dashed lines show the abundances for which $\Delta\chi^2$ is +1, +4, +9, corresponding to gas non-detections at the 1-σ, 2-σ and 3-σ significance levels, respectively.

the OSI mode,[15] and a slightly higher amount of 3×10^{-5} using the CSN recordings at a similar level.[16]

Therefore, the CIRS 3-σ upper limits derived here: 1.3×10^{-9} at (107 km, 25°S), and 1.4×10^{-8} at (224 km, 75°N) do not produce any meaningful constraint on photochemical models at the present time, nor are likely to: although CIRS measurements could perhaps provide a route to constraining the size/occurrence of ongoing outgassing events.

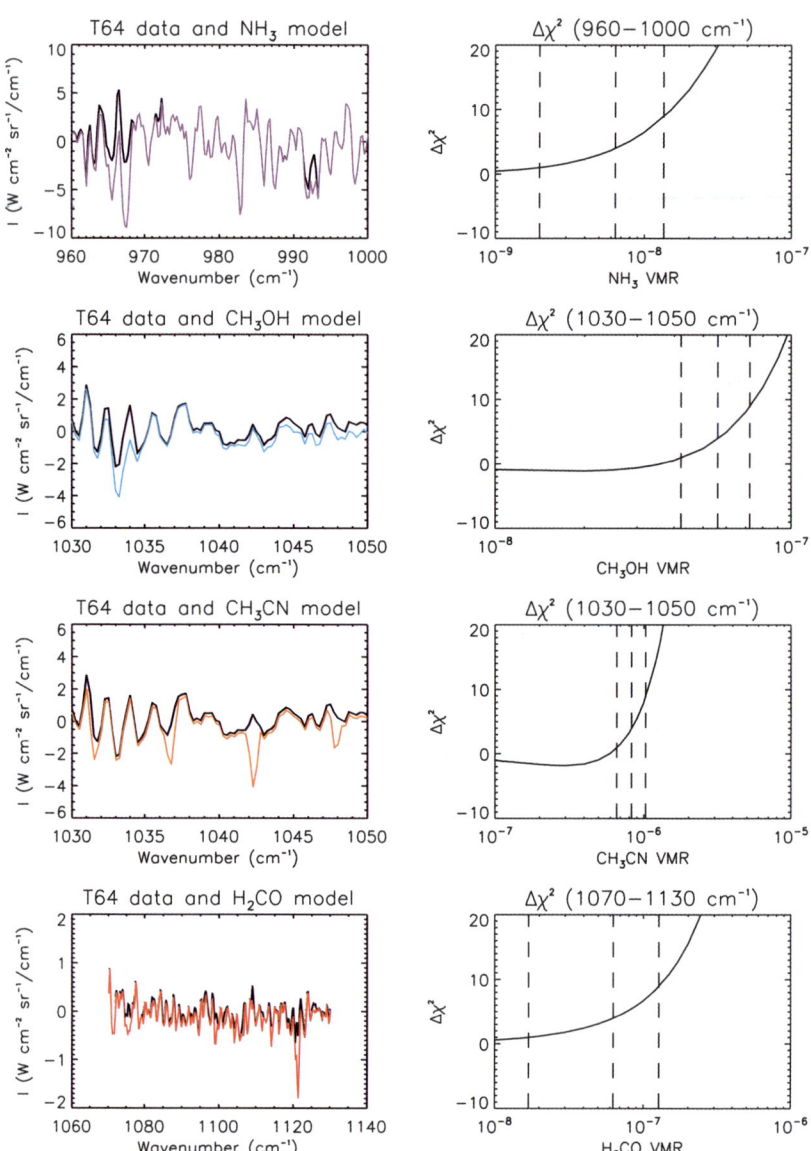

Fig. 6 Upper limits derived from T64 MIRLMPAIR data (75°N). Left column: Titan residual (data-model) spectrum in selected ranges after removal of known gases (black), and the residual found after addition of an arbitrary amount of trial gas to the model, showing location of spectral features (colored lines). Right column: the change $\Delta\chi^2$ in the goodness of fit measure χ^2 plotted over a wide range of trial gas abundances (solid line). Vertical dashed lines show the abundances for which $\Delta\chi^2$ is +1, +4, +9, corresponding to gas non-detections at the 1-σ, 2-σ and 3-σ significance levels, respectively.

5.2 Methanol

Methanol is expected to be a very minor component of Titan's stratosphere, and occurs through the combination of the CH_3 and OH radicals, the latter due to the small amount of water detected in Titan's atmosphere (about 8×10^{-9} at 400 km).[11] In the model of WA04, methanol reaches a peak value of a few 10^{-10}

Table 2 Calculated upper limits on abundances of undetected trace gases in Titan's stratosphere

Gas	Latitude	Pressure/mbar	Band	Wavenumber range for calculation Start/cm^{-1}	End/cm^{-1}	1σ NESR (nW cm^{-2} sr^{-1}/cm^{-1})	Abundance upper limits (ppb[a]) 1σ	2σ	3σ
NH$_3$	25°S	7.60	v_2	960	1000	2.34	0.59	0.88	1.3
CH$_3$OH	25°S	0.27	v_8	1030	1050	2.16	6.4	8.2	10
CH$_3$CN	25°S	0.27	v_7	1030	1050	1.56	49	78	109
H$_2$CO	25°S	0.27	v_6	1070	1130	0.39	2.2	8	16
NH$_3$	76°N	0.26	v_2	960	1000	3.47	2.0	6.4	14
CH$_3$OH	76°N	0.018	v_8	1030	1050	1.19	42	56	72
CH$_3$CN	76°N	0.018	v_7	1030	1050	0.86	660	830	1000
H$_2$CO	76°N	0.018	v_6	1070	1130	0.27	17	63	130

[a] Parts per billion.

at ~700 km, and declines steadily through the stratosphere, reaching 10^{-12} between 200–300 km.

As in the case of ammonia, our derived 3-σ upper limits for CH$_3$OH: 1×10^{-8} at (247 km, 25°S), and 7×10^{-8} at (348 km, 75°N), do not provide a strong restriction on the chemistry despite the relatively intense band available for the purpose. We note that an upper limit of 3×10^{-8} was inferred from the Cassini INMS OSI data at 1100 km.[15]

5.3 Acetonitrile

Following the discovery of this molecule *via* ground-based sub-mm spectroscopy,[10] the vertical profile was subsequently measured by the same technique (IRAM telescope) applied to the 147.6 and 220.7 GHz transitions measured at 78 kHz resolution.[52] This study revealed a disk-averaged vertical profile of a few 10^{-8} above 150 km increasingly slowly with altitude to 500 km, much more similar to the shallow HCN profile than the steep one inferred for HC$_3$N. The 1-D model of WA04 also exhibits a relatively small vertical gradient for CH$_3$CN in this range, although with somewhat lower abundances in the range $3-10 \times 10^{-9}$. ISO was unsuccessful in detecting acetonitrile at 1041 cm^{-1}, placing an upper limit of 5×10^{-10}.

Cassini CIRS measurements of the vertical profiles of HCN and HC$_3$N by limb sounding[32,53,54] reveal a more nuanced picture, with both HCN and HC$_3$N showing much steeper gradients in the equatorial regions and southern hemisphere, and shallow gradients in the north during the early Cassini epoch of Titan northern winter/spring. This results in large equator-to-north polar enhancements in abundance at 3 mbar, attributed to downward advection in a global Hadley cell,[55] as predicted by General Circulation Models (GCMs).[56] This large polar enhancement in nitriles should apply to acetonitrile as well, and therefore we might expect an enrichment of ~4 in the lower stratosphere as seen in HCN, a species with very similar photochemical lifetime,[20] and therefore abundances of up to ~10^{-7} at latitudes above 50°N.

Our 3-σ upper limit of 1×10^{-7} (247 km, 25°S) is therefore about one order of magnitude greater than the IRAM disk-average value (usually representative of low latitudes) at the same altitude, while our limit of 1×10^{-6} at (348 km, 75°N), is around two orders of magnitude greater than the IRAM value at this altitude, and does not yet provide a strong constraint on the expected enrichment factor.

Finally we note that in a pre-Cassini study,[57] the 3-σ sensitivity of CIRS to detection of various nitriles in nadir viewing was predicted. For acetonitrile, this was estimated to be 2×10^{-8} at 363 cm^{-1} (ν_8 band) and 4×10^{-8} at 717 cm^{-1} ($2\nu_8$ band), some order of magnitude less than our estimate, despite the lower intensities of these bands (4.5 and 8.5 cm^{-2} atm^{-1} respectively) compared to the ν_7 (21 cm^{-2} atm^{-1}).[58] A likely contributing factor for this difference is the non-optimal altitude probed here by CIRS FP4, as FP3 was instead targeted at the lower stratosphere where the gas pressure (and hence opacity for optically thin species) was 30× higher.

5.4 Formaldehyde

Formaldehyde is produced in Titan's atmosphere *via* the reaction of two HCO (formyl) radicals that are produced when an H atom attaches to CO.[20] WA04 predict an abundance of 1×10^{-9} at 4 mbar (130 km) in the lower stratosphere, with a declining abundance above to 400 km when the profile increases again. Note that oxygen species (CO, CO_2) do not appear to show significant polar enhancement,[31,54,59] and we presume the same may be true for H_2CO. Our 3-σ constraints of 2×10^{-8} at (247 km, 25°S) and 1×10^{-7} at (348 km, 75°N) are therefore not restrictive at the present time.

6 Conclusions and further work

This work has provided a preliminary search for NH_3, CH_3OH, CH_3CN and H_2CO in Titan's stratosphere from Cassini CIRS spectra, resulting in upper limits for these species at two (latitude, altitude) coordinates. NH_3 and CH_3OH in particular are predicted to occur at very low abundances, almost certainly out of reach of CIRS detection, whilst CH_3CN and H_2CO are predicted at abundances within 1–2 orders of magnitude of the current upper limits, holding out the possibility of detection by CIRS if the sensitivity of the search technique can be improved. There are two main routes to this goal: (i) by utilizing different spectral regions, where the combination of stronger gas bands and/or lower spectral noise produces a more favourable situation for detection, and (ii) targeting different altitudes and/or latitudes.

In the case of CH_3CN, a more intense band than the ν_7 currently used exists in the CIRS FP4 range: the ν_6 centred at 1450 cm^{-1} with an intensity nearly 6× greater. The current impediments to exploiting this band are two-fold: (i) the ν_6 (1379 cm^{-1}) and ν_8 (1468 cm^{-1}) bands of ethane which are clearly seen on the CIRS spectra, but for which we do not have a good line atlas; and (ii) aliasing in the CIRS spectrum which is predicted to become important at $\sim$1430 cm^{-1}.[28] Both these modelling issues can in principle be resolved, which may open an improved route to CH_3CN detection. In the case of H_2CO, no other exploitable bands exist in the CIRS range, however sensitivity may still be improved by targeting FP4 at a lower altitude ($\sim$125 km) where the predicted abundance is higher, and the atmosphere thicker.

Finally, future work must also expand the remit of this study to include other likely gas candidates in Titan's atmosphere, including allene (CH_2CCH_2), an isomer of the already detected propyne (CH_3C_2H); the two isomers of C_3H_6: propylene and cyclopropane, and many possible nitriles.

Acknowledgements

C. A. N. and D. E. J. were funded during this work by the NASA Cassini Data Analysis Program grant number NNX09AK55G. Research at the Jet Propulsion Laboratory (JPL), California Institute of Technology, was performed under contract with the National Aeronautics and Space Administration. I. K. thanks the Programme National de Planétologie (PNP, France) for their funding of the spectroscopy project.

References

1. G. P. Kuiper, *Astrophys. J.*, 1944, **100**, 378–383.
2. L. Trafton, *Astrophys. J.*, 1972, **175**, 295–306.
3. A. Tokunaga, S. Beck, T. Geballe and J. Lacy, *Bull. Am. Astron. Soc.*, 1980, **12**, 669.
4. R. Hanel, B. Conrath, F. M. Flasar, V. Kunde, W. Maguire, J. Pearl, J. Pirraglia, R. Samuelson, L. Herath, M. Allison, D. Gautier, P. Gierasch, L. Horn, R. Koppany and C. Ponnamperuma, *Science*, 1981, **212**, 192–200.
5. V. G. Kunde, A. C. Aiken, R. A. Hanel, D. E. Jennings, W. C. Maguire and R. E. Samuelson, *Nature*, 1981, **292**, 686–688.
6. W. C. Maguire, R. A. Hanel, D. E. Jennings, V. G. Kunde and R. E. Samuelson, *Nature*, 1981, **292**, 683–686.
7. R. E. Samuelson, W. C. Maguire, R. A. Hanel, V. G. Kunde, D. E. Jennings, Y. L. Yung and A. C. Aikin, *J. Geophys. Res.*, 1983, **88**, 8709–8715.
8. A. L. Broadfoot, B. R. Sandel, D. E. Shemansky, J. B. Holberg, G. R. Smith, D. F. Strobel, J. C. McConnell, S. Kumar, D. M. Hunten, S. K. Atreya, T. M. Donahue, H. W. Moos, J. L. Bertaux, J. E. Blamont, R. B. Ponphrey and S. Linick, *Science*, 1981, **212**, 206–211.
9. B. L. Lutz, C. de Bergh and T. Owen, *Science*, 1983, **220**, 1374–1375.
10. B. Bézard, A. Marten and G. Paubert, *Bull. Am. Astron. Soc.*, 1993, **25**, 1100.
11. A. Coustenis, A. Salama, E. Lellouch, Th. Encrenaz, G. L. Bjoraker, R. E. Samuelson, Th. de Graauw, H. Feuchtgruber and M. F. Kessler, *Astron. Astrophys.*, 1998, **336**, L85–L89.
12. A. Coustenis, A. Salama, B. Schulz, S. Ott, E. Lellouch, Th. Encrenaz, D. Gautier and H. Feuchtgruber, *Icarus*, 2003, **161**, 383–403.
13. J. H. Waite, H. Niemann, R. V. Yelle, W. T. Kasprzak, T. E. Cravens, J. G. Luhmann, R. L. McNutt, W.-H. Ip, D. Gell, V. De La Haye, I. Müller-Wodarg, B. Magee, N. Borggren, S. Ledvina, G. Fletcher, E. Walter, R. Miller, S. Scherer, R. Thorpe, J. Xu, B. Block and K. Arnett, *Science*, 2005, **308**, 982–986.
14. J. H. Waite, D. T. Young, T. E. Cravens, A. J. Coates, F. J. Crary, B. Magee and J. Westlake, *Science*, 2007, **316**, 870–875.
15. V. Vuitton, R. V. Yelle and M. J. McEwan, *Icarus*, 2007, **191**, 722–742.
16. J. Cui, R. V. Yelle, V. Vuitton, J. H. Waite, W. T. Kasprzak, D. A. Gell, H. B. Niemann, I. C. F. Müller-Wodarg, N. Borggren, G. G. Fletcher, E. L. Patrick, E. Raaen and B. A. Magee, *Icarus*, 2009, **200**, 581–615.
17. Y. L. Yung, M. Allen and J. P. Pinto, *Astrophys. J. Suppl.*, 1984, **55**, 465–506.
18. D. Toublanc, J. P. Parisot, J. Brillet, D. Gautier, F. Raulin and C. P. McKay, Photochemical modeling of Titan's atmosphere, *Icarus*, 1995, **113**, 2–26.
19. L. M. Lara, E. Lellouch, J. J. López-Moreno and R. Rodrigo, *J. Geophys. Res.*, 1996, **101**, 23261–23283.
20. E. H. Wilson and S. K. Atreya, *J. Geophys. Res.*, 2004, **109**, E06002.
21. V. G. Kunde, P. A. Ade, R. D. Barney, D. Bergman, J.-F. Bonnal, R. Borelli, D. Boyd, J. C. Brasunas, G. Brown, S. B. Calcutt, F. Carroll, R. Courtin, J. Cretolle, J. A. Crooke, M. A. Davis, S. Edberg, R. Fettig, M. Flasar, D. A. Glenar, S. Graham, J. G. Hagopian, C. F. Hakun, P. A. Hayes, L. Herath, L. Horn, D. E. Jennings, G. Karpati, C. Kellebenz, B. Lakew, J. Lindsay, J. Lohr, J. J. Lyons, R. J. Martineau, A. J. Martino, M. Matsumura, J. McCloskey, T. Melak, G. Michel, A. Morell, C. Mosier, L. Pack, M. Plants, D. Robinson, L. Rodriguez, P. Romani, W. J. Schaefer, S. Schmidt, C. Trujillo, T. Vellacott, K. Wagner and D. Yun, A Mission to the Saturnian Systems, *Proc. SPIE*, 1996, **2803**, 162–177.
22. F. M. Flasar, V. G. Kunde, M. M. Abbas, R. K. Achterberg, P. A. R. Ade, A. Barucci, B. Bézard, G. L. Bjoraker, J. C. Brasunas, S. B. Calcutt, R. Carlson, C. J. Césarsky, B. J. Conrath, A. Coradini, R. Courtin, A. Coustenis, S. Edberg, S. Edgington, C. Ferrari, T. Fouchet, D. Gautier, P. J. Gierasch, K. Grossman, P. G. J. Irwin, D. E. Jennings, E. Lellouch, A. A. Mamoutkine, A. Marten, J. P. Meyer, C. A. Nixon, G. S. Orton, T. C. Owen, J. C. Pearl, R. Prangé, F. Raulin, P. L. Read, P. N. Romani, R. E. Samuelson, M. E. Segura, M. R. Showalter, A. A. Simon-Miller, M. D. Smith, J. R. Spencer, L. J. Spilker and F. W. Taylor, *Space Sci. Rev.*, 2004, **115**, 169–297.
23. S. Vinatier, B. Bézard and C. A. Nixon, *Icarus*, 2007, **191**, 712–721.
24. B. Bézard, C. A. Nixon, I. Kleiner and D. E. Jennings, *Icarus*, 2007, **191**, 397–400.
25. C. A. Nixon, D. E. Jennings, B. Bézard, N. A. Teanby, R. K. Achterberg, A. Coustenis, S. Vinatier, P. G. J. Irwin, P. N. Romani, T. Hewagama and F. M. Flasar, *Astrophys. J.*, 2008, **681**, L101–L103.
26. D. E. Jennings, C. A. Nixon, A. Jolly, B. Bézard, A. Coustenis, S. Vinatier, P. G. J. Irwin, N. A. Teanby, P. N. Romani, R. K. Achterberg and F. M. Flasar, *Astrophys. J.*, 2008, **681**, L109–L111.

27 A. Coustenis, D. E. Jennings, A. Jolly, Y. Bénilan, C. A. Nixon, S. Vinatier, D. Gautier, G. L. Bjoraker, P. N. Romani, R. C. Carlson and F. M. Flasar, *Icarus*, 2008, **197**, 539–548.
28 C. A. Nixon, D. E. Jennings, J.-M. Flaud, B. Bézard, N. A. Teanby, P. G. J. Irwin, T. M. Ansty, A. Coustenis, S. Vinatier and F. M. Flasar, *Planet. Space Sci.*, 2009, **57**, 1573–1585.
29 M. Fulchignoni, F. Ferri, F. Angrilli, A. J. Ball, A. Bar-Nun, M. A. Barucci, C. Bettanini, G. Bianchini, W. Borucki, G. Colombatti, M. Coradini, A. Coustenis, S. Debei, P. Falkner, G. Fanti, E. Flamini, V. Gaborit, R. Grard, M. Hamelin, A. M. Harri, B. Hathi, I. Jernej, M. R. Leese, A. Lehto, P. F. Lion Stoppato, J. J. López-Moreno, T. Mäkinen, J. A. M. McDonnell, C. P. McKay, G. Molina-Cuberos, F. M. Neubauer, V. Pirronello, R. Rodrigo, B. Saggin, K. Schwingenschuh, A. Seiff, F. Simões, H. Svedhem, T. Tokano, M. C. Towner, R. Trautner, P. Withers and J. C. Zarnecki, *Nature*, 2005, **438**, 785–791.
30 H. B. Niemann, S. K. Atreya, S. J. Bauer, G. R. Carignan, J. E. Demick, R. L. Frost, D. Gautier, J. A. Haberman, D. N. Harpold, D. M. Hunten, G. Israel, J. I. Lunine, W. T. Kasprzak, T. C. Owen, M. Paulkovich, F. Raulin, E. Raaen and S. H. Way, *Nature*, 2005, **438**, 779–784.
31 A. Coustenis, R. K. Achterberg, B. J. Conrath, D. E. Jennings, A. Marten, D. Gautier, C. A. Nixon, F. M. Flasar, N. A. Teanby, B. Bézard, R. E. Samuelson, R. C. Carlson, E. Lellouch, G. L. Bjoraker, P. N. Romani, F. W. Taylor, P. G. J. Irwin, T. Fouchet, A. Hubert, G. S. Orton, V. G. Kunde, S. Vinatier, J. Mondellini, M. M. Abbas and R. Courtin, *Icarus*, 2007, **189**, 35–62.
32 S. Vinatier, B. Bézard, T. Fouchet, N. A. Teanby, R. de Kok, P. G. J. Irwin, B. J. Conrath, C. A. Nixon, P. N. Romani, F. M. Flasar and A. Coustenis, *Icarus*, 2007, **188**, 120–138.
33 B. N. Khare, C. Sagan, E. T. Arakawa, F. Suits, T. A. Callcott and M. W. Williams, *Icarus*, 1984, **60**, 127–137.
34 P. G. J. Irwin, N. A. Teanby, R. de Kok, L. N. Fletcher, C. J. A. Howett, C. Tsang, C. Wilson, S. B. Calcutt, C. A. Nixon and P. Parrish, *J. Quant. Spectrosc. Radiat. Transfer*, 2008, **109**, 1136–1150.
35 L. S. Rothman, I. E. Gordon, A. Barbe, D. Chris Benner, P. F. Bernath, M. Birk, V. Boudon, L. R. Brown, A. Campargue, J.-P. Champion, K. Chance, L. H. Coudert, V. Dana, V. M. Devi, S. Fally, J.-M. Flaud, R. R. Gamache, A. Goldman, D. Jacquemart, I. Kleiner, N. Lacome, W. J. Lafferty, J.-Y. Mandin, S. T. Massie, S. N. Mikhailenko, C. E. Miller, N. Moazzen-Ahmadi, O. V. Naumenko, A. V. Nitikin, J. Orpal, V. I. Perevalov, A. Perrin, A. Predoi-Cross, C. P. Rinsland, M. Rotger, M. Simeckova, M. A. H. Smith, K. Sung, S. A. Tashkun, J. Tennyson, R. A. Toth, A. C. Vandaele and J. Vander Auwera, *J. Quant. Spectrosc. Radiat. Transfer*, 2009, **110**, 533–572.
36 A. Perrin, F. Keller and J.-M. Flaud, *J. Mol. Spectrosc.*, 2003, **221**, 192–198.
37 R. Goody, R. West, L. Chen and D. Crisp, *J. Quant. Spectrosc. Radiat. Transfer*, 1989, **42**, 539–550.
38 R. Goody, R. West, L. Chen and D. Crisp, *J. Quant. Spectrosc. Radiat. Transfer*, 1989, **43**, 191–199.
39 C. A. Nixon, N. A. Teanby, S. B. Calcutt, S. Aslam, D. E. Jennings, V. G. Kunde, F. M. Flasar, P. G. J. Irwin, F. W. Taylor, D. A. Glenar and M. D. Smith, *Appl. Opt.*, 2009, **48**, 1912–1925.
40 C. P. Rinsland, V. Malathy Devi, D. Chris Benner, T. A. Blake, R. L. Sams, L. R. Brown, I. Kleiner, A. Dehayem-Kamadjeu, H. S. P. Müller, R. R. Gamache, D. L. Niles and T. Masiello, *J. Quant. Spectrosc. Radiat. Transfer*, 2008, **109**, 974–994.
41 C. P. Rinsland, S. W. Sharpe and R. L. Sams, *J. Quant. Spectrosc. Radiat. Transfer*, 2005, **96**, 271–280.
42 A.-M. Tolonen, M. Koivusaari, R. Paso, J. Schroderus, S. Alanko and R. Antilla, *J. Mol. Spectrosc.*, 1993, **160**, 554–565.
43 C. D. Rodgers. Inverse Methods for Atmospheric Sounding, *Series on Atmospheric, Oceanic and Planetary Physics – Vol. 2*. World Scientific, Singapore, 2000.
44 R. K. Achterberg, B. J. Conrath, P. J. Gierasch, F. M. Flasar and C. A. Nixon, *Icarus*, 2008, **194**, 263–277.
45 N. A. Teanby, P. G. J. Irwin, R. de Kok, A. Jolly, B. Bézard, C. A. Nixon and S. B. Calcutt, *Icarus*, 2009, **202**, 620–631.
46 W. H. Press, S. A. Teukolsky, W. T. Vetterling, and B. P. Flannery. *Numerical Recipes*, CUP, Cambridge, 2nd edn, 1992.
47 L.-H. Xu, R. M. Lees, P. Wang, L. R. Brown, I. Kleiner and J. W. C. Johns, *J. Mol. Spectrosc.*, 2004, **228**, 453–470.
48 N. A. Teanby, P. G. J. Irwin, R. de Kok, C. A. Nixon, A. Coustenis, B. Bézard, S. B. Calcutt, N. E. Bowles, F. M. Flasar, L. Fletcher, C. Howett and F. W. Taylor, *Icarus*, 2006, **181**, 243–255.

49 R. de Kok, P. G. J. Irwin, N. A. Teanby, E. Lellouch, B. Bézard, S. Vinatier, C. A. Nixon, L. Fletcher, C. Howett, S. B. Calcutt, N. E. Bowles, F. M. Flasar and F. W. Taylor, *Icarus*, 2007, **186**, 354–363.
50 C. A. Nixon, R. K. Achterberg, S. Vinatier, B. Bézard, A. Coustenis, P. G. J. Irwin, N. A. Teanby, R. de Kok, P. N. Romani, D. E. Jennings, G. L. Bjoraker and F. M. Flasar, *Icarus*, 2008, **195**, 778–791.
51 A. Coustenis, D. E. Jennings, C. A. Nixon, R. K. Achterberg, P. Lavvas, S. Vinatier, N. A. Teanby, G. L. Bjoraker, R. C. Carlson, L. Piani, G. Bampasidis, F. M. Flasar and P. N. Romani, *Icarus*, 2010, **207**, 461–476.
52 A. Marten, T. Hidayat, Y. Biraud and R. Moreno, *Icarus*, 2002, **158**, 532–544.
53 N. A. Teanby, P. G. J. Irwin, R. de Kok, S. Vinatier, B. Bézard, C. A. Nixon, F. M. Flasar, S. B. Calcutt, N. E. Bowles, L. Fletcher, C. Howett and F. W. Taylor, *Icarus*, 2007, **186**, 364–384.
54 N. A. Teanby, R. de Kok, P. G. J. Irwin, S. Osprey, S. Vinatier, P. J. Gierasch, P. L. Read, F. M. Flasar, B. J. Conrath, R. K. Achterberg, B. Bézard, C. A. Nixon and S. B. Calcutt, *J. Geophys. Res.*, 2008, **113**(e12), E12003.
55 N. A. Teanby, P. G. J. Irwin, R. de Kok and C. A. Nixon, *Philos. Trans. R. Soc. London, Ser. A*, 2009, **367**, 697–711.
56 S. Lebonnois, P. Rannou and F. Hourdin, *Philos. Trans. R. Soc. London, Ser. A*, 2009, **367**, 665–682.
57 A. Coustenis, Th. Encrenaz, B. Bézard, G. Bjoraker, G. Graner, M. Dang-Nhu and E. Arié, *Icarus*, 1993, **102**, 240–260.
58 F. Cerceau, F. Raulin, R. Courtin and D. Gautier, *Icarus*, 1985, **62**, 207–220.
59 S. Vinatier, B. Bézard, C. A. Nixon, A. Mamoutkine, R. C. Carlson, D. E. Jennings, E. A. Guandique, N. A. Teanby, G. L. Bjoraker, F. Michael Flasar and V. G. Kunde, *Icarus*, 2010, **205**, 559–570.

General Discussion

Dr Adams opened the discussion of the paper by Professor Sushil Atreya: With the detection of negative ions, ion–ion recombination becomes more important. Since it is an order of magnitude slower than electron–ion recombination, there is more time for build up of clusters on both the positive and negative ions. Also, ion–ion recombination is less energetic than electron–ion recombination and there will be less fragmentation and thus more rapid build up of the neutral clusters perhaps producing PAHs and PANHs.

Professor Atreya responded: This seems like a very attractive way for building really large negative ions, *via* clusters to 10 000 Da. Unfortunately the CAPS/ELS was not designed to measure ions, so identification of anything beyond the simple low mass ions is not possible. Indeed, negative as well as positive ion clusters do form in the earth's D-region as well as higher. Some discussion on this matter is included in my paper in this volume of *Faraday Discussions*.

Dr Manzanares asked: Although the solubility of water in cryogenic solvents is very small (probably below 0.01 ppm), our experience in the laboratory is that ice crystals form in solution. It is possible to see the crystals in liquid Ar, Kr, Xe, and N_2. They usually come from water impurities in the gas tank or the vacuum line before the gas condenses in the cell to form the liquid solvent. The absorption band of the ice crystals has a maximum around 3230 cm^{-1}. The crystals can be seen floating in the CH_4 and C_2H_6 solvents but the IR band of the ice crystals cannot be measured because the C–H absorption band of the hydrocarbon is too strong and wide. This C–H band overlaps any small bands near 3000 cm^{-1}. During the laboratory experiments, turbulence occurs when the gas condenses and the ice that is formed appears floating and moving in the solvent. It is possible that in Titan, perturbations of the ice layers underneath the lakes will produce pieces of ice of different sizes that will float in the liquid solution of methane and ethane.

Professor Atreya replied: This is a very interesting finding. First, the Cassini data are incapable of determining the make up of the lakes except for the identification of ethane. The presence of liquid methane, although suspected, is not yet confirmed due to the interference by the huge amount of methane gas in the atmosphere, but work is under way to fine-tune the analysis precisely for such identification in the VIMS data. The density of methane liquid is about one half that of water ice (I think ethane is similar to methane, but methane liquid is expected to dominate), so water ice may not float up, unless large pockets of gas make up its solid-state structure. However, it could be possible for pieces of water ice to be present and to be within the volume of the lakes. This is something that a future lake lander mission could attempt to look for.

Dr Vuitton asked: Is it possible to determine the acetylene-to-benzene ratio at the Huygens landing site from the GCMS data?

Professor Atreya replied: Not at this time, but there is some hope. Although the acetylene signature from the surface was unmistakable, the GCMS data cannot unambiguously confirm the presence of benzene at the surface. If present, it is too low to be separated from the benzene that was permanently present as background in the instrument throughout the probe's descent through the atmosphere. A simulation chamber is being built at Goddard, and using the flight spare there is a chance we could say something about actual benzene from the surface (amongst other things), hence the C_2H_2/C_2H_6 data

Professor Kaiser enquired: In your presentation, you outlined a critical temperature range in which the formation of molecular nitrogen *via* photolysis can happen. At lower temperature, your study comes to the conclusion that gas phase formation of molecular nitrogen is not feasible, since the hydrazine intermediate (N_2H_4) freezes out. We have shown in our group that the interaction of ionizing radiation, in the form of energetic electrons as generated in the track of galactic cosmic ray particles with solid ammonia at temperatures as low as 10 K, can lead to the formation of molecular nitrogen *via* N_2H_4 and HNNH as well as NNH_2.[1] How can this be applied to Titan? In a recent study, Molina-Cubero et al. derived an energy deposition of secondary particles from cosmic rays on Titan's surface of 4.5×10^9 eV cm^{-2} s^{-1}.[2] Therefore, in principle, molecular nitrogen could be also formed at lower temperatures *via* interaction of frozen hydrazine with secondary particles from the galactic cosmic radiation field. How easy it is to include these processes in your models? It this planned?

1 W. Zheng, D. C. Jewitt, Y. Osamura, and R. I. Kaiser, Formation of molecular hydrogen, nitrogen, hydrazine, and diazene in solid ammonia and implications to Solar System ices, *Astrophys. J.*, 2008, **674**, 1242–1250.
2 G. J. Molina-Cuberosa and J. J. Lopez-Morenoa, Ionization by cosmic rays of the atmosphere of Titan, *Planet. Space Sci.*, 1999, **47**, 1347–1354.

Professor Atreya replied: Since Titan was warm during its accretionary heating phase, plenty of ammonia, water and methane were outgassed from Titan's interior into the atmosphere. Once in the atmosphere, NH_3 was photolyzed to form nitrogen. It could not escape. Titan's purported ocean is believed to have a concentration of 5–10% dissolved ammonia, which is where the majority of the ammonia on Titan is present today. Any ammonia that is in the ice is buried under hundreds-of-metres-thick hydrocarbon–nitrile–PAH haze, so it's inaccessible even to the galactic cosmic rays (GCRs) that do make it down to the surface. There may be some exposed patches of ice, but the IR data do not show them to be ubiquitous. Ammonia could also be produced from N_2 in the ionosphere today, and some of it could make it to the ground intact. Firstly, the amounts would be so low that, even before its burial, dissociation by GCRs would make little difference to the atmospheric N_2. Secondly, this is circular reasoning—to make N_2, you need N_2 (which produced the NH_3 in the ionosphere to start with). Considering the above, although we are looking into the modification of surface material by GCRs, NH_3 is not a priority.

Professor Strobel said: As a follow-up to the question by Dr Vuitton, was benzene, at mass 78 Daltons, observed to rise in mixing ratio after the Huygens probe landed and heated the surface of Titan as observed for acetylene?

Professor Atreya replied: Within the range of uncertainty, there was no perceptible change from the permanent background of 78 Da in the GCMS, especially that which was present before the probe actually landed. If benzene had happened to be present on Titan's surface, it's concentration would have been too low to distinguish from the instrument's background.

Professor Plane opened the discussion of the paper by Dr Roger Yelle: Dr Yelle's paper emphasises the importance of dissociative electron recombination (DR) reactions in providing important synthetic pathways in planetary atmospheres. It is curious that the experimental study of DR is such a neglected field. A recent review by Florescu-Mitchell and Mitchell (*Phys. Rep.*, 2006,

430, 277–374, which contains an extensive compilation) emphasises that many classes of DR reactions have not been studied. A common response is "but they all have the same rate coefficients". This is not the case: DR rate coefficients of small positive ions vary by up to a factor of 20, and there is intriguing evidence that the DR rates of clusters grow rapidly with cluster size, which has implications for the ion-induced nucleation of particles.

Professor Mitchell replied: Dissociative recombination is generally studied using either the flowing afterglow or the merged beams method. The first method has been developed to the point where branching ratios can be measured, but the ion to be studied must be a terminal ion, *i.e.* an ion that does not react further with the parent gas. Thus this is a limitation on the method.

The modern merged beams technique is conducted with heavy ion storage rings, and there any ion that can be stored in the ring with a sufficient velocity to allow velocity matching with the electrons, can be studied. However the measurement of branching ratios requires that the fragments be separated by the energy sensitive neutral detector and this means that the beam energy must be sufficiently high to do this. This sets an upper limit on the useful ion mass. Most of the branching ratios listed in Florescu-Mitchell and Mitchell were measured using this technique.

Unfortunately of the four original storage rings used for these studies, two (CRYRING and TARN II) are now closed and a third (ASTRID) has been converted to full-time synchrotron radiation production. Only the TSR ring in Heidelberg is still active, but this is not a user facility and the concentration there is on light ions. Essentially at the current time, there is no opportunity to perform further studies using this method. This may change if the ion storage ring in China is available to such experiments.

With regard to absolute rates, these can still be measured using afterglow machines such as that of Nigel Adams or the group at Rennes. Indeed we at Rennes still have an on-going programme in this area and have developed a method for studying species with low volatility such as PAHs. Identification of specific data needs would be useful. Cluster ions of water and ammonia do indeed have high recombination rates that increase with the degree of clustering. This is not necessarily true for other clusters however although there is a scarcity of data. The mechanism of cluster ion recombination ("super-dissociative recombination") is not clear, so this could be an interesting field to study in the future. Metallic ion recombination is also a largely unexplored field.

Dr Adams commented: In this comment on the products of electron–ion recombination, an alternative to Storage Ring (SR) measurements is the recently developed Flowing Afterglow (FA) technique (C. D. Molek, V. Poterya, N. G. Adams and J. L. McLain, 2009, *Int. J. Mass Spectrom.*, **285**, 1–11). When the same ion species (H_3O^+, N_2H^+, CH_5^+) are studied by the two techniques, there are usually discrepancies. This has been explained for N_2H^+ where the SR recombining ions were misidentified and were actually $^{14}N^{15}N^+$ (W. D. Geppert and M. Larsson, *Mol. Phys.*, 2008, **106**, 2199–2226). For CH_5^+, the main FA product was CH_4 and in the SR was CH_3. A possible explanation (V. Zhaunerchyk, M. Kaminska, E. Vigren, M. Hamberg, W. D. Geppert, M. Larsson, R. D. Thomas and J. Semaniak, *Phys. Rev. A*, 2009, **79**, 030701(R)), is that in the FA, the fragmentation energy is quenched by collisions with He on a time scale of 10^{-7} s, whereas in the SR the products are detected after the longer time of 10^{-6} s allowing time for the excitation to cause fragmentation of CH_4 to CH_3 + H. The discrepancy for H_3O^+ has not been explained.

Dr Adams opened the discussion of the paper by Dr Roger Yelle: In the paper, the electron–ion recombination of $H_2CNH_2^+$ is considered. Since Titan's ion chemistry is that of protonation and proton transfer, would the recombination of protonated methylamine be important?

Dr Yelle answered: We do include CH_3NH_2 and $CH_3NH_3^+$ in our model and $CH_3NH_3^+$ has been detected in the Cassini/INMS ion spectra of Titan's upper atmosphere. According to the models $CH_3NH_3^+$ is produced primarily by proton transfer from XH^+ to CH_3NH_2 and is lost primarily through electron recombination. The products of the latter reaction are assumed to be either $CH_2NH + H_2 + H$ or $CH_4 + NH_2$, with equal probability. Recombination of $CH_3NH_3^+$ is therefore a source of CH_2NH, but it is a very minor source, with a rate of only 10^{-6} that of R42 in our paper, according to our models.

Professor Mitchell commented: I agree with Nigel Adam's comment that discrepancies between afterglow and storage ring measurements can be due to the production of fragments in excited states that can subsequently dissociate to form different products under the differing conditions of the two experimental techniques. In past models, it was usual to suppose that recombination of hydrocarbon ions would lead to the release of one or more hydrogen atoms. In fact, the breaking of carbon–carbon bonds is often the dominant channel and it should be realized that dissociative recombination is an explosive process involving high energy releases (often 10 eV or more). In one process that we have not yet published, that of the trimethylamine ion, $N(CH_3)_3^+$, we found that the release of a methyl group was the dominant dissociation channel and not the release of a hydrogen atom.

Professor Atreya inquired: In Fig. 3 of your paper, you show several pathways for the formation of NH_3 in the ionosphere, particularly those that involve NH_4^+ and NH_2 in the final steps. Could you summarize their relative importance, *i.e.* percentages of production from each?

Dr Yelle responded: Production of NH_3 from NH_2 (which is produced primarily from $CH_2NH_2^+$) dominates production of NH_3. The column integrated rate for this pathway is about 25 times larger than for the NH_x^+ pathway shown in the bottom of Fig. 3.

Professor Casavecchia (speaking on behalf of N. Balucani, A. Bergeat, L. Cartechini, G. G. Volpi, D. Skouteris and M. Rosi) asked: In relation to the paper by Yelle *et al.*,[1] we would like to make some comments on the $N(^2D)$ reactions included in model and listed in Table 1 of ref. 1. Only two reaction channels are considered for the reaction $N(^2D) + CH_4$, that is those leading to $CH_2=NH + H$ and $N + CH_3$ (R29 and R30 of Table 1, respectively). Apparently, while for the reaction R29 a temperature dependence has been considered and the rate coefficient given is $3.84 \times 10^{-11} e^{-750/T}$ cm^3 $molec^{-1}$ s^{-1}, for the reaction R30 a rate coefficient of 9.6×10^{-12} cm^3 $molec^{-1}$ s^{-1} without T dependence has been used. Which is the reason for this choice? According to the cited references,[2,3] indeed, the rate coefficient as a function of the temperature in the 223–292 K range for the $N(^2D)+CH_4$ reaction is given by $k = 7.1 \times 10^{-11} e^{-750/T}$ cm^3 $molec^{-1}$ s^{-1}.[1,2] Since in the kinetic study of ref. 3 only the disappearance rate of $N(^2D)$ was followed, it is not correct to attribute the temperature dependence only to R29. Interestingly, the room temperature absolute yields of NH (produced in conjunction with CH_3) and H (which can be produced in conjunction not only with $CH_2=NH$, but also with its isomers CH_3N and $CH-NH_2$) have been determined to be 0.3 ± 0.1 and 0.8 ± 0.2, respectively, by Umemoto *et al.*[4] Therefore, the rate coefficient used in the model by Yelle *et al.* for the channel $NH + CH_3$ is too large by a factor of five at room temperature. In addition, in our laboratory we have investigated the H-displacement channel for the reaction $N(^2D) + CH_4$ by means of the crossed molecular beam technique (CMB) with mass-spectrometric (MS) detection with the aim of understanding which of the three possible isomers ($CH_2=NH$, $CHNH_2$ and CH_3N) is actually produced in conjunction with H. To achieve that, we have performed a systematic study of the $N(^2D) + CH_4$ reaction in a wide range of collision energies (E_c, from about 20 to 60 kJ mol^{-1}).[5,6] From the experimental results we have evidence that two isomers are actually formed, that is methanimine (CH$_2$=NH) and methylnitrene, a diradical with formula CH$_3$N.[5,6] Notably, the relative branching ratio for the formation of CH$_2$=NH with respect to CH$_3$N was found to change considerably with E_c,[6] spanning from 1.3 at E_c = 20 kJ mol^{-1} to 0.03 at 60 kJ mol^{-1} (see Fig. 1, below). This is due to a competition between two different reaction mechanisms, one dominated by dynamical effects—with the formation of the products CH$_3$N (methylnitrene) + H or NH + CH$_3$—and one by the statistical redistribution of the available energy—which invariably leads to the formation of the much more stable products CH$_2$=NH + H. Since the production of the CH$_3$N isomer is dominated by the same dynamical effects that lead to the production of CH$_3$ + NH, the yields of these two channels are expected to be comparable at all collision energies (and temperatures).[6] Given the trend with E_c (see Fig. 1), the yield of NH + CH$_3$ at the temperature of the upper atmosphere of Titan is probably even lower than derived by Umemoto *et al.* at room temperature. In any case, we would like to ask Prof. Yelle whether the inclusion of the methylnitrene formation channel can alter the output of their model. CH$_3$N is a very reactive species and can open up reaction chains not considered in the models so far.

Another comment regards the N(^{2}D) + H$_2$ reaction, which has been included in the model as reactions R31 and R32 (see Table 1 of ref. 1). Again, it is not clear why the authors chose these rate coefficients. Reaction 31 is written as the physical quenching of the ^{2}D state due to collisions with H$_2$. In the model, the rate constant attributed to this process at room temperature is the one recommended by Herron[2] (2.2 × 10^{-12} cm^3 molec^{-1} s^{-1}) regarding the global disappearance rate of N(^{2}D) in the presence of H$_2$. Nevertheless, there is wide consensus on the fact that the disappearance rate measured for the N(^{2}D) + H$_2$ system is mainly due to reactive collisions.[2,7,8] Also, a collisional quenching by H$_2$ more efficient than that by N$_2$ (1.7 × 10^{14} cm^3 molec^{-1} s^{-1}) is not convincing. Finally, the excellent agreement between the

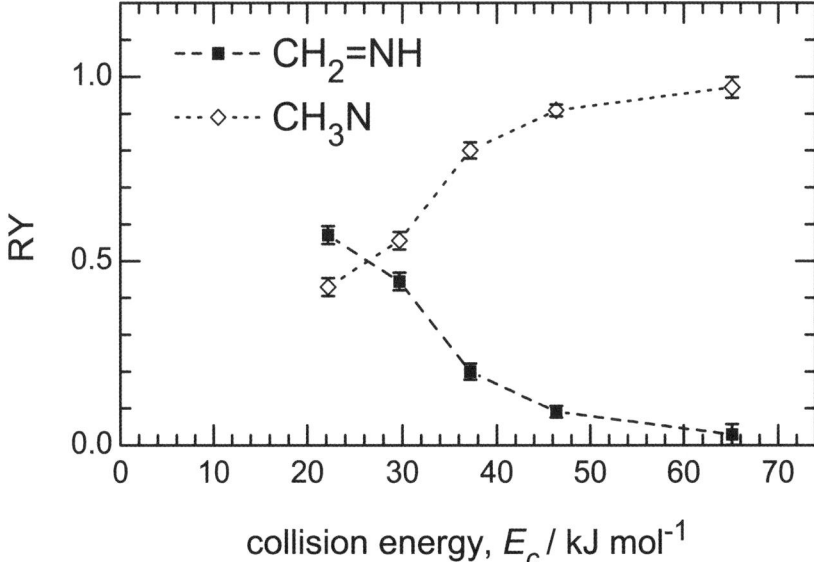

Fig. 1 Relative yields, RY, of the isomeric products CH$_2$=NH (solid squares) and CH$_3$N (open diamonds) as a function of collision energy, E_c, derived from crossed molecular beam studies of the N(^{2}D) + CH$_4$ reaction.[6] The dashed and dotted lines joining the data points are drawn to guide the eye only.

calculated rate coefficient on the NH_2 reactive potential energy surface with the measured value at 300 K definitely points to a negligible contribution of the collisional quenching to the disappearance rate of $N(^2D)$ in the $N(^2D) + H_2$ system.[9,10] Incidentally, the $N(^2D) + H_2$ reaction (and its isotopic variant $N(^2D) + D_2$) has been thoroughly investigated at the level of reaction dynamics[9,10] and is actually one of the few reactions for which a comparison between state-of-the-art quantum dynamical calculations and detailed experimental data has been possible.[9] This has allowed us to reach a complete description of the reaction mechanism and of some important reaction characteristics. For instance, the NH product is formed by the reaction in the $v' = 0$–4 vibrational levels.[9,11] Since the energy content of the vibrationally excited NH levels is quite sizeable ($v_{0-1} = 43.8$ kJ mol^{-1}; $v_{0-2} = 85.8$ kJ mol^{-1}; ($v_{0-3} = 125.9$ kJ mol^{-1}), the vibrational excitation can significantly enhance the capability of NH to undergo subsequent reactions. Also, from the determination of the product energy release, we have observed that the H atoms are produced with ~50 kJ mol^{-1} of translational energy, which translates into an average velocity of ~10 000 m s^{-1}. Is this velocity enough to escape the atmosphere of Titan? Can this type of information be of use to determine the escape probability of various species in the upper atmosphere of Titan? Other $N(^2D)$ reactions we have investigated in our laboratory include $N(^2D) + C_2H_4$, C_2H_2 and C_2H_6.[12–14] The main reaction products for the reaction $N(^2D) + C_2H_4$ are 2H-azirine (cyclic-$H_2C(N)CH$) and ketenimine ($CH_2=C=NH$).[12,13] According to the experimental determination of the product energy release, we have been able to establish that roughly 67% of the 2H-azirine molecules and 35% of the ketenimine molecules are formed with enough internal energy to spontaneously tautomerize to the most stable isomer, acetonitrile.[12] More recent results from our laboratory have also shown that ~10% of the reaction proceeds directly toward the formation of acetonitrile at $E_c = 28.2$ kJ mol^{-1}.[13] We are now performing RRKM statistical calculations to estimate the fraction of 2H-azirine and ketenimine molecules that isomerize to acetonitrile.[13] In any case, the use of a unitary branching ratio for the formation of CH_3CN in the reaction $N(^2D) + C_2H_4$, as done in the model by Yelle et al.[1] as well as in other models, is certainly too simplistic. Quite interestingly, molecules such as 2H-azirine (characterized by a strained ring) and ketenimine (with two unsaturated bonds) provide polymerization sites which might ultimately lead to the formation of the nitrogen-rich aerosols of Titan.[15] Finally, as far as the reaction $N(^2D) + C_2H_6$ is concerned, we note that in this model, as well as in some of the preceding ones, the only reaction channel considered is the one leading to $NH + C_2H_5$ (reaction R35 of Table 1 of ref. 1). As we have demonstrated in the paper presented at this discussion,[16] this is not actually the case and the main reaction products are $CH_2=NH + CH_3$ and $CH_3CH=NH + H$.

1 R. V. Yelle, V. Vuitton, P. Lavvas, S. J. Klippenstein, M. A. Smith, S. M. Horst and J. Cui, *Faraday Discuss.*, 2010, **147**, DOI: 10.1039/c004787m
2 J. T. Herron, *J. Phys. Chem. Ref. Data*, 1999, **28**, 1453.
3 T. Takayanagi, Y. Kurosaki, K. Sato, K. Misawa, Y. Kobayashi and S. Tsunashima, *J. Phys. Chem. A*, 1999, **103**, 250.
4 H. Umemoto, T. Nakae, H. Hashimoto, K. Kongo and M. Kawasaki, *J. Chem. Phys.*, 1998, **109**, 5844.
5 P. Casavecchia, N. Balucani, L. Cartechini, G. Capozza, A. Bergeat and G. G. Volpi, *Faraday Discuss.*, 2001, **119**, 27; N. Balucani and P. Casavecchia, *Orig. Life Evol. Biosphere*, 2006, **36**, 443; N. Balucani, *Int. J. Mol. Sci.*, 2009, **10**, 2304.
6 N. Balucani, A. Bergeat, L. Cartechini, G. G. Volpi, P. Casavecchia, D. Skouteris and M. Rosi, *J. Phys. Chem. A*, 2009, **113**, 11138.
7 R. J. Donovan and D. Husain, *Chem. Rev.*, 1970, **70**, 489.
8 K. Schofield, *J. Phys. Chem. Ref. Data*, 1979, **8**, 723.
9 N. Balucani, L. Cartechini, G. Capozza, E. Segoloni, P. Casavecchia, G. G. Volpi, F. J. Aoiz, L. Banares, P. Honvault and J. M. Launay, *Phys. Rev. Lett.*, 2002, **89**, 013201; N. Balucani, P. Casavecchia, L. Banares, F. J. Aoiz, T. Gonzalez-Lezana, P. Honvault and J.-M. Launay, *J. Phys. Chem. A*, 2006, **110**, 817.

10 M. Alagia, N. Balucani, L. Cartechini, P. Casavecchia, G. G. Volpi, L. A. Pederson, G. C. Schatz, G. Lendvay, L. B. Harding, T. Hollebek, T.-S. Ho and H. Rabitz, H., *J. Chem. Phys.*, 1999, **110**, 8857; N. Balucani, M. Alagia, L. Cartechini, P. Casavecchia, G. G. Volpi, L. A. Pederson and G. C. Schatz, *J. Phys. Chem. A*, 2001, **105**, 2414; P.Casavecchia, N. Balucani, M. Alagia, L. Cartechini and G. G. Volpi, *Acc. Chem. Res.*, 1999, **32**, 503.
11 H. Umemoto, N. Terada and K. Tanaka, *J. Chem. Phys.*, 2000, **112**, 5762.
12 N. Balucani, L. Cartechini, M. Alagia, P. Casavecchia and G. G. Volpi, *J. Phys. Chem. A*, 2000, **104**, 5655.
13 N. Balucani, F. Leonori, R. Petrucci, W. D. Geppert, M. Hamberg, P. Casavecchia, D. Skouteris and M. Rosi, manuscript in preparation.
14 N. Balucani, M. Alagia, L. Cartechini, P. Casavecchia, G. G. Volpi, K. Sato, T. Takayanagi and Y. Kurosaki, *J. Am. Chem. Soc.*, 2000, **122**, 4443.
15 G. Israel, C. Szopa, F. Raulin, M. Cabane, H. B. Niemann, S. K. Atreya, S. J. Bauer, J. F. Brun, E. Chassefiere, P. Coll, E. Conde, D. Coscia, A. Hauchecorne, P. Millian, M. J. Nguyen, T. Owen, W. Riedler, R. E. Samuelson, J. M. Siguier, M. Steller, R. Sternberg and C. Vidal-Madjar, *Nature*, 2005, **438**, 796.
16 N. Balucani, F. Leonori, R. Petrucci, M. Stazi, D. Skouteris, M. Rosi and P. Casavecchia, *Faraday Discuss.*, 2010, **147**, DOI: 10.1039/c004748a.

Dr Yelle responded: 1. Unfortunately, there was a typographical error in the original version of Table 1 that resulted in the omission on the temperature factor for reaction R30. We do in fact include the temperature dependence and the rate is correct.

2. We do not include CH_3N in our models. Perhaps it would be interesting to investigate this in the future, if the available reaction rate data allowed it.

3. I cannot comment about whether your interpretation of the situation is to be preferred over that in Herron,[1] but it is unlikely to be important for the model. Loss of $N(^2D)$ through collisions with H_2 is about a factor of 40 times smaller than other loss processes in the model. The dominant loss processes are collision de-excitation with N_2, reaction with C_2H_4, and radiative de-excitation.

4. We agree that our chemical scheme (and others) is likely too simplistic. This is especially true regarding isomers, including the specific example that you point out. However, at this time this is really no alternative to oversimplifying because very little is known about the chemistry of these isomers and it is not possible to include them in the model in any reasonable way. At least with these oversimplified schemes the content of the model is clear as well as its shortcomings and these can serve as a guide to future laboratory measurements or theoretical estimates of reaction rate data, which is one of our main objectives.

1 J. T. Herron, *J. Phys. Chem. Ref. Data*, 1999, **28**, 1453.

Professor Sims spoke: Professor Casavecchia, in his comments, has referred to the role of vibrational excitation in promoting reactions. Numerous studies on reactions of small radicals and molecules have shown that when the radical is vibrationally excited, and the bond where the vibrational excitation is localised is conserved during the reaction, then the vibrational excitation has little or no effect on the rate coefficient for the reaction—the reaction can be said to be 'vibrationally adiabatic', as shown for example in a series of studies that Ian Smith and I performed on reactions of the CN radical in its ground and first vibrationally-excited states.[1-3] In general, any rate enhancement will be most pronounced when the bond that is broken in the reaction is excited vibrationally, and when that reaction can be considered 'direct'.[4] The vibrational excitation sometimes observed by Professor Casavecchia and co-workers in their radical beams should not unduly influence the overall reaction cross-section if the excited bond is conserved in the reaction products. However, if the reaction proceeds *via* a collision complex, then this excess vibrational energy may become available to the system and possibly influence the subsequent product branching that is measured in these experiments.

1 I. R. Sims and I. W. M. Smith, *J. Chem. Soc., Faraday Trans. II*, 1989, **85**, 915.
2 I. R. Sims and I. W. M. Smith, *J. Chem. Soc., Faraday Trans. II*, 1988, **84**, 527.
3 I. R. Sims and I. W. M. Smith, *Chem. Phys. Lett.*, 1988, **149**, 565.
4 I. W. M. Smith, *Acc. Chem. Res.*, 1990, **23**, 101.

Professor Casavecchia added: I would like to remind this audience that nowadays it is possible for many bimolecular reactions to measure the rate constant down to very low temperatures (10 K or so), as the work of Rowe, Smith, Sims *et al.* with the CRESU technique has shown over the years[1] and also how the paper by Morales *et al.*[2] in this Discussion documents. Therefore, it is possible to have suitable information at the relevant temperature also for $N(^2D)$ reactions for use in photochemical models of Titan's atmosphere. Furthermore, reaction dynamics studies with the "universal" crossed molecular beam technique are able to provide information on primary product and branching ratios of bimolecular reactions also at quite low translational energies (temperatures),[3] which are approaching those of the atmosphere of Titan.[4] For instance, in the case of the reaction $C(^3P,^1D) + C_2H_2$ we were able to reach translational energies of 3.6 kJ mol^{-1} (corresponding to 300 K) by crossing the two reactant beams at 45°,[5] but lower energies are also possible, of the order of 1–2 kJ mol^{-1} (corresponding to about 80–160 K) by using smaller beam crossing angles. I would also like to note that an even lower translational energy, of 0.8 kJ mol^{-1} (corresponding to a temperature of about 65 K), was reached for the $C + C_2H_2$ reaction using pulsed crossed-beam techniques with laser detection at the interaction region, as done by Costes and coworkers[5,6] by employing a beam crossing angle of 22°.

1 A. Canosa, I. R. Sims, D. Travers, I. W. M. Smith and B. R. Rowe, *Astron. Astrophys.*, 1997, **323**, 644.
2 S. B. Morales, S. D. Le Picard, A. Canosa and I. R. Sims, *Faraday Discuss.*, 2010, **147**, DOI: 10.1039/c004219f
3 P. Casavecchia, F. Leonori, N. Balucani, R. Petrucci, G. Capozza and E. Segoloni, *Phys. Chem. Chem. Phys.*, 2009, **11**, 46; N. Balucani, G. Capozza, F. Leonori, E. Segoloni and P. Casavecchia, *Int. Rev. Phys. Chem.*, 2006, **25**, 109.
4 N. Balucani, A. Bergeat, L. Cartechini, G. G. Volpi, P. Casavecchia, D. Skouteris and M. Rosi, *J. Phys. Chem. A*, 2009, **113**, 11138.
5 F. Leonori, R. Petrucci, E. Segoloni, A. Bergeat, K.M. Hickson, N. Balucani and P. Casavecchia, *J. Phys. Chem. A*, 2008, **112**, 1363.
6 M. Costes, N. Daugey, C. Naulin, A. Bergeat, F. Leonori, E. Segoloni, R. Petrucci, N. Balucani and P. Casavecchia, *Faraday Discuss.*, 2006, **133**, 157.

Professor Mason asked: I would like to ask about and upon the role of heterogeneous chemistry in the production of these species *e.g.* on the surface of Titan's dust/tholins. Recent experiments we have performed using discharges to mimic the chemistry in Titan's atmosphere (see our poster at this meeting) have shown that ammonia production rates are greatly increased when there is a surface present for the radicals to act upon. Can you comment upon the role of heterogeneous chemistry in the models?

Dr Yelle responded: We did not include it in the paper, but we did perform some tests of the possible importance of heterogeneous chemistry on NH_3 production. We used the rates derived for H recombination on tholins measured by Sekine *et al.* (2008) and an aerosol density based on the density of large mass negative ions from Coates *et al.* (2006) and found no noticeable effect on the calculated NH_3 density. I think that this is because we are studying the NH_3 density in Titan's upper atmosphere, essentially the thermosphere, and the aerosols densities, though large and extremely surprising for a thermosphere, are really quite small in an absolute sense. I suspect that heterogeneous chemistry does play an important role deeper in Titan's atmosphere.

Y. Sekine, H. Imanaka, T. Matsui, B. N. Khare, E. L. O. Bakes, C. P. McKay and S. Sugita, The role of organic haze in Titan's atmospheric chemistry. I. Laboratory investigation on heterogeneous reaction of atomic hydrogen with Titan tholin, *Icarus*, 2008, **194**, 186–200.

A. J. Coates, F. J. Crary, G. R. Lewis, D. T. Young, J. H. Waite and E. C. Sittler, Discovery of heavy negative ions in Titan's ionosphere, *Geophys. Res. Lett.*, 2007, **34**, L22103.

Dr Ellinger asked two questions: 1: You present transition states with energies very close to that of the starting products. Do you think that the procedure of extrapolating energies to CBS limit is an alternative to the usual techniques generally used for correcting BSSE artifacts ?

2: The CAS for the NH_2 + H_2CN reaction (4e in 4mo) contains the two open shells together with the π and π^* orbitals of H_2CN. The CH bond that is broken is not included in the active space. Why not using a (6e in 6mo) active space?

Dr Klippenstein answered: The CBS limit provides an effective approach for removing BSSE artifacts that arise in studies of weak binding with limited basis sets. The more commonly employed approaches, based on comparisons of calculations with or without the extra basis sets of the interacting molecule, generally provide only qualitative to semiquantitative corrections. In contrast, by definition, the CBS limit provides the accurate binding energy for a given method. Our experience is that estimates of the CBS limits based on the extrapolation of explicit cc-pVTZ and cc-pVQZ calculations are generally accurate enough that BSSE artifacts are insignificant.

You are correct that the inclusion of the breaking CH bond in the active space for the abstraction reaction NH_2 + $H_2CN \rightarrow NH_3$ + HCN to yield a (6e,6o) active space would be a better calculation. However, our experience from related calculations is that for highly exothermic abstractions such as this, where the transition state is very early, the inclusion of the active CH bond has only a modest effect on the energetics (*e.g.*, a 10% effect on the predicted rate), and so we have chosen to employ the simpler (4e,4o) active space described in the paper. In such calculations, we have found that it is most important to include the two radical orbitals and that the inclusion of the full π space can also be of some kinetic significance.

Professor Plane opened the discussion of the paper by Dr Teanby: Using the gradients of species within the polar vortex is an elegant way to estimate photochemical lifetimes. Over the length of time that the CIRS instrument on Cassini has been making observations, do you have some sense of the variability of this dynamical "focusing" effect? Also, removal of a species could be through a process other than "photochemistry"—such as poymerization to form aerosols. Is this a possible reason for the much shorter lifetimes of the nitrogen-containing species in Fig. 8 in the paper?

Dr Teanby replied: The Cassini mission currently covers 6 years (1/5th of a Titan year). Some variations have indeed been seen throughout the mission—for example a decrease in abundances over the south pole (Teanby *et al.*, 2009, ref. 10). However, the equatorial region has maintained a stable composition during this time and the north polar latitudes have sparse coverage. Therefore, the dynamical focusing presented in this work represents an average for the northern winter season, but should be representative of the atmosphere as a whole. We expect that the magnitude of the effect will change soon as the atmospheric circulation changes from a single cell to two symmetric cells.

Removal processes for the chemical species could definitely involve processes other than photochemistry. In fact, this is one of the conclusions of this work—the difference in enrichment ratios of nitrile and hydrocarbon species with similar expected photochemical lifetimes indicates that additional processes are at work. Polymerisation is a definite contender. Other possibilities suggested at the meeting were loss due to formation of dimers/clusters or heterogeneous chemistry catalysed on the surface of the haze particles. Measurements from Cassini's extended mission and further chemical modelling should help shed more light on the cause.

Dr Moses commented: Do you have any ideas for the processes that could be destroying the nitriles in Titan's stratosphere that are not being included in the photochemical models?

Dr Teanby responded: Some possibilities include: nitrile polymerisation; haze particles acting as sinks; formation of dimers or clusters; or heterogeneous chemistry/loss processes catalysed on the surface of haze particles. Cosmic rays appear to be important in Titan's stratosphere, which could cause some additional effects. Uncertainties in rate coefficients could also contribute.

Dr Knox asked: Reactions taking place on the surface of the haze particles could represent a further sink for the nitriles. Have such reactions been considered to date?

Dr Teanby replied: As far as I am aware, such reactions are not considered in current photochemical schemes. This could be an important source of nitrile loss, although the processes involved would need to be nitrile-specific, so as not to produce a similar effect on the hydrocarbons.

Professor Zwier asked: Can you say anything further about the thermodynamics of dimer/trimer formation under the conditions of Titan's atmosphere? Is it plausible that small clusters of the nitriles might be stable, and hence serve as the basis for subsequent photochemistry?

Dr Teanby answered: This could be a plausible mechanism—although further lab work would be required to determine the efficiency of such reactions under the conditions in Titan's atmosphere.

Professor Suits remarked: I would like to call your attention to some recent work from my group, in collaboration with Ralf Kaiser and Alex Mebel. We have shown (P. C. Singh *et al.*, *Astrophys. J.*, 2010, **710**, 112) that dimers or complexes may play an important but overlooked role in Titan's low temperature atmosphere. In this work we showed that the diacetylene dimer underwent photodissociation quite efficiently to give H transfer and H loss processes with thresholds 2 eV lower than dissociation of the monomer. These reactions produced resonantly stabilized radicals directly in one step. We anticipate similar behavior and higher mixing ratios for other complexes involving more abundant species such as acetylene, cyanoacetylene and HCN. One other possible implication of these reactions is that such complexes may provide a "built-in" third-body that can take away excess energy, without need of a second collision, if the complex is involved in some reaction.

Dr Moses remarked: If clustering or dimer formation were responsible for the loss of HCN and HC_3N in Titan's atmosphere, and you point out that C_2H_2 clustering operates in a similar manner, then why isn't there also a model-data discrepancy with C_2H_2, as occurs with the nitriles?

Dr Klippenstein replied: Both HCN and HC_3N have strong dipole moments, which likely yield greatly increased bond strengths for their dimers as compared with C_2H_2 dimers. Thus, the concentration of C_2H_2 dimers may be substantially lower and their effect may be greatly reduced.

Professor Strobel enquired: CO in Titan's atmosphere has a very long chemical time constant on the order of 1 GY and thus should be well mixed throughout the atmosphere including the vortex region in the winter northern polar region. Could one infer the effective temperature for CO line formation in the vortex based on these facts?

Dr Teanby replied: Yes, in theory this should be possible. By assuming that the equatorially derived CO abundance, where we have good temperature control, is applicable to the entire atmosphere, the height of the far-IR CO emission peaks should be mainly controlled by the temperature of the lower stratosphere. This should give us an idea of the temperature at the 20 mbar pressure level.

Professor Plane remarked: In order to determine the abundances of species from the far infra-red, you presumably need to know the local temperature of the emitting species. Can you say more about how that is measured?

Dr Teanby replied: Yes, it is essential to know the local temperature. Because the methane abundance was measured accurately by the Huygens probe GCMS we can fix the methane mixing ratio and use the v_4 emission band (around 1300 cm^{-1} or 7.5 µm), also measured by CIRS, to determine the stratospheric temperature. This gives us temperature information for 10–0.1 mbar at the equator and 1–0.01 mbar around the north pole. The local temperature of HCN is well constrained as condensation removes it from the atmosphere below ~10 mbar at the equator and ~2 mbar at the north pole, so emission from deeper levels can be neglected. However, CO does not condense so at the north pole the temperature is poorly constrained and only equatorial abundances of CO are reliable.

Professor Plane opened the discussion of the paper by Conor A Nixon: Were there any big surprises, where you failed to detect species which should occur at concentrations above the detection limits you have estimated?

Dr Nixon answered: No—at this time our upper limits for the four species are not constraining, relative to existing photochemical model predictions and (in the case of CH$_3$CN) prior measurements. The closest we come to testing predictions at present is for acetonitrile and formaldehyde, where our constraints at (25°S, 250 km) are approximately one order of magnitude greater than the measured value (CH$_3$CN, Marten *et al.*, *Icarus*, 2002, **58**, 532–544) and predicted value (H$_2$CO, Wilson and Atreya, *J. Geophys. Res.*, 2004, **109**, 6002) respectively. In future we expect to increase our sensitivity by targeting our limb sounding path with the lowest-noise, shortest wavelength array (FP4, 7-9 µm) lower in Titan's atmosphere—at 125 km rather than 250 km (in the present work). This will increase the gas density at the tangent altitude—and hence the opacity for optically thin gas bands—by about one order of magnitude, and decrease our upper limits by a similar amount. At this stage we will either make detections of these species, or meaningfully test the previous measurements/predictions.

Dr Moses remarked: This question does not necessarily have to relate to the four species you discuss in your paper, but in terms of helping with the analysis of Cassini CIRS data of Titan and/or Saturn, what are your biggest needs now that might be resolved with new laboratory or theoretical line parameters or other data (*i.e.*, what would go on your wish list)?

Dr Nixon responded: The top priority right now is to acquire spectral line data for several of the weak bands of propane that have been detected by CIRS in the 8–12 micron region, for which no model or list currently exists. These are the v_8 (869 cm^{-1}), the v_{21} (922 cm^{-1}), the v_{20} (1053 cm^{-1}), the v_7 (1158 cm^{-1}) and the v_{25} (1192 cm^{-1}) as described in Nixon *et al.* (*Planet. Space Sci.*, 2009, **57**, 1573–1585). Second in importance is spectral line data for molecules that are predicted to be present in

Titan's atmosphere at levels that may be detectable by CIRS, including: cyanides, such as propionitrile (C_2H_5CN) and acrylonitrile (C_2H_3CN); hydrocarbons including propylene and cyclopropane (isomers of C_3H_6); methylamine (CH_3NH_2) and methanimine (CH_2NH); and small rings such as toluene (C_7H_8), naphthalene ($C_{10}H_8$) and pyridine (C_5H_5N). Aside from spectroscopy, we also critically require reaction rates and kinetic isotopic effect (KIE) fractions for the abstraction of hydrogen from the various isotopes of methane ($^{12}CH_4$, $^{13}CH_4$, $^{12}CH_3D$) by ethynyl (C_2H). This reaction is rapid and catalytic in the sense that the resulting acetylene is recycled back to ethynyl by photolysis. The KIE effect of this reaction likely dominates the evolution of methane isotopes in Titan's atmosphere.

Professor Atreya asked: Have you tried to search for formaldehyde? It does have lines in the CIRS spectral range; perhaps some clean ones in the sub-mm range.

Dr Nixon replied: Formaldehyde (H_2CO) is one of the four molecules considered in this paper; we place upper limits on its abundance that are somewhat greater than model predictions, and so are not constraining at the present time. See also our response to Professor Plane's opening question. In this work we considered only the v_6 vibrational band at 1167 cm^{-1}, which is probably best suited for CIRS (*i.e.* most sensitive) due to low instrument noise and few strong gas emissions from Titan in this part of the spectrum. We did not consider far-infrared/sub-millimetre pure rotational emissions of formaldehyde, which have been detected from astrophysical sources (*e.g.* the Orion nebula, Johnstone *et al.*, *Astron. Astrophys.*, 2003, **412**, 157–174). Although the CIRS spectral range does indeed extend to 10 cm^{-1} (1 mm), this long-wavelength region is less suited for detection of faint trace species for several reasons: (i) the NESR (noise equivalent spectral radiance) is up to two orders of magnitude higher than at 9 μm; (ii) there are many emissions of CH_4, CO and HCN in this region that occupy much of the spectrum; (iii) there are ongoing calibration problems which affect the level of the continuum. For these regions, we consider it unlikely that H_2CO will be detected in the sub-millimetre by CIRS, but a search with ground-based telescopes having much higher spectral resolution might be very worthwhile.

Dr Vuitton remarked: I notice some confusion in the introduction of your paper. You mention that "Detailed modelling of the INMS cracking patterns based on open-source ion (OSI) spectra has proposed the presence of species unseen in the stratosphere such as ammonia (NH_3), toluene (C_7H_8), tri- and tetra-cetylene (C_6H_2, C_8H_2), methylenimine (CH_2NH), and several nitriles including vinyl cyanide (acrylonitrile, C_2H_3CN) and ethyl cyanide (propionitrile, C_2H_5CN). However, a later evaluation of the closed-source neutral (CSN) mass spectra, which are less contaminated by wall reactions, found confirmation only for ammonia on the above list— beyond the species already known from the infrared." Actually, in the OSI mode, the ions are collected and directly analyzed, there are no cracking patterns, no wall reactions. The assignment of each m/z is based on an ion chemistry model, which also allows the estimation of the abundance of the neutrals required for the protonated species to be present at such a density (*i.e.* $m/z = 42$ attributed to CH_3CNH^+ is used to retrieve the density of CH_3CN). In the CSN mode, the neutrals are collected in an antechamber and subsequently ionized in the instrument. We then retrieve their density based on calibration spectra of their cracking pattern. Because the ionization is not a super-efficient process, the detection limit for the neutrals is lower than for the ions (ppm *vs.* ppb, roughly). There is also some evidence for wall reactions in the antechamber, which makes things difficult for some more sticky species, including NH_3.

Also, in the next paragraph, C_4H_6 is a perfectly stable molecule.

Dr Nixon answered: Thank you to Dr Vuitton for this clarification.

Dr Teanby asked: Is it also possible to determine upper limits for allene?

Dr Nixon answered: Yes, we also recently calculated upper limits for allene (CH_2CCH_2) from the same pair of spectral datasets. Note that allene is an isomer of the molecule methyl acetylene (propyne, CH_3C_2H), which has been previously detected in Titan's atmosphere with substantial variations from the summer to winter poles (*i.e.* 4–24 ppb at 10 mbar, Coustenis *et al.*, *Icarus*, 2010, **207**, 461–476, 2010). The model of Wilson and Atreya (*J. Geophys. Res.*, 2004, **109**, E06002) predicts an allene abundance of ~1 ppb at 200 km for low latitudes, while the model of Lavvas *et al.* (*Planet Space Sci.*, 2008, **56**, 67–99; Fig. 5) seems to show even less. A 3-σ upper limit of 5 ppb (constant stratospheric value) was determined by modeling the Voyager 1 IRIS spectrum (Coustenis *et al.*, *Icarus*, 1993, **102**, 240–260) at 841 cm^{-1} (ν_{10} band) using a line list based on the work of Chazelas *et al.* (*J. Mol. Spectrosc.*, 1985, **110**, 326–338) and Koga *et al.* (*J. Chem. Phys.*, 1979, **71**, 2404–2411). The limit was later reduced to 2 ppb by application of the same method to ISO spectral data (Coustenis *et al.*, *Icarus*, 2003, **161**, 383–403).

We have now also made a quantitative estimate of upper limits for allene from Cassini CIRS limb data, using the same data and method as described in the paper for the other moelcules, and using the line list of Coustenis *et al.* (1993) for the ν_{10} band. In this spectral region, the residual emission after fitting the spectrum with known gases (mainly the ν_9 band of ethane) appears to be dominated by systematic effects: apparently ethane emissions that are imperfectly modeled, which then determines our 'noise' level. We attempted to include allene towards the short-wavelength edge of the ethane band, in the region 845–880 cm^{-1}. The indicated upper limits (1, 2, 3-σ) are: 0.1, 0.2, 0.3 ppb at 25°S, 107 km, and: 0.5, 1.0, 1.6 ppb at 76°N, 224 km. See also Fig. 2, below. The value at low latitudes of 0.3 ppb (3-σ) is seven times lower than the ISO limit, and similar to the predicted amounts from photochemical models at that altitude.

Professor Zwier opened the general discussion: One of the themes of this first set of papers is the chemistry of nitriles, and looking ahead to other papers coming later in this discussion, this theme will continue in what follows. As a result, I thought it

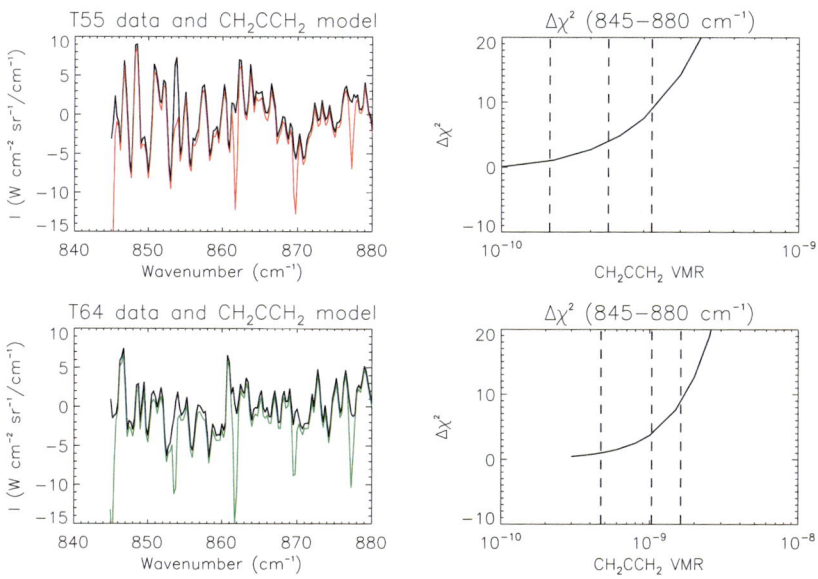

Fig. 2 T55 and T64 data with CH_2CCH_2 models

appropriate to present some data from our laboratory at Purdue on the photochemistry of cyanoacetylene, HC_3N. In the paper by Teanby et al.,[1] Fig. 8 showed an unusual enrichment for the nitriles that led them to postulate the need for an additional sink term in the photochemical models as a cause for the steeper than predicted vertical gradients observed. The reactions of metastable cyanoacetylene, HC_3N^*, may play such a role, whether in the gas phase, in clusters, or in aerosols.[2–7]

This photochemistry has been studied by others in the past, but we have used our experimental methods[8–13] to study, at least in a preliminary way, the reactions of UV-excited HC_3N with itself, C_4H_2, and 1,3-butadiene. The results are intriguing enough to present them here.

HC_3N is the nitrile analog of diacetylene, C_4H_2, and like C_4H_2, it has only strong chemical bonds. The recent theoretical study by Silva et al. places the C–H bond dissociation energy for HC_3N at 560 kJ mol^{-1}, equivalent to a 213 nm photon, calling into question previous experimental studies that had deduced a lower threshold.[14] This means that much of the UV absorption of HC_3N occurs below the lowest dissociation threshold. At the same time, previous studies of the photochemistry of HC_3N in a bulb show extensive polymerization, and long-lived excited states (either singlet or triplet in character) have been postulated as responsible for this chemistry.[2,5,7,15–16]

We have carried out our photochemical studies using the experimental apparatus shown in Fig. 1 of our paper, with the modification that the source shown as inset was used as a photochemical reaction channel rather than a discharge source, with the UV excitation laser counter-propogating the reaction channel. The work on HC_3N was carried out by Christopher Ramos during his time as a graduate student in my group.[17] A mixture of 5% HC_3N in helium was pulsed from a pulsed valve (R. M. Jordan) into the 2 mm diameter X 7 mm long reaction channel. The temperature in the constrained expansion in the channel has been determined previously to be ~75 K.[18] The doubled output of a Nd:YAG-pumped dye laser at 224 nm is used to excite HC_3N via the 5_0^1 transition,[19–20] and initiating its reactions with other molecules in the reaction mixture. This reaction mixture then expands into vacuum before being ionized ~10 cm downstream via VUV single-photon or resonant two-photon ionization (R2PI) with time-of-flight mass analysis of the ionized products. One of the particular strengths of this experimental arrangement is that reaction is initiated and subsequently quenched on a timescale of about 20 μs, so that only early-time behavior is probed, with primary products dominating.

Due to the high ionization potential of ground state HC_3N (11.62 eV), which is not ionized by the 118 nm photons (10.5 eV) used in the VUV photoionization step, the time-of-flight mass spectrum in the absence of the UV laser shows no background ion signal at m/z 51, the mass of HC_3N. This opens the possibility of directly observing the metastable HC_3N^* molecules by looking for the appearance of m/z 51 ion signal produced by UV excitation. Fig. 3a shows the time-of-flight mass spectrum over a mass range that includes m/z 51 with the VUV laser timed to intersect the UV-excited molecules when they traverse the ion source region of the time-of-flight mass spectrometer. The m/z 51 signal is present with intensity approximately ten times that of any of the detected products (e.g., m/z 101) from the HC_3N^* + HC_3N reaction, showing the early-time behavior probed under these conditions. The energy level diagram shown as an inset in Fig. 3a shows the scheme for detection of HC_3N^*. Fig. 3b is an action spectrum recorded by tuning the UV photoexcitation laser under timing conditions in which the HC_3N^* signal is detected downstream. This action spectrum shows a wavelength dependence characteristic of the $^1\Sigma$–$^1\Delta$ vibronic transition that carries the oscillator strength in the 224 nm region. The detection downstream occurs at a delay of ~50 μs, proving that, even under the collision conditions in the reaction channel, the metastable HC_3N is not electronically quenched.

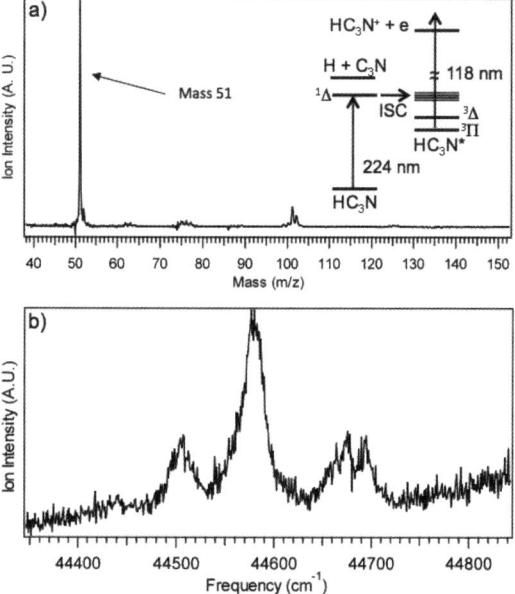

Fig. 3 (a) Difference time-of-flight mass spectrum (TOFMS) of a 5% HC$_3$N mixture in helium recorded with photoexcitation laser at 224.3 nm (44 570 cm^{-1}), and 118 nm probe laser fixed at a time delay to intercept the laser-excited HC$_3$N molecules traversing the ion source region of the time-of-flight mass spectrometer. The energy level diagram used for the experiment is shown as an inset. (b) Action spectrum monitoring the HC$_3$N* signal as the photoexcitation laser is tuned. The spectrum maps out the absorptions of HC$_3$N in this wavelength region.

Here we report preliminary measurements of the primary products formed from the reactions of HC$_3$N* with HC$_3$N, C$_4$H$_2$, and 1,3-butadiene (1,3-C$_4$H$_6$, H$_2$C=CH–CH=CH$_2$). These are carried out by incorporating into the reaction mixture known concentrations of the co-reactant, using selective UV excitation of HC$_3$N, and checking that the wavelength dependence of the action spectra for each reaction mixture reflect the absorption spectrum of HC$_3$N (as shown in Fig. 4a). Unfortunately, many of the anticipated reaction products of the reactions of HC$_3$N* with HC$_3$N and C$_4$H$_2$ (*e.g.*, HC$_5$N, C$_2$H$_2$, HCN, C$_6$N$_2$) have ionization potentials above 10.5 eV, and hence are not observed with 118 nm VUV ionization. On the other hand, hydrocarbon products (e.g., C$_6$H$_2$) or radical products, with lower ionization potentials, can be detected. Thus, for the HC$_3$N* + HC$_3$N reaction, we observe

$$HC_3N^* + HC_3N \rightarrow C_6H_2 \; (m/z \; 74) + N_2 \quad (1a)$$

$$\rightarrow HC_6N_2 \; (m/z \; 101) + H \quad (1b)$$

The low IP products of the reactions of HC$_3$N* with C$_4$H$_2$ and of C$_4$H$_2$* with HC$_3$N are not included here for brevity's sake.[17] In all these cases, the inability to detect products with ionization potentials above 10.5 eV prevents any deductions regarding product quantum yields under our short-time conditions. A more thorough study will require VUV photoionization deeper in the VUV, either *via* laser-based or synchrotron methods.

The main purpose of this comment, however, is to introduce the products of the HC$_3$N* + 1,3-butadiene reaction, since in this case, most of the anticipated products

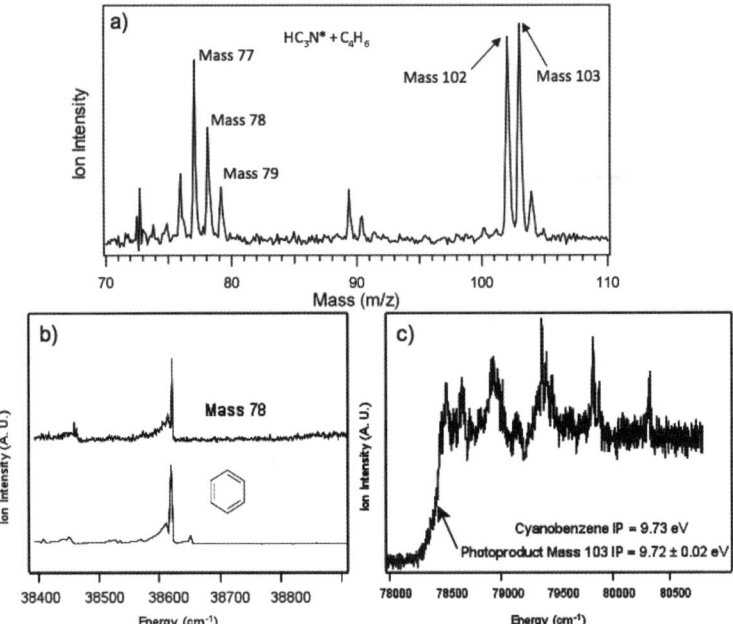

Fig. 4 (a) Difference TOFMS of the products of the HC_3N^* + 1,3-butadiene reaction. (b) R2PI spectrum of the m/z 78 photoproducts (top trace) and (bottom trace) spectrum of benzene, proving that some portion of the m/z 78 VUV ion signal from (a) is due to benzene. (c) One-color 4 two-photon ionization efficiency scan of the m/z 103 photoproduct, showing its ionization threshold at 9.72 ± 0.02 eV, consistent with the known ionization potential of cyanobenzene (9.73 eV).

have ionization potentials below 10.5 eV. Fig. 4a shows the difference VUV time-of-flight mass spectrum (taken as the difference signal with and without the UV excitation laser present) for the HC_3N^* + 1,3-C_4H_6 reaction. Primary products include those at m/z 77, 78, 102, and 103. Of these, the most intriguing are those at m/z 78 and 103. As shown in Fig. 4b and 4c, in these cases we have used one-color R2PI spectroscopy to identify the products as benzene and cyanobenzene, respectively. In the case of benzene, the photoproduct R2PI signal (top trace) is compared to the R2PI spectrum from benzene under otherwise identical conditions. The correspondence provides unambiguous identification of the m/z 78 product as benzene. In the case of mass 103, we observe a sharp two-photon ionization threshold for our photoproduct at 9.72 ± 0.02 eV, and vibronic substructure above this threshold. This is consistent with identification of the m/z 103 product as cyanobenzene, which has a known ionization potential of 9.73 eV.[21] The presence of the cyano group in this product raises its IP relative to that of benzene (9.24 eV),[22–23] so that the ionization threshold occurs well above the S_0–S_1 origin transition, and the structure observed just above threshold is likely due to vibronic structure in the S_0–S_1 transition in the first photon of the two-photon process. The near-coincidence between the observed and known values of the ionization potential for the m/z 103 product positively identifies it as cyanobenzene. Thus, the reaction of HC_3N^* + 1,3-C_4H_6 produces two aromatic products: benzene and cyanobenzene.

I'd like to finish my comments by raising the following question: Are there circumstances on Titan (and elsewhere) in which metastable HC_3N^* plays a role in producing larger more complex hydrocarbons and nitriles, including aromatics? This question has been posed in the past by others, but recent work by Silva et al.[24] on metastable C_4H_2 has called this role into question, at least in the gas phase.[3,6,15,25] However, HC_3N^* has several attributes that make this question worth considering again.

Like C_4H_2, HC_3N absorbs at longer wavelengths than C_2H_2, HCN, and other molecules known to play important photochemical roles, where the solar spectrum is more intense. Thus, even lower down in the atmosphere where shorter wavelength radiation has been removed by absorption by these molecules, HC_3N will continue to absorb. At these wavelengths, HC_3N^* will exclusively be produced, since these wavelengths are below the photodissociation threshold of HC_3N.

The efficiency of reaction will depend sensitively on the lifetime of HC_3N^* under the pressures and temperatures at a given altitude/latitude. A direct measure of this lifetime is still needed. The previous work of Silva et al.[24] included a measurement of the analogous collision-free lifetime of $C_4H_2^*$ that led these authors to conclude that $C_4H_2^*$ was unlikely to be a major source of photochemical processing of C_4H_2 in Titan's atmosphere. Nevertheless, the triplet lifetime of HC_3N (and C_4H_2) is likely to increase dramatically as its internal energy is reduced by collision, and a more complete study of the HC_3N^* lifetime as a function of temperature and pressure is needed.

If reaction does not occur, the most likely fate of the HC_3N^* after electronic quenching is to return to the ground state where it can be recycled by photoexcitation. During the course of many cycles, reaction of HC_3N^* may still be important, especially deeper in the atmosphere where the shorter wavelengths needed to photodissociate HC_3N are filtered out.

When HC_3N is in a cluster or aerosol, it will have reaction partners immediately adjacent to it, therefore rendering lifetime issues moot. In fact, Huang et al. have recently studied the photochemistry of C_4H_2 dimer, and have suggested that reaction is facile under such circumstances.[26] The same is likely to be the case for HC_3N clusters, but experiments are needed to confirm this fact. The large dipole moment of HC_3N (3.7 Debye) is likely to make clustering even more facile than in C_4H_2.

Finally, the peak at m/z 79 in Fig. 4a is due to a photoproduct with molecular formula C_5H_5N. While we have no spectroscopic evidence to date regarding its chemical structure, it is intriguing to note that one of the isomers consistent with this formula is pyridine, a nitrogen hetero-aromatic that bears a close chemical similarity with important pre-biotic molecules such as the DNA bases.

Support for this work from the NASA Planetary Atmospheres program under grant NNX10AB89G is gratefully acknowledged.

1 N. A. Teanby, P. G. J. Irwin, R. d. Kok and C. A. Nixon, Faraday Discuss., 2010, **147**, DOI: 10.1039/c001690j.
2 D. W. Clarke and J. P. Ferris, J. Geophys. Res.-Planets, 1996, **101**, 7575–7584.
3 I. Couturier-Tamburelli, A. Coupeaud, Z. Guennoun, N. Pietri and J. P. Aycard, Actual Chim., 2008, XV–XIX.
4 J. P. Ferris, B. N. Tran, V. Vuitton, J. Joseph, P. Persans, J. J. Chera, R. Briggs and M. Force, Origins Life Evol. Biosphere, 2006, **36**, 331–332.
5 J. B. Halpern, L. Petway, R. Lu, W. M. Jackson, V. R. McCrary and W. Nottingham, J. Phys. Chem., 1990, **94**, 1869–1873.
6 S. Lebonnois, E. L. O. Bakes and C. P. McKay, Icarus, 2002, **159**, 505–517.
7 K. Seki, M. Q. He, R. Z. Liu and H. Okabe, J. Phys. Chem., 1996, **100**, 5349–5353.
8 R. E. Bandy, C. Lakshminarayan, R. K. Frost and T. S. Zwier, Science, 1992, **258**, 1630–1633.
9 R. E. Bandy, C. Lakshminarayan, R. K. Frost and T. S. Zwier, J. Chem. Phys., 1993, **98**, 5362–5374.
10 C. A. Arrington, C. Ramos, A. D. Robinson and T. S. Zwier, J. Phys. Chem. A, 1998, **102**, 3315–3322.
11 C. A. Arrington, C. Ramos, A. D. Robinson and T. S. Zwier, J. Phys. Chem. A, 1999, **103**, 1294–1299.
12 A. G. Robinson, P. R. Winter, C. Ramos and T. S. Zwier, J. Phys. Chem. A, 2000, **104**, 10312–10320.
13 A. G. Robinson, P. R. Winter and T. S. Zwier, J. Phys. Chem. A, 2002, **106**, 5789–5796.

14 R. Silva, W. K. Gichuhi, V. V. Kislov, A. Landera, A. M. Mebel and A. G. Suits, *J. Phys. Chem. A*, 2009, **113**, 11182–11186.
15 D. W. Clarke and J. P. Ferris, *Icarus*, 1995, **115**, 119–125.
16 J. P. Ferris and J. C. Guillemin, *J. Org. Chem.*, 1990, **55**, 5601–5606.
17 C. Ramos, *Master's Thesis*, 1998.
18 J. A. Stearns, T. S. Zwier, E. Kraka and D. Cremer, *Phys. Chem. Chem. Phys.*, 2006, **8**, 5317–5327.
19 Y. Benilan, D. Andrieux, M. Khlifi, P. Bruston, F. Raulin, J. C. Guillemin and C. Magos, *Astrophys. Space Sci.*, 1997, **236**, 85–95.
20 P. Bruston, H. Poncet, F. Raulin, C. Cossart-Magos and R. Courtin, *Icarus*, 1989, **78**, 38–53.
21 M. Araki, S. Sato and K. Kimura, *J. Phys. Chem.*, 1996, **100**, 10542–10546.
22 L. A. Chewter, M. Sander, K. Mullerdethlefs and E. W. Schlag, *J. Chem. Phys.*, 1987, **86**, 4737–4744.
23 G. I. Nemeth, H. L. Selzle and E. W. Schlag, *Chem. Phys. Lett.*, 1993, **215**, 151–155.
24 R. Silva, W. K. Gichuhi, C. Huang, M. B. Doyle, V. V. Kislov, A. M. Mebel and A. G. Suits, *Proc. Natl. Acad. Sci. U. S. A.*, 2008, **105**, 12713–12718.
25 D. W. Clarke and J. P. Ferris, *Icarus*, 1997, **127**, 158–172.
26 C. S. Huang, F. T. Zhang, R. I. Kaiser, V. V. Kislov, A. M. Mebel, R. Silva, W. K. Gichuhi and A. G. Suits, *Astrophys. J.*, 2010, **714**, 1249–1255.

Professor Ashfold continued: Developing Professor Zwier's comment, it is of course well known that much simpler molecules, like C_2H_2[1,2] and CO,[3,4] also have metastable triplet states, which can display lifetimes in the hundreds of microseconds to millisecond range.

I also had a question for Professor Yelle, who reports 'fairly large abundances' of H_2CN radicals in the upper atmosphere of Titan (>10^4 cm^{-3} at altitudes ~1000 km) and proposes reaction of these radicals with NH_2 as a source of NH_3 (and HCN). H_2CN photolysis (even at near UV wavelengths) also leads to HCN formation;[5,6] are the authors able to estimate the relative efficiencies of H_2CN loss by photolysis, and by reaction with H atoms, as compared with loss by reaction with NH_2?

1 T. Suzuki and Y. Shi, *J. Phys. Chem. A*, 1998, **102**, 7414.
2 S. Altunata, K. L. Cunningham, M. Canagaratna, R. Thom and R. W. Field, *J. Phys. Chem. A*, 2002, **106**, 1122.
3 M. Nooney, J. M. Price, R. M. Martin and A. M. Wodtke, *Chem. Phys. Lett.*, 1995, **245**, 377.
4 J. J. Gilijamse, S. Hoekstra, S. A. Meek, M. Metsala, S. Y. T. van de Meerakker, G. Meijer and G. C. Groenenboom, *J. Chem. Phys.*, 2007, **127**, 221102.
5 E. J. Bernard, B. R. Strazisar and F. F. Davis, *Chem. Phys. Lett.*, 1999, **313**, 461.
6 A. Teslja, P. J. Dagdigian, M. Banck and W. Eisfeld, *J. Phys. Chem. A*, 2006, **110**, 7826.

Dr Yelle responded: In general, there is an absence of cross section and product data for photolysis of radicals and more laboratory measurements would be welcome; however, especially in outer solar system atmospheres, it is difficult for photolysis of radicals to compete with loss through chemical reactions. This is certainly the case for H_2CN. This radical reacts with both H to produce HCN and H_2 and with CH_3 to produce CH_4 and H_2, and these are the dominant loss processes in our model. The net H_2CN chemical loss rate at 1000 km in our model is 3.6×10^{-4} s^{-1}. Even though the H_2CN cross section is, as far as I am aware, poorly known, this chemical rate is several orders of magnitude larger than any conceivable photolysis rate. A typical large photolysis rate on Titan (*e.g.* C_2H_2) is 10^{-7} s^{-1}.

Professor Suits commented: Stimulated by the rich chemistry in the ionosphere revealed by INMS, Wilson K. Gichuhi, Alexander M. Mebel and I have begun a series of studies of the photochemistry of state-prepared ions using slice velocity map imaging methods. One of the advantages of ion imaging is we can readily study neutral or ion dissociation processes, and examine photoelectron spectra as well.

I would like to present our current results on ethylamine cation photodissociation, as these relate in particular to the species discussed in Professor Yelle's paper. I should emphasize that we are not necessarily interested in the photochemistry of

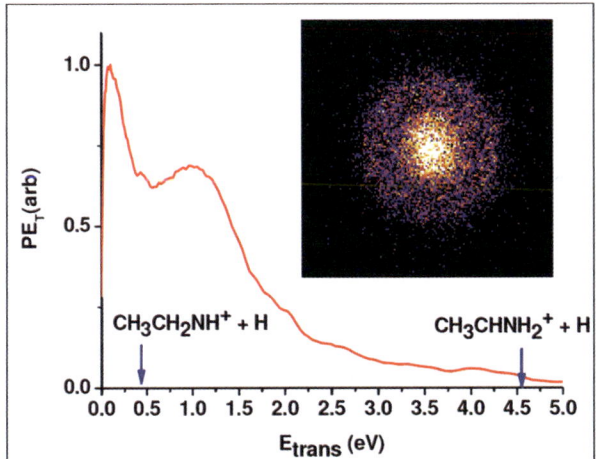

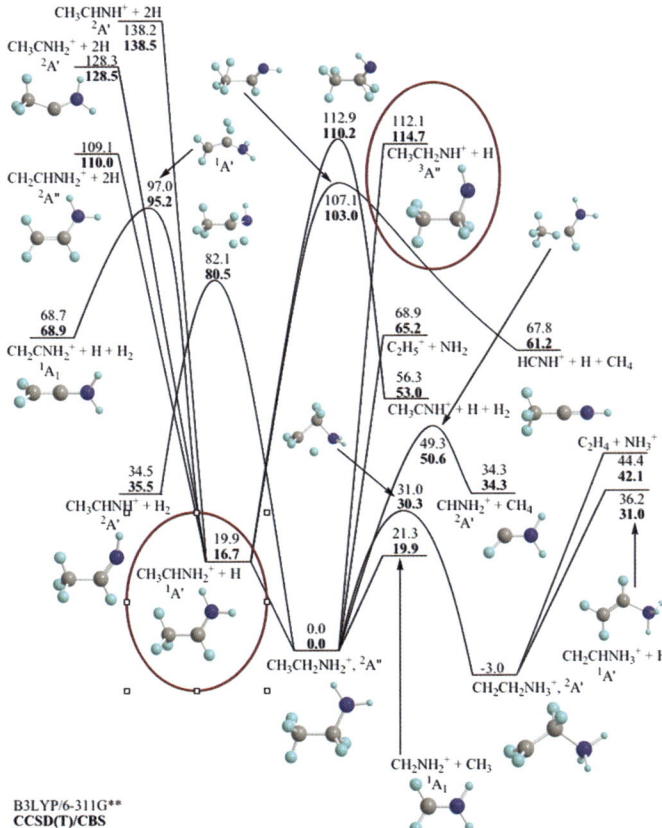

Fig. 5 Inset image is $m/z = 44$ (H loss) from photodissociation of ethylamine cation. The plot shows the total translational energy distribution obtained from the image, along with energy thresholds for formation of the indicated products.

Fig. 6 Stationary points and energies (kcal mol^{-1}) for key species on the ethylamine cation potential surface obtained at the indicated level of theory. Encircled products correspond to thresholds given in Fig. 6.

the ions as an important reaction pathway in the ionosphere—rather, these are so-called "half-collision" studies of the corresponding ion–molecule reactions. When we examine these experimental results in the context of high-level *ab initio* calculations, we gain insight into reaction pathways, barriers and product branching.

Ethylamine cation is prepared by (1+1) resonant ionization at 233 nm. Absorption of an additional photon leads to fragmentation. We see several product channels, including H loss and $HCNH^+$ formation, with the latter the dominant process. Here we discuss the H loss channel, in which the dynamics are fairly clear. The ion image and associated total translational energy distribution are shown in Fig. 5. We see two components in the image that give a bimodal translational energy distribution upon analysis. One of these extends to the limit of the available energy for formation of the CH_3CHNH^{2+} product but drops off on the low energy side at about 0.5 eV. A second slow component peaks at very low energy (0.1–0.2 eV) and drops rapidly at 0.5 eV and beyond. In Fig. 6 we show *ab initio* calculations for some of the key stationary points on this complex potential surface, with energies given in kcal mol^{-1}. Our available energy is 123 kcal mol^{-1}. One can see two possible H loss pathways circled in red: one is the lowest energy dissociation channel of all, with a threshold of about 17 kcal mol^{-1}: this clearly corresponds to the outer ring in our image. There is a "hole" in the image for that product, because any of that isomer formed with less that 0.5–1 eV in translation can undergo secondary decomposition *via* several pathways, as seen in Fig. 6. The second, very slow isomer must then correspond to the triplet $CH_3CH_2NH^+$ product at high energy. These preliminary results indicate the means by which the combination of high-level *ab initio* calculations and photofragment imaging of state-selected ions can give insight into product branching and energetics for species relevant to Titan's ionosphere.

On the abundance of non-cometary HCN on Jupiter†

Julianne I. Moses,[*a] Channon Visscher,[b] Thomas C. Keane[c] and Aubrey Sperier[d]

Received 11th March 2010, Accepted 7th April 2010
DOI: 10.1039/c003954c

Using one-dimensional thermochemical/photochemical kinetics and transport models, we examine the chemistry of nitrogen-bearing species in the Jovian troposphere in an attempt to explain the low observational upper limit for HCN. We track the dominant mechanisms for interconversion of N_2–NH_3 and HCN–NH_3 in the deep, high-temperature troposphere and predict the rate-limiting step for the quenching of HCN at cooler tropospheric altitudes. Consistent with some other investigations that were based solely on time-scale arguments, our models suggest that transport-induced quenching of thermochemically derived HCN leads to very small predicted mole fractions of hydrogen cyanide in Jupiter's upper troposphere. By the same token, photochemical production of HCN is ineffective in Jupiter's troposphere: CH_4–NH_3 coupling is inhibited by the physical separation of the CH_4 photolysis region in the upper stratosphere from the NH_3 photolysis and condensation region in the troposphere, and C_2H_2–NH_3 coupling is inhibited by the low tropospheric abundance of C_2H_2. The upper limits from infrared and submillimetre observations can be used to place constraints on the production of HCN and other species from lightning and thundershock sources.

1. Introduction

Although hydrogen cyanide (HCN) was detected in the Jovian stratosphere following the Comet Shoemaker-Levy 9 impacts,[1–9] no convincing observational evidence exists for the presence of non-cometary HCN in Jupiter's troposphere. Tentative detections of HCN from the 1960s and 1970s have all been discounted.[10] The most credible report of the detection of non-cometary HCN on Jupiter resulted from ground-based 13.5 µm observations of three spectral lines by Tokunaga *et al.*;[11] however, this detection is now considered dubious due to the lack of confirmation from subsequent infrared, sub-millimetre, and millimetre observations.[12–15] Bézard *et al.*[13] suggest that one of the purported HCN absorption lines identified by Tokunaga *et al.*[11] is actually an expected "valley" between two nearby C_2H_2 emission lines, a second absorption line is much narrower than it would be if caused by HCN (and is likely an instrument artifact), and the identification of the third line is suspect due to uncertainties in the position (frequency) of the feature. From recent

[a] *Space Science Institute, 1602 Old Orchard Ln, Seabrook, TX 77586, USA. E-mail: jmoses@spacescience.org; Tel: +281.474.9996*
[b] *Lunar and Planetary Institute, 3600 Bay Area Blvd., Houston, TX 77058, USA*
[c] *Dept. of Chemistry and Biochemistry, Russell Sage College, Troy, NY 12180, USA*
[d] *Dept. of Chemistry and Physics, University of St. Thomas, Houston, TX 77006-4626, USA*

† Electronic supplementary information (ESI) available: Jupiter photochemical model output. See DOI: 10.1039/c003954c

850-μm observations, a strict upper limit of 0.93 ppb has been placed on the Jovian tropospheric HCN mole fraction, assuming HCN condenses in the upper troposphere, or as small as 0.16 ppb if the HCN is assumed to be uniformly mixed throughout the troposphere and stratosphere.[15]

The production of hydrogen cyanide and other nitrogen-bearing organics in reducing atmospheres such as that of Jupiter has attracted considerable interest in the past half century due to prebiological chemistry implications[16] and to the long-standing puzzle of the cloud coloring agents on Jupiter, for which it has been suggested that HCN polymers could play a role.[17,18] The HCN abundance is expected to be negligible in thermochemical equilibrium at the cold atmospheric levels that can be probed by remote-sensing observations,[19] but several disequilibrium processes could supply HCN to the Jovian troposphere. These processes include rapid transport from the deep troposphere,[10,19–21] photochemical processing of CH_3NH_2 dredged up from the deep atmosphere,[10] lightning and related processes in thunderstorms,[17,22–26] coupled CH_4–NH_3 photochemistry,[27–34] and coupled C_2H_2–NH_3 photochemistry.[35–40]

Kaye and Strobel[35] and Lewis and Fegley[10] have evaluated the various disequilibrium processes for HCN production and conclude that NH_3–C_2H_2 photochemical coupling is the most plausible mechanism for producing HCN on Jupiter— HCN production from the chemistry of hot H atoms released from NH_3 (or H_2S) photolysis in the presence of methane is inhibited under Jovian conditions because of rapid hot-atom thermalization from collisions with H_2,[31,41,42] photochemical production of the HCN precursor CH_3NH_2 is inhibited due to an insufficient source of CH_3 in the NH_3 photolysis region,[29,34] and lightning and thunder shockwave production of HCN appears to be inadequate when observed production efficiencies in realistic laboratory discharge and shock-synthesis experiments are scaled to Jovian conditions.[10,23] We point out, however, that Bar-Nun and Podolak[24] and Podolak and Bar-Nun[25] continue to favor the thundershock hypothesis, and Fegley and Lodders[20] and Lodders and Fegley[21] still support a deep-atmospheric source.

The purpose of this paper is to use updated information on the kinetics of nitrogen species to reevaluate both the quenched chemistry (deep atmospheric source) and photochemistry (coupled C_2H_2–NH_3 photochemistry) hypotheses for the production of HCN on Jupiter. We use two different one-dimensional (1D) kinetic-transport numerical models for this investigation. Both are based on the Caltech/JPL KINETICS code,[43] which uses finite-difference techniques to solve the 1D continuity equations. The first model considers photochemical kinetics and molecular and eddy diffusion, and we apply that model to the stratosphere and upper troposphere of Jupiter. The second model considers thermochemical kinetics and eddy diffusion, and we apply that model to the deep Jovian troposphere. Previous investigators who have looked in detail at the possibility of HCN transport from the deep troposphere[44,20,21] have used time-constant arguments rather than full kinetic-transport models to predict the quenched HCN abundance in the upper troposphere. Recent suggested improvements to the time-constant arguments[45–48] have prompted us to reevaluate the thermochemical kinetics and quenching of the C–H–O system on Jupiter.[49] Our success at modeling the transition from thermochemical equilibrium to transport-induced quenching in that C–H–O system has led us to investigate nitrogen species thermochemistry and quenching for this study.

2. Nitrogen species kinetics

The full reaction list for our Jovian photochemical model includes 145 species and 1973 reactions. The H–C–O reactions are discussed elsewhere.[49,50] The kinetics of carbon–hydrogen–oxygen species has been well studied because of numerous combustion-chemistry applications[51–54] and terrestrial atmospheric-chemistry applications;[55–59] however, less information is available for reactions of nitrogen-bearing species, particularly in reducing environments. Some relevant experimental data are discussed in the above data compilations[51–59] and in numerous individual rate-coefficient studies. However, for our application, laboratory data must be supplemented by theory—both quantum chemistry and master-equation calculations—for many reactions of importance in our models. We therefore rely heavily on the theoretical calculations of Dean and Bozzelli[60] for the generation of our reaction mechanism. Several other investigations have also been useful for identifying important reactions and rate coefficients.[61–65] Our adopted reaction-rate coefficients are included in the full photochemical model output available in the ESI.†

We consider the kinetics of H_2, H, and 58 oxygen- and carbon-bearing species,[49,50] as well as N, N_2, NH, NH_2, NH_3, NNH, N_2H_2, H_2NN, N_2H_3, N_2H_4, CN, HCN, H_2CN, CH_2NH, CH_3NH, CH_2NH_2, CH_3NH_2 CH_2CN, CH_3CN, C_3N, HC_3N, C_2H_2CN, C_2H_3CN, NO, NO_2, N_2O, HNO, HNO_2, NCO, HNCO, CH_3NO, PN, and NH_2PH_2 in both our thermochemical and photochemical models. We also include several other phosphorus-bearing species, but a full discussion of the phosphorus kinetics and NH_3–PH_3 photochemical coupling is deferred to a later paper (see also ref. 66 and 67). We initially included HNC, C_2N, and C_2N_2 in the models, but these species were produced in trivial amounts in the photochemical models and had little effect on the kinetics of other constituents, so we dropped them from consideration. For our photochemical model, we also include C_2H_4N (*i.e.*, $CH_2=CH\dot{N}H$, $CH_3\dot{C}=NH$, and/or $CH_3CH=N\cdot$ isomers), C_2H_5N (*i.e.*, $CH_2=CHNH_2$ and $CH_3CH=NH$ isomers), $C_2H_5\dot{N}H$, $C_2H_5NH_2$, $CH_3CH=NNH_2$, $CH_3CH=NC_2H_5$, and $CH_3CH=NN=CHCH_3$ based on laboratory photolysis investigations.[36–39] However, we omit these latter species from our deep-tropospheric thermochemical model due to a lack of information on their thermodynamic parameters—information that is needed to fully reverse the kinetic reactions through the principle of microscopic reversibility. Although these species are unlikely to be significant constituents in the deep troposphere of Jupiter, they are potentially important photochemical products of coupled NH_3–C_2H_2 photochemistry, as well as precursors to HCN formation in Jupiter's upper troposphere,[36–39] and must be included in the photochemical model.

Many of our reaction rate coefficients derive from Dean and Bozzelli,[60] who validate their proposed mechanism by comparing their model predictions with experimental results from several flame studies. The expressions provided for their individual reaction rate coefficients are valid for temperatures in the range of 600–2500 K. Those temperatures are appropriate for our deep-tropospheric thermochemical modeling but not for conditions in Jupiter's upper troposphere and stratosphere, where temperatures can drop to $\lesssim$110 K at the tropopause.[68–71] Therefore, although we generally adopt the Dean and Bozzelli expressions as given in their paper,[60] we check for pathological or inconsistent behavior at low temperatures and alter the expressions, as necessary. Some of the Dean and Bozzelli rate coefficients have also been replaced due to the availability of experimentally derived rate coefficients or due to inappropriate rate coefficients calculated for the reverse reaction (*i.e.*, those in excess of kinetic collision-rate considerations). Moreover, the Dean and Bozzelli[60] mechanism does not cover the full suite of nitriles, amines, hydrazones, and other complex nitrogen-bearing organics that are expected to form in coupled NH_3–C_2H_2 photochemistry.[36–39] Some important rate coefficients for the production and loss of these organo-nitrogen compounds could be found in the literature, but many could not. We therefore apply our reaction list to simulations of laboratory photolysis investigations to help constrain uncertain reaction rate coefficients and to test our overall mechanism.

The first such simulation we perform is for the photolysis of pure ammonia, as described in Groth *et al.*[72] In this experiment, 37.5 torr of pure ammonia at room temperature is introduced to a cylindrical quartz cell 10 cm long and 5.5 cm in diameter and irradiated by 206.2-nm photons from an iodine lamp.[72] The resulting photolysis product quantum yields are derived as a function of photons absorbed. To simulate this investigation, we use the KINETICS code[43] with our reaction list described above and apply it to a 1D "box" of the appropriate length (10 cm), with the appropriate 298 K, 37.5-torr NH_3 initial conditions. Since the NH_3 photolysis rate has a gradient within the cell under these conditions, we divide our box into a 21-segment grid. We assume a lamp flux at 206.2 nm at the front of our cell of 3×10^{14} photons cm^{-2} s^{-1}, although exact knowledge of the lamp flux is not necessary because the experimental results were reported in terms of quantum yields per quanta absorbed.[72] After 1.03×10^{19} photons absorbed, Groth et al.[72] find quantum yields for N_2, H_2, and N_2H_4 production of 0.163, 0.490, 0.0005, respectively. From our box-model simulations, we derive quantum yields of 0.162, 0.487, and 0.00115 for N_2, H_2, and N_2H_4, respectively, at a corresponding number of photons absorbed. As with the Groth et al. experiment,[72] we find that the quantum yields of H_2 and N_2 level off after an initial rise, whereas the N_2H_4 quantum yield goes through an early maximum, followed by a drop off to low values. The results of our simulation are insensitive to reasonable assumptions about the diffusion coefficients within the cylinder or the lamp flux.

The photolysis of ammonia at 206.2 nm occurs exclusively through the $NH_3 + h\nu \rightarrow NH_2(\tilde{X}^2B_1) + H$ pathway.[72] Hydrazine production and loss depends critically on these photolysis products, a fact that can explain the observed quantum-yield behavior of the N_2H_4. Hydrazine is produced in our model mainly from the termolecular reaction $2\,NH_2 + M \rightarrow N_2H_4 + M$, where M is any third body, and the dominant two loss processes (of roughly equal importance at later times) are $H + N_2H_4 \rightarrow N_2H_3 + H_2$ and $NH_2 + N_2H_4 \rightarrow N_2H_3 + NH_3$. When NH_2 is released from NH_3 photolysis, hydrazine is readily synthesized and continues to increase in concentration until enough N_2H_4, NH_2, and H is built up that loss processes can compete. At that point, the N_2H_4 reaches a constant concentration (i.e., it is in steady state), while N_2 and H_2 production continues through the net equation $2NH_3 \rightarrow N_2 + 3H_2$, which goes through N_2H_4 and other N_2H_x species as intermediates. For example, the dominant pathway for N_2 and H_2 production in our simulation is the following scheme:

$$
\begin{array}{lr}
3\,(NH_3 + h\nu \longrightarrow NH_2 + H) & \text{Cheng et al.}^{73} \\
2\,NH_2 + M \longrightarrow N_2H_4 + M & \text{Fagerstrom et al.}^{74} \\
N_2H_4 + H \longrightarrow N_2H_3 + H_2 & \text{Vaghjiani}^{75} \\
NH_2 + N_2H_3 \longrightarrow NH_3 + H_2NN & \text{Dean and Bozzelli}^{60} \\
H_2NN + H \longrightarrow NNH + H_2 & \text{Dean and Bozzelli}^{60} \\
NNH \longrightarrow N_2 + H & \text{Dean and Bozzelli}^{60} \\
\underline{2\,H + M \longrightarrow H_2 + M} & \underline{\text{Baulch et al.}^{52}} \\
\text{Net}:\ 2\,NH_3 \longrightarrow N_2 + 3\,H_2, &
\end{array}
$$

where the reference at the end of each reaction represents the source of the rate coefficient or photolysis cross section. Note the importance of N_2H_x intermediate species in this scheme (see also ref. 76); these species are also likely to be important in the kinetics of nitrogen species under combustion-chemistry conditions[60] and Jovian tropospheric conditions. Reactions involving H_2NN, a singlet biradical, are speculative at this point, but H_2NN is expected to be a major product of the $NH_2 + NH_2$ reaction under combustion-chemistry conditions.[60] We strictly follow the Dean and Bozzelli theoretically derived rate coefficients[60] for the production and loss of this species and find that it can contribute to the conversion of NH_3 to N_2 under low-temperature NH_3 photolysis conditions.

A main secondary scheme in our simulation involves the $NH_2 + N_2H_4 \rightarrow NH_3 + N_2H_3$ abstraction reaction. Dean and Bozzelli[60] use a "Direct Hydrogen Transfer" (DHT) method to derive a rate coefficient of $6.1 \times 10^{-18} \, T^{1.94} \exp(-820/T) \, cm^3 \, s^{-1}$ (for T in K) for this reaction, which if extrapolated to 300 K would produce a value of $2.5 \times 10^{-14} \, cm^3 \, s^{-1}$. In contrast, experimental data have been used to estimate a 300-K rate coefficient of $5 \times 10^{-13} \, cm^3 \, s^{-1}$ for this reaction[77]—a value about 20 times larger than that derived theoretically.[60] We find that we get the best agreement with the Groth *et al.* experimentally derived N_2 and H_2 quantum yields[72] if we adopt a value that lies in between the experimental[77] and theoretical[60] values, *i.e.*, if we adopt a rate coefficient of $4.3 \times 10^{-17} \, T^{1.94} \exp(-820/T) \, cm^3 \, s^{-1}$ for this reaction (seven times the Dean and Bozzelli[60] expression). We use this expression throughout our subsequent modeling.

Although our modeled quantum yields (and their time variation) for N_2 and H_2 are in excellent agreement with experimental results,[72] our quantum yields for N_2H_4 do not exactly match the experiments. Our N_2H_4 quantum yield goes through a maximum at slightly earlier times (albeit at a similar peak magnitude of ~0.03–0.04) and falls off more quickly initially than was observed,[72] then more slowly at later times. Moreover, our N_2H_4 concentration reaches a steady state (with a quantum yield that therefore linearly decreases with photons absorbed), whereas the N_2H_4 concentration in the experiments apparently decreases after an early maximum before possibly reaching a low constant value (see Fig. 1 in Groth *et al.*[72]). Photolysis of N_2H_4 is included in our model, and although occurring, the N_2H_4 loss due to photolysis cannot compete with abstraction by hydrogen atoms and NH_2. It is unclear what the additional loss process might be. Despite this slight quantitative inconsistency with the N_2H_4 behavior, we have chosen not to tweak the reaction rate coefficients any further, as the dominant reactions (except for the one for $NH_2 + N_2H_4$, which we modified) all have literature-derived values. Keep in mind, however, that our mechanism may slightly overpredict the net production of N_2H_4 under these conditions.

The second simulation we perform is the investigation of the photochemical coupling of ammonia (NH_3) and acetylene (C_2H_2) in the presence of H_2.[35–39] As has been discussed in detail in these investigations, several complex organo-nitrogen compounds are produced when H_2–NH_3–C_2H_2 mixtures are irradiated by 184.9-nm and 206.2-nm photons. The rate coefficients for the production and loss of these compounds are generally not available in the literature. We therefore simulate the conditions in the Keane *et al.* experiments[39] and compare models to experimental results in order to help constrain the relevant kinetics.

The specific experiment we simulate is the irradiation of a mixture of 600 torr H_2, 40 torr NH_3, and 5 torr C_2H_2 at 296 K by 206.2-nm ultraviolet photons. Keane *et al.*[39] present quantum yields resulting from that investigation; however, we utilize more detailed information on this experiment, including the time variation of the photoproducts, from the thesis and laboratory notebooks of T. C. Keane.[38] During the experiment, quartz cells of 2.5-cm diameter and 10-cm length were filled with the above gas mixture, and an iodine lamp was used to irradiate the cell with 206.2 nm photons for various lengths of time. Ammonia actinometry[76] was used to determine that 4.438×10^{15} photons per second were entering the cell, for a corresponding 206.2-nm flux of 9.04×10^{14} photons $cm^{-2} \, s^{-1}$. The composition and abundance of the photoproducts were measured by 500 MHz NMR spectroscopy, and full details of the experimental and analysis procedure can be found in the original reports.[38,39]

We use KINETICS[43] to simulate this experiment in a similar manner as for the pure ammonia photolysis experiment.[72] We start with a one-dimensional 10-cm box (subdivided into a six-segment grid) filled with the appropriate 600/40/5 torr mix of $H_2/NH_3/C_2H_2$ irradiated by a 206.2-nm flux of 9.04×10^{14} photons $cm^{-2} \, s^{-1}$, and use the KINETICS model with our full 1973-reaction list to solve for the time variation in the abundances of carbon-, nitrogen-, and hydrogen-bearing species. Our model results are compared with the experimental results[38] in Fig. 1.

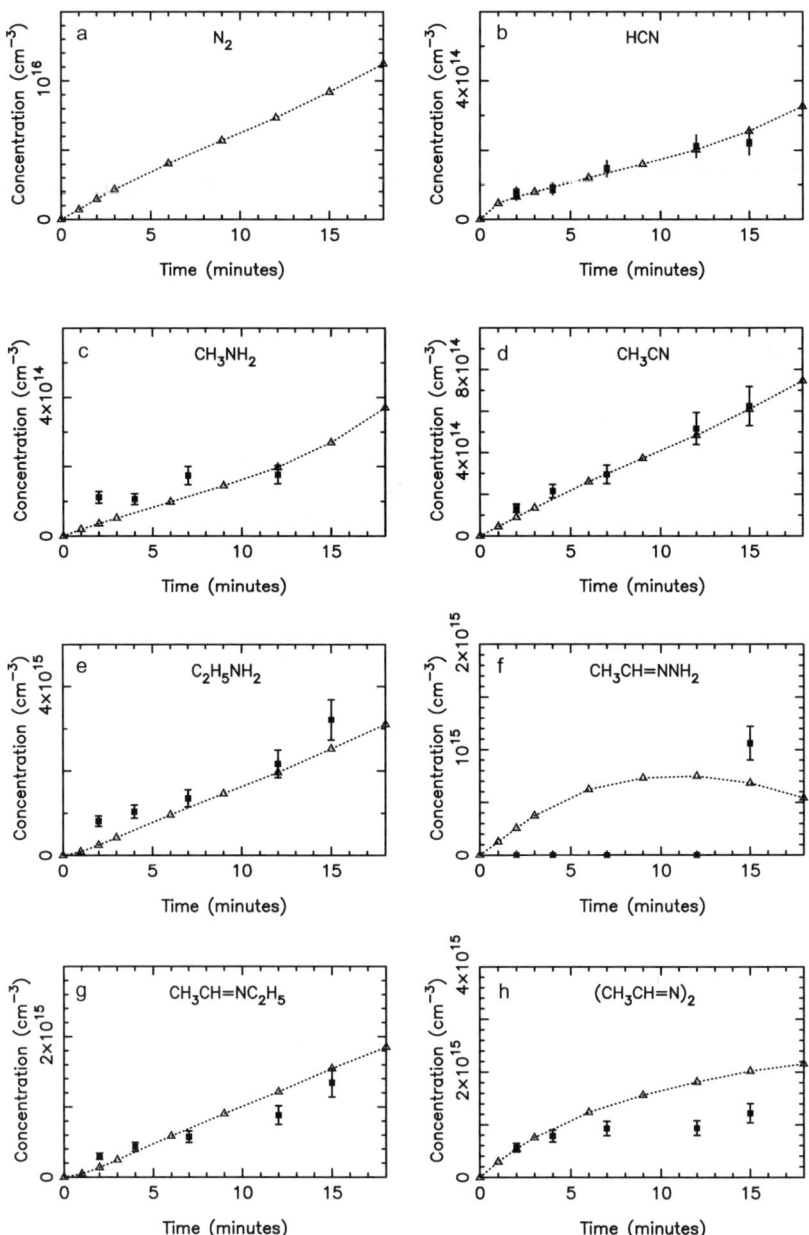

Fig. 1 Concentrations of (a) molecular nitrogen, (b) hydrogen cyanide, (c) methylamine, (d) acetonitrile, (e) ethylamine, (f) acetaldehyde hydrazone, (g) N-ethylethylideneimine, and (h) acetaldazine as a function of time in the photolysis cell. The dotted lines and open triangles correspond to the model results, whereas the solid squares correspond to the experimental results.[38] Note the change in the ordinate range for the different species. Reported measurement errors are 10–15% on species abundances, mostly due to the NMR technique.[38]

The dominant nitrogen-bearing products in our model are N_2 (not investigated by Keane 1995), acetaldazine ($CH_3CH=N-N=CHCH_3$), ethylamine ($C_2H_5NH_2$), and N-ethylethylideneimine ($CH_3CH=NC_2H_5$), with lesser amounts of the other species shown in Fig. 1. Almost no quantitative kinetic information exists for these species.

Kaye and Strobel,[35] Ferris and Ishikawa,[37] and Keane[38] all propose that the critical first step in the coupled photochemistry of NH_3–C_2H_2 is the sequence $NH_3 + h\nu \rightarrow NH_2 + H$ and $C_2H_2 + H + M \rightarrow C_2H_3 + M$, followed by $NH_2 + C_2H_3 + M \rightarrow C_2H_5N + M$. However, the identity and fate of the C_2H_5N isomer, and the resulting production sequences for the different complex organo-nitrogen species differ between these investigations. Kaye and Strobel[35] do not attempt to distinguish the main C_2H_5N isomer, although they do note that aziridine, a cyclic C_2H_5N isomer, is known to yield HCN upon photolysis,[78] and they suggest that the other isomers would, as well. Kaye and Strobel[35] stop their mechanism at the C_2H_5N photolysis stage and do not further investigate the production of complex nitrogen-bearing organics. Ferris and Ishikawa[37] and Keane[38] suggest that vinylamine ($C_2H_3NH_2$) is formed initially, followed by isomerization to ethylideneimine ($CH_3CH=NH$), a process that is common with enamines.[79–81] Keane[38] goes on to propose that the key to the formation of HCN, as well as some of the complex species is the ethylideneiminyl radical ($CH_3CH=N\cdot$), which can react with itself to form acetaldazine, or thermally decompose (or photolyze) to form HCN or CH_3CN. Although we do not explicitly distinguish between C_2H_5N and C_2H_4N isomers in our model, our reaction list implicitly follows the main $CH_3CH=NH$ and $CH_3CH=N\cdot$ pathways suggested by Keane.[38] However, some of our production mechanisms for the complex species diverge from those suggested by the earlier investigations.[37,38]

As an example, the dominant acetaldazine (see Fig. 1h) formation mechanism in our model is $2\ C_2H_4N + M \rightarrow CH_3CH=NN=CHCH_3$, as suggested by Keane[38] and Ferris et al.,[37] where their expected C_2H_4N isomer is $CH_3CH=N\cdot$. However, the main non-recycling production pathway for C_2H_4N in our model is the reaction $C_2H_2 + NH_2 + M \rightarrow C_2H_4N + M$, followed very closely in terms of importance by hydrogen abstraction from C_2H_5N by both H and NH_2, whereas only the H-atom abstraction pathway C_2H_5N (i.e., as $CH_3CH=NH) + H \rightarrow C_2H_4N + H_2$ is considered in the earlier investigations.[37,38] Note that the $C_2H_2 + NH_2 + M$ reaction has been studied experimentally[82–85] and theoretically,[86] and we adopt a rate coefficient for this reaction of $k_0 \approx 1 \times 10^{-26}$ cm^6 s^{-1} and $k_\infty = 1.3 \times 10^{-19}\ T^{2.03} \exp(-1300/T)$ cm^3 s^{-1}, where T is the temperature in kelvins. Photolysis of acetaldazine helps recycle C_2H_4N in our model and is the main effective loss process of acetaldazine.

Ethylamine (Fig. 1e) is produced in our model through the reaction $NH_2 + C_2H_5 + M \rightarrow C_2H_5NH_2 + M$, with loss due to photolysis and abstraction by H atoms to form $C_2H_5\dot{N}H$. These pathways are consistent with previous suggestions.[38] Although the rate coefficient for the reaction of C_2H_5 with NH_2 has been measured,[87] the relative roles of addition and disproportionation are not clear, and the assumed relative rates of the different product pathways for this reaction can affect our results. To prevent large quantities of ethylamine from building up in our cell, we have assumed that $NH_2 + C_2H_5 \rightarrow NH_3 + C_2H_4$ is about three times faster than $NH_2 + C_2H_5 \rightarrow C_2H_5NH_2$. This factor-of-three value may be a overestimate, as can be seen from the fact that we slightly underestimate the ethylamine abundance shown in Fig. 1e. A thorough discussion of the photolysis of ethylamine can be found elsewhere.[38]

Keane[38] suggests that N-ethylethylideneimine (Fig. 1g) is produced through the nucleophilic addition/elimination reaction $CH_3CH=NH + C_2H_5NH_2 \rightarrow CH_3CH=N–C_2H_5 + NH_3$. However, this reaction may have a significant activation energy barrier in the gas phase, given that the reactants are molecules and not radicals. We instead suggest that $CH_3CH=N–C_2H_5$ formation might occur through reactions such as (1) $CH_3CH=N\cdot + C_2H_5 \rightarrow CH_3CH=N–C_2H_5$, (2) $C_2H_5N + C_2H_5\dot{N}H \rightarrow CH_3CH=N–C_2H_5 + NH_2$, (3) $CH_3CH=N\cdot + C_2H_6 \rightarrow CH_3CH=N–C_2H_5 + H$, (4) $CH_3CH=N\cdot + C_2H_5NH_2 \rightarrow CH_3CH=N–C_2H_5 + NH_2$, (5) $C_2H_5\dot{N}H + C_2H_5 \rightarrow CH_3CH=N–C_2H_5 + H_2$, (6) $C_2H_5N + C_2H_5 \rightarrow CH_3CH=N–C_2H_5 + H$, or (7) $C_2H_5NH_2 + C_2H_3 \rightarrow CH_3CH=N–C_2H_5 + H$. Without thermodynamic parameters for N-ethylethylideneimine and some of these other species, the exothermicity and likely activation energies of these reactions cannot be determined. In our model, the first of our suggested reactions dominates

N-ethylethylideneimine production, but other potential pathways are also included in the model; see the ESI† for our estimates of these reaction rate coefficients. N-Ethylethylideneimine is destroyed in our model predominantly through reactions with H (to form either $CH_3CH=N\cdot + C_2H_6$ or $C_2H_5N + C_2H_5$) and with $CH_3CH=N\cdot$ (to form acetaldazine + C_2H_5). The production and loss mechanisms for N-ethylethylideneimine remain considerably uncertain. In particular, given the rapid observed rate of the gas-phase reaction of CH_3CHO and N_2H_4 to form $CH_3CH=N-NH_2$ and other products,[38] it is possible that $CH_3CH=NH$ participates in nucleophilic addition/elimination reactions without much of an energy barrier in the gas phase, as originally proposed by Keane.[38]

Acetaldehyde hydrazone exhibits interesting behavior in Keane's experimental data (see Fig. 1f). At room temperature, the $CH_3CH=NNH_2$ abundance is negligible until irradiation times of 15 min or longer, at which point it jumps up to become a major product of coupled $C_2H_2-NH_3$ photochemistry. The reason for this late production is unclear, and the model does not reproduce this behavior. One possibility is that slow photolysis of one of the more abundant photoproducts such as $C_2H_5NH_2$, $CH_3CH=NC_2H_5$, or $CH_3CH=NN=CHCH_3$ contributes, but the profiles of these species show no evidence for a sudden, significant loss at late times. Atomic hydrogen begins to build up in our model at later times as more and more of the C_2H_2 is destroyed, and a second possibility is that hydrogen abstraction commences as a significant loss process for some of our species as H atoms build up in the cell, with a corresponding significant increase in the production rate for $CH_3CH=NNH_2$. However, again, there are no signs of the corresponding reduction of any of the other observed species, and we are unable to find a combination of reactions that would reproduce this behavior. The late production of acetaldehyde hydrazone remains a mystery. In our model, $CH_3CH=NNH_2$ is produced predominantly through the addition reactions $NH_2 + C_2H_4N \rightarrow CH_3CH=NNH_2$ and $H_2NN + C_2H_4 \rightarrow CH_3CH=NNH_2$ and is lost mainly through $H + CH_3CH=NNH_2 \rightarrow NH_2 + C_2H_5N$, but given our poor comparisons with the experimental data,[38] we have no confidence in our adopted kinetics for $CH_3CH=NNH_2$.

Kinetic information for some of the simpler species like HCN, CH_3CN, and CH_3NH_2 exists in the literature, but that does not mean the production and loss of these species under the Keane[38] experimental conditions are well understood. Methylamine (Fig. 1c) is produced in our model largely from $NH_2 + CH_3 + M \rightarrow CH_3NH_2 + M$, for which rate-coefficient information is available.[88] Photolysis dominates the loss of CH_3NH_2, a process that has been well studied.[89–93] Acetonitrile (Fig. 1d) has four main production mechanisms in our model, all involving Keane's key intermediate, the $CH_3CH=N\cdot$ radical.[38] In order of decreasing importance, these are (1) $2CH_3CH=N\cdot \rightarrow CH_3CN + C_2H_5N$, (2) $CH_3CH=N\cdot + C_2H_5 \rightarrow CH_3CN + C_2H_6$, (3) $CH_3CH=N\cdot + H \rightarrow CH_3CN + H_2$, and (4) $CH_3CH=N\cdot + NH_2 \rightarrow CH_3CN + NH_3$, for which there are no literature values for rate coefficients. In our model, CH_3CN is destroyed largely through the abstraction reactions $CH_3CN + CH_3 \rightarrow CH_2CN + CH_4$ and $CH_3CN + H \rightarrow CH_2CN + H_2$, as well as through $CH_3CN + H \rightarrow HCN + CH_3$. Rate coefficients for the two reactions of H with CH_3CN have been reported.[63] The reaction $CH_3CN + CH_3 \rightarrow CH_2CN + CH_4$ is exothermic under our conditions but likely possesses a significant activation barrier. We estimate a rate coefficient of $1.0 \times 10^{-12} \exp(-3000/T)$ (for T in K) for this reaction, based in part on analogy with $H + CH_3CN$.

Hydrogen cyanide (Fig. 1b) is produced in multiple ways under the conditions of this experiment, including from photolysis of many of the above species.[38] Due to the rapid synthesis of C_2H_4N and H radicals with our proposed mechanism, the dominant production pathway in our model is the speculative reaction $C_2H_4N + H \rightarrow HCN + CH_4$. This reaction is likely to be highly exothermic; however, other product pathways may compete, including formation of the C_2H_5N adduct, formation of $CH_3CN + H_2$, and, less likely because of the amount of rearrangement involved and the necessity of breaking the strong $C=N$ bond, the formation of $NH_3 + C_2H_2$. We have adopted a rate coefficient of 4×10^{-11} cm^3 s^{-1} for the $C_2H_4N + H \rightarrow HCN + CH_4$ reaction. Other significant production mechanisms in our model involve H_2CN, either through self reaction,[94] reaction with H,[95] or reaction with NH_2.[60] Hydrogen cyanide is lost mainly through the reaction $HCN + C_2H_3 \rightarrow C_2H_3CN + H$.[96]

The acrylonitrile (C_2H_3CN, not shown in Fig. 1) that forms in the latter reaction also builds up in the cell in our simulation, for an overall abundance that lies in between that of acetonitrile and ethylamine. Keane[38] finds no evidence for acrylonitrile in the NMR spectra, although he does note that a liquid polymer forms on the window of the cell during the photolysis experiment. It is possible that the C_2H_3CN polymerizes or that we have neglected some other significant loss process for this molecule. The dominant loss mechanism currently in our model is the reverse of the production reaction, *i.e.*, $C_2H_3CN + H \rightarrow HCN + C_2H_3$.

Two other species that form in noticeable quantities in our model but are not detected by Keane[38] are methanimine (CH_2NH), whose concentration reaches a peak value of 5.8×10^{14} cm^{-3} after 21 min before slowly dropping off with time, and C_2H_5N, whose concentration peaks at 4.6×10^{14} cm^{-3} after 15 min before slowly dropping off with time. Methanimine is produced in our model mainly from reaction of H_2CN with C_2H_3 and C_2H_5 radicals and is lost from abstraction by C_2H_3 to form $H_2CN + C_2H_4$. Because of its suspected importance in their overall reaction mechanism, Keane[38] searched specifically for a signature that could be caused by the C_2H_5N isomer ethanimine ($CH_3CH=NH$), as well as attempted to synthesize and isolate this imine, but both attempts were unsuccessful. The primary production mechanism for forming C_2H_5N in our model, and indeed one of the top three mechanisms for forming the C–N bond in our simulation, is the reaction originally proposed by Kaye and Strobel:[35] $NH_2 + C_2H_3 + M \rightarrow C_2H_5N + M$. Reaction of $CH_3CH=NNH_2$ with H to form $C_2H_5N + NH_2$ contributes at later times. The primary loss processes for C_2H_5N in our model include abstraction by H or NH_2 to form C_2H_4N and H_2 or NH_3, and reaction with C_2H_5NH to form $CH_3CH=N-C_2H_5 + NH_2$. Aside from the $C_2H_3 + NH_2 \xrightarrow{M} C_2H_5N$ reaction proposed earlier,[35] the two other key reactions leading to the formation of the C–N bond in our model are $NH_2 + C_2H_2 \xrightarrow{M} C_2H_4N$ and $NH_2 + C_2H_5 \xrightarrow{M} C_2H_5NH_2$ (see above).

Keane[38] does not track the time dependence of the N_2H_4 abundance in their experiment. However, a quantum yield is reported[39] for N_2H_4 formation of 0.007 from the first few minutes of the irradiation (further details are not specified). We obtain that N_2H_4 quantum yield after 7 min in our model.

Although our model results compare well with the time variation observed for many of the species in the Keane[38] experiment (see Fig. 1), problems do exist, especially for $CH_3CH=NNH_2$ (Fig. 1f), and to a lesser extent for $CH_3CH=NN=CHCH_3$ (Fig. 1h). The actual mechanism involved is likely far more complex than our limited reaction list can attempt to reproduce, but without further information on the thermodynamic properties of the key species and on the rate coefficients for individual reactions, we are unlikely to implement meaningful improvements to the proposed mechanism. Despite the incomplete and cursory nature of the proposed mechanism, we at least now have in place a reaction list that includes estimates for the production and loss of the major species involved in the coupled chemistry of $C_2H_2-NH_2-H_2$ mixtures, and we can test the effectiveness of coupled $NH_3-C_2H_2$ photochemistry under Jovian conditions. We can test, in particular, whether the addition of these various complex nitrogen-bearing species contributes to the formation of HCN on Jupiter.

3. Photochemical model

The photochemistry of ammonia on Jupiter was first investigated qualitatively by Wildt[97] and quantitatively by Cadle[98] and McNesby;[99] the latter two authors both discussed the likelihood of carbon–nitrogen coupling. Strobel[100] was the first to develop a realistic model that explained the continuing presence of NH_3 on Jupiter, through a nitrogen cycle in which convection allows photochemical products like N_2H_4 to be transported to deeper, hotter levels of the troposphere, where they can be converted back into NH_3. Strobel[100] suggested that slow vertical mixing above the ammonia clouds, combined with efficient NH_3 photolysis and N_2H_4 production, would limit the abundance of NH_3 in Jupiter's stratosphere and thus inhibit carbon–nitrogen coupling; he was also the first to suggest that condensed N_2H_4 could be a major component of the Jovian upper tropospheric and lower stratospheric haze. Atreya et al.[101] further refined these earlier models and considered the possible effects of hydrazine supersaturation on the distribution of nitrogen-bearing species, Kuhn et al.[29] examined CH_4–NH_3 coupled photochemistry through the production of methylamine, and Strobel[102] examined NH_3–PH_3 photochemical coupling. Kaye and Strobel[34] refuted the suggestion[29] that methylamine production would be significant on Jupiter and demonstrated that CH_3NH_2 and HCN production would be greatly inhibited even if the catalytic destruction of CH_4 through C_2H_2 photolysis products were included in the model. Kaye and Strobel[35] suggested instead that coupled NH_3–C_2H_2 photochemistry could lead to the production of HCN and other carbon–nitrogen species on Jupiter.

After the seminal works of Kaye and Strobel,[34,35,66,67] there was a hiatus in Jovian tropospheric photochemical modeling until observations became advanced enough to provide constraints on the vertical and horizontal profiles of ammonia and other tropospheric species. The first to exploit these advances were Edgington et al.,[103,104] who created a photochemical model to investigate the latitude and altitude variation of ammonia and phosphine from ultraviolet Hubble Space Telescope observations. However, the Edgington et al. models were based on the earlier models[101,105] that did not include the coupled C_2H_2–NH_3 photochemistry suggested by theoretical[35] and experimental[36–39] studies. The models we develop for this paper represent the first attempt to include the photochemical production of the complex organo-nitrogen species observed in these NH_3–C_2H_2 photolysis experiments. Advances in our knowledge of the vertical distribution of temperatures, stratospheric hydrocarbons, and tropospheric constituents on Jupiter[106,107] greatly aid our current investigation. Details concerning the photochemical model and our assumptions and inputs to that model are described below.

3.1 Photochemical model inputs

Our photochemical model is designed to represent global-average conditions on Jupiter. We adopt a temperature profile for the upper troposphere and middle atmosphere as described by Moses et al.[50] The temperature profile at pressures less than 0.001 mbar derives from the Galileo probe ASI data,[69] whereas the profile at pressures greater than 1 mbar derives largely from global-average Infrared Space Observatory observations;[70] in between these pressure regions, a roughly isothermal atmosphere is assumed. The model extends from 6.7 bar to 2.3×10^{-8} mbar, in a grid of 111 pressure levels, with a vertical resolution of at least three levels per atmospheric scale height. This extensive vertical range allows us to encompass not only the NH_3 photolysis region in the upper troposphere, but the methane photolysis region in the upper stratosphere. Low-to-average solar ultraviolet flux values are adopted,[50] and Jupiter's orbital distance is fixed at 5.2 AU from the Sun. As with the model described in Section 2, our reaction list contains 145 C–H–O–N–P species and 1973 reactions. Diurnally averaged quantities are considered, and we run the model until steady-state conditions are achieved. Condensation of NH_3, N_2H_4, P_2H_4, HCN, CH_3CN, and H_2O (the latter from external sources) are included in a manner described elsewhere.[108] Hydrogen sulfide (H_2S) and the tropospheric water and NH_4SH clouds, with their potential effects on ammonia and other constituents,[107,109] are not considered.

At the lower boundary of the model, we assume a He mole fraction of 0.136,[110,111] a CH_4 mole fraction of 2.04×10^{-3},[112] a CO mole fraction of 1.0×10^{-9},[48] a PH_3 mole fraction of 7×10^{-7},[113] and an NH_3 mole fraction of either 1×10^{-4},[114] 3×10^{-4},[115,116] or 5.72×10^{-4}.[112] The three values for the bottom NH_3 boundary condition were chosen to reflect the differences between the Jovian belt regions, for which NH_3 is apparently dynamically depleted at the few-bar level, and zone regions, for which NH_3 might achieve its deep, well-mixed value at the few-bar level.[107,109,112,115,117,118] All other species are assumed to have a zero concentration gradient at the lower boundary such that the species are transported through the lower boundary at a maximum possible rate. Zero flux boundary conditions are adopted at the upper boundary for all species except H, H_2O, CO, and CO_2. Atomic hydrogen, which is produced in the thermosphere, is assumed to have a downward flux of 8.0×10^8 atoms cm^{-2} s^{-1} at the top boundary.[50] We also assume that H_2O, CO, and CO_2 from external sources are entering at the top of the atmosphere with fluxes of, respectively, 4×10^4, 1×10^6, and 1×10^4 cm^{-2} s^{-1}.[8,48,50,119]

Both eddy and molecular diffusion are considered in the transport terms of the continuity equations. Our adopted molecular diffusion coefficients are described elsewhere.[120] In 1D models, the "eddy" diffusion coefficient K_{zz} provides a convenient means by which to parameterize vertical motions of all scales in the atmosphere. The stratospheric values for K_{zz} are taken from Moses et al.,[50] who provide a full discussion of the various observations that were used to help constrain the eddy-diffusion-coefficient profile. All these observational constraints have model dependencies that are difficult to characterize quantitatively. Near the methane homopause region in the upper stratosphere, K_{zz} is constrained from observations of the methane density profile from spacecraft ultraviolet occultation observations,[121,122] from ground-based near-infrared stellar occultation observations,[123] and from ultraviolet airglow observations of the 121.6-nm H Lyman α line emission and the 58.4-nm He line emission.[124,125] In the middle stratosphere, near the few tenths of a millibar level, K_{zz} is constrained from observations of the evolution of the Shoemaker-Levy 9 vapor deposited during the plume splashback phase of the impacts.[5,8,119] In the lower stratosphere ($\sim$ 1–100 mbar), K_{zz} is constrained from mid-infrared ethane observations.[50] The minimum value of $K_{zz} \approx$ 300–1500 cm^2 s^{-1} in the upper troposphere derives from CO observations[48] (see also the inferences from infrared and ultraviolet observations of PH_3 and NH_3[103,104,126–128]). This minimum value affects the profiles of the species that have a production source at high altitudes; a low value (i.e., a stagnant lower stratosphere) allows a greater build up of the column abundance of photochemically stable species like CO and C_2H_6 in the stratosphere.[129] Due to the numerous observational constraints on stratospheric K_{zz} values, we keep our stratospheric K_{zz} profile fixed but allow the tropospheric value to vary as a free parameter.

Tropospheric K_{zz} values are difficult to constrain by remote-sensing observations. Within and below the water clouds, convective motions are expected to result in high effective eddy diffusion coefficients of order $K_{zz} \approx 10^8$–10^9 cm^2 s^{-1}.[130] However, when radiative processes become important and the temperature profile ceases to become purely convective (i.e., in the upper troposphere above the clouds), vertical mixing is expected to be reduced. The PH_3 and NH_3 vertical profiles derived from infrared and ultraviolet observations have been used in this region, usually in combination with photochemistry or diffusion models, to constrain K_{zz} in the upper troposphere, although the results are very model dependent.[103,104,126,128] Moreover, these observations demonstrate that the PH_3 and NH_3 profiles, and the resulting K_{zz} inferences, vary with latitude, whereas we are attempting to construct a global-average model. As we will show, our results have some sensitivity to the adopted tropospheric K_{zz} values: high values allow for rapid removal of photochemically generated species into the deep troposphere. Based on Edgington et al.,[104] we adopt a K_{zz} profile that varies slowly and linearly with pressure in the troposphere with a value of $\sim$1–2 $\times 10^4$ cm^2 s^{-1} between $\sim$180 mbar and 1 bar, and we test the sensitivity of the results to the tropospheric eddy diffusion coefficient by simply multiplying this linear K_{zz} profile by a constant value.

Rayleigh scattering of H_2, He, and CH_4 has been included in the model. Aerosol opacity in the 150–230 nm wavelength region can also influence our model results, through shielding of PH_3, NH_3, and other molecules from photolysis. West et al.[131] review our current state of knowledge of the Jovian cloud and haze properties. Unfortunately, there appears to be little consensus regarding the details of the optical and physical properties and structure of the upper tropospheric and lower stratospheric clouds and hazes on Jupiter, as different groups, using different data sets and analysis procedures, derive different results. Uniquely deriving these properties is difficult due to the large number of unknown parameters that can strongly affect the results, and the problem is exacerbated by horizontal variations in these properties across Jupiter. Because of a lack of reliable, detailed information on the aerosol properties in this wavelength region,[103,104,132,133] we simplify the problem by including aerosol absorption only and neglect aerosol scattering. For our nominal model, we assume an optically thick haze in the 300–700 mbar region with vertical optical depth 2.7 at 150–230 nm, presumed to be the NH_3 cloud itself, and an optically thin haze in the 10–150 mbar region of vertical optical depth 0.14 (see West et al.,[131] Sromovsky and Fry,[134] and references therein for further details and comparisons with other models). We test the sensitivity of these results to the assumed optical thickness of the 300–700 mbar haze. Fortunately, condensation itself (and the saturation vapor pressure variation with temperature) has the largest effect on the vertical profile of NH_3 in the upper troposphere, and our results regarding the coupled NH_3–C_2H_2 photochemistry are relatively insensitive to our aerosol-opacity assumptions.

3.2 Photochemical model results

Our model results in terms of the mole-fraction profiles for several important nitrogen-bearing species in our nominal model are shown in Fig. 2. Also shown in the Figure is the C_2H_2 profile derived from the model. Note that although C_2H_2 is abundant in the stratosphere, its mole fraction is expected to fall off significantly

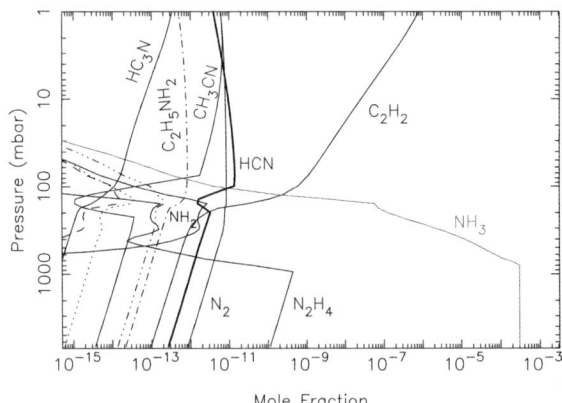

Fig. 2 Mole-fraction profiles for several nitrogen-bearing species in our nominal photochemical model (as labeled). The triple-dot-dashed profile represents CH_2NH, the dotted profile represents $CH_3CH=NH$, and the dashed profile represents CH_3NH_2. The acetylene profile is included for comparison; C_2H_2 is relatively abundant in the stratosphere but its mole fraction falls off rapidly with decreasing altitude due to photochemical loss processes. Condensation is responsible for the sharp drop off in altitude of the NH_3 and N_2H_4 profiles in the few-hundred mbar range and the more localized "bite-outs" in the HCN and CH_3CN profiles between ~100–200 mbar.

with decreasing altitude because of reaction with atomic H, photolysis, and other photochemical loss processes. Our model atmosphere contains only 0.11 ppb of C_2H_2 at the tropopause (140 mbar). Similarly, the NH_3 mole fraction in the tropopause region is also significantly reduced by condensation, photolysis, and other loss processes. Photochemical coupling of C_2H_2–NH_3 is therefore greatly inhibited compared to static photolysis experiments[38] described above, in which the initial mole fractions of NH_3 and C_2H_2 were 6.2% and 0.775%. Some NH_3–C_2H_2 coupling does occur, however, in our Jovian photochemical model, resulting in the production of small amounts of HCN, CH_3CN, $C_2H_5NH_2$, and the other complex organo-nitrogen species seen in the photolysis experiments.[36–39] Hydrogen cyanide is the dominant end product of this coupled chemistry, but its predicted abundance is well below the observational upper limits.[13–15] Our full photochemical model output, including species abundances, reaction rate coefficients, photolysis rates, production and loss rates, and chemical loss time scales can be found in the ESI.†

Because of the large amount of atomic H produced from NH_3 photolysis in and above the ammonia condensation region, the dominant scheme for producing HCN in our Jovian photochemical model is

$$NH_3 \xrightarrow{h\nu} NH_2 + H$$
$$H + C_2H_2 \xrightarrow{M} C_2H_3$$
$$H + C_2H_3 \xrightarrow{M} C_2H_4$$
$$H + C_2H_4 \xrightarrow{M} C_2H_5$$
$$H + C_2H_5 \longrightarrow 2\,CH_3$$
$$H + CH_3 \xrightarrow{M} CH_4$$
$$NH_2 + CH_3 \xrightarrow{M} CH_3NH_2$$
$$CH_3NH_2 \xrightarrow{h\nu} CH_3NH + H$$
$$H + CH_3NH \longrightarrow CH_2NH + H_2$$
$$CH_2NH \xrightarrow{h\nu} HCN + H_2$$

$$\text{Net}: NH_3 + C_2H_2 + 4\,H \longrightarrow HCN + CH_4 + + 2\,H_2.$$
(1)

The column-integrated rate of the $NH_2 + CH_3 + M \rightarrow CH_3NH_2 + M$ reaction is about an order of magnitude larger than the next most important reaction for producing carbon–nitrogen bonds, that of $NH_2 + C_2H_3 + M \rightarrow C_2H_5N + M$. This solution differs from theoretical models of Kaye and Strobel[35] for which the $NH_2 + C_2H_3 \rightarrow C_2H_5N$ reaction dominates, despite the fact that we use a slightly higher estimate for the rate coefficient for this reaction. The main difference between our model and that of Kaye and Strobel[35] in this regard appears to be the significance of CH_3 production through sequential addition of atomic H to C_2H_x hydrocarbons (i.e., the first half of the scheme (1) above). The sheer amount of H produced from NH_3 photolysis (and from C_2H_2 photolysis at higher altitudes) makes these three-body addition reactions for atomic H effective (see also[135]). Our solution here also differs from the results of our box-model simulations described in Section 2, for which the $NH_2 + C_2H_2 + M \rightarrow C_2H_4N + M$ reaction dominates the formation of carbon–nitrogen bonds, due to the large concentration of C_2H_2 and relatively high pressure in the reaction cell.

None of the carbon–nitrogen species appear to be produced in our Jovian photochemical model in large enough quantities to be observable with current technologies. After HCN, the second and third most abundant carbon–nitrogen species produced in our Jovian photochemical model are $C_2H_5NH_2$ and CH_3CN. The dominant mechanism for producing CH_3CN in our model is

$$
\begin{aligned}
NH_3 &\xrightarrow{h\nu} NH_2 + H \\
H + C_2H_2 &\xrightarrow{M} C_2H_3 \\
NH_2 + C_2H_3 &\xrightarrow{M} C_2H_5N \\
C_2H_5N + H &\longrightarrow C_2H_4N + H_2 \\
C_2H_4N &\longrightarrow CH_3CN + H \\
\hline
Net: NH_3 + C_2H_2 &\longrightarrow CH_3CN + H_2.
\end{aligned}
$$

Note that this mechanism was proposed as a way to form acetonitrile in coupled NH_3–C_2H_2 photolysis experiments.[36–38]

The dominant mechanism for producing $C_2H_5NH_2$ in our model is

$$
\begin{aligned}
NH_3 &\xrightarrow{h\nu} NH_2 + H \\
H + C_2H_2 &\xrightarrow{M} C_2H_3 \\
H + C_2H_3 &\xrightarrow{M} C_2H_4 \\
H + C_2H_4 &\xrightarrow{M} C_2H_5 \\
NH_2 + C_2H_5 &\xrightarrow{M} C_2H_5NH_2 \\
\hline
Net: NH_3 + C_2H_2 + 2H &\longrightarrow C_2H_5NH_2.
\end{aligned}
$$

This mechanism was proposed as being important for the production of ethylamine in coupled NH_3–C_2H_2 photolysis experiments.[38]

Fig. 3 shows the sensitivity of some of the species abundances to the tropospheric eddy diffusion coefficient and to the assumed haze optical depth in the ammonia condensation region. A larger tropospheric eddy diffusion coefficient allows slightly more NH_3 to be carried up into the stratosphere, where slightly more NH_2 is formed as a result. The stratospheric abundances of species that depend on NH_2 for their production, which includes all the carbon–nitrogen species, are then slightly increased for the case of the larger K_{zz}, and stratospheric mole fractions are correspondingly increased. However, the larger tropospheric K_{zz} also allows these species to diffuse more quickly through the bottom boundary of the model, so that the tropospheric mole fractions of species that are produced from the stratospheric coupled C_2H_2–NH_3 photochemistry are reduced when K_{zz} is increased. The NH_3 and N_2H_4 abundances themselves are controlled by condensation and evaporation in the troposphere and show little sensitivity to tropospheric K_{zz} values. Molecular nitrogen, on the other hand, does not condense and is produced largely in the troposphere, so increasing the tropospheric K_{zz} leads to increased transport and loss through the lower boundary, resulting in a reduced N_2 column abundance when tropospheric K_{zz} values are increased.

The dotted lines in Fig. 3 represent a model in which the eddy diffusion coefficient profile is the same as that of our nominal model (solid lines), but in which the haze vertical optical depth at 150–230 nm in the 300–700 mbar region is 8 instead of 2.7. The larger optical depth leads to increased shielding of the NH_3, with

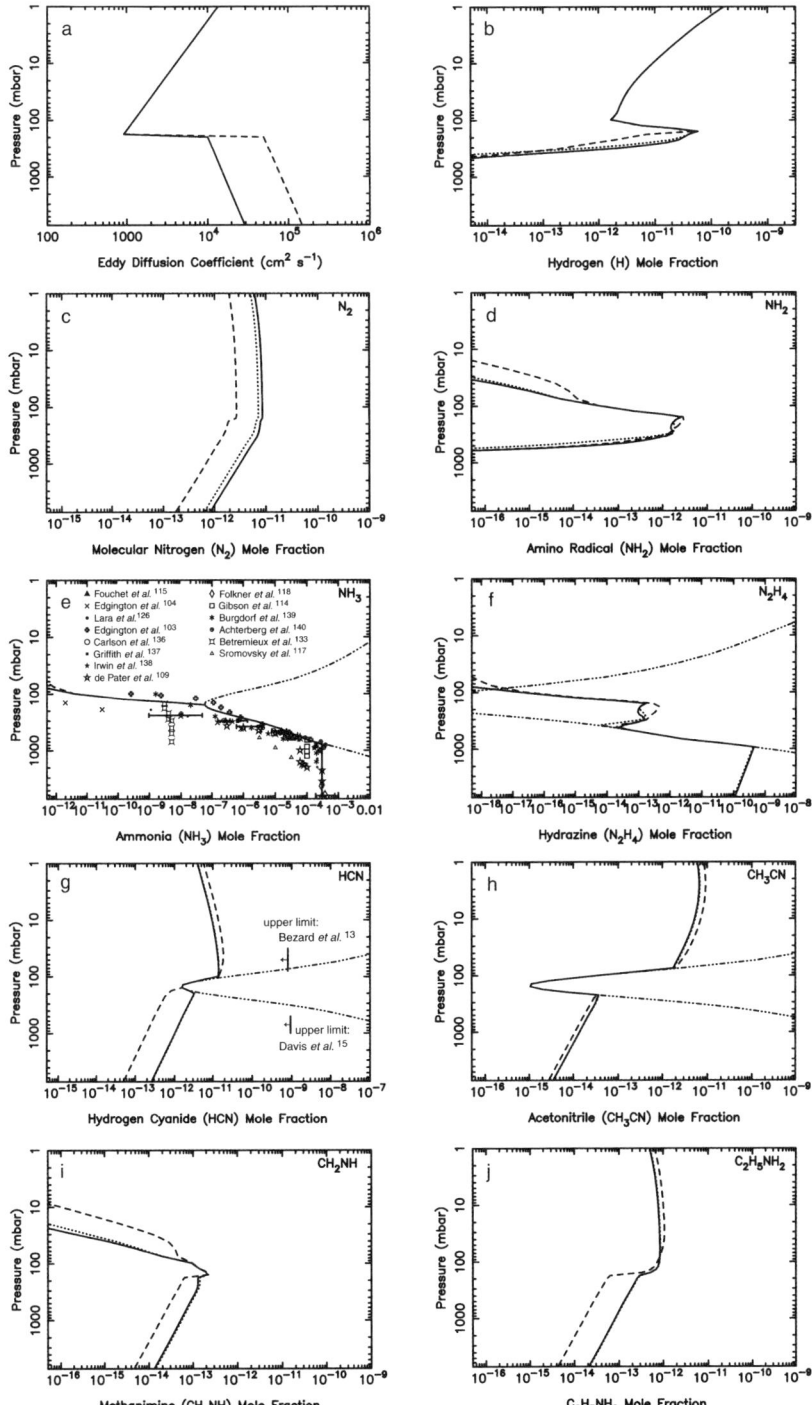

Fig. 3 The eddy diffusion coefficient profiles assumed in the photochemical model (a), along with the mole-fraction profiles for atomic hydrogen (b), molecular nitrogen (c), amino radicals (d), ammonia (e), hydrazine (f), hydrogen cyanide (g), acetonitrile (h), methanimine (i), and ethylamine (j) in our Jovian photochemical model. The solid lines represent our nominal model,

a corresponding slight reduction in the NH_2 production rate at the base of the NH_3 cloud and a very slight increase in the NH_3 mole fraction at the top of the cloud. This change in optical depth in the 300–700 mbar pressure region has almost no effect on species abundances, except for that of N_2, whose production rate slightly drops. Similarly, changing the assumed NH_3 mole fraction at the lower boundary (not shown in the Figure) has almost no effect on the species abundances other than on NH_3 itself below its condensation region. Note that we have not included chemical loss processes in the model such as reaction of NH_3 with H_2S or interaction with the NH_4SH cloud that would explain the "stair step" reduction behavior of the NH_3 mole fraction in the 1–2 bar region implied by some of the observations.[107,109]

These simple sensitivity studies show that our main conclusion about the unimportance of coupled C_2H_2–NH_3 photochemistry on Jupiter cannot be changed by adjusting uncertain free parameters in the model. The observed PH_3 profile on Jupiter puts limits on how small the tropospheric eddy diffusion coefficient can be,[103,104,126,128] and our tropospheric K_{zz} is unlikely to be much smaller than we have adopted in our nominal model. The production rate of nitriles and other organo-nitrogen compounds is mostly unaffected by the assumed optical depth within the ammonia cloud layer or the assumed NH_3 abundance below the cloud. Observations show NH_3 to be subsaturated on Jupiter, particularly in belt regions,[107] whereas our ammonia profile closely follows its saturation vapor pressure curve in its condensation region (in contrast to N_2H_4, which is produced rapidly enough in its condensation region that condensation loss cannot keep pace, and supersaturations are maintained). A subsaturated ammonia profile would lead to even less C_2H_2–NH_3 photochemical coupling and lower abundances of HCN and other carbon–nitrogen compounds.

The lack of coupled C_2H_2–NH_3 photochemistry in our model is entirely due to the low derived abundance of C_2H_2 in the NH_3 photolysis region in Jupiter's upper troposphere and lower stratosphere. Kaye and Strobel[35] assumed a much greater C_2H_2 abundance in their model—their Model A contains C_2H_2 concentrations a factor of ~70 higher than our nominal model at their 50-km altitude upper boundary and more than four orders of magnitude greater than our nominal model throughout the troposphere. That difference is the main cause of our different predictions concerning the HCN abundance on Jupiter. On the other hand, we confirm the models of Kaye and Strobel[35] and Strobel[100] in that coupled CH_4–NH_3 photochemistry is greatly inhibited due to the physical separation of the CH_4 photolysis region in the upper stratosphere from the NH_3 photolysis region in the troposphere. The only way we could increase the net production rate of HCN in our model would be to invoke an upper tropospheric source of C_2H_2[24,25,132,133] or unusual dynamical conditions that allow rapid transport of stratospheric C_2H_2 into the troposphere; we discuss these possibilities further in Section 5.2.

4. Thermochemical kinetics and transport model for the deep troposphere

At very high temperatures (≥ 1500 K) in Jupiter's deep troposphere, the atmospheric composition is controlled by thermochemical equilibrium. Equilibrium models that

---

with a tropospheric eddy diffusion coefficient similar to that derived by Edgington et al.[104] and with a haze vertical optical depth at 150–230 nm of 2.7 in the 300–700 mbar region. The dashed lines represent a model that is the same as our nominal model, except the tropospheric K_{zz} profile has been multiplied by a factor of 5. The dotted lines represent a model that is the same as our nominal model, except the haze vertical optical depth at 150–230 nm in the 300–700 mbar region is 8 instead of 2.7. The dash-triple-dot lines represent the saturation vapor density curves for the molecules in question. The model results for NH_3 are compared with various observations[103,104,114–118,126,136–140] in (e), and the HCN profiles are compared with the stratospheric[13] and tropospheric[15] upper limits.

include nitrogen species[141–143,19–21] show that NH_3 is the dominant nitrogen-bearing constituent throughout the Jovian atmosphere; HCN and N_2 are not very abundant at colder high altitudes, but the equilibrium mole fractions of N_2 and HCN are both expected to increase toward the deeper, hotter regions of the troposphere. For the observable regions in the upper troposphere, the equilibrium HCN abundance, in particular, is negligible. However, as was first discussed quantitatively by Prinn and Barshay,[144] thermochemical equilibrium may be difficult to maintain on Jupiter in the presence of rapid convective mixing. In this scenario, thermochemical equilibrium can be preserved only as long as the chemical kinetic time scale for conversion between different molecular species is shorter than the time scale for vertical atmospheric mixing. As a parcel of gas from deeper, hotter levels is transported up to cooler atmospheric regions, it will eventually encounter regions where the chemical kinetic conversion time scale becomes longer than the transport time scale, at which point the mole fractions of species like CO, N_2, CH_3NH_2, or HCN can become "quenched" or "frozen in" due to the inability of the kinetic reactions to overcome activation energy barriers. At altitudes above that quenching level (*i.e.*, at altitudes above the level where the chemical-kinetic time constant equals the convective mixing time constant), the mole fraction of the quenched species will remain fixed at the equilibrium abundances achieved at the quench point.[144] Because of this transport-induced disequilibrium process, species can be present in the upper troposphere of Jupiter in abundances much greater than their predicted equilibrium abundances. The disequilibrium quenching of N_2 and HCN on Jupiter has been discussed by several investigators[145,10,20,21] (see also the relevant discussions for other solar-system applications[146–148]).

Through time-constant and quenching-level arguments, Lewis and Fegley[10] suggest that HCN would quench at ~1200 K on Jupiter, resulting in a quenched steady-state mole fraction for HCN of only ~1×10^{-12} for a solar-composition gas. Lewis and Fegley,[10] using arguments from Prinn and Fegley,[147] suggest that the rate-limiting step responsible for quenching the HCN abundance is the reaction $H_2 + HCN \leftrightarrow CH_2 + NH$. The rate coefficient for this reaction and its reverse have never been measured; Lewis and Fegley[10] assume that the reverse reaction will proceed rapidly with a rate coefficient of 1×10^{-10} cm^3 s^{-1}. The rate coefficient in the forward direction is then estimated from the equilibrium constant of the reaction, along with the reverse rate coefficient, using the principle of microscopic reversibility. Fegley and Lodders[20] make these same assumptions in their follow-up study, but they derive a much larger quenched HCN mole fraction of 0.6–2.6 ppb for Jupiter. The differences between the two results are attributed to the larger elemental enrichment factors used in the more recent study.[20,21] We note that the most likely products of the $CH_2 + NH$ reaction could be $H_2CN + H$ and/or $N + CH_3$, rather than $HCN + H_2$, so that the overall $H_2 + HCN$ reaction as stated might be an oversimplification of a multiple-step process, but otherwise the assumptions seem reasonable. Is that reaction the only available mechanism for HCN loss, though? That premise seems unlikely. The reaction between HCN and H_2 is very endothermic and will be exceedingly slow, even at 1500 K, and we suspect there are more effective HCN destruction mechanisms under deep-tropospheric conditions on Jupiter. Moreover, as is demonstrated by Smith,[46] the adoption of the pressure scale height H for the characteristic length scale in the expression for the convective mixing time constant leads to a roughly two-order-of-magnitude overestimation of the transport time scale used in the time-constant approach[144,10,20,21] to predict quenched disequilibrium abundances on Jupiter. If Smith[46] is correct, the HCN quenching level is likely even deeper in the atmosphere than Fegley and Lodders[20] have assumed (given their rate-limiting mechanism), which in turn suggests an even higher predicted quenched HCN mole fraction—a result in clear violation of the HCN upper limits.[13–15]

The details of the original Prinn and Barshay[144] time-constant approach for CO quenching on Jupiter have been questioned.[24,45–49] Both the assumed rate-limiting step (and its estimated rate coefficient) and the assumptions regarding the transport time scale have been criticized; several groups have suggested ways in which the original assumptions could be improved.[24,45–49] Fuelled by these criticisms and suggested improvements, we have recently developed a way to bypass the back-of-the-envelope time-scale approach by directly modeling chemical kinetics and transport in the deep troposphere of Jupiter to more quantitatively investigate carbon–hydrogen–oxygen chemistry and the transport-induced quenching of disequilibrium C–H–O species. The results are presented in Visscher *et al.*[49] Using this model, we confirm the results of Smith[46] regarding the transport time constants and identify the most likely rate-limiting step for the quenching of CO on Jupiter. The new rate-limiting mechanism suggested by Visscher *et al.*[49] helps resolve a long-standing controversy regarding the Prinn and Barshay[144] scheme and the origin of tropospheric CO on Jupiter, as well as helps constrain the deep water abundance on Jupiter. Based on kinetics and diffusion only, the model is able to reproduce the equilibrium composition at deep, hot atmospheric levels, and then transitions smoothly to a quenched regime at higher altitude levels based on the rates of the reactions controlling the interconversion of the different atmospheric constituents. For this paper, we discuss the results regarding nitrogen chemistry in the deep Jovian troposphere.

4.1. Thermochemical model description

We again use the Caltech/JPL KINETICS code[43] to solve the continuity equations for the atmospheric constituents, but we focus this time on Jupiter's deep troposphere. The model extends from 12 650 bar (2500 K) to 17.4 bar (399 K) in a grid of 144 atmospheric levels, with a vertical resolution of at least twenty altitude levels per scale height. The assumed pressure–temperature profile in the 17–24 bar region is taken from the Galileo probe ASI data[69] and is extended to greater depths along an adiabat, assuming ideal-gas behavior.[148] Thermochemical equilibrium is adopted as an initial condition, with elemental abundances taken from the Galileo Probe Mass Spectrometer (GPMS) results[110,112] for carbon, nitrogen, and helium, but not for oxygen. The oxygen elemental abundance determined by the GPMS is considered to be a lower limit to the deep Jovian abundance due to the probe's entry into an anomalous "hot-spot" region.[107] The oxygen abundance in our model is set at 2.6 times the assumed protosolar value of $H_2O/H_2 = 9.61 \times 10^{-4}$,[149] where it is assumed here that a portion of the total oxygen content has already been removed by rock-forming elements (which we do not consider in the model). This level of oxygen enrichment was found by Visscher *et al.*[49] to provide a good fit to the observed tropospheric CO mole fraction.[48] The NASA CEA code[150] is used to calculate thermodynamic equilibrium, with thermodynamic parameters taken from Gurvich *et al.*,[151] Chase,[152] Burcat and Ruscic,[153] and other literature sources. Zero flux boundary conditions are adopted at the top and bottom of the model such that no mass enters or leaves the system. Transport is assumed to occur through vertical eddy diffusion, with our nominal model adopting a constant tropospheric K_{zz} of 1×10^8 cm^2 s^{-1} (see Visscher *et al.*[49] for further details).

We use a subset of ~1800 reactions and ~120 species from our photochemical model described in Sections 2 and 3. Because we must fully reverse all our reactions (using the principle of microscopic reversibility) in order to accurately reproduce equilibrium compositions with the kinetic model, we are forced to omit several of the complex organo-nitrogen compounds observed in coupled C_2H_2–NH_3 photolysis experiments[36–39] due to a lack of information on thermodynamic properties (see Section 2 for a list of these species). The top of our model (17 bar) is also deep enough that ultraviolet photons do not penetrate, so we neglect photolysis reactions. Other than these changes, the rate coefficient expressions are taken from the photochemical model. The kinetics of nitrogen species under deep-tropospheric conditions on Jupiter is less well understood than the corresponding case for C–H–O species, and our predicted quench levels for the nitrogen-bearing species will be correspondingly less certain. We deem the modeling exercise worth the effort, however, if for nothing else than to suggest reactions that could be important for the dominant quenching mechanisms and thus warrant further study. More importantly, we hope to resolve the apparent paradox mentioned above regarding the large quenched disequilibrium abundance of HCN in excess of observational limits (cf.[15,20,21]) that is predicted when the Lewis and Fegley[10] rate-limiting kinetic mechanism for HCN destruction is used in combination with the convective-mixing length scale derived by Smith.[46]

4.2. Thermochemical model results

Fig. 4 shows the results of our thermochemical kinetics and transport model for some of the major nitrogen-bearing species. Because our kinetic model considers the same species as our thermochemical-equilibrium calculations and because we fully reverse all of our kinetic reactions using that same thermodynamic data, our kinetic model will reproduce the equilibrium results in the absence of transport, provided that we allow enough time for equilibrium to be achieved at any particular temperature. At hot, deep levels in the kinetics-transport model shown in Fig. 4, the kinetic reactions are very fast, and equilibrium is maintained. However, the species profiles diverge from equilibrium (shown with dashed lines in Fig. 4) when the kinetic rates can no longer keep up with dynamical mixing (i.e., when the transport time scale falls below the kinetic loss time scale for the species in question). This divergence occurs in the cooler, upper regions of the model: many reactions have significant activation barriers that cannot readily be overcome as temperatures drop, so that the reactions proceed predominantly in one direction, and equilibrium is not preserved. At altitudes above this quench level, the mole fraction of a quenched species remains constant. This behavior was first predicted analytically by Prinn and Barshay.[144] Note that different species shown in Fig. 4 have different quench levels. Molecular nitrogen, with its strong triple bond, is difficult to destroy kinetically and so is quenched at relatively hot, deep levels, whereas HCN equilibrium is preserved even at relatively cool temperatures such that the quench level is at higher altitudes. Some species, such as HNCO and CH$_2$NH, exhibit more complex behavior due to the effects of the quenching of other constituents. For instance, HNCO first begins to diverge from equilibrium when CO quenches, but HNCO itself does not quench until it reaches higher altitudes. In the intervening regions, HNCO continues to

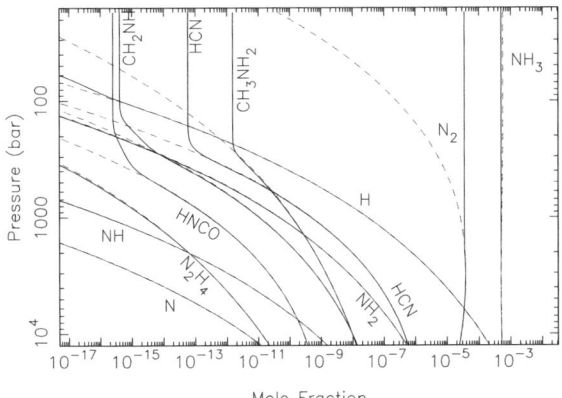

Fig. 4 Mole-fraction profiles (as labeled) for several nitrogen-bearing species in our Jovian deep-troposphere thermochemical kinetics and transport model for an assumed constant eddy diffusion coefficient of 1×10^8 cm^2 s^{-1}. The dashed lines show our thermochemical equilibrium solution. Note that kinetic reactions are so fast in the hotter, deeper regions of the model that equilibrium can be maintained. However, the species abundances diverge from equilibrium and are quenched at colder, higher levels as the transport time scale drops below the kinetic time constants for conversion between the different species.

maintain an equilibrium with the quenched CO. Similarly, CH_2NH diverges from equilibrium when HCN quenches but does not itself fully quench until it reaches higher altitudes.

Based on our adopted nitrogen reaction mechanism, we find that HCN does not quench until it reaches the ~880-K, 260-bar level. As a result, the quenched mole fraction is only $\sim 6 \times 10^{-14}$, a value well below the observational limit for HCN of 0.93 ppb in the Jovian upper troposphere.[15] The dominant mechanism for HCN loss in our model is the following scheme:

$$\begin{aligned}
H + HCN &\xrightarrow{M} H_2CN \\
H_2 + H_2CN &\longrightarrow CH_2NH + H \\
H + CH_2NH &\xrightarrow{M} CH_2NH_2 \\
H_2 + CH_2NH_2 &\longrightarrow CH_3NH_2 + H \\
CH_3NH_2 &\xrightarrow{M} CH_3 + NH_2 \\
H_2 + NH_2 &\longrightarrow NH_3 + H \\
H_2 + CH_3 &\longrightarrow CH_4 + H \\
\underline{2H &\xrightarrow{M} H_2} \\
\text{Net}: HCN + 3H_2 &\longrightarrow NH_3 + CH_4.
\end{aligned} \qquad (2)$$

The rate-limiting step in this scheme is the $H_2 + H_2CN \rightarrow CH_2NH + H$ reaction. Our rate coefficient for this reaction comes from the reverse reaction, whose rate coefficient has been calculated from the DHT method.[60] Although the rate coefficient for the $H_2 + H_2CN \rightarrow CH_2NH + H$ reaction is a relatively low $\sim 3 \times 10^{-18}$ cm^3 s^{-1} at the quench level, it is still much faster than the likely rate for the $H_2 + HCN$ reaction suggested as the rate-limiting step by Lewis and Fegley[10] and subsequent modelers[20,21] because of the relative reactivity of H_2CN *versus* HCN. Note that our scheme begins with the three-body addition of H to HCN, and continues with the products reacting with H and/or H_2, which are abundant in the Jovian troposphere, to form a hydrogen-saturated single-bonded species, which can then thermally dissociate to break the C–N bond. A very similar scheme was invoked by Visscher *et al.*[49] to explain the dominant mechanism for breaking the strong carbon–oxygen bond and destroying CO in the Jovian deep troposphere (see also ref. 45), and we find N_2 to be destroyed by a similar process.

For instance, the dominant scheme leading to N_2 destruction in our model is

$$\begin{aligned}
H + N_2 &\xrightarrow{M} NNH \\
H_2 + NNH &\longrightarrow N_2H_2 + H \\
H + N_2H_2 &\xrightarrow{M} N_2H_3 \\
H_2 + N_2H_3 &\longrightarrow NH_2 + NH_3 \\
\underline{H_2 + NH_2 &\longrightarrow NH_3 + H} \\
\text{Net}: N_2 + 3H_2 &\longrightarrow 2NH_3,
\end{aligned} \qquad (3)$$

where NNH has recently been recognized as an important intermediate in the combustion chemistry of nitrogen species, particularly in flame fronts where the concentration of atoms is high.[60] Note the similarity of the first three steps in this reaction scheme (3) to the first three steps in our dominant HCN destruction scheme (2) above. The rate-limiting step in this N_2 destruction scheme (3) is $H + N_2H_2 + M \rightarrow N_2H_3 + M$. Our rate coefficient for this reaction again comes from the reverse reaction, whose rate coefficient was derived from QRRK analysis by Dean and Bozzelli.[60]

As with the destruction of other species with strong bonds in our model, our proposed mechanism begins with H addition to the very stable N_2 molecule, followed by sequential reactions of H_2 and H to form a single-bonded N–N species, before the N–N bond is broken. This mechanism differs significantly from the $N_2 + H_2 \rightarrow 2$ NH gas-phase, rate-limiting mechanism suggested by previous investigators.[10,20,21,145–147] Our mechanism leads to a somewhat more effective conversion of $N_2 \rightarrow NH_3$ in Jupiter's deep troposphere; however, given the near-vertical slope of the N_2 equilibrium profile at depth, our prediction for the quenched disequilibrium N_2 mole fraction of 3.5×10^{-5} does not differ too much from the $2–3 \times 10^{-5}$ mole fraction predicted by Fegley and Lodders,[20] who assumed a slightly smaller nitrogen elemental abundance in their model (see also[21]). In fact, the assumptions about the deep nitrogen elemental abundances on Jupiter have the largest effect on the predictions concerning the mole fraction of N_2 dredged up from the deep atmosphere.[20] No matter what the actual rate-limiting step is, we agree with the conclusions of Prinn and Olaguer[145] and subsequent modelers that N_2 is likely to be the most abundant quenched disequilibrium species in the upper troposphere of Jupiter.

Note that the conversion of N_2 to NH_3 might also occur heterogeneously on the surface of metallic iron grains,[145] as in the industrial Haber process. If catalytic N_2 destruction on grain surfaces is occurring on Jupiter, Prinn and Olaguer[145] demonstrate that this process could be more efficient than the pure gas-phase mechanism we consider, leading to a reduced, but still significant, N_2 mole fraction of 0.3–6 ppm in their model. However, modern chemical equilibrium calculations[20,21,154] demonstrate that Fe is removed from the atmosphere by condensation at altitudes much deeper than the level where the catalysis would be occurring on Jupiter, making metallic iron unavailable as a catalyst for $N_2 \rightarrow NH_3$ conversion. The presence of another suitable catalyst remains problematic, and homogeneous gas-phase reactions are expected to dominate.[20]

Molecular nitrogen is stable in the Jovian upper troposphere: N_2 does not readily react with the photolytic products of NH_3 and hydrocarbon photochemistry, and it is shielded from photolysis by the large overlying column of H_2. We expect N_2 to survive well into the upper stratosphere, where its mole fraction will eventually be reduced due to molecular diffusion. Galactic cosmic rays may initiate interesting N_2 chemistry, as suggested for Neptune,[155] and N_2 may participate in interesting ionospheric chemistry, as on Titan.[156]

As originally noted by Lewis and Fegley,[10] the quenched mole fraction of methylamine (CH_3NH_2) is greater than that of HCN, so that the additional deep tropospheric source of CH_3NH_2 could enhance photochemical production of HCN (see the last part of scheme (1) above). However, the upper-tropospheric quenched mole fraction of CH_3NH_2 in our model is only 1.5×10^{-12}. Even if the conversion of CH_3NH_2 into HCN were 100% effective, the HCN produced from CH_3NH_2 photochemistry would still be well below the HCN upper limits.[13–15]

Fegley and Lodders,[20] Lodders and Fegley,[21] and Visscher et al.[49] demonstrate that the mole fractions of the quenched species on Jupiter are very sensitive to the adopted eddy diffusion coefficient in the vicinity of the quench level. Our adopted nominal K_{zz} value of 1×10^8 cm^2 s^{-1} is based on free-convection and mixing-length theory for a rapidly rotating atmosphere.[157,158,49] The exact value of the eddy diffusion coefficient is uncertain by about an order of magnitude. A larger value would lead to more rapid mixing and a quench level deeper in the atmosphere, with correspondingly higher values for the mole fractions of quenched disequilibrium species

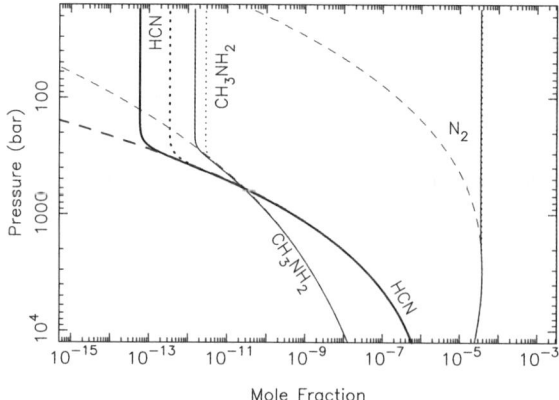

Fig. 5 Mole-fraction profiles for N_2, HCN, and CH_3NH_2 in our nominal thermochemical kinetics and diffusion model with $K_{zz} = 1.0 \times 10^8$ cm^2 s^{-1} (solid lines) compared with a model with $K_{zz} = 1.0 \times 10^9$ cm^2 s^{-1} (dotted lines). The dashed lines show the thermochemical equilibrium solution.

like HCN, CH_3NH_2, and CO. Fig. 5 illustrates how the quenched abundances of N_2, HCN, and CH_3NH_2 would change for a larger assumed K_{zz} of 1×10^9 cm^2 s^{-1}. Even with this maximum value of K_{zz}, the predicted upper-tropospheric mole fraction for HCN is well below the observational upper limit. Note that the equilibrium gradient in the deep troposphere controls how the quenched abundance will be affected by the eddy diffusion coefficient; N_2, with its nearly vertical profile near the quench levels, is relatively unaffected, whereas HCN is strongly affected.

One loss mechanism for HCN that is not in the model is reaction of HCN with NH_3 to form condensed NH_4CN salt, as was originally suggested by Lewis.[143] The formation of this species may significantly limit the possible vapor abundance of HCN at temperatures below ~160 K (*i.e.*, above 0.9 bar) on Jupiter.

5. Discussion

Our modeling suggests that neither the photochemical source nor the deep tropospheric source can provide much HCN to the Jovian upper troposphere—a result that is consistent with the low observational upper limit for non-cometary HCN on Jupiter.[13-15] Our conclusions are at odds with some of the previous modeling predictions found in the literature,[35,20,21] and we now discuss some of the reasons for the differences.

5.1. Deep tropospheric HCN source compared with previous models

The quenched-disequilibrium model that can be most directly compared with ours is that of Fegley and Lodders,[20] due to the greater-than-solar nitrogen elemental abundance assumed in both models: we assume a NH_3/H_2 ratio of 6.64×10^{-4},[112] whereas Fegley and Lodders[20] assume a NH_3/H_2 ratio of 5.2×10^{-4}. Fegley and Lodders[20] use a time-constant approach rather than thermochemical kinetics and transport models, but we can use similar arguments for our comparisons. Based on an earlier estimate from Prinn and Fegley,[147] Fegley and Lodders[20] assume that the rate-limiting step in the reduction of HCN is the reaction $HCN + H_2 \rightarrow CH_2 + NH$, with an estimated rate coefficient of $1.08 \times 10^{-8} \exp(-70,456/T)$ cm^3 s^{-1}, for T in K. The estimated chemical kinetic time constant for HCN destruction is then

$$\tau_{chem} = (1.08 \times 10^{-8} \exp(-70,456/T) [H_2])^{-1}.$$

If one then assumes that the convective mixing time scale $\tau_{mix} = L^2/K_{zz}$, with the effective convective length scale L being the atmospheric pressure scale height H, then τ_{mix} is equal to τ_{chem} at level at which the equilibrium HCN mole fraction is $\sim 1 \times 10^{-9}$ in the Fegley and Lodders model[20]—a value similar to the observed HCN upper limit of 0.93 ppb.[15] However, Smith[46] argues that the effective mixing length scale L is more like $0.11H$, such that τ_{mix} would be equal to τ_{chem} deeper in the atmosphere, near the 2300-bar, 1630-K level in our model, at which point the equilibrium HCN mole fraction would be $\sim 1.6 \times 10^{-8}$. This value is almost twenty times the observational upper limit for HCN and is clearly not supported by the infrared and submillimetre observations.[13–15] Thus, we have an apparent paradox in that the predicted HCN mole fraction from the quenched deep-tropospheric source can only be consistent with the observational upper limits if one ignores the modeling of Smith.[46]

We propose a resolution to this apparent paradox. Our model contains several more efficient pathways to HCN → NH_3 conversion than the HCN + H_2 → CH_2 + NH rate-limiting step first proposed by Prinn and Fegley.[147] Scheme (2) above shows the most effective HCN → NH_3 conversion scheme in our model, but there are also several others that are more efficient than the Prinn and Fegley mechanism.[147] The rate-limiting step in scheme (2) is the reaction

$$H_2 + H_2CN \rightarrow CH_2NH + H, \qquad (4)$$

for which the rate coefficient in our model is determined from the reverse reaction.[60] We can fit an Arhennius expression to our rate coefficient for reaction (4), as calculated from the reverse reaction: $k_4 = 1.011 \times 10^{-18} T^{1.941} \exp(-10,682.5/T)$ cm^3 s^{-1}, for T in K. If we use this reaction as our rate-limiting step, the kinetic time constant for HCN destruction is

$$\tau_{chem} = \frac{[HCN]}{d[HCN]/dt} = \frac{[HCN]}{k_4[H_2][H_2CN]}.$$

Then, using equilibrium calculations along with an effective length scale of $L \approx 0.11H$,[46] we find that τ_{chem} equals τ_{mix} at the $\sim$870-K, $\sim$250-bar level in our model, where the equilibrium HCN mole fraction is 4×10^{-14} from this time-constant approach, compared to our model-derived value of 6×10^{-14}. The similarity in the results from the time-constant approach using the Smith[46] length scale as compared with our thermochemical kinetics and transport model results suggests that (a) the Smith[46] results are indeed reasonable and should not be ignored (as they frequently are now) by those who use time-constant arguments to calculate the abundance of quenched disequilibrium species dredged up from the deep troposphere, and (b) the time-constant approach is valid (*i.e.*, one does not need a full thermochemical kinetics and transport model to predict quenched disequilibrium abundances on the giant planets). Visscher *et al.*[49] came to similar conclusions from their modeling of CO reduction on Jupiter (see also Bézard *et al.*[48]). Furthermore, the fact that our derived quenched HCN mole fraction lies comfortably below the observational upper limits suggests that there are faster mechanisms for converting HCN to NH_3 in the Jovian troposphere than the HCN + H_2 → CH_2 + NH reaction originally proposed by Prinn and Fegley.[147]

Our dominant schemes for the reduction of HCN, N_2, and CO (see Visscher *et al.*[49]) on Jupiter all start with H-atom addition, followed by reaction of the H_2CN, NNH, or HCO adducts with H_2, and subsequent reactions with H and/or H_2 to eventually form species with single C–N, N–N, or C–O bonds, before those bonds are finally broken. We thus find alternative pathways to the H_2-plus-strongly-bonded-constituent reactions that form the backbone of the mechanisms proposed by Prinn and Barshay,[144] Prinn and Fegley,[147] Prinn and Olaguer,[145] and subsequent modelers. Yung *et al.*[45] call such reactions "kinetically too ambitious"—a wonderful phrase that has been often quoted. Our mechanism suggests that other, less ambitious reactions can do the job.

Not all the reactions in our full mechanism have firmly determined rate coefficients, and we may be missing important pathways and/or species in our reaction list. As such, we cannot be completely confident in our derived abundances for the quenched disequilibrium species. However, the model development is based on the best reaction rate coefficients available today from combustion-chemistry studies and terrestrial atmospheric chemistry studies (both of which are unfortunately concerned more with the oxidation of reduced species than the reduction of oxidized species) and are likely significant improvements over the $H_2 + N_2$ and $H_2 + HCN$ mechanisms proposed 30 years ago. Given the likely importance of such processes for extrasolar giant planets, as well as for giant planets within our own solar system, we encourage further study of the dominant mechanisms for the reduction of N_2 and HCN in hydrogen-dominated atmospheres.

5.2. Photochemical HCN source and the tropospheric C_2H_2 abundance

Our modeling in Section 3 demonstrates that NH_3–C_2H_2 photochemical coupling is not a significant source of HCN on Jupiter. Photochemical destruction of C_2H_2 in the lower stratosphere limits the amount of acetylene that diffuses into the NH_3 photolysis region, and both condensation and photolysis limit the amount of NH_3 that diffuses into the stratosphere. Ammonia and acetylene are simply not present in large enough quantities together to provide a source of HCN. In contrast, Kaye and Strobel[35] predict that as much as a few ppb HCN could form in the upper troposphere from coupled NH_3–C_2H_2 photochemistry. As we mention above, the main difference in our models relates to the C_2H_2 abundance rather than to major differences in the kinetic reaction-rate coefficients adopted for the first critical pathway for formation of carbon–nitrogen bonds. For their model in which a few ppb of HCN is formed, Kaye and Strobel[35] assume a fixed acetylene distribution in which C_2H_2 is uniformly mixed with altitude at a mixing ratio of 3×10^{-8}, based on observations available at the time,[159] whereas our model predicts C_2H_2 mole fractions of 1 ppb at ~80 mbar, with rapidly increasing mole fractions at higher altitudes and decreasing mole fractions at lower altitudes (see Fig. 2). However, even with the older observations,[159] the mixing-ratio profiles that provided the best fit to the observed spectra required C_2H_2 to be depleted below the 100-mbar level, suggesting little C_2H_2 in the troposphere. More recent ultraviolet and infrared observations clearly indicate that the C_2H_2 mole fraction decreases with decreasing altitude in Jupiter's stratosphere,[13,106,115,160,161] firmly pointing to an upper stratospheric photochemical source for the acetylene. Moreover, a severe constraint on the tropospheric C_2H_2 mole fraction is indicated by the lack of absorption wings observed for the resolved C_2H_2 line profiles in thermal-infrared spectra.[133,161] Thus, the fact that all recent Jovian hydrocarbon photochemical models predict a rapidly decreasing C_2H_2 mole fraction in the lower stratosphere and into the troposphere[50,106,125,135,161–167] appears consistent with present-day observations.[106]

In contrast, two sets of observations have been used to suggest that the C_2H_2 mole fraction is relatively large in the Jovian troposphere. The first observation consists of ultraviolet spectra from the Jovian equatorial region taken with the Faint Object Spectrograph (FOS) onboard the Hubble Space Telescope (HST).[132,133] The acetylene profile derived from the Bétrémieux and Yelle[132] and Bétrémieux *et al.*[133] analyses of this data exhibit a mole fraction that decreases with decreasing altitude in the stratosphere but then increases again in the troposphere, for a best fit mole fraction of 1.5×10^{-7} in the 120–700 mbar region. This large increase in the troposphere has been used as evidence that lightning and thundershock sources of acetylene production exist in the Jovian troposphere.[168] The second observation is the Galileo probe mass spectrometer data[110,112] that are interpreted as being consistent with ethane and other non-methane hydrocarbons going through a mole-fraction minimum near the 1-bar level, followed by an increased mole fraction at pressures greater than 16 bar.[169] Hunten[169] interprets the GPMS signature as being caused by efficient ethane and acetylene adsorption on stratospheric and tropospheric aerosols, which then rain down through the atmosphere until they reach temperatures high enough for desorption to occur.

We find these reports of large tropospheric mole fractions of C_2H_2 and/or C_2H_6 to be unconvincing, at least for the bulk of the planet. The HST/FOS ultraviolet spectra are affected by scattering and absorption from many gas-phase and aerosol species, all of which have poorly constrained parameters. Both NH_3 and C_2H_2 are clearly detected in the HST/FOS ultraviolet spectra,[132,133] but deriving mole fractions from the spectra may be problematic. The strongest argument against the 1.5×10^{-7} tropospheric C_2H_2 mole fraction derived by Bétrémieux et al.[133] is that such a large amount would generate absorption wings around observed mid-infrared C_2H_2 emission lines, which Bétrémieux et al. recognize is inconsistent with their own ground-based infrared data,[133] as well as with other mid-infrared observations.[115,161] The NH_3 profile derived from the HST/FOS dataset[133] also differs considerably from profiles inferred from other observations (see Fig. 3e) for reasons that are unclear. Bétrémieux et al.[133] suggest that their unexpectedly large derived C_2H_2 tropospheric mole fraction, as well as their unexpectedly small derived NH_3 mole fraction, could result from uncertain NH_3 ultraviolet absorption cross sections at relevant Jovian temperatures, from some kind of dynamical situation where the temperature profile masks the C_2H_2 infrared absorption wings, or from tropospheric NH_3 and C_2H_2 profiles that vary with location and/or time on Jupiter. We note that the spatially resolved Cassini Composite Infrared Spectrometer (CIRS) data of Nixon et al.[161] preclude such large C_2H_2 mole fractions over all the latitude regions that were investigated by CIRS, although given the size of the HST/FOS footprint, the explanation of an anomalous localized atmospheric region may still be possible.

In any case, the global-average Infrared Space Observatory data of Fouchet et al.[115] and the spatially resolved Cassini/CIRS data analyzed by Nixon et al.[161] demonstrate that the spectral shape of the thermal-infrared C_2H_2 features is inconsistent with large tropospheric C_2H_2 mole fractions being present over the bulk of Jupiter.

The increase in the mole fraction of heavy hydrocarbons in the GPMS data described by Hunten[169] occurs too deep to affect infrared spectra, and no remote-sensing observations currently exist to test this claim. However, the suggestion[169] that the GPMS data are consistent with the adsorption of non-methane hydrocarbons onto fluffy stratospheric and upper-tropospheric aerosols that then sediment into the troposphere to evaporate at high temperatures has been questioned by Wong,[170] based on the extensive GMPS calibrations from his thesis.[171] Wong[170] states that the GPMS data "do not support the vertical variation of ethane mixing ratio" that is key to the Hunten[169] aerosol adsorption/desorption model. In particular, the GPMS data do not support an upper-tropospheric minimum in the ethane mole fraction, nor do they indicate that the mole fraction increases with depth in the troposphere. Wong[170] provides evidence, including the relative abundances of the measured species, that suggests that the non-methane hydrocarbons measured in the troposphere in the 8.5–12 bar region by the GPMS were instrumentally generated.

We suggest additional problems with the adsorption/desorption hypothesis. In the Hunten[169] model, a mole fraction of 3 ppm of non-methane hydrocarbons (mostly ethane) must be removed from ~30 mbar (see his Fig. 3) through adsorption. For a 30-mbar atmospheric density of 1.6×10^{18} cm^{-3}, that means 4.8×10^{12} ethane molecules cm^{-3} must be removed, for a column density of 9.6×10^{18} ethane molecules cm^{-2} (or really only ~70% of this value since some ethane gas remains at 30 mbar), compared with an estimated stratospheric haze column density of $(3–8) \times 10^8$ particles cm^{-2} (see Fig. 12 of West et al.[172]). These haze particles must be fluffy indeed to accommodate $\sim 10^{10}$ adsorbed ethane molecules per particle. The situation is even worse at 1 bar, where an ethane mole fraction of $\sim 4 \times 10^{-7}$ must be lost, corresponding to an ethane column density of $\sim 4.4 \times 10^{19}$ cm^{-2} that must removed from the 1-bar region. The total mass that must be adsorbed per particle is likely greater than the mass of the particle itself, and layer upon layer of ethane molecules must adsorb on top of each other, which does not typically happen in adsorption processes. In fact, Curtis et al.[173] show that significant ethane adsorption (i.e., a monolayer or greater) does not occur on tholin particles unless the ethane concentration is at or above the ethane saturation vapor density, which never occurs in the Jovian atmosphere. Although the stratospheric hazes on Jupiter may be composed of solid H_2O (from an external source), benzene, and/or butane,[50] or perhaps P_2H_4 and N_2H_4 particles that have been transported through the tropopause from below, rather than tholins, the Curtis et al.[173] study suggests that ethane and probably acetylene condensation under the greatly subsaturated conditions on Jupiter is not likely to be as efficient as Hunten[169] have assumed. Therefore, because of problems with the physics of the adsorption process itself, as well as to the interpretation of the GPMS data, we deem it unlikely that the Hunten aerosol adsorption/desorption mechanism[169] could operate to release a significant amount C_2H_2 into the Jovian troposphere.

Although it is clear that C_2H_2 is not abundant in the troposphere of Jupiter in a global sense, the possibility that localized regions could contain enhanced tropospheric acetylene due to lightning[24,25] or dynamical effects has not been ruled out. We therefore investigate a photochemical model for which the bottom boundary condition for C_2H_2 has been changed to a mole fraction of 1.5×10^{-7}.[132,133] We also change the bottom boundary conditions for N_2 and CH_3NH_2 to mole fractions of 3.52×10^{-5} and 1.52×10^{-12}, respectively, to reflect the deep-tropospheric quenched source from our nominal thermochemical kinetics and transport model (see Fig. 4). Species like NH_3, NH_2, and N_2H_4 are relatively unaffected by the increased C_2H_2 and CH_3NH_2 abundances (N_2 is effectively inert in the Jovian upper troposphere), but CH_2NH, CH_3NH_2, HCN, CH_3CN, $C_2H_5NH_2$, C_2H_3CN, and all the nitrogen-bearing organic species observed in the Keane et al.[39] $C_2H_2/NH_3/H_2$ photolysis experiments are significantly enhanced when the C_2H_2 abundance is increased. Ethylamine, in particular, shows a particularly large increase and approaches 1 ppb in our model; however, ethylamine condensation, which is not currently included in the model, will eventually limit the column abundance. Fig. 6 shows the enhancement in the HCN abundance that would result from coupled C_2H_2–NH_3 photochemistry when more C_2H_2 is available in the Jovian troposphere. The resulting increase in the tropospheric HCN mole fraction of two orders of magnitude is not as large as one might expect because significant loss processes for C_2H_2 still exist in the lower stratosphere and tropopause region that convert C_2H_2 to C_2H_6 and heavier hydrocarbons rather than to nitriles and other nitrogen-bearing organics. The C_2H_2 mole fraction remains low in the NH_3 photolysis region, with a minimum mole fraction of only 1.4 ppb at 124 mbar. The HCN mole fraction therefore still remains below the observational upper limits, even with the increased C_2H_2 lower boundary condition. The only way to increase the HCN abundance from coupled NH_3–C_2H_2 coupled chemistry would be to invoke rapid dynamical mixing near the tropopause, such that more C_2H_2 and NH_3 are available in the ammonia photolysis region.

5.3. Thunderstorm source for HCN

Bar-Nun and Podolak[24] and Podolak and Bar-Nun[25] advocate lightning, and the resulting shock waves produced from lightning, as a source of HCN (and C_2H_2 and CO) for the Jovian troposphere. A discussion of the relative effectiveness of such a source is beyond the scope of this paper. The Bar-Nun and Podolak[24] and Podolak and Bar-Nun[25] model calculations depend on several unknown parameters such as production efficiencies per energy released in the lightning event, the location of lightning within the cloud (which must be in the top portion of the H_2O cloud to have HCN and C_2H_2 as significant products), the assumed attenuation of the

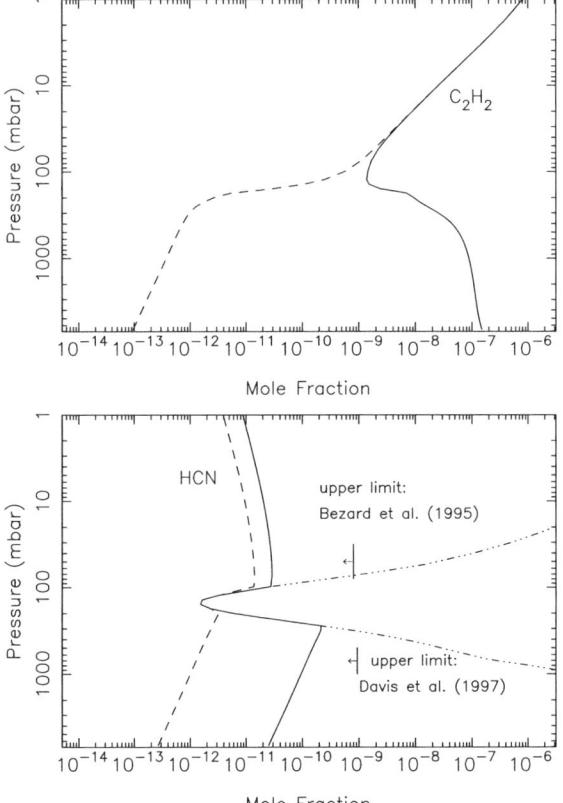

Fig. 6 Mole-fraction profiles for acetylene (top) and hydrogen cyanide (bottom) from our Jovian photochemical model. The dashed line represents our nominal model, as discussed in Section 3. The solid line represents a model in which the bottom boundary conditions for C_2H_2, N_2, and CH_3NH_2 have been increased to mole fractions of 1.5×10^{-7}, 3.52×10^{-5}, and 1.52×10^{-12}.

observed visible light within the cloud, the optical efficiency of the lightning, and the fate of the products once they are generated. The fact that their derived HCN mole fraction is greater than the observational upper limits[13–15] suggests that some of the terms in their calculation may need reevaluation. Note that although lightning production of C_2H_2 could lead to enhanced photochemical production of HCN, the resulting HCN mole fraction would not necessarily be observable (see Fig. 6).

Several observational tests should be available to evaluate the likelihood of lightning contributing to the global production of disequilibrium species on Jupiter. If production is as effective as the models suggest,[24,25] disequilibrium species should be greatly enhanced within active thunderstorms, and the dispersion to other latitudes would not be instantaneous. Therefore, investigators could look for local enhancements in disequilibrium tropospheric species (*e.g.*, C_2H_2 from thermal infrared or ultraviolet observations, CO and HCN from infrared, millimetre, or sub-millimetre observations) at latitudes where active lightning storms are known to be prevalent.[174] Such observational tests might be more readily available for Saturn, from Cassini data already obtained or from the extended mission, but should also be possible for Jupiter from ground-based observations and from the Juno mission. We encourage studies of the spatial distribution of tropospheric disequilibrium constituents on Jupiter and Saturn to help evaluate the lightning source.

6. Conclusions

We have developed two theoretical models to investigate the production and loss of HCN and other nitrogen-bearing organics in the atmosphere of Jupiter. The first model covers the upper troposphere and stratosphere, for which we use the Caltech/JPL KINETICS code[43] to track coupled NH_3–C_2H_2 and NH_3–CH_4 photochemistry. The second model covers the deep troposphere, for which we again use the KINETICS code, but this time to track the thermochemical kinetics and transport of nitrogen species in the hot, high-pressure troposphere. We use simulations of the laboratory photolysis experiments,[38,39] along with theoretical calculations from the combustion-chemistry literature,[60] to help constrain uncertain rate coefficients in our reaction mechanism. We find that the photochemical production of HCN and other organo-nitrogen compounds is greatly inhibited in Jupiter's atmosphere. As was first discussed by Strobel,[100] our models suggest that ammonia condensation in the upper troposphere, combined with efficient photolysis and the resulting generation of the condensible N_2H_4 photoproduct, limit the availability of NH_3 in the Jovian stratosphere. Ammonia then does not diffuse into the upper atmosphere to the methane photolysis region to participate in coupled NH_3–CH_4 photochemistry. Contrary to several suggestions in the literature,[35–39] we find that coupled NH_3–C_2H_2 photochemistry is inefficient in Jupiter's troposphere when realistic C_2H_2 mixing-ratio profiles are considered. As is apparent from both photochemical models and observations, acetylene has a large mixing-ratio gradient in the stratosphere, leading to low abundances of C_2H_2 in the tropopause region. The main factor inhibiting the photochemical production of HCN and other nitrogen-bearing organics on Jupiter is the low acetylene abundance in the region where NH_3 is being photolyzed. Our predicted HCN mole fraction from our photochemical model is well below the upper limits derived from infrared and sub-millimetre observations.[13–15]

Consistent with other investigations that were based solely on time-constant arguments,[10,20,21,145] our thermochemical kinetics and transport models suggest that transport-induced quenching of equilibrium abundances in Jupiter's deep troposphere leads to large predicted mole fractions of N_2 and small predicted mole fractions of hydrogen cyanide in Jupiter's upper troposphere. However, our mechanisms for HCN and N_2 destruction differ considerably from those suggested by previous investigators. Our models confirm the results of Smith,[46] who demonstrates that the effective length scale for atmospheric mixing has been overestimated in the time-scale arguments of the above investigators. As a result, the suggested rate-limiting step for HCN destruction originally suggested by Fegley and Prinn[147] and used by later investigators is too slow to be consistent with the observed HCN upper limits—HCN would quench too deep in the atmosphere, where the equilibrium HCN mole fraction is large, if the suggested $HCN + H_2 \rightarrow CH_2 + NH$ reaction is the rate-limiting step. We instead find that HCN in our thermochemical kinetics and transport model is destroyed through a series of reactions that begin with H-atom addition to HCN, followed by reactions with H_2 and H to eventually form a single-bonded C–N species (CH_3NH_2) that can then thermally decompose to break the C–N bond (see scheme (2) above). The rate-limiting step for HCN reduction in our mechanism is the reaction $H_2 + H_2CN \rightarrow CH_2NH + H$, where we have calculated the rate coefficient of this reaction from that of the reverse reaction.[60] This scheme, and others like it, are much more efficient than the proposed $HCN + H_2$ destruction reaction,[147] and our predicted quenched disequilibrium HCN mole fraction is therefore comfortably below the HCN observational upper limits.[13–15]

Reduction of N_2 in our model follows a similar scheme (see scheme (3) above) with the reaction $N_2H_2 + H + M \rightarrow N_2H_3 + M$ being the rate-limiting step. Molecular nitrogen is likely to be very abundant ($\sim$30 ppm) on Jupiter from this quenched deep-tropospheric source (see also Prinn and Olaguer[147] and subsequent modelers), and N_2 should survive to be transported up into the upper atmosphere, where its mole fraction will eventually be reduced due to molecular diffusion. The interaction of galactic cosmic rays with N_2 might initiate interesting N_2 chemistry on Jupiter (see the equivalent Neptune study[155]), although the stratospheric HCN upper limit of 0.8 ppb[13] may place constraints on the effectiveness of this process. Molecular nitrogen may also participate in Jovian ionospheric chemistry, as on Titan,[156,175–181] with interesting consequences for the composition, structure, and time-variability of the lower ionosphere, as well as for the production of polycyclic aromatic hydrocarbons and other neutral species. The chemistry of N_2 in the auroral regions might be particularly interesting. The suggested large N_2 mole fraction on Jupiter needs observational confirmation, however.

One way to investigate the hypothesis of $\sim$30 ppm N_2 on Jupiter would be to take a more detailed look at the Galileo Probe Mass Spectrometer data. The N_2 abundance has never been definitively determined from the GPMS data because of concerns over an internal source of CO_2 from the instrument itself,[110] which could contaminate the signal at 28 dalton/e$^-$ due to the CO^+ daughter ion of CO_2.[182] Wong[171] calculated a combined CO + N_2 mixing ratio relative to H_2 of 2.3×10^{-7} in the 8.5–12 bar pressure region of the probe entry site. An N_2 mixing ratio greater than this value can only be supported if the contribution from the internal instrumental CO_2 source was overestimated in the Wong[171] study;[182] this uncertainty highlights the need for further calibration studies to better characterize the instrument-generated CO_2 signal (if any) in the GPMS data.

The observational upper limits for tropospheric HCN[13–15] and the lack of C_2H_2 absorption wings in thermal-infrared data[133,161] can be used help constrain the effectiveness of other disequilibrium production processes, such as that of lightning-induced chemistry.[24,25] Upper limits to the tropospheric C_2H_2 mole fraction from thermal-infrared observations are seldom provided in the literature, although such information would aid our understanding of the role of lightning on Jupiter and the other giant planets. Spatially resolved observations that compare the tropospheric composition of latitude regions known to have active thunderstorms with the composition of quiescent regions would also provide important information on this topic.

The mechanisms proposed in our investigation are still speculative due to the limited experimental data available for individual reactions of interest in the models. A better understanding of the pathways for reduction of HCN and N_2 is of importance for studies of the composition and chemical behavior of extrasolar giant planets and brown dwarfs, as well as for giant planets within our own solar system, and further investigation into these processes is warranted. Our thermochemical kinetics and transport model in particular can be applied to studies of the "hot Jupiters" that are being discovered at an astonishing rate around other stars. The photochemical model may also be of use to extrasolar-giant-planet studies. Although we find that coupled carbon–nitrogen photochemistry is not important on Jupiter, that result is largely due to the removal of NH_3 from the upper atmosphere due to condensation, and we may anticipate alternative scenarios for warmer giant planets. If an extrasolar giant planet were located closer to its parent star (such that NH_3 does not condense) but not so close that it is being intensely bombarded by ultraviolet radiation (such that strongly bonded species are not the only surviving molecules), then coupled NH_3–C_2H_2 and NH_3–CH_4 photochemistry could be very important indeed (see the theoretical planetary classes of Sudarsky et al.[183]). Coupled NH_3–C_2H_2 photochemistry might also be important in cometary comae or any other astronomical environment where ammonia and acetylene are brought together in the presence of ultraviolet radiation. We therefore encourage further investigation into the thermodynamic and kinetic properties of the organo-nitrogen compounds observed in the photolysis experiments of Ferris and Ishikawa,[36,37] Keane,[38] and Keane et al.[39]

Acknowledgements

The Caltech/JPL KINETICS code was developed jointly by Yuk L. Yung and Mark Allen, with assistance from many people over the years, and we thank them for letting us continue to use this powerful and flexible code. Veronica LaMothe helped with initial forays into Jovian tropospheric photochemistry. We thank Michael H. Wong for useful discussions, and we gratefully acknowledge financial support from the NASA Planetary Atmospheres Program (NNX08AF05G for the photochemistry portion, NNX09AB55G for the thermochemistry portion).

References

1. A. Marten, D. Gautier, M. J. Griffin, H. E. Matthews, D. A. Naylor, G. R. Davis, T. Owen, G. Orton, D. Bockelée-Morvan, P. Colom, J. Crovisier, E. Lellouch, I. de Pater, S. Atreya, D. Strobel, B. Han and D. B. Sanders, *Geophys. Res. Lett.*, 1995, **22**, 1589–1592.
2. E. Lellouch, in *The Collision of Comet Shoemaker-Levy 9 and Jupiter*, ed. K. S. Noll, H. A. Weaver and P. D. Feldman, Cambridge University Press, Cambridge, 1996, pp. 213–242.
3. B. Bézard, C. A. Griffith, D. M. Kelly, J. H. Lacy, T. Greathouse and G. Orton, *Icarus*, 1997, **125**, 94–120.
4. H. E. Matthews, A. Marten, R. Moreno and T. Owen, *Astrophys. J.*, 2002, **580**, 598–605.
5. R. Moreno, A. Marten, H. E. Matthews and Y. Biraud, *Planet. Space Sci.*, 2003, **51**, 591–611.
6. C. A. Griffith, B. Bézard, T. Greathouse, E. Lellouch, J. Lacy, D. Kelly and M. J. Richter, *Icarus*, 2004, **170**, 58–69.
7. V. G. Kunde, F. M. Flasar, D. E. Jennings, B. Bézard, D. F. Strobel, B. J. Conrath, C. A. Nixon, G. L. Bjoraker, P. N. Romani, R. K. Achterberg, A. A. Simon-Miller, P. Irwin, J. C. Brasunas, J. C. Pearl, M. D. Smith, G. S. Orton, P. J. Gierasch, L. J. Spilker, R. C. Carlson, A. A. Mamoutkine, S. B. Calcutt, P. L. Read, F. W. Taylor, T. Fouchet, P. Parrish, A. Barucci, R. Courtin, A. Coustenis, D. Gautier, E. Lellouch, A. Marten, R. Prangé, Y. Biraud, C. Ferrari, T. C. Owen, M. M. Abbas, R. E. Samuelson, F. Raulin, P. Ade, C. J. Césarsky, K. U. Grossman and A. Coradini, *Science*, 2004, **305**, 1582–1587.
8. E. Lellouch, B. Bézard, D. F. Strobel, G. L. Bjoraker, F. M. Flasar and P. N. Romani, *Icarus*, 2006, **184**, 478–497.
9. R. Moreno, M. Gurwell, A. Marten and E. Lellouch, *Bull. Am. Astron. Soc.*, 2007, **39**, 423.
10. J. S. Lewis and M. B. Fegley Jr, *Space Sci. Rev.*, 1984, **39**, 163–192.
11. A. T. Tokunaga, S. C. Beck, T. R. Geballe, J. H. Lacy and E. Serabyn, *Icarus*, 1981, **48**, 283–289.
12. E. Lellouch, F. Combes and T. Encrenaz, *Astron. Astrophys.*, 1984, **140**, 216–219.
13. B. Bézard, C. Griffith, J. Lacy and T. Owen, *Icarus*, 1995, **118**, 384–391.
14. E. W. Weisstein and E. Serabyn, *Icarus*, 1996, **123**, 23–36.
15. G. R. Davis, D. A. Naylor, M. J. Griffin, T. A. Clark and W. S. Holland, *Icarus*, 1997, **130**, 387–403.
16. S. L. Miller and H. C. Urey, *Science*, 1959, **130**, 245–251.
17. F. Woeller and C. Ponnamperuma, *Icarus*, 1969, **10**, 386–392.
18. C. N. Matthews, *Origins Life Evol. Biosphere*, 1991, **21**, 421–434.
19. S. S. Barshay and J. S. Lewis, *Icarus*, 1978, **33**, 593–611.
20. B. Fegley, Jr. and K. Lodders, *Icarus*, 1994, **110**, 117–154.
21. K. Lodders and B. Fegley Jr., *Icarus*, 2002, **155**, 393–424.
22. A. Bar-Nun and A. Shaviv, *Icarus*, 1975, **24**, 197–210.
23. J. S. Lewis, *Icarus*, 1980, **43**, 85–95.
24. A. Bar-Nun and M. Podolak, *Icarus*, 1985, **64**, 112–124.
25. M. Podolak and A. Bar-Nun, *Icarus*, 1988, **75**, 566–570.
26. R. Stribling and S. L. Miller, *Icarus*, 1987, **72**, 48–52.
27. J. P. Ferris and C. T. Chen, *Nature*, 1975, **258**, 587–588.
28. J. P. Ferris, C. Nakagawa and C. T. Chen, *Life Sci. Space Res.*, 1977, **15**, 95–99.
29. W. R. Kuhn, S. K. Atreya and S. Chang, *Geophys. Res. Lett.*, 1977, **4**, 203–206.
30. G. Toupance, A. Bossard and F. Raulin, *Origins Life*, 1977, **8**, 259–266.
31. F. Raulin, A. Bossard, G. Toupance and C. Ponnamperuma, *Icarus*, 1979, **38**, 358–366.
32. A. Bossard and G. Toupance, *Nature*, 1980, **288**, 243–245.
33. E. P. Gardner and J. R. McNesby, *J. Photochem.*, 1980, **13**, 353–356.
34. J. A. Kaye and D. F. Strobel, *Icarus*, 1983, **55**, 399–419.
35. J. A. Kaye and D. F. Strobel, *Icarus*, 1983, **54**, 417–433.
36. J. P. Ferris and Y. Ishikawa, *Nature*, 1987, **326**, 777–778.
37. J. P. Ferris and Y. Ishikawa, *J. Am. Chem. Soc.*, 1988, **110**, 4306–4312.
38. T. C. Keane, The Coupled Photochemistry of Ammonia and Acetylene: Applications to the Atmospheric Chemistry on Jupiter, Ph.D. Thesis, Rensselaer Polytechnic Institute, Troy, NY, 1995.

39 T. C. Keane, F. Yuan and J. P. Ferris, *Icarus*, 1996, **122**, 205–207.
40 J. P. Ferris, R. R. Jacobson and J. C. Guillemin, *Icarus*, 1992, **95**, 54–59.
41 J. S. Lewis and M. B. Fegley, Jr, *Astrophys. J.*, 1979, **232**, L135–L137.
42 J. P. Ferris and J. Y. Morimoto, *Icarus*, 1981, **48**, 118–126.
43 M. Allen, Y. L. Yung and J. W. Waters, *J. Geophys. Res.*, 1981, **86**, 3617–3627.
44 B. Fegley, Jr. and R. G. Prinn, *Astrophys. J.*, 1988, **324**, 621–625.
45 Y. L. Yung, W. A. Drew, J. P. Pinto and R. R. Friedl, *Icarus*, 1988, **73**, 516–526.
46 M. D. Smith, *Icarus*, 1998, **132**, 176–184.
47 C. A. Griffith and R. V. Yelle, *Astrophys. J.*, 1999, **519**, L85–L88.
48 B. Bézard, E. Lellouch, D. Strobel, J.-P. Maillard and P. Drossart, *Icarus*, 2002, **159**, 95–111.
49 C. Visscher, J. I. Moses and S. A. Saslow, *Icarus*, 2010, submitted.
50 J. I. Moses, T. Fouchet, B. Bézard, G. R. Gladstone, E. Lellouch and H. Feuchtgruber, *J. Geophys. Res.*, 2005, **110**, E08001, DOI: 10.1029/2005JE002411.
51 D. L. Baulch, C. J. Cobos, R. A. Cox, C. Esser, P. Frank, T. Just, J. A. Kerr, M. J. Pilling, J. Troe, R. W. Walker and J. Warnatz, *J. Phys. Chem. Ref. Data*, 1992, **21**, 411–734.
52 D. L. Baulch, C. J. Cobos, R. A. Cox, P. Frank, G. Hayman, T. Just, J. A. Kerr, M. J. Pilling, J. Troe, R. W. Walker and J. Warnatz, *J. Phys. Chem. Ref. Data*, 1994, **23**, 847–1033. Errata: *J. Phys. Chem. Ref. Data*, 1995, **24**, 1609.
53 D. L. Baulch, C. T. Bowman, C. J. Cobos, R. A. Cox, T. Just, J. A. Kerr, M. J. Pilling, D. Stocker, J. Troe, W. Tsang, R. W. Walker and J. Warnatz, *J. Phys. Chem. Ref. Data*, 2005, **34**, 757–1397.
54 G. P. Smith, D. M. Golden, M. Frenklach, N. W. Moriarty, B. Eiteneer, M. Goldenberg, C. T. Bowman, R. K. Hanson, S. Song, W. C. Gardiner, Jr., V. V. Lissianski and Z. Qin, GRI Mech 3.0, 2000, http://www.me.berkeley.edu/gri_mech/.
55 R. Atkinson, D. L. Baulch, R. A. Cox, R. F. HampsonJr., J. A. Kerr, M. J. Rossi and J. Troe, *J. Phys. Chem. Ref. Data*, 1997, **26**, 1329–1499.
56 R. Atkinson, D. L. Baulch, R. A. Cox, R. F. Hampson, Jr., J. A. Kerr, M. J. Rossi and J. Troe, *J. Phys. Chem. Ref. Data*, 1999, **28**, 191–393.
57 R. Atkinson, D. L. Baulch, R. A. Cox, J. N. Crowley, R. F. Hampson, R. G. Hynes, M. E. Jenkin, M. J. Rossi and J. Troe, *Atmos. Chem. Phys.*, 2006, **6**, 3625–4055.
58 W. B. DeMore, S. P. Sander, D. M. Golden, R. F. Hampson, M. J. Kurylo, C. J. Howard, A. R. Ravishankara, C. E. Kolb and M. J. Molina, *Chemical Kinetics and Photochemical Data for Use in Stratospheric Modeling*. Evaluation Number 12, JPL Publication 97-4, Jet Propulsion Laboratory, Pasadena, 1997.
59 S. P. Sander, R. R. Friedl, D. M. Golden, M. J. Kurylo, R. E. Huie, V. L. Orkin, G. K. Moortgat, A. R. Ravishankara, C. E. Kolb, M. J. Molina and B. J. Finlayson-Pitts, *Chemical Kinetics and Photochemical Data for Use in Atmospheric Studies*. Evaluation Number 14, JPL Publication 02-25, Jet Propulsion Laboratory, Pasadena, 2003.
60 A. M. Dean and J. W. Bozzelli, in *Gas Phase Combustion Chemistry*, ed. W. C. Gardiner, Jr., Springer-Verlag, New York, 2000, ch. 2, pp. 125–341.
61 J. A. Miller and C. T. Bowman, *Prog. Energy Combust. Sci.*, 1989, **15**, 287–338.
62 D. F. Davidson, K. Kohse-Höinghaus, A. Y. Chang and R. K. Hanson, *Int. J. Chem. Kinet.*, 1990, **22**, 513–535.
63 B. A. Williams and J. W. Fleming, *Combust. Flame*, 1997, **110**, 1–13.
64 D.-Y. Hwang and A. M. Mebel, *J. Phys. Chem. A*, 2003, **107**, 2865–2874.
65 P. Dagaut, P. Glarborg and M. U. Alzueta, *Prog. Energy Combust. Sci.*, 2008, **34**, 1–46.
66 J. A. Kaye and D. F. Strobel, *Geophys. Res. Lett.*, 1983, **10**, 957–960.
67 J. A. Kaye and D. F. Strobel, *Icarus*, 1984, **59**, 314–335.
68 G. F. Lindal, G. E. Wood, G. S. Levy, J. D. Anderson, D. N. Sweetnam, H. B. Hotz, B. J. Buckles, D. P. Holmes, P. E. Doms, V. R. Eshleman, G. L. Tyler and T. A. Croft, *J. Geophys. Res.*, 1981, **86**, 8721–8727.
69 A. Seiff, D. B. Kirk, T. C. D. Knight, R. E. Young, J. D. Mihalov, L. A. Young, F. S. Milos, G. Schubert, R. C. Blanchard and D. Atkinson, *J. Geophys. Res.*, 1998, **103**, 22857–22889.
70 E. Lellouch, B. Bézard, T. Fouchet, H. Feuchtgruber, T. Encrenaz and T. de Graauw, *Astron. Astrophys.*, 2001, **370**, 610–622.
71 A. Simon-Miller, B. J. Conrath, P. J. Gierasch, G. S. Orton, R. K. Achterberg, F. M. Flasar and B. M. Fisher, *Icarus*, 2006, **180**, 98–112.
72 W. E. Groth, U. Schurath and R. N. Schindler, *J. Phys. Chem.*, 1968, **72**, 3914–3920.
73 B.-M. Cheng, H. –C. Lu, H.-K. Chen, M. Bahou, Y.-P. Lee, A. M. Mebel, L. C. Lee, M. –C. Liang and Y. L. Yung, *Astrophys. J.*, 2006, **647**, 1535–1542.
74 K. Fagerström, J. T. Jodkowski, A. Lund and E. Ratajczak, *Chem. Phys. Lett.*, 1995, **236**, 103–110.

75 G. L. Vaghjiani, *Int. J. Chem. Kinet.*, 1995, **27**, 777–790.
76 W. E. Groth and H. J. Rommel, *Z. Phys. Chem. Neue Folge*, 1965, **45**, 96–116.
77 V. M. Gehring, K. Hoyermann, H. G. Wagner and J. Wolfrum, *Ber. Bunsen-Ges. Phys. Chem.*, 1971, **75**, 1287.
78 A. A. Scala and D. Salomon, *J. Chem. Phys.*, 1976, **65**, 4455–4461.
79 F. J. Lovas, F. O. Clark and E. Tiemann, *J. Chem. Phys.*, 1975, **62**, 1925–1931.
80 V. G. Granik, *Russ. Chem. Rev.*, 1984, **53**, 383–400.
81 K. Lammertsma and B. V. Prasad, *J. Am. Chem. Soc.*, 1994, **116**, 642–650.
82 W. Hack, H. Schacke, M. Schröter and H. G. Wagner, *Symp. (Int.) Combust., [Proc.]*, 1979, **17**, 505.
83 S. R. Bosco, D. F. Nava, W. D. Brobst and L. J. Stief, *J. Chem. Phys.*, 1984, **81**, 3505–3511.
84 R. Lesclaux, B. Veyret and P. Roussel, *Ber. Bunsen.-Ges. Phys. Chem.*, 1985, **89**, 330–335.
85 G. Hennig and H. G. Wagner, *Ber. Bunsen.-Ges. Phys. Chem.*, 1995, **99**, 989–994.
86 L. V. Moskaleva and M. C. Lin, *J. Phys. Chem. A*, 1998, **102**, 4687–4693.
87 M. Demissy and R. Lesclaux, *Int. J. Chem. Kinet.*, 1982, **14**, 1–12.
88 J. T. Jodkowski, E. Ratajczak, K. Fagerström, A. Lund, N. D. Stothard, R. Humpfer and H.-H. Grotheer, *Chem. Phys. Lett.*, 1995, **240**, 63–71.
89 J. V. Michael and W. A. Noyes, Jr., *J. Am. Chem. Soc.*, 1963, **85**, 1228–1233.
90 H. Nishi, H. Shinohara and I. Hanazaki, *Chem. Phys. Lett.*, 1980, **73**, 473–477.
91 E. P. Gardner and J. R. McNesby, *J. Phys. Chem.*, 1982, **86**, 2646–2651.
92 G. C. G. Waschewsky, D. C. Kitchen, P. W. Browning and L. J. Butler, *J. Phys. Chem.*, 1995, **99**, 2635–2645.
93 K. M. Dunn and K. Morokuma, *J. Phys. Chem.*, 1996, **100**, 123–129.
94 D. G. Horne and R. G. W. Norrish, *Proc. R. Soc. London, Ser. A*, 1970, **315**, 301–322.
95 F. L. Nesbitt, G. Marston and L. J. Stief, *J. Phys. Chem.*, 1990, **94**, 4946–4951.
96 P. S. Monks, P. N. Romani, F. L. Nesbitt, M. Scanlon and L. J. Stief, *J. Geophys. Res.*, 1993, **98**, 17115–17122.
97 R. Wildt, *Astrophys. J.*, 1937, **86**, 321–326.
98 R. D. Cadle, *J. Atmos. Sci.*, 1962, **19**, 281–285.
99 J. R. McNesby, *J. Atmos. Sci.*, 1969, **26**, 594–599.
100 D. F. Strobel, *J. Atmos. Sci.*, 1973, **30**, 1205–1209.
101 S. K. Atreya, T. M. Donahue and W. R. Kuhn, *Icarus*, 1977, **31**, 348–355.
102 D. F. Strobel, *Astrophys. J.*, 1977, **214**, L97–L99.
103 S. G. Edgington, S. K. Atreya, L. M. Trafton, J. J. Caldwell, R. F. Beebe, A. A. Simon, R. A. West and C. Barnet, *Icarus*, 1998, **133**, 192–209.
104 S. G. Edgington, S. K. Atreya, L. M. Trafton, J. J. Caldwell, R. F. Beebe, A. A. Simon and R. A. West, *Icarus*, 1999, **142**, 342–356.
105 S. K. Atreya, *Atmospheres and Ionospheres of the Outer Planets and Their Satellites*, Springer-Verlag, Berlin, 1986.
106 J. I. Moses, T. Fouchet, R. V. Yelle, A. J. Friedson, G. S. Orton, B. Bézard, P. Drossart, G. R. Gladstone, T. Kostiuk and T.A. Livengood, in *Jupiter: The Planet, Satellites and Magnetosphere*, ed. F. Bagenal, T. E. Dowling and W. B. McKinnon, Cambridge University Press, Cambridge, 2004, ch. 7, pp. 129–157.
107 F. W. Taylor, S. K. Atreya, T. Encrenaz, D. M. Hunten, P. G. J. Irwin and T. C. Owen, in *Jupiter: The Planet, Satellites and Magnetosphere*, ed. F. Bagenal, T. E. Dowling and W. B. McKinnon, Cambridge University Press, Cambridge, 2004, ch. 4, pp. 59–78.
108 J. I. Moses, E. Lellouch, B. Bézard, G. R. Gladstone, H. Feuchtgruber and M. Allen, *Icarus*, 2000, **145**, 166–202.
109 I. dePater, D. Dunn, P. N. Romani and K. Zahnle, *Icarus*, 2001, **149**, 66–78.
110 H. B. Niemann, S. K. Atreya, G. R. Carignan, T. M. Donahue, J. A. Haberman, D. N. Harpold, R. E. Hartle, D. M. Hunten, W. T. Kasprzak, P. R. Mahaffy, T. C. Owen and S. H. Way, *J. Geophys. Res.*, 1998, **103**, 22831–22845.
111 U. von Zahn, D. M. Hunten and G. Lemacher, *J. Geophys. Res.*, 1998, **103**, 22815–22829.
112 M. H. Wong, P. R. Mahaffy, S. K. Atreya, H. B. Niemann and T. C. Owen, *Icarus*, 2004, **171**, 153–170.
113 G. L. Bjoraker, H. P. Larson and V. G. Kunde, *Icarus*, 1986, **66**, 579–609.
114 J. Gibson, W. J. Welch and I. de Pater, *Icarus*, 2005, **173**, 439–446.
115 T. Fouchet, E. Lellouch, B. Bézard, T. Encrenaz, P. Drossart, H. Feuchtgruber and T. de Graauw, *Icarus*, 2000, **143**, 223–243.
116 I. de Pater and S. T. Massie, *Icarus*, 1985, **62**, 143–171.
117 L. A. Sromovsky, A. D. Collard, P. M. Fry, G. S. Orton, M. T. Lemmon, M. G. Tomasko and R. S. Freedman, *J. Geophys. Res.*, 1998, **103**, 22929–22977.
118 W. M. Folkner, R. Woo and S. Nandi, *J. Geophys. Res.*, 1998, **103**, 22847–22855.
119 E. Lellouch, B. Bézard, J. I. Moses, G. R. Davis, P. Drossart, H. Feuchtgruber, E. A. Bergin, R. Moreno and T. Encrenaz, *Icarus*, 2002, **159**, 112–131.

120 J. I. Moses, B. Bézard, E. Lellouch, G. R. Gladstone, H. Feuchtgruber and M. Allen, *Icarus*, 2000, **143**, 244–298.
121 R. V. Yelle, L. A. Young, R. J. Vervack, Jr., R. Young, L. Pfister and B. R. Sandel, *J. Geophys. Res.*, 1996, **101**, 2149–2161.
122 T. K. Greathouse, G. R. Gladstone, J. I. Moses, S. A. Stern, K. D. Retherford, R. J. Vervack, Jr., D. C. Slater, M. H. Versteeg, M. W. Davis, L. A. Young, A. J. Steffl, H. Throop and J. W. Parker, *Icarus*, 2010, in press.
123 P. Drossart, B. Sicardy, F. Roques, T. Widemann, G. R. Gladstone, J. H. Waite and M. Vincent, *Bull. Am. Astron. Soc.*, 2000, **32**, 1013.
124 J. Vervack, Jr., B. R. Sandel, G. R. Gladstone, J. C. McConnell and C. D. Parkinson, *Icarus*, 1995, **114**, 163–173.
125 G. R. Gladstone, M. Allen and Y. L. Yung, *Icarus*, 1996, **119**, 1–52.
126 L.-M. Lara, B. Bézard, C. A. Griffith, J. H. Lacy and T. C. Owen, *Icarus*, 1998, **131**, 317–333.
127 M. –C. Liang, B.-M. Cheng, H. –C. Lu, H.-K. Chen, M. S. Alum, Y.-P. Lee and Y. L. Yung, *Astrophys. J.*, 2007, **657**, L117–L120.
128 L. N. Fletcher, G. S. Orton, N. A. Teanby and P. G. J. Irwin, *Icarus*, 2009, **202**, 543–564.
129 B. Landry, M. Allen and Y. L. Yung, *Icarus*, 1991, **89**, 377–383.
130 P. H. Stone, in *Jupiter*, ed. T. Gehrels, Univ. Arizona Press, Tucson, 1976, pp. 586–618.
131 R. A. West, K. H. Baines, A. J. Friedson, D. Banfield, B. Ragent and F. W. Taylor, in *Jupiter: The Planet, Satellites and Magnetosphere*, ed. F. Bagenal, T. E. Dowling and W. B. McKinnon, Cambridge University Press, Cambridge, 2004, ch. 5, pp. 79–128.
132 Y. Bétrémieux and R. V. Yelle, *Icarus*, 1999, **142**, 324–341.
133 Y. Bétrémieux, R. V. Yelle and C. A. Griffith, *Icarus*, 2003, **163**, 414–427.
134 L. A. Sromovsky and P. M. Fry, *Icarus*, 2002, **157**, 373–400.
135 M. Allen, Y. L. Yung and G. R. Gladstone, *Icarus*, 1992, **100**, 527–533.
136 B. E. Carlson, A. A. Lacis and W. B. Rossow, *J. Geophys. Res.*, 1993, **98**, 5251–5290.
137 C. A. Griffith, B. Bézard, T. Owen and D. Gautier, *Icarus*, 1992, **98**, 82–93.
138 P. G. J. Irwin, A. L. Weir, S. E. Smith, F. W. Taylor, A. L. Lambert, S. B. Calcutt, P. J. Cameron-Smith, R. W. Carlson, K. Baines, G. S. Orton, P. Drossart, T. Encrenaz and M. Roos-Serote, *J. Geophys. Res.*, 1998, **103**, 23001–23021.
139 M. J. Burgdorf, G. S. Orton, T. Encrenaz, G. R. Davis, S. D. Sidher, E. Lellouch and B. M. Swinyard, *Planet. Space Sci.*, 2004, **52**, 379–383.
140 R. K. Achterberg, B. J. Conrath and P. J. Gierasch, *Icarus*, 2006, **182**, 169–180.
141 E. R. Lippincott, R. V. Eck, M. O. Dayhoff and C. Sagan, *Astrophys. J.*, 1967, **147**, 753–764.
142 J. A. Greenspan and T. Owen, *Science*, 1967, **156**, 1489–1494.
143 J. S. Lewis, *Icarus*, 1969, **10**, 393–409.
144 R. G. Prinn and S. S. Barshay, *Science*, 1977, **198**, 1031–1034.
145 R. G. Prinn and E. P. Olaguer, *J. Geophys. Res.*, 1981, **86**, 9895–9899.
146 J. S. Lewis and R. G. Prinn, *Astrophys. J.*, 1980, **238**, 357–364.
147 R. G. Prinn and B. Fegley, Jr., *Astrophys. J.*, 1981, **249**, 308–317.
148 B. Fegley Jr. and R. G. Prinn, *Astrophys. J.*, 1985, **299**, 1067–1078.
149 K. Lodders, H. Palme and H.-P. Gail, 2009, chapter submitted to Landolt-Börnstein, New Series, *Astronomy and Astrophysics*, Springer-Verlag, Berlin, ArXiv: 0901.1149.
150 S. Gordon and B. J. McBride, 1994, NASA Reference Publication 1311.
151 L. V. Gurvich, I. V. Veyts and C. B. Alcock, Thermodynamic Properties of Individual Substances, Hemisphere Publishing, New York, fourth ed., 3 vols., 1989-1994.
152 M. W. Chase, *J. Phys. Chem. Ref. Data*, 1998, 28, Monograph 9.
153 A. Burcat and B. Ruscic, Third millenium ideal and condensed phase thermochemical database for combustion with updates from active thermochemical tables, *TAE 960, ANL-05/20, Argonne National Laboratory*, 2005.
154 C. Visscher, K. Lodders and B. Fegley Jr., *Astrophys. J.*, 2010, arXiv:1001:3639, submitted.
155 E. Lellouch, P. N. Romani and J. Rosenqvist, *Icarus*, 1994, **108**, 112–136.
156 V. Vuitton, R. V. Yelle and P. Lavvas, *Philos. Trans. R. Soc. London, Ser. A*, 2009, **367**, 729–741.
157 F. M. Flasar and P. J. Gierasch, in Proc. Nineteenth Symp. Roy. Soc. Canada, ed. A. V. Jones, Royal Society of Canada, Ottowa, pp. 85–87.
158 F. M. Flasar and P. J. Gierasch, *Geophys. Astrophys. Fluid Dyn.*, 1978, **10**, 175–212.
159 G. S. Orton and H. H. Aumann, *Icarus*, 1977, **32**, 431–436.
160 R. V. Yelle, C. A. Griffith and L. A. Young, *Icarus*, 2001, **152**, 331–346.
161 C. A. Nixon, R. K. Achterberg, B. J. Conrath, P. G. J. Irwin, N. A. Teanby, T. Fouchet, P. D. Parrish, P. N. Romani, M. Abbas, A. LeClair, D. Strobel, A. A. Simon-Miller, D. J. Jennings, F. M. Flasar and V. M. Kunde, *Icarus*, 2007, **188**, 47–71.

162 M.-C. Liang, R.-L. Shia, A. Y.-T. Lee, M. Allen, A. J. Friedson and Y. L. Yung, *Astrophys. J.*, 2005, **635**, L177–L180.
163 S. Lebonnois, *Planet. Space Sci.*, 2005, **53**, 486–497.
164 A.-S. Wong, A. Y.-T. Lee, Y. L. Yung and J. M. Ajello, *Astrophys. J.*, 2000, **534**, L215–L217.
165 A.-S. Wong, Y. L. Yung and A. J. Friedson, *Geophys. Res. Lett.*, 2003, **30**, 1447, DOI: 10.1029/2002GL016661.
166 A. Y. T. Lee, Y. L. Yung and J. I. Moses, *J. Geophys. Res.*, 2000, **105**, 20207–20225.
167 P. N. Romani, *Icarus*, 1996, **122**, 233–241.
168 S. J. Desch, W. J. Borucki, C. T. Russell and A. Bar-Nun, *Rep. Prog. Phys.*, 2002, **65**, 955–997.
169 D. M. Hunten, *Icarus*, 2008, **194**, 616–622.
170 M. H. Wong, *Icarus*, 2009, **199**, 231–235.
171 M. H. Wong, Hydrocarbons and Condensible Volatiles of Jupiter's Galileo Probe Entry Site, Ph.D. Thesis, Univ. Michigan, Ann Arbor, MI, 2001.
172 R. A. West, D. F. Strobel and M. G. Tomasko, *Icarus*, 1986, **65**, 161–217.
173 D. B. Curtis, C. D. Hatch, C. A. Hasenkopf, O. B. Toon, M. A. Tolbert, C. P. McKay and B. N. Khare, *Icarus*, 2008, **195**, 792–801.
174 B. Little, C. D. Anger, A. P. Ingersoll, A. R. Vasavada, D. A. Senske, H. H. Breneman, W. J. Borucki and the Galileo SSI Team, *Icarus*, 1999, **142**, 306–323.
175 J. H. Waite, Jr., H. Niemann, R. V. Yelle, W. T. Kasprzak, T. E. Cravens, J. G. Luhmann, R. L. McNutt, W.-H. Ip, D. Gell, V. De La Haye, I. Müller-Wodarg, B. Magee, N. Borggren, S. Ledvina, G. Fletcher, E. Walter, R. Miller, S. Scherer, R. Thorpe, J. Xu, B. Block and K. Arnett, *Science*, 2005, **308**, 982–986.
176 J. H. Waite, Jr., D. T. Young, T. E. Cravens, A. J. Coates, F. J. Crary, B. Magee and J. Westlake, *Science*, 2007, **316**, 870–875.
177 V. Vuitton, R. V. Yelle and V. G. Anicich, *Astrophys. J.*, 2006, **647**, L175–L178.
178 V. Vuitton, R. V. Yelle and M. J. McEwan, *Icarus*, 2007, **191**, 722–742.
179 H. Imanaka and M. A. Smith, *Geophys. Res. Lett.*, 2007, **34**, L02204, DOI: 10.1029/2006GL028317.
180 V. De La Haye, J. H. WaiteJr., T. E. Cravens, I. P. Robertson and S. Lebonnois, *Icarus*, 2008, **197**, 110–136.
181 T. E. Cravens, I. P. Robertson, J. H. Waite, Jr., R. V. Yelle, V. Vuitton, A. J. Coates, J.-E. Wahlund, K. Agren, M. S. Richard, V. de La Haye, A. Wellbrock and F. M. Neubauer, *Icarus*, 2009, **199**, 174–188.
182 Michael H. Wong, personal communication.
183 D. Sudarsky, A. Burrows and I. Hubeny, *Astrophys. J.*, 2003, **588**, 1121–1148.

PAPER

Photochemical modeling of Titan atmosphere at the "10 percent uncertainty horizon"

Zhe Peng,[a] Michel Dobrijevic,[b] Eric Hébrard,[b] Nathalie Carrasco[c] and Pascal Pernot[*a]

Received 23rd February 2010, Accepted 6th April 2010
DOI: 10.1039/c003366a

Titan's atmospheric chemistry modeling is presently limited by the lack of knowledge about many reaction rate coefficients at low temperature (50–200 K). Considering the difficulty of measuring such data, the only way to improve this situation is to identify key reactions as the ones for which better estimations of reaction rates is guaranteed to have a strong influence on the precision of model predictions. This is a slow iterative process, the limit of which has never been clearly defined in terms of model precision. The fact is that this limit is not a fully deterministic simulation, since one should not expect all reaction rate coefficients ever to become available with null uncertainty. The present study considers a quite optimistic scenario, in which all reaction rate coefficients in the chemical model are assumed to be known with a 10% relative uncertainty. The implications for chemical growth modeling are discussed.

1 Introduction

The overall precision of photochemical models of planetary atmospheres has unambiguously been shown to be highly sensitive to the uncertainty in the rates of involved chemical reactions.[1-8] Monte Carlo uncertainty propagation enabled Hébrard *et al.*[9] to assess the effect of these uncertainties on the computed abundances of major chemical species predicted by a 1D photochemical model of Titan's atmosphere. Strikingly, the uncertainties of most of the computed abundances could be much larger than the estimated uncertainty of the abundances gathered from observations, even for basic hydrocarbons like CH_4, C_2H_2, C_2H_4 and C_2H_6.

A major obstacle to precise prediction is the lack of data on the reactivity of neutral species at low temperature (low-T); for instance, in state-of-the-art photochemical models of Titan's atmosphere, less than 10% of the reaction rates have been measured in the relevant temperature range. In consequence, photochemical models of Titan's atmosphere are based mostly on low-T extrapolations of Arrhenius-type laws, which are known to be often inappropriate in this context.[10,11] Until reliable extrapolation models are made available, low-T extrapolation of reaction rates is to be treated with great care and considered as highly uncertain.[12]

Sensitivity analysis can be used to identify *key reactions*, responsible for large uncertainties in model prediction of some target property.[13] This approach guarantees that the reduction of the uncertainty on the rates of key reactions will have the strongest impact on the precision of the target property. This is particularly

---
[a] *Laboratoire de Chimie Physique, UMR 8000, CNRS, Université Paris-Sud 11, 91405 Orsay cedex, France. E-mail: pascal.pernot@lcp.u-psud.fr*
[b] *Université de Bordeaux, Laboratoire d'Astrophysique de Bordeaux, CNRS/INSU, UMR 5804, BP 89, 33271 Floirac Cedex, France*
[c] *Laboratoire Atmosphères, Milieux, Observations Spatiales, Université de Versailles Saint-Quentin, UMR 8190, 91371 Verrières le Buisson cedex, France*

important to assist in designing new rate constant measurement campaigns or in prioritizing the review, by experts, of existing data.

The improvement of model precision by key reaction identification/reevaluation is an iterative process (new key reactions are eventually revealed following an update of the previous ones[12]), which can take a very long time to achieve a prescribed precision level. The goal of the present paper is to extrapolate this process and observe what could be expected in the limit where all reaction rates are well determined. We want to emphasize here that an absolute accuracy of reaction rates will probably never be achieved, and we retained here a very optimistic limit of 10% relative uncertainty for all known reaction rates. Model improvement might also come from the addition of missing processes when they are discovered. Although model complexity is a salient issue in the present study, prediction of a model completion effect is beyond the scope of this paper.

Monte Carlo uncertainty propagation was performed for 0D and 1D photochemical models of Titan's atmosphere and we present here the results and their analysis in order to better understand uncertainty patterns in chemical networks with regard to molecular complexification.

2 Methods

The results of this article are based on three types of chemical models. A 1D photochemical model is used to study the global uncertainty patterns appearing in a complex network, whereas a simpler 0D model, without transport, is better fit to study local uncertainty (at the species level). Elementary models based on simplified reaction networks, are introduced in the course of the analysis to illustrate various observations.

2.1 1D and 0D photochemical models

The main lines of the 1D photochemical model and statistical procedures for uncertainty propagation and sensitivity analysis are presented here. More details can be found in Hébrard et al.[9] and Dobrijevic et al.[13,14]

In our 1D photochemical model extending from Titan's surface to 1300 km, the species densities are governed by the altitude-dependent continuity–diffusion equation. A detailed description of hydrocarbon, nitriles and oxygen coupled photochemistry, vertical eddy diffusion, molecular diffusion, and radiative transfer (including Rayleigh scattering by N_2 and aerosols absorption) are included in this model. Ions are not considered, and the loss and productions are due to photodissociations, bimolecular and termolecular reactions between neutral species. The model calculates abundances for 127 hydrocarbons, nitriles and oxygenated species, involved in 676 chemical reactions and 69 photodissociation processes.

A preliminary study is based on a simplified 0D model of Titan's hydrocarbon chemistry at 800 km, proposed by Dobrijevic et al.[13] as a benchmark for sensitivity analysis methods. This model contains reactions between H, H_2 and hydrocarbons with less than three carbon atoms, i.e. 15 species involved in 48 reactions. This 0D model had no stationary state in Dobrijevic et al.[13] and has been completed in the present study with additional production and loss processes, tuned to provide stationary densities close to those of the original model at the representative time $t = 10^7$ s

$$\begin{aligned} \varnothing &\rightarrow CH_4 &&; k = 175 \text{ cm}^{-3} \cdot \text{s}^{-1} \\ H &\rightarrow \varnothing &&; k = 8.68 \times 10^{-8} \text{ s}^{-1} \\ H_2 &\rightarrow \varnothing &&; k = 8.17 \times 10^{-8} \text{ s}^{-1} \\ C_2H_2 &\rightarrow \varnothing &&; k = 1.33 \times 10^{-7} \text{ s}^{-1} \\ C_2H_6 &\rightarrow \varnothing &&; k = 7.98 \times 10^{-8} \text{ s}^{-1} \end{aligned} \quad (1)$$

Those rates have no attached uncertainty.

Elementary Model 1. Linear chain of (quasi-)unimolecular reactions

Scheme

$$\emptyset \xrightarrow{K} A_1 \xrightarrow{k_1} A_2 \xrightarrow{k_2} \ldots \xrightarrow{k_{n-1}} A_n \xrightarrow{k_n} \emptyset$$

Stationary state concentration

$$a_i = \frac{K}{k_i}$$

Relative variance

$$\frac{\sigma_{a_i}^2}{a_i^2} = \frac{\sigma_K^2}{K^2} + \frac{\sigma_{k_i}^2}{k_i^2}$$

Notes
There is no uncertainty accumulation along the chain. The uncertainty of a given species depends only on the initial production rate K and the loss rate of this species k_i. The intermediate steps have no influence on the relative uncertainty of species i. This offers an important simplification rule in the analysis of more complex networks. Note also that if such chains occur in a photochemical model, the photolysis rates (corresponding to K) will systematically appear as key reactions.

2.2 Elementary models

To analyze the results of uncertainty propagation in more complex networks, we introduce in the course of this paper a set of basic chemical networks, presented as "Elementary Models" or EMs. For each elementary model (see *e.g.* EM 1), we provide expressions for the stationary state concentration of species of interest $a_i = [A_i]_{t=\infty}$ and the relative variance $\sigma_{a_i}^2/a_i^2$, obtained by the standard law of uncertainty propagation by combination of variances.[15] The corresponding uncertainty factor is $F_{a_i} = 1 + \sqrt{\sigma_{a_i}^2/a_i^2}$.

2.3 Uncertainty propagation and sensitivity analysis

Vertical structure, solar irradiance and diffusion coefficients (eddy and molecular) were kept fixed throughout the calculations. Uncertain values of photodissociation and reaction rates are represented by lognormal probability distributions

$$p(x) = \frac{1}{\sqrt{2\pi}x\sigma}\exp\left(-\frac{(\ln x - \mu)^2}{\sigma^2}\right) \quad (2)$$

with two parameters $\mu = \ln k(T)$, the logarithm of the nominal value of the reaction rate at temperature T, and $\sigma = \ln F(T)$, where $F(T)$ is the geometric standard uncertainty of the lognormal distribution. With these notations, the 67% confidence interval for a reaction rate at a given temperature is $[k(T)/F(T), k(T) \times F(T)]$. For small uncertainties, one can write $F \simeq 1 + \frac{\Delta k}{k}$: for instance, a 10% relative uncertainty on k corresponds to $F = 1.1$. The reaction rate coefficients and the photodissociation coefficients used in the present study were extracted from the review by Hébrard *et al.*,[16] with some important revisions detailed in a recent article,[12] but a global uncertainty factor $F = 1.1$ was assumed for all processes.

For Monte Carlo uncertainty propagation, random reactions rates are generated from their pdf and model outputs are computed for each draw. For the 1D model, long computation times required to reach the stationarity of the species densities

Table 1 Densities and uncertainty factors for all species of the 0D model. Simulation results for uncertainty factors, F(simul), are compared to estimations by Elementary Models, F(EM), when available. The last column reports uncertainty factors for the same species as simulated with the 1D model at 900 km. All simulations are run with a uniform uncertainty factor $F = 1.1$ on reaction rate constants

Species	0D model				1D model
	Density/cm^{-3}	F (simul)	F (EM)	EM #	F(simul)
H	2.6×10^9	1.03	1.03	2	1.06
H$_2$	1.7×10^9	1.03	1.03	2	1.03
CH	2.3×10^2	1.13	1.12	2	1.11
^{3}CH$_2$	2.6×10^1	1.13	1.14	3	1.08
^{1}CH$_2$	9.2×10^1	1.08	1.09	2	1.11
CH$_3$	4.4×10^7	1.08	—	—	1.11
CH$_4$	2.2×10^9	1.07	1.06	2	1.02
C$_2$H	6.0×10^3	1.24	—	—	1.08
C$_2$H$_2$	6.3×10^8	1.01	1.03	2	1.05
C$_2$H$_4$	2.4×10^8	1.07	1.07	2	1.05
C$_2$H$_5$	2.6×10^1	1.20	—	—	1.13
C$_2$H$_6$	4.1×10^7	1.31	—	—	1.09

limit the number of Monte Carlo samples: typically about 500 independent samples are generated. This provides a convergence of average values and correlation coefficients to better than 5%. For the smaller 0D model, the number of Monte Carlo runs is not limited, and we are able to estimate output uncertainty factors with better than 1% accuracy (10 000 runs).

For each run, one records the reaction rate coefficients (inputs) and neutral mole fractions (outputs) at different altitudes, which are used for statistical uncertainty and sensitivity analysis. Input-output correlation coefficients provide sensitivity measures well adapted to key reaction search.[13] They are easy to estimate within the Monte Carlo uncertainty propagation framework and do not require dedicated sampling schemes.[17,18] The input and output samples recorded for uncertainty evaluation can be directly used for the sensitivity analysis.[12–14,19]

3 Results and discussion

3.1 Analysis of the 0D simulations

Stationary densities and the associated uncertainty factors are given in Table 1. We observe that many species densities are simulated with uncertainty factors smaller than $F = 1.1$, as low as $F = 1.01$ for C$_2$H$_2$ or $F = 1.03$ for H and H$_2$. By contrast, two species have remarkably enhanced uncertainty factors: C$_2$H ($F = 1.24$) and C$_2$H$_6$ ($F = 1.31$).

We will show now how these observations can be interpreted through elementary models, mostly based on unimolecular or pseudo-unimolecular reactions.

3.1.1 C$_2$H$_4$, H$_2$, H, CH$_4$, CH and ^{1}CH$_2$. These species have an uncertainty factor smaller than the nominal value ($F = 1.1$) and are affected by numerous production and/or loss processes. For instance, C$_2$H$_4$ has 12 production pathways and 2 loss reactions. This pattern can be linked to EM 2, in the hypothesis of independent pathways. Observing that in this case all pathways have almost equivalent contributions, we have

Elementary Model 2. m productions $-$ n losses

Scheme

$$\emptyset \xrightarrow{K_1} A_1 \xrightarrow{k_{10}} \quad \xrightarrow{k_{11}} A_1' \xrightarrow{k_{12}} \emptyset$$
$$\emptyset \xrightarrow{K_2} A_2 \xrightarrow{k_{20}} A \xrightarrow{k_{21}} A_2' \xrightarrow{k_{22}} \emptyset$$
$$\cdots \quad \cdots$$
$$\emptyset \xrightarrow{K_m} A_m \xrightarrow{k_{m0}} \quad \xrightarrow{k_{n1}} A_n' \xrightarrow{k_{n2}} \emptyset$$

Stationary state concentration

$$a = \frac{\sum_{i=1}^m K_i}{\sum_{i=1}^n k_{i1}}$$

Relative variance

$$\frac{\sigma_a^2}{a^2} = \sum_{i=1}^m \left(\frac{K_i}{\sum_{j=1}^m K_j}\right)^2 \frac{\sigma_{K_i}^2}{K_i^2} + \sum_{i=1}^n \left(\frac{k_{i1}}{\sum_{j=1}^n k_{j1}}\right)^2 \frac{\sigma_{k_{i1}}^2}{k_{i1}^2}$$

Notes
- The results are independent of the rates k_{i0} of direct production processes of A.
- In the case of equal values of $\{K_i\}$ and $\{k_{j1}\}$, one gets

$$\frac{\sigma_a^2}{a^2} = \frac{1}{m}\frac{\sigma_{K_i}^2}{K_i^2} + \frac{1}{n}\frac{\sigma_{k_{j1}}^2}{k_{j1}^2}$$

The production and loss contributions are independent, and in both cases, the relative variance is inversely proportional to the number of processes, which follows from the standard law for the sum of independent random variables.

$$F_{C_2H_4} = 1 + \sqrt{\left(\frac{1}{12}+\frac{1}{2}\right) \times 0.1^2} \tag{3}$$

$$F_{C_2H_4} \simeq 1.08 \tag{4}$$

which is very close to the value obtained by simulation ($F = 1.07$). When pathways have different contributions, we use the general expression for EM 2, which provides also favorable comparisons for H_2, H, CH_4, CH and $^1CH_2$ (see Table (1)), validating the initial hypothesis of pathway independence. Indeed, almost all species densities in this model are correlated by reactions, but we observe that, due to this general correlation, individual correlations are typically weak, even in the case where two species are directly linked by a bimolecular reaction (Fig. (1)).

3.1.2 $^3CH_2$. This species has an uncertainty factor $F = 1.13$, larger than the nominal value. Its loss is dominated by reaction

$$H + {}^3CH_2 \rightarrow H_2 + CH \tag{5}$$

and the main production pathways are the deactivation processes of $^1CH_2$, mainly

$${}^1CH_2 + H_2 \rightarrow {}^3CH_2 + H_2 \tag{6}$$

Elementary Model 3. Parallel pathways

Scheme

$$\emptyset \xrightarrow{K} A_0 \begin{array}{c} \nearrow A_1 \searrow \\ \xrightarrow{k_{10}} \\ \xrightarrow{k_{20}} A_2 \xrightarrow{k_{21}} \\ \vdots \\ \xrightarrow{k_{n0}} A_n \xrightarrow{k_{n1}} \end{array} A \xrightarrow{k_2} \emptyset$$

Stationary solution
Solving the system

$$\sum_{i=1}^{n} k_{i0} a_0 = K$$
$$k_{11} a_1 = k_{10} a_0$$
$$k_{21} a_2 = k_{20} a_0$$
$$\vdots$$
$$k_{n1} a_n = k_{n0} a_0$$
$$k_2 a = \sum_{i=1}^{n} (k_{i1} a_i)$$

one obtains

$$a = \frac{K}{k_2}$$

Relative variance

$$\frac{\sigma_a^2}{a^2} = \frac{\sigma_K^2}{K^2} + \frac{\sigma_{k_2}^2}{k_2^2}$$

Notes
As there is no flux loss or production along the chain, this model is equivalent to EM 1.

$$^1CH_2 + CH_4 \rightarrow {}^3CH_2 + CH_4 \qquad (7)$$

These reactions, being limited by the small concentration of $^1CH_2$ with regard to CH_4 and H_2, can be considered as pseudo-unimolecular. This system is an analog of EM 3. Having the same origin, the production pathways can be considered as a single pathway, and we recover the simple linear chain, EM 1. The uncertainty factor is thus $F_{3CH_2} \approx 1 + \sqrt{(F_k - 1)^2 + (F_k - 1)^2} \approx 1.14$, to be compared with the simulation result $F = 1.13$.

3.1.3 Other species. C_2H, C_2H_2, C_2H_6, C_2H_5 and CH_3 are involved in complex networks with loops[13] and bimolecular reactions and/or they are strongly correlated to each other (Fig. 1). Although we had some success with various other Elementary Models to reproduce the observed tendencies, the basic hypotheses of pathways independence and unimolecularity find here their limits. Nevertheless, uncertainty enhancement for C_2H, C_2H_5 and C_2H_6 have been traced back to bimolecular effects (EM 5).

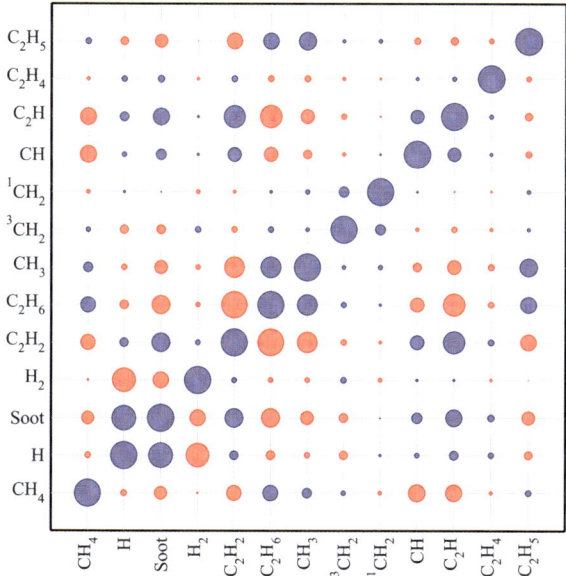

Fig. 1 Correlation matrix between stationary densities of species in the 0D model: (blue) positive correlation; (red) negative correlation. Linear scale of symbol size from 1 (on the diagonal) to 0.

For C_2H_2, there is an equilibrium with C_2H

$$C_2H_2 \underset{+H_2\cdots,\, k_2}{\overset{h\nu,\, k_1}{\rightleftarrows}} C_2H \qquad (8)$$

According to EM 4, equilibrium (8) has no influence on the stationary density of C_2H_2. Moreover, this species has a loss reaction without uncertainty and 14 other production pathways, which is relevant to EM 2

$$F_{C_2H_2} \approx 1 + \sqrt{\frac{1}{14} \times 0.1^2} \approx 1.027$$

This value is larger than the simulation result ($F = 1.01$), but the uncertainty attenuation is fairly well reproduced. For a better estimation, one should take explicitly into account the weak uncertainty factors of species involved in C_2H_2 formation, such as H_2, C_2H_4, $^1CH_2$, etc.

3.1.4 Intermediate conclusion. The analysis of the 0D model shows that the relative uncertainty of many species can be explained simply by counting their direct production–loss pathways. For these species, the relative variance is inversely proportional to the number of production–loss pathways. A remarkable result, which confirms previous observations in the sensitivity analysis of such systems,[12,13] is that the rates of direct production pathways of these species do not contribute to the uncertainty (as in EM 2); instead, initiation processes (photodissociation) play a major role in the uncertainty budget of the whole system.

Some species are nevertheless affected by bimolecular processes, which can have various effects on the uncertainty, depending on the correlation between the reactants densities (EM 5). This correlation is determined by the overall reaction network (Fig. 1). Uncertainty enhancement is maximal for reactions between species

Elementary Model 4. Lateral equilibrium

Scheme

$$\emptyset \to \to \to A_1 \underset{k_2}{\overset{k_1}{\rightleftarrows}} \rightleftarrows \rightleftarrows A_2$$
$$ \downarrow$$
$$ k_3 \downarrow$$
$$ \downarrow$$
$$ \emptyset$$

with rate K entering A_1.

Stationary solution

$$(k_1 + k_3)a_1 = K + k_2 a_2$$
$$k_2 a_2 = k_1 a_1$$
$$a_1 = \frac{K}{k_3}$$

Relative variance

$$\frac{\sigma_{a_1}^2}{a_1^2} = \frac{\sigma_K^2}{K^2} + \frac{\sigma_{k_3}^2}{k_3^2}$$

Notes
Another analog of EM 1.

---

with strongly positively correlated densities, as for instance in the formation of C_2H_6 from two CH_3 radicals.

3.2 Analysis of the 1D simulations

Considering the large prediction uncertainties observed for simulations based on evaluated reaction rate databases, it is interesting to assess the effect of reducing the uncertainty on rate coefficients to a very small value, $F = 1.1$, to mimic what one could expect from photochemical models when reaction rates will be measured with this kind of precision.

We compare outputs of the present simulation with the results of a state-of-the-art simulation with evaluated uncertainty factors, as presented in Hébrard et al.[12] A comparison of the mole fractions for three representative species is reported in Fig. 2 and the uncertainty factors for all species are reported for an altitude of 1200 km in Fig. 3. For CH_4, the reduction of dispersion is remarkable (the uncertainty factor F_{CH_4} at 1200 km is reduced from 1.28 to 1.02); for C_2H_6, all outlier profiles have disappeared and $F_{C_2H_6}$ is contracted from 3.3 to 1.1; and for one of the heavier species in the model C_6H_{14}, one has a reduction of $F_{C_6H_{14}}$ from 6.2 to 1.5, which is the largest uncertainty factor in the present $F = 1.1$ scenario.

The uncertainty reduction for the densities of all species of the model at 1200 km is presented in Fig.3. Globally, all points lie below the curve $F_{10\%} = F_{SA}^{0.25}$, which means that if the density of a species had 68% chances to lie within the interval $[x/F_{SA}, x * F_{SA}]$, this probability is now higher than 99.99% (the analog on a linear scale would be for a 1σ interval to become a 4σ interval or better). Some species have a spectacular uncertainty reduction: for instance, $F_{C_3H_8}$ went from 6.4 to 1.2. In the bottom left corner we observe a set of species, containing H, H_2 and CH_4, having low uncertainty factors, as was also observed in the 0D model.

3.2.1 Small species.
For basic species, such as H, H_2, CH_4, or C_2H_4 the results agree with the results of the 0D model (cf. Table 1). However, some differences are noticeable, which can be related to different factors:

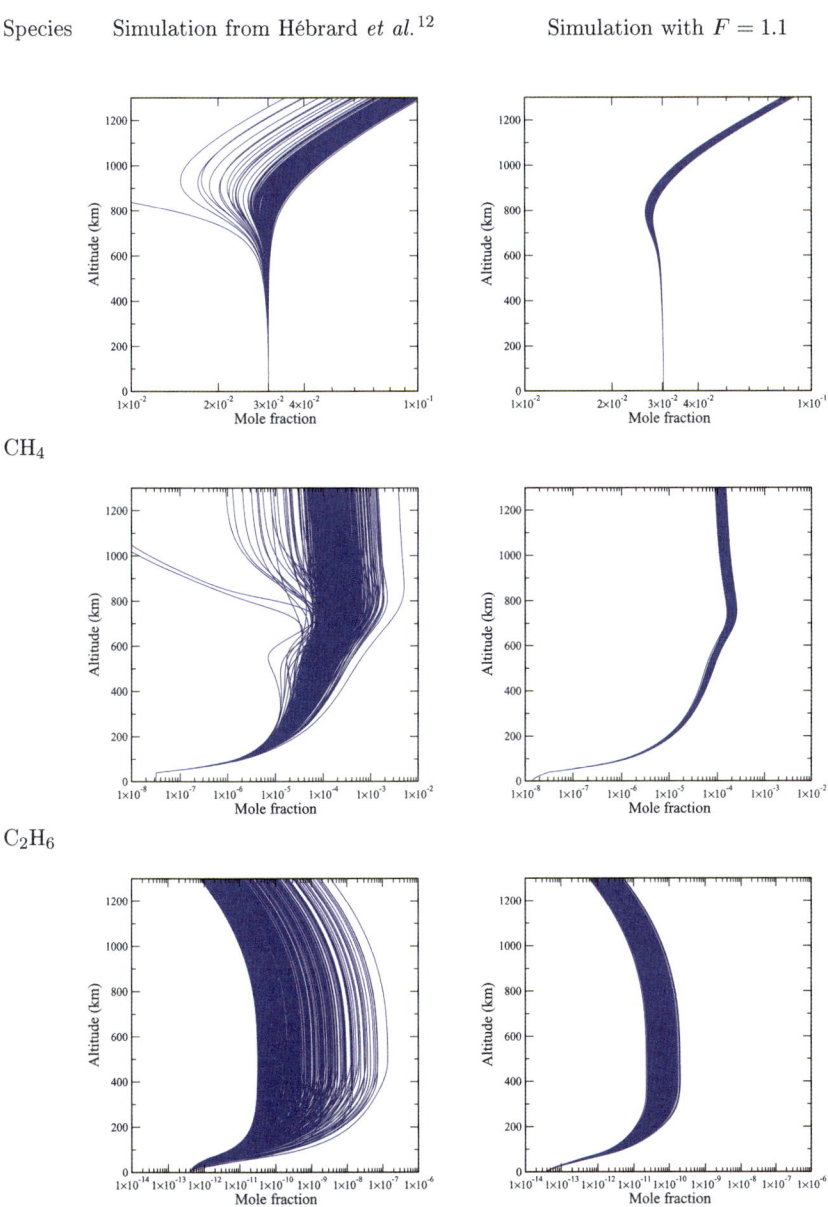

Fig. 2 Comparison of Monte Carlo samples of density profiles for representative hydrocarbons: (left) current *state-of-the-art* reaction rate database;[12] (right) simulation with all uncertainty factors set to $F = 1.1$. A slight shift in density at low altitudes is due to an update of boundary conditions between both simulations.

Elementary Model 5. Bimolecular production

Scheme

$$A_1 + A_2 \xrightarrow{k_1} A_3 \xrightarrow{k_2} \emptyset$$

Stationary solution

$$a_3 = \frac{k_1 a_1 a_2}{k_2}$$

Relative variance

In the limit where k_1 has a weak influence on a_1 and a_2 (as in complex systems) we can assume a null correlation between these variables ($\mathrm{Corr}(k_1, a_1) = \mathrm{Corr}(k_1, a_2) = 0$), and one gets

$$\frac{\sigma_{a_3}^2}{a_3^2} = \frac{\sigma_{k_1}^2}{k_1^2} + \frac{\sigma_{k_2}^2}{k_2^2} + \frac{\sigma_{a_1}^2}{a_1^2} + \frac{\sigma_{a_2}^2}{a_2^2} + 2\mathrm{Corr}(a_1, a_2)\frac{\sigma_{a_1}}{a_1}\frac{\sigma_{a_2}}{a_2}$$

which involves the correlation coefficient between stationary densities $\mathrm{Corr}(a_1, a_2)$, as shown in Fig. 1.

Notes

Depending on the correlation coefficient, the relative variance takes its values between the minimal ($\mathrm{Corr}(a_1, a_2) = -1$)

$$\frac{\sigma_{a_3}^2}{a_3^2} = \frac{\sigma_{k_1}^2}{k_1^2} + \frac{\sigma_{k_2}^2}{k_2^2} + \left(\frac{\sigma_{a_1}}{a_1} - \frac{\sigma_{a_2}}{a_2}\right)^2 \geq \frac{\sigma_{k_1}^2}{k_1^2} + \frac{\sigma_{k_2}^2}{k_2^2}$$

and maximal ($\mathrm{Corr}(a_1, a_2) = 1$) value

$$\frac{\sigma_{a_3}^2}{a_3^2} = \frac{\sigma_{k_1}^2}{k_1^2} + \frac{\sigma_{k_2}^2}{k_2^2} + \left(\frac{\sigma_{a_1}}{a_1} + \frac{\sigma_{a_2}}{a_2}\right)^2$$

In the case of a dimer formation ($A_1 = A_2$), the relative uncertainty corresponds to the maximal value (full positive correlation between reactants)

$$\frac{\sigma_{a_3}^2}{a_3^2} = \frac{\sigma_{k_1}^2}{k_1^2} + \frac{\sigma_{k_2}^2}{k_2^2} + 4\frac{\sigma_{a_1}^2}{a_1^2}$$

• **Diffusion.** A major difference between the 0D and 1D models is transport. In these simulations, the diffusion coefficients have fixed values, which introduces in the system a production–loss process without uncertainty. When compared to 0D results, this contributes to decrease globally the uncertainty on stationary densities (*cf.* Table 1), an effect which is modulated by relative values of the chemical and diffusion lifetimes for each species. The fact that we used in the 0D model a few input/output processes with fixed rates to ensure stationarity, has also an influence on the differences between the 0D and 1D results: the effect is large for those species not involved in the input/output reactions with fixed rates introduced in the 0D model, *e.g.* C_2H and C_2H_6, whereas it is very small for species such as H_2, directly concerned by these reactions.

• **Number of production and loss processes for a given species.** According to EM 2, relative uncertainty is decreased by a large number of production/loss processes. The 1D model being more complex than the 0D model, this can contribute to reduce relative uncertainty for node species as CH_4, $^3CH_2$, C_2H and C_2H_6.

Elementary Model 6. Branching chain

Scheme

$$\emptyset \xrightarrow{K} A_1 \xrightarrow{k_1} A_2 \xrightarrow{k_2} \ldots \xrightarrow{k_{n-1}} A_n \xrightarrow{k_n} \emptyset$$
$$\quad\quad\; \downarrow k_{p1} \quad\; \downarrow k_{p2} \quad\quad\quad\quad\quad\; \downarrow k_{pn}$$
$$\quad\quad\; \emptyset \quad\quad\; \emptyset \quad\quad\quad\quad\quad\quad\; \emptyset$$

Stationary solution

$$a_i = \frac{K}{k_i}\left[\prod_{j=1}^{i}\frac{k_j}{k_j + k_{pj}}\right]$$

Relative variance

$$\frac{\sigma_{a_i}^2}{a_i^2} = \frac{\sigma_K^2}{K^2} + \left(\frac{k_i}{k_i + k_{pi}}\right)^2 \frac{\sigma_{k_i}^2}{k_i^2}$$
$$+ \sum_{j=1}^{i-1}\left(\frac{k_{pj}}{k_j + k_{pj}}\right)^2 \frac{\sigma_{k_j}^2}{k_j^2}$$
$$+ \sum_{j=1}^{i}\left(\frac{k_{pj}}{k_j + k_{pj}}\right)^2 \frac{\sigma_{k_{pj}}^2}{k_{pj}^2}$$

Notes
For identical rates and uncertainty factors for all reactions, one has a square root dependence of the relative uncertainty as a function of the position of the species of interest in the chain

$$\frac{\sigma_{a_i}}{a_i} = (F_k - 1)^*\sqrt{1 + i/2}$$

- **Reactions involving reactants with very uncertain densities.** Very reactive species, such as C, CH, $^1CH_2$, C_2 and C_2H_3, can react with a lot of heavy species having large uncertainty. This is a source of uncertainty increase, when compared with less reactive species, such as $^3CH_2$ et C_2H. Moreover, CH_3 et C_2H_5 are the main products of complex species metathesis, and they also suffer from larger uncertainty factors.

3.2.2 **Uncertainty patterns for chemical families.** To apprehend the pattern of uncertainty distribution amongst the species, notably amongst hydrocarbons, we plotted the uncertainty factor against the mole fraction (Fig. 4, left). Globally, the uncertainty factors increase as the mole fractions decrease, for all compounds. Except for a few points, the values lie between the curves $F_{min}(y) \simeq 1 - 0.01 * \log(y)$ and $F_{max}(y) \simeq 1 - 0.05 * \log(y)$, where y is the mole fraction. A similar plot of the uncertainty factor vs. the molecular mass is shown for hydrocarbons (Fig. 4, right). There is a positive correlation between the uncertainty factor and molecular mass and also an increasing dispersion of uncertainty factors. We have shown through the Elementary Models that molecular complexification was not necessarily associated with uncertainty growth, notably when pseudo-unimolecular processes are dominant. A square-root law for growth can however be obtained in the case of a branching reaction chain (EM 6). The linear uncertainty growth observed for the 1D simulations is thus at least in part due to bimolecular reactions.

It is noteworthy that alkanes, alkenes and alkynes present identical quasi-linear trends. For alkanes, one has $F(M) \simeq 1.012 + 0.007 * (M - M_{CH_4})$, where M_{CH_4} is the molecular mass of CH_4. This provides a rule of thumbs in terms of number

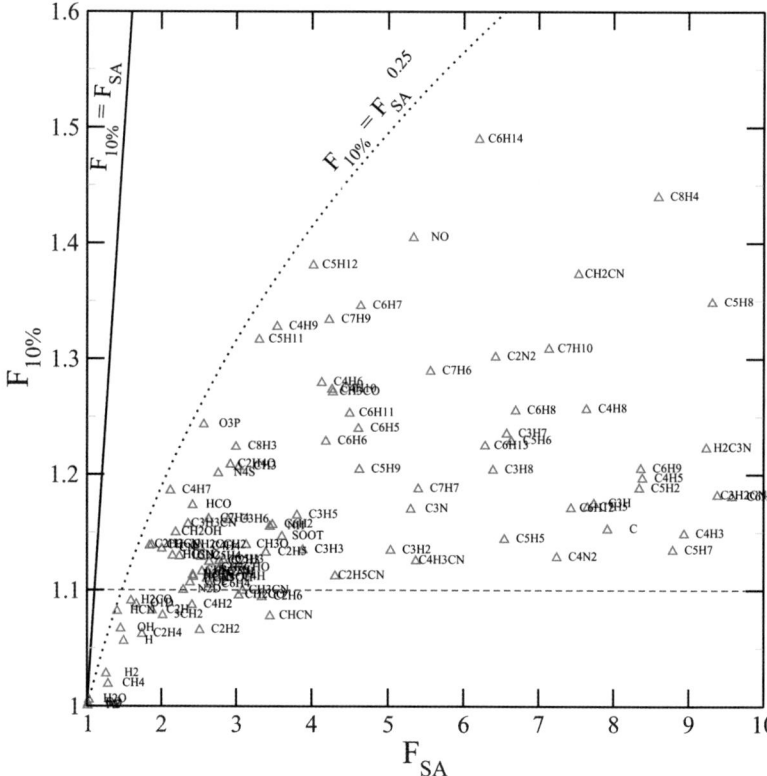

Fig. 3 Reduction of uncertainty factors on stationary mole fractions at 1200 km. F_{SA} refers to the simulation with current state-of-the-art kinetics database,[12] and $F_{10\%}$ refers to the present simulation with the uncertainty factors of all processes set to $F = 1.1$.

of carbon atoms $F(n_C) \simeq 1.01 + 0.093 * (n_C - 1)$, *i.e.* almost a 10% relative uncertainty increase by additional carbon atom. A similar linearity is observed for alkanes on the F *vs.* mole fraction plot, with $F(y) \simeq 1 - 0.05 * \log(y/y_{CH_4})$. The pattern is less regular for alkenes and alkynes. Other families, as for instance C_nH and C_nH_2, present also regular, albeit nonlinear, uncertainty growth pattern with mass (Fig. 4, right).

At the moment, we have no full explanation for this linearity of alkane uncertainty factors. As molecular mass and mole fraction within the family are strongly anticorrelated,[20] it is not clear which one should be retained as an explanatory variable for this trend. Notwithstanding, we see that, when accounting only for reaction rates uncertainty, there are limits to the prediction precision by a photochemical model of the mole fractions of complex hydrocarbons.

3.2.3 Most influential reactions.

The present simulations with reduced reaction rates uncertainty are assumed to probe the ultimate photochemical accuracy for a given model. Identification of key reactions is therefore not aiming at model precision improvement; instead, we use it here as a tool to detect influential reactions, *i.e.* reactions that affect the densities of many species.

Identification of key reactions is performed by analyzing input–output correlations (Table 2). The influence of a reaction is quantified by the number of species having input–output absolute rank correlation coefficients larger than 0.2 with this reaction. As in Hébrard *et al.*,[12] reactions with at least one score larger than 15 are reported.

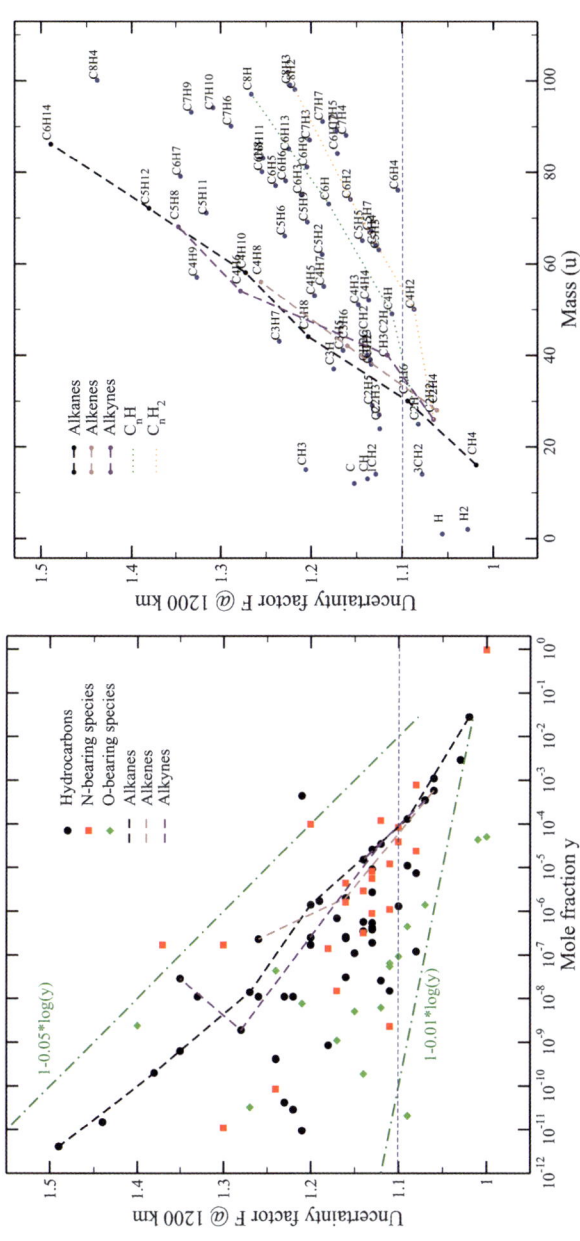

Fig. 4 Correlation plots between the uncertainty factor of stationary densities at 1200 km and (left) the mole fractions, and (right) the masses. Lines are visual guides to depict some chemical families with regular pattern.

Table 2 Key reactions with the number of species they influence at a set of representative altitudes. The total number of species in the model is 127. Reactions with at least one score larger than 15 are shown; scores below 10 are not reported

Reaction	300 km	600 km	900 km	1200 km
$N_2 + h\nu$	25	31	57	63
$CH_3 + h\nu$	37	46	46	49
$CH_4 + h\nu$	48	48	63	64
$C_2H_2 + h\nu$	27	11	—	—
$C_2H_4 + h\nu$	13	35	20	18
$C_6H_6 + h\nu$	69	63	31	32
$CH + CH_4$	17	30	45	42
$C_2H + CH_4$	26	13	—	—
$C_2H + C_2H_6$	18	12	11	11
$C_4H + C_2H_4$	—	23	16	16
$CH_3 + CH_3$	17	14	—	—
$H + C_2H_5$	14	14	15	15

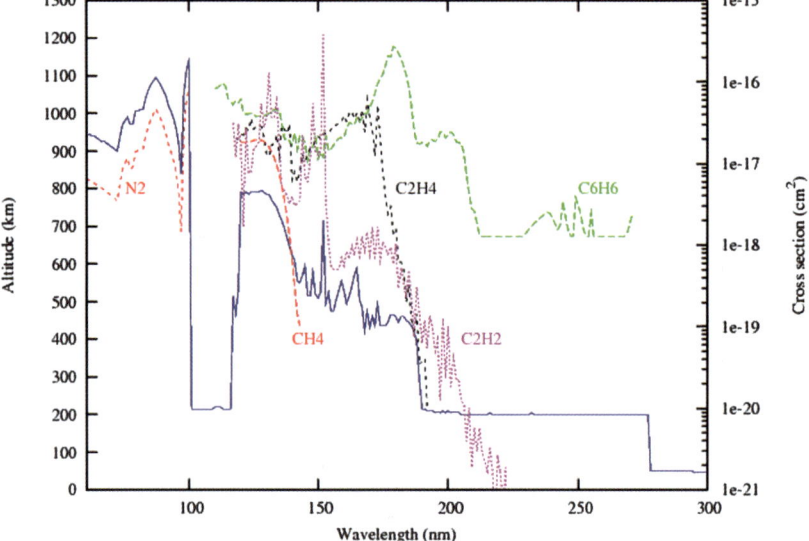

Fig. 5 The depth of penetration of solar radiation as a function of wavelength in Titan's atmosphere. The blue solid line represents the altitude at which the total optical depth is unity in our 1D model. Cross sections of main absorbers are given. N_2 is mainly photodissociated at 1000 km, CH_4 at 800 km, C_2H_2 around 600 km, C_2H_4 around 400–500 km and C_6H_6 around 200 km.

Very few reactions are selected by this procedure. Unsurprisingly, photodissociation rates play a dominant role. This would confirm the patterns outlined for the 0D model, *i.e.* that many species are involved in quasi-unimolecular reaction chains. In terms of influence, the photodissociation of N_2 and CH_4 significantly affect about half of the 127 modeled species at high altitudes. This high score is equaled by the photolysis of C_6H_6 at lower altitudes. Indeed, most photodissociation rates see their influence decrease with altitude, except for C_2H_2 and C_6H_6. A plot comparing the cross sections of these processes (Fig. 5) shows that C_6H_6 has a residual absorption in the 220–270 nm range, where the other absorbers have no impact. Similarly, C_2H_2 absorbs weakly around 190–210 nm, where CH_4 and C_2H_4 have negligible cross-sections, and it is sufficiently more abundant than C_6H_6 to have some influence.

Amongst the set of reactions, only one (CH + CH$_4$) has a strong influence, almost at all altitudes. This reaction was identified in recent work on the bimodality in the density profiles of some species[14] and on the effect of low-T measurements on model predictivity.[12] In the latter study, it was shown that updating the rate constant of this reaction with low-T measured data[21,22] had the effect of getting this reaction out of the list of key reactions. The rate constants of several others reactions identified here (C$_2$H + CH$_4$, C$_2$H + C$_2$H$_6$, C$_4$H + C$_2$H$_6$) have also been recently updated at low-T.[23–26] The current remaining uncertainty regarding the rates of these identified reactions, apart from some possible systematic effects, concerns mainly the nature of their products which has not been investigated thoroughly until now, even at room temperature. Other reactions rates, with much larger uncertainty factors, are currently more in need of improved accuracy. It is interesting to observe that, when all uncertainty factors are fictitiously set to a minimum achievable value, CH + CH$_4$ and these other reactions, though to a somewhat lower extent, stand out as cornerstones of Titan's photochemistry. Many of both experimental and theoretical studies have been published in order to investigate the rate constant of the three-body recombination reaction CH$_3$ + CH$_3$ (see Klippenstein *et al.*[27] for a quite exhaustive review). As most of the three-body recombination reactions, very few of these studies have however been performed in conditions appropriate for planetary atmospheres. Most of the rate expressions available in the literature still have to be extrapolated down to the lowest temperatures encountered in the outer planets' atmospheres. This ever-existing scarcity reflects both laboratory limitations and the importance of this reaction in hydrocarbon combustion chemistry. Likewise, there is no direct measurement of the rate constant nor of the products channels of the reaction H + C$_2$H$_5$ which occurs as a secondary process in the combustion studies of the H + C$_2$H$_4$ and H + C$_2$H$_6$ reactions;[28] information on any temperature dependence at conditions representative of Titan's atmosphere is thus very limited.

Future photochemical models of Titan's atmosphere would thus greatly benefit from a much closer investigation of the reactions identified here as it would help to improve significantly their predictivity.

4 Conclusion

Simulation of the atmospheric photochemistry of Titan by a 1D model and optimistically small uncertainty of reaction rates ($F = 1.1$), reveal interesting features for the precision of predicted density profiles. It has to be noted that these values correspond to a lower limit, as uncertainty sources other than chemical rates, such as diffusion coefficients, temperature, thermodynamics... have not been taken into account.

On the positive side, all modeled species densities have uncertainty factors below $F = 1.5$, which is a huge improvement in comparison to the present state of affairs. On a more pessimistic side, we have observed a linear increase of F with molecular mass of about 10% per additional carbon atom for the three major hydrocarbon families. This questions the possibility to achieve predictive detailed models of molecular complexification for these species. Other chemical families present a milder uncertainty increase with mass. We have still to elucidate the origins of these regular patterns in terms of reaction network structure.

Keeping in mind that the present model, as all photochemical models of Titan's atmosphere, is not complete, the present simulations help to define limits to the predictivity/interpretation level of such models. To be deemed significant, any relative variation of the mole fraction of a compound should be larger than the relative uncertainty due to photochemistry. For a given species, let's say C$_3$H$_8$ ($F \simeq 1.2$), a photochemical model would therefore not enable us to identify the origin of temporal/latitudinal/longitudinal mole fraction variations smaller than 20%. Similarly, an additional process introduced in the model should induce a change in mole fraction of C$_3$H$_8$ larger than 20% to be considered as important for this species.

This statistical concept has been used by Carrasco et al.[19] to reduce the ion–molecule reaction set for Titan's ionosphere.

An interesting outcome of this study is that the initial processes (e.g. photodissociation) are always important in unimolecular reaction chains; intermediate reactions leading to a product can often be neglected, meaning that they play no role in the stationary concentration of species further in the chain. This sheds some light on the key role of photodissociations revealed by previous sensitivity analyses of this system, in apparent contradiction with the fact that photodissociation rates have modest uncertainty factors (typically estimated to be $F = 1.5$) when compared to many neutral–neutral reaction rates.[12] Of course, Titan's atmospheric chemistry cannot be reduced to a unimolecular reactions network, and bimolecular reactions play a major role on the observed uncertainty factors of a number of species. It is difficult to predict the amplitude of their effect, because it depends strongly on the level of correlation between the reactants' densities. A consequence is that it is practically impossible to ascertain beforehand the effect of the addition of new reactions into the model on the uncertainty factors of most species densities. Uncertainty propagation remains a necessary tool to solve this kind of question.

Sensitivity analysis enabled us to identify a list of reactions (Table 2), which comes as a complement to the list previously published by Hébrard et al.[12] The latter list defines *key reactions*, for which a more accurate estimation of low-T rate constants and branching ratios would impact significantly the precision of predictions with the present photochemical models; we would assign them highest priority for the improvement of model predictivity. The new list highlights a core of *influential reactions*, which would appear ultimately as key reactions (*i.e.* when all the reaction rates will be known at low-T with a precision better than 10% and if the reaction network does not undergo drastic modifications). From our precision-oriented point of view, these reactions are of lower priority, but the accurate determination of their low-T rate constants and branching ratios is nevertheless a safe investment for the future of Titan's photochemical modeling.

Acknowledgements

Z. P. thanks the Laboratoire de Chimie Physique for financial support during his Master's thesis.

References

1 R. S. Stolarski, D. M. Butler and R. D. Rundel, *J. Geophys. Res.*, 1978, **83**, 3074–3078.
2 A. Thompson and R. Stewart, *J. Geophys. Res.*, 1991, **96**, 13089–13108.
3 R. Stewart and A. Thompson, *J. Geophys. Res.*, 1996, **101**, 20953–20964.
4 M. Dobrijevic and J. Parisot, *Planet. Space Sci.*, 1998, **46**, 491–505.
5 N. S. Smith and F. Raulin, *J. Geophys. Res.*, 1999, **104**, 1873–1876.
6 M. Dobrijevic, J. Ollivier, F. Billebaud, J. Brillet and J. Parisot, *Astron. Astrophys.*, 2003, **398**, 335–344.
7 E. Hebrard, Y. Bénilan and F. Raulin, *Adv. Space Res.*, 2005, **36**, 268–273.
8 G. P. Smith and D. Nash, *Icarus*, 2006, **182**, 181–201.
9 E. Hébrard, M. Dobrijevic, Y. Bénilan and F. Raulin, *Planet. Space Sci.*, 2007, **55**, 1470–1489.
10 H. Sabbah, L. Biennier, I. R. Sims, Y. Georgievskii, S. J. Klippenstein and I. W. M. Smith, *Science*, 2007, **317**, 102–105.
11 I. W. M. Smith, *Chem. Soc. Rev.*, 2008, **37**, 812–826.
12 E. Hébrard, P. Pernot, M. Dobrijevic, N. Carrasco, A. Bergeat, K. M. Hickson, A. Canosa, S. D. L. Picard and I. R. Sims, *J. Phys. Chem. A*, 2009, **113**, 11227–11237.
13 M. Dobrijevic, E. Hébrard, S. Plessis, N. Carrasco, M. Bruno-Claeys and P. Pernot, *Adv. Space Res.*, 2010, **45**, 77–91.
14 M. Dobrijevic, N. Carrasco, E. Hébrard and P. Pernot, *Planet. Space Sci.*, 2008, **56**, 1630–1643.
15 J. Tellinghuisen, *J. Phys. Chem. A*, 2001, **105**, 3917–3921.

16 E. Hébrard, M. Dobrijevic, Y. Bénilan and F. Raulin, *J. Photochem. Photobiol., C*, 2006, **7**, 211–230.
17 A. Saltelli, M. Ratto, S. Tarantola and F. Campolongo, *Chem. Rev.*, 2005, **105**, 2811–2827.
18 J. C. Helton, J. D. Johnson, C. J. Sallabery and C. B. Storlie, *Reliab. Eng. Syst. Saf.*, 2006, **91**, 1175–1209.
19 N. Carrasco, S. Plessis, M. Dobrijevic and P. Pernot, *Int. J. Chem. Kinet.*, 2008, **40**, 699–709.
20 M. Dobrijevic and I. Dutour, *Planet. Space Sci.*, 2006, **54**, 287–295.
21 A. Canosa, I. R. Sims, D. Travers, I. W. M. Smith and B. R. Rowe, *Astron. Astrophys.*, 1997, **323**, 644–651.
22 N. Daugey, P. Caubet, B. Retail, M. Costes, A. Bergeat and G. Dorthe, *Phys. Chem. Chem. Phys.*, 2005, **7**, 2921–2927.
23 B. J. Opansky and S. R. Leone, *J. Phys. Chem.*, 1996, **100**, 19904–19910.
24 B. J. Opansky and S. R. Leone, *J. Phys. Chem.*, 1996, **100**, 4888–4892.
25 J. E. Murphy, A. B. Vakhtin and S. R. Leone, *Icarus*, 2003, **163**, 175–181.
26 C. Berteloite, S. D. Le Picard, P. Birza, M.-C. Gazeau, A. Canosa, Y. Bénilan and I. R. Sims, *Icarus*, 2008, **194**, 746–757.
27 S. Klippenstein, Y. Georgievskii and L. Harding, *Phys. Chem. Chem. Phys.*, 2006, **8**, 1133–1147.
28 D. L. Baulch, C. T. Bowman, C. J. Cobos, R. A. Cox, T. Just, J. A. Kerr, M. J. Pilling, D. Stocker, J. Troe, W. Tsang, R. W. Walker and J. Warnatz, *J. Phys. Chem. Ref. Data*, 2005, **34**, 757–1397.

Experimental measurements of low temperature rate coefficients for neutral–neutral reactions of interest for atmospheric chemistry of Titan, Pluto and Triton: Reactions of the CN radical

Sébastien B. Morales, Sébastien D. Le Picard,* André Canosa and Ian R. Sims

Received 12th March 2010, Accepted 22nd April 2010
DOI: 10.1039/c004219f

The kinetics of the reactions of cyano radical, CN ($X^2\Sigma^+$) with three hydrocarbons, propane ($CH_3CH_2CH_3$), propene ($CH_3CH=CH_2$) and 1-butyne ($CH\equiv CCH_2CH_3$) have been studied over the temperature range of 23–298 K using a CRESU (Cinétique de Réaction en Ecoulement Supersonique Uniforme or Reaction Kinetics in Uniform Supersonic Flow) apparatus combined with the pulsed laser photolysis–laser induced fluorescence technique. These reactions are of interest for the cold atmospheres of Titan, Pluto and Triton, as they might participate in the formation of nitrogen and carbon bearing molecules, including nitriles, that are thought to play an important role in the formation of hazes and biological molecules. All three reactions are rapid with rate coefficients in excess of 10^{-10} cm^3 molecule^{-1} s^{-1} at the lowest temperatures of this study and show behaviour characteristic of barrierless reactions. Temperature dependences, different for each reaction, are compared to those used in the most recent photochemical models of Titan's atmosphere.

1 Introduction

Titan, Pluto and Triton are the three bodies of the Solar System that are surrounded by a permanent atmosphere containing predominantly nitrogen, with methane in significant amounts.[1] The dissociation of these two major components is the starting point for a complex photochemistry. On Titan, and presumably in a similar way on Pluto and Triton, the N_2 bond is predominantly broken by dissociative ionization at high altitude following impact of magnetospheric electrons and solar EUV photons, forming $N(^2D)$, N and N^+ as final products.[2] Reactions involving these three atoms can lead to the formation of various nitriles, such as HCN. CH_4 on the other hand, is predominantly photodissociated by Lyman α photons at 121.6 nm,[2] forming essentially CH_3, $CH_2(a\ ^1A_1)$ and CH, with branching ratios that are still not well known, as reviewed recently by Romanzin et al.[3] Subsequent chemical reactions undergone by these primary products lead to the formation of many small hydrocarbons including C_2H_2, C_2H_4, C_2H_6, C_3H_4, C_3H_6 and C_4H_2[1] which have been observed on Titan during the recent Cassini–Huygens mission.[4] Carbon and nitrogen chemistries must therefore be coupled in these atmospheres and, according to the latest photochemical models,[5–8] this will lead to the formation of condensable species that are likely to participate to the formation of hazes as those observed for instance in

Institut de Physique de Rennes, Equipe Astrochimie Expérimentale, UMR 6251 du CNRS - Université de Rennes 1, Campus de Beaulieu, 35042 Rennes Cedex, France. E-mail: sebastien.le-picard@univ-rennes1.fr

Titan's atmosphere.[9] Among the species issuing from the carbon and nitrogen chemistries, nitriles have been the object of particular attention as they are thought to be precursors for the formation of amino acids, the building blocks of biological molecules.[10,11]

If little is still known about Pluto and Triton, our knowledge of Titan has increased significantly in the last few decades because of the success of space missions, such as Cassini–Huygens, and the improvement of Earth based observations and photochemical models. In order to model as accurately as possible the chemistry occurring in these environments, data on the numerous chemical and photochemical processes involved are needed, among which the rate coefficients of a large number of chemical reactions play a central role.[12] The kinetics of chemical reactions and their branching ratios can however, be very dependent on the physical conditions. While the pressure dependence of gas phase reactions is relatively well understood, the effect of temperature remains highly unpredictable. Accurate kinetics measurements therefore, performed under the relevant physical conditions, enable significant improvement in the predictability of the photochemical models.[12–14] Titan, Pluto and Triton for instance have cold atmospheres as their surface temperatures are 94 K, 35–50 K and 38 K respectively, while the dayside temperatures of their exospheres are 150 K for Titan, and 100 K for Pluto and Triton.[1] Reaction kinetics data measured at these low temperatures are required therefore, in order to model properly their atmospheric chemistry.

Among the primary species formed in the various processes involved in the photochemistry of these atmospheres rich in nitrogen and methane, the CN radical, originating from HCN photodissociation,[15] is likely to play an important role because of its high reactivity with a variety of hydrocarbons that has been demonstrated experimentally even at the very low temperatures prevailing in Titan, Pluto or Triton.[16–20] Almost all of the reactions of CN with hydrocarbons studied to date have been found to be rapid, with rate coefficients $\sim 10^{-10}$ cm^3 molecules^{-1} s^{-1}, corresponding to reaction on essentially every collision. Their temperature dependences however, show a wide variety of behaviours as will be illustrated by the present work.

In this paper, we present a kinetic study at temperatures down to 23 K of the reactions between the cyano radical, CN ($X^2\Sigma^+$), and three hydrocarbons, propane ($CH_3CH_2CH_3$), propene ($CH_3CH=CH_2$), both observed on Titan,[4] and 1-butyne ($CH\equiv CCH_2CH_3$), the presence of which is predicted by photochemical models of Titan.[6,7,21] Overall rate coefficients for these reactions have been measured using the CRESU (Cinétique de Réaction en Ecoulement Supersonique Uniforme) apparatus coupled with the Pumped Laser Photolysis–Laser Induced Fluorescence technique, which is currently the only experimental setup available which enables reaction kinetics measurements of neutral–neutral reactions at temperatures down to less than 10 K.[22] Temperature dependences for these three reactions are also compared to those used in the most recent photochemical models of Titan.

2 Experimental

In the CRESU technique, low temperatures are achieved by the isentropic expansion of a gas mixture, consisting predominantly (typically >98%) of a buffer gas (He, Ar or N_2), through an axisymmetric convergent–divergent Laval nozzle. The method and its adaptation to the study of neutral–neutral reactions have been described in detail elsewhere.[23] Briefly, a supersonic flow of gas is generated in which the Mach number, the temperature, the total density and the mole fraction of the components of the mixture are uniform along the flow. All these properties are conserved in the core of the flow over a distance of a few tens of centimetres corresponding to a few hundred microseconds. The relatively high density of the supersonic flow (typically, 10^{16}–10^{17} molecule cm^{-3}) ensures that frequent collisions maintain thermal equilibrium among the translational and rotational degrees of freedom of the flow constituents. In this study we used seven different Laval nozzles, each of them providing an axially uniform supersonic flow at a particular temperature and density for a given buffer gas. In addition, the rate coefficients were determined at room temperature (298 K) by increasing the pressure in the main chamber up to the same pressure of the reservoir of about 10 mbar; in this case, a subsonic flow is generated. Cyano radicals, CN, were produced coaxially to the flow by pulsed laser photolysis of cyanogen iodide (ICN; Aldrich, 95%) at 266 nm using the fourth harmonic of a Nd:YAG laser (Spectra Physics GCR 190, 10 Hz) at a fluence in the reaction zone of about 80 mJ cm^{-2}. A small flow of a few standard cm^3 min^{-1} of helium was passed over crystals of cyanogen iodide before entering the gas reservoir upstream of the Laval nozzle. We estimate the density of CN radicals introduced in the supersonic flow to be *ca.* 10^8 cm^{-3}. Buffer and reactant gases were taken directly from cylinders and regulated by means of separately calibrated Tylan mass flow controllers. The flow controllers were calibrated by standard pressure rise/drop techniques using an independently calibrated volume. Knowledge of the total gas density along the flow, achieved by means of impact pressure measurements using a Pitot tube, and of each gas flow rate enabled the calculation of the concentration of the reactants in the supersonic expansion. The purities of the gases used are given by the manufacturer (Air Liquide) as follows: argon (Ar) ≥ 99.997%, helium (He) and nitrogen (N$_2$) ≥ 99.995%, propane (C$_3$H$_8$) > 99.95%, propene (C$_3$H$_6$) > 99.5% and 1-butyne (C$_4$H$_6$) > 95%.

The photodissociation of cyanogen iodide at 266 nm produces cyano radicals that are overwhelmingly in the vibrational ground state ($v = 0$) of the $X^2\Sigma^+$ electronic ground state, but distributed over a wide range of rotational levels.[24] Cyano radicals were detected by laser induced fluorescence by exciting them in the (0,0) band of the $B^2\Sigma^+$–$X^2\Sigma^+$ system at about 388 nm using the doubled idler output of a tunable optical parametric oscillator (Spectra Physics MOPO) pumped by the 355 nm tripled output of another Nd:YAG laser (Spectra Physics GCR 230, 10 Hz). This probe laser beam propagated coaxially to the supersonic flow. Fluorescence in the (0,1) band at about 420 nm was collected at right angles *via* an appropriate combination of a ultraviolet (UV)-enhanced mirror and fused silica lenses onto the photocathode of a UV sensitive photomultiplier tube (EMI) through a narrow band interference filter centred at 420 nm with a 10 nm bandwidth (Ealing Optics). Decay traces, such as that shown in Fig. 1, were obtained by recording the laser induced fluorescence (LIF) intensity as the delay time between the photolysis and probe pulses was

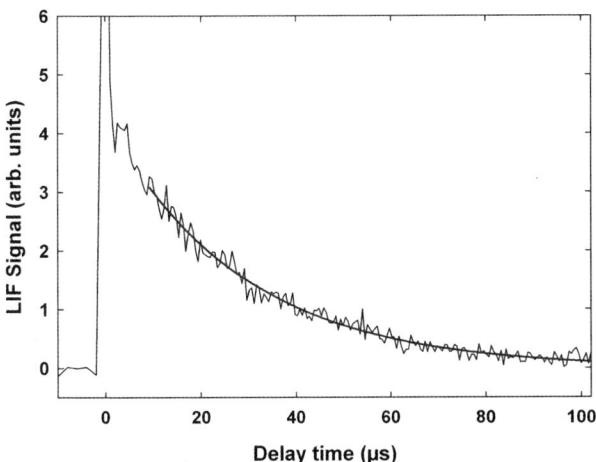

Fig. 1 Decay of CN ($B^2\Sigma^+$–$X^2\Sigma^+$) LIF signal at 23 K in the presence of propene ([C$_3$H$_6$] = 7.5 × 10^{13} molecule cm^{-3}) and He buffer ([He] = 4.73 × 10^{16} molecule cm^{-3}), fit to a single-exponential decay function.

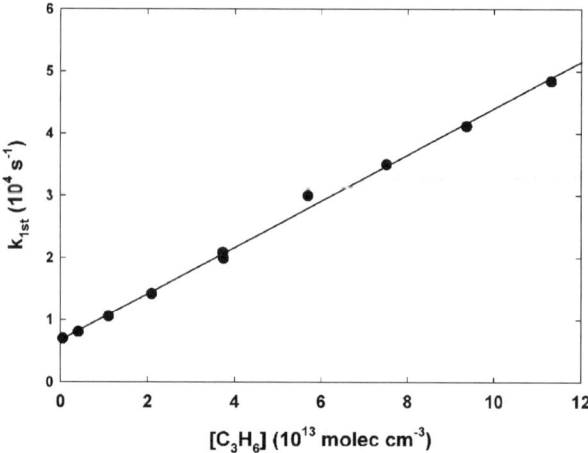

Fig. 2 Second order plot for the reaction of CN(X$^2\Sigma^+$) with propene, C$_3$H$_6$, at 23 K in He buffer, leading to a value for the second-order rate coefficient of $k = (3.73 \pm 0.40) \times 10^{-10}$ cm^3 molecule^{-1} s^{-1}.

systematically varied between zero and typically a few hundred microseconds. These traces of the LIF signal *versus* the time delay were fitted to single exponential functions, starting the fit at time delays long enough to allow for complete rotational relaxation, typically 20 μs. This procedure yielded pseudo-first-order rate coefficients, k_{1st}, related to the rate of loss of cyano radicals. Second order rate coefficients, k, were obtained by varying the concentration of reactants, propane, [C$_3$H$_8$], propene [C$_3$H$_6$], or 1-butyne, [C$_4$H$_6$], and then plotting k_{1st} against these concentrations as illustrated in Fig. 2 for the reaction of CN with propene at 23 K.

3 Results

Our experimental results for the reaction of CN(X$^2\Sigma^+$) with propane (C$_3$H$_8$), propene (C$_3$H$_6$), and 1-butyne (C$_4$H$_6$), are summarized in Table 1. For each measurement, the main flow conditions (buffer gas, total density, range of reactant gas density) are detailed, as well as the number of experimental points used in the second order plots. Quoted uncertainties include both statistical and systematic errors and are evaluated as the square root of the sum of the squares of these two independent errors. Statistical errors are calculated as the standard errors from the fit of the second order plot multiplied by the appropriate Student's t factor (depending upon the number of experimental points in the second order plot) for the 95% confidence limit. Systematic errors were more difficult to estimate and are likely to arise from inaccuracies in the calibration of the flow controllers, affecting the determination of the buffer and reactant densities. Every effort was made to minimize these systematic errors and we estimate that they do not exceed 10%. We also performed for each reaction studies of reactivity as a function of the total density. Measurements were performed at 52 K using two different nozzles functioning with argon as a buffer gas but with two different densities: 5.1×10^{16} molecule cm^{-3} and 1.03×10^{17} molecule cm^{-3}. At 298 K two different pressure conditions were used to obtain total densities of 10^{16} molecule cm^{-3} and 11.8×10^{16} molecule cm^{-3} for propane and propene, and 10^{16} molecule cm^{-3} and 10.2×10^{16} molecule cm^{-3} for 1-butyne (see Table 1). For both temperatures, 52 K and 298 K, rate coefficients measured at these different pressures were indistinguishable within the error bars. The lack of pressure dependence observed does not necessarily indicate that the reactions in question are bimolecular, as it can not be excluded that high pressure limits

Table 1 Rate coefficients and experimental conditions for the reaction between CN($X^2\Sigma^+$) and propane (C_3H_8), propene (C_3H_6) and 1-butyne (C_4H_6)

T/K	M	[M]/10^{16} molecule cm^{-3}	[Reactant]/10^{13} molecule cm^{-3}	No of Points	k/10^{-10} cm^3 molecule^{-1} s^{-1}
CN + propane					
23	He	4.73	0–7.7	8	2.90 ± 0.35^a
39	N$_2$	3.2	0.-2.3	12	2.99 ± 0.42
52	Ar	10.3	0.2– 9.6	11	1.62 ± 0.17
52	Ar	5.15	0.1–4.8	11	1.71 ± 0.19
83	N$_2$	4.88	0– 2.8	11	1.38 ± 0.16
123	He	12.7	0.1–21.9	15	0.93 ± 0.10
200	N$_2$	5.57	0– 39.3	11	0.77 ± 0.08
298	N$_2$	11.8	0.6– 49.9	11	0.76 ± 0.08
298	N$_2$	1	1.3–66.6	12	0.81 ± 0.08
CN + propene					
23	He	4.73	0–11.3	10	3.73 ± 0.40
39	N$_2$	3.2	0.-2.6	13	3.94 ± 0.53
52	Ar	10.3	1.1– 10.9	11	3.48 ± 0.38
52	Ar	5.15	0.1–5.4	10	3.71 ± 0.41
83	N$_2$	4.88	0.8– 3.6	7	3.50 ± 0.51
123	He	12.7	0.1–5.6	11	3.23 ± 0.36
200	N$_2$	5.57	0– 22.2	14	3.11 ± 0.32
298	N$_2$	11.8	0.5– 43.3	11	3.03 ± 0.31
298	N$_2$	1	1.5–14.5	10	3.12 ± 0.33
CN + 1-butyne					
23	He	4.73	0–5.5	9	4.60 ± 0.60
39	N$_2$	3.2	0.-1.2	12	4.09 ± 0.50
52	Ar	10.3	0.1– 2.1	7	3.59 ± 0.46
52	Ar	5.15	0.2–2.5	11	3.23 ± 0.48
83	N$_2$	4.88	0– 2.8	12	4.03 ± 0.49
120	He	12.7	0–7.8	8	4.46 ± 0.49
200	N$_2$	5.57	0.1–3.9	11	4.22 ± 0.45
298	N$_2$	10.2	0.5–35.9	14	3.94 ± 0.40
298	N$_2$	1	1–9.5	10	4.24 ± 0.44

^{a} quoted uncertainty (here and throughout the table) corresponds to statistical error (95% uncertainty) plus 10% likely systematic error.

of a termolecular reaction can be reached under the experimental conditions of this study, at least in the case of propene and 1-butyne, given the size of the potential collision complexes formed.

Temperature dependences of the rate coefficients measured in this study are displayed on a log–log plot of k (cm^3 molecule^{-1} s^{-1}) *versus* T/K over the temperature range 23–298 K in Fig. 3 to 5. Fig. 3 and 4 also display the results obtained by other authors for propane[25–27] and propene[16,28–31] and show a fair agreement with our results in the overlapping temperature range. In all three cases, the rate coefficients are high, around 10^{-10} cm^3 molecule^{-1} s^{-1}, the reactions with unsaturated hydrocarbons, C_3H_6 and C_4H_6 being faster however than with propane, the only alkane of this study. Consequently, there must be no barrier along the minimum energy path leading from reactants to products the potential energy surface(s) that correlate with reagents and products.

As can be seen in Fig. 3 to 5, the temperature dependences show different behaviours for each reaction. The rate coefficient of the reaction between CN and 1-butyne (Fig. 5) shows essentially no temperature dependence, with an average value of $(4.0 \pm 0.7) \times 10^{-10}$ cm^3 molecule^{-1} s^{-1}. The reaction between CN and propene

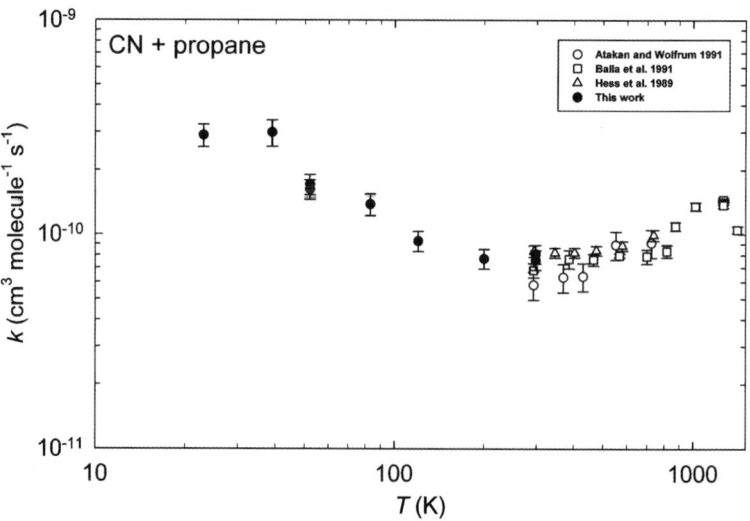

Fig. 3 Rate coefficients for the reaction of CN($X^2\Sigma^+$) with propane, C_3H_8, as a function of temperature, displayed on a log–log scale. The filled circles represent the experimental results obtained in this work with the CRESU apparatus. The open circles show the results of Atakan and Wolfrum[26] obtained between 294 and 1260 K. The open squares show the results of Balla et al.[27] obtained between 292 and 1500 K. The open triangles show the results of Hess et al.[25] obtained between 294 and 736 K.

(Fig. 4) shows a slightly negative temperature dependence that can be reproduced by the following equation:

$$k_{CN+C_3H_8}(T) = (3.06 \pm 0.14) \times 10^{-10} \left(\frac{T}{298\ K}\right)^{-0.094 \pm 0.030} \text{cm}^3\ \text{molecule}^{-1}\ \text{s}^{-1}$$

that we recommend to use only over the temperature range, 23–298 K, of the present experimental study. Finally, as can be seen on Fig. 3, the temperature dependence of the rate coefficient of the reaction with propane is more complex and will be discussed in the following section.

4 Discussion

Rate coefficients and temperature dependences

One of the most surprising results revealed by the first CRESU experiments performed in the early 1990s, has been to show that some reactions between a radical and a saturated molecule could apparently proceed by H-atom transfer.[16,23,32–34] Of particular interest was the kinetics study of the reaction between CN radical and ethane,[16] C_2H_6 which was shown to proceed via an H abstraction mechanism. CRESU measurements performed between 23 K and 295 K indeed, combined with results obtained by Balla et al.[27] at temperatures up to 1140 K showed a very unusual temperature dependence of the rate coefficient: it is high (>10^{-10} cm^3 molecule^{-1} s^{-1}) at 23 K, decreases monotonically until reaching a minimum value of ca. 8×10^{-11} cm^3 molecule^{-1} s^{-1} at 295 K, where it starts to increase monotonically to reach a value of ca. 1.5×10^{-10} cm^3 molecule^{-1} s^{-1} at about 1000 K.[16] This behaviour is very similar to the behaviour of the CN + propane reaction as shown in Fig. 3. It is likely therefore, that the mechanism is similar for these two reactions. In the case of CN + C_2H_6, Georgievskii and Klippenstein[35] were able to predict the temperature dependence using high level quantum chemical and transition state theory methods.[35] They explained this unusual behaviour of the rate coefficient of this reaction by a two transition state model.[36] At low temperatures, the dominant transition state lies at large separation (the outer transition state) where only long range interactions dominate, while at higher temperatures, the dominant transition state lies in the neighbourhood of the saddle point (the inner transition state). The zero point energy of the inner transition state is lower than in the outer one, but the energy levels are more widely spaced. In other words, the inner transition state is "tight" with respect to the outer, "loose" transition state. At high temperatures, the passage through the inner transition state is rate determining, whereas at low temperatures, the passage through the outer transition state becomes rate determining. The rate coefficient of the reaction depends therefore on passage through both transition states with the balance changing with temperature. A detailed theoretical kinetic study of the reaction between CN and propane using this two transition state model would be therefore very interesting, but it necessitates an *ab initio* calculation of the potential energy surface.

Reactions between CN and unsaturated hydrocarbons, that is species with double or triple bonds, have also been the object of several studies at low temperatures.[16,18,19,27-29,31,37] They are likely to proceed *via* the formation of a strongly bound adduct which may dissociate exothermically to form product species or undergo pressure dependent association. In many cases the rate coefficients rise to a limiting value as the temperature is lowered, and remain constant (or fall slightly) when the temperature is further lowered. The limiting value can be estimated on the basis of capture by a long-range potential arising from dispersion forces.[38] The expression for the capture rate coefficient is given by:[38]

$$k_{\text{capture}}(T) = 2\sqrt{\frac{2\pi}{\mu}}\Gamma\left(\frac{2}{3}\right)(2C_6)^{1/3}(k_\text{b}T)^{1/6}$$

where μ is the reduced mass, $\Gamma(x)$ the gamma function, C_6 is the dispersion coefficient and k_b is the Boltzmann constant. The positive temperature dependence $(T)^{1/6}$, given by this formula does not reproduce our measurements. This expression can be used however, to estimate the order of magnitude of the rate coefficient at very low temperatures. As recommended by Faure *et al.*,[39,40] we used the following expression of C_6:

$$C_6 = 2 \times C_6^{\text{London}} = 2 \times \frac{3}{2}\frac{I_1 I_2}{I_1 + I_2}\alpha_1\alpha_2$$

where C_6^{London} is a well-known approximation of the London formula, I_1 and I_2 are the ionisation potentials and α_1 and α_2 the polarizabilities of the reactants. Using ionisation energies[41] of 13.6 eV for CN, 9.73 eV for C_3H_6, 10.2 eV for 1-butyne, and polarizabilities[42] of 1.97×10^{-24} cm^3 for CN, 6.26×10^{-24} cm^3 for propane and 7.41×10^{-24} cm^3 for 1-butyne, we derived the following values for the rate coefficients at 20 K: $k_{\text{capture}}(20\text{ K}) = 3.8 \times 10^{-10}$ cm^3 molecule^{-1} s^{-1} for CN + propene and $k_{\text{capture}}(20\text{ K}) = 4.4 \times 10^{-10}$ cm^3 molecule^{-1} s^{-1} for CN + 1-butyne. These two values are very close to those obtained at 23 K in the present experimental work, indicating that the rates of both reactions are dominated by long range dispersion interactions.

It can be seen from Fig. 4 that for the reaction between CN and propene, Butterfield *et al.*'s results[28] for the rate coefficient at higher temperatures are lower than might be expected, and that these results are also in disagreement with the high temperature study of Herbert *et al.*[29] Butterfield *et al.*'s article[28] does not specify the pressure at which these kinetic studies were carried out, and so it is difficult to comment further on possible reasons for these discrepancies.

Interestingly, very recently we measured the rate coefficient of the reaction of $CN(X^2\Sigma^+)$ with 1,3-butadiene, an isomer of 1-butyne, at temperatures from 298 K down to 23 K. The value of the rate coefficient at 298 K was found to be very close

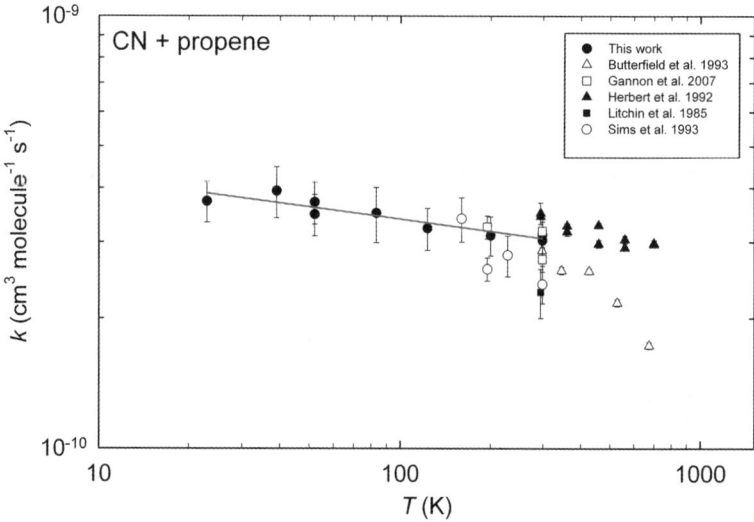

Fig. 4 Rate coefficients for the reaction of CN($X^2\Sigma^+$) with propene, C_3H_6, as a function of temperature, displayed on a log–log scale. The filled circles represent the experimental results obtained in this work with the CRESU apparatus. The open triangles show the results of Butterfield et al.[28] obtained between 297 and 673 K. The open squares show the results of Balla et al.[27] obtained between 292 and 1500 K. The open triangles show the results of Hess et al.[25] obtained between 294 and 736 K. The open squares show the results of Gannon et al.[31] at 195 and 298 K. The black triangles show the results of Herbert et al.[29] obtained between 294 and 698 K. The black square shows the result of Litchin et al.[30] at 294 K. The open circles show the results of Sims et al. between 160 and 298 K. The solid line represents the fit of the present CRESU data with the equation:

$$k_{CN+C_3H_6}(T) = (3.06 \pm 0.14) \times 10^{-10} \left(\frac{T}{298 \text{ K}}\right)^{-0.094 \pm 0.030} \text{ cm}^3 \text{ molecule}^{-1} \text{ s}^{-1}.$$

to that obtained here for reaction with 1-butyne. The temperature dependence was found, however, to be slightly positive, following an Arrhenius behaviour.[19] This example illustrates the variety of temperature dependences that can be observed, even for reactions involving the same radical with isomeric reactants.

Products of reactions

It is worth noting that our kinetics measurements are obtained from the total rate of removal of CN radical and do not provide any information on the products of the reactions studied. As mentioned above, the reaction of CN with propane is likely to proceed *via* hydrogen abstraction, C_3H_7 and H being the products of this reaction. For reaction with propene and 1-butyne, the dissociation of the strongly bound adduct formed during the reaction can give rise to various products, as long as these various channels are exothermic, given the low temperatures involved in these experiments. In recent years Leone and co-workers[17,37,43–45] at Berkeley and Sandia (USA), and Pilling and Seakins and co-workers[31,46] at the University of Leeds (UK) have determined branching ratios of some reactions of interest for combustion and planetary atmospheres, almost exclusively at room temperature. Very recently for instance, Gannon et al.[31] have performed H atom yield measurements at 298 K and 195 K for reactions involving CN and various unsaturated hydrocarbons, including propene. In these experiments, H atom product yield is determined by using a reference reaction, $CN + H_2$, giving 100% H atoms. Comparison of the VUV laser induced fluorescence of H atoms for this test reaction to the signal obtained for a given reaction in the same experimental conditions, enables the

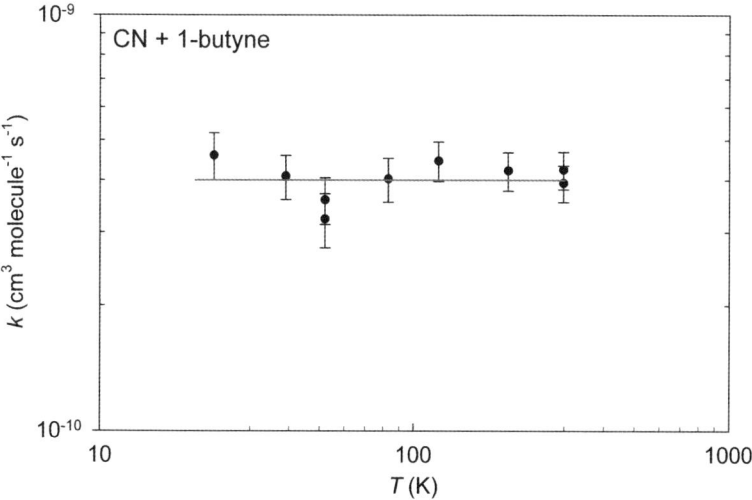

Fig. 5 Rate coefficients for the reaction of $CN(X^2\Sigma^+)$ with 1-butyne, C_4H_6, as a function of temperature, displayed with a log–log scale. The solid line represents the fit of the data with a constant value equals to the average of the present results: $k_{CN+C_4H_6} = (4.0 \pm 0.7) \times 10^{-10}$ cm^3 molecule^{-1} s^{-1}.

determination of its branching ratio, *i.e.* the percentage of the overall reaction proceeding *via* the channel producing H atoms. The three exothermic channels accessible at the temperatures of this study are:

$$CN + C_3H_6 \rightarrow C_4H_5N \text{ (and isomers)} + H$$

$$CN + C_3H_6 \rightarrow C_3H_3N + CH_3, \Delta_r H^\circ_{298\ K} = -130 \text{ kJ mol}^{-1}\ ^{47}$$

$$CN + C_3H_6 \rightarrow C_3H_5 + HCN, \Delta_r H^\circ_{298\ K} = -149 \text{ kJ mol}^{-1}\ ^{47}$$

Interestingly, in the case of CN + C_3H_6, H atom production was found by Gannon *et al.*[31] to be pressure dependent at room temperature, with a H atom yield of 0.5 at the low pressure (2 Torr) that decreases to approach a value corresponding to the limit of detection of the experiment at 120 Torr, as the fluorescence signal disappeared. Complementary Master Equation calculations were also performed, indicating that at low temperatures, H atom and CH_3 production were both significant and pressure dependent channels. More recently, Trevitt *et al.*[37] performed product detection of the reaction of CN with propene at room temperature using tuneable vacuum ultraviolet (VUV) synchrotron photoinization coupled with time resolved multiplexed mass spectrometry. This technique also enables the identification of specific product isomers. These results confirmed the two dominant channels to be H and CH_3 elimination. 1-Cyanopropene and 2-cyanopropene were found to be the products of H elimination, whereas only cyanoethene was formed with CH_3 elimination. From this study Trevitt *et al.*[37] determined the following branching ratios at room temperature: 75($\pm$15)% cyanoethene + CH_3, and 25($\pm$15)% to the C_4H_5N + H channel with the C_4H_5N signal found to comprise 57($\pm$15)% 1-cyanopropene, 43($\pm$15)% 2-cyanopropene and <15% 3-cyanopropene.

The ratio of H atom production is found to be significantly smaller than the one derived by Gannon et al.[31] Furthermore, no pressure dependence of the H atom production was found over the range 1–6 torr in this study, which is in disagreement with the measurements of Gannon et al.[31] These contradictory results illustrate the difficulty in determining branching ratios accurately, and show the need for a combination of state of the art experiments and high level quantum calculations.

There have, to our knowledge, been no studies of product formation at low temperature for the reaction between CN and 1-butyne and no quantitative branching ratio measurements performed at any temperature. About ten years ago, Chastaing et al.[48] pointed out the correlation between the rate of addition of C_2H radical to an unsaturated hydrocarbon, and the ionisation energy of this unsaturated hydrocarbon. Addition of two co-reactants is promoted by electron withdrawal, by the approaching radical having a high electronic affinity, and will be easier, therefore the lower the ionisation energy of the unsaturated hydrocarbon.[49] Consequently, the rate coefficient for a reaction involving a radical such as CN and an unsaturated hydrocarbon is expected to be larger for reactions involving higher alkenes or alkynes. Carty et al.,[18] using a CRESU apparatus, performed a kinetics study similar to this work at temperatures from 295 K down to 15 K of the reactions of CN with two unsaturated hydrocarbons, the isomeric allene, CH_2=C=CH_2 and methyacetylene, $CH_3C{\equiv}CH$. For both reactions the rate constants were found to be independent of the temperature from 15 K to 295 K, with an identical average value of $(4.1 \pm 0.5) \times 10^{-10}$ cm^3 $molecule^{-1}$ s^{-1} very close to the value obtained in this work for CN with $CH{\equiv}CCH_2CH_3$. 1-Butyne however, contains one atomic carbon more. As can be seen in Table 2 however, it seems that when very high values (here $>3 \times 10^{-10}$ cm^3 $molecule^{-1}$ s^{-1}) of the rate coefficients have been reached, the value of the ionization energy ceases to influence the rate of the reaction as the rate coefficient is essentially driven by the rate of capture as seen from the estimation of the capture rate coefficient presented in the previous section for CN + propene and CN + 1-butyne.

More recently, Smith and co-workers[50,51] have correlated the height of the barrier corresponding to the inner transition state in radical–molecule reactions to the difference (I.P. − E.A.) between the ionisation potential of the hydrocarbon molecular co-reactant (I.P.) and the electron affinity (E.A.) of the radical reagent, proposing a threshold value of 8.75 eV for this difference, below which the reaction is likely not to possess any absolute barrier and to remain rapid down to very low temperatures. Employing a value for the E.A. of CN of 3.86 eV and I.P. values taken from the NIST Chemistry Webbook[42] shown in Table 2, the values of (I.P − E.A.) for the reactions for CN + propene and CN + 1-butyne are 5.86 eV and 6.32 eV, respectively. This corresponds well with the experimental results reported here, where both reactions display behaviour characteristic of barrierless reactions as described above.

Table 2 Comparison of rate coefficients for reactions between CN and various alkenes and alkynes at $T \approx 25$ K. Ionisation energies of the unsaturated hydrocarbons are also listed

	Rate constant/10^{-10} cm^3 $molecule^{-1}$ s^{-1}	Ionisation energy/eV[42]	Ref.
HC≡CH	4.6 ± 0.2	11.4	Sims et al.[16]
H_2C=CH_2	4.4 ± 0.2	10.51	Sims et al.[16]
$CH_3C{\equiv}CH$	3.6 ± 0.7	10.36	Carty et al.[18]
$CH{\equiv}CCH_2CH_3$	4.6 ± 0.6	10.18	This work
CH_3CH=CH_2	3.7 ± 0.1	9.73	This work
CH_2=C=CH_2	3.9 ± 0.6	9.69	Carty et al.[18]

Photochemical models

The experimental results obtained in this work have some potential importance for the chemistry of Titan, Pluto and Triton, as they all have atmospheres rich in nitrogen (N_2) and methane (CH_4), and these rate constants should be included in their photochemical models. For reactions of CN with unsaturated hydrocarbons for instance, it is likely that the multiple bond of the hydrocarbon remains in the nitrile compounds formed during the reaction. Further photochemical reactions therefore, should lead to the formation of more complex species, among which are potential precursors of the haze present on Titan or Pluto,[1,4,52,53] as well as intermediate species necessary for the formation of biological molecules.[11] In the photochemical models of Lavvas et al.[6] and of Wilson and Atreya,[7] the reactions studied in this work are included, and we compare below the rate coefficients used in these models to our experimental kinetic results. For the reaction between CN and C_3H_8, both models cite the work by Hess et al.[25] obtained between 298 and 736 K but use very different temperature dependences of the rate coefficient. As shown in Fig. 6, the rate coefficients used by Lavvas et al. ($k(T) = 3.58 \times 10^{-15} (T/K)^{1.14} \exp(284\,K/T)$ cm^3 molecule^{-1} s^{-1}) are about a factor of ten lower than those obtained in this work for the temperatures prevailing on Titan. The constant value used by Wilson and Atreya (8.34×10^{-11} cm^3 molecule^{-1} s^{-1} at any temperature) is closer to our results, but is lower by a factor of two at the lowest temperatures of Titan's atmosphere. For the rate coefficient of the reaction CN + C_3H_6, Lavvas et al.[6] used the measurements of Sims et al.[16] obtained in a tubular reactor cryogenically cooled down to 160 K. Results of this study performed at temperatures from 160 K to 298 K are shown in Fig. 4 (open circles). Lavvas et al.[6] as well as Wilson and Atreya[7] fitted these experimental values with the following equation: $k(T) = 1.73 \times 10^{-10} \exp(101\,K/T$ cm^3 molecule^{-1} s^{-1}) and presumably used this equation at the lower temperatures prevailing in Titan's atmosphere. Unfortunately, the present results show a different temperature dependence, as displayed on Fig. 7, illustrating the risk of extrapolating experimental results obtained even over a quite small range of temperatures. Finally, for the reaction between CN and C_4H_6, Lavvas et al. cited the high temperature measurements for the butadiene isomer of C_4H_6 of Butterfield

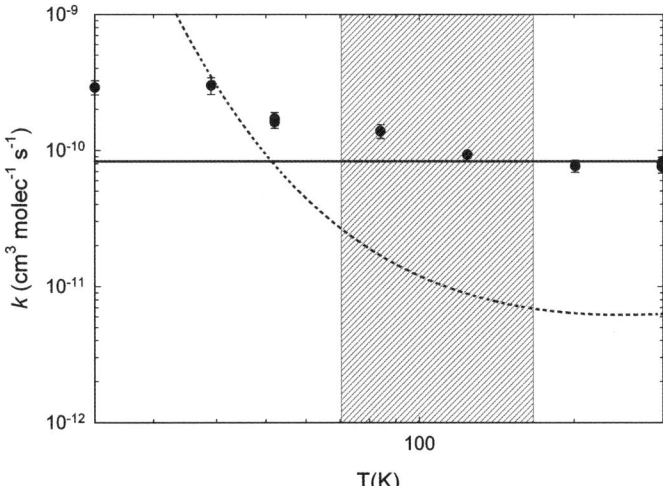

Fig. 6 Comparison of the rate coefficients for the reaction of CN($X^2\Sigma^+$) with propane, C_3H_8, obtained in this work (filled circles), with the fit used by Lavvas et al.[6] (dashed line) and Wilson et al.[7] (solid line), see text. The hatched area indicates the temperature range prevailing on Titan's atmosphere (∼70–170 K).

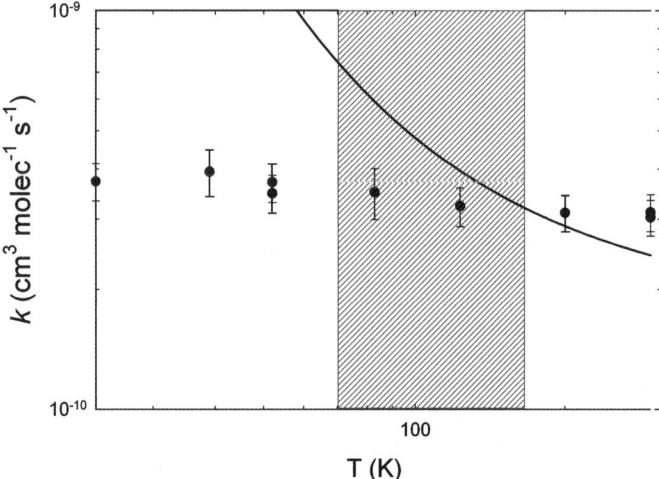

Fig. 7 Comparison of the rate coefficients for the reaction of CN($X^2\Sigma^+$) with propene, C_3H_6, obtained in this work (filled circles), with the identical fit used by Lavvas et al.[6] and Wilson et al.[7] (see text). The hatched area indicates the temperature range prevailing on Titan's atmosphere ($\sim$70–170 K).

et al.[28] and used a temperature dependence that slightly differed[6] ($k(T) = 2.57 \times 10^{-10} \exp(-171\ K/T)$ cm^3 molecule^{-1} s^{-1}) from the equation given by the fit of the authors:[28] $k(T) = 10^{-9.59} \exp(-169.2\ K/T)$ cm^3 molecule^{-1} s^{-1}, giving rate coefficients about ten times lower than those measured in this work for the 1-butyne isomer, as shown on Fig. 8. It is worth noting however, that as mentioned above for the reaction CN + propene, the temperature dependence found by Butterfield et al.[28] for CN + 1,3-butadiene is not in line with very recent results we obtained for the reaction of CN with butadiene between 298 K and 23 K.[19] While for that reaction there is no

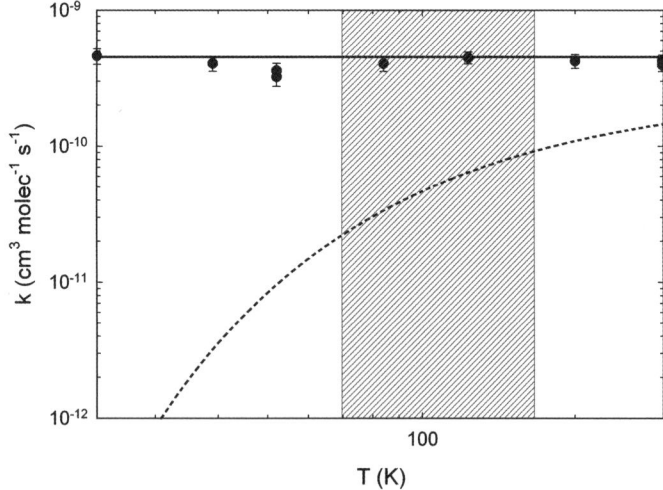

Fig. 8 Comparison of the rate coefficients for the reaction of CN($X^2\Sigma^+$) with 1-butyne, C_4H_6, obtained in this work (filled circles), with the fit by Lavvas et al.[6] (dashed line) and Wilson et al.[7] (solid line), see text. The hatched area indicates the temperature range prevailing on Titan's atmosphere ($\sim$70–170 K).

overlap in temperature range (Butterfield *et al.*'s measurements cover from 297 to 740 K while our measurements cover the range from 23 to 298 K), the very mild positive temperature dependence that we found is in marked contrast to the strong fall in the rate coefficient at higher temperatures reported by Butterfield *et al.*[28] Wilson and Atreya[7] on the other hand used the value obtained by Butterfield *et al.*[28] at 295 K for CN + butadiene, $k = 4.52 \times 10^{-10}$ cm^3 molecule^{-1} s^{-1}, which is very close to the value we obtained here for CN + 1-butyne, with no temperature dependence. As can be seen on Fig. 8, the values used by Wilson and Atreya[7] are very close to our experimental measurements for CN + 1-butyne. Finally, it should be mentioned that in both photochemical models, for the reactions of this study, in some cases the rate coefficients were assigned to only one exit channel of the reactions, whereas in other cases they were considered as overall values. The virtual absence of any insights concerning the expected products for most of these reactions, especially at the low temperatures prevailing in Titan's atmosphere, makes the choice of favouring some channels very risky.[12,14,54,55]

Finally we note that in the most recent photochemical model of Titan, by Krasnopolsky,[21] none of the reactions presented in this work are included in the chemical reaction scheme proposed.

Conclusion and perspectives

We have measured the rate coefficients for the reaction of CN radical with propane ($CH_3CH_2CH_3$), propene ($CH_3CH=CH_2$) and 1-butyne ($CH{\equiv}CCH_2CH_3$), at temperatures between 23 and 298 K. These reactions are of interest for the atmospheres of Titan, Pluto and Triton, containing predominantly nitrogen and a significant amount of methane, as they might participate in the formation of nitrogen and carbon bearing molecules, including nitriles, that are thought to play an important role in the formation of hazes and biological molecules. All three reactions are rapid with rate coefficients in excess of 10^{-10} cm^3 molecule^{-1} s^{-1} at the lowest temperatures of this study and show a behaviour characteristic of barrierless reactions. Temperature dependences were found however, to be different for each reaction: the reaction of CN with 1-butyne does not show any significant temperature dependence, the reaction of CN with propene displays a slight positive temperature dependence, whereas the reaction between CN and propane shows a remarkable temperature dependence when combined with results obtained at higher temperatures.[25–27] The rate coefficient shows a minimum ($\sim 0.80 \times 10^{-10}$ cm^3 molecule^{-1} s^{-1}) around 250 K, reaching values in excess of 10^{-10} cm^3 molecule^{-1} s^{-1} at high temperatures (~ 1000 K) and at the lowest temperatures of this study. The values obtained here have been compared with those used in the most recent models and this showed that in some cases, the difference can reach an order of magnitude. It should be noted, however, that in the recent years, because more experimental results have become available, and also because of more frequent meetings between kineticists and modellers, often initiated by multidisciplinary research networks, significant progress has been achieved in terms of the rate coefficient values used in photochemical models.

Much remains to be accomplished in this area. First, it is obvious that we will not be able to study experimentally at low temperature all the reactions involved in the chemical schemes of cold planetary atmospheres. For instance, less than 10% of the chemical reaction rate coefficients used in recent photochemical models of Titan have been measured below 150 K.[14] With about a hundred reactions studied using the CRESU apparatus however, considerable progress has been made in the understanding of the mechanisms that control the rates of these reactions,[51] a prerequisite to predicting the rate coefficients for other reactions that might play a role in the chemistry of a given atmosphere. Several authors have pointed out that imprecisions in reaction and photolysis rates used in photochemical models introduce significant uncertainties on computed concentrations. Early works on this topic were devoted to the Earth's atmosphere,[56–59] and in recent years, Michel Dobrijevic, Pascal Pernot and co-workers have improved and extended this work to the photochemistry of other planets, such as Saturn,[13] Neptune[60] and Titan.[12,55,61] These authors worked hard to first give a proper definition of what is a *a key reaction* in a photochemical model:[14,62] "it is a reaction that has a strong influence on the uncertainty of the model outputs";[14] and second to determine the influence of uncertainties in rate coefficients on the outputs of photochemical models, enabling the identification of reactions that should be measured at low temperature with high accuracy.[12,54,63] It is worth noting that comparable work has been performed for Titan's ionosphere.[61,64] The recent work by Hébrard[12] *et al.* performed on Titan, illustrates the difficulty of low temperature extrapolation of reaction rates and was based on reactions already studied at low temperatures. It does not include therefore the reactions of this kinetic study. Hébrard *et al.*[12] and more recently Dobrijevic *et al.*[14] have also made a first attempt to give a list of key reactions that includes essentially chemical reactions responsible for the formation of hydrocarbon molecules. Little has been done so far, on the nitrogen chemistry or on the chemistry coupling nitrogen and hydrocarbon compounds. It is difficult therefore, to give any insight on the influence or sensitivity of the three reactions studied here on the photochemistry of Titan.

Maybe the most challenging work for the future will be to determine the branching ratios of reactions, that is the proportion of each open output channel, at the low temperatures of these atmospheres. Currently, two kinds of methods have been used to tackle this problem. Crossed-beam experiments, where reactions are studied under single collision conditions, gives insights into the mechanism of a reaction at the microscopic level and identifies the primary products. These experiments are usually performed at collision energies varying from a few kJ mol^{-1} to a few tens of kJ mol^{-1}, corresponding to temperatures higher or much higher than those of cold planetary atmospheres. It is worth noting, however, that some efforts have been made in recent years to reduce the collision energies in some crossed beam apparatuses.[65,66] Moreover, they cannot easily determine the temperature dependences of the rate coefficients and have some difficulty in identifying kinetically less-favourable channels. Furthermore, they are not able to determine absolute values of the reactive cross sections. They have to be combined therefore, with kinetics experiments at low temperatures and electronic structure calculations to give some information on the potential products formed in the conditions of cold atmospheres. However, uncertainties in electronic structure calculations are typically in excess of 1 kJ mol^{-1} (corresponding to 120 K) which precludes any prediction of the branching ratio at low temperature, as the presence of even very small barriers would prevent the formation of some products of reaction at low temperature. Alternatively, as described above, Leone and co-workers at Berkeley and Sandia,[17,37,43–45] and Pilling and Seakins and co-workers at Leeds,[31,46] have undertaken different experiments to identify and quantify the formation of some products of reactions of importance for combustion or atmospheric chemistry. Only one of these experiments has been, however, performed below room temperature (195 K),[31,46] and both groups find different results for the only common reaction studied, between CN and propene at room temperature,[31,37] underlining the difficulty of such measurements that also need to be coupled to state of the art quantum calculations. Recently, the Rennes group has developed a new pulsed CRESU system,[67] that should enable us to perform experiments on branching ratio determination at temperatures down to 20 K over the next few years.

The importance of the potential role of the ion chemistry in the formation of polycyclic aromatics hydrocarbons (PAHs), nitriles and other complex molecules as suggested by the recent data from the Ion Neutral Mass Spectrometer (INMS) aboard the Cassini spacecraft[68] has been highlighted by a number of authors.[68–72] Only a small fraction of the gases present in the atmospheres, however, are ionized by solar radiation or charged particle impact, and neutral species potentially formed in the ionosphere have to migrate to the deep atmosphere to explain the formation of complex molecules and hazes by these processes in Titan's atmosphere, for instance. Photochemical models need therefore to couple ion–molecule chemistry with the nitrogen and carbon chemistries of the neutral atmospheres.

Finally, chemical models currently used are largely inspired by combustion models, and we suggest that other schemes leading to the formation of complex molecules should be considered, as low temperature environments are more sensitive to the presence of any small energy barrier along the minimum energy path leading from reactants to products. Recently for instance, combining kinetics experiments at low temperatures, crossed beam experiments and electronic structure calculations, we have for the first time provided evidence that the aromatic pyridine molecule (C_5H_5N; X^1A_1) can be synthesized in low temperature environments under single collision conditions of CN with 1,3-butadiene.[19]

Acknowledgements

We are grateful for support from the U.S. National Science Foundation "Collaborative Research in Chemistry Program" (NSF-CRC; CHE-0627854), the French "Programme National de Planétologie", the *Région de Bretagne* and *Rennes Métropole*. V. Wakelam is warmly thanked for promoting fruitful discussions between experimentalists and modellers in the frame of the KInetics Database for Astrochemistry (KIDA) project.

References

1 I. C. F. Mueller-Wodarg, D. F. Strobel, J. I. Moses, J. H. Waite, J. Crovisier, R. V. Yelle, S. W. Bougher and R. G. Roble, *Space Sci. Rev.*, 2008, **139**, 191–234.
2 D. F. Strobel, *Space Sci. Rev.*, 2005, **116**, 155–170.
3 C. Romanzin, Y. Benilan, A. Jolly and M. C. Gazeau, *Adv. Space Res.*, 2008, **42**, 2036–2044.
4 S. Vinatier, B. Bezard, C. A. Nixon, A. Mamoutkine, R. C. Carlson, D. E. Jennings, E. A. Guandique, N. A. Teanby, G. L. Bjoraker, F. M. Flasar and V. G. Kunde, *Icarus*, 2010, **205**, 559–570.
5 V. Krasnopolsky and D. P. Cruikshank, *J. Geophys. Res.*, 1995, **100**, 21271–21286.
6 P. P. Lavvas, A. Coustenis and I. M. Vardavas, *Planet. Space Sci.*, 2008, **56**, 27–66.
7 E. H. Wilson and S. K. Atreya, *J. Geophys. Res. [Planets]*, 2004, **109**.
8 P. P. Lavvas, A. Coustenis and I. M. Vardavas, *Planet. Space Sci.*, 2008, **56**, 67–99.
9 C. P. McKay, *Planet. Space Sci.*, 1996, **44**, 741–747.
10 A. Eschenmoser and E. Loewenthal, *Chem. Soc. Rev.*, 1992, **21**, 1–16.
11 N. Balucani, *Int. J. Mol. Sci.*, 2009, **10**, 2304–2335.
12 E. Hébrard, M. Dobrijevic, P. Pernot, N. Carrasco, A. Bergeat, K. M. Hickson, A. Canosa, S. D. Le Picard and I. R. Sims, *J. Phys. Chem. A*, 2009, **113**, 11227–11237.
13 M. Dobrijevic, J. L. Ollivier, F. Billebaud, J. Brillet and J. P. Parisot, *Astron. Astrophys.*, 2003, **398**, 335–344.
14 M. Dobrijevic, E. Hébrard, S. Plessis, N. Carrasco, P. Pernot and M. Bruno-Claeys, *Adv. Space Res.*, 2010, **45**, 77–91.
15 S. Lebonnois, E. L. O. Bakes and C. P. McKay, *Icarus*, 2002, **159**, 505–517.
16 I. R. Sims, J. L. Queffelec, D. Travers, B. R. Rowe, L. B. Herbert, J. Karthauser and I. W. M. Smith, *Chem. Phys. Lett.*, 1993, **211**, 461–468.
17 A. J. Trevitt, F. Goulay, C. A. Taatjes, D. L. Osborn and S. R. Leone, *J. Phys. Chem. A*, 2010, **114**, 1749–1755.
18 D. Carty, V. Le Page, I. R. Sims and I. W. M. Smith, *Chem. Phys. Lett.*, 2001, **344**, 310–316.
19 S. B. Morales, C. J. Bennett, S. D. Le Picard, A. Canosa, I. R. Sims, B. J. Sun, P. H. Chen, A. H. H. Chang, A. M. Mebel, X. Gu, F. Zhang and R. I. Kaiser, 2010, submitted.
20 C. J. Bennett, S. B. Morales, S. D. Le Picard, A. Canosa, I. R. Sims, Y. H. Shih, A. H. H. Chang, X. Gu, F. Zhang and R. I. Kaiser, *Phys. Chem. Chem. Phys*, 2010, **12**(31), 8737–8749.
21 V. A. Krasnopolsky, *Icarus*, 2009, **201**, 226–256.
22 C. Berteloite, M. Lara, A. Bergeat, S. D. Le Picard, F. Dayou, K. M. Hickson, A. Canosa, C. Naulin, J.-M. Launay, I. R. Sims and M. Costes, 2010, submitted.
23 I. R. Sims, J. L. Queffelec, A. Defrance, C. Rebrion-Rowe, D. Travers, P. Bocherel, B. R. Rowe and I. W. M. Smith, *J. Chem. Phys.*, 1994, **100**, 4229–4241.

24 A. P. Baronavski and J. R. McDonald, *Chem. Phys. Lett.*, 1977, **45**, 172–176.
25 W. P. Hess, J. L. Durant and F. P. Tully, *J. Phys. Chem.*, 1989, **93**, 6402–6407.
26 B. Atakan and J. Wolfrum, *Chem. Phys. Lett.*, 1991, **186**, 547–552.
27 R. J. Balla, K. H. Casleton, J. S. Adams and L. Pasternack, *J. Phys. Chem.*, 1991, **95**, 8694–8701.
28 M. T. Butterfield, T. Yu and M. C. Lin, *Chem. Phys.*, 1993, **169**, 129–134.
29 L. Herbert, I. W. M. Smith and R. D. Spencersmith, *Int. J. Chem. Kinet.*, 1992, **24**, 791–802.
30 D. A. Lichtin and M. C. Lin, *Chem. Phys.*, 1985, **96**, 473–482.
31 K. L. Gannon, D. R. Glowacki, M. A. Blitz, K. J. Hughes, M. J. Pilling and P. W. Seakins, *J. Phys. Chem. A*, 2007, **111**, 6679–6692.
32 C. Berteloite, S. D. Le Picard, P. Birza, M. C. Gazeau, A. Canosa, Y. Benilan and I. R. Sims, *Icarus*, 2008, **194**, 746–755.
33 I. R. Sims, I. W. M. Smith, D. C. Clary, P. Bocherel and B. R. Rowe, *J. Chem. Phys.*, 1994, **101**, 1748–1751.
34 A. Canosa, I. R. Sims, D. Travers, I. W. M. Smith and B. R. Rowe, *Astron. Astrophys.*, 1997, **323**, 644–651.
35 Y. Georgievskii and S. J. Klippenstein, *J. Phys. Chem. A*, 2007, **111**, 3802–3811.
36 E. E. Greenwald, S. W. North, Y. Georgievskii and S. J. Klippenstein, *J. Phys. Chem. A*, 2005, **109**, 6031–6044.
37 A. J. Trevitt, F. Goulay, G. Meloni, D. L. Osborn, C. A. Taatjes and S. R. Leone, *Int. J. Mass Spectrom.*, 2009, **280**, 113–118.
38 T. Stoecklin, C. E. Dateo and D. C. Clary, *J. Chem. Soc., Faraday Trans.*, 1991, **87**, 1667–1679.
39 A. Faure, V. Vuitton, R. Thissen and L. Wiesenfeld, *J. Phys. Chem. A*, 2009, **113**, 13694–13699.
40 A. Faure, C. Rist and P. Valiron, *Chem. Phys.*, 1999, **241**, 29–42.
41 S. G. Lias, *Ionization Energy Evaluation*, http://webbook.nist.gov, accessed March 2010, 2010.
42 H. Y. Afeefy, J. F. Liebman and S. E. Stein, *Neutral Thermochemical Data*, http://webbook.nist.gov, accessed March 2010, 2010.
43 F. Goulay, D. L. Osborn, C. A. Taatjes, P. Zou, G. Meloni and S. R. Leone, *Phys. Chem. Chem. Phys.*, 2007, **9**, 4291–4300.
44 F. Goulay, A. J. Trevitt, G. Meloni, T. M. Selby, D. L. Osborn, C. A. Taatjes, L. Vereecken and S. R. Leone, *J. Am. Chem. Soc.*, 2009, **131**, 993–1005.
45 D. L. Osborn, P. Zou, H. Johnsen, C. C. Hayden, C. A. Taatjes, V. D. Knyazev, S. W. North, D. S. Peterka, M. Ahmed and S. R. Leone, *Rev. Sci. Instrum.*, 2008, **79**, 104103.
46 K. L. Gannon, M. A. Blitz, M. J. Pilling, P. W. Seakins, S. J. Klippenstein and L. B. Harding, *J. Phys. Chem. A*, 2008, **112**, 9575–9583.
47 NIST Chemistry WebBook, Standard Reference, Database Number 69, National Institute of Standards and Technology, ed. P. J. Linstrom and W. G. Mallard, Gaithersburg, MD, p. 20899.
48 D. Chastaing, P. L. James, I. R. Sims and I. W. M. Smith, *Faraday Discuss.*, 1998, **109**, 165–181.
49 E. Martinez, B. Cabanas, A. Aranda, J. Albaladejo and R. P. Wayne, *J. Chem. Soc., Faraday Trans.*, 1997, **93**, 2043–2047.
50 H. Sabbah, L. Biennier, I. R. Sims, Y. Georgievskii, S. J. Klippenstein and I. W. M. Smith, *Science*, 2007, **317**, 102–105.
51 I. W. M. Smith, A. M. Sage, N. M. Donahue, E. Herbst and D. Quan, *Faraday Discuss.*, 2006, **133**, 137–156.
52 L. A. Young, S. A. Stern, H. A. Weaver, F. Bagenal, R. P. Binzel, B. Buratti, A. F. Cheng, D. Cruikshank, G. R. Gladstone, W. M. Grundy, D. P. Hinson, M. Horanyi, D. E. Jennings, I. R. Linscott, D. J. McComas, W. B. McKinnon, R. McNutt, J. M. Moore, S. Murchie, C. B. Olkin, C. C. Porco, H. Reitsema, D. C. Reuter, J. R. Spencer, D. C. Slater, D. Strobel, M. E. Summers and G. L. Tyler, *Space Sci. Rev.*, 2008, **140**, 93–127.
53 D. F. Strobel, X. Zhu, M. E. Summers and M. H. Stevens, *Icarus*, 1996, **120**, 266–289.
54 E. Hébrard, M. Dobrijevic, Y. Benilan and F. Raulin, *Planet. Space Sci.*, 2007, **55**, 1470–1489.
55 E. Hébrard, M. Dobrijevic, Y. Benilan and F. Raulin, *J. Photochem. Photobiol., C*, 2006, **7**, 211–230.
56 R. D. Rundel, D. M. Butler and R. S. Stolarski, *J. Geophys. Res.*, 1978, **83**, 3063–3073.
57 R. S. Stolarski, D. M. Butler and R. D. Rundel, *J. Geophys. Res.*, 1978, **83**, 3074–3078.
58 A. M. Thompson and R. W. Stewart, *J. Geophys. Res.*, 1991, **96**, 13089–13108.
59 R. W. Stewart and A. M. Thompson, *J. Geophys. Res.*, 1996, **101**, 20953–20964.

60 M. Dobrijevic and J. P. Parisot, *Planet. Space Sci.*, 1998, **46**, 491–505.
61 N. Carrasco, E. Hébrard, M. Banaszkiewicz, M. Dobrijevic and P. Pernot, *Icarus*, 2007, **192**, 519–526.
62 M. Dobrijevic, N. Carrasco, E. Hébrard and P. Pernot, *Planet. Space Sci.*, 2008, **56**, 1630–1643.
63 N. Carrasco and P. Pernot, *J. Phys. Chem. A*, 2007, **111**, 3507–3512.
64 N. Carrasco, C. Alcaraz, O. Dutuit, S. Plessis, R. Thissen, V. Vuitton, R. Yelle and P. Pernot, *Planet. Space Sci.*, 2008, **56**, 1644–1657.
65 P. Casavecchia, F. Leonori, N. Balucani, R. Petrucci, G. Capozza and E. Segoloni, *Phys. Chem. Chem. Phys.*, 2009, **11**, 46–65.
66 M. Costes and C. Naulin, *Phys. Chem. Chem. Phys.*, 2010, **12**, 9154–9164.
67 S. B. Morales, *Doctor of Philosophy*, December 2009.
68 T. E. Cravens, I. P. Robertson, J. H. Waite, R. V. Yelle, W. T. Kasprzak, C. N. Keller, S. A. Ledvina, H. B. Niemann, J. G. Luhmann, R. L. McNutt, W. H. Ip, V. De La Haye, I. Mueller-Wodarg, J. E. Wahlund, V. G. Anicich and V. Vuitton, *Geophys. Res. Lett.*, 2006, **33**, L07105.
69 V. Vuitton, R. V. Yelle and V. G. Anicich, *Astrophys. J.*, 2006, **647**, L175–L178.
70 J. Cui, R. V. Yelle, V. Vuitton, J. H. Waite, W. T. Kasprzak, D. A. Gell, H. B. Niemann, I. C. F. Muller-Wodarg, N. Borggren, G. G. Fletcher, E. L. Patrick, E. Raaen and B. A. Magee, *Icarus*, 2009, **2009**, 581–615.
71 R. Thissen, V. Vuitton, P. Lavvas, J. Lemaire, C. Dehon, O. Dutuit, M. A. Smith, S. Turchini, D. Catone, R. V. Yelle, P. Pernot, A. Somogyi and M. Coreno, *J. Phys. Chem. A*, 2009, **113**, 11211–11220.
72 I. P. Robertson, T. E. Cravens, J. H. Waite, R. V. Yelle, V. Vuitton, A. J. Coates, J. E. Wahlund, K. Agren, K. Mandt, B. Magee, M. S. Richard and E. Fattig, *Planet. Space Sci.*, 2009, **57**, 1834–1846.

PAPER

An experimental and theoretical investigation of the competition between chemical reaction and relaxation for the reactions of $^1CH_2$ with acetylene and ethene: implications for the chemistry of the giant planets

Kelly L. Gannon,[a] Mark A. Blitz,[a] Chi-Hsiu Liang,[a] Michael J. Pilling,[a] Paul W. Seakins,*[a] David R. Glowacki[b] and Jeremy N. Harvey[b]

Received 10th March 2010, Accepted 12th April 2010
DOI: 10.1039/c004131a

The temperature dependence of the branching ratios for H atom production from the reactions of the first excited state of methylene (a^1A_1 $^1CH_2$) with acetylene and ethene have been measured at ~1 Torr total pressure and temperatures of 195, 250 and 298 K by monitoring the production of H atoms using laser induced fluorescence, comparing the signal to that observed from a calibration reaction. For the reaction with acetylene the yield of H increases from 0.28 (195 K) to 0.53 (250 K) to 0.88 at 298 K. The H atom yield from the reaction of $^1CH_2$ with ethene shows similar behaviour, the yields being 0.35 (195 K), 0.51 (250 K) and 0.71 (298 K). The co-products, propargyl (C_3H_3) and allyl (C_3H_5) are formed from the dissociation of chemically activated C_3H_4 and C_3H_6 intermediates respectively, and are important species in the formation of higher hydrocarbons, including benzene, in the atmospheres of the outer planets and Titan. H atom production is in competition with electronic relaxation to form ground state methylene (X^3B_1, $^3CH_2$) and collisional stabilization to form C_3H_4 and C_3H_6. Master equation calculations have been carried out to demonstrate that for the reaction of $^1CH_2$ with acetylene, collisional stabilization is insignificant under experimental conditions and hence the balance of reaction is due to electronic relaxation. Non-adiabatic transition state theory has been applied to the reaction of $^1CH_2$ with acetylene. The calculations show reasonable agreement with experiment, generally being within the combined errors, and reproduce the negative temperature dependence for electronic relaxation. The implications of the temperature dependence of the absolute rate coefficients for $^1CH_2$ reactions with inert gases, hydrogen, acetylene and ethene and of the branching ratios between chemical reaction and electronic relaxation are discussed.

Introduction

Small hydrocarbon radicals such as CH, CH_2, C_2H play important roles in the growth of larger, more complex hydrocarbons from methane in the atmospheres of the outer planets including Saturn's satellite Titan.[1] The first electronically excited

---
[a]School of Chemistry, University of Leeds, Leeds, LS2 9JT, UK. E-mail: p.w.seakins@leeds.ac.uk
[b]Centre for Computational Chemistry, School of Chemistry, University of Bristol, Bristol, BS8 1TS, UK

state of methylene, a^1A_1 $^1CH_2$, singlet methylene, is one such species. Lavvas et al.[2] state that the main loss of acetylene in the upper atmosphere of Titan is reaction with $^1CH_2$ to form propargyl (C_3H_3). The current understanding is that $^1CH_2$ is formed both from the Lyman-α (121.6 nm) photolysis of methane (the quantum yield for $^1CH_2$ production, $\Phi_{^1CH_2}$, is 0.58[3]) and methyl radicals.[4,5] $^1CH_2$ is a reactive molecule and can take part in addition reactions with unsaturated species (e.g. $^1CH_2 + C_2H_2$) or insertion reactions into saturated species such as H_2 or CH_4. The initial complex formed following addition or insertion can either be stabilised by collisions with a bath gas, M, or dissociate into products; in the above reaction of $^1CH_2$ with acetylene, the products would be C_3H_4 (allene, propyne or cyclopropene formed by stabilization) or C_3H_3 (propargyl) + H.

$$^1CH_2 + C_2H_2 \rightarrow C_3H_4^* \rightarrow C_3H_3 + H \quad (R1a)$$

$$\downarrow + M$$

$$C_3H_4 \quad (R1b)$$

In conjunction with chemical reaction, collision of $^1CH_2$ with species can lead to electronic relaxation to the triplet ground state (X^3B_1, $^3CH_2$).

$$^1CH_2 + C_2H_2 \rightarrow {}^3CH_2 + C_2H_2 \quad (R1c)$$

The fraction of reaction 1 proceeding via relaxation at room temperature has been determined to be ~20%[6] but little is known about the branching ratio at low temperatures. $^3CH_2$ is far less reactive than the first excited singlet state despite the energy splitting (37.65 kJ mol^{-1}), being small compared to the high heat of formation of methylene. For example the rate coefficient for the reaction of $^3CH_2$ with H_2 is at least four orders of magnitude smaller than that for $^1CH_2$ with H_2.[8,9] The lower reactivity also applies to $^3CH_2$ reactions with unsaturated species.[8]

For collisions with rare gas and inert species such as nitrogen, electronic relaxation is the only possible fate for $^1CH_2$. The reaction has been proposed to take place via a gateway state mechanism, GSM,[10] and recent studies in this laboratory[11] have demonstrated that the ortho and para quantum states of $^1CH_2$ have different collision efficiencies with light species (e.g. He, Ne) at temperatures below 300 K, consistent with such a mechanism. The rate coefficients for relaxation increase with temperature.

$^1CH_2$ reactions with reactive species have previously been shown to demonstrate quite different behaviour. The rate coefficient of the overall loss of $^1CH_2$ is either temperature independent (H_2)[9] or increases with decreasing temperature (C_2H_2,[12,13] C_2H_4[12,14]). For H_2 the efficiency of the electronic relaxation process increases with decreasing temperature[9,15] in contrast to the rare gas species. The primary objective of this paper is to measure and model the branching ratio between reaction and electronic relaxation for the reactions of $^1CH_2$ with acetylene and ethene (reaction 2) at temperatures approaching those which are relevant for modelling outer planetary atmospheres.

$$^1CH_2 + C_2H_4 \rightarrow C_3H_5 + H \quad (R2a)$$

$$^1CH_2 + C_2H_4 + M \rightarrow C_3H_6 + M \quad (R2b)$$

$$^1CH_2 + C_2H_4 \rightarrow {}^3CH_2 + C_2H_4 \quad (R2c)$$

Conventional laser flash photolysis techniques are generally limited to temperatures of ~200 K, somewhat higher than the temperatures (~70–~180 K) required for planetary modelling. The development of the Laval expansion technique in either continuous[16,17] or pulsed[18–20] forms has allowed the measurement of removal rate coefficients for a range of radical species, however, there has been limited success in probing the branching ratios of reactions. For $^1CH_2$ studies the relatively fast relaxation of $^1CH_2$ with inert gases limits the total pressure at which reactions can be studied to ~1 Torr using thermal distributions of $^1CH_2$ and time resolved studies. Whilst this is relevant to the pressures present in the upper atmospheres of relevant planets, it does hinder our understanding of $^1CH_2$ chemistry in other higher pressure regions. There is clearly a need to develop models to allow extrapolation of experimental data to the temperature and pressure conditions of relevant applications.

In this paper, we have combined results from two different theoretical models – the energy grained master equation (EGME), and the non-adiabatic transition state theory (NA-TST) – in order to understand the experimentally obtained results and extrapolate the experimental results to planetary conditions. The EGME has been described in detail previously.[21] Using a combination of Rice–Ramsperger–Kassel–Marcus (RRKM) theory and collisional energy transfer models, it permits calculation of the time evolution of molecules on multidimensional potential energy surfaces (PESs).

The microcanonical version of NA-TST utilized in this work is similar to more conventional statistical theories such as RRKM theory and TST. However, whereas the latter are used for calculating rate coefficients over energy bottlenecks on an adiabatic surface, NA-TST is used for calculating rate coefficients between different non-adiabatic PESs.[22] The NA-TST analogue to the transition state is the minimum energy crossing point (MECP) between the two non-adiabatic surfaces. Because it is a statistical theory, NA-TST is most appropriate to large molecules with very high state densities, and it generally agrees with experiment with roughly the same range of accuracy as TST provides from adiabatic reactions.[22] In this paper, we report new calculations on the electronic relaxation of $^1CH_2$ with C_2H_2 to compare with the experimental results. We also include calculations on the reaction with Xe, as a further test of the method and to investigate the role of processes with negative temperature dependence in the electronic relaxation of $^1CH_2$ by inert gases. The positive temperature dependence of the electronic relaxation rate coefficients for reaction of $^1CH_2$ by inert gases decreases in magnitude from He to Xe in a way that is not predicted by the GSM.

Our motivation for the theoretical study of reaction 1 lies in the fact that existing collision induced intersystem crossing (CIISC) models, particularly the GSM are unable to explain the negative temperature dependences observed in this work. We compare our results and the NA-TST physical picture with experiment and the GSM picture.

Current comprehensive chemical models of planetary atmospheres treat many $^1CH_2$ reactions as temperature independent, both in terms of the overall removal of $^1CH_2$ and the branching ratios between various chemical products and electronic relaxation to $^3CH_2$.[2,23,24] The results presented in this paper, and other studies from this group and elsewhere, suggest that such an approach is unrealistic and in the final section of this paper we discuss some possible implications of temperature dependence $^1CH_2$ kinetics and branching ratios.

Experimental

All product studies were carried out by laser flash photolysis with pulsed laser induced fluorescence detection (LIF) of the H atom product within a slow flow reaction cell. The cell was a six way cross design with the photolysis and probe lasers being introduced at 90° and the fluorescence being observed with a photomultiplier tube (PMT) mounted perpendicularly to the plane of the laser beams. To obtain sub-ambient cooling the cell could be immersed into a bath containing coolant maintained at the appropriate temperature by the use of a cooling probe (Thermocoax). Temperatures were measured just above and below the observation region by type K Chromel Alumel thermocouples.

The substrate (C_2H_2, C_2H_4) and methylene precursor (ketene, diluted in helium) were flowed through calibrated mass flow controllers and then mixed with the main helium bath gas flow in a mixing manifold before flowing into the reaction cell. The total pressure in the cell (as measured by a capacitance manometer) was controlled by throttling the exhaust line. Flows were maintained such that all the gases in the photolysis zone were replaced between photolysis laser pulses.

Ketene, the $^1CH_2$ precursor, was generated by the pyrolysis of acetic anhydride,[25] purified by trap-to-trap distillation and the purity checked by IR spectroscopy.[26] The substrate gases used: C_2H_2 (99.9%, Air Products), C_2H_4 (99% Aldrich) were further purified by trap-to-trap distillations before being stored in darkened glass bulbs.

$^1CH_2$ was produced by the pulsed photolysis of ketene at 308 nm (Questek $\nu\beta$2000, 10 Hz, 60 mJ cm^{-2} pulse^{-1}). The photolysis mechanism is well established at this wavelength with approximately 95% of the methylene produced being in the required singlet state.[27,28] The minor component of ground triplet state would not influence these studies as the time resolved detection of H atom product allows separation of signal from $^1CH_2$ reactions which occurs orders of magnitude faster than $^3CH_2$ reactions.

Product studies were carried out monitoring the H atom signal from the reaction using pulsed LIF using resonance fluorescence from the Lyman-α transition at 121.6 nm. VUV radiation was generated by frequency tripling the output of a pulsed dye laser (Lambda Physik LPX 100 pumping a Lambda Physik FL3002) at 364.8 nm in a glass cell containing krypton and argon phased matched to promote production of Lyman-α radiation. The dye laser was focused into the tripling cell ($\sim$800 Torr with Kr : Ar = 1 : 2.5) by a lens with a focal length of 3.0 cm and the resulting VUV radiation was coupled into the cell through a magnesium fluoride window. Resonance fluorescence was directly observed perpendicular to the photolysis and probe beams using a solar blind photomultiplier (Electron Tube). The VUV radiation was monitored by a second photomultiplier mounted at exactly the same distance from the reaction zone as the primary detection PMT. The output from this second PMT could then be used to normalize the fluorescence signal for laser power and absorption by the substrate gases. Temporal profiles of the H atom signal were obtained by delaying the delay time between the photolysis and probe pulses.

LIF is an extremely sensitive but relative technique and the H atom signal needs to be put on an absolute scale in order to extract the branching ratio. This is best done by the use of a calibration reaction; a reaction of $^1CH_2$ that generates a known fraction of H atoms.[29] $^1CH_2$ was generated in the presence of an excess of the calibrant reagent and the H atom signal recorded. The calibrant was then replaced by the substrate under test, the identical concentration of $^1CH_2$ was generated and the H atom signal was recorded. Comparison of the final H atom signals yields the branching ratio. In reality the analysis was slightly more complex as the traces are biexponential and corrections have to the made for the absorption of the Lyman-α radiation by the substrate gases.[30]

Table 1 Branching ratios of H atom production, Γ_{3H}[9,15]

T/K	Γ_{3H}
195	0.71 ± 0.07
250	0.78 ± 0.11
298	0.85 ± 0.08

For these studies the reaction of $^1CH_2$ with H_2 was used as the calibration reaction.

$$^1CH_2 + H_2 \rightarrow CH_3 + H \tag{R3}$$

The branching ratio to H atoms for reaction 3, Γ_{3H}, has previously been studied in this laboratory[9,15] and values of Γ_{3H} as a function of temperature are given in Table 1.

Theoretical calculations

Initial geometry optimization of reactants, adducts, and minimum energy crossing points (MECPs) was carried out with the B3LYP density functional theory (DFT) method and the aug-cc-pVTZ basis set. For Xe, we represented the core electrons using a 28 electron aug-cc-pVTZ spin–orbit pseudo-potential.[31] These DFT calculations used Gaussian 03,[32] combined with the code developed by one of us[33] to locate the MECP. Vibrational frequencies for the minima and transition states (TSs) were obtained at the same level of theory. Vibrational analysis at the MECPs was carried out using the method described by Koga and Morokuma[34,35] wherein the Hessians corresponding to each of the non-adiabatic states are combined to obtain an effective Hessian describing motion within the seam of state crossing. For this purpose, the gradients and Hessians on the two potential energy surfaces at the MECP were computed in Gaussian. Then, using a new program†, this information was extracted from the Gaussian output, the effective Hessian constructed, and the eigenvectors corresponding to overall rotational motion, translational motion, and atomic motion orthogonal to the MECP crossing seam were projected out.[36] Finally, this program was used to diagonalize the resulting effective Hessian to provide vibrational frequencies and eigenvectors. These were used to confirm that the MECPs located in this work are stationary points within the non-adiabatic crossing seam, and to provide zero-point energy corrections and input for the non-adiabatic transition state theory.

Calculation of the spin–orbit coupling matrix element was carried out at the geometry of the B3LYP MECPs, using the CASSCF(6,6) method for $^1CH_2$ + C_2H_2 and CASSCF(2,2) for $^1CH_2$ + Xe, together with the aug-cc-pVTZ basis. For the reactants, $^3CH_2$, and MECPs only, re-optimization at the QCISD(T) level of theory was carried out, again using the aug-cc-pVTZ basis as above (except that the hydrogen atoms were described by the aug-cc-pVDZ basis). These CASSCF and QCISD(T) calculations were carried out using MOLPRO.[37]

All non-adiabatic transition state theory (NA-TST) calculations were carried out using the recently developed MESMER program,[38] into which the NA-TST[39] has recently been implemented. Treating all molecules with the rigid rotor, harmonic oscillator approximation, $k(E)$s for surface crossing are described by an RRKM expression:

$$k(E) = \frac{N_{MECP}(E)}{h\rho(E)} \tag{E1}$$

where $\rho(E)$ is the reactant state density and h is Planck's constant. Compared to the usual equation for $k(E)$, eqn (1) replaces the transition state sum of states with the MECP sum of states, $N_{MECP}(E)$, which is a convolution of the density of states at the MECP geometry, $\rho_{MECP}(E)$, and the spin forbidden hopping (SH) probabilities, $p_{SH}(E)$:

---

† The code for performing the Vibrational analysis at the MECP recently developed by Glowacki and Harvey is available on request from Jeremy Harvey.

$$N_{\text{MECP}}(E) = \int_0^E \rho^{\text{MECP}}(E-E_H) p_{\text{SH}}(E) dE_E \qquad (E2)$$

The SH probabilities are calculated according to Landau–Zener theory as follows:

$$p_{\text{SH}}(E_H) = (1+P)(1-P)$$

$$P = \exp\left(\frac{-2\pi V_{\text{ST}}^2}{h\Delta F} \sqrt{\frac{\mu}{2(E - E_{\text{MECP}})}}\right) \qquad (E3)$$

where V_{ST} is the electronic matrix element for coupling between the two surfaces, which was determined from CAS calculations; μ is the reduced mass for movement along the gradient orthogonal to the crossing seam, and ΔF is the norm of the difference in the gradients between the two surfaces.

Canonical rate coefficients, $k(T)$ are then determined using the usual equation:

$$k(T) = \frac{1}{Q(T)} \int k(E)\rho(E)\exp(-\beta E) dE \qquad (E4)$$

where $Q(T)$ is the reactant partition function.

Experimental results

The H atom profiles (Fig. 1) were biexponential in nature with the growth corresponding to the $^1CH_2$ decay (obtained monitoring $^1CH_2$ by laser induced fluorescence[12]) and the longer time decay being mainly due to diffusive loss of H and a slow reaction of H with the substrate or precursor. As can be seen from Fig. 1, the pseudo-first-order rate coefficients obtained by monitoring either $^1CH_2$ removal or H atom production were identical confirming that the observed H atom signal did come from the target reaction.

The yields of H from the reaction of $^1CH_2$ with acetylene and ethene are given in Table 2. Also included in Table 2 are the yield data from our earlier high temperature study.[12] The yield of H, the surrogate for chemical reaction as opposed to electronic relaxation, clearly increases with temperature and appears to reach a limiting

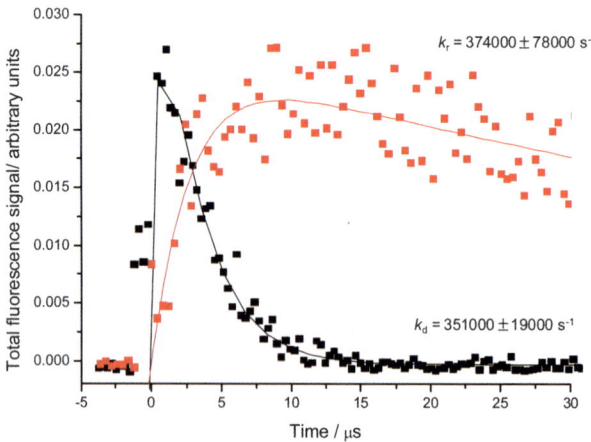

Fig. 1 Typical H atom (growth) and $^1CH_2$ (decay) time profiles for the reaction of $^1CH_2$ with H_2 at 298 K, 0.5 Torr total pressure, $[H_2] = 1.75 \times 10^{15}$ molecule cm^{-3}, together with the corresponding pseudo-first-order rate coefficients of $^1CH_2$ decay (k_d) and H atom growth (k_r).

Table 2 Absolute H atom yields for reactive removal of $^1CH_2$ with acetylene and ethene

$^1CH_2$ +	Γ_H				
	195 K	250 K	298 K	398 K	498 K
C_2H_2	0.28 ± 0.11^a	0.53 ± 0.15	0.88 ± 0.09	1.1 ± 0.16^b	1.1 ± 0.42^b
C_2H_4	0.35 ± 0.09	0.51 ± 0.13	0.71 ± 0.08	0.86 ± 0.16^b	1.08 ± 0.19^b

^{a} Reported errors represent statistical uncertainty (2σ) in the experimental data. ^{b} High temperature values from our earlier study.[12]

value of close to 100% at 398 K. In the subsequent sections of this paper we discuss possible rationalizations of this observation and the potential implications for planetary atmospheres.

At room temperature, Böhland and Temps[6] have reported a yield for electronic relaxation of 20% by observing the production of $^3CH_2$ by laser magnetic resonance spectroscopy. Therefore at room temperature the yields of channels 1a and 1c sum to unity (within experimental error) suggesting that there is no stabilization of the C_3H_4 intermediate.

Master equation calculations

Below 300 K there are no experimental determinations of electronic relaxation and hence we are forced to rely on theoretical models of collisional stabilization to determine whether the balance of reaction 1 that does not lead to H atom production is solely due to electronic relaxation or to a combination of electronic relaxation and collisional stabilization to a C_3H_4 compound (cyclopropene, allene or propyne). We have used the MESMER (Master Equation Solver for Multi-Energy well Reactions) to model the competition between reaction 1a and 1b at temperatures of 150–300 K.[38] This analysis does not take into account electronic relaxation (channel 1c), and investigates only the competition between collisional stabilization $C_3H_4^*$ and its dissociation to form C_3H_3 + H. Master equation modeling[21] of reaction 1 at low temperatures is particularly challenging due to the deep potential wells corresponding to the stable allene and propyne intermediates.[13] As described in our previous publication,[12] we have used a reservoir approximation to reduce the size

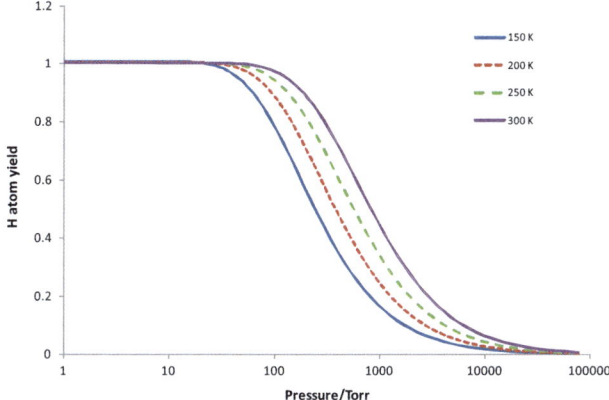

Fig. 2 H atom yield from reaction 1 calculated from master equation analysis. This analysis does not include electronic relaxation, so the balance of the reaction is collisional stabilization to C_3H_4 isomers.

of the collision matrix allowing the use of increased precision arithmetic libraries that overcome machine precision issues caused by the deep wells. The reservoir approximation falls into a wider category of steady state master equation methods as discussed by Green and Bhatti.[40]

Fig. 2 shows the yield of channel 1a as a function of temperature between 150 and 300 K. The predicted yield of channel 1a at 300 K is unity in agreement with the observation that the experimental measurements of H and $^3CH_2$ sum to give unity, *i.e.* both experiment and calculation agree that there is no collisional stabilization of the adduct. At the lower temperature, comparable to our experimental study at 195 K, the yield of H under experimental pressures in the master equation calculations is also unity, suggesting that the balance of the reaction is still electronic relaxation.

Whilst the reservoir approximation allows us to extend our calculations for reaction 1 down to 200 K, we cannot currently extend them beyond 150 K into the temperature region relevant for modeling planetary atmospheres. We are currently investigating other implementations of the master equation[41] that may allow access to relevant temperatures for chemical systems containing deep wells.

Non-adiabatic transition state theory (NA-TST) calculations

The singlet potential energy surface for the reaction with C_2H_2 is complex, as it includes a reactive channel leading to stable complexes such as cyclopropene, allene, and propyne. The stationary points on this PES are described in another publication,[12] and master equation analysis shows that the addition process is effectively irreversible. Fig. 3 shows a schematic of the initial portion of the $^1CH_2 + C_2H_2$ PES. The first new feature of this work is location of a kinetically significant

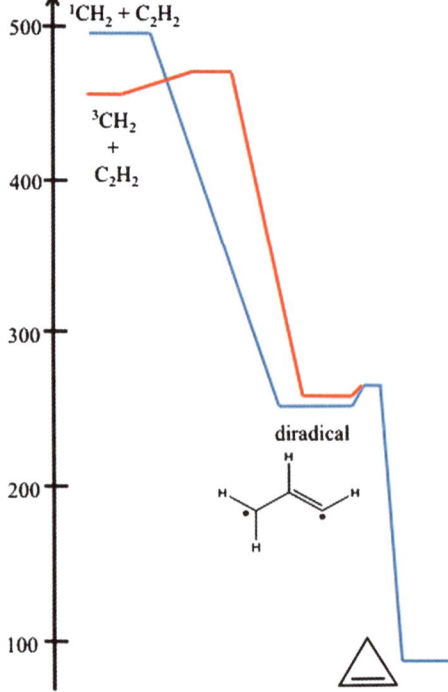

Fig. 3 Schematic of stationary points and MECPS on the $^1CH_2 + C_2H_2$; energy scale is kJ mol^{-1}.

Table 3 Zero point corrected energies of reactants, MECPS, and products (units are kJ mol^{-1}, and energies are relative to reactants)

System	Stationary point	B3LYP	QCISD(T)
CH_2–C_2H_2	$^1CH_2 + C_2H_2$	0.0	0.0
	$^3CH_2 + C_2H_2$	−45.8	−40.3
	MECP	−17.2	−15.7
CH_2–Xe	$^1CH_2 + $ Xe	0.0	0.0
	$^3CH_2 + $ Xe	−45.8	−40.3
	MECP	−8.9	−9.9

minimum energy crossing point (MECP) that occurs earlier than the addition barrier on the triplet surface. At this MECP, the separation between the C–C methylene–acetylene distance is 2.64 Å, and it lies 17.2 kJ mol^{-1} (QCISD(T), 15.7 kJ mol^{-1}) below the energy of the reactants. There is another MECP occurring in the region of the 1,3-diradical adduct of CH_2 to acetylene, which lies well below the energy of the reactants. However, because of the low isomerization barriers on the singlet surface, the rate coefficients for isomerization of the singlet are several orders of magnitude larger than those for non-adiabatic hopping. Hence this MECP does not play a role in formation of $^3CH_2$, and we did not consider it in this kinetic analysis. Energies of the MECPs, reactants, and products are given in Table 3 and details on the properties of the MECP are in Table 4.

As a test of the NA-TST methodology, we also carried out calculations on the simpler collisional electronic relaxation by Xe. Fig. 4 shows a relaxed PES along the methylene–acetylene C–C bond distance and the CH_2–Xenon C–Xe bond distance using the B3LYP functional with the aug-cc-pVTZ basis set. The crossing shown in Fig. 4 for Xe + $^1CH_2$ is not a true crossing because the optimization of orthogonal degrees of freedom on each PES yields different geometries; nevertheless the Figure shows that the non-adiabatic crossing occurs at larger separations than a non-covalent $^1CH_2$–Xe adduct lying 18.4 kJ mol^{-1} below the reactants. The MECP has a C–Xe separation of 3.12 Å, and lies 8.9 kJ mol^{-1} (QCISD(T), 9.9 kJ mol^{-1}) below the reactants.

NA-TST calculations of the electronic relaxation rate coefficient using the MECP properties in Table 3 are shown in Fig. 5 and 6. In Fig. 5 the 'experimental' data are obtained by multiplying the measured experimentally determined rate coefficient for

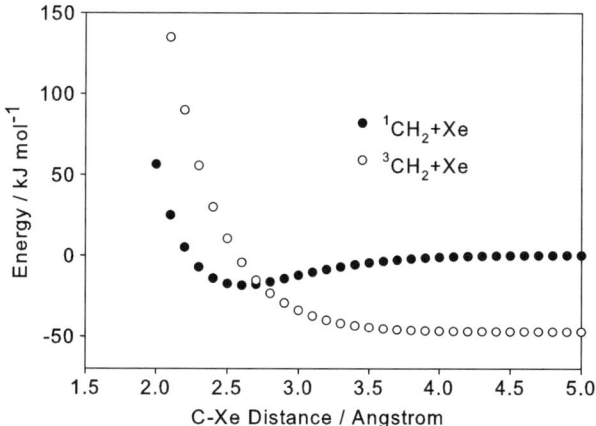

Fig. 4 Relaxed B3LYP/aug-cc-pVTZ scan along the C–Xe coordinate.

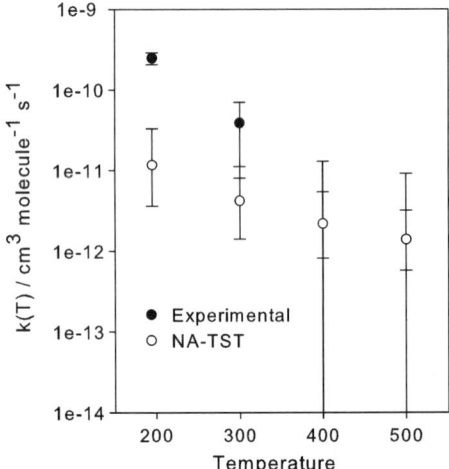

Fig. 5 Comparison of experimental and NA-TST rate coefficients for $^1CH_2 + C_2H_2$. Error bars on the NA-TST values correspond to rate coefficients obtained with an MECP energy of ± 4 kJ mol^{-1} with respect to the QCISD(T) value.

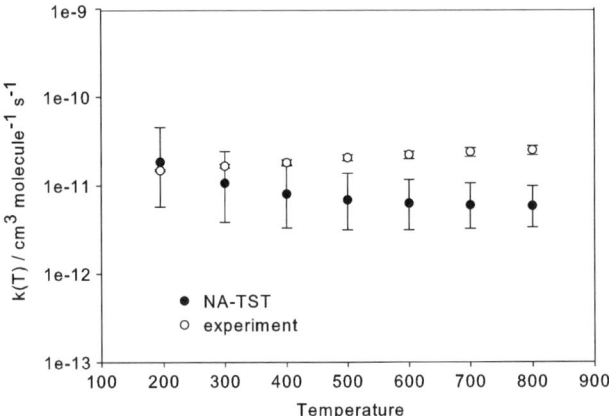

Fig. 6 Comparison of experimental[11] and NA-TST rate coefficients for $^1CH_2 + $ Xe. Error bars on the NA-TST values correspond to rate coefficients obtained with an MECP energy of ± 4 kJ mol^{-1} with respect to the QCISD(T) value.

$^1CH_2$ removal determined previously ($k_{^1CH_2+C_2H_2} = (3.06 \pm 0.11) \times 10^{-10}$ $(T/298)^{-0.39\pm0.07}$ cm^3 molecule^{-1} s^{-1}[12]) by $(1 - \Gamma_{^1H})$. For both systems, the NA-TST rate coefficients are in reasonable agreement with the measured values over the experimental temperature range, and reproduce the observed negative temperature dependence of electronic relaxation in the $^1CH_2 + C_2H_2$ system, despite the simplicity of the NA-TST theory. For $^1CH_2 + C_2H_2$, the rate coefficient at 298 K is smaller than the experimental error bound; however, as in conventional TST, the NA-TST rate coefficients are very sensitive to the MECP energy, and adjustment of this MECP by a few kJ mol^{-1}, which is well within the uncertainty of the electronic structure methods utilized in this work, brings the NA-TST rate coefficient within the experimental error bounds. The error bars in Fig. 5 are generated by varying the energy of the MECP by ± 4 kJ mol^{-1}. The competition between the capture transition state and the MECP suggests a type of two transition state

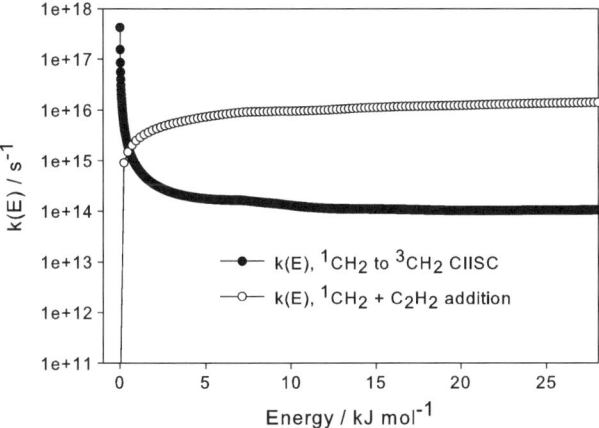

Fig. 7 Comparison of the energy dependent (microcanonical) rate coefficients for electronic relaxation compared to capture on the singlet addition surface.

model,[42] where the MECP operates as a sort of 'spin forbidden' submerged tight barrier. The competition between electronic relaxation and capture on the addition surface is shown in Fig. 7 where the cross over in the microcanonical rate coefficients for the two processes at low energies can be seen. The microcanonical rate coefficients shown in this Figure for capture on the adiabatic surface were obtained using inverse Laplace transform of the experimental rate coefficients. The difference between this system and more conventional two transition state models is that the outcome is different for the two transition states – i.e., electronic relaxation of $^1CH_2$ to $^3CH_2$ vs. chemical reaction on the singlet PES. In qualitative agreement with the laboratory measurements,[11] the NA-TST also correctly predicts a more negative temperature dependence for electronic relaxation for $^1CH_2 + C_2H_2$ than for $^1CH_2 + Xe$.

The most significant discrepancy between the NA-TST and the experimental data is that NA-TST predicts a small negative temperature dependence for $^1CH_2 + Xe$ compared to the small positive temperature dependence observed experimentally. This derives from the fact that the MECP lies lower in energy than the reactants. While the MECP energy may be subject to errors in the electronic structure theory calculations, the error in the calculations is not likely to bring the MECP to a higher energy than the reactants. The disagreement could also derive from the fact that the actual electronic relaxation with Xe takes place via both a gateway and MECP mechanism. The NA-TST mechanism emphasizes the occurrence of discrete spin-state changes occurring only upon seam-crossing in the presence of colliders, and assumes a continuum of rovibrational states that may be treated using statistical theory. Because of this latter assumption, which is most appropriate in the large molecule regime,[43] NA-TST predicts a rate coefficient for deactivation of $^1CH_2$ to $^3CH_2$ in the absence of collisions, which is not physically realistic. In the GSM, the overall electronic relaxation process may be described as: (a) collisionally induced transition to a state which has mixed singlet–triplet character, and (b) selective collisional deactivation of this mixed state into a lower rovibrational state of purely triplet character. The statistical NA-TST description views these processes as occurring in a single step, with MECP passage representing the bottleneck to state-hopping. The NA-TST model is less appropriate for the deactivation processes where the state density is far from a continuum (e.g., $^1CH_2 + Xe$); however, it has the advantage that it describes – at least roughly – the way in which the rovibrational energy levels of $^1CH_2$ and $^3CH_2$ are perturbed during collision, which is particularly important where the potential energy is more attractive and the state density is larger, as with $^1CH_2 + C_2H_2$ and $^1CH_2 + C_2H_4$. Hence, unlike the GSM, NA-TST has no particular difficulty obtaining a negative temperature dependence.

The negative temperature dependence obtained in NA-TST may be expressed in terms of the GSM: during strongly perturbative collisions with more attractive partners that have a higher state density (such as C_2H_2 or C_2H_4), steps (a) and (b) can be concerted, and short-lived 'gateway' rovibrational states of the collision complex are formed in which spin-state change can occur. As these short-lived collisional states may be lower in energy than the ground rovibrational state of $^1CH_2$, it becomes possible to observe a negative temperature dependence – unlike in the simple GSM.

The relatively simple NA-TST model has successfully reproduced the temperature dependence of the branching ratio for reaction 1. Further refinements are required to allow for quantitative predictions for planetary conditions and these would then ideally need to be tested against experimental data generated from Laval expansion systems. Once validated the model could be used to predict the temperature dependence of electronic relaxation in the reaction of $^1CH_2$ with other important reagents such as H_2, CH_4, C_2H_4 and C_2H_6.

NA-TST may provide some rationalisation of the temperature dependence of $^1CH_2$ electronic relaxation by inert gases. The reaction of $^1CH_2$ with N_2 would be particularly relevant for Titan (see below) but are more complex than for $^1CH_2$ + Xe. Calculations are currently in progress to investigate this system.

Implications

In a previous paper[12] we reported the rate coefficients for reactions 1 and 2 as a function of temperature from 195–798 K. The overall temperature dependence of the rate coefficients is relevant for the discussions below. The parameterizations of the rate coefficients are:

$$k_{^1CH_2+C_2H_2} = (3.06 \pm 0.11) \times 10^{-10} \, (T/298)^{-0.39 \pm 0.07} \, \text{cm}^3 \, \text{molecule}^{-1} \, \text{s}^{-1\,12}$$

$$k_{^1CH_2+C_2H_4} = (2.10 \pm 0.18) \times 10^{-10} \, (T/298)^{-0.84 \pm 0.18} \, \text{cm}^3 \, \text{molecule}^{-1} \, \text{s}^{-1\,12}$$

Reactions of $^1CH_2$ with rare gases and nitrogen

A number of studies[14,44] including recent work from this laboratory[11] have shown a clear positive temperature dependence for electronic relaxation of $^1CH_2$ by N_2. However, many models do not utilize this information. For example Hebrard et al.[23] and Lavvas et al.[2] use a temperature independent value of 1.0×10^{-11} cm^3 molecule^{-1} s^{-1} in their model of Titan. This value is consistent with the lowest temperature rate coefficient: $8.8 \pm 0.5 \times 10^{-12}$ cm^3 molecule^{-1} s^{-1} at 195 K, measured by Gannon et al.[11] but significantly higher than the extrapolation of the temperature

Table 4 MECP data used for carrying out NA-TST calculations. All units are cm^{-1}, with the exception of ΔF and μ_H, whose respective units are a.u./Bohr and a.m.u

MECP	Rotational constants	Frequencies	V_{ST}	ΔF	μ_H
CH$_2$–C$_2$H$_2$	1.121, 0.228, 0.197	131, 137, 199, 314, 524, 641, 666, 758, 773, 2037, 2966, 3209, 3410, 3509	6.6	0.078	8.23
CH$_2$–Xe	7.28, 0.136, 0.135	116, 335, 444, 2883, 3092	11.45	0.079	11.43

dependent measurements to the relevant temperature ranges of 70–180 K ($k(T) = (10.95 \pm 0.33) \times 10^{-12} \left(\frac{T}{298}\right)^{0.81}$ [11]). At 70 K the lifetime of $^1CH_2$ with respect to electronic relaxation by N_2 would be a factor of three greater than that predicted using a temperature independent rate coefficient for electronic relaxation. Wilson and Atreya[24] use a temperature dependent rate coefficient ($k(T) = 2.36 \times 10^{-14}T$) which under predicts the measured rate coefficient at 195 K by approximately a factor of two.

The reaction of $^1CH_2$ with H_2

Recent studies have shown that whilst the temperature dependence of the overall rate coefficient is temperature independent, the branching ratio between reaction and electronic relaxation does show temperature dependence with electronic relaxation becoming more efficient at lower temperatures.[9,15] Evidence for the temperature dependence of the ratio of reaction to electronic relaxation has appeared only relatively recently and hence models have only included a temperature independent ratio. Hebrard et al.,[23] Wilson and Atreya[24] and Lavvas et al.[2] all use a branching ratio of ~10% to relaxation, comparable to the measured room temperature value, but considerably lower than the measured value of ~30% at 195 K. Actual branching ratios to relaxation over the temperature range 70–180 K are expected to be even higher given the observed temperature dependence of the branching ratio.

In Jovian type atmospheres with high concentrations of hydrogen, chemical reaction of $^1CH_2$ with H_2 promotes recycling of $^1CH_2$ to CH_4. Under these conditions greater efficiency of electronic relaxation is likely to lead to more complex chemistry *via* radical–radical reactions of $^3CH_2$ such as:

$$^3CH_2 + CH_3 \rightarrow C_2H_4 + H$$

or *via* formation of CH radicals from

$$^3CH_2 + H \rightarrow CH + H_2$$

The reaction of $^1CH_2$ with C_2H_2

A number of studies have demonstrated a negative temperature dependence for the overall rate coefficient for the removal of $^1CH_2$ by acetylene.[12,13] The temperature independent values for the overall rate coefficient used by Hebrard et al.,[23] 3.7 × 10^{-10} cm^3 molecule^{-1} s^{-1}, Wilson and Atreya 3.6 × 10^{-10} cm^3 molecule^{-1} s^{-1} and Lavvas et al.[2] 4.2 × 10^{-10} cm^3 molecule^{-1} s^{-1} are similar to that measured at 195 K by Gannon et al.[12] ($(3.43 \pm 0.06) \times 10^{-10}$ cm^3 molecule^{-1} s^{-1}), however, extrapolation of the measured temperature dependence using the parameterization above yields values of 3.7 and 5.4 × 10^{-10} cm^3 molecule^{-1} s^{-1} at temperatures of 180 and 70 K respectively.

The results of this work demonstrate that the use of a temperature independent branching ratio between reaction and electronic relaxation is inappropriate for this reaction. Hebrard et al.[23] and Lavvas et al.[2] use a temperature independent branching ratio of 22% which is in good agreement with previous room temperature studies[6] and this work, however our measurements and calculations show a strong increase in the efficiency of relaxation as the temperature is lowered; at 195 K the fraction of electronic relaxation rises to 72%. Wilson and Atreya only include channel 1a in their model.[24]

Another major difference between this work and models relates to the products of the chemical reaction. Hebrard et al.[23] assign equal probability to three chemical reactions: the formation of propargyl + H, allene and propyne. Master equation studies demonstrate that at the pressures relevant for the upper atmospheres of Titan and other planets, collisional stabilization of the initially formed C_3H_4 intermediate is not likely and hence that the formation of propargyl + H will be the dominant chemical channel.

The propargyl radical is an important intermediate in the formation of benzene, the step in the production of polycyclic aromatic hydrocarbons and then aerosols. The simulation studies of Lebonnois[45] and Wilson et al.[1] conclude that benzene abundance depends primarily on reactions that effect the propargyl radical concentration indicating the importance of the propargyl pathway to benzene. Reaction 1 is an important route to propargyl formation.[1] In comparison to the model of Hebrard et al., the observation of decreased efficiency of reaction with decreasing temperature is offset by calculations which show that propargyl + H is likely to be the only chemical product at low pressures. It is therefore possible that predictions of the yield of benzene may not be greatly altered.

The reaction of $^1CH_2$ with C_2H_4

The behaviour of this reaction is similar to that of reaction 1 with a negative temperature dependence of the overall rate coefficient and a decrease in the yield of H atoms with decreasing temperature. The room temperature yield of H atoms is consistent with only two reaction channels being relevant at low pressures:

$$^1CH_2 + C_2H_4 \rightarrow {}^3CH_2 + C_2H_4$$

$$^1CH_2 + C_2H_4 \rightarrow H + C_3H_5$$

As no master equations could be carried out on this reaction we cannot definitively assign the non H atom production solely to relaxation; it is possible that collisional stabilization of the C_3H_6 intermediate (as either cyclopropane or propene) could also contribute. However, as the H atom yield shows similar behaviour to reaction 1, we believe that a majority of the remaining reaction occurs *via* electronic relaxation.

The observation of a significant H atom channel for reactive loss is in contrast to models where the formation of propene, C_3H_6, is the sole or dominant product. Lavvas et al.[2] treat formation of propene as a pressure dependent process; at 180 K and 1 Torr total pressure propene formation would be the dominant process in contrast to the experimental observations of H atom production. The observation of a significant H atom product, and by implication allyl radical (C_3H_5) formation, will have implications on predicted propene concentrations and on the relative importance of other minor compounds formed from the allyl radical.

Summary

The experimental work presented in this and other studies demonstrates temperature dependence in both overall $^1CH_2$ rate coefficients and in the branching ratios of reaction products. These effects need to be taken into account in modelling studies. Experimental studies are complex and there is an important role for sensitivity analysis of models to identify important reactions for study.[5,46] The relevant temperatures of the upper atmospheres of the outer planets lie outside conditions that are easily accessed *via* conventional flash photolysis type experiments. Laval expansion systems open up lower temperature regions, but to date have mainly only focused on the measurement of overall rate coefficients. The above discussion emphasises the importance of determining temperature dependent data (rate coefficients and branching ratios), but comprehensive experimental data will not be available from current techniques. Clearly there is an important role for theory to underpin extrapolation of experimental data to the relevant conditions.

Transition state theory is well established as a technique for modelling rate coefficients for reactions with well defined barriers. However, for the low temperatures prevalent in the upper atmospheres of the outer planets, barrierless reactions of radical species will be important. The temperature dependence of such reactions and modelling by variational transition state theory has recently been reviewed by Smith[47] and Klippenstein et al.[48] In this paper we have demonstrated that models based on transition state theory can be applied to electronic surface crossing allowing us to model the temperature dependence of the competition between reaction and electronic relaxation. These calculations provide insight into the difference between the reaction of $^1CH_2$ with Xe, where the MECP is only slightly lower in energy than reactants, and acetylene, where it is notably lower. The calculations are able to reproduce the rate constants in a qualitative way, and somewhat improved agreement might be obtained with more accurate potential energy surfaces. However, for more quantitative results, a more sophisticated approach incorporating (1) the non-continuous rovibrational state density, (2) corrections to the rigid rotor harmonic oscillator approximation, and (3) treatment of the coupling between the spin-allowed singlet channel and the spin-forbidden channel would be needed.

Acknowledgements

We acknowledge the support of EPSRC for funding *via* a studentship for K. L. G. and through Grant GR/T28560/01. D. R. G. and J. N. H. acknowledge EPSRC Grant EP/G00224X/1, as well as previous work by Kittusamy Senthilkumar for assistance with the Hessian projection code. We are grateful for discussions with Dr Luc Vereecken regarding master equation modelling.

References

1 E. H. Wilson, S. K. Atreya and A. Coustenis, *J. Geophys. Res. [Planets]*, 2003, **108**, 10.
2 P. P. Lavvas, A. Coustenis and I. M. Vardavas, *Planet. Space Sci.*, 2008, **56**, 27–66.
3 J. H. Wang, K. P. Liu, Z. Y. Min, H. M. Su, R. Bersohn, J. Preses and J. Z. Larese, *J. Chem. Phys.*, 2000, **113**, 4146–4152.
4 S. H. S. Wilson, J. D. Howe, K. N. Rosser, M. N. R. Ashfold and R. N. Dixon, *Chem. Phys. Lett.*, 1994, **227**, 456–460.
5 M. Dobrijevic, E. Hebrard, S. Plessis, N. Carrasco, P. Pernot and M. Bruno-Claeys, *Adv. Space Res.*, 2010, **45**, 77–91.
6 T. Bohland, F. Temps and H. G. Wagner, *Ber. Bunsen-Ges. Phys. Chem. Chem. Phys.*, 1985, **89**, 1013–1018.
7 P. Jensen and P. R. Bunker, *J. Chem. Phys.*, 1988, **89**, 1327–1333.
8 D. C. Darwin and C. B. Moore, *J. Phys. Chem.*, 1995, **99**, 13467–13470.
9 K. L. Gannon, M. A. Blitz, M. J. Pilling, P. W. Seakins, S. J. Klippenstein and L. B. Harding, *J. Phys. Chem. A*, 2008, **112**, 9575–9583.
10 U. Bley and F. Temps, *J. Chem. Phys.*, 1993, **98**, 1058–1072.
11 K. L. Gannon, M. A. Blitz, T. Kovacs, M. J. Pilling and P. W. Seakins, *J. Chem. Phys.*, 2010, **132**, 024302.
12 K. L. Gannon, M. A. Blitz, D. R. Glowacki, C.-H. Liang, M. J. Pilling and P. W. Seakins, *J. Phys. Chem. A*, 2010, submitted.
13 M. A. Blitz, M. S. Beasley, M. J. Pilling and S. H. Robertson, *Phys. Chem. Chem. Phys.*, 2000, **2**, 805–812.
14 R. Wagener, *Z. Naturforsch., A: Phys. Sci.*, 1990, **45a**, 649–656.
15 M. A. Blitz, N. Choi, T. Kovacs, P. W. Seakins and M. J. Pilling, *Proc. Combust. Inst.*, 2005, **30**, 927–933.
16 I. W. M. Smith, *Angew. Chem., Int. Ed.*, 2006, **45**, 2842–2861.

17 N. Daugey, P. Caubet, B. Retail, M. Costes, A. Bergeat and G. Dorthe, *Phys. Chem. Chem. Phys.*, 2005, **7**, 2921–2927.
18 D. B. Atkinson and M. A. Smith, *Rev. Sci. Instrum.*, 1995, **66**, 4434–4446.
19 S. Lee, D. A. Samuels, R. J. Hoobler and S. R. Leone, *J. Geophys. Res.*, 2000, **105**, 15085–15090.
20 S. E. Taylor, A. Goddard, M. A. Blitz, P. A. Cleary and D. E. Heard, *Phys. Chem. Chem. Phys.*, 2008, **10**, 422–437.
21 M. J. Pilling and S. H. Robertson, *Annu. Rev. Phys. Chem.*, 2003, **54**, 245–275.
22 J. N. Harvey, *Phys. Chem. Chem. Phys.*, 2007, **9**, 331–343.
23 E. Hebrard, M. Dobrijevic, Y. Benilan and F. Raulin, *J. Photochem. Photobiol., C*, 2006, **7**, 211–230.
24 E. H. Wilson and S. K. Atreya, *J. Geophys. Res. [Planets]*, 2004, 109.
25 G. J. Fisher, A. F. Maclean and A. W. Schnizer, *J. Org. Chem.*, 1953, **18**, 1055–1057.
26 W. F. Arendale and W. H. Fletcher, *J. Chem. Phys.*, 1957, **26**, 793–797.
27 C. G. Morgan, M. Drabbels and A. M. Wodtke, *J. Chem. Phys.*, 1996, **104**, 7460–7474.
28 E. A. Wade, H. Clauberg, S. K. Kim, A. Mellinger and C. B. Moore, *J. Phys. Chem. A*, 1997, **101**, 732–739.
29 P. W. Seakins, *Annu. Rep. Prog. Chem., Sect. C*, 2007, **103**.
30 K. McKee, M. A. Blitz, K. J. Hughes, M. J. Pilling, H. B. Qian, A. Taylor and P. W. Seakins, *J. Phys. Chem. A*, 2003, **107**, 5710–5716.
31 K. A. Peterson, D. Figgen, E. Goll, H. Stoll and M. Dolg, *J. Chem. Phys.*, 2003, **119**, 11113–11123.
32 M. J. Frisch, G. W. Trucks, H. B. Schlegel, G. E. Scuseria, M. A. Robb, J. R. Cheeseman, J. A. Montgomery, Jr., T. Vreven, K. N. Kudin, J. C. Burant, J. M. Millam, S. S. Iyengar, J. Tomasi, V. Barone, B. Mennucci, M. Cossi, G. Scalmani, N. Rega, G. A. Petersson, H. Nakatsuji, M. Hada, M. Ehara, K. Toyota, R. Fukuda, J. Hasegawa, M. Ishida, T. Nakajima, Y. Honda, O. Kitao, H. Nakai, M. Klene, X. Li, J. E. Knox, H. P. Hratchian, J. B. Cross, V. Bakken, C. Adamo, J. Jaramillo, R. Gomperts, R. E. Stratmann, O. Yazyev, A. J. Austin, R. Cammi, C. Pomelli, J. Ochterski, P. Y. Ayala, K. Morokuma, G. A. Voth, P. Salvador, J. J. Dannenberg, V. G. Zakrzewski, S. Dapprich, A. D. Daniels, M. C. Strain, O. Farkas, D. K. Malick, A. D. Rabuck, K. Raghavachari, J. B. Foresman, J. V. Ortiz, Q. Cui, A. G. Baboul, S. Clifford, J. Cioslowski, B. B. Stefanov, G. Liu, A. Liashenko, P. Piskorz, I. Komaromi, R. L. Martin, D. J. Fox, T. Keith, M. A. Al-Laham, C. Y. Peng, A. Nanayakkara, M. Challacombe, P. M. W. Gill, B. G. Johnson, W. Chen, M. W. Wong, C. Gonzalez and J. A. Pople, *GAUSSIAN 03 (Revision B.03)*, Gaussian, Inc., Wallingford, CT, 2004.
33 J. N. Harvey, M. Aschi, H. Schwarz and W. Koch, *Theor. Chem. Acc.*, 1998, **99**, 95–99.
34 N. Koga and K. Morokuma, *Chem. Phys. Lett.*, 1985, **119**, 371–374.
35 J. N. Harvey and M. Aschi, *Phys. Chem. Chem. Phys.*, 1999, **1**, 5555–5563.
36 W. H. Miller, N. C. Handy and J. E. Adams, *J. Chem. Phys.*, 1980, **72**, 99–112.
37 R. D. Amos, A. Bernhardsson, A. Berning, P. Celani, D. L. Cooper, M. J. O. Deegan, A. J. Dobbyn, F. Eckert, C. Hampel, G. Hetzer, P. J. Knowles, T. Korona, R. Lindh, A. W. Lloyd, S. J. McNicholas, F. R. Manby, W. Meyer, M. E. Mura, A. Nicklass, P. Palmieri, R. Pitzer, G. Rauhut, M. Schütz, U. Schumann, H. Stoll, A. J. Stone, R. Tarroni, T. Thorsteinsson and H.-J. Werner, *MOLPRO, a package of* ab initio *programs designed by H.-J. Werner and P. J. Knowles, Version 2002.1*, 2002.
38 S. H. Robertson, D. R. Glowacki, C. H. Liang, C. M. Morley and M. J. Pilling, *MESMER*, http://sourceforge.net/projects/mesmer, 2009.
39 J. N. Harvey and M. Aschi, *Faraday Discuss.*, 2003, **124**, 129–143.
40 N. J. B. Green and Z. A. Bhatti, *Phys. Chem. Chem. Phys.*, 2007, **9**, 4275–4290.
41 L. Vereecken, G. Huyberechts and J. Peeters, *J. Chem. Phys.*, 1997, **106**, 6564–6573.
42 E. E. Greenwald, S. W. North, Y. Georgievskii and S. J. Klippenstein, *J. Phys. Chem. A*, 2005, **109**, 6031–6044.
43 K. F. Freed, *Acc. Chem. Res.*, 1978, **11**, 74–80.
44 F. Hayes, W. D. Lawrance, W. S. Staker and K. D. King, *J. Phys. Chem.*, 1996, **100**, 11314–11318.
45 S. Lebonnois, *Planet. Space Sci.*, 2005, **53**, 486–497.
46 G. P. Smith and D. Nash, *Icarus*, 2006, **182**, 181–201.
47 I. W. M. Smith, *Chem. Soc. Rev.*, 2008, **37**, 812–826.
48 S. J. Klippenstein, Y. Georgievskii and L. B. Harding, *Phys. Chem. Chem. Phys.*, 2006, **8**, 1133–1147.

PAPER

Formation of nitriles and imines in the atmosphere of Titan: combined crossed-beam and theoretical studies on the reaction dynamics of excited nitrogen atoms N(^{2}D) with ethane

Nadia Balucani,[a] Francesca Leonori,[a] Raffaele Petrucci,[a] Massimiliano Stazi,[a] Dimitris Skouteris,[b] Marzio Rosi[c] and Piergiorgio Casavecchia*[a]

Received 30th March 2010, Accepted 14th April 2010
DOI: 10.1039/c004748a

The dynamics of the H-displacement channels in the reaction N(^{2}D) + C$_2$H$_6$ have been investigated by the crossed molecular beam technique with mass spectrometric detection and time-of-flight analysis at two different collision energies (18.0 and 31.4 kJ mol^{-1}). From the derived center-of-mass product angular and translational energy distributions the reaction micromechanisms and the product energy partitioning have been obtained. The interpretation of the scattering results is assisted by new *ab initio* electronic structure calculations of stationary points and product energetics for the C$_2$H$_6$N ground state doublet potential energy surface. C–C bond breaking and NH production channels have been theoretically characterized and the statistical branching ratio derived at the temperatures relevant for the atmosphere of Titan. Methanimine plus CH$_3$ and ethanimine plus H are the main reaction channels. Implications for the atmospheric chemistry of Titan are discussed.

1 Introduction

Titan, the largest moon of Saturn, is the only solar system body to have a dense atmosphere mainly composed of molecular nitrogen like our own planet.[1–3] The second most abundant component is methane (~1.4% in the stratosphere rising up to 4.9% in the proximity of the surface[3]), while observations by Voyager/IRIS (InfraRed Interferometer Spectrometer),[1] confirmed by the more recent ones by the Composite InfraRed Spectrometer (CIRS) aboard Cassini,[2] have also revealed the presence of a set of larger hydrocarbons in trace amounts.[1,2] A complex chemistry initiated by the photodissociation of CH$_4$ at high altitude is alleged to lead to the formation of larger hydrocarbons,[1,2] amongst which the most abundant is ethane with several ppm above 120 km.[2] Ethane is believed to be formed from the self combination of methyl (CH$_3$) radicals according to the scheme[4–10]

$$CH_3 + CH_3 + M \rightarrow C_2H_6 + M$$

[a] Dipartimento di Chimica, Università degli Studi di Perugia, 06123, Perugia, Italy. E-mail: nadia.balucani@unipg.it; piero@dyn.unipg.it
[b] Dipartimento di Matematica e Informatica and Dipartimento di Chimica, Università degli Studi di Perugia, 06123, Perugia, Italy. E-mail: dimitris@dyn.unipg.it
[c] Dipartimento di Ingegneria Civile e Ambientale and ISTM-CNR, c/o Dipartimento di Chimica, Università degli Studi di Perugia, 06123, Perugia, Italy. E-mail: marzio@thch.unipg.it

CH$_3$ is a radical relatively abundant in the upper atmosphere of Titan since it can be produced[4,6–10] directly by the photodissociation of methane or by methane catalytic dissociation[4–10] induced by several hydrocarbon radicals such as

$$CH_2(a^1A_1) + CH_4 \rightarrow CH_3 + CH_3$$

$$C_2H(X^2\Sigma^+) + CH_4 \rightarrow C_2H_2 + CH_3$$

As CH$_2$(a^{1}A$_1$) and C$_2$H(X$^2\Sigma^+$) are generated by the photodissociation of methane and acetylene (which, in turn, is generated in a sequence of steps that start with methane photodissociation) we can say that the energy deposited by the UV solar photons converts methane into ethane and higher hydrocarbons.[4–10] C$_2$H$_6$ is considered to be quite stable in the conditions of the atmosphere of Titan with a chemical lifetime of $\sim$200 y[10] and to be mainly lost by condensation in the lower stratosphere. As a consequence, ethane is predicted to be uniformly mixed in the stratosphere with similar mixing ratio profiles over the disk.[9,10] Various observations confirm that the ethane mixing ratio profile is almost constant with altitude, while a weak global increase towards high northern latitudes has been observed.[11,12]

The possible chemical loss processes of C$_2$H$_6$ are UV photolysis and chemical reactions with radical species.[6–10] Regarding the photolysis process, the UV absorption spectrum of ethane overlaps with that of the much more abundant methane, so that methane acts as a shield.[6–10] This means that chemical loss proceeds mainly through chemical reactions with reactive radicals. The reaction of ethane with an active form of nitrogen, such as N(^{2}D), can certainly contribute to the chemical loss of ethane in the upper atmosphere. Nitrogen atoms in the first, metastable electronic ^{2}D state can be produced in the upper atmosphere of Titan by several processes involving molecular nitrogen (such as photodissociation, electron impact dissociation or dissociative photoionization) or the molecular ion N$_2^+$ (such as dissociative recombination).[13] These processes lead to atomic nitrogen in the ground, ^{4}S, and metastable electronically excited, ^{2}D$_{5/2,3/2}$, states (energy content: 230.0 kJ mol^{-1}; radiative lifetimes of ^{2}D$_{3/2}$ and ^{2}D$_{5/2}$ are 6.1 $\times$ 10^4 and 1.4 $\times$ 10^5 s, respectively[14]). Since collisional deactivation of N(^{2}D) by N$_2$ is a slow process ($k_{298} \sim$ 10^{-14} cm^3 molecule^{-1} s^{-1}),[15] the main fate of N(^{2}D) above 800 km is chemical reaction with other constituents of Titan's atmosphere.[7] The production of N atoms in the ^{2}D state is especially relevant, because it is well known that ground state N atoms exhibit very low reactivity with closed shell molecules and the probability of collision with a radical is small.

The paucity of experimental information on the N(^{2}D) + C$_2$H$_6$ reaction has so far prevented the correct inclusion of this reaction in the chemical models of Titan. For instance, this reaction was not considered in the models by Yung et al.[5] and Toublanc et al.,[6] even though in the same models the reaction of N(^{2}D) with acetylene (less abundant than ethane) was included.[5b,6] In the models by Lara et al.[7] and Wilson and Atreya,[8] the reaction N(^{2}D) + C$_2$H$_6$ has been included with a room temperature rate constant of 3.0 $\times$ 10^{-12} cm^3 s^{-1}, estimated by analogy with the reaction N(^{2}D) + CH$_4$. Nevertheless, the measured rate contant for N(^{2}D) + C$_2$H$_6$ is larger by roughly one order of magnitude, being 2.12 $\pm$ 0.16 $\times$ 10^{-11} cm^3 s^{-1} according to the measurements by Umemoto et al.[16] In the more recent models by Lavvas et al.[9] and Krasnopolsky,[10] the rate constant value has been set to be 1.9 $\times$ 10^{-11} cm^3 s^{-1}, as recommended by a critical review article[15] on the reactions of atomic and molecular nitrogen, where the results by Umemoto et al.[16] have been averaged over previous data. Yet, uncertainty about the nature of the reaction products remains. In particular, while in the models by Lara et al.,[7] Wilson and Atreya,[8] and Krasnopolsky[10] the only reaction channel considered is the one leading to NH + C$_2$H$_5$, Lavvas et al.[9] have put forward the suggestion that the main products are aziridine (a closed-shell cyclic molecule with formula c-CH$_2$(NH)CH$_2$) and atomic hydrogen. In the model by Lavvas *et al.*[9] the alleged production of aziridine has some interesting implications because this molecule is a possible precursor of ammonia *via* UV photolysis.[17] As a matter of fact, according to a photodissociation study at λ = 147 and 123.6 nm the channel leading to C_2H_2 + NH_3 is open, even though its quantum yield is as small as 3%.[17] In this way, Lavvas *et al.* have been able to account for the presence of ammonia in the Titan's thermosphere (1100 km), albeit the predicted mixing ratio is one order of magnitude lower than that inferred by the analysis of the spectra recorded by INMS on board Cassini.[18] It should be noted that, according to the prediction of the models by Lavvas *et al.*, aziridine should be present in the thermosphere in such amounts as to be observable,[9] while the instruments on board Cassini have not identified it yet. In the assumption that NH and C_2H_5 are the only (or main) products of the title reaction, as suggested by Lara *et al.*,[7] Wilson and Atreya,[8] and Krasnopolsky,[10] the principal role of the $N(^2D)$ + C_2H_6 reaction would be that of contributing to the formation of larger hydrocarbons through the formation of the ethyl radical. Remarkably, in no model has the possibility of reaction channels proceeding through C–C bond breaking been considered. Clearly, to establish the role of the $N(^2D)$ + C_2H_6 reaction in the chemical evolution of the atmosphere of Titan, the nature of the primary products has to be assessed in laboratory experiments.

In our laboratory, we have successfully investigated the reactions involving atomic nitrogen in its first electronically excited state $N(^2D)$ with H_2 and several hydrocarbons[13,19–22] – namely CH_4, C_2H_2 and C_2H_4 – by means of the crossed molecular beam (CMB) method with mass-spectrometric (MS) detection and time-of-flight (TOF) analysis.[22–25] In CMB experiments bimolecular reactions are studied under well-defined conditions. Differently to a bulk experiment, the reacting species are confined in separate beams which cross each other at a specific angle and collision energy. The products are formed at the collision center and then fly undisturbed towards the detector because of the large mean free path achieved by operating at a very low pressure. In this way, the results of well-defined molecular collisions are observed and secondary/wall collisions, that could alter the nature of the primary products, are avoided.[22–25] The use of MS detection makes the method universal, as any species can be ionized in the ionizer which precedes the mass filter and it is possible to determine the mass and the gross formula of all primary products of a bimolecular reaction.[22,24] As they allow determining the reaction mechanism, the nature of the primary products and, in favourable cases, their branching ratio, CMB-MS experiments are complementary to gas kinetics studies that normally furnish the global disappearance rate of the reactants without providing information on the products. In recent years the CMB-MS method has been successfully applied to the study of elementary reactions of relevance in the atmospheric chemistry of our own and other planets, as well as in the chemistry of the interstellar medium.[13,19–22,25–30] In the cases of the reactions $N(^2D)$ + CH_4, C_2H_2 and C_2H_4, we have observed[13,20–22] the formation of molecular species containing a novel C–N bond (CH_2=NH, CH_3N, HCCN, CH_2=CNH, c-CH(N)CH_2 and CH_3CN), thus demonstrating that the reactions of an active form of nitrogen with relatively abundant hydrocarbons are viable routes of formation of species containing a CN group in the atmosphere of Titan, as well as in that of other bodies where both molecular nitrogen – the precursor of $N(^2D)$ – and hydrocarbons are present.

In a recent paper on the reaction $N(^2D)$ + CH_4, of great relevance in the atmospheric chemistry of Titan because of the abundance of methane, we have complemented the CMB experimental results with a theoretical study of the relevant CH_4N potential energy surface (PES) and statistical prediction of the product branching ratio (BR) based on the Rice–Ramsperger–Kassel–Marcus (RRKM) method.[13] RRKM BRs were determined at the conditions of the CMB experiments for comparison and at the conditions (particularly the temperature) of the atmosphere of Titan.[13] An interesting conclusion of that study is that the $N(^2D)$ + CH_4 reaction proceeds *via* the insertion of $N(^2D)$ into one of the methane C–H bonds thus forming the CH$_3$NH bound intermediate. The excess energy released during the formation of the intermediate CH$_3$NH causes its fragmentation into products. Interestingly, there has been experimental evidence that there is a competition between the fragmentation of one of the two new bonds formed by N(^{2}D) insertion (that is, N–H and C–N of CH$_3$NH) and of one of the preexisting C–H bonds, not directly involved in the insertion process. The C–H bonds are weaker than the newly formed N–H and C–N bonds, but the energy released after N(^{2}D) insertion is initially concentrated in the "activated" N–H and C–N bonds and there is a significant probability that they break apart before sufficient intramolecular vibrational energy redistribution (IVR) occurs.[13,34a] As a result of this competition, the yield of the channels leading to the products CH$_2$=NH + H, CH$_3$N + H and NH + CH$_3$ has been seen to vary with the available energy, because the lifetime of the CH$_3$NH insertion intermediate decreases with the available energy and so does the extent of IVR.[13] Notably, the RRKM BRs are not in line with the experimental ones because the RRKM method, as well as all the other statistical methods, relies on the assumption that IVR is complete and that the available energy is randomly distributed amongst all degrees of freedom of the decomposing intermediate. The conclusion of that study was that part of the reaction is dominated by dynamical effects – with the formation of the products CH$_3$N(methylnitrene) + H and NH + CH$_3$ – and part by the statistical redistribution of the available energy – leading to the formation of the much more stable products CH$_2$=NH + H.[13] Those conclusions were in line with the characteristics of the internal population of the NH product, as was determined in a previous spectroscopic study.[34a]

In this contribution, we extend the same combined experimental and theoretical approach to the investigation of the reaction of N(^{2}D) with the second most abundant hydrocarbon in the atmosphere of Titan, that is, C$_2$H$_6$. In particular, the focus of the experimental investigation has been on the possible H-displacement channels. As we are going to see, in fact, in addition to the aziridine molecule suggested by Lavvas et al.[9] there are many molecular products which can be formed in an H-displacement reaction. Differently from other cases,[31–33] in the present study our experimental technique could not be applied to the characterization of the NH formation or C–C bond breaking channels because of some interfering signals (see below). All the possible channels, however, have been characterized at the theoretical levels and their importance discussed, also in the light of a previous dynamical experiment on the N(^{2}D) + C$_2$H$_6$ → NH + C$_2$H$_5$ channel.[34b]

The reaction N(^{2}D)+C$_2$H$_6$ (1) is indeed quite complex. According to the present ab initio calculations, the thermodynamically allowed channels (listed in order of decreasing reaction exothermicity) are:

$$N(^2D) + C_2H_6 \rightarrow CH_2=NH(^1A') + CH_3(^2A_2'') \ \Delta H_0° = -366.2 \text{ kJ mol}^{-1} \quad (1a)$$

$$N(^2D) + C_2H_6 \rightarrow C_2H_4(X^1A_g) + NH_2(^2B_1) \ \Delta H_0° = -363.9 \text{ kJ mol}^{-1} \quad (1b)$$

$$N(^2D) + C_2H_6 \rightarrow CH_3CH=NH(^1A') + H(^2S) \ \Delta H_0° = -338.2 \text{ kJ mol}^{-1} \quad (1c)$$

$$N(^2D) + C_2H_6 \rightarrow CH_2=CHNH_2(^1A) + H(^2S) \ \Delta H_0° = -321.3 \text{ kJ mol}^{-1} \quad (1d)$$

$$N(^2D) + C_2H_6 \rightarrow CH_3N=CH_2(^1A') + H(^2S) \ \Delta H_0° = -299.3 \text{ kJ mol}^{-1} \quad (1e)$$

$$N(^2D) + C_2H_6 \rightarrow c\text{-}CH_2(NH)CH_2(^1A') + H(^2S) \ \Delta H_0° = -249.8 \text{ kJ mol}^{-1} \quad (1f)$$

$$N(^2D) + C_2H_6 \rightarrow CHNH_2(^1A') + CH_3(^2A_2'') \quad \Delta H_0° = -218.6 \text{ kJ mol}^{-1} \quad (1g)$$

$$N(^2D) + C_2H_6 \rightarrow CH_3CNH_2(^1A') + H(^2S) \quad \Delta H_0° = -194.5 \text{ kJ mol}^{-1} \quad (1h)$$

$$N(^2D) + C_2H_6 \rightarrow CH_3N(^3A_2) + CH_3(^2A_2'') \quad \Delta H_0° = -149.9 \text{ kJ mol}^{-1} \quad (1i)$$

$$N(^2D) + C_2H_6 \rightarrow CH_3CH_2(^2A') + NH(^3\Sigma^-) \quad \Delta H_0° = -135.6 \text{ kJ mol}^{-1} \quad (1j)$$

$$N(^2D) + C_2H_6 \rightarrow CH_3CH_2N(^3A'') + H(^2S) \quad \Delta H_0° = -98.8 \text{ kJ mol}^{-1} \quad (1k)$$

$$N(^2D) + C_2H_6 \rightarrow CH_2NH(^3A'') + CH_3(^2A_2'') \quad \Delta H_0° = -94.4 \text{ kJ mol}^{-1} \quad (1l)$$

$$N(^2D) + C_2H_6 \rightarrow CH_3CH(^3A'') + NH_2(^2B_1) \quad \Delta H_0° = -73.0 \text{ kJ mol}^{-1} \quad (1m)$$

$$N(^2D) + C_2H_6 \rightarrow CH_2NH_2(^2A') + CH_2(^3B_1) \quad \Delta H_0° = -67.7 \text{ kJ mol}^{-1} \quad (1n)$$

$$N(^2D) + C_2H_6 \rightarrow CH_3CH(^1A) + NH_2(^2B_1) \quad \Delta H_0° = -58.3 \text{ kJ mol}^{-1} \quad (1o)$$

$$N(^2D) + C_2H_6 \rightarrow CH_2CHNH_2(^3A) + H(^2S) \quad \Delta H_0° = -56.2 \text{ kJ mol}^{-1} \quad (1p)$$

$$N(^2D) + C_2H_6 \rightarrow CH_3NH(^2A'') + CH_2(^3B_1) \quad \Delta H_0° = -43.3 \text{ kJ mol}^{-1} \quad (1q)$$

$$N(^2D) + C_2H_6 \rightarrow CH_3CNH_2(^3A'') + H(^2S) \quad \Delta H_0° = -42.9 \text{ kJ mol}^{-1} \quad (1r)$$

$$N(^2D) + C_2H_6 \rightarrow CH_2NHCH_2(^3A) + H(^2S) \quad \Delta H_0° = -38.1 \text{ kJ mol}^{-1} \quad (1s)$$

$$N(^2D) + C_2H_6 \rightarrow CH_3NCH_2(^3A'') + H(^2S) \quad \Delta H_0° = -37.6 \text{ kJ mol}^{-1} \quad (1t)$$

$$N(^2D) + C_2H_6 \rightarrow CH_2CH_2NH\ (^3A') + H(^2S) \quad \Delta H_0° = -31.6 \text{ kJ mol}^{-1} \quad (1u)$$

$$N(^2D) + C_2H_6 \rightarrow CH_3NH(^2A'') + CH_2(^1A_1) \quad \Delta H_0° = -28.5 \text{ kJ mol}^{-1} \quad (1v)$$

where the enthalpies of the reaction channels reported are those calculated in the present work at the CCSD(T) level of calculations (see below). Notably, all the above mentioned reaction channels correlate with the reactant asymptote directly or through one or more isomerization steps. Some molecular products can be formed either in the ground singlet or excited triplet states (see, for instance, channels (1d) and (1p) or (1e) and (1t)). The variety of products that can be generated is the result of the versatility of carbon and nitrogen in forming bonds, of the relatively large number of atoms involved and of the high energy content of the reactants. The species $CH_2=NH$ (methanimine), $CH_3CH=NH$ (ethanimine), $CH_2=CHNH_2$ (ethenamine or vinyl amine), $CH_3N=CH_2$ (N-methylmethanimine) and *cyclic*-CH$_2$(NH)CH$_2$ (aziridine) are closed-shell species characterized by some stability.

The implication for the atmospheric chemistry of Titan, as well as that of other solar bodies where both nitrogen and ethane might be present, such as Uranus and Neptune, will also be addressed.

2 Experimental

The scattering experiments were carried out by using a CMB apparatus that has been described in detail elsewhere.[23–25] Briefly, two well-collimated, in angle and velocity, continuous supersonic beams of the reactants are crossed at 90° in a large scattering chamber with background pressure in the 10^{-6} hPa range, which assures the single collision conditions. The beam crossing angle can also be set to 45° or 135°.[24,25] The detection system consists of an electron impact ionizer, a quadrupole mass filter and an off-axis (90°) secondary electron multiplier. The ionizer is located in the innermost region of a triply-differentially-pumped ultra-high-vacuum chamber which is maintained in the 10^{-11} hPa pressure range in operating conditions by extensive turbo- and cryo-pumping. The whole detector unit can be rotated in the collision plane around an axis passing through the collision center and the velocities of the particles can be derived from TOF measurements.

The study of reaction (1) has been possible following the development in our laboratory of a continuous supersonic beam of nitrogen atoms containing, in addition to the electronic ground state ^{4}S, a sizeable amount of the excited, metastable state ^{2}D.[35] Atomic nitrogen beams have been generated by the high-pressure radio-frequency discharge beam source[35] successfully used in our laboratory over a number of years to generate intense supersonic beams of atoms and radicals.[13,19–26,28–33] In the present series of experiments, we have used a 0.43 mm dia. quartz nozzle and a boron nitride skimmer (dia. 0.8 mm) located at a distance of 6.5 mm from the nozzle. Starting from mixtures of N$_2$ (2.5%) in He or Ne(30%)/He, a high degree of molecular dissociation ($\sim$60%) was achieved. Atomic nitrogen was produced in a distribution of electronic states which has been characterized by Stern–Gerlach magnetic analysis:[35] 72% of the N atoms were found in the ground ^{4}S state, and 21% and 7% in the metastable excited ^{2}D and ^{2}P states (the latter lying 343.5 kJ mol^{-1} above the ground state[14]). The use of nitrogen atom beams which contain also N(^{4}S) and N(^{2}P), does not represent a complication in the present experiments because the rate constants for the N(^{4}S) reactions with saturated hydrocarbons are immeasurably small[36] while the N(^{2}P) decay rate constant is smaller (5.3 × 10^{-13} cm^3 s^{-1}) than that for N(^{2}D) and is believed to be due essentially to physical quenching.[37]

In the experiment at the higher collision energy, the atomic nitrogen beam was generated by expanding the N$_2$(2.5%)/He mixture at 90 hPa and 250 W of RF power; the resulting beam has a peak velocity of 2425 m s^{-1} and a speed ratio of 5.7. In the experiment at the lower collision energy, the atomic nitrogen beam was generated by expanding the N$_2$(2.5%)/Ne(30%)/He mixture at 75 hPa and 260 W of RF power; the resulting beam has a peak velocity of 1747 m s^{-1} and speed ratio 4.0. The beam of C$_2$H$_6$ was produced by supersonic expansion through a 100 µm stainless-steel nozzle of pure C$_2$H$_6$ with a stagnation pressure of 500 hPa. The beam was skimmed by a 1 mm stainless-steel skimmer. The resulting collision energies were 31.4 and 18.0 kJ mol^{-1}.

Reaction products could be detected at mass-to-charge ratios, *m/z*, of 43 (C$_2$H$_5$N$^+$), 42 (C$_2$H$_4$N$^+$) and 41 (C$_2$H$_3$N$^+$). We have also attempted to measure laboratory distributions at *m/z* = 28–30 and 14–16 to characterize the channels (1a), (1b), (1g), (1i), (1j), (1l), (1m), (1n), (1o) and (1q). This attempt failed because in the *m/z* range of interest there were strong interfering signals caused by both atomic (^{14}N and ^{15}N) and molecular nitrogen (^{14}N^{14}N, ^{14}N^{15}N) and ethane (also producing, through dissociative ionization, the daughter ions C$_2$H$_5^+$, C$_2$H$_4^+$, CH$_3^+$, CH$_2^+$).

The distributions recorded at *m/z* = 43, 42 and 41 were superimposable, thus suggesting that they are generated by the same molecular products with gross formula

C$_2$H$_5$N that partially undergoes dissociative ionization in the electron impact ionizer. Because of the better signal-to-noise ratio, all the final measurements were carried out at m/z = 42. The m/z = 42 laboratory angular distributions, N(Θ), were obtained by taking at least 4 scans of 50 s counts at each angle. The nominal angular resolution of the detector for a point collision zone is 1°. The secondary target beam (C$_2$H$_6$ beam) was modulated at 160 Hz with a tuning fork chopper for background subtraction. Velocity analysis of the beams was carried out by conventional "single-shot" TOF techniques, using a high-speed multichannel scaler and a CAMAC data acquisition system controlled by a personal computer. Velocity distributions of the products were obtained at selected different angles using the cross-correlation TOF technique with four 127-bit pseudorandom sequences. High-time resolution was achieved by spinning the TOF disk, located at the entrance of the detector, at 393.7 Hz corresponding to a dwell time of 5 μs/channel. The flight length was 23.6 cm. Counting times varied from 120 to 360 min depending upon signal intensity.

The scattering measurements have been carried out in the laboratory (LAB) system of coordinates, while for the physical interpretation of the scattering process it is necessary to transform the data (angular, N(Θ), and time-of-flight, N(Θ,t) distributions) to a coordinate system which moves with the center-of-mass (CM) of the colliding system.[23] Because of the finite resolution of experimental conditions, i.e., finite angular and velocity spread of the reactant beams and angular resolution of the detector, the LAB-CM transformation is not single-valued and, therefore, analysis of the laboratory data is carried out by the usual forward convolution procedure.[23] Trial CM angular and velocity distributions are assumed, averaged and transformed to the LAB for comparison with the experimental data until the best fit of the LAB distributions is achieved.

3. Computational details

The potential energy surface of the system N(^{2}D) + C$_2$H$_6$ has been investigated by locating the lowest stationary points at the B3LYP[38] level of theory in conjunction with the correlation consistent valence polarized set aug-cc-pVTZ.[39] At the same level of theory we have computed the harmonic vibrational frequencies in order to check the nature of the stationary points, i.e. minimum if all the frequencies are real, saddle point if there is one, and only one, imaginary frequency. The assignment of the saddle points was performed using intrinsic reaction coordinate (IRC) calculations.[40] The energy of all the stationary points was computed at the higher level of calculation CCSD(T)[41] using the same basis set aug-cc-pVTZ. Both the B3LYP and the CCSD(T) energies were corrected to 0 K by adding the zero point energy correction computed using the scaled harmonic vibrational frequencies evaluated at B3LYP/aug-cc-pVTZ level. The energy of N(^{2}D) was estimated by adding the experimental[14] separation N(^{4}S) − N(^{2}D) of 230.0 kJ mol^{-1} to the energy of N(^{4}S) at all levels of calculation. Thermochemical calculations were performed at the W1[42] level of theory for selected systems. All calculations were done using Gaussian 03[43] while the analysis of the vibrational frequencies was performed using Molekel.[44]

4. RRKM calculations

We have performed RRKM calculations on the N(^{2}D) + C$_2$H$_6$ system, using a code recently developed for this purpose.[13,33a] In accordance with the RRKM scheme,[45] the microcanonical rate constant for a specific reaction at a specific total energy (hereby denoted by k(E)) is given by the expression

$$k(E) = \frac{N_{TS}(E)}{\rho_r(E)}$$

where $N_{TS}(E)$ stands for the number of states (orthogonal to the reaction coordinate) open at the transition state at an energy E, and $\rho_r(E)$ denotes the reactant density of states at the same energy. Regarding the total angular momentum (J), the system was assumed to be described by a distribution of J states between 0 and 50, where each value of J is weighted by a factor proportional to its corresponding density of states at the energy concerned. We have noted that, even though the J distribution can influence considerably the absolute value of the rate constants, it only has a minimal effect on the branching ratio.

The rotational densities of states, both for the reactants and for the transition states, were calculated using an inverse Laplace transform of the corresponding partition functions. Subsequently, the rotational densities of states were convoluted with the corresponding vibrational ones using a direct count algorithm. Finally, the density of states for the transition state was appropriately integrated with respect to the energy in order to produce the sum of states required.

In the exact expression for the microcanonical rate constant, it is the cumulative reaction probability that should be written in the expression instead of the sum of states. Among other factors, this includes the possibility of tunnelling from classically unavailable states as well as antitunnelling (a reaction probability less than 1) from classically available ones. Since many reactions in this system involve displacement of H atoms, we have included tunnelling corrections in the RRKM calculations using the imaginary vibrational frequency of the tight transition states and simulating the transition state as an Eckart barrier. In other words, in order to obtain the appropriate 'sum of states', the TS density of states was not multiplied with a Heaviside step function before integration but with an appropriately smoothed function given the imaginary frequency of the transition state.

In the cases of 'loose' transition states (monotonic dissociation channels), there is no well defined transition state and one has to utilise the variational RRKM approach. We performed *ab initio* and RRKM calculations at various points along the reaction coordinate (typically at a step of 0.5–1.0 Å), choosing as a transition state the point yielding the minimum value of the rate constant (alternatively, sum of states) in accordance with the variational (VTST) approach.[46] However, intermediate points were not available for all loose dissociation channels due to difficulties in the electronic structure calculations. In these cases, we have assumed as transition states for the system the products at infinite separation.

5. Results and analysis of reactive scattering experiments

In Fig. 1 and 2 are shown the $m/z = 42$ angular distributions in the LAB system obtained at the collision energies of 31.4 kJ mol^{-1} and 18.0 kJ mol^{-1}, respectively, with the corresponding Newton diagrams. The error bars (representing ±1 standard deviation) are also reported when they exceed the size of the dots indicating the intensity averaged over the different scans. In the Newton diagrams of Fig. 1 and 2 are also shown the Newton circles relative to the possible isomers $CH_3CH=NH$ (dotted line), $CH_2=CHNH_2$ (continuous line) and CH_3CH_2N (dashed line) in the assumption that all the available energy is converted into product translational energy. The Newton circles delimit the LAB angular range within which each specific isomer can be scattered. The exothermicity of the H-displacement channels implies a different extension of the Newton circles and of the relative scattering angular ranges. $CH_3CH=NH$ and $CH_2=CHNH_2$ are produced in channels (1c) and (1d) which are characterized by a very similar reaction enthalpy. For this reason, the two circles are almost overlapping each other, while the circle associated with ethylnitrene is quite smaller. We have not reported the Newton circles associated with the other C_2H_5N isomers to avoid too many lines and in consideration of their negligible contribution to the global reaction (see Discussion).

TOF spectra at $m/z = 42$ were measured at selected LAB angles at $E_c = 31.4$ kJ mol^{-1} and only in the proximity of the CM angle at the lower E_c of 18.0 kJ mol^{-1} because of the much lower signal-to-noise ratio of this set of data.

Both LAB angular distributions reported in Fig. 1 and 2 appear to be quite symmetric with respect to the CM angle, with similar intensity to its right and left. This feature by itself might suggest that the reaction mechanism is dominated by the formation of bound intermediates. The TOF spectra show some structure, with a fast peak and a slow shoulder. Together with the width of the LAB angular distributions, the bimodality of the TOF is a clear indication that the peak of the product translational energy distributions is far from $E_T' = 0$.

The solid lines in Fig. 1–4 represent the curves calculated with the best-fit functions depicted in Fig. 5 and 6. The grey areas delimit the range of CM functions which still afford an acceptable fit of the data, *i.e.*, they represent the error bars of the present determination. At both E_c, the best CM angular distributions are almost isotropic with a best fit $T(90°)/T(0°)$ value of 0.90 at both E_cs. This ratio can vary by about ±0.15 within the error bars. Such a shape is consistent with the formation of a bound intermediate with a lifetime longer than its rotational period. Also the best-fit product translational energy distributions exhibit about the same characteristics for the two experiments. The fit of both angular and TOF distributions was particularly sensitive to the rise and peak shape of $P(E_T')$, while it was less sensitive to its high energy cut-off, as clearly visible from the extension of the grey areas. The low sensitivity to the high-energy cut-off does not allow the attribution of reactive scattering signal to any specific H-displacement channel, as many of them are compatible with the largest value of E_T' derived by the best fit analysis (namely, channels (1c), (1d), (1e), (1f) and (1h)). Other, much less exothermic channels – such as (1p), (1r), (1s), (1t) and (1u) – can be ruled out, while the channel (1k) leading to the diradical ethylnitrene (CH_3CH_2N) can still make an important contribution because the main part of the $P(E_T')$ is comprised within the maximum E_T' achievable by this isomeric product.

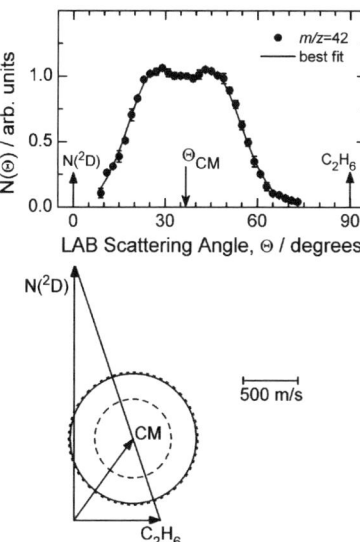

Fig. 1 Laboratory angular distributions recorded at $m/z = 42$ for the reaction $N(^2D) + C_2H_6$ at $E_c = 31.4$ kJ mol^{-1}. Error bars, when visible outside the dots, represent ±1 standard deviation from the mean. The circles in the Newton diagram delimit the maximum velocity that the $CH_3CH=NH$ (dotted), $CH_2=CHNH_2$ (continuous) and CH_3CH_2N (dashed) products from channels (1c), (1d) and (1k), respectively, can attain if all the available energy is channelled into product translational energy. The solid line is the total $N(\Theta)$ calculated when using the best-fit CM angular and translational energy distributions of Fig. 5.

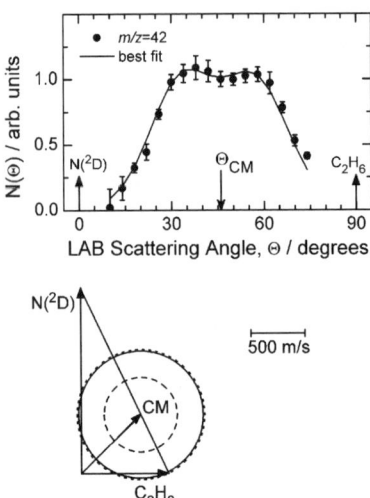

Fig. 2 As in Fig. 1, but at $E_c = 18.0$ kJ mol^{-1}. The solid line is the total $N(\Theta)$ calculated when using the best-fit CM angular and translational energy distributions of Fig. 6.

The average product translational energy, defined as $<E_T'> = \sum P(E_T')E_T'/\sum P(E_T')$ corresponds to a fraction, f_T, of the total available energy ($E_{tot} = E_c - \Delta H_0°$ where the theoretical value of the most exothemic H-displacement channel (1c) is used) of 0.20 at both E_c, even though the peak is quite displaced from $E_T' = 0$.

By analogy with the similar N(^{2}D) + CH$_4$ system, during the best-fit procedure we have tried to simulate the experimental results by using two different contributions,

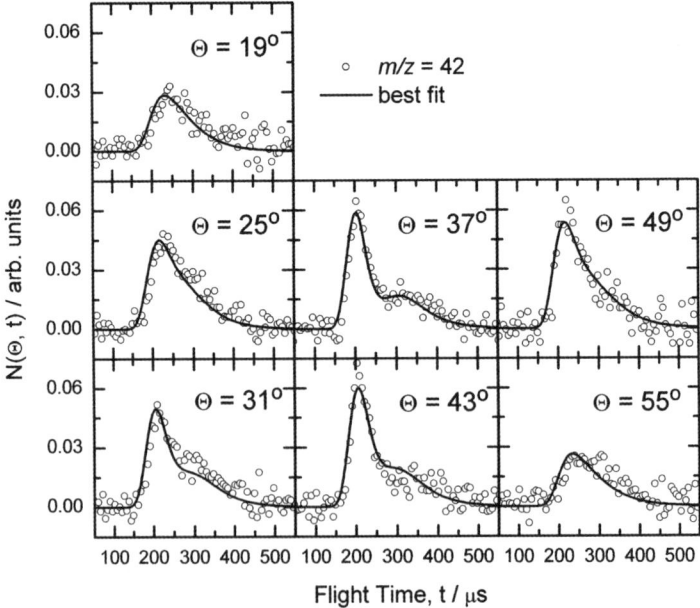

Fig. 3 Time-of-flight distributions of the products (open circles) detected at $m/z = 42$ for the reaction N(^{2}D) + C$_2$H$_6$ at $E_c = 31.4$ kJ mol^{-1} at the indicated LAB angles. Solid lines represent the TOF distributions calculated from the best fit CM functions reported in Fig. 5.

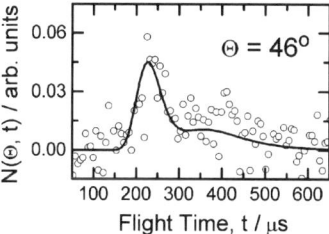

Fig. 4 Time-of-flight distribution of the products (open circles) detected at $m/z = 42$ for the reaction $N(^2D) + C_2H_6$ at $E_c = 18.0$ kJ mol^{-1} at $\Theta = 46°$. Solid lines represent the TOF distribution calculated from the best fit CM functions reported in Fig. 6.

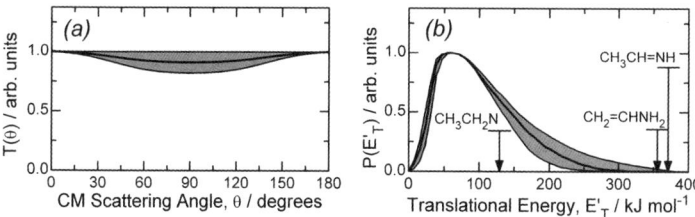

Fig. 5 Best fit CM product (a) angular and (b) translational energy distributions for the reaction $N(^2D) + C_2H_6$ at $E_c = 31.4$ kJ mol^{-1}. The arrows in panel (b) indicate the total energy available to the product isomers $CH_3CH=NH$, $CH_2=CHNH_2$ and CH_3CH_2N. The shaded areas delimit the range of CM functions which still afford an acceptable fit of the data.

one associated with the most exothermic *H*-displacement channel, (1c), which should be the favoured one in the case of complete IVR, and one associated with the formation of ethylnitrene (channel 1k) formed by the break-up of the newly formed N–H bond before an efficient IVR takes place. Nonetheless, since a satisfying fit could also be obtained using a single set of CM functions, we have preferred not to report the double channel fit.

6. Computational results

The lowest stationary points localized on the $N(^2D) + C_2H_6$ PES have been reported in Fig. 7, where the main geometrical parameters (Å and °) are shown together with the energies computed at B3LYP/aug-cc-pVTZ and CCSD(T)/aug-cc-pVTZ (in parentheses) level, relatively to that of CH_3CHNH_2 which is the most stable species at both levels of calculation. In Fig. 8 we have reported also the structure and the relative energies of the main transition states, while in Fig. 9 the geometries of the main dissociation products are shown. The energy changes and barrier heights

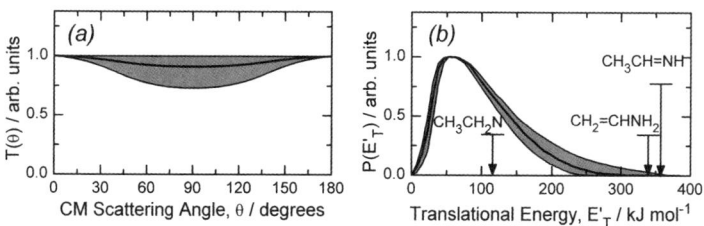

Fig. 6 As in Fig. 5, but for $E_c = 18.0$ kJ mol^{-1}.

computed at 0 K with inclusion of the zero point energy correction for the main isomerisation and dissociation processes are reported in Table 1, where for comparison purposes, we have reported also some ΔH_0^0 values computed at the W1 level. As expected, CCSD(T) relative energies are in better agreement with the W1 values than with the B3LYP ones; for this reason we will consider only CCSD(T) values in the following discussion. A schematic representation of the potential energy surface of the system $N(^2D) + C_2H_6$ is shown in Fig. 10 and 11. In order to have a clearer picture of the exit channels we have reported them in two different Figures, of which Fig. 10 represents all the exit channels characterized by an exit barrier, while Fig. 11 represents those without an exit barrier. The energy scale of Fig. 11 has been expanded in the upper part to make visible the numerous channels originating from the fragmentation of the five reaction intermediates. For the sake of simplicity in Fig. 10 and 11 we have reported only the relative energies computed at CCSD(T) level, while in Table 1 we have reported the values computed at all levels of calculation.

As is visible in Fig. 10 and 11, the insertion of $N(^2D)$ into one of the C–H bonds of C_2H_6 gives rise to the species CH_3CH_2NH which is more stable than the reactants by 447.0 kJ mol^{-1} at CCSD(T) level. For this insertive approach we have located a saddle point which is, however, below the reactants both at B3LYP and CCSD(T) level of calculation. The presence of saddle points below the reactants has already been observed for very exothermic reactions, in particular involving small hydrocarbons.[13,47] Nonetheless, since we were not successful in localizing a stable initial complex, the energy of this saddle point does not seem to be very meaningful.

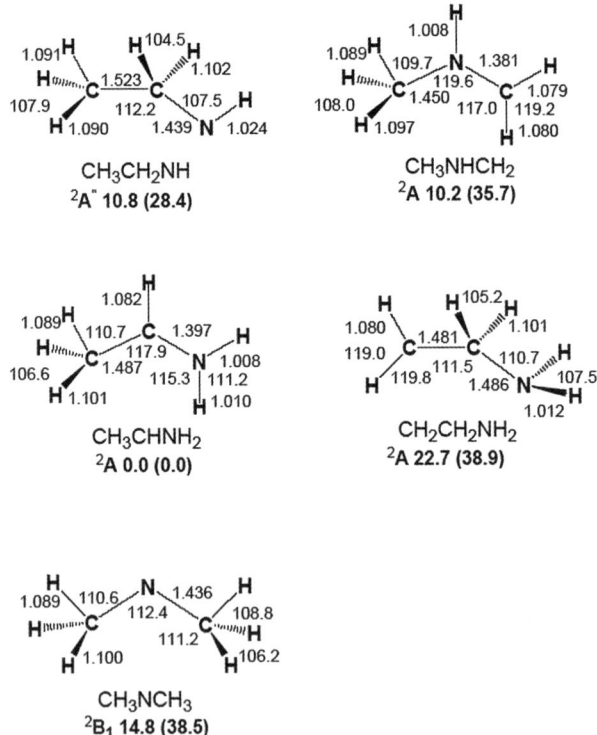

Fig. 7 B3LYP optimized geometries (Å and °) and relative energies (in kJ mol^{-1}; the zero of energy is associated with CH_3CHNH_2, the most stable species of the PES) at 0 K of minima localized on the PES of $N(^2D) + C_2H_6$; CCSD(T) relative energies are reported in parentheses.

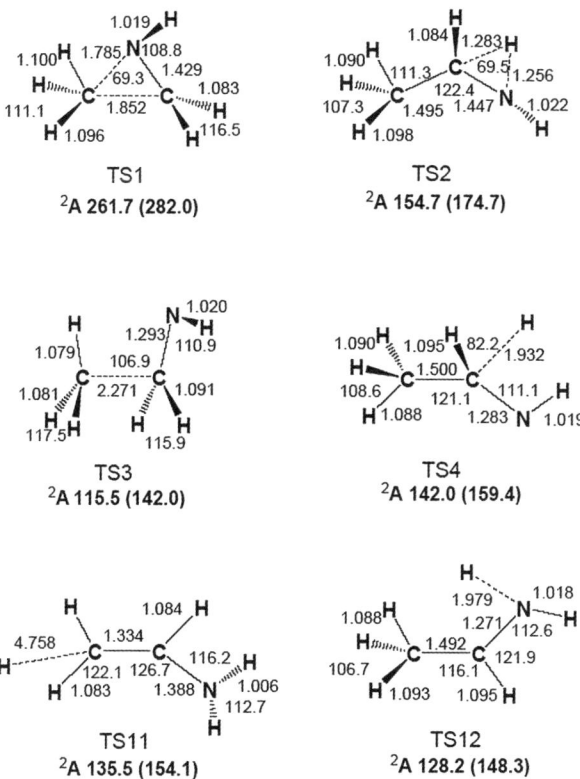

Fig. 8 B3LYP optimized geometries (Å and °) and relative energies (kJ mol^{-1}) at 0 K of main saddle points on the the PES of N(^{2}D) + C$_2$H$_6$; CCSD(T) relative energies are reported in parentheses.

Once formed, the CH$_3$CH$_2$NH intermediate can rearrange to the isomer CH$_3$NHCH$_2$, with a relatively high barrier of 253.6 kJ mol^{-1} or to the isomer CH$_3$CHNH$_2$, which is the most stable intermediate on the N(^{2}D) + C$_2$H$_6$ PES, with a barrier of 146.3 kJ mol^{-1}. Alternatively, CH$_3$CH$_2$NH can directly dissociate to CH$_3$CH=NH and H with a slightly lower barrier of 131.0 kJ mol^{-1} or to the more stable products CH$_3$ and CH$_2$=NH with an even lower barrier of 113.6 kJ mol^{-1}. The other channels, *i.e.* the dissociation of CH$_3$CH$_2$NH to CH$_3$CH$_2$ + NH, CH$_3$CH$_2$N + H, CH$_3$ + CH$_2$NH in its excited triplet state, and CH$_2$CH$_2$NH in its excited triplet state + H are also barrierless, but are characterized by a higher endothermicity, the dissociation energies being 311.4, 348.2, 352.6, and 415.4 kJ mol^{-1}, respectively. The intermediate CH$_3$NHCH$_2$ can isomerize to CH$_3$NCH$_3$ with a barrier of 172.2 kJ mol^{-1} or dissociate to CH$_3$ + CH$_2$NH, CH$_3$NCH$_2$ + H, c-CH$_2$(NH)CH$_2$ + H, or CH$_2$NHCH$_2$ + H with barriers of 125.2, 140.2, 268.7, 276.7 kJ mol^{-1}, respectively. CH$_3$NHCH$_2$ can dissociate also to CH$_3$ + CH$_2$NH in its excited triplet state, CH$_3$NH + CH$_2$, CH$_2$NHCH$_2$ in its excited triplet state + H, and CH$_3$NCH$_2$ in its excited triplet state + H. All these processes are barrierless, but they are very endothermic, the dissociation energies being 345.4, 396.4, 401.6, and 402.1 kJ mol^{-1}, respectively. The most stable isomer CH$_3$CHNH$_2$ can isomerize to CH$_2$CH$_2$NH$_2$ with a relatively low barrier of 193.6 kJ mol^{-1} or can dissociate to CH$_3$CHNH + H, CH$_2$CHNH$_2$ + H, CHNH$_2$ + CH$_3$, CH$_3$CNH$_2$ + H with barriers of 148.3, 154.1, 258.9, 280.9 kJ mol^{-1}, respectively. The intermediate CH$_3$CHNH$_2$ can dissociate also to CH$_3$CNH$_2$ in its excited triplet state + H, CH$_3$ +

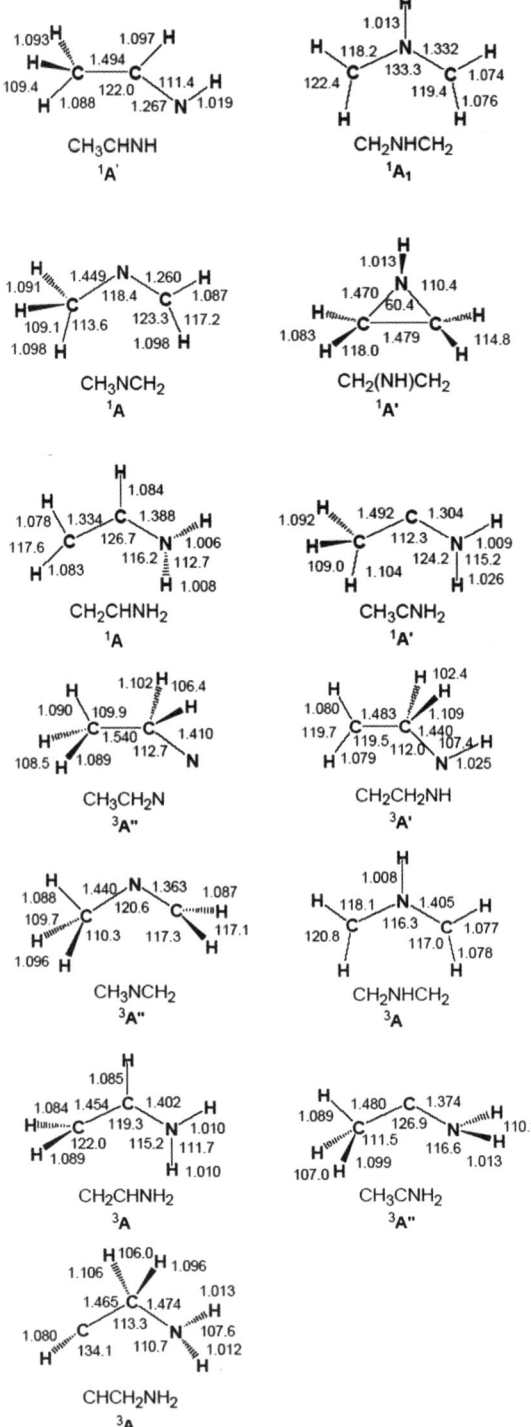

Fig. 9 B3LYP optimized geometries (Å and °) of main dissociation products on the PES of N(^{2}D) + C$_2$H$_6$.

Table 1 Enthalpy changes and barrier heights (kJ mol^{-1}, 0 K) computed at the B3LYP/aug-cc-pVTZ and CCSD(T)/aug-cc-pVTZ levels of theory for selected dissociation and isomerization processes for the system N(^{2}D) + C$_2$H$_6$. Selected enthalpy changes have also been computed at W1 level for comparison purposes

	$\Delta H_0°$			Barrier height	
	B3LYP	CCSD(T)	W1	B3LYP	CCSD(T)
N(^{2}D) + C$_2$H$_6$ → CH$_3$CH$_2$NH	−482.9	−447.0	−465.1	−113.4	−65.6
CH$_3$CH$_2$NH → CH$_3$ + ^{1}CH$_2$NH	72.7	80.8	80.1	104.7	113.6
CH$_3$CH$_2$NH → CH$_3$ + ^{3}CH$_2$NH	323.7	352.6	357.6		
CH$_3$CH$_2$NH → ^{1}CH$_3$CHNH + H	114.6	108.8	105.9	131.2	131.0
CH$_3$CH$_2$NH → CH$_3$CH$_2$ + ^{3}NH	309.0	311.4	323.1		
CH$_3$CH$_2$NH → ^{3}CH$_3$CH$_2$N + H	350.9	348.2	356.7		
CH$_3$CH$_2$NH → ^{3}CH$_2$CH$_2$NH + H	408.5	415.4	419.8		
CH$_3$CH$_2$NH → CH$_3$NHCH$_2$	−0.6	7.3		250.9	253.6
CH$_3$NHCH$_2$ → CH$_3$ + ^{1}CH$_2$NH	73.3	73.5		117.5	125.2
CH$_3$NHCH$_2$ → CH$_3$ + ^{3}CH$_2$NH	324.3	345.4			
CH$_3$NHCH$_2$ → CH$_3$NH + ^{3}CH$_2$	396.0	396.4			
CH$_3$NHCH$_2$ → ^{1}CH$_2$(NH)CH$_2$ + H	209.6	189.9		274.4	268.7
CH$_3$NHCH$_2$ → ^{1}CH$_2$NHCH$_2$ + H	273.4	276.6		273.7	276.7
CH$_3$NHCH$_2$ → ^{3}CH$_2$NHCH$_2$ + H	397.3	401.6			
CH$_3$NHCH$_2$ → ^{1}CH$_3$NCH$_2$ + H	152.5	140.4		152.7	140.2
CH$_3$NHCH$_2$ → ^{3}CH$_3$NCH$_2$ + H	391.3	402.1			
CH$_3$CH$_2$NH → CH$_3$CHNH$_2$	−10.8	−28.4		143.9	146.3
CH$_3$CHNH$_2$ → ^{1}CH$_2$CHNH$_2$ + H	134.3	154.1		135.5	154.1
CH$_3$CHNH$_2$ → ^{3}CH$_2$CHNH$_2$ + H	388.4	419.2			
CH$_3$CHNH$_2$ → CH$_3$ + ^{1}CHNH$_2$	227.4	257.1		231.2	258.9
CH$_3$CHNH$_2$ → CH$_3$ + ^{3}CHNH$_2$	357.9	395.8			
CH$_3$CHNH$_2$ → ^{1}CH$_3$CHNH + H	125.4	137.2		128.2	148.3
CH$_3$CHNH$_2$ → ^{1}CH$_3$CNH$_2$ + H	262.6	280.9		262.8	280.9
CH$_3$CHNH$_2$ → ^{3}CH$_3$CNH$_2$ + H	406.5	376.6			
CH$_3$CHNH$_2$ → ^{3}CH$_3$CH + NH$_2$	375.9	402.4			
CH$_3$CHNH$_2$ → ^{1}CH$_3$CH + NH$_2$	393.2	416.7			
CH$_3$NHCH$_2$ → CH$_3$NCH$_3$	4.6	2.8		176.1	172.2
CH$_3$NCH$_3$ → ^{1}CH$_3$NCH$_2$ + H	147.9	137.6	136.1	154.0	148.6
CH$_3$NCH$_3$ → ^{3}CH$_3$NCH$_2$ + H	386.7	399.3	403.1		
CH$_3$NCH$_3$ → ^{3}CH$_3$N + CH$_3$	277.5	287.0			
CH$_3$CHNH$_2$ → CH$_2$CH$_2$NH$_2$	22.7	38.9		166.8	193.6
CH$_2$CH$_2$NH$_2$ → ^{3}CH$_2$ + CH$_2$NH$_2$	355.3	368.8			
CH$_2$CH$_2$NH$_2$ → ^{1}CH$_2$ + CH$_2$NH$_2$	401.0	408.0			
CH$_2$CH$_2$NH$_2$ → C$_2$H$_4$ + NH$_2$	63.8	72.6		82.5	93.7
CH$_2$CH$_2$NH$_2$ → ^{1}CH$_2$CHNH$_2$ + H	111.6	115.2		126.3	140.1
CH$_2$CH$_2$NH$_2$ → ^{3}CH$_2$CHNH$_2$ + H	365.7	380.3			
CH$_2$CH$_2$NH$_2$ → ^{3}CH$_2$CH$_2$NH + H	396.6	404.9			
CH$_2$CH$_2$NH$_2$ → ^{3}CHCH$_2$NH$_2$ + H	450.2	445.7			

CHNH$_2$ in its excited triplet state, NH$_2$ + CH$_3$CH both in its ground state or in its excited singlet state, and CH$_2$CHNH$_2$ in its excited triplet state + H. All these processes are barrierless but very endothermic, the dissociation energies being 376.6, 395.8, 402.4, 416.7, and 419.2 kJ mol^{-1}, respectively. The intermediate CH$_2$CH$_2$NH$_2$ can dissociate to C$_2$H$_4$ + NH$_2$ or to CH$_2$CHNH$_2$ + H with relatively low barriers of 93.7 and 140.1 kJ mol^{-1}, respectively. CH$_2$CH$_2$NH$_2$ can dissociate also to CH$_2$ + CH$_2$NH$_2$, CH$_2$CHNH$_2$ in its excited triplet state + H, CH$_2$CH$_2$NH + H, and CHCH$_2$NH$_2$ + H; all these processes are barrierless but very endothermic being the dissociation energies 368.8, 380.3, 404.9, and 445.7 kJ mol^{-1}, respectively.

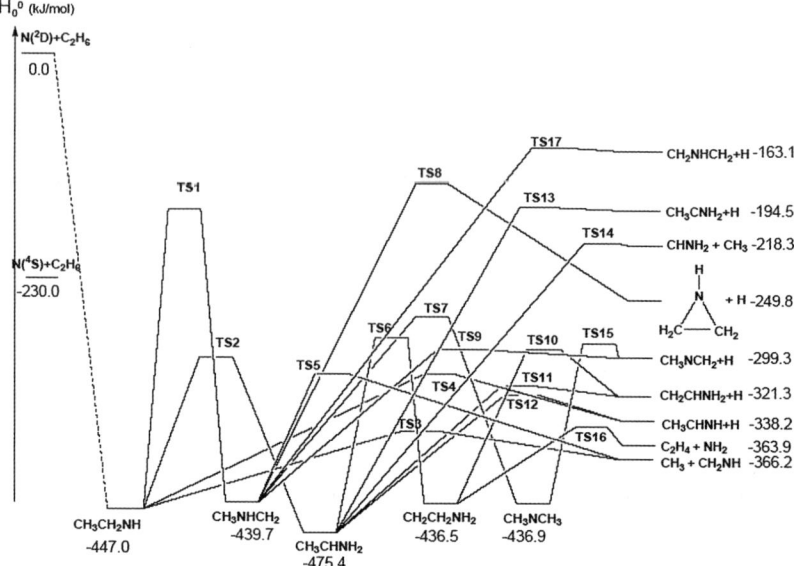

Fig. 10 Schematic representation of the N(^{2}D) + C$_2$H$_6$ potential energy surface illustrating the reaction channels which are characterized by an exit barrier. For simplicity, only the CCSD(T) relative energies (kJ mol^{-1}) are reported.

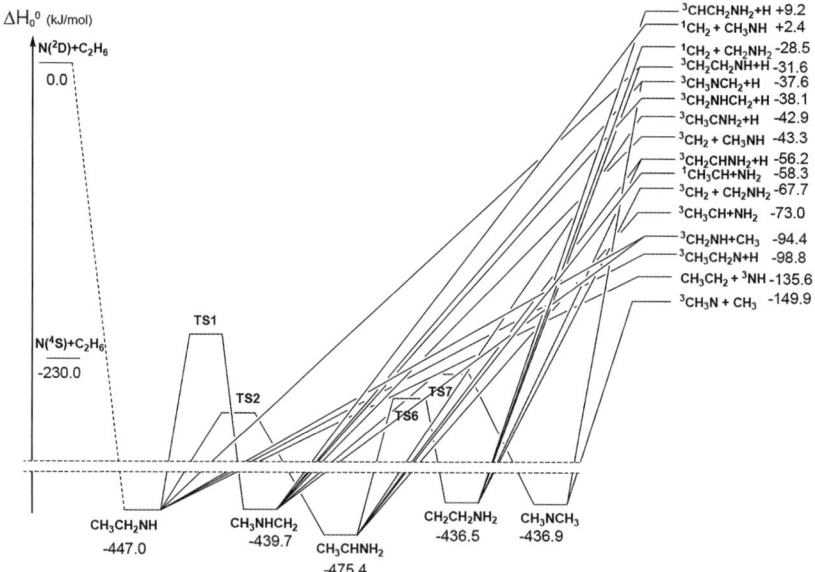

Fig. 11 Schematic representation of the N(^{2}D) + C$_2$H$_6$ potential energy surface illustrating the reaction channels which are not characterized by an exit barrier. These channels are much less exothermic than those in Fig. 10 and the energy scale has been expanded in the upper part. For simplicity, only the CCSD(T) relative energies (kJ mol^{-1}) are reported.

Finally, the intermediate CH$_3$NCH$_3$ can dissociate to CH$_3$NCH$_2$ + H with a barrier of 148.6 kJ mol^{-1} or to CH$_3$N + CH$_3$ in a barrierless process endothermic by 287.0 kJ mol^{-1}. The dissociation of CH$_3$NCH$_3$ to CH$_3$NCH$_2$ in its excited triplet state + H is much more endothermic (399.3 kJ mol^{-1}).

Finally we have been unable to locate a saddle point relative to the H-abstraction channel.

7. RRKM results

RRKM calculations have been carried out for five values of the total energy (the zero of energy is assumed to be the reactants $N(^2D) + C_2H_6$ at infinite separation). Three of these (0.8, 1.5 and 2.5 kJ mol^{-1}) correspond to the most probable collision energy at the surface temperature of Titan (94 K), at the stratospheric temperature of Titan (175 K) and at room temperature, while the other two (18.0 and 31.4 kJ mol^{-1}) correspond to the collision energies used in the present CMB experiments.

According to the present PES calculations, the favourite initial approach is $N(^2D)$ insertion into a C–H bond of ethane forming a CH_3CH_2NH intermediate. This first intermediate, subsequently, can rearrange to form CH_3NHCH_2 or CH_3CHNH_2 or, alternatively, dissociate. As we are going to see, all subsequent important channels of the reaction proceed from one of these intermediates. Even though the 'second generation' intermediates can further rearrange to other intermediates, these latter ones play a negligible role in the final branching ratios. Nevertheless, for the sake of completeness, we provide rate constants for the reactions of these intermediates as well. Because of the complexity of the system, we have found it convenient to report the rate constants distinctly for the various intermediates. In Tables 2–4 are shown, respectively, the rate constants for the reactions originating from the initial insertion intermediate CH_3CH_2NH and the 'second generation' intermediates CH_3NHCH_2 and CH_3CHNH_2. In Tables 5–6 are shown the rate constants for the reactions of the 'third generation' intermediates $CH_2CH_2NH_2$ and CH_3NCH_3 while, in Table 7, are shown the overall final BRs for all dissociation channels, obtained by combining all the possible rearrangement (including back-isomerization) and dissociation rate constants.

It can be seen that, at all energies, the dominant channel is dissociation into $CH_2=NH + CH_3$ (channel 1a). The branching ratio for this channel changes from 0.789 at the lowest energy to 0.740 at the highest one. This dominance is easily explained by the fact that this is by far the fastest dissociation channel from the initial intermediate, an immediate consequence of the fact that the transition state for this dissociation (TS3) is the one at a minimum energy. Moreover, there is a dissociation pathway to the same channel from the CH_3NHCH_2 intermediate and the latter one can be formed by direct isomerisation of CH_3CH_2NH. However, as the rate constant for this isomerisation is much smaller than rate constants of competing channels (the transition state for this isomerisation, TS1, lies at a prohibitively high energy), this latter contribution is not important. In fact, the $CH_3NCH_2 + H$ dissociation channel (obtainable, for all purposes, only from the CH_3NHCH_2 intermediate) has branching ratios of the order of 10^{-5}–10^{-4} at all energies, even though it is the fastest reaction this intermediate undergoes.

The second most important reaction of the CH_3CH_2NH intermediate is dissociation into $CH_3CH=NH + H$ (channel 1c) and, indeed, this is the second most important channel at all energies (with branching ratios around 0.125) and the corresponding transition state (TS4) the second lowest in energy. An alternative pathway to this channel is through the CH_3CHNH_2 intermediate which also can be formed from CH_3CH_2NH by direct isomerisation. This is a more important contribution than in the previous case, as the rate constant for isomerisation to CH_3CHNH_2 is three orders of magnitude higher than the one for isomerisation to CH_3NHCH_2. Clearly, the reason for this is the much lower energy of the TS2 transition state with respect to TS1. This effect is also manifested by the appreciable branching ratios of the $CH_2=CHNH_2 + H$ channel (around 0.02 at all energies), which is the dominant dissociation channel of CH_3CHNH_2 and, again, practically obtainable only by this intermediate. It is worth mentioning that this apparent paradox (the predominance of the $CH_2=CHNH_2 + H$ channel from CH_3CHNH_2

Table 2 Microcanonical rate constants for reaction channels of the CH_3CH_2NH intermediate, in units of s^{-1}. (var.) implies that the transition state has been obtained through the variational method. (inf.) implies that, in the absence of relevant data for a variational treatment, the products at infinite separation have been used as the transition state

Energy/ kJ mol^{-1}	Isomerization to CH_3NHCH_2 ($\times 10^8$)	Isomerization to CH_3CHNH_2 ($\times 10^{11}$)	$CH_3 + CH_2NH$ ($\times 10^{12}$)	$CH_3CHNH + H$ ($\times 10^{11}$)	$^3CH_2CH_2NH + H$ (var.) ($\times 10^8$)	$^3CH_3CH_2N + H$ (var.) ($\times 10^6$)	$CH_3CH_2 + ^2NH$ (var.) ($\times 10^{11}$)	$^3CH_2NH + CH_3$ (inf.) ($\times 10^{10}$)
0.8	7.18	1.37	5.27	8.19	5.31	1.97	3.63	9.06
1.5	7.30	1.37	5.29	8.25	5.70	2.03	3.71	9.27
2.5	7.47	1.39	5.33	8.33	6.30	2.13	3.82	9.57
18.0	10.5	1.60	5.92	9.64	26.2	4.33	5.89	15.3
31.4	13.7	1.80	6.45	10.9	75.7	7.57	8.30	22.2

Table 3 Microcanonical rate constants for reaction channels of the CH_3NHCH_2 intermediate, in units of s^{-1}. (var.) and (inf.) as above

Energy/ kJ mol^{-1}	Isomerization to CH_3CH_2NH ($\times 10^6$)	Isomerization to CH_3NCH_3 ($\times 10^{10}$)	$CH_3 + CH_2NH$ ($\times 10^{12}$)	Aziridine + H ($\times 10^7$)	$CH_3NCH_2 + H$ ($\times 10^{12}$)	$CH_2NHCH_2 + H$ ($\times 10^{10}$)	$^3CH_2NH + CH_3$ (inf.) ($\times 10^{10}$)	$^3CH_2 + CH_3NH$ (var.) ($\times 10^5$)	$^3CH_2NHCH_2 + H$ (inf.) ($\times 10^9$)	$^3CH_3NCH_2 + H$ (inf.) ($\times 10^8$)
0.8	4.91	2.65	3.02	8.67	4.42	6.17	6.19	8.52	2.15	1.99
1.5	4.99	2.68	3.03	8.84	4.45	6.28	6.33	8.93	2.26	2.09
2.5	5.10	2.71	3.06	9.09	4.49	6.45	6.54	9.53	2.43	2.23
18.0	7.15	3.27	3.45	13.8	5.21	9.50	10.5	24.7	6.83	5.73
31.4	9.41	3.82	3.80	19.3	5.88	13.0	15.2	51.7	15.1	11.9

Table 4 Microcanonical rate constants for reaction channels of the CH_3CHNH_2 intermediate, in units of s^{-1}. (var.) and (inf.) as above

Energy/ kJ mol^{-1}	Isomerization to CH_3CH_2NH ($\times 10^{10}$)	Isomerization to $CH_2CH_2NH_2$ ($\times 10^{10}$)	CH_2CHNH_2 + H ($\times 10^{13}$)	CH_3CHNH + H ($\times 10^{11}$)	CH_3CNH_2 + H ($\times 10^{10}$)	$CHNH_2$ + CH_3 ($\times 10^{11}$)	$^3CH_3CH$ + NH_2 (var.) ($\times 10^{7}$)	$^3CH_2CHNH_2$ + H (inf.) ($\times 10^{6}$)	$^3CH_3CNH_2$ + H (inf.) ($\times 10^{7}$)
0.8	2.76	3.23	1.41	9.06	3.66	4.18	9.96	2.20	4.17
1.5	2.78	3.26	1.42	9.13	3.72	4.24	10.3	2.30	4.38
2.5	2.81	3.30	1.43	9.22	3.81	4.32	10.9	2.44	4.69
18.0	3.33	3.99	1.64	10.8	5.44	5.82	24.4	5.88	12.8
31.4	3.83	4.66	1.83	12.3	7.25	7.40	45.8	11.7	27.8

Table 5 Microcanonical rate constants for reaction channels of the $CH_2CH_2NH_2$ intermediate, in units of s^{-1}, (var.) and (inf.) as above

Energy/ kJ mol^{-1}	Isomerization to CH_3CHNH_2 ($\times 10^{11}$)	$CH_2CHNH_2 + H$ ($\times 10^{11}$)	$C_2H_4 + NH_2$ ($\times 10^{12}$)	$^3CH_2CH_2NH + H$ (var.) ($\times 10^5$)	$^3CH_2 + CH_2NH_2$ (var.) ($\times 10^7$)	$^3CH_2CHNH_2 + H$ (inf.) ($\times 10^8$)	$^3CHCH_2NH_2 + H$ (var.) ($\times 10^5$)
0.8	1.01	8.13	3.32	4.02	2.63	6.85	3.25
1.5	1.02	8.19	3.34	4.19	2.72	7.14	3.88
2.5	1.03	8.27	3.36	4.45	2.86	7.57	4.96
18.0	1.20	9.59	3.65	10.7	5.89	17.6	99.8
31.4	1.36	10.8	3.91	21.1	10.4	34.0	656.0

Table 6 Microcanonical rate constants for reaction channels of the CH_3NCH_3 intermediate, in units of s^{-1}. (var.) and (inf.) as above

Energy/ kJ mol^{-1}	Isomerization to CH_3NHCH_2 ($\times 10^{10}$)	$CH_3NCH_2 + H$ ($\times 10^{11}$)	$^3CH_3N + CH_3$(var.) ($\times 10^{10}$)	$^3CH_3NCH_2 + H$ (inf.) ($\times 10^8$)
0.8	2.26	2.93	9.81	1.69
1.5	2.28	2.95	9.96	1.77
2.5	2.30	2.99	10.2	1.89
18.0	2.77	3.59	14.0	4.84
31.4	3.22	4.17	18.1	10.00

Table 7 Global branching ratios for the principal reaction channels

Energy/ kJ mol^{-1}	(1a)	(1b) ($\times 10^{-5}$)	(1c)	(1d)	(1e) ($\times 10^{-5}$)	(1g) ($\times 10^{-4}$)	(1h) ($\times 10^{-5}$)	(1j)	(1l)
0.8	0.789	3.33	0.124	0.0186	6.30	5.51	4.82	0.0544	0.0136
1.5	0.788	3.34	0.124	0.0186	6.36	5.55	4.87	0.0552	0.0138
2.5	0.786	3.35	0.124	0.0186	6.45	5.61	4.94	0.0564	0.0141
18.0	0.760	3.48	0.125	0.0186	7.91	6.58	6.16	0.0756	0.0197
31.4	0.740	3.59	0.126	0.0185	9.30	7.49	7.34	0.0892	0.0255

despite the fact that it has a slightly more energetic transition state) is explained by the fact that the TS11 transition state has large amplitude motions which augment its sum of states to the expense of the relatively tight TS12 transition state.

The third and fourth most important reactions of the CH_3CH_2NH intermediate are dissociation into $CH_3CH_2 + NH$ (1j) and into $CH_2NH(^3A'')+CH_3$ (1l), respectively. Even though the former channel can only be obtained from CH_3CH_2NH, it is more dominant than the latter one which can also be obtained from CH_3NHCH_2. Again, this is an effect of the very low rate of isomerisation to CH_3NHCH_2. Since both of these dissociation channels are barrierless, their rate constants have been calculated using the variational RRKM method. For both of them, the minimum sum of states has been located relatively near the reactant configuration (at a bond length of the bond undergoing fission of around 2.5–3.0 Å).

As the energy increases, the branching ratio of the dominant $CH_2=NH + CH_3$ channel drops from 0.79 to 0.74 while the corresponding one of the $CH_3CH_2 + NH$ and, to a smaller degree, that of the $CH_2NH(^3A'') + CH_3$ increases. The values of the rate constants confirm that the rate constant of the dominant channel rises much more slowly with energy than the other two. This is a result of the 'tighter' nature of the transition state for the first dissociation as opposed to the much more pronounced presence of large amplitude motions in the transition states for the other two dissociation channels.

Minor dissociation products include $CHNH_2 + CH_3$, $CH_3CNH_2 + H$, $C_2H_4 + NH_2$ and $CH_3N=CH_2 + H$. Of these, the first two are obtained from the dissociation of CH_3CHNH_2 (they are the third and fourth fastest ones, besides the two more dominant ones mentioned earlier). The third one is obtained from the dissociation of the 'third generation' intermediate $CH_2CH_2NH_2$ to which CH_3CHNH_2 rearranges at a high rate while the fourth one, as already mentioned, is the result of the dissociation of the CH_3NHCH_2 intermediate to which, however, the original intermediate rearranges only slowly.

A final consideration is to be made regarding the effects of tunnelling. As far as the reactions of the first intermediate are concerned, the one most influenced by tunnelling is isomerisation to CH_3CHNH_2, where tunnelling makes a contribution of around 15% to the total rate constant. This is to be expected since this reaction involves the migration of an H atom across a relatively narrow barrier. As expected, tunnelling contributions to isomerisation to CH_3NHCH_2 are lower (around 8%) and the ones to dissociation channels are practically negligible. The lower contributions of tunnelling to dissociation as opposed to isomerisation are due to the narrower barriers associated with the latter and the consequent higher imaginary frequencies to be considered.

Notably, the channel (1k) leading to ethylnitrene is by far less important, at the statistical level, than those leading to CH_3CH_2 + NH (1j) and $CH_2NH(^3A'')$ + CH_3 (1l), even though these channels are characterized by similar energetics. This is due to the fact that the (variationally determined) transition state leading to this channel is much tighter than the ones for the other two. The main reason for this is that the larger amplitude motions for this transition state involve primarily the motion of the light departing H particle as opposed to heavier units such as NH or CH_3. The much lower mass of the H atom brings about a drastic lowering of the TS density of states and hence of the rate constant.

It is also seen that the yield for the production of aziridine (channel 1f) is negligibly small. The main factor responsible for this is the low rate constant for the production of the CH_3NHCH_2 intermediate (the only one aziridine can be derived from). Moreover, the transition state for the dissociation step (TS8) is very high in energy and therefore, given the intermediate, this is seen to be one of its channels of minor importance.

8. Discussion

Even if limited to the H-displacement channels, the CMB experimental results clearly indicate that the reaction proceeds through the formation of one or more bound intermediates. This view is supported by the electronic structure calculations of the C_2H_6N PES, according to which $N(^2D)$ inserts into one of the C–H bonds of ethane without an entrance barrier leading to the formation of the intermediate CH_3CH_2NH. The insertion intermediate is quite stable with respect to both reactants and dissociation products, but it is formed with a high internal energy content so that it undergoes either a direct fission to the products or isomerization to other intermediates, which, in turn, fragment into other products. The fragmentation of a bond of the initial insertion intermediate leads to: (1) CH_2=NH(methanimine)+CH_3 (channel 1a) if the weakest C–C bond breaks apart and the unpaired electron which is left on the C atom after the homolytic fission of the C–C bond couples with the uncoupled electron on nitrogen, or $CH_2NH(^3A'')$+CH_3 if the C=N double bond is not formed; (2) if one of the pre-existing C–H bonds of the –CH_2–group, not directly involved in the insertion process, breaks apart, CH_3CH=NH and H are formed (channel 1c), with the unpaired electron which is left on the central atom after the homolytic fission of the C–H bond coupling with the uncoupled electron on nitrogen and forming a double C=N bond; (3) if one of the pre-existing C–H bonds of the –CH_3 group, also not directly involved in the insertion process, breaks apart, the triplet biradical CH_2CH_2NH (channel 1u) is formed; (4) finally, if one of the new bonds, formed by $N(^2D)$ insertion into a C–H bond, breaks apart we can have the formation of NH + CH_3CH_2 (if the new C–N bond fragments, channel (1j)) or CH_3CH_2N + H (if the new N–H splits, channel (1k)). The two bond energies are comparable and so are the reaction enthalpies associated with the two channels (1j) and (1k). All the mentioned fission processes compete with each other. In addition, as illustrated in Fig. 10 and 11 and already commented on in Section 6, the initial addition intermediate can rearrange to isomers CH_3CHNH_2 and CH_3NHCH_2 which open up other reaction channels. Nevertheless, according to RRKM calculations, the isomerization processes are not competitive and make a minor contribution to the reaction. The only exception comes from the fragmentation of CH_3CHNH_2 to CH_2=$CHNH_2$ + H which accounts for roughly 2% of the global reaction in the statistical prediction. As already pointed out, the RRKM method relies on the basic assumption that the lifetimes of the involved intermediates are long enough to allow a complete IVR. Therefore the RRKM BR are not representing the other possible channels where dynamical effects, rather than statistics, are at play. For the analogous reaction $N(^2D) + CH_4$ a competition between the channel favoured on statistical grounds (CH_2=NH + H) and those resulting from a relatively fast fragmentation of the new, strongly excited N–H and C–N bonds (CH_3N + H and NH + CH_3) has been observed. In particular, the non-statistical channels have been seen to be favoured with the increase of the E_c (and therefore of the internal energy of the insertion intermediate). The trend was confirmed in that case also by an increase in the forward bias of the CM angular distribution associated with the statistical channel, which is a symptom of the shortening of the lifetime of the insertion intermediate. In the present case as well, it is reasonable to expect that both mechanisms – a "statistical" one leading mostly to CH_2=NH + CH_3 or CH_3CH=NH + H and a "dynamical" one leading to CH_3CH_2N + H or NH + C_2H_5 – are at play. In this respect it is instructive to analyse the results of a previous dynamical experiment on the NH + C_2H_5 channel by Umemoto et al.[34b] According to their laser-induced-fluorescence study at room temperature, the observed rovibrational population of NH is consistent with an insertive mechanism proceeding through the formation of a short-lived intermediate complex which dissociates before IVR. They exclude a contribution from the H-abstraction mechanism or the fragmentation of a long-lived complex. That because the NH rovibrational distribution is unimodal (thus implying that only one reaction mechanism is operative) and is similar to that associated with other reactions involving the formation of short-lived complexes. Regarding the total yield of the NH channel, Umemoto et al. were not able to give an absolute yield but could furnish the relative yield with respect to that of the reaction $N(^2D) + CH_4$. In this way they have established that the NH yield is roughly one third of the one relative to the reaction $N(^2D) + CH_4$. Since the absolute yield of NH in the reaction $N(^2D) + CH_4$ is 0.3 ± 0.1,[48] the channel leading to NH + C_2H_5 accounts for ~10% of the global reaction. Therefore, the fraction of the reaction that proceeds via a short-lived complex is smaller in the case of $N(^2D) + C_2H_6$ with respect to $N(^2D) + CH_4$. This is certainly not surprising because of the increased number of the degrees of freedom in the N + C_2H_6 system. Several implications for the H-displacement reaction also follow. Here, as well, we expect a competition between the channel favoured on statistical grounds, i.e. CH_3CH=NH + H, and that originated by the fission of the activated N–H bond, i.e. CH_3CH_2N + H. Unfortunately, the sensitivity of our data has not allowed us to discriminate between the two contributions, as in the case of $N(^2D)$ + CH_4. Nevertheless, in the light of the above reasoning, we expect an essentially minor contribution from the CH_3CH_2N channel because of the increased capability of the insertive CH_3CH_2NH intermediate to randomize its high internal energy content, which unavoidably will cause the fragmentation of the weakest C–C bond or, to a minor extent, will lead to the formation of CH_3CH=NH. In conclusion, for the title reaction we expect a mostly statistical behaviour, with a minor contribution (up to a few percent according to the estimate by Umemoto et al.) from the two "dynamical" channels. Incidentally, the symmetry of the best fit CM angular distributions with respect to $\theta = 90°$ (Fig. 5 and 6), which is retained also at the higher E_c investigated, witnesses a long intermediate lifetime. All the other channels associated with H-displacement give a minor contribution: the dissociation of one of the pre-existing C–H bonds of the –CH_3 group (case 3, channel 1u) is not favoured either on statistic grounds or by dynamical effects; CH_3N=CH_2, c-$CH_2(NH)CH_2$, CH_3CNH_2, CH_2CHNH_2, CH_3CNH_2(triplet), CH_2NHCH_2 and CH_3NCH_2(triplet) can only be formed after extensive rearrangements of the initial insertive intermediate take place and therefore they fall in the statistical description which establishes their minor role. CH_2=$CHNH_2$ is the only channel which makes a sizeable – though small – contribution to the total reaction amongst the channels that require the isomerization of the first intermediate.

9. Implications for the atmosphere of Titan

An important conclusion of the dynamical and theoretical investigation of the title reaction is that, mostly, it behaves statistically. This assertion is even more valid when we move to low collision energies because of the increasing lifetime of the insertion intermediate. That implies that for all purposes concerning the modelling of the atmosphere of Titan we can refer to the RRKM BRs derived at the energy corresponding to the stratospheric temperature (see Table 7). The dominant reaction channel is the one leading to methanimine and CH_3 (78.8%), followed by the one leading to ethanimine and H (12.4%) with minor contributions from the channels leading to NH + C_2H_5 (5.5%), ethenamine plus H (1.9%) and excited triplet CH_2NH plus CH_3 (1.4%). Therefore, neither the assumption[7,8,10] that NH + C_2H_5 is the sole reaction channel nor the recent suggestion[9] that c-$CH_2(NH)CH_2$ + H is the dominant reaction channel are correct. Methanimine is also produced by the reaction $N(^2D)$ + CH_4. Methane is more abundant than ethane by roughly three orders of magnitude, but the rate constant (at room temperature) of the reaction $N(^2D)$ + C_2H_6 is one order of magnitude larger than that of methane. Therefore, the reaction of $N(^2D)$ + C_2H_6 should contribute by a few percent to the global budget of methanimine. The only information on the presence of methanimine in the atmosphere of Titan comes from the analysis of the ion spectra recorded by Cassini Ion Neutral Mass Spectrometer (INMS).[18] Previous attempts to search for CH_2=NH in Titan's atmosphere at millimetre wavelengths failed.[7] The recent model by Lavvas et al.[9] derived a larger quantity than that inferred by INMS and, therefore, there is no need of an additional source of CH_2=NH, as our new results suggest. However, there is a lot of uncertainty on the possible fate of CH_2=NH in the upper atmosphere of Titan, because of a severe lack of knowledge on the possible chemical loss pathways of this species. CH_2=NH is known to absorb in the UV region[49] and the possible photodissociation products are HCN/HNC + H_2 or H_2CN + H.[50] Also, having a C = N double bond, methanimine is a highly reactive molecule that can react with atomic/radical species and also ions, present in the upper atmosphere of Titan. For instance, a theoretical investigation on the reaction CN + CH_2=NH has revealed the absence of an entrance barrier, thus suggesting that this reaction can have a rate constant in the gas kinetic range.[51] Growing evidence suggests that nitrogen chemistry contributes to the formation of the haze aerosols in the Titan upper atmosphere.[52–54] In this respect, it should be noted that imines are well-known for their capability of polymerizing (as a matter of fact, even though methanimine is a closed-shell species, it is unstable and cannot be stored as such). In this respect, CH_2=NH is an excellent candidate to account for the nitrogen-rich aerosols of Titan[52] through polymerization and copolymerization with other unsaturated nitriles or unsaturated hydrocarbons.[9] Finally, the production of CH_3 from a reaction destroying ethane partially recycles it.

Ethanimine, the second most important molecular product of the title reaction, can also be a source of nitrogen-rich molecules and aerosols *via* addition reactions. Alternatively, it might undergo UV photodissociation, thus forming reactive radicals that can further enhance the formation of nitrogen-rich complex species. A theoretical study[55] has suggested that activated ethanimine can directly decompose to CH_3CN + H_2 and CH_4 + HCN/HNC. In this respect, activated ethanimine could be a source of acetonitrile and HCN/HNC.

Ethenamine (a minor product of the title reaction) can also represent a building block of nitrogen polymers (polyvinylamine is a commercial polymer). Also, having an amino group ($–NH_2$) ethenamine can possibly be involved in the formation of ammonia. Notably, at this stage, both mechanisms proposed by Lavvas et al.[9] based on the formation of aziridine, as well as the ionic mechanism proposed by Wilson and Atreya[8] have proved to be not effective.[56]

A last mention goes to the importance of methanimine (and possibly ethanimine) in prebiotic chemistry: methanimine can form the simplest amino acid, glycine, by reacting with HCN and then H_2O, or by reacting with formic acid (HCOOH). Therefore, the simplest amino acid can be formed starting from simple molecules relatively abundant in extraterrestrial environments, such as the interstellar medium and, possibly, primitive Earth.[30,57]

In conclusion, the chemistry that controls the formation of N-containing organic compounds in the atmosphere of Titan, as well as that of other bodies where both molecular nitrogen and hydrocarbons are present, is slowly being unveiled. Much information, however, is still missing on many processes responsible for the nitriles/imines budget. In particular, to assess the real role of the $N(^2D)$ reactions with hydrocarbons there are still important needs, such as the measurement of $k(T)$ at the relevant temperature. The rate constant temperature dependence employed in the models has been determined in a T range not typical of the atmosphere of Titan, which can strongly affect the accuracy of the models.[58] Also, the reaction products of many reactions remain undetermined and this can also strongly affect the accuracy of the models. For instance, the reactions $N(^2D)$ + CH_3CCH, C_3H_8, C_4H_8 and C_4H_{10} have been included in the recent model by Lavvas et al.,[9] but with unknown products. Since the use in the models of unidentified products corresponds to an irreversible loss of the reactants, laboratory studies that identify the primary reaction products are crucial to build realistic models of the chemistry of planetary atmospheres. Future work on the above mentioned systems is being planned in our laboratory.

Acknowledgements

We acknowledge financial support from the Italian MIUR (Ministero Istruzione Università Ricerca) under projects PRIN 2007H9S8SW-004 and 2007WLBXX9-004. R. Petrucci thanks FSE (Fondo Sociale Europeo) – Regione Umbria – Ministero del Lavoro e delle Politiche Sociali for a fellowship. D. S. thanks Alexander Mebel for useful suggestions on RRKM calculations.

References

1 A. Coustenis, E. Lellouch and B. Sicardy, in *Titan from Cassini–Huygens*, ed. R. H. Brown, J.-P. Lebreton and J. H. Waite, Springer, Dordrecht, 2009, ch. 2, pp. 9–34.
2 D. F. Strobel, S. K. Atreya, B. Bezard, F. Ferri, F. M. Flasar, M. Fulchignoni, E. Lellouch and I. M. Muller-Wodarg, in *Titan from Cassini–Huygens*, ed. R. H. Brown, J.-P. Lebreton and J. H. Waite, Springer, Dordrecht, 2009, ch. 10, pp. 235–257.
3 H. B. Niemann, S. K. Atreya, S. J. Bauer, G. R. Carignan, J. E. Demick, R. L. Frost, D. Gautier, J. A. Haberman, D. N. Harpold, D. M. Hunten, G. Israel, J. I. Lunine, W. T. Kasprzak, T. C. Owen, M. Paulkovich, F. Raulin, E. Raaen and S. H. Way, *Nature*, 2005, **438**, 779.
4 D. F. Strobel, *Space Sci. Rev.*, 2005, **116**, 155.
5 (a) Y. Yung, M. Allen and J. Pinto, *Astrophys. J. Suppl.*, 1984, **55**, 465; (b) Y. Yung, *Icarus*, 1987, **72**, 468.
6 D. Toublanc, J. Parisot, J. Brillet, D. Gautier, F. Raulin and C. McKay, *Icarus*, 1995, **113**, 2.
7 L. M. Lara, E. Lellouch, J. J. López-Moreno and R. Rodrigo, *J. Geophys. Res.*, 1996, **101**, 23261.
8 E. H. Wilson and S. Atreya, *J. Geophys. Res.*, 2004, **109**, E06002.
9 P. P. Lavvas, A. Coustenis and I. M. Vardavas, *Planet. Space Sci.*, 2008, **56**, 27; P. P. Lavvas, A. Coustenis and I. M. Vardavas, *Planet. Space Sci.*, 2008, **56**, 67.
10 V. A. Krasnopolsky, *Icarus*, 2009, **201**, 226.
11 S. Vinatier, B. Bézard, C. A. Nixon, A. Mamoutkine, R. C. Carlson, D. E. Jennings, E. A. Guandique, N. A. Teanby, G. L. Bjoraker, F. M. Flasar and V. G. Kunde, *Icarus*, 2010, **205**, 559.
12 A. Coustenis, D. E. Jennings, C. A. Nixon, R. K. Achterberg, P. Lavvas, S. Vinatier, N. A. Teanby, G. L. Bjoraker, R. C. Carlson, L. Piani, G. Bampasidis, F. M. Flasar and P. N. Romani, *Icarus*, 2010, **207**, 461, DOI: 10.1016/j.icarus.2009.11.027.
13 N. Balucani, A. Bergeat, L. Cartechini, G. G. Volpi, P. Casavecchia, D. Skouteris and M. Rosi, *J. Phys. Chem. A*, 2009, **113**, 11138.

14 K. Schofield, *J. Phys. Chem. Ref. Data*, 1979, **8**, 723.
15 J. T. Herron, *J. Phys. Chem. Ref. Data*, 1999, **28**, 1453.
16 H. Umemoto, N. Hachiya, E. Matsunaga, A. Suda and M. Kawasaki, *Chem. Phys. Lett.*, 1998, **296**, 203.
17 A. A. Scala and D. Salomon, *J. Chem. Phys.*, 1976, **65**, 4455; M. Kawasaki, M. Iwasaki, T. Ibuki and Y. Takezaki, *J. Chem. Phys.*, 1973, **59**, 6321.
18 V. Vuitton, R. V. Yelle and V. G. Anicich, *Astrophys. J.*, 2006, **647**, L175.
19 N. Balucani, L. Cartechini, G. Capozza, E. Segoloni, P. Casavecchia, G. G. Volpi, F. J. Aoiz, L. Banares, P. Honvault and J. M. Launay, *Phys. Rev. Lett.*, 2002, **89**, 013201; N. Balucani, P. Casavecchia, L. Banares, F. J. Aoiz, T. Gonzalez-Lezana, P. Honvault and J. M. Launay, *J. Phys. Chem. A*, 2006, **110**, 817.
20 N. Balucani, M. Alagia, L. Cartechini, P. Casavecchia, G. G. Volpi, K. Sato, T. Takayanagi and Y. Kurosaki, *J. Am. Chem. Soc.*, 2000, **122**, 4443.
21 N. Balucani, L. Cartechini, M. Alagia, P. Casavecchia and G. G. Volpi, *J. Phys. Chem. A*, 2000, **104**, 5655.
22 P. Casavecchia, N. Balucani, L. Cartechini, G. Capozza, A. Bergeat and G. G. Volpi, *Faraday Discuss.*, 2001, **119**, 27.
23 M. Alagia, N. Balucani, P. Casavecchia, D. Stranges and G. G. Volpi, *J. Chem. Soc., Faraday Trans.*, 1995, **91**, 575.
24 N. Balucani, G. Capozza, F. Leonori, E. Segoloni and P. Casavecchia, *Int. Rev. Phys. Chem.*, 2006, **25**, 109.
25 P. Casavecchia, F. Leonori, N. Balucani, R. Petrucci, G. Capozza and E. Segoloni, *Phys. Chem. Chem. Phys.*, 2009, **11**, 46.
26 M. Alagia, N. Balucani, P. Casavecchia, D. Stranges and G. G. Volpi, *J. Chem. Phys.*, 1993, **98**, 2459; M. Alagia, N. Balucani, P. Casavecchia, D. Stranges and G. G. Volpi, *J. Chem. Phys.*, 1993, **98**, 8341.
27 R. I. Kaiser, *Chem. Rev.*, 2002, **102**, 1309; N. Balucani, O. Asvany, L. C. L. Huang, Y. Y. Lee, R. I. Kaiser, Y. Osamura and H. F. Bettinger, *Astrophys. J.*, 2000, **545**, 892; R. I. Kaiser and N. Balucani, *Acc. Chem. Res.*, 2001, **34**, 699; X. B. Gu, Y. Guo, F. T. Zhang, A. M. Mebel and R. I. Kaiser, *Faraday Discuss.*, 2006, **133**, 245; R. I. Kaiser, C. Ochsenfeld, D. Stranges, M. Head-Gordon and Y. T. Lee, *Faraday Discuss.*, 1998, **109**, 183; R. I. Kaiser, D. Stranges, Y. T. Lee and A. G. Suits, *Astrophys. J.*, 1997, **477**, 982; R. I. Kaiser, C. Ochsenfeld, M. Head Gordon, Y. T. Lee and A. G. Suits, *Science*, 1996, **274**, 1508.
28 M. Costes, N. Daugey, C. Naulin, A. Bergeat, F. Leonori, E. Segoloni, R. Petrucci, N. Balucani and P. Casavecchia, *Faraday Discuss.*, 2006, **133**, 157; F. Leonori, R. Petrucci, E. Segoloni, A. Bergeat, K. M. Hickson, N. Balucani and P. Casavecchia, *J. Phys. Chem. A*, 2008, **112**, 1363; F. Leonori, R. Petrucci, K. M. Hickson, E. Segoloni, N. Balucani, S. D. Le Picard, P. Foggi and P. Casavecchia, *Planet. Space Sci.*, 2008, **56**, 1658.
29 F. Leonori, K. M. Hickson, S. D. Le Picard, X. Wang, R. Petrucci, P. Foggi, N. Balucani and P. Casavecchia, *Mol. Phys.*, 2010, **108**, 1097.
30 N. Balucani, *Int. J. Mol. Sci.*, 2009, **10**, 2304.
31 G. Capozza, E. Segoloni, F. Leonori, G. G. Volpi and P. Casavecchia, *J. Chem. Phys.*, 2004, **120**, 4557; P. Casavecchia, G. Capozza, E. Segoloni, F. Leonori, N. Balucani and G. G. Volpi, *J. Phys. Chem. A*, 2005, **109**, 3527.
32 F. Leonori, N. Balucani, G. Capozza, E. Segoloni, D. Stranges and P. Casavecchia, *Phys. Chem. Chem. Phys.*, 2007, **9**, 1307.
33 (*a*) F. Leonori, R. Petrucci, N. Balucani, P. Casavecchia, M. Rosi, D. Skouteris, C. Berteloite, S. Le Picard, A. Canosa and I. R. Sims, *J. Phys. Chem. A*, 2009, **113**, 15328; (*b*) F. Leonori, R. Petrucci, N. Balucani, P. Casavecchia, M. Rosi, C. Berteloite, S. Le Picard, A. Canosa and I. R. Sims, *Phys. Chem. Chem. Phys.*, 2009, **11**, 4701.
34 (*a*) H. Umemoto, Y. Kimura and T. Asai, *Chem. Phys. Lett.*, 1997, **264**, 215; (*b*) H. Umemoto, Y. Kimura and T. Asai, *Bull. Chem. Soc. Jpn.*, 1997, **70**, 2951.
35 M. Alagia, V. Aquilanti, D. Ascenzi, N. Balucani, D. Cappelletti, L. Cartechini, P. Casavecchia, F. Pirani, G. Sanchini and G. G. Volpi, *Isr. J. Chem.*, 1997, **37**, 329.
36 J. T. Herron and R. E. Huie, *J. Phys. Chem.*, 1968, **72**, 2538.
37 H. Umemoto, K. Sugiyama, S. Tsunashima and S. Sato, *Bull. Chem. Soc. Jpn.*, 1985, **58**, 3076.
38 (*a*) A. D. Becke, *J. Chem. Phys.*, 1993, **98**, 5648; (*b*) P. J. Stephens, F. J. Devlin, C. F. Chablowski and M. J. Frisch, *J. Phys. Chem.*, 1994, **98**, 11623.
39 (*a*) T. H. Dunning, Jr, *J. Chem. Phys.*, 1989, **90**, 1007; (*b*) D. E. Woon and T. H. Dunning, Jr, *J. Chem. Phys.*, 1993, **98**, 1358; (*c*) R. A. Kendall, T. H. Dunning, Jr and R. J. Harrison, *J. Chem. Phys.*, 1992, **96**, 6796.

40 (a) C. Gonzales and H. B. Schlegel, *J. Chem. Phys.*, 1989, **90**, 2154; (b) C. Gonzales and H. B. Schlegel, *J. Phys. Chem.*, 1990, **94**, 5523.
41 R. J. Bartlett, *Annu. Rev. Phys. Chem.*, 1981, **32**, 359; K. Raghavachari, G. W. Trucks, J. A. Pople and M. Head-Gordon, *Chem. Phys. Lett.*, 1989, **157**, 479; J. Olsen, P. Jorgensen, H. Koch, A. Balkova and R. J. Bartlett, *J. Chem. Phys.*, 1996, **104**, 8007.
42 J. M. L. Martin and G. de Oliveira, *J. Chem. Phys.*, 1999, **111**, 1843; S. Parthiban and J. M L. Martin, *J. Chem. Phys.*, 2001, **114**, 6014.
43 M. J. Frisch, G. W. Trucks, H. B. Schlegel, G. E. Scuseria, M. A. Robb, J. R. Cheeseman, J. A. Montgomery, Jr., T. Vreven, K. N. Kudin, J. C. Burant, J. M. Millam, S. S. Iyengar, J. Tomasi, V. Barone, B. Mennucci, M. Cossi, G. Scalmani, N. Rega, G. A. Petersson, H. Nakatsuji, M. Hada, M. Ehara, K. Toyota, R. Fukuda, J. Hasegawa, M. Ishida, T. Nakajima, Y. Honda, O. Kitao, H. Nakai, M. Klene, X. Li, J. E. Knox, H. P. Hratchian, J. B. Cross, V. Bakken, C. Adamo, J. Jaramillo, R. Gomperts, R. E. Stratmann, O. Yazyev, A. J. Austin, R. Cammi, C. Pomelli, J. Ochterski, P. Y. Ayala, K. Morokuma, G. A. Voth, P. Salvador, J. J. Dannenberg, V. G. Zakrzewski, S. Dapprich, A. D. Daniels, M. C. Strain, O. Farkas, D. K. Malick, A. D. Rabuck, K. Raghavachari, J. B. Foresman, J. V. Ortiz, Q. Cui, A. G. Baboul, S. Clifford, J. Cioslowski, B. B. Stefanov, G. Liu, A. Liashenko, P. Piskorz, I. Komaromi, R. L. Martin, D. J. Fox, T. Keith, M. A. Al-Laham, C. Y. Peng, A. Nanayakkara, M. Challacombe, P. M. W. Gill, B. G. Johnson, W. Chen, M. W. Wong, C. Gonzalez and J. A. Pople, *GAUSSIAN 03 (Revision D.01)*, Gaussian, Inc., Wallingford, CT, 2004.
44 MOLEKEL 4.3, P. Flükiger, H. P. Lüthi, S Portmann and J. Weber, Swiss Center for Scientific Computing, Manno (Switzerland), 2000–2002; S. Portmann and H. P. Lüthi, *Chimia*, 2000, **54**, 766.
45 R. G. Gilbert and S. C. Smith, *Theory of Unimolecular and Recombination Reactions*, 1990, Blackwell Scientific Publications, UK.
46 S. J. Klippenstein, *J. Chem. Phys.*, 1992, **96**, 367.
47 A. A. Fokin, T. E. Shubina, P. A. Gunchenko, S. D. Isaev, A. G. Yurchenko and P. R. Schreiner, *J. Am. Chem. Soc.*, 2002, **124**, 10718; P. Schreiner, A. A. Fokin, P. v. R. Schleyer and H. F. Schaefer, III, in *Fundamental World of Quantum Chemistry*; ed. E. J. Bründas and E. S. Kryachko, Kluwer, Netherlands, 2003, Vol. II, 349–375.
48 H. Umemoto, T. Nakae, H. Hashimoto, K. Kongo and M. Kawasaki, *J. Chem. Phys.*, 1998, **109**, 5844.
49 A. Teslja, B. Nizamov and P. J. Dagdigian, *J. Phys. Chem. A*, 2004, **108**, 4433.
50 C. Larson, Y. Y. Ji, P. Samartzis, A. M. Wodtke, S. H. Lee, J. J. M. Lin, C. Chaudhuri and T. T. Ching, *J. Chem. Phys.*, 2006, **125**, 133302.
51 V. A. Basiuk, *J. Phys. Chem. A*, 2001, **105**, 4252.
52 G. Israel, C. Szopa, F. Raulin, M. Cabane, H. B. Niemann, S. K. Atreya, S. J. Bauer, J. F. Brun, E. Chassefiere, P. Coll, E. Conde, D. Coscia, A. Hauchecorne, P. Millian, M. J. Nguyen, T. Owen, W. Riedler, R. E. Samuelson, J. M. Siguier, M. Steller, R. Sternberg and C. Vidal-Madjar, *Nature*, 2005, **438**, 796.
53 H. Imanaka and M. A. Smith, Astrobiology Science Conference 2010, League City, TX (USA) Abstract No. 5186.
54 X. Gu, R. I. Kaiser, A. M. Mebel, V. V. Kislov, S. J. Klippenstein, L. B. Harding, M. C. Liang and Y. L. Yung, *Astrophys. J.*, 2009, **701**, 1797.
55 J. F. Arenas, J. I. Marcos, I. Lopez-Tocon, J. C. Otero and Juan Soto, *J. Chem. Phys.*, 2000, **113**, 2282.
56 S.-H. Ge, X.-L. Cheng, X.-D. Yang, Z.-J. Liu and W. Wang, *Icarus*, 2006, **183**, 153.
57 N. Balucani and P. Casavecchia, *Origins Life Evol. Biosphere*, 2006, **36**, 443.
58 E. Hebrard, M. Dobrijevic, P. Pernot, N. Carrasco, A. Bergeat, K. M. Hickson, A. Canosa, S. D. Le Picard and I. R. Sims, *J. Phys. Chem. A*, 2009, **113**, 11227.

PAPER

Structural and spectroscopic characterization of mixed planetary ices

Nuria Plattner, Myung Won Lee and Markus Meuwly*

Received 24th February 2010, Accepted 20th April 2010
DOI: 10.1039/c003487h

Mixed ices play a central role in characterizing the origin, evolution, stability and chemistry of planetary ice surfaces. Examples include the polar areas of Mars, the crust of the Jupiter moon Europa, or atmospheres of planets and their satellites, particularly in the outer solar system. Atomistic simulations using accurate representations of the interaction potentials have recently shown to be suitable to quantitatively describe both, the mid- and the far-infrared spectrum of mixed H_2O/CO amorphous ices. In this work, molecular dynamics simulations are used to investigate structural and spectroscopic properties of mixed and crystalline ices containing H_2O, CO and CO_2. Particular findings include: (a) the sensitivity of the water bending mode to the local environment of the water molecules which, together with structural insights from MD simulations, provides a detailed picture for the relationship between spectroscopy and structure; and (b) the sensitivity of the low-frequency spectrum to the structure of the mixed CO_2/H_2O ice. Specifically, for mixed H_2O/CO_2 ices with low water contents isolated water molecules are found which give rise to a band shifted by only 12 cm^{-1} from the gas-phase value whereas for increasing water concentration (for a 1 : 1 mixture) the band progressively shifts to higher frequency because water clusters can form. More generally it is found that changes in the ice structure due to the presence of CO_2 are larger compared to changes induced by the presence of CO and that this difference is reflected in the shape of the water bending vibration. Thus, the water bending vibration appears to be a suitable diagnostic for structural and chemical aspects of mixed ices.

1 Introduction

Multicomponent ices play a central role in characterizing the origin, evolution, stability and chemistry of planetary ice surfaces. Examples include the polar areas of Mars, the crust of the Jupiter moon Europa, or atmospheres of planets and their satellites, particularly in the outer solar system. The chemical composition of these ices includes H_2O, CO_2, CO, CH_4 and other small molecules. The properties of these ices are key for understanding chemical processes on planetary surfaces, including determining the chemical inventory, physicochemical properties of the planet, the equilibrium and matter exchanges between surface and atmosphere of many planets[1] or planetary climate change.[2] Phase equilibria and thermodynamic properties of these ices have been studied in order to understand surface properties, including the stability of different structures of a given ice composition as well as mass and heat transport.[3,4]

Department of Chemistry, University of Basel, Klingelbergstrasse 80, CH-4056 Basel, Switzerland. E-mail: m.meuwly@unibas.ch; Fax: +41 61 267 38 55; Tel: +41 61 267 38 21

For ices with given molecular composition a range of structures, including regular or random (amorphous) mixtures, layers of pure amorphous or crystalline structures as well as ordered two- or three-component structures, including clathrate hydrates, are possible. Depending on the three-dimensional structure, the intermolecular vibrations appear in characteristic frequency ranges which can be probed by infrared (IR) spectroscopy. As an example, the infrared spectra of CO incorporated in H_2O ice at 10 K show two features at about 2138 and 2152 cm^{-1}. Upon heating, the high-frequency band gradually disappears and the main band shifts from 2138 to 2136 cm^{-1}.[5,6] The 2152 cm^{-1} band appears to be absent in astronomical spectra which was related to the fact that in interstellar ices dangling-OH binding sites are either not accessible to CO molecules or do not exist.[7] Further studies[5,8,9] have established that the details of composition and thermodynamic conditions influence the shape, width and splitting of the CO absorption band.

The composition of interstellar grains and dust has been characterized in the region of the fundamental vibrations (mid-infrared, 3 to 30 μm or ≈ 3300 to 330 cm^{-1}) using, *e.g.*, the international space observatory (ISO) or Spitzer. Molecules such as H_2O, CO, or CO_2 have been identified through their near-infrared absorptions on planets including Pluto, Charon, Triton, and Europa.[10] However, other frequency ranges also provide valuable information. In particular for homonuclear diatomics (O_2, N_2), which are not infrared active, the low frequency (typically below ≈ 150 cm^{-1}) lattice vibrations offer alternative ways of characterizing mixed ices. Typical frequencies for such far-infrared transitions in pure ices have been measured for CO, N_2, CO_2 and CH_4.[11] To complement the extraterrestrial studies, laboratory experiments have been undertaken.[8,12–14] Particularly valuable amongst these experiments are those which aim at growing layered ices in a controlled fashion because correlating spectroscopy and structural features of such ices is in general not possible. It should be noted that not all IR active bands are equally suitable to infer structural characteristics from spectroscopic features. For example, laboratory studies of solid H_2O/CO_2 and CH_3OH/CO_2 mixtures found that the band corresponding to the dangling OH at 3650 cm^{-1} is probably no diagnostic for the presence of CO_2 because it appears whenever guest molecules are present in pure water ices.[15] On the other hand, other peaks (such as the near-IR $2\nu_3$ overtone of CO_2 or the water-bending vibration) split and change shapes in characteristic fashion depending on the amount of CO_2 present.[13,15] Also, it was found that spectral features of mixed H_2O, CO_2 and CO ices depend on temperature changes.[12,13,16] This is most likely related to structural changes and MD simulations can be used to analyze this relationship.

To relate IR data with structural properties it is important to be able to compute spectral features with sufficient detail and resolution. In the present study, atomistic molecular dynamics (MD) simulations with accurate intermolecular potentials are used to relate structural properties of ices to spectral features and compare them to experimental observations for ices containing CO, CO_2 and H_2O. The present study focuses primarily on the mid-IR frequency range since it is suitable to characterize composition and structure of ices and experimental data is available for a range of different ice mixtures. For the near IR range, it has been found in previous studies that spectral features are difficult to obtain by atomistic MD simulations in particular for water ice.[17] On the other hand, for the mid- and far-IR regions, atomistic simulations using accurate representations of the interaction potentials have recently shown to be suitable to quantitatively describe experimental spectra of mixed H_2O/CO amorphous ices and CO/CH_4-containing water clathrates, respectively, and to relate spectroscopy and structure.[18,19] Electrostatics including higher multipole moments on the CO together with the spectroscopically accurate stretching potential (see Section 2.1) have recently been shown to quantitatively explain splittings and shifts for mixed H_2O/CO ices. Experimentally, the CO stretching spectrum was found to consist of a red- and a blue-shifted band (by −5 cm^{-1} and +9 cm^{-1}) relative to the gas-phase which compares with shifts of −4 cm^{-1} and +11 cm^{-1} from simulations, *i.e.* splittings of 14 cm^{-1} and 15 cm^{-1}, respectively.[5,6,18] It is remarkable that the same stretching potential and multipole moments for CO are also able to quantitatively describe splittings and shifts of the CO fundamental in myoglobin.[20] Additional validations of the simulation approaches presented here show that the radial distribution functions (RDFs) of CO reproduce those from previous investigations[21] whereas compared to experiment the present approach provides more accurate spectroscopic properties. Finally, it is found that an anharmonic water potential with distributed multipole moments (see Section 2.1) can be used to reproduce experimentally observed shifts in the water bending frequency. Furthermore, validated molecular dynamics simulations can be used to observe and characterize structural changes, analyze diffusion processes inside mixed ices and assess properties of equilibrated structures.

This work is structured as follows. First, the computational methods are presented. Then, structural and spectroscopic properties of different $H_2O/CO_2/CO$ ices are analyzed. Finally, the results are discussed in view of applications in planetary sciences.

2 Computational methods

2.1 Simulations and interaction potentials

All simulations are carried out with the CHARMM program[22] with provisions for evaluating electrostatic interactions involving higher distributed multipoles (DMs, see below). The CO bond potential is the spectroscopically accurate anharmonic RRKR (rotational Rydberg–Klein–Rees) potential,[23,24] which is a distorted Morse potential. Electrostatic interactions are described by atomic multipole moments based on the distributed multipole analysis (DMA),[25,26] which accurately represent the molecular moments up to hexadecapole.[20] For water the flexible Kumagai, Kawamura and Yokokawa (KKY) potential[27] model was used. As was previously reported, the KKY potential parameters need to be reparametrized to reproduce the water gas phase frequencies.[28] For the electrostatic interactions distributed multipoles including point charges plus additional atomic moments up to quadrupole are used (see Table 1). For CO_2 a new distributed multipole model was developed (see Table 2). The model is based on electronic structure calculations at the

Table 1 Electrostatic moments for the water multipole model. The values correspond to spherical tensor notation

	Charge [e]	Dipole [ea_0]	Quadrupole [ea_0^2]	
		$q10$	$q20$	$q22c$
Oxygen	−0.280016	−0.418741	0.028448	−1.687540
Hydrogen	0.140008	—	—	—

Table 2 Electrostatic moments for the CO_2 multipole model. The values correspond to spherical tensor notation

	Charge [e]	Dipole [ea_0] $q10$	Quadrupole [ea_0^2] $q20$
Carbon	0.031976	—	0.509873
Oxygen	−0.015988	−0.673825 ^{a}	0.848992

^{a} Opposite signs on the two oxygens.

MP2 level with a D95 basis set.[29,30] It was found that parameters based on the D95 basis set better describe the thermodynamic stability of solid crystalline CO_2 in simulations than parameters derived from other basis sets. The atomic multipole moments were then obtained using GDMA (Gaussian Distributed Multipole Analysis).[26] For the bonded interactions, parameters for harmonic potential functions were fitted to reproduce the IR active CO_2 bond and bending frequencies at 2349 and 667 cm^{-1}, respectively.[31] Conformationally dependent multipole moments are only used for CO in this work. The change of moments with bond length is important for the calculation of the CO IR spectrum, since the molecular dipole moment changes sign with bond length. For water and CO_2, the dependence of the molecular moments on changes of bond lengths and angles are reproduced qualitatively correctly by the conformationally varying charges and moments on the atoms.

In all simulations the time step was 0.4 fs which allows flexibility of the water hydrogen atoms. All production simulations were carried out in the isothermal–isobaric (*NPT*) ensemble using the Nosé–Hoover thermostat. The non-bonded interactions were truncated at a distance of 12 Å using a shift function for the electrostatic terms and a switch algorithm for the van der Waals terms. For the evaluation of the IR spectra, simulation temperatures of 20, 50 and 100 K were chosen. For generating and equilibrating the ice structures, simulations were also run at higher temperatures (see Section 2.3).

2.2 Calculation of IR spectra

The infrared (IR) spectrum is calculated from the Fourier transform of the dipole moment autocorrelation function $C(t)$ which is accumulated over 2^n time origins, where n is an integer such that 2^n corresponds to between 1/3 and 1/2 of the trajectory, with the time origins separated by 0.8 fs. $C(t)$ is then transformed to yield $C(\omega)$ using a Fast Fourier Transform with a Blackman filter to minimize noise.[32] The final infrared absorption spectrum $A(\omega)$ is then calculated from

$$A(\omega) = \omega\{1 - \exp[-\omega/(kT)]\} C(\omega) \quad (1)$$

where k is the Boltzmann constant and T is the temperature in Kelvin.

The absolute frequencies obtained using this method depend on different factors. With a spectroscopically accurate potential (*e.g.* RRKR potential for CO), the frequency calculated from MD simulations differs from the experimental value (2143 cm^{-1} for CO) for two different reasons. First, the finite time step used in the velocity-Verlet algorithm leads to a systematic shift which is 16 cm^{-1} for $\Delta t = 1$ fs and decreases with decreasing Δt.[33] Second, the classical treatment of a Morse oscillator leads to an energy-dependent fundamental frequency ω_E[34]

$$\omega_E = \alpha\sqrt{\frac{2(D_e - E_{tot})}{\mu}} \quad (2)$$

where E_{tot} is the total energy of the particular simulation, D_e is the dissociation energy and α describes the steepness of the repulsive wall. To compare the band positions and their shifts with experiment, reference stretching frequencies for gas phase CO using the RRKR potential were calculated. The calculation included ten trajectories for free CO. The free CO frequency is at 2172 cm^{-1} for a time step of 0.4 fs. Deviations between the ten trajectories were small (0.2 cm^{-1}).

For water and CO_2, potentials have been fit in order to reproduce the experimental gas phase frequencies. In all cases, the absolute values of the frequencies are not of interest. All IR spectra will be discussed in terms of relative frequencies in different environments. Important reference numbers used for comparison in the following are therefore the calculated CO gas phase frequency at 2172 cm^{-1} as explained above and the water bending gas phase frequency 1595 cm^{-1}.

2.3 The systems investigated

In the following the different compositions, structures and simulation characteristics of the systems considered will be briefly described. At the typical temperatures relevant to observations (≈ 100 K) water and CO_2 are in a crystalline state whereas CO ice melts above 50 K.[35,36] Thus, simulations for pure CO ice were carried out at 20 K. Amorphous structures were generated by starting from perfectly ordered structures and heating them to 300 K and then cooling down to the desired temperature at which equilibrium MD simulations were run.

CO_2. For pure CO_2 ices simulations were started from two different initial structures: (A) ordered CO_2 arranged in $4 \times 4 \times 4$ unit cells of dry ice cubic crystals of space group Pa3[37] and (B) randomly arranged CO_2 molecules in a box of initial dimensions $38 \times 35 \times 35$ Å. The systems contain 2688 and 3072 atoms, respectively.

H_2O. For pure water, also two different initial arrangements were investigated: (A) a hexagonal ice structure of dimensions $35 \times 31 \times 29$ Å with 3072 atoms, and (B) a random arrangement in a box with dimensions $35 \times 35 \times 35$ Å containing 3000 atoms.

CO. For pure CO ices, two different structures were investigated: (A) amorphous ice and (B) ordered ice. For the amorphous structure, the CO molecules were initially in a cubic arrangement containing 864 molecules; the system was heated to 100 K, which is above the boiling point of CO.[38] Then, the system was cooled to 50 K and 20 K for the simulations of ices. To simulate the ordered structure, the face-centered cubic structure (α-CO) with the lattice constant of 5.652 Å obtained at 20 K[39] was used as the initial structure. The CO interatomic axis was directed along the four cube body diagonals and the orientation of the molecule at each site was chosen randomly to be either CO or OC. Then, the system containing 864 CO molecules ($6 \times 6 \times 6$ cells) was heated to 20 K, and NPT simulations were carried out at this temperature.

CO_2/H_2O. For the layered, mixed CO_2/H_2O structure the hexagonal water ice structure (see above) was combined with $3 \times 4 \times 4$ dry ice unit cells. The xy-plane was chosen as the interface between the two structures. The system contains 5088 atoms in a box of dimensions $35 \times 66 \times 28$ Å. For the random structures, CO_2 molecules were inserted randomly in the water crystal structure and *vice versa*. Three different structures were generated: (I) 100 CO_2 molecules inserted in hexagonal ice; (II) 100 water molecules inserted in dry ice; (III) a 1 : 1 ratio by number of water and CO_2. In all cases, directly overlapping molecules of the more abundant species were removed. The total number of molecules in each of the three structures is about 1000. The ice was then equilibrated by initial heating to 300 K in the canonical (NVT) ensemble and equilibration during 50 ps, followed by equilibration at 200 and 100 K during 50 ps at each temperature in the NPT ensemble.

CO/H_2O. The random two-component systems of CO and H_2O were generated using the same insertion procedure as for CO_2/H_2O: CO molecules were inserted randomly into a hexagonal ice structure, generating three systems of (N_{wat}, N_{CO}) = (992,60), (912,100) and (581,500).

$CO_2/H_2O/CO$. Random structures for this three component-system were constructed by inserting CO as an additional component into hexagonal ice, in addition to 10, 20 and 100 CO_2 molecules following the same procedure as described above. Structures with (N_{wat}, N_{CO}, N_{CO_2}) = (811, 100, 100), (970, 30, 20) and (983, 30, 10) are investigated in the following. The layered CO_2/H_2O structure described above was modified by adding 60 CO molecules randomly to the hexagonal ice of the layered CO_2/H_2O system described above, with (N_{wat}, N_{CO}, N_{CO_2}) = (992, 60, 672).

Clathrate systems. CO_2 is known to form a structure type I clathrate.[40] Since methane forms the same structure type, initial water coordinates were taken from a previous study on methane clathrate[19] and CO_2 molecules were inserted into the cavities, resulting in a system of 368 water molecules and 64 CO_2 molecules. The structure was then equilibrated in the *NPT* ensemble at 100 K during 100 ps in order to adapt the volume to the presence of the guest molecule.

3 Results

3.1 H_2O/CO_2

For ices containing H_2O and CO_2, results from different structures have been compared. The most simple are random mixtures of the two components using different $H_2O : CO_2$ ratios (see Fig. 1). The strongest bimolecular interaction in such a mixture are interactions between two water molecules. Therefore in each of the mixtures, the system attempts to maximize the number of water–water contacts. Consequently, for high water concentration it is expected that local ice structures are mainly preserved and rearrange around the disturbing CO_2 molecules. For low water concentration, the formation of water clusters inside the CO_2 block can be observed. The crystalline CO_2 structure is disrupted by the presence of the water molecules, while the hexagonal ice preserves some of its local structure. This can clearly be shown by comparing the oxygen–oxygen ($g_{O_wO_w}$) and carbon–carbon (g_{CC}) radial distribution functions (RDFs) shown in Fig. 2 (A) and (B), where O_W denotes oxygen atoms of water. It is found that $g_{O_wO_w}$ loses its structuring with increasing CO_2 content, while g_{CC} shows increasing structuring with increasing CO_2 content, but the change is rather small compared to the change in $g_{O_wO_w}$.

The infrared spectra at 100 K for amorphous water/CO_2 ices in the region of the "water libration/CO_2 bending vibration" (between 600 and 900 cm^{-1}) and in the region of the "water bending vibration" (between 1550 and 1700 cm^{-1}) are shown in Fig. 2 (C) and (D). For the libration, the spectrum obtained from the simulation can be explained by the superposition of the CO_2 bending vibration at 667 cm^{-1} and the water libration which forms a broad background between 600 and 850 cm^{-1}. A superposition of the sharply peaked CO_2 bending vibration and the broad water libration band can also be seen in the experimental spectra for different CO_2/H_2O mixtures at 20 K, whereas at 100 K the spectrum is dominated by the water libration, even if the mixture contains substantial amounts of CO_2.[13] At 100 K the intensity of the water libration is larger than the CO_2 bending, therefore upon lowering the water concentration the overall intensity decreases.

The water bending vibration provides more information since a shift towards higher frequencies with increasing water content can be observed. The gas phase water bending vibration is at 1595 cm^{-1} whereas at 100 K the centers of the calculated spectral lines for low, medium and high water content are at 1607, 1614 and 1620 cm^{-1}, respectively. Such frequency shifts are consistent with experimental

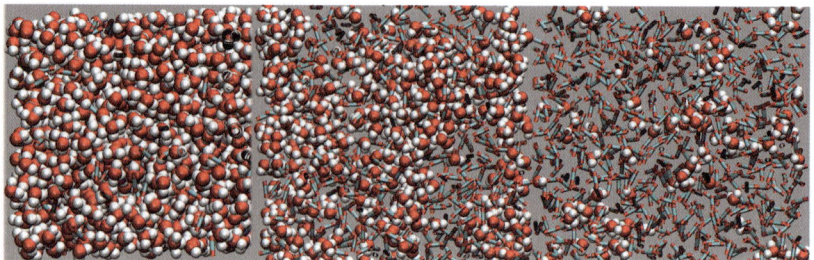

Fig. 1 Structures of different mixed $CO_2 : H_2O$ ices from water-rich (left) to CO_2-rich (right).

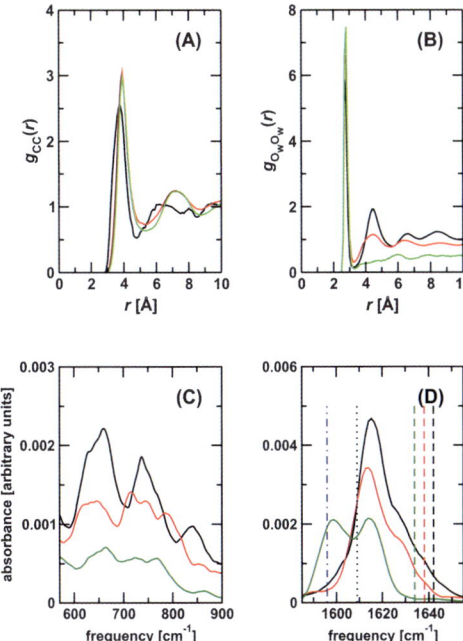

Fig. 2 Radial distribution functions (RDFs) and IR spectra for mixed CO_2/H_2O-ices with different compositions at 100 K. (A) Carbon–carbon RDFs (g_{CC}) for the C atoms of CO_2; (B) oxygen–oxygen ($g_{O_wO_w}$) RDFs for the water-O atoms; (C) IR spectra in the region of the lattice vibration; and (D) IR spectra in the water bending vibration region. Water-rich ice (black), 1 : 1 ratio (red), CO_2-rich ice (green). For comparison, experimental IR peak positions of mixed ices from ref. 13 are shown in (D) as vertical dashed lines: 1642 cm^{-1} for water-rich ice (black); 1638 cm^{-1} for 1 : 1 ratio (red); 1634 cm^{-1} for CO_2-rich ice (green); and 1609 cm^{-1} for all three cases (black). The experimental frequency of the gas-phase water bending mode is also shown in (D) at 1595 cm^{-1} (blue).

measurements of IR spectra for different CO_2/H_2O ratios which show a decrease in intensity for the low-frequency band at 1609 cm^{-1} with increasing H_2O content combined with an increase in the higher frequency bands above 1610 cm^{-1}.[13] The same is observed at 20 K (see left panel of Fig. 3). For low water content (green line), the most intense band is observed at 1609 cm^{-1}, with a large shoulder at about 1626 cm^{-1} and a small shoulder at about 1595 cm^{-1}. The two bands at 1609 and 1626 cm^{-1} are also observed for the 1 : 1 ratio of water and CO_2. For high water content, this band is only visible as a small shoulder of the main band. The observation of two bands for the water bending vibration in CO_2/H_2O mixtures at low temperatures and the increase of the intensity of the lower band with decreasing water ratio are also consistent with experiment.[13] Analysis of the corresponding structures from the simulations show that at low water contents isolated water molecules are found in the mixture which can be related to the lowest band. The low-water content mixture and the 1 : 1 mixture contain, furthermore, small water clusters which lead to the prominent middle band at 1610 cm^{-1}. The highest frequency band can be associated with larger water assemblies, found in the 1 : 1 mixture and at high water content.

Furthermore, mixtures of H_2O and CO_2 can also form layered structures which can, *e.g.*, be formed by subsequent precipitation of each of the two components. The simulations of the layered ice show that the layers are destabilized. With the interaction potentials used here, the melting points of pure water and CO_2 are around 250 and 300 K and between 150 and 200 K, respectively. In the combined, layered systems their individual structures at 100 K are predominantly conserved,

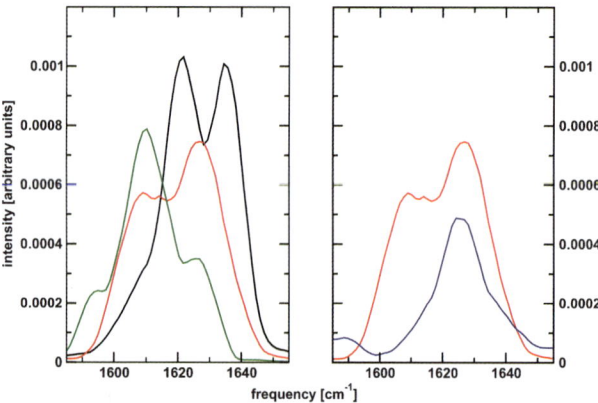

Fig. 3 IR spectra in the water bending region at 20 K. Left hand side: different compositions of CO_2/H_2O; water-rich ice (black); 1 : 1 ratio (red); CO_2-rich ice (green). Right hand side: Comparison of CO_2/H_2O (red) to CO/H_2O (blue) using 1 : 1 ratios of the two components.

showing only slight distortions at the interface. At higher temperature (150 K), the CO_2-phase loses its structure, while the water ice is only perturbed at the interface. At 200 K, both parts lose their structure and individual molecules start to diffuse into the other layer. This is illustrated by the RDFs shown in Fig. 4 (A) and (B), where a structural change is observed for g_{CC} between 100 and 150 K and for $g_{O_wO_w}$ between 150 and 200 K.

IR spectra calculated for individual layers of the layered water/CO_2 system provide information about differences between boundary- and bulk-regions. As an example, IR spectra of the CO_2 bending mode have been analyzed (not shown) for all CO_2 molecules of the layered structure and for the first and second layers of CO_2 molecules closest to the water interface and compared to spectra of pure CO_2 and the 1 : 1 mixture of H_2O and CO_2. The spectra of pure CO_2 show a splitting of the bending mode, while for the mixture with H_2O this splitting disappears and the spectrum broadens. This finding is in agreement with experiment.[16] For the different CO_2 layers in the layered structure, the splitting is not visible for the first and second layer, at the interface, but can be observed for layers farther away from the water ice. The broadening is reduced with increasing distance to the interface.

A third possibility of a H_2O/CO_2 structure is a clathrate hydrate, for which simulations were carried out in clathrate type I. The RDF $g_{O_wO_w}$ of clathrate type I at 100 K (not shown) approaches that of hexagonal ice for small r. From g_{CC} it is found that in type I clathrates the distance between CO_2 molecules is larger compared to those in other CO_2/water systems. To further investigate the motional freedom available to the CO_2 molecules, the rotational correlation function (RCF) of CO_2 in clathrate type I hexagonal water ice and in a 1 : 1 mixture of CO_2 and H_2O were computed. As the data suggested multiple time scales, the RCFs were fitted to a sum of two exponential functions, $a_1 \exp(-t/\tau_1) + a_2 \exp(-t/\tau_2) + c$ with $\tau_1 < \tau_2$. This yields $\tau_1 = 0.47$, 0.70, and 1.05 ps for CO_2 in the clathrate, CO_2 in hexagonal ice, and 1 : 1 mixture of CO_2 and H_2O, respectively. For the slower decay $\tau_2 = 1.6$, 269, and 69 ps was found. This shows that CO_2 in clathrates is relatively free to rotate whereas in the other two systems its angular motion is heavily constrained. The CO_2 reorientational correlation time was experimentally obtained from measuring the ^{17}O longitudinal relaxation time $T_1(^{17}O)$ by Nuclear Magnetic Resonance spectroscopy (NMR), under gaseous, liquid, and supercritical conditions at temperatures between 293 and 351 K over pressures ranging from 1 to 20 MPa.[41] The experimentally observed reorientation

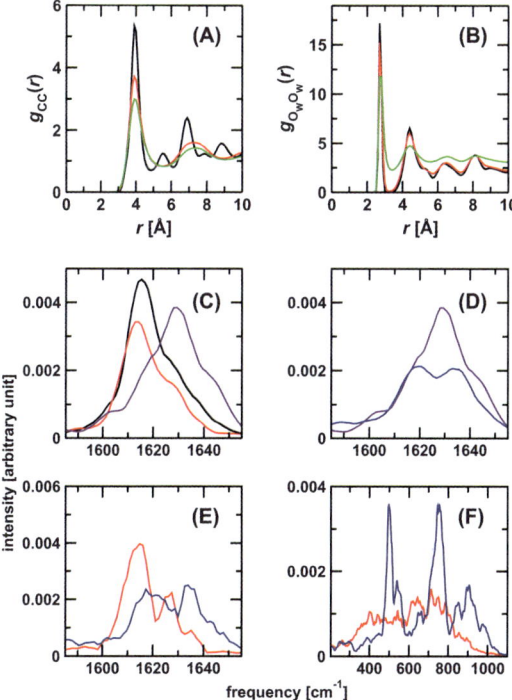

Fig. 4 Radial distribution functions (RDFs) and IR spectra for various H_2O/CO_2 ices. (A) RDFs from the C atoms of CO_2, and (B) RDFs from the O atoms of H_2O, for layered H_2O/CO_2 system at 100 K (black), 150 K (red), and 200 K (green). (C) IR spectra in the water bending mode region of CO_2 clathrate (violet), random mixtures of 100 CO_2 in ice (black) and 1 : 1 H_2O/CO_2 (red). (D) IR spectra in the water bending region of CO_2 clathrate (violet) and layered H_2O/CO_2 ice (blue). (E) IR spectra in the water bending vibration, and (F) IR spectra in the lattice vibration region, for ices with a 1 : 1 ratio of CO_2 and H_2O, with random mixture in red and layered ice in blue. Panels (E) and (F) show unaveraged data whereas panels (C) and (D) report running averages over a few neighboring points.

times τ_{2r} were between 0.24 and 0.40 ps at various conditions which qualitatively agrees with calculated reorientation times in a clathrate type I. It is expected that more sterically demanding environments, such as hexagonal ice, restrict rotational motion and thus leads to longer reorientation times which is what the calculations find.

The IR spectra of CO_2 clathrate in the water bending mode region is compared to those in random (100 CO_2 in ice and 1 : 1) and layered H_2O/CO_2 ices at 100 K in Fig. 4 (C) and (D). The comparison shows that in both random mixtures the water bending frequency is significantly lower compared to that in clathrate type I. On the other hand, the two bands essentially overlap for the layered structure and the clathrate. It is, however, notable, that in the layered ice the water bending vibration shows two distinct peaks which are not found for the CO_2 clathrate. Since from experiment the composition of an ice can be controlled, but the structure of the ice is difficult or even impossible to determine, it is interesting to compare the spectra of the layered CO_2/H_2O structure to the random 1 : 1 mixture. Fig. 4 (E) and (F) compare the two spectra and show that the H_2O bending frequency is shifted to the blue in the layered structure compared to the random mixture. Furthermore, the lower-frequency vibrations exhibit three distinct bands for the layered structure instead of one broad, unstructured band for the random mixture. This analysis shows that the different structures of the same $H_2O : CO_2$ ratio can clearly be distinguished based on their IR spectra.

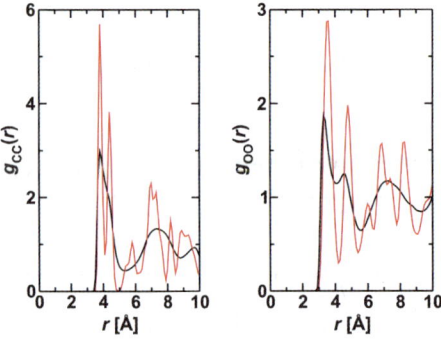

Fig. 5 Radial distribution functions (RDFs) from the carbon atoms (left) and from the oxygen atoms (right) of CO at 20 K. RDFs of amorphous structure are shown in black and RDFs of ordered structure in red.

3.2 Pure CO

As a validation of the computational strategies, the simulation for pure CO in two different structures (amorphous and ordered) were carried out. The RDFs for these two systems from simulations at 20 K are shown in Fig. 5. For the ordered structure in which the centers of the CO molecules were arranged regularly with head-to-tail disorder in the initial setup, the RDFs are reminiscent of those from previous work.[21] There, a fully ordered antiferroelectric structure was used to take short-range order into account. These earlier simulations were carried out at 31 K and used three point charge model for CO and the intermolecular potential composed of $1/R^6$, $1/R^8$, and $1/R^{10}$ terms combined with exponential functions, while in the present case the spectroscopically accurate RRKR potential for CO molecule and full multipoles up to quadrupoles for the intermolecular interactions were used. The RDFs for the amorphous system on the other hand, resemble those of liquid CO in ref. 21. A difference between the two simulations is the intensity of the peaks in the RDFs for the amorphous structure, which are slightly stronger in the present simulations. This implies that the structure of our amorphous system is similar to the liquid system, while it retains some features of solid-state as it exists as solid at the simulation temperature of 20 K.

3.3 H_2O/CO

The CO band shape in mixed H_2O/CO ices has been investigated in a previous study.[18] Using computer simulations, it was found that the experimentally observed splitting of the CO absorption band into a red- and a blue-shifted part below 50 K is primarily related to the local structure around the CO molecules and can be correlated to the number of nearby hydrogen atoms. In the present work, the analysis will focus on the water spectrum of H_2O/CO mixed ice compared to H_2O/CO_2 ice. The CO band shape is then used to analyze $H_2O/CO_2/CO$ mixtures and compare the results to the previous findings for H_2O/CO mixtures. Compared to H_2O/CO_2 systems, mixed system of H_2O and CO are stable only at lower temperatures. Due to the low boiling point of CO, already at 100 K the two components of the randomly mixed 1 : 1 systems start to separate. In the mixed ice, the CO perturbs the water ice structure less due to its smaller volume.

The spectra of H_2O/CO mixtures are compared to spectra of H_2O/CO_2 mixtures in the water angle band region. At 20 K, the low-frequency band found for the H_2O/CO_2 1 : 1 mixture is not observed for the 1 : 1 ratio of H_2O/CO (see Fig. 3). This is consistent with the experimental findings for these two systems.[13,42] A comparison of the spectra at 100 K is shown in Fig. 6 for 1 : 1 ratios and for 100 CO or CO_2

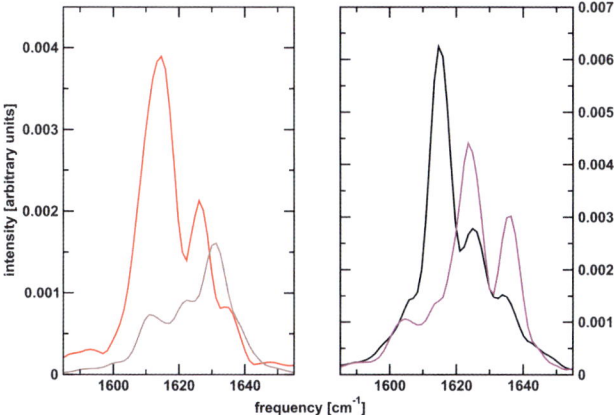

Fig. 6 IR spectra for the water bending mode for H_2O/CO and H_2O/CO_2. Left hand side: 1 : 1 mixtures with water; Right hand side: 100 CO or CO_2 molecules in ice. Color coding: H_2O/CO_2 1 : 1 (red); H_2O/CO 1 : 1 (brown); 100 CO_2 in water ice (black); 100 CO in water ice (magenta).

molecules in ice, *i.e.* high water/low CO/low CO_2 concentration. For both systems the H_2O bending frequency is at higher frequencies for mixtures with CO compared to mixtures with CO_2, which means it is closer to the frequency in pure water ice. Analysis of the simulations leads to two reasons for this finding: for low CO contents, CO disturbs the ice structure less than CO_2. For the 1 : 1 mixture of CO and water, the two components are separated, which also leads to a water bending frequency closer to that of pure water.

3.4 $H_2O/CO_2/CO$

The water absorption bands of H_2O/CO_2 with additional CO as a trace molecule resemble those of H_2O/CO_2 described above, so they are not discussed separately

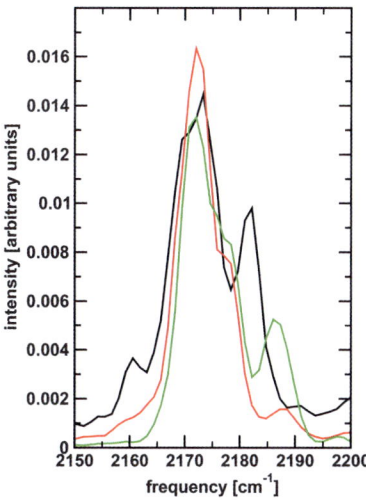

Fig. 7 IR spectra for the CO absorption band for three different $H_2O/CO_2/CO$ systems: 30 CO/10 CO_2 in water ice (black), 100 CO/100 CO_2 in water ice (red), CO in the water ice phase of a layered H_2O/CO_2 structure (green). For the potentials and methods used here, the free CO frequency is located at 2172 cm^{-1} for comparison.

here. The CO absorption band for different $H_2O/CO_2/CO$ mixtures is shown in Fig. 7. The CO stretching band is similar for 30 CO/10 CO_2 in water ice and for the layered structure of H_2O/CO_2 with CO as a trace molecule in the water phase. The absorption is split into a red- and a blue-shifted part with respect to free CO, as was found in previous work.[18] For 100 CO/100 CO_2 in water ice the intensity of the blue-shifted part is minor which indicates that the presence of CO_2 changes the water ice structure that surrounds the CO molecules. It is known experimentally that adding an additional component to H_2O/CO ice can modify the shape of the CO absorption band.[12] The modification depends on the additional molecule: in some cases, the blue-shifted band between 2143 and 2160 cm^{-1} in experiment (corresponding to 2174 to 2190 cm^{-1} in the simulations, see Section 2.2) has enhanced intensity, in others its intensity is reduced. For the mixture analyzed here with CO_2 as an additional component, no spectra have been published so far.

4 Conclusions

Experimentally observed mid-IR features, including the position and splitting of particular far- and mid-IR spectral lines in different mixed ices have been analyzed from atomistic MD simulations using accurate intermolecular interaction potentials. The potential energy functions for the bonded interactions are able to describe most of the fundamental vibrations. For CO a spectroscopically accurate potential is available whereas for water an anharmonic potential is used which is suitable for describing the water bending mode and the water libration, but not for the O–H stretch, which is generally difficult to describe with classical potentials.[17] For CO_2 the force field is able to reproduce the fundamental vibrations although the anharmonicities are probably less well captured. However, for the present work where spectral features relative to each other or splittings induced by the environment are of particular interest, accurate electrostatic interactions are more important. For this, distributed multipole expansions have been used which have repeatedly shown to provide quantitatively correct changes in spectral patterns induced by the environment.[20,43–45]

To relate spectroscopy and structure, the water bending vibration has been found to be particularly suitable in differentiating between mixed and layered H_2O/CO_2 ices. In contrast to the water libration region where the spectra can be explained by superpositions of the pure water ice spectrum with the CO_2 bending mode, the water bending depends on the local environment of each water molecule in the spectrum. Experimentally observed frequency shifts for different ice mixtures[13] can be reproduced and related to the structure of the ice. In a previous study, it had been shown that the splitting of the CO absorption band in mixed H_2O/CO ices is related to the local ice environment. Here, random structures of $H_2O/CO_2/CO$ ices were compared to layered H_2O/CO_2 ice with CO as a trace molecule in the water layer. It was found that the presence of CO_2 reduces the blue-shifted part of the CO absorption band, whereas for the layered structure, the absorption band is similar to H_2O/CO mixtures. Thus, astronomical observations in the CO stretching band region may provide insight into the chemistry and structure of $H_2O/CO_2/CO$ mixed ices. A last finding concerns the frequency range below 1000 cm^{-1} which, for a layered and randomly organized 1 : 1 H_2O : CO_2 mixed ice, leads to distinctly different spectra. Thus, for given chemical composition and spectroscopic information in one or better several IR frequency domains, simulations can provide assistance to rule out or corroborate proposed or inferred structures of the investigated ices. As is shown here, the analysis of spectra and structural details from MD simulations with accurate and validated potentials help to establish relationships between spectra and structures in cases where direct structure determination is not possible, in particular when the specimen is not physically available such as for planetary ices.

Acknowledgements

This work was supported by the Schweizerischer Nationalfonds grant 200021-117810.

References

1. N. Fray and B. Schmitt, Sublimation of ices of astrophysical interest: A bibliographic review, *Planet. Space Sci.*, 2009, **57**, 2053–2080.
2. T. Yokohata, M. Odaka and K. Kuramoto, Role of H_2O and CO_2 ices in Martian climate changes, *Icarus*, 2002, **159**, 439–448.
3. J. Longhi, Phase equilibrium in the system CO_2–H_2O: Application to Mars, *J. Geophys. Res.*, 2006, **111**, E06011.
4. X. Lu and S. W. Kieffer, Thermodynamics and mass transport in multicomponent, multiphase H_2O systems of planetary interest, *Annu. Rev. Earth Planet. Sci.*, 2009, **37**, 449–477.
5. S. A. Sandford and L. J. Allamandola, The condensation and vaporization behavior of H_2O: CO ices and implications for interstellar grains and cometary activity, *Icarus*, 1988, **76**, 201–224.
6. A. Al-Halabi, H. J. Fraser, G. J. Kroes and E. F. van Dishoeck, Adsorption of CO on amorphous water-ice surface, *Astron. Astrophys.*, 2004, **422**, 777–791.
7. H. J. Fraser, M. P. Colling, J. W. Dever and M. R. S. McCoustra, Using laboratory studies of CO–H_2O ices to understand the non-detect ion of a 2152 cm^{-1} (4.647 μm) band in the spectra of interstellar ices, *Mon. Not. R. Astron. Soc.*, 2004, **353**, 59–68.
8. M. E. Palumbo and G. Strazzulla, The 2140 cm^{-1} band of frozen CO: laboratory experiments and astrophysical applications, *Astron. Astrophys.*, 1993, **269**, 568–580.
9. B. Schmitt, J. Greenberg and R. Grim, The temperature dependence of the CO infrared band strength in CO: H_2O ices, *Astrophys. J.*, 1989, **340**, L33–L36.
10. T. J. Roush, Physical state of ices in the outer solar system, *J. Geophys. Res.*, 2001, **106**, 33315–33323.
11. A. Ron and O. Schnepp, Lattice vibrations of the solids N_2, CO_2, and CO, *J. Chem. Phys.*, 1967, **46**, 3991–3997.
12. S. A. Sandford, L. J. Allamandola, A. G. G. M. Tielens and G. J. Valero, Laboratory studies of the infrared spectral properties of CO in astrophysical ices, Astrophys. J., 1988, **329**, pp. 498–510.
13. K. I. Öberg, H. J. Fraser, A. C. A. Boogert, S. E. Bisschop, G. W. Fuchs, E. F. van Dishoeck and H. Linnartz, Effects of CO_2 on H_2O band profiles and band strengths in mixed H_2O : CO_2 ices, *Astron. Astrophys.*, 2007, **462**, 1187–1198.
14. B. Maté, O. Gálvez, B. Martín-Llorente, M. A. Moreno, V. J. Herrero, R. Escribano and E. Artacho, Ices of CO_2/H_2O mixtures. reflection-absorption IR spectroscopy and theoretical calculations, *J. Phys. Chem. A*, 2008, **112**, 457–465.
15. M. P. Bernstein, D. P. Cruikshank and S. A. Sandford, Near-infrared laboratory spectra of solid H_2O/CO_2 and CH_3OH/CO_2 ice mixtures, *Icarus*, 2005, **179**, 527–534.
16. D. W. White, P. A. Gerakines, A. M. Cook and D. C. B. Whittet, Laboratory spectra of the CO_2 bending-mode feature in interstellar ice analogues subject to thermal processing, *Astrophys. J. Suppl.*, 2009, **180**, 182–191.
17. C. J. Burnham, G. F. Reiter, J. Mayers, T. Abdul-Redah, H. Reichert and H. Dosch, On the origin of the redshift of the OH stretch in ice Ih: evidence from the momentum distribution of the protons and the infrared spectral density, *Phys. Chem. Chem. Phys.*, 2006, **8**, 3966–3977.
18. N. Plattner and M. Meuwly, Atomistic simulations of CO vibrations in ices relevant to astrochemistry, *ChemPhysChem*, 2008, **9**, 1271–1277.
19. N. Plattner, T. Bandi, J. D. Doll, D. L. Freeman and M. Meuwly, MD simulations using distributed multipole electrostatics: Structural and spectroscopic properties of CO- and methane-containing clathrates, *Mol. Phys.*, 2008, **106**, 1675–1684.
20. N. Plattner and M. Meuwly, The role of higher CO-multipole moments in understanding the dynamics of photodissociated carbonmonoxide in myoglobin, *Biophys. J.*, 2008, **94**, 2505–2515.
21. P. F. Fracassi, G. Cardini, S. O'Shea, R. W. Impey and M. L. Klein, Solid and liquid carbon monoxide studied with the use of constant-pressure molecular dynamics, *Phys. Rev. B: Condens. Matter*, 1986, **33**, 3441–3447.
22. B. R. Brooks, R. E. Bruccoleri, B. D. Olafson, D. J. States, S. Swaminathan and M. Karplus, CHARMM: A program for macromolecular energy, minimization, and dynamics calculations, *J. Comput. Chem.*, 1983, **4**, 187–217.

23 J. N. Huffaker, Diatomic molecules as perturbed Morse oscillators. I. Energy levels, *J. Chem. Phys.*, 1976, **64**, 3175–3181.
24 J. N. Huffaker, Diatomic molecules as perturbed Morse oscillators. II. Extension to higher-order parameters, *J. Chem. Phys.*, 1976, **64**, 4564–4570.
25 A. J. Stone. *The Theory of Intermolecular Forces*. Clarendon Press, Oxford, 1996.
26 A. J. Stone, Distributed multipole analysis: Stability for large basis sets, *J. Chem. Theory Comput.*, 2005, **1**, 1128–1132.
27 N. Kumagai, K. Kawamura and T. Yokokawa, An interatomic potential model for H_2O: Applications to water and ice polymorphs, *Mol. Simul.*, 1994, **12**, 177–186.
28 C. J. Burnham, J. C. Li and M. Leslie, Molecular dynamics calculations for ice Ih, *J. Phys. Chem. B*, 1997, **101**, 6192–6195.
29 M. J. Frisch, G. W. Trucks, H. B. Schlegel, G. E. Scuseria, M. A. Robb, J. R. Cheeseman, J. A. Montgomery, Jr., T. Vreven, K. N. Kudin, J. C. Burant, J. M. Millam, S. S. Iyengar, J. Tomasi, V. Barone, B. Mennucci, M. Cossi, G. Scalmani, N. Rega, G. A. Petersson, H. Nakatsuji, M. Hada, M. Ehara, K. Toyota, R. Fukuda, J. Hasegawa, M. Ishida, T. Nakajima, Y. Honda, O. Kitao, H. Nakai, M. Klene, X. Li, J. E. Knox, H. P. Hratchian, J. B. Cross, V. Bakken, C. Adamo, J. Jaramillo, R. Gomperts, R. E. Stratmann, O. Yazyev, A. J. Austin, R. Cammi, C. Pomelli, J. Ochterski, P. Y. Ayala, K. Morokuma, G. A. Voth, P. Salvador, J. J. Dannenberg, V. G. Zakrzewski, S. Dapprich, A. D. Daniels, M. C. Strain, O. Farkas, D. K. Malick, A. D. Rabuck, K. Raghavachari, J. B. Foresman, J. V. Ortiz, Q. Cui, A. G. Baboul, S. Clifford, J. Cioslowski, B. B. Stefanov, G. Liu, A. Liashenko, P. Piskorz, I. Komaromi, R. L. Martin, D. J. Fox, T. Keith, M. A. Al-Laham, C. Y. Peng, A. Nanayakkara, M. Challacombe, P. M. W. Gill, B. G. Johnson, W. Chen, M. W. Wong, C. Gonzalez and J. A. Pople, *GAUSSIAN 03 (Revision B.01)*, Gaussian, Inc., Wallingford, CT, 2003.
30 T. H. Dunning and P. J. Hay. *Modern Theoretical Chemistry*. Plenum, New York, 1976.
31 G. Herzberg. *Molecular Spectra and molecular Structure Vol. II Infrared and Raman Spectra of Polyatomic Molecules*. Van Nostrand Co, New York, 1945.
32 M. P. Allen and D. J. Tildesley. *Computer Simulation of Liquids*. Clarendon Press, Oxford, 1987.
33 P. Berens and K. Wilson, Molecular-dynamics and spectra. I. Diatomic rotation and vibration, *J. Chem. Phys.*, 1981, **74**, 4872–4882.
34 J. Danielsson and M. Meuwly, Energetics and dynamics in MbCN: CN^--vibrational relaxation from molecular dynamics simulations, *J. Phys. Chem. B*, 2007, **111**, 218–226.
35 E. K. Gill and J. A. Morrison, Thermodynamic properties of condensed CO, *J. Chem. Phys.*, 1966, **45**, 1585–1590.
36 W. H. Keesom and J. W. L. Köhler, The lattice constant and expansion coefficient of solid carbon dioxide, *Physica*, 1934, **1**, 655–658.
37 A. Simon and K. Peters, Single crystal refinement of the structure of carbon dioxide, *Acta Crystallogr., Sect. B: Struct. Crystallogr. Cryst. Chem.*, 1980, **36**, 2750–2751.
38 *CRC Handbook of Chemistry and Physics*, ed. R. C. Weast, D. R. Lide, M. J. Astle, and W. H. Beyer, CRC Press, Inc., Boca Raton, Florida, 70th edn, 1989.
39 I. N. Krupskii, A. I. Prokhvatilov, A. I. Erenburg and L. D. Yantsevich, Structure and thermal expansion of α-CO, *Phys. Status Solidi A*, 1973, **19**, 519–527.
40 S. Circone, L. A. Stern, S. H. Kirby, W. B. Durham, B. C. Chakoumakos, C. J. Rawn, A. J. Rondinone and Y. Ishii, CO_2 hydrate: Synthesis, composition, structure, dissociation behavior, and a comparison to structure I CH_4 hydrate, *J. Phys. Chem. B*, 2003, **107**, 5529–5539.
41 T. Umecky, M. Kanakubo and Y. Ikushima, Experimental determination of reorientational correlation time of CO_2 over a wide range of density and temperature, *J. Phys. Chem. B*, 2003, **107**, 12003–12008.
42 J. Bouwman, W. Ludwig, Z. Awad, K. I. Öberg, G. W. Fuchs, E. F. van Dishoeck and H. Linnartz, Band profiles and band strengths in mixed H_2O: CO ices, *Astron. Astrophys.*, 2007, **476**, 995–1003.
43 D. R. Nutt and M. Meuwly, Theoretical investigation of infrared spectra and pocket dynamics of photodissociated carbonmonoxy myoglobin, *Biophys. J.*, 2003, **85**, 3612–3623.
44 D. R. Nutt and M. Meuwly, Migration in native and mutant myoglobin: Atomistic simulations for the understanding of protein function, *Proc. Natl. Acad. Sci. U. S. A.*, 2004, **101**, 5998–6002.
45 N. Plattner and M. Meuwly, The role of higher CO-multipole moments in understanding the dynamics of photodissociated carbonmonoxide in myoglobin, *Biophys. J.*, 2008, **94**, 2505–2515.
46 S. Lutz, K. Nienhaus, G. U. Nienhaus and M. Meuwly, Ligand migration between internal docking sites in photodissociated carbonmonoxy neuroglobin, *J. Phys. Chem. B*, 2009, **113**, 15334–15343.

Isomer specific spectroscopy of $C_{10}H_n$, $n = 8$–12: Exploring pathways to naphthalene in Titan's atmosphere

Joshua A. Sebree,[a] Vadim V. Kislov,[b] Alexander M. Mebel[b] and Timothy S. Zwier[*a]

Received 26th February 2010, Accepted 8th April 2010
DOI: 10.1039/c003657a

Laboratory investigations of the isomer-specific spectroscopy of several $C_{10}H_n$ isomers with $n = 8$–12 are described, focusing on structures of relevance to the formation or subsequent reaction of naphthalene. The photochemical models of Titan's atmosphere have now progressed to the point that further development of the large-molecule end of the model must recognize and explicitly incorporate the unique spectroscopy, photochemistry, and reactivity of structural isomers. Mass-resolved, resonant two-photon ionization (R2PI) was used to record ultraviolet spectra of specific $C_{10}H_n$ composition, while hole-burning methods were used to resolve the spectra of different structural and conformational isomers under jet-cooled conditions. The R2PI spectrum of a new $C_{10}H_8$ isomer, 1-phenyl-1-butyne-3-ene, is described and contrasted with other $C_{10}H_8$ isomers. The anticipated role for resonance-stabilized radicals is illustrated by studies of the visible spectroscopy of two hydronaphthyl radical isomers, 1-$C_{10}H_9$ and 2-$C_{10}H_9$, and the trihydronaphthyl radical 1,2,3-$C_{10}H_{11}$. Conformation-specific spectra of an anticipated $C_{10}H_{12}$ recombination product of benzyl and allyl radicals is also reported. A reaction scheme that fleshes out the experimental data surrounding naphthalene and its hydrogenated radicals and ions is proposed as a basis for future modeling under Titan's conditions.

Introduction

Among the planetary bodies in our solar system, Titan's atmosphere is arguably one of the most intriguing. The high methane content of its photochemically driven atmosphere has produced a complex mixture of hydrocarbons and nitriles, leading ultimately to polymeric materials ('tholins') that are constituents of its aerosols.[1]

The Cassini mission is providing detailed, quantitative data on the existence and concentrations of larger molecules in Titan's atmosphere, including aromatics, polyynes, and nitriles.[1,2] The observation by INMS of large mass ions at 1000 km altitude, including benzene and toluene,[1] has led to a re-consideration of the photochemical models of Titan's atmosphere, with the intention of including both ion and neutral pathways. In anticipation of this data, Wilson and Atreya[3] in 2004 extended the photochemical model of Titan to include processes leading to haze formation (often referred to simply as 'polymer'), including some ion chemistry. More recently, Krasnopolsky[4] has explicitly included the interplay between ion and neutral chemistry in his model, showing a strong coupling between the two and significant

---

[a]*Department of Chemistry, Purdue University, West Lafayette, IN, 47907-2084, U.S.A. E-mail: zwier@purdue.edu*
[b]*Department of Chemistry and Biochemistry, Florida International University, Miami, FL, 33199, USA*

contributions to 'polymer' formation from both. Three general families of 'polymer' are included in the models: poly-yne, nitrile, and polyaromatic hydrocarbons (PAH), with phenyl radicals implicated as the neutral species leading to larger PAH.[3,4]

Due to a lack of experimental data, these models of Titan's atmosphere lack detailed chemical specificity in steps beyond the first aromatic ring, and use generic reactions with effective rates for PAH formation.[3,4] In fleshing out the large-molecule end of the photochemical model of Titan, laboratory studies must determine the products formed, provide temperature-dependent reaction rates, and measure the product quantum yields under the conditions relevant to the atmosphere. When photochemical steps are included, photochemical quantum yields are needed as a function of excitation wavelength.[3,4] The challenge is that the chemistry is becoming increasingly complex, with the number of pathways from phenyl and benzyl radicals that lead towards even larger products growing exponentially with size. Under such circumstances, it will be increasingly important not simply to study all possible reactions by 'brute force', but rather to explore particular pathways that show promise of leading efficiently to a particular product of interest.

Naphthalene is the smallest polyaromatic hydrocarbon ($C_{10}H_8$), consisting of two fused rings. As such, it is especially important to establish pathways from benzene and its simple derivatives to naphthalene and its derivatives, since similar routes are likely to operate in forming larger PAH. In this paper, we describe spectroscopic and thermochemical results for a series of $C_{10}H_n$ isomers with $n = 8-12$ that are likely intermediates on the pathway to naphthalene, or could be involved in reactive pathways from naphthalene once formed.

The work described in this paper has been guided by the following hypotheses. First, as the photochemical models are extended to include molecules in this size regime, explicit account must be taken of structural isomers, which will play an increasingly pervasive role. For instance, while it is tempting to ascribe all $C_{10}H_8$ in the model to naphthalene, many other $C_{10}H_8$ structural isomers exist that must be considered as well. The product distributions of reactions that form $C_{10}H_8$ products are anticipated to be extraordinarily sensitive to temperature and pressure, since the chemically activated adducts formed by initial attack are born and subsequently stabilized on a multi-dimensional potential energy surface replete with many isomeric minima, often with significant barriers separating them.[5,6] Motivated by this fact, we are currently exploring the ultraviolet spectroscopy and photochemistry of a series of $C_{10}H_8$ isomers of naphthalene, each of which possesses unique spectral signatures and reactive pathways. The *cis* and *trans* isomers of 4-phenyl-1-butene-3-yne (phenylvinylacetylene, labeled hereafter as Z-PVA and E-PVA) are one pair of $C_{10}H_8$ isomers recently explored in some detail.[7] Photoisomerization of each of these isomers has also been studied using the newly developed method of ultraviolet population transfer spectroscopy.[8] Here, we report on the ultraviolet spectroscopy of a close analog of PVA in which the order of the ethynyl and vinyl groups relative to the phenyl ring are swapped, 1-phenyl-1-butyne-3-ene, labeled as PAV in order to highlight the re-ordering of the conjugated substituents. As we shall see, this re-ordering produces a wholesale change in the ultraviolet spectrum relative to the isomers of PVA (Scheme 1).

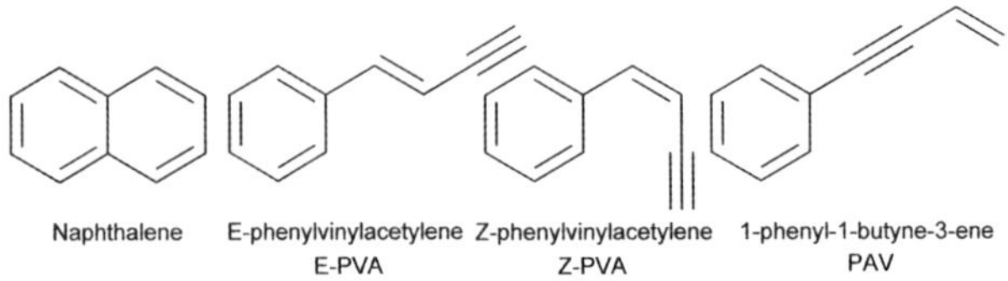

| Naphthalene | E-phenylvinylacetylene E-PVA | Z-phenylvinylacetylene Z-PVA | 1-phenyl-1-butyne-3-ene PAV |

Scheme 1

Second, resonance-stabilized radicals (RSR) are anticipated to play an especially important role in the neutral chemistry in Titan's atmosphere, due to their accumulation in the atmosphere resulting from their unusual stability.[9] Propargyl radical (C_3H_3) and benzyl radical ($C_6H_5CH_2$) are two notable examples of RSRs that are likely to play key roles in Titan's atmospheric chemistry, but the list must be expanded beyond this pair. It is thus important to characterize the spectroscopy and thermochemistry of a wider array of RSRs, which also could come in more than one isomeric form. In this paper, we focus particular attention on a series of RSRs that involve partially hydrogenated naphthalenes: 1-$C_{10}H_9$ and 2-$C_{10}H_9$ hydronaphthyl radical isomers, and 1,2,3-trihydronaphthyl radical (1,2,3-$C_{10}H_{11}$) (Scheme 2).

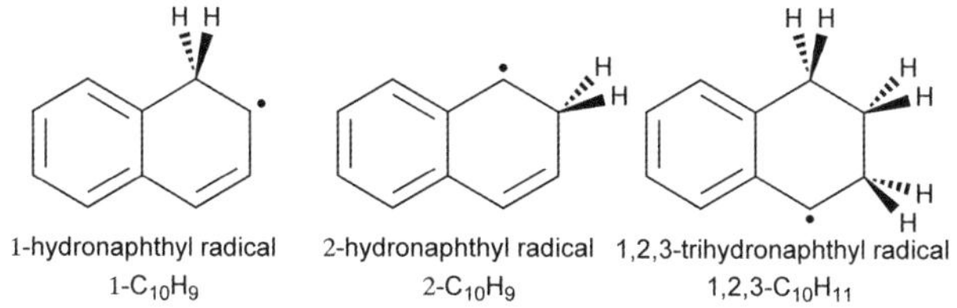

Scheme 2

These radicals involve addition of one or more hydrogen atoms to one of the aromatic rings of naphthalene, forming RSRs in which the ring not under attack retains its aromaticity, with the radical site delocalized over several sites on both rings.

Third, the recombination of two RSRs is likely to be an important means of forming larger closed-shell products, among them $C_{10}H_n$ neutrals with $n = 10, 12$. If such radical recombination reactions could also cyclize the second ring, they could serve as important pathways to formation of naphthalene. The "head-to-head" recombination of benzyl and propargyl radicals would produce 4-phenyl-1-butyne, a $C_{10}H_{10}$ isomer with several conformational isomers that has been studied recently by conformation-specific methods (Scheme 3).[10]

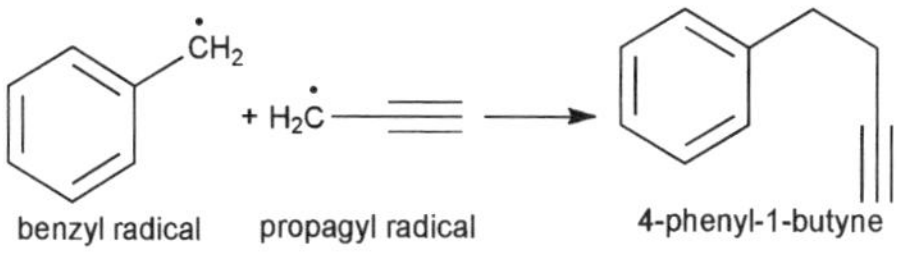

Scheme 3

The analogous "head-to-head" recombination of benzyl and allyl radicals will produce 4-phenyl-1-butene ($C_{10}H_{12}$), whose single-conformation spectroscopy is reported here (Scheme 4).

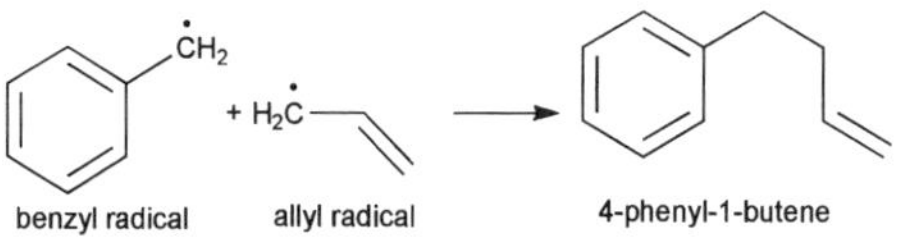

Scheme 4

The initial adducts of benzyl + propargyl and benzyl + allyl reactions will have sufficient energy to undergo isomerization, including the possibility of cyclizing to form partially hydrogenated naphthalenes such as 1,2-dihydronaphthalene

(1,2-DHN), 1,4-dihydronaphthalene (1,4-DHN), or tetralin (1,2,3,4-tetrahydronaphthalene) (Scheme 5).

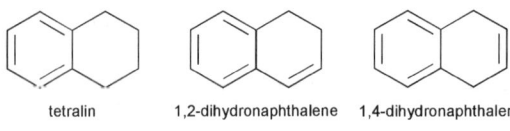

tetralin   1,2-dihydronaphthalene   1,4-dihydronaphthalene

Scheme 5

These hydrogenated naphthalenes are likely intermediates in many of the pathways that lead to naphthalene. Alternatively, they may be formed by subsequent reaction of naphthalene with other atmospheric constituents.

As a part of our studies of the hydronaphthyl radicals, we have also recorded unsaturated, jet-cooled ultraviolet spectra of these three molecules, which are reported here. This body of data on the hydrogenated naphthalene series, $C_{10}H_n$, $n = 8-12$ has been explored computationally with an eye towards its incorporation into future models of Titan's atmosphere.

Finally, an accurate account of the close coupling between neutral and ion chemistry in Titan's atmosphere requires laboratory data that explicitly connects the two, whether *via* photoionization, electron–ion recombination, or ion–molecule reactions. In the case of the hydronaphthyl radicals, we have recorded photoionization efficiency scans that determine the ionization potentials of the radicals to high accuracy.[11] The low ionization potentials of these RSRs (~6.5 eV) make them prime candidates to couple neutral and ion chemistry, producing protonated naphthalene ($C_{10}H_9^+$) or protonated dihydronaphthalene ($C_{10}H_{11}^+$) ions, the former of which is particularly important as a stable protonated PAH. Here we propose applying to Titan's atmosphere an expanded version of a reaction scheme originally proposed by LePage *et al.* to account for PAH-mediated H_2 formation in the interstellar medium.[12] The model explicitly couples $C_{10}H_n$ neutrals and $C_{10}H_n^+$ ions with $n = 8-12$, setting a foundation for future modeling of the altitude-dependent concentrations of these potentially important constituents of Titan's atmosphere.

Experimental details

Conformational studies were carried out in a vacuum chamber described previously[13] (Fig. 1). The molecule of interest, diluted in helium at a backing pressure

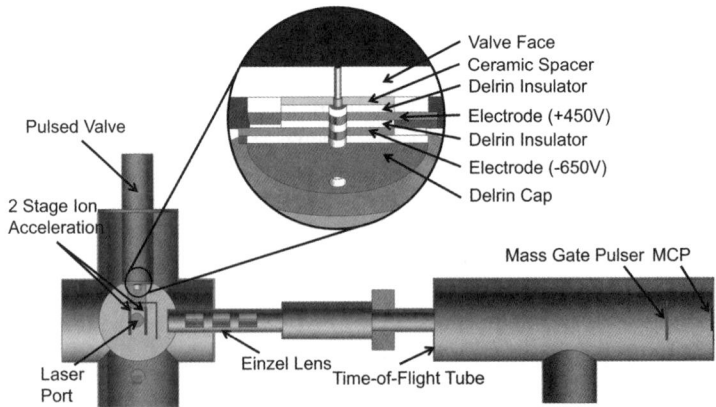

Fig. 1 Schematic of time-of-flight spectroscopy chamber with (cutaway) discharge source attached.

of 3 bar, was introduced to the chamber *via* a pulsed valve (R. M. Jordon Company) operating at 20 Hz. The gas mixture underwent a supersonic expansion upon entering the chamber and was collisionally cooled to a few degrees Kelvin prior to entering the ion-source region of a Wiley-McLaren time-of-flight (TOF) chamber.[14] To generate hydronaphthyl radicals, samples of 1,4-DHN, 1,2-DHN or tetralin were entrained in argon at a backing pressure of 1 bar and pulsed into a discharge channel (2 mm dia. × 7 mm long). Voltages of +450 V and −650 V were applied in 100 μs pulses to the inner and outer electrodes respectively and timed to intersect the gas pulse. The resulting species were then cooled *via* supersonic expansion.

When the expansion entered the ion-source region a laser, intersecting the expansion, irradiated the expansion. If the laser was resonant with a transition of a molecule in the expansion, the compound would be ionized though a process known as two-color resonance-two-photon-ionization (2C-R2PI). Once ionized, the molecule was accelerated down the time-of-flight chamber where it was detected *via* a microchannel plate (MCP) ion detector. The MCP signal was recorded on a digital oscilloscope as a function of time. An advantage of using a TOF setup is mass resolution, allowing for the removal of spectral overlap that may be generated from other molecules present in the ionization region. To increase mass resolution and remove interference from the radical precursors, a mass gate pulser[13] (MGP) could be switched on to pulse away ions that would otherwise washout signal from the species of interest.

Excitation spectra were obtained using two techniques: two-color resonance-two-photon-ionization (2C-R2PI) and holeburning (HB). To accomplish these techniques, the harmonics of an Nd:YAG (532 and 355 nm) were used to pump a tunable dye laser. The fundamental of the dye laser was then either used as is for RSR studies or doubled by passing it though a BBO crystal to give final wavelengths in the mid-UV range providing the first photon for the excitation step. A second photon, held at a fixed frequency, then ionized the excited molecules. The R2PI spectrum is recorded by monitoring the ion signal in the mass spectrum while tuning through different wavelengths. When the first laser is resonant with a vibronic or pure electronic transition, the ion signal registered by the MCP is resonantly enhanced. By plotting the change in the ion signal *versus* the wavelength of the laser, the UV-Vis spectrum is obtained.

To obtain individual spectra for each conformer or isomer, holeburning is used. In HB, the holeburn laser at 10 Hz is spatially overlapped and fired ~200 ns prior to the probe laser and tuned to a transition in the R2PI spectrum. The holeburn laser partially or totally saturates the transition. The probe laser at 20 Hz is then tuned through the R2PI spectrum and the difference in the signal between the successive laser pulses is recorded using active baseline subtraction. Whenever both the holeburn laser and the probe laser are resonant with a transition with the same ground state, there is a dip in the ion signal from the probe laser. In this way, the spectra of individual conformers can be recorded and assigned.

Rotational band contour (RBC) spectroscopy was done using 2C-R2PI. An intracavity Etalon was placed in the probe dye laser to increase the resolution. The probe laser at low power (~10–50 μJ/pulse) was then scanned over the origin of each conformer to excite the molecule without saturation effects prior to being ionized by the second laser. Band contours were then simulated using JB95[15] as described previously.[10]

Results and analysis

Isomer-specific spectroscopy of $C_{10}H_8$ isomers: 1-phenyl-1-butyne-3-ene

Fig. 2a) presents the R2PI spectrum of 1-phenyl-1-butyne-3-ene (PAV) in the 34900–36360 cm^{-1} region, while monitoring the *m/z* 128 mass channel in the time-of-flight mass spectrum. Most aromatic derivatives have spectra in which the most intense

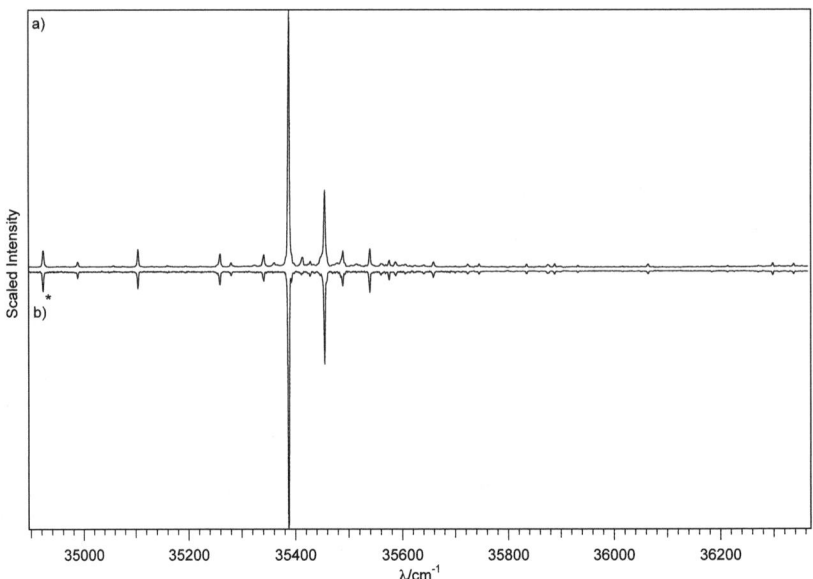

Fig. 2 (a) R2PI and (b) holeburning scans of jet-cooled 1-phenyl-1-butyne-3-ene (PAV). The asterisk (*) denotes the holeburning transition used in recording the scan in (b).

transition is the S_0–S_1 origin at the red edge of the spectrum, with Franck–Condon factors that reflect a small geometry change between ground and excited states of the π–π* transitions. The spectrum shown in Fig. 2a) is thus unusual in possessing a series of weak transitions on its red edge, followed by strong transitions well to the blue. One possibility is that the weak transitions are not associated with the S_0–S_1 transitions out of the ground state zero-point level of PAV, but rather arise from an impurity in the sample or residual hot bands due to population in vibrationally excited ground state levels of PAV not removed completely by the collisional cooling in the supersonic expansion. To test this possibility, a UVHB scan was recorded with hole-burning laser fixed on the transition marked by an asterisk in the Figure, recording a difference spectrum that produces a non-zero signal from all transitions that arise from the same ground state level tagged by the hole-burn laser. All transitions are present in the UVHB spectrum of Fig. 2b), proving that both the weak and strong transitions come from the same ground state level, the zero-point level of PAV. Thus, the S_0–S_1 origin is assigned to the transition at 34924 cm^{-1}, with all transitions to the blue of the origin assigned as vibronic transitions terminating in higher-lying vibrational levels in the S_1 manifold. The strong transitions are likely gaining their intensity from vibronic coupling to higher-lying singlet states, a hypothesis that we are presently testing using dispersed fluorescence spectroscopy. Currently the strongest vibronic mode at 464 cm^{-1} has been assigned to an in-plane bending mode of the acetylenic moiety.

Table 1 reports the S_0–S_1 origin of PAV and the other $C_{10}H_8$ isomers that have been studied to date. As this Table clearly shows, these isomers each possess unique ultraviolet absorptions that can be used as markers for their presence.

Recombination of resonance-stabilized radicals

The anticipated build-up in the concentrations of resonance-stabilized hydrocarbon radicals in Titan's atmosphere makes it likely that recombination of these radicals will be a significant pathway to more complex hydrocarbons. In a previous study of 4-phenyl-1-butyne, it was noted that this $C_{10}H_{10}$ isomer is the product of the

Table 1 S_0–S_1 origins of $C_{10}H_8$ structural isomers

Isomer	S_0–S_1 origin/cm^{-1}
trans-o-Ethynylstyrene[a]	32369
cis-m-Ethynylstyrene[a]	32672
trans-m-Ethynylstyrene[a]	32926
p-Ethynylstyrene[a]	33407
trans-Phenylvinylacetylene (E-PVA)[b]	33578
cis-Phenylvinylacetylene (Z-PVA)[b]	33838
1-Phenyl-1-butyne-3-ene (PAV)	34924

[a] Ref. 16. [b] Ref. 7.

head-to-head recombination of the benzyl and propargyl radicals. Similarly, 4-phenyl-1-butene (C_6H_5–CH_2–CH_2–CH=CH_2, 4PB) is the direct recombination product of benzyl ($C_6H_5CH_2$) and allyl radicals ($\cdot CH_2$–CH=CH_2). One of the interesting aspects of the spectroscopy of 4-phenyl-1-butene is that the alkenyl tail is sufficiently long that it can form spectroscopically distinguishable conformational isomers, much as occurs in 4-phenyl-1-butyne.[10] Thus we are faced with the spectroscopic challenge of observing and assigning the single-conformation ultraviolet spectra of the low-lying conformational isomers (often referred to via the contraction 'conformers') of 4PB. In addition to providing the spectroscopic signatures of these conformational isomers, they highlight the fact that in molecules this size, conformational isomerization may need to precede structural isomerization or intermolecular reaction. The timescales and rates of these reactions will therefore be affected by the number of conformers present, their relative energies, and the magnitudes of the energy barriers that separate them.

Fig. 3 shows the R2PI and UVHB spectra of 4PB under jet-cooled conditions. The three transitions labeled A–C served as hole-burning transitions in the UVHB spectra shown below. The resulting spectra prove that there are three conformers of 4PB, with S_0–S_1 origins for conformers A–C at 37525, 37528, and 37580 cm^{-1}, respectively. The vibronic spectra of each are closely similar, but shifted from one

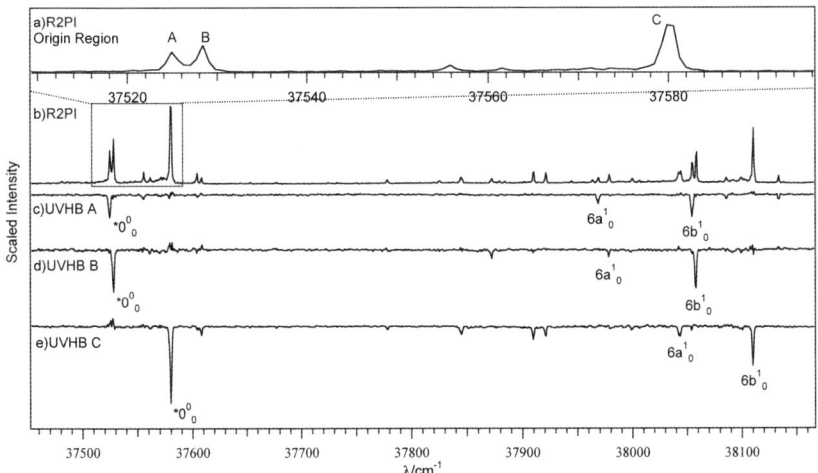

Fig. 3 (a) R2PI spectrum of 4-phenyl-1-butene expanded in the origin region, (b) R2PI of PB, (c–e) UVHB spectra of conformers A, B, and C respectively. The 0^0_0, $6a^1_0$, and $6b^1_0$ transition are labeled. The asterisk (*) indicates the transitions used for holeburning.

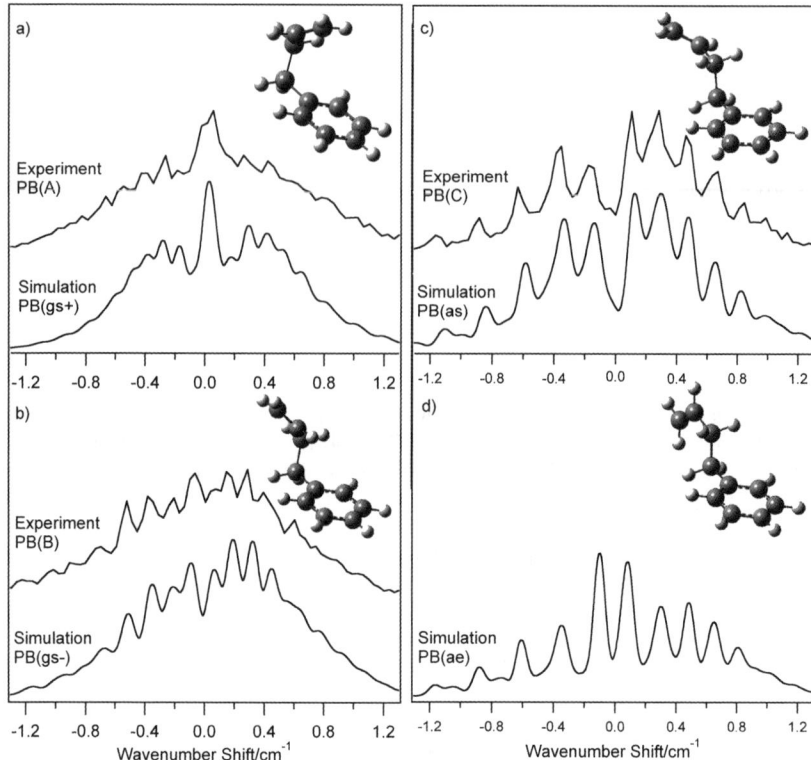

Fig. 4 (a–c) Rotational band contours of PB origins. Experimental data (top traces) were fit using JB95[15] to give simulated contours (bottom traces) for three low-energy conformers. (d) Simulated spectrum for fourth unobserved conformer of PB. Ground state rotational constants were taken from HF/6-311+G(d,p) calculations. Excited state rotational constants and transition dipole moments were taken from CIS/6-311+G(d,p) calculations.

another, producing three overlapping spectra that are revealed cleanly in the hole-burning spectra. The $6a^1_0$ and $6b^1_0$ transitions involve phenyl ring deformations, carrying large Franck–Condon intensity much as they do in many other phenyl derivatives.

Fig. 4 shows the experimental rotational band contours of conformers A–C, recorded at a resolution of 0.08 cm^{-1}. The experimental contours show very different profiles associated with changes in the direction of the transition dipole moment with conformation. This unusual circumstance is a consequence of the sensitivity of the phenyl ring transition dipole moment (TDM) to its local environment, first noted by Simons and co-workers.[17,18] It provides a robust method to obtain structural assignments in phenyl derivatives, since calculated TDM directions have been shown to match experiment closely. Indeed, Fig. 4 compares the experimental contours with simulated contours calculated using the TDM projections along the inertial axes indicated in the Figure caption. The structures shown as an inset in Fig. 4a–c are those assigned to conformers A–C, respectively. Naming of conformers is based on the dihedral angles of the butene side chain (Scheme 6).

CαCβ: 60/180 gauche/anti
CαCγ: 0/±120 eclipsed/±sync

Scheme 6

Oddly, the calculations predict the presence of a fourth low-energy conformer (Fig. 4d), which is not observed experimentally. It may be that this conformer possesses a low barrier to isomerization to one or more of the other conformers, so that its population is cooled out of this well during the collisional cooling process in the expansion.

Hydrogenated naphthalenes

Of the $C_{10}H_{10}$ and $C_{10}H_{12}$ isomers, some of the most intriguing and potentially important are the hydrogenated naphthalenes. These isomers are bicyclic, and thus are likely near-final steps on the way to forming naphthalene. Furthermore, since at most altitudes in Titan's atmosphere, hydrogen atoms are the most abundant radical, hydrogenation of naphthalene could be a significant pathway for further chemical processing of naphthalene once formed. To that end, we have undertaken a study of the vibronic spectroscopy and thermochemistry of the partially hydrogenated naphthalenes $C_{10}H_n$, $n = 9$–12.

When n is even (*i.e.*, $n = 10, 12$), the hydronaphthalenes have closed-shell electronic structures and are stable molecules, with two isomeric forms of dihydronaphthalene (1,2- and 1,4-), and one of tetrahydronaphthalene (1,2,3,4-, tetralin). In both cases the additional hydrogens are on the same ring, breaking its aromaticity, but retaining aromaticity in the ring not under attack. These molecules were used as precursors in forming the hydronaphthyl and trihydronaphthyl radicals. However, given their relevance to the chemistry of Titan's atmosphere, we also recorded jet-cooled R2PI spectra of all three molecules under similar conditions as a preliminary to our experiments on the radicals.

Fig. 5a presents the two-color R2PI spectrum of the first 1800 cm^{-1} of the S_0–S_1 transition of tetralin while monitoring the $C_{10}H_{12}^+$ mass channel. The analogous spectra of 1,4-DHN and 1,2-DHN in the $C_{10}H_{10}^+$ mass channel are shown in Fig. 5b and 5c, respectively. All three spectra were recorded under laser power the S_0–S_1 origins dominating the spectrum. The two molecules have S_0–S_1 origins in conditions in which the S_0–S_1 vibronic transitions are unsaturated. The R2PI spectra so obtained are similar to the fluorescence excitation spectra from past studies of these molecules.[19-23] We have extended the spectra of all three molecules out to ~2000 cm^{-1} above their respective S_0–S_1 origins in order to provide an overview of the Franck–Condon activity in each and to allow a direct comparison between them. The spectra of both tetralin (Fig. 5a) and 1,4-DHN (Fig. 5b) are quite simple,

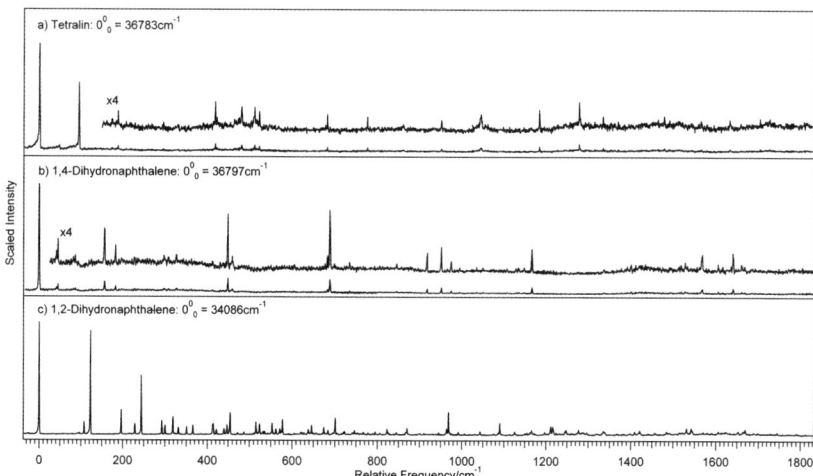

Fig. 5 R2PI spectra of (a) tetralin, (b) 1,4-dihydronaphthylene, (c) 1,2-dihydronaphthylene.

with close proximity to one another (36797 cm^{-1} for tetralin and 36783 cm^{-1} for 1,4-DHN) and to *ortho*-xylene (37310 cm^{-1}), as might be anticipated since all three molecules have two saturated carbons attached to the ring *ortho* to one another. In tetralin, the single strong vibronic band 94 cm^{-1} above the S_1 origin has been assigned previously[23] to the fundamental of a ring twist vibration of the saturated ring (ν_{31}).

Further to the blue, a series of weak transitions with intensities a few percent of the origin are observed, extending out to at least 1400 cm^{-1} above the S_1 origin. It seems likely that the intensities of these transitions are reduced due to the turn-on of a non-radiative process.

The spectrum of 1,2-DHN (Fig. 5c) shows an S_0–S_1 origin transition at 34086 cm^{-1}, flanked by much more extensive vibronic structure that reaches almost 1700 cm^{-1} above the S_1 origin. Much of the low-frequency structure can be ascribed to ring puckering modes, as has been analyzed in some detail by Autrey *et al.*[20]

Fig. 6 presents the vibronic spectra of 1-hydronaphthyl, 2-hydronaphthyl, and 1,2,3-trihydronaphthyl radicals, plotted on relative wavenumber scale (cm^{-1}) to facilitate direct comparison between them. Fig. 6a shows the 2C-R2PI spectrum of the 1-hydronaphthyl radical (1-$C_{10}H_9$) formed when 1,4-DHN was used as radical precursor in a pulsed electric discharge expansion in argon. The radical discharge source shown as an inset in Fig. 1 was used to create the free radicals of interest. The observed transitions are ascribed almost entirely to 1-$C_{10}H_9$, whose calculated ground state structure is shown in Fig. 7a. When 1,2-DHN is used as precursor, a series of new transitions appears, which are assigned to the 2-$C_{10}H_9$ radical. VHB (Fig. 6b) confirms that these transitions are due to a second $C_{10}H_9$ isomer, and provides a spectrum for this isomer free from interference from the transitions due to 1-$C_{10}H_9$. Details of the assignment process for these two radicals have been discussed elsewhere.[11] Here, we compare and contrast their spectra and properties, emphasizing those aspects that are most directly relevant to Titan's atmosphere.

The D_0–D_1 origin of 1-$C_{10}H_9$ occurs at 18949 cm^{-1} (5277 Å), in the middle of the visible. This drop in D_0–D_1 separation relative to the D_0–D_1 transitions of dihydro- and tetrahydronaphthalenes is due to the open-shell character of the resonance-stabilized radical. TDDFT B3LYP/6-311+G(d,p) calculations of the D_1 excited state vibrations match with the observed low-frequency vibronic structure remarkably well, accounting for the observed low-frequency vibronic structure as even overtones and combination bands of out-of-plane skeletal vibrations of the fused-ring system.

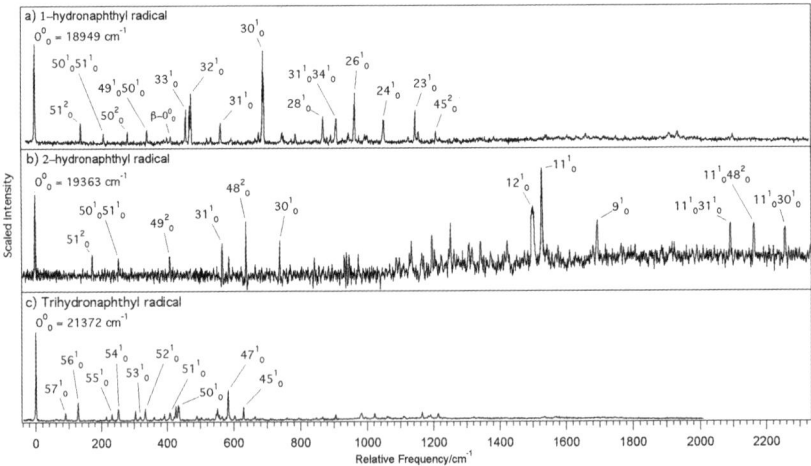

Fig. 6 (a) 2C-R2PI of 1-hydronaphthyl radical, (b) VHB of 2-hydronaphthyl, and (c) 2C-R2PI of 1,2,3-trihydronaphthylene with selected vibronic transitions labelled.

Table 2 Tentative assignments of select vibrational modes of hydronaphthyl radicals

1-$C_{10}H_9$			2-$C_{10}H_9$			1,2,3- $C_{10}H_{11}$		
Exp. Freq.[a]	Assignment[b]	D_1 Calc.[c]	Exp. Freq.[a]	Assignment[b]	D_2 Calc.[c]	Exp. Freq.[a]	Assignment[b]	D_1 Calc.[c]
1146	23^1_0	1144	2257	$11^1_0 30^1_0$	2258	629	45^1_0	641
1050	24^1_0	1086	2163	$11^1_0 48^2_0$	2167	584	47^1_0	583
965	26^1_0	965	2092	$11^1_0 31^1_0$	2099	433	50^1_0	447
908	$31^1_0 34^1_0$	933	1688	9^1_0	1688	427	$52^1_0 57^1_0$	428
869	28^1_0	859	1527	11^1_0	1521	410	51^1_0	418
689	30^1_0	691	1497	12^1_0	1494	392	$56^1_0 57^3_0$	413
564	31^1_0	581	1252	20^1_0	1251	360	$55^1_0 56^1_0$	362
475	32^1_0	487	1195	22^1_0	1170	334	52^1_0	332
472	$48^1_0 50^1_0$	480	1133	23^1_0	1147	318	53^1_0	316
460	33^1_0	478	738	30^1_0	731	305	$56^1_0 57^2_0$	317
401	49^2_0	410	637	48^2_0	640	251	54^1_0	253
342	$49^1_0 50^1_0$	354	566	31^1_0	572	233	55^1_0	237
283	50^2_0	298	407	49^2_0	382	216	$56^1_0 57^1_0$	221
209	$50^1_0 51^1_0$	211	253	$50^1_0 51^1_0$	253	129	56^1_0	125
139	51^2_0	150	173	51^2_0	160	89	57^1_0	96

[a] Experimental frequencies measured from the origin (cm^{-1}). [b] Numbering using the Mulliken[25] numbering scheme. [c] Excited state frequencies calculated with TDDFT B3LYP/6-311+G(d,p).

On this basis, 1-$C_{10}H_9$ radical is deduced to retain a plane of symmetry in its D_1 excited state. Table 2 compares the observed and calculated positions of several of the tentatively assigned bands. Dispersed fluorescence spectra are still needed to strengthen these assignments and quantify the structural changes responsible for the vibronic structure. The observed vibronic activity also contains contributions from in-plane ring deformation modes (32^1_0 and 30^1_0).

The D_0-D_2 origin of 2-$C_{10}H_9$ (Fig. 6b) occurs at 19363 cm^{-1} (5164 Å), 414 cm^{-1} to the blue of that due to 1-$C_{10}H_9$. The D_0–D_1 transition in 2-$C_{10}H_9$ has been assigned by studies of irradiated naphthalene crystals to an absorption at 6355 Å.[24] This is consistent with the predictions of the TDDFT calculations. Instead, the observed transitions assigned to 2-$C_{10}H_9$ are assigned to the D_0-D_2 transition, with an oscillator strength 20 times greater than the D_0–D_1 transition. A search for the very weak D_0–D_1 transitions of 2-$C_{10}H_9$ was unsuccessful, presumably due to this weakness. The TDDFT calculations again provide an accurate prediction of the vibrational frequencies of the out-of-plane skeletal modes of 2-$C_{10}H_9$, leading to predicted positions for their even overtones and combination bands that fall within a few wavenumbers of their observed positions (Table 2). This confirms our assignment to the D_0–D_2 transition, since the analogous predictions for D_1 are not consistent with experiment. The vibronic structure in 2-$C_{10}H_9$ extends out to more than 2200 cm^{-1} above the electronic origin. Strong vibronic bands appear 1497 and 1527 cm^{-1} above the electronic origin. These are tentatively assigned to the 12^1_0 and 11^1_0 fundamentals involving C=C stretch vibrations of the aromatic ring, which are expected to be highly Franck–Condon active in the π–π* transition. Since all three radicals are anticipated to experience similar structural changes in the rings, the lack of such Franck–Condon activity in either of the other two radicals suggests the turn-on of non-radiative processes in these regions for 1-$C_{10}H_9$ and 1,2,3-$C_{10}H_{11}$. The calculated ground state structure for 2-$C_{10}H_9$ is shown in Fig. 7b.

The 2C-R2PI spectrum of 1,2,3-trihydronaphthyl radical is shown in Fig. 6c. This spectrum was recorded with tetralin as radical precursor, monitoring the m/z 131 mass channel in the time-of-flight mass spectrum. Assignment of the observed

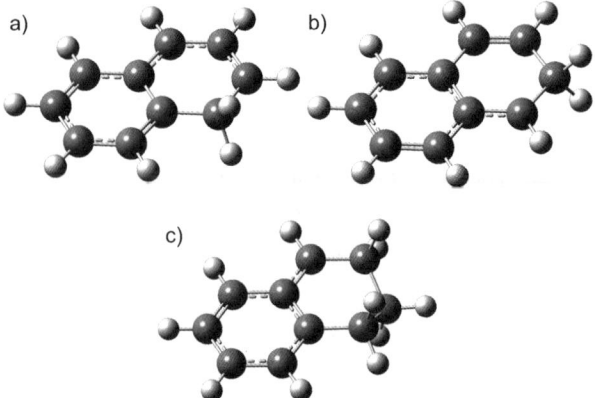

Fig. 7 Ground state structures of (a) 1-hydronaphthyl radical, (b) 2-hydronaphthyl, and (c) 1,2,3-trihydronaphthylene (B3LYP/6-311+G(d,p)).

transitions to the 1,2,3-trihydronaphthyl radical (1,2,3-$C_{10}H_{11}$, structure in Fig. 7c) is based largely on the close proximity of the observed band system, with D_0–D_1 origin at 21372 cm^{-1}, to that in the benzyl radical (22 000 cm^{-1}), with the radical site conjugated with the benzene ring. The other possible trihydronaphthyl isomer (1,2,4-$C_{10}H_{11}$) would have the radical site remote from the benzene ring, and would have electronic structure of an alkyl radical. The 1,2,3-$C_{10}H_{11}$ radical is also calculated to be more stable than 1,2,4-$C_{10}H_{11}$ by 46 kJ mol^{-1}. The spectrum of 1,2,3-$C_{10}H_{11}$ has a more congested appearance in the low-frequency region, due both to the greater intensity and number of transitions in this region. The vibronic band assignments shown in the spectrum reflect the fact that out-of-plane vibrational fundamentals are now allowed due to the non-planar carbon framework in the trihydrated ring. Once again, the calculations provide remarkably accurate predictions for the out-of-plane vibrational frequencies in the D_1 state (Table 2). This vibronic structure is reminiscent of that in 1,2-dihydronaphthalene (Fig. 5c), as might be expected based on a similar puckering of the hydrogenated ring in both. In both ground and D_1 states of 1,2,3-$C_{10}H_{11}$, C(2) is puckered out-of-plane by ~0.41 Å, while the rest of the hydrogenated ring carbons are nearly planar with the aromatic ring (Fig. 7c).

The adiabatic ionization potentials of each of these radicals has been determined using two-color photoionization efficiency scans, with the D_n excited state origin serving as intermediate state.[11] The small geometry changes between neutral ground state, neutral excited state, and ion lead to nearly vertical Franck–Condon factors to the ion, which provide sharp, highly-accurate ionization onsets. The ionization potentials of 1-$C_{10}H_9$, 2-$C_{10}H_9$, and 1,2,3-$C_{10}H_{11}$ radicals are 6.570 eV, 6.487 eV, and 6.620 eV, respectively, with errors of ±0.001 eV.

Discussion

$C_{10}H_8$ isomers

One of the particular challenges facing modelers of the rich hydrocarbon and nitrile chemistry in Titan's atmosphere is the anticipation that many of the molecules anticipated to play an especially important role (*e.g.*, naphthalene as the prototypical PAH) have many structural isomers whose chemistry and photochemistry must be considered. Naphthalene ($C_{10}H_8$) is the most stable $C_{10}H_8$ structural isomer, but the $C_{10}H_8$ potential energy surface contains many other stable minima, many of them composed of conjugated benzene derivatives.

In this paper, we have presented the ultraviolet spectrum of one such $C_{10}H_8$ isomer, PAV. The rearrangement of acetylenic and vinylic moieties in the conjugated chain relative to PVA produces wholesale changes in the position of the S_0–S_1 origin and the intensities of the vibronic structure observed. While the S_0–S_1 origin is weak, vibronic bands above the origin gain significant intensity *via* vibronic coupling to higher-lying states. This is reflected in $\Delta v = \pm 1$ emission from the vibronically-induced $v' = 1$ levels.[26] Excited state calculations of PAV locate higher-lying excited states that carry significant oscillator strength, and are thus likely candidates to be responsible for the observed vibronic activity in the S_0–S_1 transition of PAV.

Table 1 presents the observed S_0–S_1 origins of several of the $C_{10}H_8$ isomers studied to date. Each of these isomers has unique excited state properties, leading to absorptions that shift position in the UV, providing a spectroscopic means for their identification. Future studies of the photodissociation and photoisomerization pathways of these isomers is warranted. The competition between photoisomerization (which should favor naphthalene as the lowest energy $C_{10}H_8$ isomer) and photodissociation (to produce smaller free radical products) is likely to depend on the starting isomeric structure, the excited state accessed by photoexcitation, and the collisional and temperature regime in which excitation occurs.

Recombination of resonance-stabilized radicals (RSRs)

The role played by RSRs in Titan's atmosphere is likely to be significant, as it is in many other circumstances, including hydrocarbon combustion.[27] The build-up of concentration of RSRs and their barrierless recombination, leads to the prediction that such reactions will be among the important pathways for forming larger, stable hydrocarbon and nitrile products.

In this paper, we studied the single-conformation spectroscopy of 4-phenyl-1-butene, the recombination product of benzyl and allyl radicals. We have spectroscopically identified three conformer products, with S_0–S_1 origins separated by 60 cm^{-1} from one another, and near to the corresponding transitions of toluene and many other alkylbenzenes. It would be useful to carry out a more detailed first-principles investigation of the ground state surface of $C_{10}H_{12}$ in order to model the distribution of products expected from recombining benzyl and allyl radicals under the conditions of Titan's atmosphere. Photochemical studies of the 4-phenyl-1-butene can also provide experimental data on the anticipated products, if internal conversion to the ground state dominates the photoexcitation pathways.

Hydronaphthalenes and hydronaphthyl radicals

In the introduction, we posed the series of hydronaphthalenes $C_{10}H_n$, $n = 9$–12, as a potentially important set of molecules in the chemistry occurring in Titan's atmosphere, formally associated with successive addition of H-atoms to naphthalene. The experimentally determined adiabatic ionization potential of 1-$C_{10}H_9$ (6.570 eV), can be combined in a thermochemical cycle with the proton affinity of naphthalene[28] and the ionization potential of the hydrogen atom to derive the bond dissociation energy of naphthalene at the 1-carbon site.

$$1\text{-}C_{10}H_9 \rightarrow H + C_{10}H_8 \quad D_0 = 121.2 \pm 2 \text{ kJ mol}^{-1}$$

This value is sufficiently large that the 1-$C_{10}H_9$ radical is a viable intermediate worth including in future models of Titan's atmosphere. Since the corresponding proton affinities of naphthalene at the 2-site and of 1,2-dihydronaphthalene at the 3-position have not been experimentally measured, the corresponding bond dissociation energies of 2-$C_{10}H_9$ and 1,2,3-$C_{10}H_{11}$ cannot be directly extracted from the measured ionization potentials of these radicals. In order to derive these values, we have chosen instead to use state-of-the-art quantum chemical calculations of the proton affinities instead. Confidence in this procedure is derived in part from the close correspondence between the recently re-evaluated proton affinity of naphthalene at the 1-site[28] (799.4 ± 2 kJ mol^{-1}) and the value predicted by calculations at the G3(MP2, CC)//B3LYP/6-311G** level of theory (797 kJ mol^{-1}). The corresponding calculated proton affinities for naphthalene at the 2-site and for 1,2-dihydronaphthalene at the 3-position are 787 kJ mol^{-1} and 842 kJ mol^{-1}, respectively, leading to the following bond dissociation energies,

$$2\text{-}C_{10}H_9 \rightarrow H + C_{10}H_8 \quad D_0 = 103.6 \pm 2 \text{ kJ mol}^{-1}$$

$$1,2,3\text{-}C_{10}H_{11} \rightarrow H + 1,2\text{-}C_{10}H_{10} \quad D_0 = 168 \pm 3 \text{ kJ mol}^{-1},$$

where we have assumed errors of similar size in the calculated proton affinities used in the thermochemical cycles.

In order to incorporate this chemistry into future photochemical models of Titan's atmosphere, a reasonable reaction scheme must be developed, combining the experimental data with the predictions of quantum chemical calculations where experimental data is lacking. Fig. 8 summarizes the predictions of calculations for the reaction energetics (both exoergicities and transition state energies) for the successive addition of up to four H-atoms to naphthalene, forming various isomeric forms of the hydronaphthalene series $C_{10}H_n$, $n = 9$–12. All calculations were carried out at the G3(MP2, CC)//B3LYP/6-311G** level of theory. Within this computational scheme, geometries of all local minima and transition states were optimized using the hybrid density functional B3LYP[29,30] method with the 6-311G** basis set. Vibrational frequencies calculated at the same level were utilized to compute zero-point energy (ZPE) corrections. High-level single-point energy calculations were carried out employing the G3(MP2, CC) modification[31,32] of the original Gaussian 3 (G3) scheme,[33] where final energies at 0 K were obtained as

$$E0[G3(MP2,CC)] = E[CCSD(T)/6\text{-}311G(d,p)] + \Delta EMP2 ++ \Delta E(SO) + \Delta E(HLC) + E(ZPE),$$

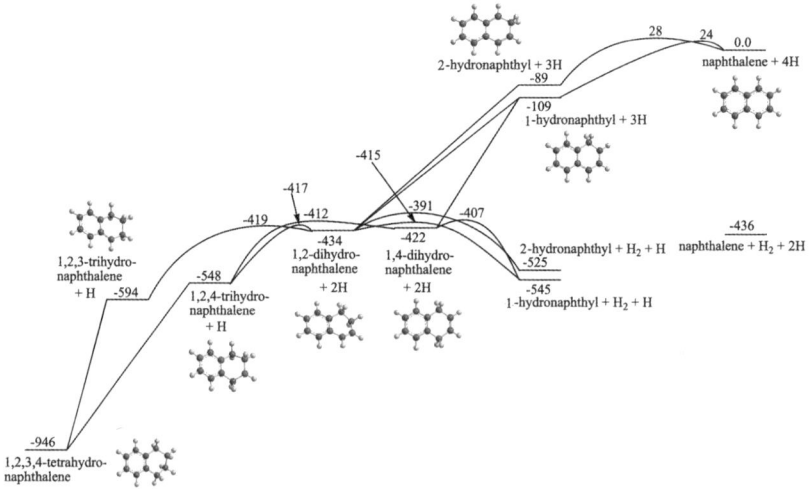

Fig. 8 Potential energy diagram of H addition and abstraction reactions in the $C_{10}H_x$ ($x = 8$–12) system calculated at the G3(MP2, CC)//B3LYP/6-311G** level. Numbers specify relative energies in kJ mol^{-1}.

Table 3 Proposed reaction scheme for modeling reactions of naphthalene and partially hydrogenated naphthalene in Titan's atmosphere. Reaction energetics are calculated at the G3(MP2,CC)// B3LYP/6-311G** level of theory. Values in parentheses are experimental results

Reaction		
A. H-Atom addition		
1. Neutrals	$\Delta E/\text{kJ mol}^{-1}$	$E_a/\text{kJ mol}^{-1}$
(1) $C_{10}H_7 + H + M \rightarrow C_{10}H_8 + M$	−461	
(2a) $C_{10}H_8 + H + M \rightarrow 1\text{-}C_{10}H_9 + M$	−109 (−121 ± 2^a)	+24
(2b) $\rightarrow 2\text{-}C_{10}H_9 + M$	−89 (−103 ± 2^a)	+28
(3a) $1\text{-}C_{10}H_9 + H + M \rightarrow 1,4\text{-}C_{10}H_{10} + M$	−313	0
(3b) $\rightarrow 1,2\text{-}C_{10}H_{10} + M$	−325	0
(4) $2\text{-}C_{10}H_9 + H + M \rightarrow 1,2\text{-}C_{10}H_{10} + M$	−345	0
(5a) $1,2\text{-}C_{10}H_{10} + H + M \rightarrow 1,2,3\text{-}C_{10}H_{11} + M$	−160	+15
(5b) $\rightarrow 1,2,4\text{-}C_{10}H_{11} + M$	−114	+17
(6) $1,4\text{-}C_{10}H_{10} + H + M \rightarrow 1,2,4\text{-}C_{10}H_{11} + M$	−126	+10
(7) $1,2,4\text{-}C_{10}H_{11} + H + M \rightarrow 1,2,3,4\text{-}C_{10}H_{12} + M$	−398	0
(8) $1,2,3\text{-}C_{10}H_{11} + H + M \rightarrow 1,2,3,4\text{-}C_{10}H_{12} + M$	−352	0
2. Ions^b	$\Delta E/\text{kJ mol}^{-1}$	$E_a/\text{kJ mol}^{-1}$
(9a) $C_{10}H_8^+ + H + M \rightarrow 1\text{-}C_{10}H_9^+ + M$	−268 (−272.7 ± 2^a)	0
(9b) $\rightarrow 2\text{-}C_{10}H_9^+ + M$	−258 (−263 ± 2^a)	0
(10a) $1\text{-}C_{10}H_9^+ + H + M \rightarrow 1,2\text{-}C_{10}H_{10}^+ + M$	−179	~2.9^c
(10b) $\rightarrow 1,4\text{-}C_{10}H_{10}^+ + M$	−116	~2.9^c
(11) $2\text{-}C_{10}H_9^+ + H + M \rightarrow 1,2\text{-}C_{10}H_{10}^+ + M$	−189	~2.9^c
(12a) $1,2\text{-}C_{10}H_{10}^+ + H + M \rightarrow 1,2,3\text{-}C_{10}H_{11}^+ + M$	−305	0^c
(12b) $\rightarrow 1,2,4\text{-}C_{10}H_{11}^+ + M$	−209	0^c
(13) $1,4\text{-}C_{10}H_{10}^+ + H + M \rightarrow 1,2,4\text{-}C_{10}H_{11}^+ + M$	−272	0^c
(14) $1,2,3\text{-}C_{10}H_{11}^+ + H + M \rightarrow 1,2,3,4\text{-}C_{10}H_{12}^+ + M$	−167	0^c
(15) $1,2,4\text{-}C_{10}H_{11}^+ + H + M \rightarrow 1,2,3,4\text{-}C_{10}H_{12}^+ + M$	−263	0^c
(16) $C_{10}H_8^+ + H_2 \rightarrow$	No Reaction	
(17) $C_{10}H_9^+ + H_2 \rightarrow$	No Reaction	
(18a) $H^+ + C_{10}H_8 + M \rightarrow 1\text{-}C_{10}H_9^+ + M$	−797 (−799.4 ± 2^a)	0
(18b) $\rightarrow 2\text{-}C_{10}H_9^+ + M$	−787 (−789 ± 2^a)	0

^{a} Ref 11. ^{b} Built upon a reaction scheme proposed by Le Page et al.[38] ^{c} Ref 39. ^{d} Ref 40. ^{e} Ref 41. ^{f} Ref 42. ^{g} Estimated rate constant of $k(T) = 2 \times 10^{-6}$ cm^3 molecule^{-1} s^{-1} $(T/300K)^{-0.8}$ for all $C_{10}H_n^+$, $n = 8\text{--}12$ electron–ion recombination reactions based on rate data on other aromatic rings from N. Adams and co-workers.[43] ^{h} Based on the theoretical calculation of product branching ratios in photodissociation of naphthalene and azulene at 193 nm in ref. 44. As the available energy in electron–ion recombination, 8.14 eV, is higher than in photodissociation, 6.42 eV, a minor amount of phenylacetylene + C_2H_2 may be also produced. ^{i} A value of 50% for H-atom loss and 50% for H_2 loss was assumed in the model of LePage et al.[38] ^{j} The energy available after electron–ion recombination is sufficient and largely exceeds the energy needed for two consecutive H atom losses.

B. H-Atom abstraction		
1. Neutrals	$\Delta E/\text{kJ mol}^{-1}$	$E_a/\text{kJ mol}^{-1}$
(19) $1\text{-}C_{10}H_9 + H \rightarrow C_{10}H_8 + H_2$	−327	0
(20) $2\text{-}C_{10}H_9 + H \rightarrow C_{10}H_8 + H_2$	−347	0
(21) $1,4\text{-}C_{10}H_{10} + H \rightarrow 1\text{-}C_{10}H_9 + H_2$	−123	+15
(22a) $1,2\text{-}C_{10}H_{10} + H \rightarrow 1\text{-}C_{10}H_9 + H_2$	−111	+19
(22b) $\rightarrow 2\text{-}C_{10}H_9 + H_2$	−91	+43
(23a) $1,2,4\text{-}C_{10}H_{11} + H \rightarrow 1,2\text{-}C_{10}H_{10} + H_2$	−322	0
(23b) $\rightarrow 1,4\text{-}C_{10}H_{10} + H_2$	−310	0
(24) $1,2,3\text{-}C_{10}H_{11} + H \rightarrow 1,2\text{-}C_{10}H_{11} + H_2$	−276	0
2. Ions	Rate	

Table 3 (*Contd.*)

B. H-Atom abstraction

(25) $C_{10}H_9^+ + H \rightarrow C_{10}H_8^+ + H_2$	Very slow[b]	

C. Photoionization	λ_{thresh}/nm	IP/eV
(26) $C_{10}H_8 + h\nu \rightarrow C_{10}H_8^+ + e$	153 (152[d])	8.09 (8.144[d])
(27) 1-$C_{10}H_9 + h\nu \rightarrow$ 1-$C_{10}H_9^+ + e$	192 (189[a])	6.47 (6.570[a])
(28) 2-$C_{10}H_9 + h\nu \rightarrow$ 2-$C_{10}H_9^+ + e$	195 (191[a])	6.37 (6.487[a])
(29) 1,2-$C_{10}H_{10} + h\nu \rightarrow$ 1,2-$C_{10}H_{10}^+ + e$	156 (153[e])	7.98 (8.14[e])
(30) 1,4-$C_{10}H_{10} + h\nu \rightarrow$ 1,4-$C_{10}H_{10}^+ + e$	146	8.50
(31) 1,2,3-$C_{10}H_{11} + h\nu \rightarrow$ 1,2,3-$C_{10}H_{11}^+ + e$	191 (188[a])	6.49 (6.620[a])
(32) 1,2,4-$C_{10}H_{11} + h\nu \rightarrow$ 1,2,4-$C_{10}H_{11}^+ + e$	177	7.00
(33) 1,2,3,4-$C_{10}H_{12} + h\nu \rightarrow$ 1,2,3,4-$C_{10}H_{12}^+ + e$	147 (147[f])	8.41 (8.46[f])

D. Photodissociation	ΔE/kJ mol^{-1}	λ_{diss}/nm
(34) $C_{10}H_8 + h\nu \rightarrow C_{10}H_7 + H$	461	≤259
(35) 1-$C_{10}H_9 + h\nu \rightarrow C_{10}H_8 + H$	109 (121[a])	≤825
(36) 2-$C_{10}H_9 + h\nu \rightarrow C_{10}H_8 + H$	89 (103[a])	≤913
(37a) 1,2-$C_{10}H_{10} + h\nu \rightarrow$ 1-$C_{10}H_9 + H$	325	≤368
(37b) $\rightarrow$ 2-$C_{10}H_9 + H$	345	≤347
(37c) $\rightarrow C_{10}H_8 + H_2$	Roaming reaction: similar to H loss	
(38a) 1,4-$C_{10}H_{10} + h\nu \rightarrow$ 1-$C_{10}H_9 + H$	313	≤382
(38b) $\rightarrow C_{10}H_8 + H_2$	Roaming reaction: similar to H loss	
(39) 1,2,3-$C_{10}H_{11} + h\nu \rightarrow$ 1,2-$C_{10}H_{10} + H$	160 (168[a])	≤708
(40) 1,2,4-$C_{10}H_{11} + h\nu \rightarrow$ 1,2-$C_{10}H_{10} + H$	Roaming reaction: similar to H loss	
(41) 1-$C_{10}H_9^+ + h\nu \rightarrow C_{10}H_8^+ + H$	268 (272.7 ± 2[a])	≤438
(42) 2-$C_{10}H_9^+ + h\nu \rightarrow C_{10}H_8^+ + H$	258 (263 ± 2[a])	≤455
(43) $C_{10}H_{11}^+ + h\nu \rightarrow C_{10}H_9^+ + H_2$	Roaming reaction: similar to H loss	

E. Electron–ion recombination[g]	Yield
(44) $C_{10}H_8^+ + e \rightarrow C_{10}H_7 + H$	~100%[h]
(45) $C_{10}H_9^+ + e \rightarrow C_{10}H_8 + H$	100%[i]
(46) $C_{10}H_{10}^+ + e \rightarrow C_{10}H_9 + H$ / $C_{10}H_8 + 2H$	??[j]
(47) $C_{10}H_{11}^+ + e \rightarrow C_{10}H_{10} + H$	100%[i]
(48) $C_{10}H_{12}^+ + e \rightarrow C_{10}H_{11} + H$ / $C_{10}H_{10} + 2H$	??[j]

where ΔEMP2 = E[MP2/G3large] – E[MP2/6-311G(d,p)] is the basis set correction, ΔE(SO) is a spin–orbit correction (not included in our calculation), ΔE(HLC) is a high level correction, and E(ZPE) is the zero-point energy. All DFT and MP2 calculations were carried out using the *Gaussian 98*[34] package, whereas the *Molpro 2002*[35] program package was used to calculate spin-restricted (R) CCSD(T) energies.

The G3-calculated bond dissociation energies for 1-$C_{10}H_9$ (109 kJ mol^{-1}) and 2-$C_{10}H_9$ (89 kJ mol^{-1}) are smaller than experiment by ~12 kJ mol^{-1}, but their calculated relative energy difference (E(2) – E(1) = 20 kJ mol^{-1}) is close to that derived from experiment (18 ± 3 kJ mol^{-1}). The calculated BDE for 1,2,3-$C_{10}H_{11}$ to form 1,2-$C_{10}H_{10}$ (160 kJ mol^{-1}) is also in good agreement with experiment (168 kJ mol^{-1}), but about 8 kJ mol^{-1} smaller. The calculated transition states for addition of H-atom to naphthalene (24 kJ mol^{-1} to form 1-$C_{10}H_9$ and 28 kJ mol^{-1} to form 2-$C_{10}H_9$) are large enough that direct H-atom addition is unlikely to be important at the temperatures in Titan's atmosphere. The corresponding energy barrier for 1,2-$C_{10}H_{10}$ + H $\rightarrow$ 1,2,3-$C_{10}H_{11}$ is 15 kJ mol^{-1}. By contrast, H-atom addition to the $C_{10}H_9$ and $C_{10}H_{11}$ radicals is barrierless, as might be expected of radical–radical recombination reactions.

One of the intriguing aspects of the properties of the hydronaphthyl and trihydronaphthyl radicals is their low ionization potentials (~6.5 eV), forming two isomeric forms of protonated naphthalene (1-$C_{10}H_9^+$, 2-$C_{10}H_9^+$), and protonated 1,2- dihydronaphthalene (1,2,3-$C_{10}H_{11}^+$). This means that photoionization in Titan's atmosphere will occur at longer wavelengths than for stable molecules, where the solar flux is greater.

The neutral $C_{10}H_n$, $n = 9$–12 series is thus likely to be closely coupled to ion chemistry involving protonated naphthalene and $C_{10}H_{11}^+$. These stable, closed-shell ions are prototypical protonated PAHs, which have been postulated to play an important role in interstellar chemistry.[36,37] Much as occurs with resonance-stabilized radicals, protonated PAH ions are likely to build in concentration in the ionosphere of Titan, and cycle back to neutral $C_{10}H_n$ species *via* electron–ion recombination. Recently, Ricca *et al.* postulated a model for processing of PAH *via* successive H-atom addition reactions to the $C_{10}H_n^+$ series,[39] much as we are considering for neutrals. This model serves as a useful starting point in developing the combined neutral and ion PAH chemistry that is likely to be involved in Titan's atmosphere.

In preparing for such modeling, we present here the expanded reaction scheme shown in Table 3 using the combined results from experiment and theory. The set of 55 reactions are divided into sub-schemes involving photoionization, H-atom addition to neutrals and ions, H-atom abstraction reactions to form H_2, photodissocation, and electron–ion recombination. For most of the reactions, calculated reaction exoergicities and activation energies (in the form of zero-point corrected transition state energies) are listed, all at the G3(MP2,CC)//B3LYP/6-311G** level of theory. Where experimental data is available, these are listed alongside for comparison. Future work will need to combine the reaction energetics with statistical predictions[45] for the temperature-dependent rate constants which can then be used directly as input to model the steady-state concentrations of the $C_{10}H_n$ neutral and $C_{10}H_n^+$ ions as a function of altitude in Titan's atmosphere.

Conclusions

This contribution to the *Faraday Discussion* has presented a series of laboratory investigations that illustrate some of the key challenges awaiting the modeling of neutral and ion chemistry on the large-molecule end of current photochemical models of Titan's atmosphere. The pervasive role of structural isomers will require isomer-specific data on their spectroscopy, photochemistry, and radical/ion chemistry, with neutral $C_{10}H_8$ isomers of naphthalene used as an example. Experimental investigations will need to focus attention on key molecules and radical intermediates, seeking to understand the important pathways that can be translated to even larger molecules with some confidence. Here we have focused on naphthalene and hydronaphthyl radicals. The resonance stabilization of 1-$C_{10}H_9$, 2-$C_{10}H_9$, and 1,2,3-$C_{10}H_{11}$ make them potential intermediates of some importance, but further understanding of the close-coupled neutral and ion chemistry is still needed before a quantitative kinetic reaction scheme is available. Stable molecules that deserve attention include those that are produced as a result of the recombination of two resonance-stabilized radicals. The photodissociation and photoisomerization pathways of such recombination products are likely candidates for future work.

Acknowledgements

J. A. S. and T. S. Z. gratefully acknowledge support from the NASA Planetary Atmospheres program under grants NNG06GC57G and NNX10AB89G for this research. V. V. K. and A. M. M. thank Chemical Sciences, Geosciences and Biosciences Division, Office of Basic Energy Sciences, Office of Sciences of the U. S. Department of Energy (Grant No. DE-FG02-04ER15570) for their support.

References

1 J. H. Waite, D. T. Young, T. E. Cravens, A. J. Coates, F. J. Crary, B. Magee and J. Westlake, *Science*, 2007, **316**, 870–875.
2 V. Vuitton, R. V. Yelle and P. Lavvas, *Philos. Trans. R. Soc. London, Ser. A*, 2009, **367**, 729–741.
3 E. H. Wilson and S. K. Atreya, *J. Geophys. Res. [Planets]*, 2004, **109**, E06002.
4 V. A. Krasnopolsky, *Icarus*, 2009, **201**, 226–256.
5 X. B. Gu, F. T. Zhang, R. I. Kaiser, V. V. Kislov and A. M. Mebel, *Chem. Phys. Lett.*, 2009, **474**, 51–56.
6 H. Ismail, J. Park, B. M. Wong, W. H. Green and M. C. Lin, in *30th International Symposium on Combustion*, Chicago, IL, 2004, pp. 1049–1056.
7 C.-P. Liu, J. J. Newby, C. W. Muller, H. D. Lee and T. S. Zwier, *J. Phys. Chem. A*, 2008, **112**, 9454–9466.
8 J. J. Newby, C. W. MuÃàller, C.-P. Liu and T. S. Zwier, *J. Am. Chem. Soc.*, 2010, **132**, 1611–1620.
9 Y. Georgievskii, J. A. Miller and S. J. Klippenstein, *Phys. Chem. Chem. Phys.*, 2007, **9**, 4259–4268.
10 T. M. Selby and T. S. Zwier, *J. Phys. Chem. A*, 2005, **109**, 8487–8496.
11 J. A. Sebree, V. V. Kislov, A. M. Mebel and T. S. Zwier, *J. Phys. Chem. A*, 2010, **114**, 6255–6262.
12 V. LePage, Y. Keheyan, V. M. Bierbaum and T. P. Snow, *J. Am. Chem. Soc.*, 1997, **119**, 8373–8374.
13 J. J. Newby, J. A. Stearns, C.-P. Liu and T. S. Zwier, *J. Phys. Chem. A*, 2007, **111**, 10914–10927.
14 W. C. Wiley and I. H. McLaren, *Rev. Sci. Instrum.*, 1955, **26**, 1150–1157.
15 D. F. Plusquellic, R. D. Suenram, B. Mate, J. O. Jensen and A. C. Samuels, *J. Chem. Phys.*, 2001, **115**, 3057.
16 T. M. Selby, J. R. Clarkson, D. Mitchell, J. A. J. Fitzpatrick, H. D. Lee, D. W. Pratt and T. S. Zwier, *J. Phys. Chem. A*, 2005, **109**, 4484–4496.
17 J. A. Dickinson, P. W. Joireman, R. T. Kroemer, E. G. Robertson and J. P. Simons, *J. Chem. Soc., Faraday Trans.*, 1997, **93**, 1467–1472.
18 J. A. Dickinson, P. W. Joireman, R. W. Randall, E. G. Robertson and J. P. Simons, *J. Phys. Chem. A*, 1997, **101**, 513–521.
19 T. Chakraborty and M. Chowdhury, *Spectrochim. Acta, Part A*, 1992, **48**, 1683–1691.
20 D. Autrey, Z. Arp, J. Choo and J. Laane, *J. Chem. Phys.*, 2003, **119**, 2557–2568.
21 M. Z. M. Rishard, M. Wagner, J. Yang and J. Laane, *Chem. Phys. Lett.*, 2007, **442**, 182–186.
22 T. Chakraborty, J. E. Delbene and E. C. Lim, *J. Chem. Phys.*, 1993, **98**, 8–13.
23 J. Yang, M. Wagner and J. Laane, *J. Phys. Chem. A*, 2007, **111**, 8429–8438.
24 S. J. Sheng, K. Nakagawa, T. Nakayama, Y. Kumazawa and N. Itoh, *Radiat. Phys. Chem.*, 1980, **15**, 577–578.
25 J. B. Hopkins, D. E. Powers and R. E. Smalley, *J. Chem. Phys.*, 1980, **72**, 5039–5048.
26 J. A. Sebree, D. F. Plusquellic and T. S. Zwier, manuscript in preparation.
27 C. S. McEnally, L. D. Pfefferle, B. Atakan and K. Kohse-Hoinghaus, *Prog. Energy Combust. Sci.*, 2006, **32**, 247–294.
28 M. Mautner, Virginia Commonwealth University, personal communication, 2010.
29 A. D. Becke, *J. Chem. Phys.*, 1993, **98**, 5648–5652.
30 C. Lee, W. Yang and R. G. Parr, *Phys. Rev. B: Condens. Matter*, 1988, **37**, 785.
31 A. G. Baboul, L. A. Curtiss, P. C. Redfern and K. Raghavachari, *J. Chem. Phys.*, 1999, **110**, 7650–7657.
32 L. A. Curtiss, K. Raghavachari, P. C. Redfern, A. G. Baboul and J. A. Pople, *Chem. Phys. Lett.*, 1999, **314**, 101–107.
33 L. A. Curtiss, K. Raghavachari, P. C. Redfern, V. Rassolov and J. A. Pople, *J. Chem. Phys.*, 1998, **109**, 7764–7776.
34 M. J. T. Frisch, G. W.; H. B. Schlegel; G. E. Scuseria; M. A. Robb; J. R. Cheeseman; V. G. Zakrzewski; J. A. Montgomery; R. E. Stratmann; J. C. Burant; S. Dapprich; J. M. Millam; A. D. Daniels; K. N. Kudin; M. C. Strain; O. Farkas; J. Tomasi; V. Barone; M. Cossi; R. Cammi; B. Mennucci; C. Pomelli; C. Adamo; S. Clifford; J. Ochterski; G. A. Petersson; P. Y. Ayala; Q. Cui; K. Morokuma; D. K. Malick; A. D. Rabuck; K. Raghavachari; J. B. Foresman; J. Cioslowski; J. V. Ortiz; A. G. Baboul; B. B. Stefanov; G. Liu, A.L.; P. Piskorz; I. Komaromi; R. Gomperts; R. L. Martin; D. J. Fox; T. Keith; M. A. Al-Laham; C. Y. Peng; A. Nanayakkara; C. Gonzalez; M. Challacombe; P. M. W. Gill; B. Johnson; W. Chen; M. W. Wong;

J. L. Andres; C. Gonzalez; M. Head-Gordon; E. S. Replogle; J. A. Pople, Gaussian, Inc., Pittsburgh, PA, 1998.
35 R. D. Amos, A. Bernhardsson, A. Berning, P. Celani, D. L. Cooper, M. J. O. Deegan, A. J. Dobbyn, F. Eckert, C. Hampel, G. Hetzer, P. J. Knowles, T. Korona, R. Lindh, A. W. Lloyd, S. J. McNicholas, F. R. Manby, W. Meyer, M. E. Mura, A. Nicklass, P. Palmieri, R. Pitzer, G. Rauhut, M. Schütz, U. Schumann, H. Stoll, A. J. Stone, R. Tarroni, T. Thorsteinsson and H.-J. Werner, *MOLPRO, a package of* ab initio *programs designed by H.-J. Werner and P. J. Knowles, Version 2002.6*, 2003.
36 A. M. Ricks, G. E. Douberly and M. A. Duncan, *Astrophys.J.*, 2009, **702**, 301–306.
37 L. W. Beegle, T. J. Wdowiak and J. G. Harrison, *Spectrochim. Acta, Part A*, 2001, **57**, 737–744.
38 V. Le Page, T. P. Snow and V. M. Bierbaum, *Astrophys. J.*, 2009, **704**, 274–280.
39 A. Ricca, E. L. O. Bakes and C. W. Bauschlicher, *Astrophys. J.*, 2007, **659**, 858–861.
40 M. C. R. Cockett, H. Ozeki, K. Okuyama and K. Kimura, *J. Chem. Phys.*, 1993, **98**, 7763–7772.
41 C. Dass and M. L. Gross, *J. Am. Chem. Soc.*, 1983, **105**, 5724–5729.
42 C. Dass and M. L. Gross, *Org. Mass Spectrom.*, 1985, **20**, 34–40.
43 N. Adams, personal correspondence, , 2010.
44 Y. A. Dyakov, C. K. Ni, S. H. Lin, Y. T. Lee and A. M. Mebel, *J. Phys. Chem. A*, 2005, **109**, 8774–8784.
45 A. M. Mebel and V. V. Kislov, *J. Phys. Chem. A*, 2009, **113**, 9825–9833.

General Discussion

Professor Miller opened the discussion of the paper by Julianne Moses: Your explanation of why CH_4 and NH_3 do not react on Jupiter reminds me of the reason that polar bears do not eat penguins—they don't mix. That may the case for Jupiter, but one can imagine that the situation may be different on warm Jupiters. Could you indicate what you think might happen in those situations where CH_4 and NH_3 do mix?

Dr Moses answered: Your analogy is indeed appropriate. If Jupiter were a bit closer to the Sun, such that the planet were warm enough that ammonia does not condense in the troposphere, then coupled ammonia–hydrocarbon photochemistry could play a larger role than it does on the real Jupiter. The main photochemical products would then be amines like ethylamine and methylamine, nitriles like HCN and acetonitrile, and more exotic species like N-ethylethylideneimine and acetaldehyde, as were seen in the Ferris and Ishikawa (1988) and Keane *et al.* (1996) photolysis experiments. The coupled NH_3–hydrocarbon photochemistry could be particularly interesting on extrasolar giant planets, and I would encourage the derivation of thermodynamic parameters for the complex species discussed in the paper, so that the kinetics of these species can be included in extrasolar giant planet models. On the other hand, the fact that NH_3 is photolyzed at longer wavelengths than CH_4 will lead to a relatively large vertical separation between the CH_4 and NH_3 photolysis regions, which will inhibit but not prevent CH_4–NH_3 photochemical coupling, especially in H_2-dominated atmospheres where hot H atoms are effectively quenched.

Dr Klippenstein noted: The reaction $H_2CN + H_2 \rightarrow CH_2NH + H$ is the rate-limiting step in your proposed resolution of an apparent paradox related to the HCN mole fractions in the deep tropospheric source in the Jovian atmosphere. There is very limited information of either experimental or theoretical nature for this reaction. Thus, we have performed *ab initio* transition state theory calculations of this rate constant and the results are plotted in Fig. 1. These predictions employ QCISD(T)/cc-pVTZ rovibrational analyses and QCISD(T)/CBS (from cc-pVTZ and

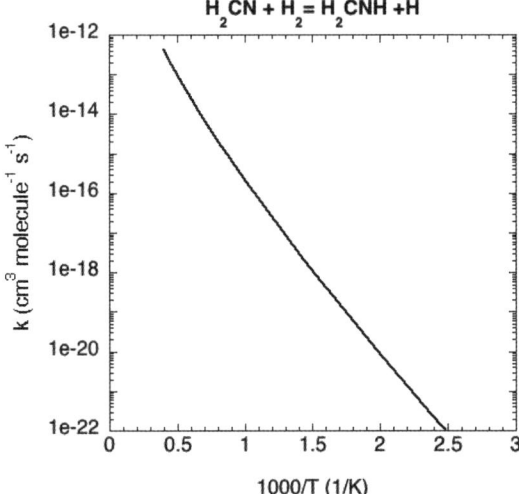

Fig. 1 Plot of the temperature dependence of the rate constant for the $H_2CN + H_2 \rightarrow CH_2NH + H$ abstraction reaction.

cc-pVQZ single point calculations) energies for the stationary points. Eckart tunneling corrections are included. The predicted rate coefficient is well represented by the modified Arrhenius expression $4.40 \times 10^{-22} T^{3.051} \exp(-7990/T)$ cm^3 molecule^{-1} s^{-1} over the 100 to 2500 K temperature range. These predictions are expected to be accurate to within about a factor of 2 for temperatures of about 800 K and higher.

As a side point, I also note that we have recently published a study of the NH$_2$OH decomposition system[1] that provides rate constants that may be of some utility for your modeling. In particular, for the NH$_2$ + NH$_2$ → N$_2$H$_4$ recombination, this study predicts Troe parameters for N$_2$ as a bath gas of $k_0 = 4.48 \times 10^{-14} T^{-5.49} \exp(-1000/T)$ cm^6 molecule^{-2} s^{-1}, $k^\infty = 9.33 \times 10^{-10} T^{-0.414} \exp(33/T)$ cm^3 molecule^{-1} s^{-1}, and $F_c = 0.31$ for the 300–2500 K temperature range. For the NH$_2$ + NH → HNNH + H reaction the collisionless limit rate coefficient k_0 is predicted to be $7.07 \times 10^{-10} T^{-0.272} \exp(39/T)$ cm^3 molecule^{-1} s^{-1} for temperatures ranging from 200 to 2500 K. I have also used the data from that study to predict the high pressure limit rate coefficient k^∞ for the H + N$_2$H$_2$ → N$_2$H$_3$ reaction to be $3.35 \times 10^{-16} T^{1.780} \exp(-790/T)$ cm^3 molecule^{-1} s^{-1} for temperatures ranging from 200 to 2500 K. Support for this work from the NASA Planetary Atmospheres program under grant NNH09AK24I is gratefully acknowledged.

1 S. J. Klippenstein, L. B. Harding, B. Ruscic, R. Sivaramakrishnan, N. K. Srinivasan, M.-C. Su, and J. V. Michael, The thermal decomposition of NH$_2$OH and subsequent reactions: *ab initio* transition state theory and reflected shock tube experiments, *J. Phys. Chem. A*, 2009, **113**, 10241–10259.

Dr Moses responded: Excellent! Thank you. I'll look into implementing these changes, and I hope you continue to be interested in the kinetics of nitrogen-bearing species under reducing conditions.

Professor Plane remarked: One has to be careful using a 1D model with a parameterised vertical "eddy" diffusion coefficient (K_{zz}) to look at the transport of species from one region of an atmosphere to another at a different height. If the species are relatively short-lived then this is usually a safe procedure. However, species which have lifetimes of weeks or months are often advected vertically by the mean residual circulation, rather than diffusion. Such vertical transport can also involve significant horizontal transport. Examples in the terrestrial atmosphere are aurorally-produced NO$_x$, meteoric smoke, and ozone, which are transported by the mesospheric meridional circulation and the stratospheric Brewer–Dobson circulation to a much greater extent than by eddy diffusion.

Dr Moses responded: While I agree that care must be exercised in these cases, observations of long-lived tracer species can be used to help derive empirical profiles of the eddy diffusion coefficient for 1D models. This technique was used successfully for years in terrestrial atmospheric studies before the onset of GCMs. Even GCMs contain parameterizations for subgrid exchanges associated with atmospheric mixing, and these processes often dominate the behavior. Atmospheric waves, for instance, can force the circulation responsible for global transport of constituents. The Brewer–Dobson circulation is a particularly apt example here, as mixing by planetary waves provides an important contribution to meridional transport. In any case, no Jovian GCMs exist that extend through the hydrocarbon photolysis region in the upper stratosphere, and we rely on observations of C$_2$H$_6$, CO, and Shoemaker–Levy 9 debris to help define the stratospheric eddy diffusion coefficient profile and on observations of PH$_3$ to help define the tropospheric profile for our 1D models.

Professor Miller added: There is always a trade-off between using global circulation models, which can transport a variety of chemical species, but only compute a limited range of chemical reactions, and 1-D models, for which it is possible to compute several thousand reactions. In that sense, it is a question of available computing power. This problem is made worse when there is a dependency between the chemistry and the dynamics of the atmosphere: this is certainly the case for the upper atmospheres of giant planets, where photochemically produced H_3^+ plays a key role in cooling, circulation and the overall energy balance of the planet. You can do the chemistry in "real-time" there, as it is fairly simple.

Dr Moses continued: This situation also arises for the middle atmospheres of the giant planets, where photochemical products like acetylene and ethane are major atmospheric coolants. Strong coupling of photochemistry, radiative transport, and dynamics will therefore occur in the middle atmospheres of the giant planets; the resulting behavior is complex and difficult to model accurately.

Dr Yelle noted: My main concern is that GCM models for atmospheres other than the Earth's are not sufficiently accurate to predict precisely vertical mixing rates. This may be a reflection of the nature of GCMs. Vertical mixing in many parts of the atmosphere is driven by waves, which are not well described by GCMs, because the forcing is not well known. Thus, in this circumstance, the simple approach of parameterizing the vertical mixing with an eddy diffusion coefficient may be best, because the complicated, physically based approach is unlikely to be accurate.

Professor Atreya asked: How do you explain HCN in the atmosphere of Neptune, but not Uranus? I would also like to note that comets and Strobel's idea of Triton's nitrogen are interesting, but my own feeling is that N_2 is involved, which would solve two problems at the same time—forming HCN and reducing the He/H_2 ratio. I included a little calculation of this in the Neptune book (chapter of Gautier *et al.*). Accordingly, HCN is produced in the N_2–CH_4 chemistry in the atmosphere of Neptune, but not Uranus. N_2 could upwell from the interior of Neptune because of its very vigorous convection, unlike Uranus where the convection is sluggish. The three orders of magnitude difference in their atmospheric eddy diffusion coefficient is an indicator of this. The presence of N_2 in Neptune's atmosphere can also solve another problem. It would reduce the He mole fraction in Neptune's atmosphere so that it is consistent with that of Uranus, as it should be.

Dr Moses answered: Given the shape of the CO line profile from both the Hesman *et al.* (2007) and Lellouch *et al.* (2005) Neptune observations (and apparently the new Herschel results, although I haven't seen the latter yet), I think the hypothesis of a "recent" comet has significant merit as the source of the stratospheric CO on Neptune, and it's possible that HCN could have been delivered during the same impact event(s). However, I agree that there are other reasons to believe that N_2 could be an important constituent on Neptune (as outlined in that University of Arizona Press book and elsewhere), in contrast with Uranus. Galactic cosmic rays could then dissociate the N_2 and trigger HCN formation (*e.g.*, Marten *et al.* 1993, Lellouch *et al.* 1994) and/or Triton could also be an external source of N for the HCN (*e.g.*, Summers and Strobel 1991). I think that it will be difficult to distinguish between these different scenarios without deep atmospheric probes.

Professor Strobel added: Based on archival data from Hubble Space Telescope observations of Neptune [R. Courtin, D. Gautier and D. F. Strobel, The CO abundance on Neptune from HST observations, *Icarus*, 1996, **123**, 37–55] we marginally detected two CO Cameron absorption bands (1,0) and (0,0) at 199.2 and 206.3 nm and inferred a CO mixing ratio of ~3 ppmv, but failed to detect N_2 Vegard–Kaplan bands in absorption. Summers and Strobel (unpublished) found that nitrogen atoms escaping Triton's atmosphere, subsequently undergoing ionization in Neptune's magnetosphere and then precipitating into Neptune's atmosphere were a sufficient source to account for the observed abundance of HCN.

Professor Plane opened the discussion of the paper by Sébastien Le Picard: In the absence of product studies, the increase of a reaction rate constant at very low temperatures can be due to the formation of a weakly-bound complex which is stable on the timescale of the CRESU flight time. Can you cite evidence for cases where a reaction has definitely occurred?

Dr Le Picard responded: Please refer first to the comments by Ian Sims, below. In addition to these comments, the very good agreement between our measurements of reaction rate coefficients at low temperatures and high level calculations of bimolecular reactions yielding products over potential energy surfaces possessing no absolute barriers by, for example, Klippenstein and co-workers,[1] where these exist, reinforce our assertion that we are measuring bimolecular reactivity rather than association to weakly-bound van der Waals complexes. We can also cite the work of your colleagues at Leeds, Gannon et al.,[2] who measured product yields for some of the CN reactions previously studied by the CRESU technique including CN + C_2H_2 and CN + C_2H_4 down to 195 K, as well as studies from the group of Michel Costes who measured the excitation function down to very low collision energies (ca. 5 meV, equivalent to ca. 40 K), again by monitoring the H-atom products, for some reactions also studied by the CRESU technique, including $C(^3P)$ + C_3H_4 (propyne and allene).[3] The energy dependence of the observed integral cross-sections agreed perfectly with what would be predicted from the observed temperature dependences of the low temperature rate coefficients. The shallow van der Waals wells that exist between these relatively small species cannot be deep enough to explain the fast rate coefficients observed from room temperature down to very low temperatures. This may, however, not be the case for species forming more strongly-bound (e.g. hydrogen-bonded) complexes where a barrier also exists to formation of subsequent bimolecular products, such as the OH + acetone system.[4]

1 Y. Georgievskii and S. J. Klippenstein, *J. Phys. Chem. A*, 2007, **111**, 3802.
2 K. L. Gannon, D. R. Glowacki, M. A. Blitz, K. J. Hughes, M. J. Pilling and P. W. Seakins, *J. Phys. Chem. A*, 2007, **111**, 6679.
3 D. Chastaing, S. D. Le Picard, I. R. Sims, I. W. M. Smith, W. D. Geppert, C. Naulin and M. Costes, *Chem. Phys. Lett.*, 2000, **331**, 170.
4 R. Shannon, S. Taylor, M. Blitz, A. Goddard and D. E. Heard, results presented at the 21st International Symposium on Gas Kinetics, Leuven, Belgium, 2010.

Professor Sims addressed Professor Plane: It has been suggested that, as our measurements of rate coefficients for CN radical reactions at low temperatures using the CRESU technique do not determine the products of reaction, collisionally assisted association of the radical and the molecular co-reagent could be responsible for the observed rates and in particular for the increases in rate coefficients observed at low temperatures, for example in our measurements on the reaction CN + propane reported in this Discussion.[1] However, the very large values for the measured rate coefficients—well in excess of 10^{-10} cm^3 molecule^{-1} s^{-1} for CN + propene and CN + but-1-yne, and >8 x 10^{-11} cm^3 molecule^{-1} s^{-1} for CN + propane—would not seem to be explicable by such a mechanism at the relatively low pressures of these measurements (around 1 mbar or below at low temperature). Calculations on similar reactions have repeatedly shown exothermic product channels accessible over submerged barriers,[2] and the minimum that is observed in $k(T)$ for the reaction CN + propane here strongly resembles that observed for CN + ethane,[3] and which has been explained by Georgievskii and Klippenstein[4] using a two transition state model with the inner transition state corresponding to such a submerged barrier.

1 S. B. Morales, S. D. Le Picard, A. Canosa and I. R. Sims, *Faraday Discuss.*, 2010, **147**, DOI 10.1039/C004219F.
2 H. Sabbah, L. Biennier, I. R. Sims, Y. Georgievskii, S. J. Klippenstein and I. W. M. Smith, *Science*, 2007, **317**, 102.

3 I. R. Sims, J. L. Queffelec, D. Travers, B. R. Rowe, L. B. Herbert, J. Karthäuser and I. W. M. Smith, *Chem. Phys. Lett.*, 1993, **211**, 461.
4 Y. Georgievskii and S. J. Klippenstein, *J. Phys. Chem. A*, 2007, **111**, 3802.

Dr Klippenstein addressed Professor Sims and Professor Plane: A rate constant that increases with increasing temperature at low temperatures is well understood by the dynamics on the long-range potential. A summary of the predicted low temperature behaviour for barrierless recombinations under various interaction potentials is provided in our paper on long-range transition state theory (TST).[1] Typically, the long-range TST predicted rate constant increases with increasing temperature due to terms such as the dispersion interaction. As noted by Ian Sims, a two transition state model provides a quantitative explanation for the observed temperature dependence over a wide range of temperature (*e.g.*, 20 to 1000 K) for a number of such reactions. In this model, at some point as the temperature increases from low temperature, the submerged barrier begins to be an important bottleneck to the reaction due to its lower entropy. The transformation from a long-range transition state to the low entropy submerged barrier transition state correlates with a rate constant that decreases with increasing temperature. Ultimately, when the submerged barrier has become the dominant bottleneck, the rate constant again increases with increasing temperature. This increase is in line with the typical behaviour for the prefactor arising from transition state theory.

1 Y. Georgievskii and S. J. Klippenstein, Transition state theory for long-range potentials, *J. Chem. Phys.*, 2005, **122**, 194103.

Professor Casavecchia (speaking on behalf of also F. Leonori, R. Petrucci, E. Segoloni, A. Bergeat, K. M. Hickson and N. Balucani) said: In relation to the paper by Morales *et al.*,[1] we wish to comment on several aspects.

First of all, we would like to stress that the crossed molecular beam (CMB) method with mass-spectrometric (MS) detection actually represents a powerful approach not only to determine the nature of the primary products and their mechanism of formation, but also their branching ratios (see, for instance, ref. 2–6). In our laboratory, we have also explored the energy dependence of the branching ratio over a range of collision energies (E_c)[3-6] and this approach has been recently empowered by the development of a variable beam crossing angle set-up.[2] As a matter of fact, in a CMB experiment the relative E_c is given by $1/2\mu v_r^2$, where μ is the reduced mass of the system and v_r is the relative velocity. The relative velocity is given by $v_r^2 = v_1^2 + v_2^2 - 2v_1v_2\cos\gamma$, where v_1 and v_2 are the two reactant beam velocities in the laboratory frame and γ is the beam crossing angle. Classical CMB instruments with MS detection feature a beam crossing angle γ of 90°. In this case the E_cs achievable are limited by the velocities with which the two beams can be produced during the expansion and, consequently, the branching ratio can be obtained in a limited E_c range. To overcome that, we have implemented a variable beam crossing angle set-up,[2] which allows crossing the two reactant beams at three different angles of 45°, 90° and 135°. With this new setting, we have succeeded in varying E_c over a much wider range than previously possible: with the beam velocities remaining constant, the $\gamma = 135°$ configuration allows us to reach higher E_c while the $\gamma = 45°$ arrangement is intended for reaching very low E_c. In particular, we have investigated the multichannel reaction

$$C(^3P, ^1D) + C_2H_2 \rightarrow c\text{-}C_3H\,(X^2B_2) + H(^2S_{1/2}) \quad \Delta H_0° = -14.1 \text{ kJ mol}^{-1} \quad (1a)$$

$$\rightarrow l\text{-}C_3H(X^2\Pi_{1/2}) + H(^2S_{1/2}) \quad \Delta H_0° = -3.1 \text{ kJ mol}^{-1} \quad (1b)$$

$$\rightarrow C_3(X1\Sigma_g^+) + H_2(X^1\Sigma_g^+) \quad \Delta H_0° = -106 \text{ kJ mol}^{-1} \quad (1c)$$

at four different E_cs spanning from 3.6 kJ mol^{-1} to 49.1 kJ mol^{-1}.[3] The ratios of cross sections $\sigma(c\text{-}C_3H)/\sigma(l\text{-}C_3H)$ and the branching ratio $\sigma(C_3)/[\sigma(C_3) + \sigma(C_3H)]$ have been determined as a function of E_c and both have been found to increase with decreasing E_c. The experimental trend has been compared with the available theoretical predictions (approximate quantum scattering calculations[7] and quasi-classical trajectory[8] calculations on the triplet PES without nonadiabatic effects, and statistical calculations including nonadiabatic effects[9]) and with the results of another crossed molecular beam study using pulsed beams[10] (see Fig. 2). The $c\text{-}C_3H/l\text{-}C_3H$ ratio is in good agreement with the results of statistical calculations on an *ab initio* PES,[9] while it is at variance with the results of the dynamical calculations on different *ab initio* triplet PESs. The statistical calculations corroborate also the trend with E_c of the experimental branching ratio $\sigma(C_3)/[\sigma(C3) + \sigma(C_3H)]$ (see Fig. 2). The comparison with $\sigma(C_3)/[\sigma(C_3) + \sigma(C_3H)]$ determined in previous CMB experiments[10] is satisfactory between E_c of 20 and 30 kJ mol^{-1}, while there is some deviation at low E_c (see Fig. 2). We are now considering to change the collision geometry and move to a crossing angle $\gamma = 25-30°$. The experimental E_c will reduce further and will become comparable to the average collision energy corresponding to the atmospheric temperatures of Titan. It will be interesting, therefore, to reinvestigate some important reactions for the chemical evolution of the atmosphere of Titan, such as $N(^2D) + CH_4$ and $N(^2D) + C_2H_6$, already investigated in our laboratory at higher E_c.[6,11]

1 S. B. Morales, S. D. Le Picard, A. Canosa and I. R. Sims, *Faraday Discuss.*, 2010, DOI: 10.1039/c004219f
2 P. Casavecchia, F. Leonori, N. Balucani, R. Petrucci, G. Capozza and E. Segoloni, *Phys. Chem. Chem. Phys.*, 2009, **11**, 46; N. Balucani, G. Capozza, F. Leonori, E. Segoloni and P. Casavecchia, *Int. Rev. Phys. Chem.*, 2006, **25**, 109.
3 M. Costes, N. Daugey, C. Naulin, A. Bergeat, F. Leonori, E. Segoloni, R. Petrucci, N. Balucani and P. Casavecchia, *Faraday Discuss.*, 2006, **133**, 157; F. Leonori, R. Petrucci, E. Segoloni, A. Bergeat, K.M. Hickson, N. Balucani and P. Casavecchia, *J. Phys. Chem. A*, 2008, **112**, 1363.
4 F. Leonori, R. Petrucci, N. Balucani, P. Casavecchia, M. Rosi, D. Skouteris, C. Berteloite, S. Le Picard, A. Canosa and I. R. Sims, *J. Phys. Chem. A*, 2009, **113**, 15328; F. Leonori,

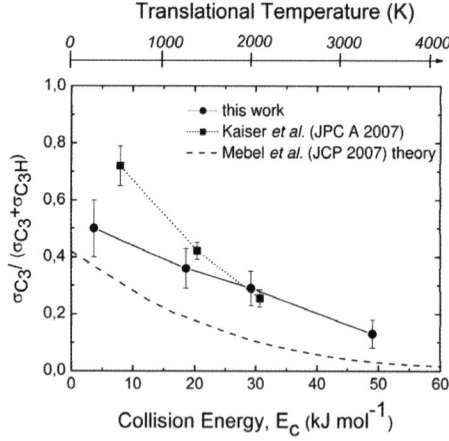

Fig. 2 C_3 product branching ratio as a function of collision energy, E_c, for the $C(^3P) + C_2H_2$ reaction. The related translational temperatures are indicated on the top abscissa. The lines joining the data points are drawn to guide the eye only. The branching ratio from statistical calculations on *ab initio* potential surfaces including ISC is reported with a dashed line for comparison.[9] The results of Kaiser and coworkers[10] obtained at three E_cs are also shown (solid squares). Reproduced with permission from *J. Phys. Chem. A*, 2008, **112**, 1363–1379. Copyright 2008 American Chemical Society.

R. Petrucci, N. Balucani, P. Casavecchia, M. Rosi, C. Berteloite, S. Le Picard, A. Canosa and I. R. Sims, *Phys. Chem. Chem. Phys.*, 2009, **11**, 4701.
5 P. Casavecchia, G. Capozza, E.Segoloni, F. Leonori, N. Balucani and G. G. Volpi, *J. Phys. Chem. A*, 2005, **109**, 3527.
6 N. Balucani, A. Bergeat, L. Cartechini, G. G. Volpi, P. Casavecchia, D. Skouteris and M. Rosi, *J. Phys. Chem. A*, 2009, **113**, 11138.
7 E. Buonomo and D. C. Clary, *J. Phys. Chem. A*, 2001, **105**, 2694; T. Takayanagi, *J. Phys. Chem. A*, 2006, **110**, 361.
8 W. K. Park, J. Park, S. C. Park, B. J. Braams, C. Chen and J. M. Bowman, *J. Chem. Phys.*, 2006, **125**, 081101.
9 A. M. Mebel, V. V. Kislov and M. Hayashi, *J. Chem. Phys.*, 2007, **126**, 204310.
10 Y. Guo, A. M. Mebel, X. Gu and R. I. Kaiser, *J. Phys. Chem. A*, 2007, **111**, 2980.
11 N. Balucani, F. Leonori, R. Petrucci, M. Stazi, D. Skouteris, M. Rosi and P. Casavecchia, *Faraday Discuss.*, 2010, DOI: 10.1039/c004748a.

Dr Le Picard asked: The results on branching ratios that you give cover a large range of collision energies and you converted them to translational temperatures. Could you comment on the rotational and vibrational temperatures of the reactants in your experiments? This could be of major importance for using these results in photochemical models of planetary atmospheres, as branching ratios could be influenced by the rotational and vibrational energies of the reactants, especially if a collision complex is not formed during the reaction.

Professor Casavecchia responded: In general, the rotational and vibrational energies of the reactants of a bimolecular reaction can influence the product branching ratios and therefore it is appropriate to consider them. In particular, the vibrational excitation can play an important role in "direct" bimolecular reactions, as the recent work of Liu and coworkers,[1] Crim and coworkers[2] and Zare and coworkers[3] on reactions of partially deuterated methane molecules have demonstrated. For long-lived complex-forming reactions the effect of vibrational excitation on product branching ratios has been much less explored.[4] In any case, in the $C(^3P) + C_2H_2$ reaction mentioned in our comment above to your paper,[5] the internal energy that can affect the branching ratios is the ro-vibrational energy of the acetylene molecule. In our experiments[6,7] the acetylene beam is produced in a continuous supersonic expansion of the pure gas and therefore it is expected to be translationally quite cold (a few tens of degrees K) and therefore collisionally relaxed to the lowest few rotational levels (assuming an equilibration of rotational and translational degrees of freedom). Low bending vibrations are not expected to relax efficiently in the supersonic expansion, but the fraction of vibrationally excited acetylene in the beam is expected to be minor. The internal temperature of acetylene molecules in our experiments is then comparable to the one that can be expected in low-temperature environments, such as those which occur in astrochemistry (from molecular clouds to planetary atmospheres). A similar situation occurs for supersonic beams of methane and ethane reactants, as those which were used in our laboratory in molecular beam studies of the $N(^2D) + CH_4$[8] and $N((^2D) + C_2H_6$ reactions.[9]

1 W. Zhang, H. Kawamata and K. Liu, *Science*, 2009, **325**, 303; and references therein.
2 F. F. Crim, *Proc. Natl. Acad. Sci. U. S. A.*, 2008, **105**, 12654; and references therein.
3 J. P. Camden, H. A. Bechtel, D. J. A. Brown and R. N. Zare, *J. Chem. Phys.*, 2006, **124**, 034311; and references therein.
4 D. L. Proctor, H. F. Davis, *Proc. Natl. Acad. Sci. U. S. A.*, 2008, **105**, 12673; and references therein.
5 S. B. Morales, S.D. Le Picard, A. Canosa and I.R. Sims, *Faraday Discuss.*, 2010, **147**, DOI: 10.1039/C004219F.
6 M. Costes, N. Daugey, C. Naulin, A. Bergeat, F. Leonori, E. Segoloni, R. Petrucci, N. Balucani and P. Casavecchia, *Faraday Discuss.*, 2006, **133**, 157.
7 F. Leonori, R. Petrucci, E. Segoloni, A. Bergeat, K. M. Hickson, N. Balucani and P. Casavecchia, *J. Phys. Chem. A*, 2008, **112**, 1363.
8 N. Balucani, A. Bergeat, L. Cartechini, G. G. Volpi, P. Casavecchia, D. Skouteris and M. Rosi, *J. Phys. Chem. A*, 2009, **113**, 11138.

9 N. Balucani, F. Leonori, R. Petrucci, M. Stazi, D. Skouteris, M. Rosi and P. Casavecchia, *Faraday Discuss.*, 2010, **147**, DOI: 10.1039/C004748A.

Professor Casavecchia (speaking on behalf of also N. Balucani, F. Leonori, R. Petrucci, K. M. Hickson, S. D. Le Picard and P. Foggi) then addressed the meeting: In relation to the paper by Morales *et al.*,[1] another point we would like to make regards our recent results on several CN reactions with unsaturated hydrocarbons. We have recently reported on the reactions CN + C_2H_2 (1) and CN + CH_3CCH (2) by the CMB-MS technique,[2] while more recent results have been obtained on the reaction CN + C_2H_4 (4).[3] In these experiments we have generated continuous supersonic beams of cyano radicals by a high-pressure radio-frequency discharge beam source[4] starting from dilute mixtures of $CO_2(0.8\%)$/$N_2(2.5\%)$ in He. The mechanism leading to formation of CN is not clear, but presumably involves reactions of C atoms with molecular nitrogen (with one or both species in an electronically excited state), electronically excited atomic/molecular nitrogen with CO (or CO_2) and/or C + N recombination in the plasma before the supersonic expansion is completed. Clearly, in a case like this, the characterisation of the internal rovibrational levels of the diatomic radicals becomes crucial for a correct interpretation of the scattering results, because the limited number of relaxing collisions during the supersonic expansion does not allow the internal distributions to be efficiently relaxed. For this reason we have pursued the characterization of the CN beams by laser-induced-fluorescence *via* the (0,0), (1,1), (2,2) and (3,3) bands of the $B^2\Sigma^+$–$X^2\Sigma^+$ system at 387.7, 386.5, 385.5 and 384.7 nm, respectively. The rotational distributions are bimodal: a rotational temperature of ~250 K accounts for low rotational quantum numbers ($N < 10$) with a peak in the population at $N = 6$; the rotational populations in higher quantum numbers, instead, largely exceed those of a thermalised distribution (at 250 K, the rotational population falls to zero at $N \sim 24$) with a minimum in the population at $\sim N = 22$–28 and a second maximum around $N = 39$–44 (peak intensity ~1/3 of the first maximum). It is worth noting that bimodal rotational distributions have also been obtained for CN radicals generated from photolysis of precursor molecules such as $(CN)_2$, ClCN and BrCN.[5] The vibrational populations were found to be reasonably well described by a thermalised distribution with a vibrational temperature of 6500 K.

Quite interestingly, reactions (1) and (2) appear to be not affected by the vibrational excitation of CN. As a matter of fact, the results of our study are perfectly in line with those obtained in a previous pulsed CMB study, where CN is known to be formed mainly in $v' = 0$. In addition to that, in the fit of the experimental distributions there is no need to consider the extra amount of energy added by CN in $v =$ 1–4 even though the energy content of the vibrationally excited CN levels is quite sizable (*e.g.* $\Delta v_{1-0} = 24.4$ kJ mol^{-1}, $\Delta v_{2-0} = 48.6$ kJ mol^{-1}, $\Delta v_{3-0} = 72.4$ kJ mol^{-1}). The lack of visible effects in the scattering functions due to the vibrational excitation of CN is not surprising, as the vibrational excitation is expected to promote a reaction when it affects a bond which is going to break during the reaction itself, while the CN moiety is not directly involved in the bond rearrangements leading from the reactants to the products and acts as a spectator. In other words, reactions (1) and (2) are vibrationally adiabatic and the initial vibrational excitation of CN is not converted into product translational energy.

Less clear is the role of the CN rotational excitation. We expect that a fast rotating CN will smear the preference for the C-side attack to the π bond of unsaturated hydrocarbons, possibly rendering less unfavourable the channels leading to isonitrile isomers. At least in the case of the reaction CN + C_2H_4, this appears to be the case,[3] as with increasing E_c a best fit of the laboratory data can only be obtained by using two contributions, of which one is associated to the dominant channel leading to C_2H–CN and one (minor) is associated to the slightly endothermic channel leading to vinyl isocyanide, C_2H_3NC.

These findings have added further insights into the role of CN rotational and vibrational excitation which are in line with those based on kinetic experiments[6] for the reaction CN(v,N) + C_2H_2 and on time-resolved IR absorption studies[5] for the reactions CN(v,N) + C_2H_6 and CH_4.

A similar LIF characterisation has been performed to characterize the internal population of beams of C2($X^1\Sigma^+_g$) and $C_2(a^3\Pi_u)$ also employed in reactive scattering experiments.[7]

1 S. B. Morales, S. D. Le Picard, A. Canosa and I. R. Sims, *Faraday Discuss.*, 2010, **147**, DOI: 10.1039/C004219F
2 F. Leonori, K. M. Hickson, S. D. Le Picard, X. Wang, R. Petrucci, P. Foggi, N. Balucani and P. Casavecchia, *Mol. Phys.*, 2010, **108**, 1097.
3 In preparation
4 M. Alagia, V. Aquilanti, D. Ascenzi, N. Balucani, D. Cappelletti, L. Cartechini, P. Casavecchia, F. Pirani, G. Sanchini and G. G. Volpi, *Isr. J. Chem.*, 1997, **37**, 329.
5 G. A. Bethardy, F. J. Northrup and R. G. Macdonald, *J. Chem. Phys.*, 1996, **105**, 4533.
6 R. V. Olkhov and I. W. M. Smith, *J. Chem. Phys.*, 2007, **126**, 134314.
7 F. Leonori, R. Petrucci, K. M. Hickson, E. Segoloni, N. Balucani, S. D. Le Picard, P. Foggi and P. Casavecchia, *Planet. Space Sci.*, 2008, **56**, 1658.

Professor Suits commented: Armando Estillore, and Laura M. Visger and I have been studying reactions of the cyano radical in crossed molecular beams using the ion imaging technique. In the results we show here, we have studied the reaction of CN with 1-pentene at a collision energy of ~4 kcal mol^{-1}. Crossed-beam imaging has several advantages: there are no kinematic limitations, so we can readily study any mass combination and kinetic energy release in the products. In addition, we obtain the contour map for the reaction directly, without any fitting or assumptions about the dynamics.

In the CN + pentene reaction, we detect the H abstraction channel, *i.e.*, we study the following reaction: CN + 1-C_5H_{10} → HCN + C_5H_9. We detect the hydrocarbon radical by single photon ionization at 157 nm. This channel has largely been neglected in studies of analogous systems such as CN + propene, even though it is the most exoergic pathway, at least in the case that it forms the resonance stabilized allyl radical. The scattering image in Fig. 3A shows the velocity distribution of the radical product, and also intense background from photodissociation of the pentene centered on that beam. This precludes determination of the scattering in that region, so it is omitted from the angular distribution shown in Fig. 3B. The sideways and backward scattered translational energy distributions are shown in Fig. 3C. Together these distributions suggest, but do not prove, reaction through an intermediate complex. They also show unequivocally that the H abstraction channel is occurring in this reaction. Although it has not been seen directly in the reaction of CN with propene or methyl acetylene, these observations suggest that it will be worthwhile to investigate this more deeply, to discern whether the additional carbon chain length is somehow key to the presence of this channel in the pentene reaction.

Professor Seakins asked: The mechanism of the reactions of CN with alkanes is of great interest. Bill Jackson's group has done a lot of work on CN reactions (*e.g.* Copeland *et al.*, *J. Chem. Phys.*, 1992, **96**, 5817). They observe IR chemiluminescence from the HCN product that both identifies a product and gives information on the mechanism (vibrational population inversion observed for most abstractions). Is there any potential for carrying out such studies in the CRESU apparatus? For propane one would expect the abstraction to occur and for C_2H_4 and C_2H_2, addition/elimination is the only viable mechanism (see Choi *et al.*, *Chem. Phys. Lett.*, 2004, **384**, 68; Vereecken *et al. Phys. Chem. Chem. Phys.*, 2003, **5**, 5070) but for propene abstraction could be competitive with addition elimination due to the formation of the relatively stable allyl co-product. Our experiments and calculations on the CN + C_3H_6 reaction (Gannon *et al.*, *J. Phys. Chem. A*, 2007, **111**, 6679) are

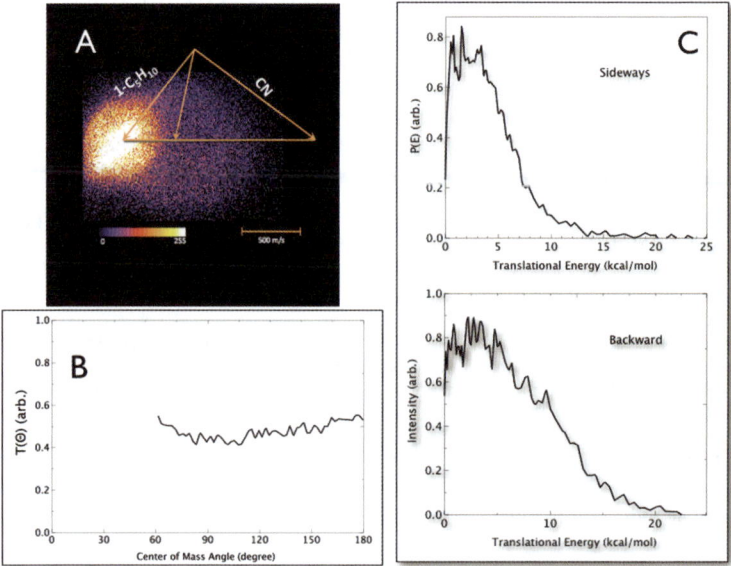

Fig. 3 (A) DC sliced image of C_5H_9 radical from reaction of CN with 1-pentene. Nominal Newton diagram is superimposed. (B) Angular distribution derived from image in A. (C) Total translational energy distributions for sideways (60–120°) and backward (120–180°) regions of the image.

consistent with just an addition/elimination mechanism but relatively low-level calculations showed that the abstraction reaction to be barrierless. Studies on the HCN product from CN + propene would be particularly interesting.

CN is isoelectronic with Cl and is often referred to as a pseudo halogen. Have you tried Cl studies in the CRESU apparatus? If so does Cl behave similarly to CN? The HCl product might be slightly easier to study and interpret, although PES calculations involving halogens are likely to be harder.

Dr Le Picard replied: Your suggestion to look at IR chemiluminescence from CN + hydrocarbon reactions in the CRESU is most interesting and would certainly be feasible given the right equipment.

In reply to your last question, we have not yet studied Cl atom reactions in the CRESU apparatus. Our colleagues from Bordeaux, however, have recently studied the kinetics of the reactions of Cl atoms with ethane and propane over the range 47–167 K.[1] The rate coefficients of the reaction between Cl and C_2H_6 show a slightly positive temperature dependence down to 47 K and display however, a non-Arrhenius behaviour below 120 K. The rate coefficients for the reaction between Cl and C_3H_8 present no temperature dependence from 48 K up to 700 K. Both temperature dependences seem, therefore, to be different from those observed for the corresponding reactions with CN.

1 K. M. Hickson, A. Bergeat and M. Costes, *J. Phys. Chem. A*, 2010, **114**, 3038.

Dr Klippenstein commented: I agree very much with your comments on the importance of developing and applying experimental methods to the measurement of branching ratios at low temperatures. Indeed, I am excited to hear about the experiments that you and the Rennes group intend to perform. Such predictions are indeed difficult to obtain from theory, but the difficulty in such calculations is somewhat different from what you describe. In particular, you describe the difficulties as being related to the problems in obtaining barrier heights to better than 1 kJ mol^{-1}. It is true that current state of the art electronic structure methods generally cannot predict barrier heights to better than about 1 kJ mol^{-1}. However, the product branching ratios generally are not extremely sensitive to the barrier heights in the exit channel. The product branching often arises from a competition between various alternative isomerizations and dissociations from the entrance channel chemical adduct. The saddle points for the channels that lead to significant product formation generally lie well below the reactant energy. As a result, there is not a Boltzmann factor sensitivity to errors in the saddle point energies. Instead, uncertainties of a few kJ mol^{-1} may correlate with uncertainties of only a few percent for the product branching. The difficulties in making accurate theoretical predictions arise instead from: (i) the difficulty in mapping out the whole potential energy surface for the dynamics that can be accessed from the entrance adduct; and (ii) the possibility of non-statistical effects arising in the case of highly exothermic reactions where the decay of the initially formed adduct is more rapid than intramolecular vibrational energy redistribution rates. Paper 22 by Landera and Mebel demonstrates that such potential energy surfaces can be effectively explored, but such calculations are rather complex and one is often concerned that all significant pathways have been discovered. For highly exothermic channels it is possible to employ direct classical trajectory simulations to predict the branching (*cf.* our study of the branching in the CH$_3$ + O reaction[1]), but such calculations are also quite time consuming.

1 T. P. Marcy, R. R. Diaz, D. Heard, S. R. Leone, L. B. Harding and S. J. Klippenstein, Theoretical and experimental investigation of the dynamics of the production of CO from the CH$_3$ + O and CD$_3$ + O reactions, *J. Phys. Chem. A*, 2001, **105**, 8361–8369.

Dr Klippenstein then asked: In your presentation you comment on an apparent discrepancy between your low temperature results and other higher temperature results for the reaction of CN with butadiene. I wonder whether a relatively sharp decrease in rate coefficient with increasing temperature at relatively high T might be explained by the two transition state model, but with somewhat different parameters than in some of our prior studies. In particular, the barrier is likely to be more submerged in this case, and this would lead to a later and perhaps sharper drop from the long-range collision rate.

Dr Faure remarked: The rate coefficient for the CN–butadiene reaction, as plotted in your talk, decreases with temperature below about 100 K. Do you have any possible explanation for this non-usual behaviour (rates for fast CN–neutral reactions are generally found to saturate below about 30 K)?

What are your current plans to measure products and branching ratios for neutral–neutral reactions with CRESU?

Dr Le Picard answered: As shown in this paper, a wide variety of behaviours for the temperature dependence of fast CN–neutral reactions has been observed at low temperatures. Not all reactions display the 'saturation' you refer to, illustrating the danger of extrapolating high temperature rate coefficients down to very low temperatures in the absence of experimental measurements or reliable *ab initio* calculations of the potential energy surface. For the CN + 1,3-butadiene reaction, we have performed a joint study with the crossed molecular beams group of Kaiser and co-workers, and the theoretical group of Mebel *et al.*[1] In the behaviour of the reaction of cyano radicals with 1,3-butadiene we reported a very fast reaction down to low temperatures which nonetheless displays a slight positive temperature dependence to its rate coefficient—maybe due to passage over a very slightly submerged barrier on the electronic potential energy surface corresponding to a tight transition state resulting in a positive temperature dependence down to very low temperatures.

To address your second question, we plan to set-up in the next few years a new, transportable, pulsed version of the CRESU[2] apparatus associated with a mass spectrometer to probe reactants and products of reaction after threshold photoionisation by the VUV DESIRS beamline of the synchrotron SOLEIL. A first version of a pulsed nozzle working at 23 K has been characterized and used for kinetics measurements using the standard PLP-LIF technique at Rennes. Results obtained are in perfect agreement with those previously measured using the continuous version of the CRESU.[3]

1 S. B. Morales, C. J. Bennett, S. D. Le Picard, A. Canosa, I. R. Sims, B. J. Sun, P. H. Chen, A. H. H. Chang, A. M. Mebel, X. Gu, F. Zhang and R. I. Kaiser, 2010, submitted.
2 S. B. Morales, B. R. Rowe, 2009, Patent 09/03663.
3 S. B. Morales, DPhil. thesis, 2009, University of Rennes 1, France.

Dr Ellinger noted: The fact that the rate constant increases when lowering the temperature below, say 100 K, is typical of potential energy surfaces with an attractive well before an activation barrier on the reaction path. Provided the potential well is deep enough to allow the formation of a stable complex and contains one vibrationnal level (at least), the decrease in temperature gives more time for tunneling through the barrier, which explains the increase of the rate constant below 100 K.

Professor Sims remarked: It has been proposed that the minimum and subsequent increase with decrease in temperature of the rate coefficient for the reaction CN + propane, reported by us in this Discussion,[1] could have been due to the effect of quantum mechanical tunnelling. Given the very high values of the observed rate coefficients (on the order of 10^{-10} cm^3 molecule^{-1} s^{-1}) it would appear more probable that the reaction takes place on potential energy surface possessing no absolute barrier to reaction, and that the increase in rate coefficient with temperature observed at higher temperatures is due to the presence of an 'inner' and relatively 'tight' submerged transition state as alluded to above.

1 S. B. Morales, S. D. Le Picard, A. Canosa and I. R. Sims, Faraday Discuss., 2010, **147**, DOI: 10.1039/C004219F.

Professor Mason opened the discussion of the paper by Dr Pernot: This paper is a very nice review of the current database of the chemical reaction reactions important to Titan's atmospheric chemistry and reveals a common problem in planetary science and astrochemistry (and indeed many other areas of science and technology); that of the accumulation and archiving of databases. Different models use different data sets and therefore many of the differences in the results of the models may be due to the differences in the datasets used. It is therefore necessary to test the sensitivity of any model to the input data sets, and these may lead to errors larger than the 10% the author mentions in his paper. I would ask him to comment upon his experience in such databases.

I would also note that there is a new major EU funded project that aims to address these issues the Virtual Atomic and Molecular Data Centre http://www.vamdc.org/. The VAMDC project aims at building an infrastructure for the exchange of atomic and molecular data that will allow, through a single portal, access to a range of databases. It brings together on the one hand scientists from a wide spectrum of disciplines in atomic and molecular (AM) Physics with a strong coupling to the users of their AM data (astrochemistry, atmospheric physics, planetary science) and on the other hand scientists and engineers from the ICT community used to deal with deploying interoperable e-infrastructure.

Dr Pernot replied: I see three main issues related to the use of rate coefficients databases for the modeling of complex chemical systems:

(1) The **collection** of rate coefficients for all the relevant processes in the relevant conditions (pressure, temperature…); (2) the **assessment** of the reliability of the collected data; and (3) the **comparison** of the model outputs with other models using possibly different input datasets.

At the collection stage, it is exceedingly important to be able to find *evaluated data*, *i.e.* data that have been reviewed and evaluated by experts. Some databases provide the modeler with a lot of references (a literature search) without pointing out the "right" ones. Other databases provide you with evaluated data without a trace of the evaluation process, which is not much better. This is what motivated the development of KIDA (http://kida.obs.u-bordeaux1.fr) for astro- and planeto-chemists. KIDA is designed to provide evaluated rate constants for specific conditions/environments, with uncertainty statements, and data sheets ensuring the traceability of the expert's review. KIDA is linked to the VAMDC. The assessment stage is crucial, as it defines confidence intervals on the rate coefficients, and thus enables quantification of the precision of model outputs (for instance by Monte Carlo Uncertainty Propagation). Here again, few databases provide reliable uncertainty statements, and there is an enormous overload of work for the modeller willing to account for uncertainties. Finally, comparison of the results between models, generally based on different chemical schemes, is not straightforward. It is enlightening to compare side-by-side the predictions of a set of models of Titan atmosphere (see for instance http://pagesperso.lcp.u-psud.fr/pernot/ISSI_1D/Runs_2), based on different physical and/or chemical datasets. A recent ISSI workshop gathered different modelers of Titan's atmosphere with the aim of checking if their models produced identical outputs when provided identical inputs. The main conclusion is yes, apart from some residual differences related to implementation schemes of the equations. It would be a huge step towards modeling consistency, if the chemical schemes and datasets used in publications were archived and reusable by others. KIDA is also aimed at providing such a repository.

Some communities are more advanced than ours regarding evaluated databases and data/models sharing, *e.g.* combustion (http://primekinetics.org) or systems biology (http://www.ebi.ac.uk/biomodels-main/), and it is very positive that EU funds are devoted to such issues.

Dr Klippenstein said: Your uncertainty analysis indicates that the reaction of CH with CH_4 is a singularly important reaction for the chemistry of Titan's atmosphere. For this reason, I have chosen to perform a detailed theoretical study of this reaction implementing our two transition state model. The plot of the minimum energy path potential in Fig. 4 illustrates that the submerged inner transition state for the insertion reaction to form C_2H_5 is fairly low lying; certainly well below reactants. Interestingly, at our highest level of theory (CASPT2/CBS//CASPT2/cc-pVTZ) the saddle point in the inner transition state region for this insertion reaction is actually absent. Nevertheless, this inner transition state region still provides a significant bottleneck and is responsible for the interesting temperature dependence of this rate coefficient.

A comparison of the theoretical predictions with the available experimental results[1-6] for this reaction is provided in Fig. 5. The near degeneracy of the $^2\Pi$ states in CH yields some uncertainty in the theoretical predictions. The curves labeled adiabatic employ an adiabatic assumption for the electronic state dynamics. In contrast, the electronic states are assumed to be completely randomized for the curve labeled statistical. The curve labeled adiabatic;E-0.2 employs an empirical 0.2 kcal mol^{-1} lowering of the potential surface in the region of the inner saddle point energy. This adjustment is well within the uncertainties of the theoretical predictions and yields a curve that smoothly interpolates the full set of experimental data. This curve is reproduced to within about 5% over the 20 to 500 K temperature range by the expression $3.76 \times 10^{-8}\ T^{-1.057} \exp(-29.2/T)$ cm^3 $molecule^{-1}$ s^{-1}. Remarkably, this expression may provide a rate coefficient that is accurate to within the "optimistic uncertainty" of 10% that was the focus of your study.

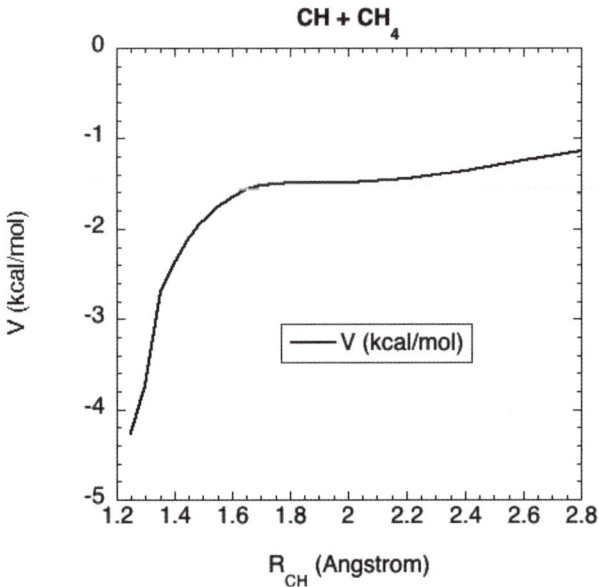

Fig. 4 Plot of the CASPT2(5,5)/cc-pVTZ predicted minimum energy path potential for the interaction of CH_4 with CH as a function of the constrained H3CH⋯CH distance.

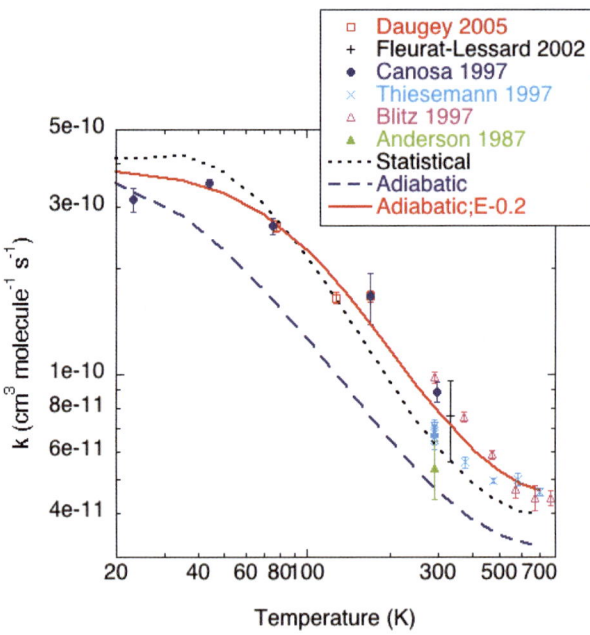

Fig. 5 A comparison of the theoretical predictions with the available experimental results.

For reference purposes we provide here some details of these theoretical evaluations. The E and J resolved number of states for the inner transition state is evaluated from variational transition state theory (TST) employing rigid rotor harmonic oscillator energies for all but one torsional mode, for which a hindered rotor treatment is implemented. The rovibrational properties along the minimum energy path for this inner transition state are obtained from constrained optimizations and vibrational analyses employing the CASPT2/cc-pVTZ method. The energies along this path are obtained from basis set extrapolations of explicit CASPT2 calculations employing the cc-pVTZ and cc-pVQZ basis sets. These calculations all employ a 5-electron 5-orbital active space consisting of the HOMO, radical, and LUMO orbitals (3e,3o) in CH and the σ–σ* orbitals for the CH bond in CH_4 that is being inserted into. The corresponding E and J resolved number of states for the outer transition state was evaluated with direct CASPT2 variable reaction coordinate (VRC) TST.[7] The VRC-TST treatment for the outer transition state employs CASPT2/CBS estimates based on explicit cc-pVDZ and cc-pVTZ evaluations. The active space consists of the (3e,3o) HOMO, radical, and LUMO orbitals in CH. A dynamical correction factor of 0.85 is appended to the TST number of states evaluations.

Support for this work from the NASA Planetary Atmospheres program under grant NNH09AK24I is gratefully acknowledged.

1 S. M. Anderson, A. Freedman, and C. E. Kolb, Fast flow studies of CH radical kinetics at 290 K, *J. Phys. Chem.*, 1987, **91**, 6272–6277.
2 M. A. Blitz, D. G. Johnson, M. Pesa, M. J. Pilling, S. H. Robertson, and P. W. Seakins, Reaction of CH radicals with methane isotopomers, *J. Chem. Soc., Faraday Trans.*, 1997, **93**, 1473–1479.
3 H. Thiesemann, J. MacNamara, and C. A. Taatjes, Deuterium kinetic isotope effect and temperature dependence in the reactions of CH($^2\Pi$) with methane and acetylene, *J. Phys. Chem. A*, 1997, **101**, 1881–1886.
4 A. Canosa, I. R. Sims, D. Travers, I. W. M. Smith, and B. R. Rowe, Reactions of the methylidyne radical with CH_4, C_2H_2, C_2H_6, and but-1-ene studied between 23 and 295 K with a CRESU apparatus, *Astron. Astrophys.*, 1997, **323**, 644–651.
5 P. Fleurat-Lessard, J.-C. Rayez, A. Bergeat, and J.-C. Loison, Reaction of methylidyne CH($X^2\Pi$) radical with CH_4 and H_2S: overall rate constant and absolute atomic hydrogen production, *Chem. Phys.*, 2002, **279**, 87–99.
6 N. Daugey, P. Caubet, B. Retail, M. Costes, A. Bergeat, and G. Dorthe, Kinetic measurements on methylidyne radical reactions with several hydrocarbons at low temperatures, *Phys. Chem. Chem. Phys.*, 2005, **7**, 2921–2927.
7 S. J. Klippenstein, Y. Georgievskii, and L. B. Harding, Predictive theory for the combination kinetics of two alkyl radicals, *Phys. Chem. Chem. Phys.*, 2006, **8**, 1133–1147.

Dr Moses remarked: Do you assume that the rate-coefficient uncertainties are uncorrelated in your analysis? In reality, there will often be correlations between the rate coefficients such that if you increase k for one particular reaction, the rate coefficient for another reaction will need to be decreased (or increased) to remain consistent with concentrations observed in laboratory experiments. Would you expect your overall model uncertainties to decrease if these correlations were taken into account?

Dr Pernot responded: We introduce as many correlations as we can get a track of. You are right that if a set of reaction rates is issued from the analysis of a lab experiment, the rate coefficients can be expected to be highly correlated. To our knowledge, this type of correlation is never published nor archived in kinetics databases. On the other hand, this might concern a few pairs of reactions only, as experimentalists make their best to measure single processes. This kind of correlation would certainly contribute to decrease prediction uncertainty, but I would expect the effect would be rather minor within a set of hundreds of extrapolated reaction rates. Very important correlations we consider in our models are: correlations between parameters of Arrhenius-type rate laws, which are crucial for low-temperature extrapolation [Hébrard *et al.*, *J. Phys Chem. A*, 2009, **113**, 11227–11237]; branching ratios (sum to one) correlations that are to be specifically accounted for when their have been measured separately from the global rate coefficient [Carrasco and Pernot, J. Phys. Chem. A, 2007, **111**, 3507–3512; Carrasco *et al.*, *Planet. Space Sci.*, 2008, 56, 1644–1657.

Dr Bézard asked: In your article, you conclude that any relative variation (temporal/latitudinal/longitudinal) of a constituent's mole fraction should be larger than the uncertainty due to photochemistry to be deemed significant. However, it seems to me that, even if the *absolute* value of a constituent's abundance may be subject to large model (*e.g.*, systematic) uncertainties, a smaller observed *relative* variation (*e.g.*, latitudinal) is still significant, as it is to the first order not affected by these systematic uncertainties. For example, in my opinion, the latitudinal variations of all photochemical compounds on Titan can be reliably interpreted by dynamics, even if they are smaller than the photochemical model uncertainties.

Dr Pernot answered: The key word in your question is "systematic". Before discussing it in the 1D modeling context, let us see its implications in a simple but representative statistical estimation setup.

The statistical setup. Let us consider a series of M simulated/observed/inferred values y_i at positions x_i (latitude, longitude, time...) , ($i = 1, M$), with uniform uncertainty u_y and correlation coefficient c.

For simplicity, the positions are shifted by their mean value (*i.e.* $x = 0$).

The best value of the slope of a regression line $y = a + b \times x$ through these points in the least squares sense is $b_o = \overline{xy}u_x^2$, and the variance of this estimate is $u_b^2 = u_y^2 \times (1 - c)/(M \times u_x^2)$, where u_x^2 is the variance of the set of positions, $u_x^2 = \overline{x^2}$. There is thus 0.95 probability that the slope is strictly positive (*i.e.* we can reject the null slope hypothesis) if $b_o > 1.96 \times (u_y/u_x) \times ((1- c)/M)^{1/2}$. The variation d$y$ of y over a characteristic interval (standard deviation) u_x has thus to satisfy d$y > 1.96 \times u_y \times ((1 - c)/M)^{1/2}$, from which we can now discuss the effect of correlation.

If the uncertainty on y is fully correlated (purely systematic uncertainty, $c = 1$), the test becomes d$y > 0$, *i.e.* any variation of y can be considered as significant. Outside of this singular case, the conclusion depends on u_y, c and M. As an example, let us assume that we have 4 values of y ($M = 4$), and a mixture of systematic and random uncertainties ($c = 0.75$). The 95% confidence test becomes d$y > 0.5 \times u_y$. For purely random uncertainties ($c = 0$), one should have d$y > u_y$, which is what we had in mind in our conclusion.

Systematic, or not systematic? Focusing on the results of 1D photochemical models (*e.g.* concentration of a given species at a given altitude) the question is now: "if a change in lat/long is large enough to induce an observable change, does this position change induce also a variation in the chemical processes or not, through a change in actinic flux, vertical transport and/or temperature ?". If the chemical processes are not affected, one could assume a very strong correlation between the simulation uncertainties, and we would agree with your remark. Otherwise, we have no guarantee that the correlation is large enough to counteract the effect of u_y (for instance, strong decorrelation effects are observed in our simulations between different altitudes).

We thus agree that if systematic uncertainties are dominant; the statement you referred to in our conclusion is too strong. At this stage however, we find it difficult to conclude on this specific condition without performing test simulations. Nevertheless, this discussion emphasizes that before asserting the statistical significance of a latitude/longitude/time variation of a simulated observable, it is highly desirable to estimate the uncertainty on the observable and the correlation coefficient between the uncertainties at different positions. This is very easy to do in our uncertainty-aware Monte Carlo setup, but out of reach from nominal/deterministic simulations.

Observations. If we now focus on observations instead of modeling, for instance the concentrations of minor species inferred from CIRS spectra, the values are mostly affected by random uncertainties related to the signal to noise ratio in the observed spectra, and the significance of lat/long variations has certainly to be tested, as proposed above.

We thank you for this question, which has enabled us to clarify our statement about an interesting issue in photochemical modeling.

Professor Plane addressed Dr Klippenstein: In the example you show, your theoretical prediction over a wide range of temperature appears to agree well with the general trend from several different studies. However, on closer inspection the experimental data sets tend not to be in good agreement with each other—certainly within their quoted error bars—at the temperatures where they overlap.

Dr Klippenstein replied: This observation is indeed correct. In fact, it is quite commonplace for even the best experiments to disagree by somewhat more than their stated error bars. As an ignorant theoretician, I have made an empirical observation that increasing the stated error bars by a factor of two often correlates quite well with the increase in the uncertainty required to bring about complete consistency between different sets of experimental data. Thus, I generally consider theory and experiment to be in good agreement when they agree to within a factor of two of their joint error bars.

Professor Sims addressed Professor Plane: The proper incorporation of the effects of experimental errors on the reliability of model predictions in the excellent work reported by Dr Pernot and colleagues[1] requires clear information on the magnitude of these errors to be provided in publications by experimental kinetics groups such as our own. In the past,[2] we have provided uncertainties at the 95% confidence level derived from the standard errors observed in the gradients of the 'second-order plots', multiplied by the appropriate value of the Student's t distribution (which depends on the number of experimental points). In the text we have noted that to this uncertainty should be added an estimated likely systematic error of 10% (resulting from uncertainties in flow and density calibrations *etc.*). However, experience has shown that it is better to incorporate this likely systematic uncertainty in the values quoted in the abstract and data tables, and so more recently this is what we have done.[3] The most important point is to state clearly what is being quoted.

1 Z. Peng, M. Dobrijevic, E. Hébrard, N. Carrasco and P. Pernot, *Faraday Discuss.*, 2010, **147**, DOI: 10.1039/C003366A.
2 See for example: A. Canosa, I. R. Sims, D. Travers, I. W. M. Smith and B. R. Rowe, *Astron. Astrophys.*, 1997, **323**, 644.
3 See for example: C. Berteloite, S. D. Le Picard, N. Balucani, A. Canosa and I. R. Sims, *Phys. Chem. Chem. Phys.*, 2010, **12**, 3666; C. Berteloite, S. D. Le Picard, N. Balucani, A. Canosa and I. R. Sims, *Phys. Chem. Chem. Phys.*, 2010, **12**, 3677.

Dr Manzanares opened the general discussion to address Dr Le Picard: At a particular temperature, the measured rate constants of CN reactions increase in the order: k (propane) < k (propene) < k (1-butyne). At the same time, temperature dependences are different for each reaction. What is unknown is the possible mechanism for each reaction. The identification of the C–H bond where the abstraction occurs could be obtained from experimental studies of deuterated samples or theoretical calculations. The theoretical method could be used to obtain the stability of the molecular fragment produced.

Dr Le Picard answered: In our paper we point out that the rate coefficients obtained at the lowest temperatures (23 K) are in good agreement with those predicted by a simple capture theory, indicating that the reactivity is essentially governed by long range electrostatic forces. The increase in rate coefficients with the size of the hydrocarbon is consistent with this theory in which the polarizabilities of the reactant molecules are involved. Concerning your last remark, in these neutral–neutral reactions no molecular ions are produced.

Professor Kaiser opened the discussion of the paper by Paul W Seakins: You have observed the atomic hydrogen loss pathway. In our crossed beam experiments, we often conduct experiments with partially deuterated chemicals to pin down the actual position of the hydrogen loss. So to elucidate if the hydrogen atom comes from the CH_2 or from C_2H_2 (or from both), you could conduct the experiemnts $CD_2 + C_2H_2$ and $CH_2 + C_2D_2$, and possibly $CH_2 + C_2HD$ and $CD_2 + C_2HD$. This would provide explicit evidence on the position of the H atom loss and could also help to elucidate the reaction pathway(s). Is this planned?

Professor Seakins responded: In the past we have used partially deuterated systems to investigate the mechanism of the reaction, for example probing the H : D ratio of decomposition of the vibrationally excited CH_2D_2 molecule formed from the insertion of $^1CH_2$ into D_2.[1] It would certainly be possible and potentially interesting to undertake the experiments that you suggest, however, it is not clear how one would interpret such data.

The expected mechanism is the initial formation of cyclopropene followed by the formation of the biradical, e.g. $\cdot CH_2CD=C\cdot D$ from the reaction of $^1CH_2$ with C_2D_2. Transfer of the middle D to either terminal carbon yields either allene or propyne, both of which can eliminate H or D to give propargyl + H/D. We would therefore expect a mixture of both H and D whether the dominant C_3H_4 intermediate was allene or propyne. Theory might be able to predict whether there should be a measurable difference in the H/D ratio depending on whether elimination occurs from propyne or allene, but similarity of the energies of propyne and allene suggests that both would be formed. Additionally, in a long-lived complex there may be further isotopic scrambling during allene/propyne interconversions via the biradical.

1 K. L. Gannon et al., Kinetics and product branching ratios of the reaction of $^1CH_2$ with H_2 and D_2, J. Phys. Chem. A, 2008, **112**(39), 9575–9583.

Dr Vuitton commented: The lowest temperature at which your experiments were done is 195 K. Titan's upper atmosphere is at 150 K. Are you planning to acquire data at lower temperature? If not, can you suggest a value to use in the models?

Professor Seakins answered: Experiments in our conventional slow-flow laser-flash-photolysis cell are limited to 195 K, however, we do have plans to carry out such experiments in a pulsed Laval system constructed in this laboratory.[1] The Laval systems such as those described in paper 7 have primarily been used solely to observe the removal of radicals—our experiments are more challenging both in requiring the observation of product growth and detecting H at Lyman-α.

At the moment all we can do is to recommend extrapolation of the data recorded at 298, 250 and 195 K, but recognise that there will be some additional uncertainty in such extrapolations. The qualitative agreement between experiment and theory and the potential developments such as those described in the response to Dr Meuwly (below) suggests that, in the future, theory may be able to determine the nature of such an extrapolation.

1 S. E. Taylor et al., Pulsed Laval nozzle study of the kinetics of OH with unsaturated hydrocarbons at very low temperatures., Phys. Chem. Chem. Phys., 2008, **10**(3), 422–437.

Professor Atreya commented: How do you extrapolate your results to the much lower temperatures? For example, the temperature in the regime of methylene reactions at Saturn can be as low as 140 K and even lower in Titan's atmosphere.

Professor Seakins replied: We would not claim to be able to extrapolate our results beyond the experimental range and, given that our theory does not produce quantitative agreement with experiment, we have no basis for recommending any method of extrapolation.

As mentioned in other responses, developments in either theory or experiment are non-trivial and it would be helpful to have some modelling studies to assess whether such work would generate useful results for the community.

Professor Meuwly remarked: In Fig. 5, your computations do not really reflect the good agreement we saw on several other occasions during this meeting. Can you comment on the differences (electronic structure methods, entropic effects, *etc.*) between other TST-based rate coefficients and those presented here in view of their ability to reproduce/explain experimental data?

As a comment: in the printed version it is not quite clear where the experimental (filled circles) data points are for $T = 400$ K and 500 K and some of the error bars look strikingly large.

Professor Seakins responded: The calculations carried out for relaxation process of $^1CH_2$ to $^3CH_2$ are significantly more challenging than a standard TST calculation. The *ab initio* component requires the calculation of two potential energy surfaces and the non-adiabatic TST then requires further calculations to determine the surface hopping probabilities. It should also be noted that these are absolute calculations—some TST calculations are tuned to reproduce an absolute rate coefficient at a certain temperature, constraining the structure of the transition state. This constrained model can then be used to predict the temperature dependence of the reaction, often with excellent results.

We should also emphasise that these are the first such calculations and there is potential of improvements in the methodology. For example, the 'bath' of vibrational modes surrounding a single reaction coordinate as used now could be replaced by a smaller number of bath modes, with more degrees of freedom treated exactly (*e.g.* the CH_2-X distance AND the H–C–H bend). Additionally, instead of using simplistic Landau–Zener theory, the Schrödinger equation could be solved exactly within the space of those dimensions. This would potentially be much more accurate. However, such calculations would not be trivial and it would be useful to have a modelling evaluation of the importance of such reactions to assess the potential return on the investment in coding development.

The comparisons with the experimental data at higher temperature data do have the potential to mislead. In Table 2 we show that for the reaction of $^1CH_2$ with acetylene the yield of H above 398 K is unity (*i.e.* no relaxation). However, the experiments are also challenging and there is a significant error in the determinations. So for Fig. 5, which plots the rate coefficient for relaxation, the experimental data points would be zero, but with significant errors on the log scale.

Professor Zwier asked: When I look at your low-temperature value for the rate constant, it looks to me to be near gas-kinetic. If you were to tell the modelers how to extrapolate to even lower temperature, wouldn't it be best to simply tell them to use a gas kinetic value?

Professor Seakins answered: Certainly the overall rate coefficient for $^1CH_2$ removal could be approximated in this way, although an experimental determination of such a value should be undertaken in the relatively near future using a pulsed Laval system (80–140 K).

The critical issue is the temperature dependence of the branching ratio and these are much more challenging experiments in the pulsed Laval experiment. Clearly, simply setting the rate coefficient for relaxation to the gas kinetic limit, *i.e.* 100% relaxation, would have important implications on the chemistry. Given that further developments of both the theory and experiment to allow quantitative determinations of branching ratios at lower temperature would be time consuming and challenging, it would be worth modellers carrying out some evaluations to assess the sensitivity of important parameters (*e.g.* benzene formation) to the relaxation/reaction branching ratio to determine whether such an investment would be worthwhile.

Professor Sims noted: Use of the so-called gas-kinetic limit is not very useful for extrapolation of temperature dependent rate coefficients of fast barrierless reactions to lower temperatures as essentially this corresponds to simple capture theory which yields positive temperature dependences (*e.g.* $k \propto T^{1/6}$ where the interaction is dominated by dispersion) while experimentally we more frequently observe negative temperature dependences.

Dr Moses remarked: The relative roles of $^1CH_2$ relaxation *versus* chemical reaction are also critical in defining the relative abundances of unsaturated versus saturated hydrocarbons on the giant planets. Here, the key reactions are singlet methylene with H_2 and with CH_4. Reaction of $^1CH_2$ with H_2 to form CH_3 + H will lead to the production of C_2H_6 (and CH_4) via termolecular reactions with CH_3 (and atomic H) [and it's worth noting here that your suggested $CH_3 + H_2 \rightarrow CH_4$ + H reaction will not be very important at the cold stratospheric temperatures of the giant planets], whereas relaxation would be followed by $^3CH_2 + H \rightarrow CH + H_2$, with subsequent CH insertion in to methane or other hydrocarbons to form unsaturated hydrocarbons on the giant planets. You mention that you have studied $^1CH_2 + H_2$ in previous investigations. Have you ever taken a look at $^1CH_2 + CH_4$, or do you have any plans to do so?

Professor Seakins replied: We have studied the reaction of $^1CH_2$ with H_2 and D_2 and have observed the competition between relaxation and reaction. This reaction shows the same qualitative trend with temperature as those studied in this work, with relaxation being favoured at low temperatures.

The reaction of $^1CH_2 + CH_4$ would indeed be of interest, however we would need to be able to detect either the products of the chemical reaction ($^2CH_3$) or the $^3CH_2$ formed from relaxation, neither of which we can do with the current experimental set up. A photoionization mass spectrometer system has been constructed in our laboratory[2] and would be capable of detecting methyl radicals, so such studies would be possible in the future although we would need to find either a way of determining absolute concentrations of $^1CH_2$ and CH_3 or finding an alternative calibration reaction. Your comments on the importance of the $CH_3 + H_2$ reaction following production of CH_3 from the reaction of $^1CH_2$ with H_2 are noted and appreciated, but as you point out there would still be different products formed from $^1CH_2$ and $^3CH_2$ chemistry.

1 K. L. Gannon *et al.*, Kinetics and product branching ratios of the reaction of $^1CH_2$ with H_2 and D_2, *J. Phys. Chem. A*, 2008, **112**(39), 9575–9583.
2 M. A. Blitz *et al.*, Time-of-flight mass spectrometry for time-resolved measurements, *Rev. Sci. Instrum.*, 2007, **78**(3), 034103.

Professor Miller asked: CH_2 is going to be a pretty anharmonic oscillator and a not very rigid rotator. How much importance is there in having and using accurate ro-vibrational states calculated from first principles? Will these make much difference to your reaction rates?

Professor Seakins answered: This is something that we plan to look into, however, we feel that the potential developments outlined in our response to Dr Meuwly (above) will be more important in determining the accuracy of the calculations.

Dr Klippenstein noted: Your nonadiabatic TST treatment of the crossing dynamics should be reasonably accurate and the examination of the effect of varying the crossing energy by ±4 kJ mol^{-1} is useful. As suggested by the authors, anharmonicities in the potential may be of some importance. They would affect both the capture rate and the electronic relaxation rate. The use of the inverse Laplace transform of the experimental rate should properly treat the anharmonicities in the capture rate. It is not clear to me how significant anharmonicities would be for your nonadiabatic TST treatment of the electronic relaxation rate. I would guess that their effect would not be dramatic. It is unclear to me whether or not (and how) you have combined the treatments for the capture rate and the electronic relaxation rate. Are the results in Fig. 5 and 6 based on some two TS model for the competition between these two reactions (as alluded to on p. 10) or are they simply an implementation of eqn (E1) to (E4) for the electronic relaxation channel? The extremely large magnitude of the rate coefficients in Fig. 7 may imply some important nonstatistical effects. However, it is not clear to me what process is being considered in that plot. The units in Fig. 7 imply you are considering the dissociation from some complex rather than a bimolecular association, but it is not clear to me what the complex is that provided the density of states for the RRKM rate coefficients.

Professor Seakins replied: With respect to Fig. 5 and 6, the theoretical results for both plots are simply the NA-TST rate coefficients. The experimental results shown in Fig. 5 were obtained by subtracting the rate coefficient for $^1CH_2 + C_2H_2 \rightarrow H + C_3H_3$ from the total loss rate coefficient for $^1CH_2 + C_2H_2$. Because our master equation results suggest that the only significant channel on the singlet surface leads to quantitative H production, this difference should give the rate coefficient for electronic relaxation to the triplet surface. In the case of $^1CH_2 + Xe$, the only loss channel is relaxation to the triplet surface.

For a $[C_2H_2] = 2.25 \times 10^{15}$ molecule cm^{-3}, Fig. 7 shows: (1) pseudo first order rate coefficients for $^1CH_2 + C_2H_2 \rightarrow$ cyclopropene using an ILT of the experimental rate coefficient; and (2) pseudo first order rate coefficients for $^1CH_2 + C_2H_2 \rightarrow {}^3CH_2 + C_2H_2$ using NA-TST. For both reactions, the reactant density of states (DOS) is the convoluted DOS for $^1CH_2 + C_2H_2$, and the TS DOS is that obtained using ILT, and at the MECP. We are simply noting that the shape of the k_{ES} resembles those in a two TS model. In order to explore this effect further, we would need to know the location of the variational TS for $^1CH_2 + C_2H_2$ capture with respect to the MECP location.

Dr Dobrijevic opened the discussion of the paper by Professor Casavecchia: It is very important for modelers to know what are the branching ratios of all the reactions included in photochemical models. In order to perform uncertainty propagation studies and sensitivity analysis in models it is of prime interest to know what are the uncertainty factors of the branching ratios (see for instance ref. 1–3).

Consequently, could you estimate the uncertainty factors of the branching ratios you have determined?

1 E. Hébrard, P. Pernot, M. Dobrijevic, N. Carrasco, A. Bergeat, K. M. Hickson, A. Canosa, S. D. Le Picard, and I. R. Sims, How measurements of rate coefficients at low temperature increase the predictivity of photochemical models of Titan's atmosphere, *J. Phys. Chem. A*, 2009, **113**, 11227–11237
2 E. Hébrard, M. Dobrijevic, Y. Bénilan, F. Raulin, Photochemical kinetics uncertainties in modeling Titan's atmosphere: First consequences, *Planet. Space Sci.*, 2007, **55**, 1470–1489.
3 E. Hébrard, M. Dobrijevic, Y. Bénilan, F. Raulin, Photochemical kinetics uncertainties in modeling Titan's atmosphere: a review, *J. Photochem. Photobiol. C: Photochem. Rev.*, 2006, **7**(4), 211–230.

Professor Casavecchia responded on behalf of also N. Balucani, F. Leonori, D. Skouteris and M. Rosi: We agree with you about the importance of the uncertainty factors in the branching ratios. In our comment (to Dr Yelle's paper) containing branching ratios data for the reaction N(^{2}D) + CH$_4$ we have reported error bars in the graph of the relative importance of the two channels examined. These are, however, relative branching ratios because we don't have data for the 3rd significant channel, that leading to NH + CH$_3$. Regarding the branching ratios derived from RRKM calculations, we remark that the RRKM estimates result directly from the energy level pattern resulting from the electronic structure calculations. In order to obtain the relevant uncertainties, one has to take into account: (i) the uncertainty in the heights of the relevant barriers; (ii) the uncertainty in the PES curvatures (which determines the vibrational frequencies); and (iii) the uncertainty in the moments of inertia (which determines the rotational structure). The most conservative possibility would be to produce two RRKM estimates corresponding to the two extremes of the uncertainty spectrum of the electronic structure calculations. In principle, further uncertainties in the RRKM estimates result from the fact that the vibrational modes of sufficiently high frequency are considered harmonic, whereas the rotational level structure is assumed to be continuous (permitting a simple inverse Laplace transformation without the need to count energy levels). However, such uncertainties are expected to be of much lower importance. In any case, for the N(^{2}D) + C$_2$H$_6$ reaction discussed in our paper the branching ratio uncertainties, obtainable along the lines outlined above, have not been calculated.

Dr Thissen asked: This is a question about the "soft ionization" method, by electron ionization, that is used in your method to identify the products. It appears to "saturate" for this C$_2$NH$_6$ system because of: (1) dissociative ionization; and (2) overlap with other products. My question is: "Do you have a strategy for improving this in order to evaluate more complex systems?"

Professor Casavecchia replied: Soft ionization (by low energy electrons or, even better, by VUV tunable photons as from a synchrotron source) is the key to reduce (or even suppress) the problem of dissociative ionization in crossed-beam experiments with mass spectrometric detection. However, in systems containing many hydrogen atoms, like N(^{2}D) + C$_2$H$_6$, where a large variety of possible products that can be formed as a consequence of the versatility of C and N atoms in forming bonds, and a variety of isomeric product species exist, things become very complicated and particularly difficult. Certainly, overlap with other products may also be a serious problem, in particular with this system, because the N atom has mass 14 (like CH$_2$) and the natural isotopic abundance of ^{15}N is a disturbance if you want to detect the methyl radical (CH$_3$). The higher energy resolution afforded by VUV synchrotron photons may help considerably with the identification of isomeric species, but still considerable difficulties remain. I don't have a general strategy for improving soft electron ionization detection. The extent of its success depends on the system under study.

Dr Klippenstein said: Your paper, and particularly the plots in Fig. 10 and 11, suggest that the overall reaction rate for the N(^{2}D) + C$_2$H$_6$ reaction will be determined by the apparently barrierless capture rate. Thus, I thought I could provide a nice complement to your crossed beam determination of the branching ratios by predicting this capture rate with direct variable reaction coordinate transition state theory. However, my initial attempt to study this process suggested that there is actually a significant barrier to the insertion process. This was somewhat surprising to me, given the isoelectronic nature of N(^{2}D) and CH together with the barrierless nature of the insertion reactions of CH with alkanes such as CH$_4$ and C$_2$H$_6$.

The presence or absence of a barrier height for this reaction is of some significance to the modeling of Titan's atmospheric chemistry. Thus, in an attempt to learn more about the saddle point energies I undertook a fairly thorough electronic structure study of the N(^{2}D) + CH$_4$ and N(^{2}D) + C$_2$H$_6$ reactions. To begin I evaluated the N(^{4}S) to N(^{2}D) splitting with various levels of theory. At the QCISD(T), CASPT2(3e,3o), and CAS+1+2+QC(3e,3o) levels this splitting is predicted to be 60.9, 54.5, and 54.6 kcal mol^{-1} respectively, while the experimental value is 55.0 kcal mol^{-1}. Each of these calculations employ a CBS extrapolation based on the cc-pVQZ and cc-pVTZ basis sets. The error for the QCISD(T) prediction is indicative of the multireference nature of the wavefunction for the N(^{2}D) state. This suggests that QCISD(T) or CCSD(T) based estimates of the saddle point energies for the alkane insertion reactions are likely to be inaccurate. For the N(^{2}D) + CH$_4$ reaction the barrier height relative to reactants is predicted to be 4.7, 0.7, and 1.9 kcal mol^{-1} at the CASPT2(5e,5o), CAS+1+2+QC(5e,5o)//CASPT2(5e,5o) and CAS+1+2+Q-C(5e,5o) levels, respectively. These calculations employ optimizations and frequency analyses with the cc-pVTZ basis set, CBS extrapolations with the pVQZ and cc-pVTZ basis sets, and harmonic oscillator zero-point energy corrections. The (5e,5o) active space employed is analogous to that employed in our earlier comments on the CH + CH$_4$ reaction. These results are in reasonable agreement with earlier multireference electronic structure results of Takayanagi et al.[1] The difference between the CASPT2 and CAS+1+2+QC results is relatively large. The lower, but still positive, value for the CAS+1+2+QC barrier suggests that TST calculations based on it should give a rough upper bound to the predicted rate constant. The plot in Fig. 6 shows that the TST predicted rate coefficient at the CAS+1+2+Q-C(5e,5o)/CBS//CAS+1+2+QC/cc-pVTZ level (including Eckart tunneling) is in quite reasonable agreement with the corresponding experimental measurement from Takayanagi et al.[2]

These results suggest that related calculations for the N(^{2}D) + C$_2$H$_6$ reaction should provide quite accurate estimates of the barrier height for that reaction. In this case, the CASPT2 and CAS+1+2+QC//CASPT2 zero-point corrected barrier heights are 3.4 and 1.2 kcal mol^{-1}, respectively. The full CAS+1+2+QC barrier would be expected to fall between these two limits as it did for the methane case. These results suggest that the room temperature rate constant for the reaction of N(^{2}D) with ethane should be quite similar to that for its reaction with methane. In contrast, the study of Umemoto et al.[3] indicates that it is 6 times faster. It is unclear to us the reason for this discrepancy, but there is some possibility that the measured rate constant includes some contribution from relaxation in addition to insertion, and that this contribution is significantly greater for C$_2$H$_6$ due to increased long-range interactions. Alternatively, the electronic structure theory may be over-estimating the barrier heights or there is some alternative lower energy pathway to the ones that we explored. We did make some attempt to locate various insertion and abstraction pathways but this search may not have been exhaustive.

Notably, the present theoretically estimated lower bound of ~1.2 kcal mol^{-1} for the barrier height would yield a rather small rate coefficient for the N(^{2}D) + C$_2$H$_6$

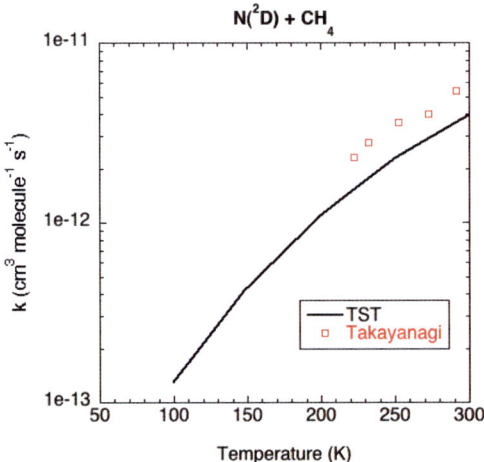

Fig. 6 Plot of the predicted and observed rate coefficient for the insertion of N(^{2}D) into CH$_4$.

reaction at low temperature. It would be valuable to have some experimental confirmation of the smallness of this rate coefficient at low temperature.

Support for this work from the NASA Planetary Atmospheres program under grant NNH09AK24I is gratefully acknowledged.

1 T. Takayanagi, Y. Kurosaki, and K. Yokoyama, Ab initio calculations for the $N(^2D) + CH_4$ reaction: Does the $N(^2D)$ atom really insert into CH bonds of alkane molecules, *Int. J. Quantum Chem.*, 2000, **79**, 190–197.
2 T. Takayanagi, Y. Kurosaki, K. Sato, K. Misawa, Y. Kobayashi, and S. Tsunashima, Kinetics studies on the $N(^2D,^2P) + CH_4$ and CD_4 reactions: the role of nonadiabatic transitions on thermal rate constants, *J. Phys. Chem. A*, 1999, **103**, 250–255.
3 H. Umemoto, N. Hachiya, E. Matsunaga, A. Suda, and M. Kawasaki, Rate constants for the deactivation of $N(^2D)$ by simple hydride and deuteride molecules, *Chem. Phys. Lett.*, 1908, **296**, 203–207.

Professor Casavecchia replied: The comment by S. Klippenstein is interesting and appropriate. Yes, I agree that CCSD(T) calculations tend to underestimate barrier heights. Higher level calculations as carried out by you certainly are expected to improve on barrier heights. I am wondering, though, if in the case of $N(^2D) + CH_4$ and $N(^2D) + C_2H_6$ your best estimates tend to overestimate the barrier height. In fact, your calculations of rate constants as a function of temperature in Fig. 1 for $N(^2D) + CH_4$ underestimate by a noticeable amount the experimental values, and for $N(^2D) + C_2H_6$ your predicted values using the calculated barrier height would underestimate the experimental value at 300 K even more than in the case of methane. In the light of the underestimation for $N(^2D) + CH_4$, one tends to think that the electronic structure theory may be overestimating the barrier heights, rather that unaccounted for relaxation in the experiment being the cause of the disagreement between experiment and theory. Of course, both things could be at play, *i.e.*, a slight overestimate of the barrier height and some relaxation occurring in the experiment. In this regard, it would certainly be desirable to have new rate constant measurements, also down to low temperatures, as possible by exploiting the CRESU technique (and possibly by detecting the NH product, for instance, by LIF), to shed further light on this disagreement.

Regarding our work on $N(^2D) + C_2H_6$ presented at this Discussion, I would like to point out that the presence of a small barrier in the entrance channel of the $N(^2D) + C_2H_6$ reaction does not affect any conclusions that we have derived.

Professor Seakins remarked: ^2N is isoelectronic with CH. For the CH reaction with ethane there is potential to insert into the C–C bond of ethane as well as the C–H bonds.[1,2] Insertion of CH into the C–C bond produces isopropyl which would decompose to H + propene. The analogous process in your system would be the production of CH_3NCH_3 following the insertion. In your theoretical calculations did you look for a direct pathway between the reagents and CH_3NCH_3? In the PES shown in Fig. 10 this is only accessible through TS1 and TS7, with the first transition state looking particularly 'tight'. Experimentally would you be able to distinguish the resulting CH_3NCH_2 product from other isomers?

1 K. McKee, K *et al.*, H atom branching ratios from the reactions of CH with C_2H_2, C_2H_4, C_2H_6, and neo-C_5H_{12} at room temperature and 25 torr, *J. Phys. Chem. A*, 2003, **107**(30), 5710–5716.
2 J. C. Loison *et al.*, Rate constants and H atom branching ratios of the gas-phase reactions of methylidyne CH((XII)-I-2) radical with a series of alkanes, *J. Phys. Chem. A*, 2006, **110**(50), 13500–13506.

Professor Casavecchia responded on behalf of also N. Balucani, F. Leonori, D. Skouteris and M. Rosi: At the level of our calculations we were unable, despite considerable search, to locate a saddle point for insertion of $N(^2D)$ into the C–C bond of ethane. Therefore, we do not have evidence for a direct pathway from reactants to CH_3NCH_3.

Regarding the question whether we would be able to distinguish the resulting CH_3NCH_2 product from other C_2H_5N isomers, the answer is no. In fact, the exoergicity of channel (1e) leading to CH_3NCH_2 is 299.3 kJ mol^{-1}, and this is only slightly lower than the exoergicity of the other two H forming channels (1c) (-338.2 kJ mol^{-1}) and (1d) (-321.3 kJ mol^{-1}) indicated in the Fig. 5(b) and 6(b) of our paper,[1] which report the product translational energy distributions at the two investigated collision energies. The differences between the total available energy for the above three channels is relatively small and well within the uncertainty of our experimental determination.

1 N. Balucani, F. Leonori, R. Petrucci, M. Stazi, D. Skouteris, M. Rosi and P. Casavecchia, *Faraday Discuss.*, 2010, **147**, DOI: 10.1039/C004748A.

Professor Mason opened the discussion of the paper by Professor Meuwly: On a planetary surface the ices may be irradiated and, hence, both chemically and physically altered; how might these effects be allowed for in your model and effect the IR signatures (*e.g.* break up of water clusters). The morphology of the ice may also be set by the long time scales of deposition; is a layered structure therefore truly representative of the type of ice you may see on a planetary surface? The random ice would seem a better mimic.

Professor Meuwly responded: The irradiation of planetary ice can trigger the formation of radicals and therefore promote the formation of new species or distort the ice structure locally. In the present contribution reactive processes were not of interest. Rather, we compare spectroscopic characteristics of different structures such as layered, random and others, since the nature of ice structures on planetary surfaces is largely unknown. Therefore quantitative atomistic simulations of specific spectral features (*e.g.* in the infrared) and comparison with experiment provide insights into the structures which are responsible for the spectroscopic observations and which can not be determined from experiment because—as was correctly pointed out—the structure that is most likely present is a random ice. For planetary ices, structural data are unavailable and the only possibility to obtain insight into these structures is *via* structure–spectroscopy relationships. While it is straightforward to determine the structure of pure crystalline ices experimentally, structure determination is much more difficult, if not impossible, for mixed and amorphous ices.

Professor Zwier asked: Can you comment a bit more on the extent to which the intramolecular vibrations (*e.g.*, CO stretch, H_2O bend) are copuled to one another in the mixed ices to form delocalized (*i.e.*, coupled) vibrations that extend over many molecules?

Professor Meuwly responded: The most likely region where coupling between CO stretching and H_2O bending/stretching modes are visible is in the far infrared region. Such spectra were calculated and have been found to be in good agreement with experiment for CH_4 in water (see Plattner *et al.*, *Mol. Phys.*, 2008, **106**, 1675). In the same manuscript, predictions for spectra in the acoustic region for CO in water have been made. Further, avoided crossings of rattling modes involving guest molecules (Xe) with the host lattice acoustic phonon branches of the same symmetry have been shown to result in vibrational couplings in clathrate hydrates.[1]

1 J. S. Tse, V. P. Shpakov, V. V. Murashov, V. R. Belosludov, *J. Chem. Phys.*, 1997, **107**, 9271–9274.

Dr Adams opened the discussion of the paper by Professor Zwier: In your Table 3, you consider the neutral products of some $C_{10}H_n^+$ electron–ion recombinations as requiring ejection of one or two H atoms. What is your justification for this considering that many storage ring studies show more fragmentation?

Professor Zwier answered: Some of the entries in Table 3 included two H-atoms as products because the calculated exothermicity for the electron–ion recombination (equivalent to the ionization potential of the neutral molecule) is sufficient to cause dissociative recombination in two steps, first losing a single H-atom, but still retaining sufficient internal energy in the product to lose a second. Thus, the magnitude of the initial internal energy of the neutral ion–electron recombination determined whether this was possible.

Dr Biennier asked: Do you observe the production of species much larger than precursor molecules in your pulsed discharge nozzle source?

Professor Zwier answered: In a recent paper on the discharge chemistry of 1,3-butadiene (J. J. Newby, J. A. Stearns, C. P. Liu, and T. S. Zwier, *J. Phys. Chem. A*, 2007, **111**, 10914–109127) we explored these issues to some extent. In that case, we concentrated primarily on aromatic products that we could positively identify structurally *via* resonant two-photon ionization spectroscopy. There we identified products that included benzene, toluene, styrene, phenylacetylene, and almost a dozen other aromatic derivatives as large as phenylcyclopentadiene, indene, and naphthalene. We also have data on an analogous study on vinylacetylene, which shows a detectable signal from molecules as large as naphthalene. So, the discharge definitely processes the organic mixture much more thoroughly than photolysis, even under the short time and flow conditions used in this study, in which the molecules are exposed to the discharge for only ~20 microseconds as they flow down the discharge tube.

Professor Ashfold commented: Professor Zwier has highlighted differences in the spectral signatures of different structural (and even conformational) isomers, but it is also worth reminding ourselves that subtle changes in structure can result in different photochemistry. For example, we have recently been investigating the near-UV photochemistry of 4- and 5-hydroxyindole molecules in the gas phase.[1] Photo-exciting the former isomer at wavelengths around its S_1–S_0 origin (284.893 nm) results in O–H bond fission and formation of fast H atoms and partner 4-hydroxyindolyl radicals in selected vibrational levels as revealed by the structure at TKER $\sim$7000 cm^{-1} in Fig. 7(a)—*i.e.* a mechanism involving eventual dissociation on the $\pi\sigma^*$ potential energy surface, reminiscent of that exhibited by bare phenol.[2] Excitation of 5-hydroxyindole, in contrast, yields no analogous structured features when excited at its S_1–S_0 origin (303.881 nm, Fig. 7(b)). Dissociation in this case is deduced to involve internal conversion to high vibrational levels of the ground state and subsequent 'statistical' unimolecular decay.

1 T. A. A. Oliver, G. A. King and M. N. R. Ashfold, unpublished results.
2 M. N. R. Ashfold, G. A. King, D. Murdock, M. G. D. Nix, T. A. A. Oliver and A. G. Sage, *Phys. Chem. Chem. Phys.*, 2010, **12**, 1218 and references therein.

Professor Suits asked: These are very powerful experiments. I wonder if you have begun incorporating nitrogen into these studies yet?

Professor Zwier replied: No we haven't, but this is an obvious an important extension that deserves detailed exploration in the future. We hope soon to take steps in that direction. In this regard, it is interesting to note that some of the nitrile-containing aromatics worth studying share the same mass as important hydrocarbon aromatics. For instance, the di-cyano benzene isomers (*ortho*-, *meta*-, and *para*-) have *m/z* 128, identical to that of naphthalene. Until we learn more about this

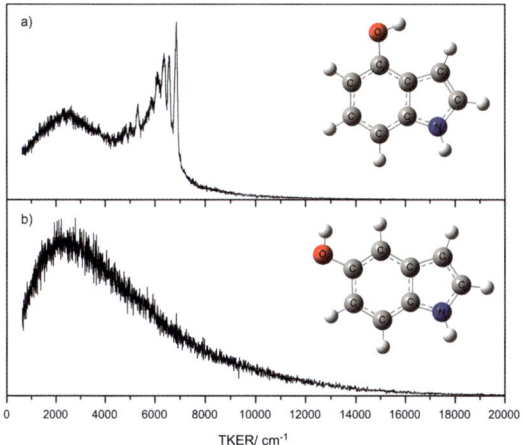

Fig. 7 Total kinetic energy release (TKER) spectra derived from H atom time-of-flight measurements following photolysis of (a) 4-hydroxyindole and (b) 5-hydroxyindole at their respective S_1–S_0 origins.

chemistry, we as a community should be cautious in assuming that ion signals that appear at *m/z* 128 are necessarily naphthalene. In fact, they may not even be pure hydrocarbons, if the nitrile chemistry turns out to be facile.

Dr Thissen asked: You mention in the paper that a very low ionization potential (IP) was found for molecules of the type $C_{10}H_9$.

I would like to mention the papers by the group of D. Schroder in Prague (ref. 1 & 2) where a possible boost of Titan's ionic chemistry is proposed that could be induced by the presence of molecular dications, resulting from photoionization in the atmosphere. Did you consider this issue, and what would be your prognosis in terms of IP for similar molecules but of the type C_9NH_8?

1 C. L. Ricketts, D. Schroder, C. Alcaraz, J. Roithova, *Chem.–Eur. J.*, 2008, **14**(16), 4779–4783.
2 E. L. Zins and D. Schroder, *J. Phys. Chem. A*, 2010, **114**(19), 5989–5996.

Dr Moses opened a general discussion of the papers and addressed Professor Casavecchia: Professor Casavecchia has criticized photochemical modelers for not always tracking the correct products from specific reactions, and Dr Yelle has responded that the fate of these products is not always known (*i.e.*, the subsequent kinetic pathways of the products have not been studied or reported in the literature), which prevents the inclusion of those species in the model. Ethanimine is one such example. I have a general question for the group. What do you recommend that modellers do in these situations? Obviously, it would be best to perform new laboratory or theoretical investigations to study the fate of those species, but that's not always feasible. It is a particular problem when thermodynamic or detailed structural data are unavailable for the species in question. Should we include these species in the model anyway and try to estimate the subsequent reactions? Should we omit the species from the model? One technique mentioned by Dr Yelle is to assume that a particular less-stable product transforms quickly to a more stable isomer. Is that assumption acceptable? Although I realize that each case is unique, and it will depend on the potential importance of the original reaction and products, I'm wondering if you have any general advice for these situations.

Professor Casavecchia answered: In response to the first part of Dr Moses' question, I would like to clarify that I am not suggesting the inclusion of the subsequent kinetic pathways of primary products of a polyatomic multichannel reaction in the photochemical models. I was simply saying that if one elementary reaction has, let's say, five thermodynamically allowed channels, it is not sufficient to know and include in the model the overall rate constant, because this refers to the disappearance of the reagents. In particular, what should not be done is to use the global rate constant and assign it to only one specific reaction channel without knowing its relative importance, because this channel may be negligible, or not be the most important one. For this reason it is important to know for each elementary reaction, besides the overall rate constant, the nature of the primary products and their relative importance (branching ratios). Unfortunately, this is information that standard kinetic studies do not usually provide. Dynamic studies, assisted by theoretical calculations of relevant potential energy surfaces and possibly by also statistical and/or dynamical computations on these surfaces, may be able to provide such information, as I have discussed in my comment, and in this respect they are complementary to kinetic studies.

Coming to the second part of the question, it is worth emphasizing that chemical reactions form products starting from reagents. I believe that in the models you would like to include reactions of which you know both the reactants and the products. This is the point. Certainly the primary products of an elementary reaction are the reactants of subsequent reactions, and these may ultimately form the products that one actually observes in a macroscopic environment. If you know the products, you may also think to consider possible (and likely) secondary reactions, but the story at this point starts over again, because to include secondary reactions means to include new reactions in the model. Regarding the technique that Dr. Yelle mentioned, that of assuming that a particular less-stable product transforms quickly to a more stable isomer, this depends on the type of product; there is no general advice for these situations.

Professor Plane commented: One concern I have with the application of lab experiments and theory to elucidating reaction pathways in Titan's atmosphere is that there need to be observations against which detailed chemical models can be tested in a reasonably unambiguous way. Models containing mechanisms and rates for pathways from N_2 and CH_4 to small unsaturated and saturated hydrocarbons and nitriles, and larger molecules such as benzene and naphthalene fall into this category. But after that there is a very long way to the large negative ions detected in the upper atmosphere.

Dr Yelle replied: I agree completely. It is essential to validate these models in as many ways as is possible. Certainly any theories for the production of the large negative ions and aerosols will remain speculative until the models are carefully validated. I would like to make two additional points however. First, at least for Titan's upper atmosphere, the chemistry is far more complicated than expected. Thus, Cassini did not carry adequate instrumentation to fully characterize the chemistry. The INMS and CAPS results have been extraordinary, but the INMS mass range and the resolution of both CAPS and INMS is not sufficient for a full characterization of the upper atmosphere composition. These instruments have been very successful at discovering things, but it follows that they were not well designed to characterize this undiscovered chemistry. So, given this situation we must look elsewhere to validate the models. One approach, which probably has not been exploited to its full potential, is modeling of Titan simulation experiments carried out in terrestrial laboratories. In this regard, there is a very interesting paper by Imanaka and Smith[1] on the link between ion chemistry and benzene production and another on the first stages of ion chemistry by Thissen *et al.*[2] Second, it is essential to return to Titan with instruments designed to measure the ion and neutral composition out to masses of several hundred amu, with a resolution sufficient to uniquely identify the molecules.

1 H. Imanaka and M. Smith, Role of photoionization in the formation of complex organic molecules in Titan's upper atmosphere, Geophys. Res. Lett., 2006, **34**, L02204.
2 R. Thissen, V. Vuitton, P. Lavvas, J. Lemaire, C. Dehon, O. Dutuit, M. Smith, S. Turchini, D. Catone, R. V. Yelle, P. Pernot, A. Somogyi, and M. Coreno, Laboratory studies of molecular growth in Titan's ionosphere, J. Phys. Chem. A, 2009, **113**, 11211–11220.

Professor Zwier noted: I agree that there is a vast chemistry separating individual reactions leading to naphthalene from the highly processed polymers present in tholins created by discharge of a mixture over many hours. My personal preference is to study reactions from the molecular viewpoint, choosing reactions for detailed study that are anticipated to be important or that are representative of reactions that will occur repeatedly as larger and larger products are formed. For example, if we understand the processes that lead from benzene and its simple derivatives to naphthalene and its simple derivatives, analogous processes are likely to govern formation of larger multi-ring PAHs.

Professor Ashfold noted: Following Professor Mitchell's observation, I have been struck by the parallels (and the differences) in the C/N coupling schemes used in modelling the atmospheres of Jupiter and Titan and those we have used previously in modelling the gas phase chemistry involved in the chemical vapour deposition (CVD) of nitrogen-doped diamond from $N_2/CH_4/H_2$ and $NH_3/CH_4/H_2$ gas mixtures. Photochemistry plays no role in the CVD process, the pressures are orders of magnitude higher, the peak gas temperatures often exceed 2500 K and the necessary H atoms are formed by thermal dissociation of H_2. Nonetheless, we rely on the same H-shifting reactions to interchange between the various C_2H_x ($x \leq 6$) and C_1H_x ($x \leq 4$) species, and end up with C_2H_2 and HCN as stable reaction products. The NH_x ($x \leq 3$) species are also linked by a sequence of H-shifting reactions; one key difference between the two environments is that, under typical CVD conditions, N atoms are the most abundant NH_x radical species and the $N + CH_3$ reaction is the dominant source of HCN.[1]

1 J. A. Smith, J. B. Wills, H. S. Moores, A. J. Orr-Ewing, M. N. R. Ashfold, Y. A. Mankelevich and N. V. Suetin, J. Appl. Phys., 2002, **92**, 672.

Professor Ashfold spoke: As a follow up to Professor Adams' question regarding the proposed electron–ion recombination steps liberating two H atoms in Table 3 of paper 11,[1] I highlight several examples of molecules (e.g. H_2O,[2,3] H_2S,[4] CH_4[5]) where excitation at high (e.g. Lyman-α) photon energies can result in the loss of two H atoms. The mechanism in each case involves prompt fission of one X–H (X = O, S, C) bond. In each of these cases, some of the radical partners are formed with such high levels of internal excitation that they undergo further (slower) unimolecular decay—yielding a second H atom. Such behaviour is not restricted to hydrides. For example, acetone photolysis at 193 nm yields one prompt CH_3 radical.[6,7] The CH_3CO co-fragments are formed with substantial internal excitation which, again, can be sufficient to allow further decay—yielding a second (slower) CH_3 radical. The continued extension of such fragmentation studies to higher photon energies will surely encourage recognition of many more sequential three-body fragmentation processes, and thereby help to inform the interpretation of electron–ion recombination processes (which, at one level, are simply another route to forming highly excited neutral molecules).

1 J. A. Sebree, V. V. Kislov, A. M. Mebel and T. S. Zwier, this volume.
2 D. H. Mordaunt, M. N. R. Ashfold and R. N. Dixon, J. Chem. Phys., 1994, **100**, 7360.
3 S. A. Harich, D. W. H. Hwang, X. F. Yang, J. J. Lin, X. M. Yang and X. M. Yang, J. Chem. Phys., 2000, **113**, 10073.
4 P. A. Cook, S. R. Langford, R. N. Dixon and M. N. R. Ashfold, J. Chem. Phys., 2001, **114**, 1672.
5 D. H. Mordaunt, I. R. Lambert, G. P. Morley, M. N. R. Ashfold, R. N. Dixon, C .M. Western, L. Schnieder and K. H. Welge, J. Chem. Phys., 1993, **98**, 2054.

6 E. L. Woodbridge, T. R. Fletcher and S. R. Leone, *J. Phys. Chem.*, 1988, **92**, 5387 and references therein.
7 S. K. Kim, S. Pedersen and A. H. Zewail, *J. Chem. Phys.*, 1995, **103**, 477.

Dr Adams responded: I agree with the comment that there are similarities between recombination and photoexcitation in producing highly excited neutral species and was interested to hear that sequential H-atom ejection is observed to occur in the latter process.

Professor Mebel commented: The dominant dissociation channel of the $C_{10}H_n^+$ ($n = 8$–12) ions after their recombination with an electron is elimination of a hydrogen atom. This conclusion is supported by our *ab initio* calculations of the potential energy surfaces for the neutral $C_{10}H_n$ species, including naphthalene, hydronaphthyl radicals, dihydronaphthalenes, and so on. The loss of molecular hydrogen, H_2, is never competitive on these potential energy surfaces. Moreover, the recombination of the ion with an electron forms the neutral molecule with so much internal energy that after splitting a first H atom, the molecule still may have enough energy to lose a second atomic hydrogen. In such a case, the products of dissociative recombination of the $C_{10}H_n^+$ ion with an electron will be $C_{10}H_{n-2}$ + H + H.

Professor Schram noted: An interesting area of possible comparison are low pressure highly activated plasmas used for deposition of carbon/nitrogen compounds. Densities are such that three particle association forming molecules is improbable. Still strong conversion is observed (forming e.g. HCN in $N_2/CH_4/H_2$ plasmas), likely at surfaces of surrounding walls and possibly clusters formed in the plasma.

There is strong evidence of cluster and powder formation in hydrocarbon plasmas. This is even so for the outside of a Tokamak plasma: very high electron and low neutral densities. It could be interesting to confront models for Titan chemistry with low neutral density plasmas used for deposition.

Professor Mitchell contributed: There have been discussions concerning the recombination of PAH ions with electrons and the feeling is that this will results in the loss of one or more hydrogen atoms. We have studied PAH ion recombination at the University of Rennes using the FlAPI technique (Novotny *et al.*, *J. Chem. Phys.*, 2005, **123**, 104303) though we have not measured branching ratios in these experiments. In separate experiments however, performed at the ASTRID storage ring, we studied the recombination of $C_3H_3^+$ and $C_4H_5^+$ ions, both of which are believed to have a closed triangular ring structures in their most stable states. We found that the production of C_2 + C products (ring opening?) accounted for 9.3% of the dissociations in the former case and while in the latter, the $C_2 + C_2$ channel accounted for 47% of the dissociations (Mitchell *et al.*, *Int. J. Mass Spectrom.*, 2004, **235**, 7; *ibid.* 2003, **227**, 273). I believe therefore that it is not necessarily the case that the PAH carbon skeleton will stay intact during the electron recombination process.

Dr Loison stated: KIDA (http://kida.obs.u-bordeaux1.fr/) is a database of kinetic data of interest for astrochemical (interstellar medium and planetary atmospheres) studies. In addition to the available reference data, KIDA provides recommendations over a number of important reactions. The purpose of the database is not to make a copy of the NIST database but to propose in the future ONE reliable value for each reaction used in astrochemical (interstellar medium and planetary atmospheres) models. The end of the process, which should involve the majority of the community, will be the writing of a datasheet with recommended rate constants in the same spirit of the datasheet of *J. Phys. Chem. Ref. Data.*

Professor Meuwly and S. Lutz communicated: A recurrent theme at this Faraday Discussion was the calculation of reaction rates and branching ratios for elementary reactions. Many details are elusive experimentally, because "the reaction" itself is a transient process, the transition state is unstable and thus, the most interesting regions along a reaction path can not be investigated experimentally in a direct fashion. To shed light on such questions, theoretical and computational work has become invaluable companions to experimental efforts in understanding particular reaction schemes.

One viable approach are methods based on transition state theory. They are computationally feasible and can be carried out with high-level *ab initio* methods. One disadvantage of such an approach is that typically a pre-defined reaction coordinates is required which sometimes is quite difficult to guess and may influence the outcome of the reaction. Another limitation is that TST assumes no recrossing of the reactive seam which separates products from products.

A more natural approach to treating chemical reactions where quantum effects are less important is to use time as the reaction coordinate. Adiabatic reactive molecular dynamics (ARMD) is one such variant which has been presented recently and has been applied to reactions in the condensed phase.[1-4] The algorithm involves two (or several) potential energy surfaces described with a set of force-field parameters corresponding to reactant (V_R) and product (V_P) states. The dynamics of the system is then initiated and propagated in the reactant state and the energy difference $\Delta E = V_R - V_P$ is monitored at each time step. When ΔE changes sign, a crossing is detected and the two states involved in the reaction are mixed over a finite time window to conserve total energy. In this way the system smoothly approaches the product state.[3] Given the possibility to describe multiple reaction outcomes, ARMD is able to describe branching ratios through a final-state analysis. Furthermore, each individual reactive process can be re-investigated by using rigorous electronic structure methods to ensure that the force field parametrization is meaningful. Finally, as the rates can be directly calculated from the trajectories, entropic and enthalpic effects are automatically included.

As an example for the current state of ARMD for small molecules,[5] a final state analysis for the photodissociation of ClCN → Cl + CN is briefly presented. This reaction has been investigated in the past[6-9] and serves as a test of our implementation in CHARMM. Fig. 8 shows the calculated rotational state distributions of the dissociating CN molecule for photo excitation at different wavelengths ranging from 190.6 to 208.7 nm. Each of these distributions $p(j)$ is built from 5000 individual trajectories. The maximum j_{max} of these quantum-state distributions ranges between $<j> = 47$–57 which is slightly lower than the experimentally observed peak distribution of $<j> = 50$–65.[9]

Classical[7] and semiclassical[9] simulations were able to correctly describe j_{max} but consistently overestimated the population at low j compared with experiment. Comparing $p(j)$ in Fig. 8 between the different excitation wavelengths it is found that j_{max} is shifted towards smaller j for longer wavelengths and that $p(j)$ become narrower. Both observations qualitatively reflect the experimental observations.[9] Narrowing of $p(j)$ can be attributed to a decreased population which is consistent with the experiments.[9] In fact, only 28% of the simulations with excitation at 208.7 nm showed a transition to the excited state within 500 fs of simulation time. This supports the proposition of a second PES to be involved in experiments above 206 nm.[9] Fig. 9 shows the time dependence of $<j>$ after dissociation. The upper graph compares the time course for different wavelengths at 300 K while the lower one compares it for different temperatures. All time series show a nearly linear increase of $<j>$ until it reaches its final value. In an equivalent calculation[10] convergence was obtained within comparable simulation time.

We wanted to bring such methods to the attention of this *Faraday Discussion* because we believe that classical MD approaches with provisions to describe reactive processes, such as ARMD, may assist in better understanding chemical reactions at a molecular level including enthalpic and entropic effects and with the possibility to also assess branching ratios.

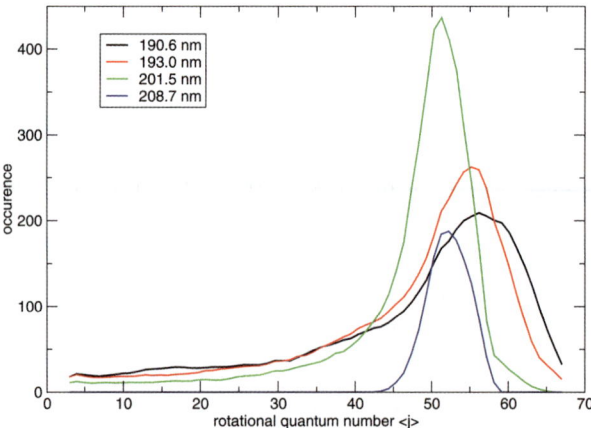

Fig. 8 Rotational state distributions calculated at 300 K for different excitation wavelengths ranging from 190.6 to 208.7 nm.

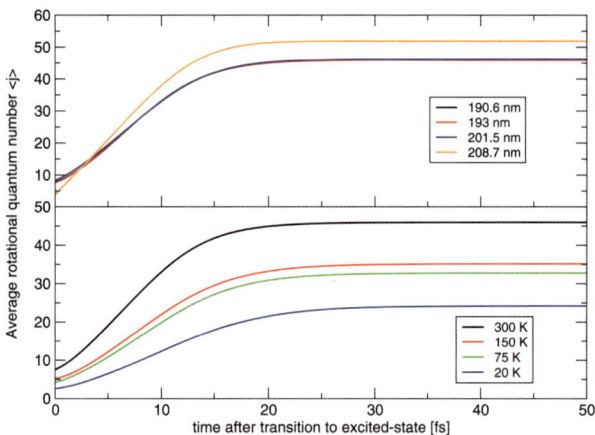

Fig. 9 Time series of $<j>$ at 300 K for different wavelengths (top graph) and for different temperatures (lower graph) at 193 nm.

1 M. Meuwly, O. M. Becker, R, Stone and M. Karplus, NO rebinding to myoglobin: A reactive molecular dynamics study, *Biophys. Chem.*, 2002, **98**, 183–207.
2 D. Nutt and M. Meuwly, Studying reactive processes with classical dynamics: Rebinding dynamics in MbNO, *Biophys. J.*, 2006, **90**, 1191–1201.
3 J. Danielsson and M. Meuwly, Atomistic simulation of adiabatic reactive processes based on multi-state potential energy surfaces, *J. Chem. Theory Comput.*, 2008, **4**, 1083–1093.
4 S. Mishra and M. Meuwly, A study of nitric oxide detoxification in truncated hemoglobin with reactive molecular dynamics, *J. Am. Chem. Soc.*, 2010, **132**, 2968.
5 S. Lutz and M. Meuwly, in preparation.
6 J. Halpern and W. M. Jackson, Partitioning of excess energy in the photolysis of cyanogen chloride and cyanogen bromide at 193 nm, *J. Phys. Chem.*, 1982, **86**, 3528–3533.
7 B. A. Waite and N. I. Dunlap, The photodissociation of ClCN: A theoretical determination of the rotational state distribution of the CN fragment, *J. Chem. Phys.*, 1986, **84**, 1391–1396.
8 R. Schinke and V. Engel, The rotational reflection principle in photodissociation dynamics, *Faraday Discuss.*, 1986, **82**, 111.
9 S. A. Barts and J. B. Halpern, Photodissociation of ClCN between 190 and 213 nm, *J. Phys. Chem.*, 1989, **93**, 7346–7351.
10 R. Schinke, *Photodissociation Dynamics: Spectroscopy and Fragmentation of Small Polyatomic Molecules*, 1995, Cambridge University Press, UK.

H_3^+ cooling in planetary atmospheres

Steve Miller,[*a] Tom Stallard,[b] Henrik Melin[b] and Jonathan Tennyson[a]

Received 10th March 2010, Accepted 8th April 2010
DOI: 10.1039/c004152c

We review the role of H_3^+ in planetary atmospheres, with a particular emphasis on its effect in cooling and stabilising, an effect that has been termed the "H_3^+ thermostat" (see Miller *et al.*, *Philos. Trans. R. Soc. London, Ser. A*, 2000, **58**, 2485). In the course of our analysis of this effect, we found that cooling functions that make use of the partition function, $Q(T)$ based on the calculated H_3^+ energy levels of Neale and Tennyson (*Astrophys. J.*, 1995, **454**, L169) may underestimate just how much energy this ion is radiating to space. So we present a new fit to the calculated values of $Q(T)$ that is accurate to within 2% for the range 100 K to 10 000 K, a very significant improvement on the fit originally provided by Neale and Tennyson themselves. We also present a fit to $Q(T)$ calculated from only those values Neale and Tennyson computed from first principles, which may be more appropriate for planetary scientists wishing to calculate the amount of atmospheric cooling from the H_3^+ ion.

1 Introduction

Infrared emission from H_3^+ has been detected in Jupiter,[1,2] Saturn[3] and Uranus,[4] but not—so far—in Neptune. For Jupiter, most of the emission comes from the auroral/polar regions, although there is a planet-wide glow: at mid-to-low latitudes, this cannot be explained by EUV ionisation alone,[3] but the exact cause(s) and their relative importance are not fully understood. Total H_3^+ emission from Jupiter is $\sim 10^{13}$ W.[4] Saturn's emission is a few percent of that of Jupiter, and is—again—concentrated around the poles as auroral activity.[5] For Uranus the situation is rather different: auroral emission is probably not more than 20% of the total, planetwide, and there is a glow that covers the entire disk. Again, uranian emission is a few percent that of Jupiter.[6] Taken together, however, these observations demonstrate that H_3^+ is an important constituent of giant planet atmospheres and ionospheres. An outstanding problem for the Solar System's giant planets is how to account for their high exospheric temperatures, all of which are hundreds of degrees hotter than can be accounted for by the absorption of sunlight alone.[7] Amongst the leading candidates to explain these temperatures are gravity wave heating from the lower atmosphere[8] and the distribution of energy from the auroral/polar regions.[9] Neither explanation is without serious problems, however.[10,11]

The upper atmosphere of a planet such as Jupiter is an important interface between its space environment and the denser atmosphere below. In the Solar System, all planets are irradiated by the Sun and impacted upon by the Solar Wind, a stream of rarified plasma travelling at several hundred kilometres per

---

[a]*Department of Physics and Astronomy, University College London, Gower Street, London, WC1E 6BT, U.K. E-mail: s.miller@ucl.ac.uk; j.tennyson@ucl.ac.uk; Fax: +44 20 7679 7155; Tel: +44 20 7679 2000*
[b]*Department of Physics and Astronomy, University of Leicester, University Road, Leicester, LE1 7RH, U.K. E-mail: tss@ion.le.ac.uk; hm@ion.le.ac.uk; Fax: +44 0116 252 2770; Tel: +44 0116 252 3575*

second through interplanetary space with particle densities that range from a few million per cubic metre, at Earth, to ~100 times less than that at Saturn. For magnetised planets, the interaction with the Solar Wind is mediated by a *magnetosphere*, a "teardrop-shaped" region of space dominated by the planetary magnetic field. For Earth, this magnetosphere stretches about 60 000 km in the direction of the Sun, and for about ten times this distance in the anti-Sun direction. For Jupiter, the corresponding distances are several million km and several hundred million. The precise values for both planets depend on the prevailing Solar Wind conditions, as densities and velocities—hence the pressure exerted on the planet's magnetic field—can vary considerably. For Earth, most of the plasma that fills the magnetosphere itself comes from the Solar Wind. But for Jupiter and, to a lesser extent, Saturn, internal sources of plasma are also important: Io's volcanoes for Jupiter; the rings and the emissions of Enceladus for Saturn.

2 H_3^+ in giant planet atmospheres

Since its first detection in the auroral regions of Jupiter,[1,2] the H_3^+ molecular ion has been shown to play an important role in the upper atmospheres of giant planets.[4,12] These upper atmospheres are hydrogen dominated, with molecular hydrogen, H_2, predominating at lower levels, giving way to H atoms at the top. They are not convectively mixed, so individual species settle out with their own scale height, which is inversely proportional to their atomic or molecular weight. Giant planet atmospheres have ~10% of He atoms, which are concentrated toward the bottom since they are twice as heavy as H_2. In the lower reaches of the upper atmosphere there may also be small amounts of other heavier species, such as hydrocarbons (especially methane, ethane and acetylene). Typically, pressures range from a microbar or so at the lower levels of the upper atmosphere, to picobars at the top. The corresponding temperature range can start from as low as 120 K, and rise to around 1500 K at the top. The pressure range between a few microbars and a few tenths of a nanobar, a region in which the temperature of the atmosphere rises monotonically with altitude, is known as the *thermosphere*. Above that is the *exosphere*, a region that blends into interplanetary space as atmospheric species escape the planet's gravitational field. The exosphere is characterised by a (more or less) uniform temperature throughout.

In the upper atmosphere, solar extreme ultraviolet (EUV) radiation can cause ionisation of the atmospheric constituents. In the regions around the poles, various mechanisms can cause particles, such as electrons with energies of several kilovolts, to be accelerated along the magnetic field lines. Given enough energy in the direction towards the planet, these can precipitate into the atmosphere, creating further ionisation. Moreover, high-energy particles can often penetrate deeper into the atmosphere than EUV photons. The locations at which such precipitation occurs are often marked by bright auroral emissions: for Jupiter, these were first detected by Voyager spacecraft[13] and the International Ultraviolet Explorer satellite[14] in the EUV; the emissions were from H Lyman-α at 121.6 nanometres, and the Lyman and Werner bands of H_2. In hydrogen-rich atmospheres, H_3^+ may be formed by a series of reactions:

$H_2 + h\nu/e^- \rightarrow H_2^+ + e^- [+ e^-]$—photo/electron impact ionisation
$H_2^+ + H_2 \rightarrow H_3^+ + H$—proton hopping
$H + h\nu/e^- \rightarrow H^+ + e^- [+ e^-]$—photo/electron impact ionisation
$H^+ + H_2(v \geq 4) \rightarrow H + H_2^+$—charge exchange
$H_2^+ + H_2 \rightarrow H_3^+ + H$—proton hopping

Note that in the charge exchange, the difference in ionisation potential between H (13.6 eV) and H_2 (15.4 eV) is made up by having the H_2 molecule excited to the fourth vibrational energy level or above. These reaction sequences show that H_3^+ is readily formed when hydrogen-rich atmospheres are ionised, so long as *molecular* hydrogen is also sufficiently abundant. The H_3^+ infrared emission first detected for

Jupiter was auroral,[2] although it was being formed higher in the thermosphere than the corresponding EUV emission.[15] The overall impact of EUV radiation and particle precipitation is to produce a weakly ionised plasma that coexists with the thermosphere: this is known as the ionosphere, and there is strong coupling between the two.[16]

The exact composition of the ionosphere depends on a large number of factors, including the composition of the thermosphere and the depth to which photons and particles can penetrate, and the local time. Clearly, solar irradiation is only a factor during the hours of dawn until dusk, and during the nighttime photoionised ions will be expected to be lost to recombination with electrons. For a hydrogen rich atmosphere, H_3^+ recombines dissociatively, at a rate of some 10^{-8} to 10^{-7} cm^3 s^{-1},[17] which is much faster than radiative recombination, the mechanism that neutralises H^+. During the course of the night, therefore, the ionosphere produced by photoionisation will become less molecular and more atomic in character. At higher latitudes, however, where precipitation is the main ionising agent, much less change might be expected, at least at lower altitudes, where high energy particles tend to expend most of their energy.

As an ion, H_3^+ is a charge carrier and it is affected by magnetic fields that do not directly affect the neutral thermosphere. Millward et al.[18] showed that H_3^+ is the main charge carrier and provider of conductivity for the jovian auroral atmosphere: depending on the exact flux and individual energy of the electrons precipitating into the atmosphere, the height integrated Pedersen conductivity, Σ_P, varied from a few tenths of a mho to nearly 8 mho. This means H_3^+ plays the major role in generating upper atmosphere Joule heating.[19] The same team also showed that H_3^+ ion winds,[20] generated through the coupling of electric and magnetic fields, would create heating through ion drag.[21] At the same time, these ion winds also generated *neutral* winds that inhibited the transfer of energy from the auroral/polar regions to lower latitudes.[11] In auroral "events" the energy generated can be considerable,[22] however, and it may be that energy is transferred to lower latitudes under non-equilibrium conditions. Recent observations of Saturn's auroral activity by Cassini show major changes on timescales of minutes to hours,[23] faster than the atmospheric response time, begging the question as to whether such planetary upper atmospheres are ever in "equilibrium".

Melin et al.'s[22] study of a jovian auroral "event" in 1998 showed that the Joule heating and ion drag inputs to the overall energy budget increased by a factor of four over two days, leading to an excess energy in the atmospheric column of 175 mW m^{-2}. Integrated across the auroral/polar region, this amounts to an excess energy of ~10 TW, roughly ten times the total EUV absorbed by the atmosphere planetwide. Melin's study also showed that the heating due to increased precipitation associated with the event was balanced by an increase in the amount of infrared radiation emitted by H_3^+, which moderated the local temperature increase. This thermostat effect demonstrates the importance of H_3^+ cooling in planetary atmospheres dominated by hydrogen.

3 H_3^+ cooling

Atomic hydrogen has only (relatively) energetic electronic transitions and therefore cannot cool planetary atmospheres where the temperature is much below 8000 K;[24] and even then is rather inefficient. Thus it plays virtually no part in the thermal stability of planetary atmospheres in the Solar System. Molecular hydrogen is a homo-nuclear diatomic molecule, and its vibrational mode results in no change of dipole, so it is infrared inactive. H_2 quadrupole transitions are known, but are weak, with Einstein A_{if} values between 10^{-5} and 10^{-7} s^{-1}.[25] H_3^+ is a stable though reactive molecule. It has two vibrational modes: a symmetric "breathing" stretch, ν_1 centred on 3.1463 μm; a doubly degenerate asymmetric stretch-bend, ν_2 centred on 3.9662 μm. Of these, only ν_2 is infrared allowed. The $\nu_2 = 1 \to 0$ ν_2 fundamental has an A_{if} coefficient of 129 s^{-1}.[26] The large molecular anharmonicity associated with this "floppy" molecule, however, means that transitions that are classically infrared inactive are also associated with large A_{if} values.[27] Thus the $v_2 = 2 \rightarrow 0\ 2v_2(2)$ overtone has $A_{if} = 145$ s^{-1}, making it $\sim 10^9$ times more intense than the H$_2$ ($v = 1 \rightarrow 0$) quadrupole transitions that are found in the same spectral region, around 2 μm. Even difference bands, such as the ($v_1 = 1 \rightarrow 0\ v_2 = 0 \rightarrow 1$) ν_1–ν_2 band centred on 15.219 μm, have A_{if} values of ~ 1 s^{-1}.[26] This means that H$_3^+$ is capable of cooling throughout a wide temperature range from as low as 200 K up to several thousands of degrees, after which thermal dissociation becomes significant.

The H$_3^+$ thermostat effect[28] noted above may not help with explaining the *higher* than expected thermospheric temperatures of the Solar System giant planets, but it has recently been shown to be extremely important in controlling the atmospheric stability of giant exoplanets. Using a three-dimensional global circulation model (3D GCM) developed for Jupiter and Saturn, Koskinen *et al.*[29] showed that it was possible to bring a Jupiter-like planet in to some 0.16 Astronomical Units (AU; 1 AU = 149 598 000 km) from the Sun, and—whilst the exospheric temperature increased from $\sim$1300 K to $\sim$3800 K—the planet remained Jupiter-like, insofar as the atmosphere was stable and its extent was $\sim$6000 km, less than 10% of the overall planetary radius. At this distance, H$_3^+$ cooling was able to offset the increased heating due to EUV insolation. Bringing the planet just 3 million km closer to the Sun, however, to 0.14 AU results in the atmosphere becoming unstable, with temperatures at the top exceeding 20 000 K, inflating to more than 75 000 km (greater than a normal jovian radius), and escaping hydrodynamically. This process can be summarised as follows:

1) As a Jupiter-like planet orbits closer to a Sun-like star, the increased EUV flux has a tendency to heat the upper atmosphere;

2) But increased EUV also produces increased H$_3^+$ densities by photoionisation;

3) The increased H$_3^+$ cools the atmosphere, maintaining stability ...

4) ...until the planet is so close that increasing EUV heating overwhelms the H$_3^+$ thermostat, whereupon...

5) ...the temperature increases sufficiently to dissociate H$_2$, and the "feedstock" for forming H$_3^+$ is removed;

6) So a vicious spiral of increasing temperature, dissociating H$_2$, and non-formation of H$_3^+$ occurs, leading to a rapidly heating and rapidly inflating atmosphere, with hydrodynamic escape.

The scenario outlined above explains why some of the closely orbiting exo-giants, such as HD189733B and HD209458B, have been found with very extended atmospheres[30,31] that show signs of escaping the planetary gravitational field, although escape rates are not considered large enough for the planet to evaporate entirely over the lifetime of a stellar system[32] (some 9 billion years, in the case of the Solar System).

The amount of H$_3^+$ cooling will depend on the number of molecules and the average cooling per molecule. This latter may be calculated by computing the energy radiated in all the possible transitions between populated upper and lower states, weighted by some factor to take into account the fraction of molecules that are in the upper state at any time. Under conditions of Local Thermodynamical Equilibrium (LTE), the energy emitted in a single line is given by:

$$I(\omega_{if}, T) = g_f (2J_f + 1) \times hc\omega_{if} \times A_{if} \exp[-hcE_f/kT]/4\pi Q(T) \quad (1)$$

where g_f is the nuclear spin degeneracy, J_f the angular momentum, and E_f the energy, of the upper state. ω_{if} and A_{if} are the energy and Einstein A coefficient of the transition, and the factor hc comes in if we wish to convert energies in wavenumbers to SI units. $Q(T)$ is the partition function. The total cooling in this instance is then:

$$E(H_3^+, T) = \Sigma_{if}\ I(\omega_{if}, T) \quad (2)$$

Neale et al.[33] have produced an extensive and highly accurate linelist for the H_3^+ molecular ion, and Melin and coworkers[34,35] have fitted the values of $E(H_3^+,T)$ that result from Neale et al.'s[33] results for temperatures between 400 K and 3000 K; this temperature range covers limits below which H_3^+ cooling is fairly insignificant ($<10^{-21}$ W sterradian^{-1} molecule^{-1}) or above which H_3^+ formation is inhibited by the thermal dissociation of H_2. The fit is given by:

$$E(H_3^+,T) = -6.11904 \times 10^{-21} + 4.96694 \times 10^{-23} T - 1.443608 \times 10^{-25} T^2 + 1.60926 \times 10^{-28} T^3 - 3.87932 \times 10^{-32} T^4 \quad (3a)$$

for the range $T = 400$–900 K, and

$$E(H_3^+,T) = -8{,}24045 \times 10^{-21} + 3.54583 \times 10^{-23} T - 8.66296 \times 10^{-26} T^2 + 9.76608 \times 10^{-29} T^3 - 1.61317 \times 10^{-32} T^4 \quad (3b)$$

for the range $T = 900$–3000 K.

The values of $E(H_3^+,T)$ thus produced are fully valid for LTE conditions. To a first approximation, they may be weighted by a non-LTE factor based on the population of vibrational levels:[35] for the gas (exo-)giants, this factor has been calculated by a detailed balance calculation making use of the method of Oka and Epp[36] and the vibration-only energy levels and A_{if} values of Dinelli et al.[26]

4 New fits of the partition function

Eqn (1) shows that the partition function $Q(T)$ is an important input to the calculation of H_3^+ cooling: $E(H_3^+,T)$ depends inversely on $Q(T)$. Neale and Tennyson[37] calculated a partition function which was aimed particularly at studies at elevated temperatures and, in particular, for work on cool white dwarfs.[38] They published the values of $Q(T)$ derived from their calculated ro-vibrational levels given by:

$$Q(T) = \Sigma_n \, g_n \, (2J_n + 1) \, \exp[-hcE_n/kT] \quad (4)$$

Neale and Tennyson (NT)[37] calculated states for angular momentum values of $J = 0$–20, with energies up to 15 000 cm^{-1} above the ground state from first principles. For higher energy states, they augmented their list by making use of an effective Hamiltonian. This increased the total number of energy levels that went into the summation by a factor of nearly 35, although this made only 0.4% difference to the value of $Q(T)$ at 2000 K and just over 6% at 3000 K. Although more reliable theoretical models for the energy levels of H_3^+ are now available,[39] the biggest problem with the NT partition function is actually due to the fit to Q(T).

Since computing $Q(T)$ over and over again for use in a variety of circumstances is time-consuming, these workers then fitted $\log_{10}(Q(T))$. This fit was used in the subsequent fits to $E(H_3^+,T)$ by Melin and coworkers.[36] In Table 1, however, we show that the Neale and Tennyson (Q(NT)) functional form of $Q(T)$ is a poor fit to their own values calculated from eqn (4) (Q(all levels)) at the lower temperatures important for planetary studies.

From Table 1 it is clear that the NT fit gets to within 10% of the actual values of $Q(T)$ only for temperatures above ~800 K, and to within 5% for temperatures above ~1400 K. This means that Q(NT) significantly overestimates $Q(T)$ for much of the range of interest in planetary atmospheres. As a result, the cooling will be proportionately underestimated. To deal with this we have therefore refitted the values of $Q(T)$ obtained from eqn (3), paying particular attention to the range from 400 K up to 3000 K. Our fit does not provide a universal set of parameters, but instead is divided into three regions: 100–1800 K; 1800–5000 K; 5000–10 000 K. Moreover, whereas Neale and Tennyson[37] fitted to $\log_{10}[Q(T)]$, our fit is to $Q(T)$ itself, approximated by:

Table 1 Partition function for H_3^+: all levels—first principles + effective Hamiltonian

T/K	Q(all levels)	Q(NT)	Q(NT)/Q(all levels)	Q(MSMT)	Q(MSMT)/Q(all levels)
100	7.360	18.516	2.516	7.465	1.014
120	9.756	20.402	2.091	9.778	1.002
150	13.599	24.516	1.803	13.573	0.998
200	20.726	33.054	1.595	20.704	0.999
300	37.608	52.096	1.386	37.630	1.001
500	80.579	98.362	1.221	80.580	1.000
1000	245.762	264.504	1.076	245.781	1.000
1400	473.751	497.496	1.050	473.763	1.000
2000	1106.162	1151.482	1.041	1102.279	0.996
2400	1833.777	1911.053	1.042	1825.934	0.996
3000	3654.411	3824.257	1.046	3717.796	1.017
4000	10005.957	10450.312	1.044	10141.354	1.013
5000	23731.476	24389.597	1.028	23964.559	1.010
7000	91465.603	91740.611	1.003	91989.372	1.006
10 000	337371.359	341484.481	1.012	337087.391	0.999

$$Q(T) = A_0 + A_1 T + A_2 T^2 + A_3 T^3 + A_4 T^4 + A_5 T^5 + A_6 T^6 \quad (5)$$

The values obtained for the As are:
$T = 100$–1800 K
$A_0 = -1.11391$
$A_1 = +0.0581076$
$A_2 = +0.000302967$
$A_3 = -2.83724 \times 10^{-7}$
$A_4 = +2.31119 \times 10^{-10}$
$A_5 = -7.15895 \times 10^{-14}$
$A_6 = +1.00150 \times 10^{-17}$
$T = 1800$–5000 K
$A_0 = -22125.5$
$A_1 = +51.1539$
$A_2 = -0.0472256$
$A_3 = +2.26131 \times 10^{-5}$
$A_4 = -5.85307 \times 10^{-9}$
$A_5 = +7.90879 \times 10^{-13}$
$A_6 = -4.28349 \times 10^{-17}$
$T = 5000$–$10\,000$ K
$A_0 = -654293.0$
$A_1 = +617.630$
$A_2 = -0.237058$
$A_3 = +4.74466 \times 10^{-5}$
$A_4 = -5.20566 \times 10^{-9}$
$A_5 = +3.05824 \times 10^{-13}$
$A_6 = -7.45152 \times 10^{-18}$

Our results are also shown in Table 1 as Q(MSMT). Comparison with the values of Q(T) calculated from the full level summation (eqn (3)) shows that our three-region fit matches Q(T) to within 2% at all temperatures. This is a significant improvement over the original Neale and Tennyson fit, particularly in the 100–3000 K region of importance in planetary atmospheres; apart from a small temperature domain around 7000 K, our new fit is better at all temperatures from 100 K to 10 000 K.

5 Corrections to the H_3^+ cooling function, $E(H_3^+)$

The line list calculated by Neale et al.,[33] of necessity, includes only transitions between the levels that were calculated from "first principles": while energy levels *may* be extended using effective Hamiltonians, it is not so straightforward to apply this to line strengths for which accidental degeneracies between levels, leading to level mixing, can cause otherwise infrared inactive transitions to "light up". Moreover, the number of individual levels increases by a factor of 35 when states calculated from the effective Hamiltonian are included, leading to a potential increase in transitions by a factor of 1225. But this raises the question as to whether a partition function based on eqn (3) should be limited to just a sum over the first principles levels for use in calculating $E(H_3^+)$; the logic is that if the sum over transitions is limited to a sum over first principles transitions, the population weighting in eqn (1), $(g_f(2J_f + 1) \exp[-hcE_f/kT]/Q(T))$, which feeds into the sum in eqn (2), should take this into account, at least as far as $E(H_3^+)$ is a parameter to be fed into model atmospheres.

Table 1I shows the value of $Q(T)$ calculated from eqn (3) using just the levels calculated from first principles. Comparison with the Neale and Tennyson fit to $\log_{10}Q(T)$ shows clearly how this underestimates the full partition function above 3000 K, even allowing for the fact that the values of $Q(NT)$ overestimate the value of $Q(\text{all levels})$. We have also fitted the values of $Q(FP$ levels) to the same form as eqn (5). The resulting A parameters are:

$T = 100-1800$ K
$A_0 = -1.11391$
$A_1 = +0.0581076$
$A_2 = +0.000302967$
$A_3 = -2.83724 \times 10^{-7}$
$A_4 = +2.31119 \times 10^{-10}$
$A_5 = -7.15895 \times 10^{-14}$
$A_6 = +1.00150 \times 10^{-17}$
$T = 1800-5000$ K
$A_0 = -378.621$
$A_1 = +0.839719$
$A_2 = -0.000349567$

Table 2 Partition function for H_3^+: first principle levels only

T	Q(FP levels)	Q(NT)	Q(NT)/Q(FP levels)	Q_{FP}(MSMT)	Q_{FP}(MSMT)/Q(FP levels)
100	7.360	18.516	2.516	7.465	1.014
120	9.756	20.402	2.091	9.778	1.002
150	13.599	24.516	1.803	13.573	0.998
200	20.726	33.054	1.595	20.704	0.999
300	37.608	52.096	1.386	37.630	1.001
500	80.579	98.362	1.221	80.580	1.000
1000	245.762	264.504	1.076	245.781	1.000
1400	473.751	497.496	1.050	473.763	1.000
2000	1102.988	1151.482	1.044	1102.760	1.000
2400	1808.507	1911.053	1.042	1808.319	1.000
3000	3438.277	3824.257	1.057	3438.590	1.000
4000	7871.175	10450.312	1.044	7871.126	1.000
5000	14259.796	24389.597	1.112	14260.253	1.000
7000	30660.898	91740.611	2.992	30661.035	1.000
10 000	57504.066	341484.481	5.938	57503.198	1.000

$A_3 = +5.17514 \times 10^{-8}$
$A_4 = +7.79447 \times 10^{-11}$
$A_5 = -1.63248 \times 10^{-14}$
$A_6 = +9.60597 \times 10^{-19}$
$T = 5000–10\,000$ K
$A_0 = +6200.41$
$A_1 = -4.55558$
$A_2 = +0.000805172$
$A_3 = +2.53004 \times 10^{-7}$
$A_4 = -4.69402 \times 10^{-11}$
$A_5 = +3.06177 \times 10^{-15}$
$A_6 = -7.34376 \times 10^{-20}$

Table 2 shows that this fit is 0.2% or better for the range from 120 K to 10 000 K. Since Melin's fit[34] for the $E(H_3^+)$ values is based on dividing by $Q(NT)$, it may be that it underestimates the actual H_3^+ cooling by as much as 30% at 400 K, falling to a few percent at higher temperatures. This will tend, for instance, to alter atmospheric temperature profiles, with lower temperatures where H_3^+ cooling is important than originally anticipated, and to move the exoplanet stability limit closer to the Sun, albeit by a few percent of the 21–24 million km proposed by Koskinen et al.[29]

6 Conclusions: The Europlanet modelling and data analysis facility

Since it has widespread applicability for planetary atmospheres and other astrophysical applications, a suite of programmes to calculate H_3^+ cooling is now being grid-enabled as part of the European Modelling and Data Analysis Facility (EMDAF) under construction by Europlanet RI, the European Union funded network of planetary scientists (see http://www.europlanet-ri.eu/research/jra3). This programme suite will allow for non-LTE effects, and have the facility to take model atmospheres as inputs, as well as enabling grids of datapoints to be calculated. In this way, those computing the properties of hydrogen-rich environments may specify their own requirements to be provided "off the shelf" for incorporation in models as required. Given the discussion in Section 3, we can consider the impact of H_3^+ cooling on the formation of our own Solar System. Although Jupiter is currently well outside of the Sun's "stability limit", as far as the jovian atmosphere is concerned, solar EUV fluxes were much higher at earlier epochs. The "Sun-in-Time" project, which has looked at current exemplars of how our early Sun would have behaved, has produced a graph that shows the EUV flux of the Sun can be represented by:

$$\log_{10} [F_{EUV}(t)/F_{EUV}(4.58)] \sim 1.23 \times \log_{10} [4.58/t] \qquad ((6))$$

where t is the age of the Solar System in gigayears.[40] This would put Jupiter within the "stability" limit when the Sun was 15 million years old or less, during a critical period of atmospheric accretion. To date, the role that H_3^+ cooling may have played in stabilising Jupiter around the 15 Myr period has not been modelled.

Finally we note that Sochi and Tennyson[41] have recently calculated a comprehensive linelist, partition function and cooling function for deuterated H_3^+, H_2D^+. Cooling by H_2D^+ could play an important role at lower temperatures where cooling by H_3^+ becomes very inefficient: this may have an important bearing in the physics of stellar discs and planet formation,[42] and on star formation in the early universe.[43]

References

1 L. M. Trafton, J. Carr, D. Lester and P. Harvey, in *Time Variable Phenomena in the Jovian System*, ed. M. J. S. Belton, R. A. West and J. Rahe, 1987, p. 229.
2 P. Drossart, et al., *Nature*, 1989, **340**, 539.

3 D. Rego, S. Miller, N. Achilleos, T. S. Stallard, R. Prangé, M. Dougherty and R. D. Joseph, *Icarus*, 2000, **147**, 366.
4 S. Miller, *et al.*, *Philos. Trans. R. Soc. London, Ser. A*, 2000, **358**, 2485.
5 T. R. Geballe, M. F. Jagod and T. Oka, *Astrophys. J.*, 1993, **408**, L109.
6 L. M. Trafton, T. R. Geballe, S. Miller, J. Tennyson and G. E. Ballester, *Astrophys. J.*, 1993, **405**, 761.
7 R. V. Yelle and S. Miller, in *Jupiter: the Planet, Satellites and Magnetosphere*, ed. F. Bagenal, T. Dowling and W. McKinnon, 2004, p. 185.
8 L. A. Young, R. V. Yelle, R. E. Young, A. Sieff and D. B. Kirk, *Science*, 1997, **276**, 108.
9 J. H. Waite Jr., T. Cravens, J. Kozyra, A. F. Nagy and S. K. Atreya, *J. Geophys. Res.*, 1983, **88**, 6143.
10 K. I. Matcheva and D. F. Strobel, *Icarus*, 1999, **140**, 328.
11 C. G. A. Smith, A. D. Aylward, G. H. Millward, S. Miller and L. E. Moore, *Nature*, 2007, **445**, 399.
12 S. Miller, T. Stallard, C. Smith, G. Millward, H. Melin, M. Lystrup and A. Aylward, *Philos. Trans. R. Soc. London, Ser. A*, 2006, **A364**, 3121.
13 A. L. Broadfoot, *et al.*, *Science*, 1979, **204**, 979.
14 J. T. Clarke, H. W. Moos, S. K. Atreya and A. L. Lane, *Astrophys. J.*, 1980, **241**, L179.
15 J. T. Clarke *et al.*, in *Jupiter: the Planet, Satellites and Magnetosphere*, ed. F. Bagenal, T. Dowling and W. McKinnon, 2004, p. 639.
16 G. Millward, S. Miller, T. Stallard, A. Aylward and N. Achilleos, *Icarus*, 2005, **173**, 200.
17 M. Larsson, *Philos. Trans. R. Soc. London, Ser. A*, 2000, **A358**, 2433.
18 G. Millward, S. Miller, T. Stallard, A. Aylward and N. Achilleos, *Icarus*, 2002, **160**, 95.
19 S. Miller, A. Aylward and G. Millward, in *The Outer Planets and their Moons*, ed. T. Encrenaz, R. Kallenbach, T. C. Owen and C. Sotin, 2005, p. 319.
20 D. Rego, N. Achilleos, T. Stallard, S. Miller, R. Prange, M. Dougherty and R. D. Joseph, *Nature*, 1999, **399**, 121.
21 C. G. A. Smith, S. Miller and A. D. Aylward, *Ann. Geophys.*, 2005, **23**, 1379.
22 H. Melin, S. Miller, T. Stallard, C. Smith and D. Grodent, *Icarus*, 2006, **181**, 256.
23 T. Stallard, *et al.*, *Nature*, 2008, **456**, 214.
24 S. C. O. Glover and D. W. Slavin, *Mon. Not. R. Astron. Soc.*, 2009, **393**, 911.
25 J. Turner, K. Kirby-Docken and A. Dalgarno, *Astrophys. J. Suppl.*, 1977, **35**, 281.
26 B. M. Dinelli, S. Miller and J. Tennyson, *J. Mol. Spectrosc.*, 1992, **153**, 718; B. M. Dinelli, S. Miller and J. Tennyson, *J. Mol. Spectrosc.*, 1992, **156**, 243, erratum.
27 S. Miller, J. Tennyson and B. T. Sutcliffe, *J. Mol. Spectrosc.*, 1990, **141**, 104.
28 S. Miller, H. A. Lam and J. Tennyson, *Can. J. Phys.*, 1994, **72**, 760.
29 T. T. Koskinen, A. D. Aylward and S. Miller, *Nature*, 2007, **450**, 845.
30 A. Vidal Madjar, *et al.*, *Nature*, 2003, **422**, 143.
31 A. Vidal Madjar, *et al.*, *Astrophys. J.*, 2004, **604**, L69.
32 R. V. Yelle, *Icarus*, 2004, **170**, 167.
33 L. Neale, S. Miller and J. Tennyson, *Astrophys. J.*, 1996, **464**, 516.
34 H. Melin, PhD thesis, 2006, University of London.
35 H. Melin, S. Miller, T. Stallard and D. Grodent, *Icarus*, 2005, **178**, 97.
36 T. Oka and E. Epp, *Astrophys. J.*, 2004, **613**, 349.
37 L. Neale and J. Tennyson, *Astrophys. J.*, 1995, **454**, L169.
38 P. Bergeron, M. T. Ruiz and S. K. Leggett, *Astrophys. J. Suppl.*, 1997, **108**, 339.
39 O. L. Polyansky and J. Tennyson, *J. Chem. Phys.*, 1999, **110**, 5056.
40 I. Ribas, E. F. Guinan, M. Gudel and M. Audard, *Astrophys. J.*, 2005, **622**, 680.
41 T. Sochi and J. Tennyson, *Mon. Not. R. Astron. Soc.*, DOI: 10.1111/j.1365-2966.2010.16665.x.
42 C. Ceccarelli and C. Dominik, *Philos. Trans. R. Soc. London, Ser. A*, 2006, **364**, 3091.
43 D. R. Flower and G. Pineau des Forets, *Mon. Not. R. Astron. Soc.*, 2001, **323**, 672.

PAPER

Negative ions at Titan and Enceladus: recent results

Andrew J. Coates,[*ab] Anne Wellbrock,[ab] Gethyn R. Lewis,[ab] Geraint H. Jones,[ab] David T. Young,[c] Frank J. Crary,[c] J. Hunter Waite,[c] Robert E. Johnson,[d] Thomas W. Hill[e] and Edward C. Sittler Jr.[f]

Received 30th March 2010, Accepted 14th April 2010
DOI: 10.1039/c004700g

The detection of heavy negative ions (up to 13 800 amu) in Titan's ionosphere is one of the tantalizing new results from the Cassini mission. These heavy ions indicate for the first time the existence of heavy hydrocarbon and nitrile molecules in this primitive Earth-like atmosphere. These ions were suggested to be precursors of aerosols in Titan's atmosphere and may precipitate to the surface as tholins. We present the evidence for and the analysis of these heavy negative ions at Titan. In addition we examine the variation of the maximum mass of the Titan negative ions with altitude and latitude for the relevant encounters so far, and we discuss the implications for the negative ion formation process. We present data from a recent set of encounters where the latitude was varied between encounters, with other parameters fixed. Models are beginning to explain the low mass negative ions, but the formation process for the higher mass ions is still not understood. It is possible that the structures may be chains, rings or even fullerenes. Negative ions, mainly water clusters in this case, were seen during Cassini's recent close flybys of Enceladus. We present mass spectra from the Enceladus plume, showing water clusters and additional species. As at Titan, the negative ions indicate chemical complexities which were unknown before the Cassini encounters, and are indicative of a complex balance between neutrals and positively and negatively charged ions.

1. Introduction

The detection of heavy negative ions (up to 13 800 amu q^{-1}) in Titan's ionosphere is one of the tantalizing new results from the Cassini mission.[1–3] These heavy ions indicate for the first time the existence of heavy hydrocarbon and nitrile molecules in this primitive Earth-like atmosphere. These ions were suggested[1,2,4] to be precursors of aerosols in Titan's atmosphere and may precipitate to the surface as tholins.[5] We present the evidence for and the analysis of these heavy negative ions at Titan. In addition we examine the variation of the maximum mass[6] of the Titan negative ions with altitude and latitude for the relevant encounters so far, and we discuss

[a]Mullard Space Science Laboratory, University College London, Holmbury St Mary, Dorking, RH5 6NT, UK
[b]Centre for Planetary Science at UCL/Birkbeck, Gower Street, London, WC1E 6BT, UK
[c]Southwest Research Institute, Division of Space Science and Engineering, 6220 Culebra Road, San Antonio, TX 78228, USA
[d]University of Virginia, Engineering Physics, Thornton Hall, Charlottesville, VA 22904, USA
[e]Physics and Astronomy Department, Rice University, MS 108, Houston, TX 77251-1892, USA
[f]NASA/Goddard Space Flight Center, 8800 Greenbelt Road, Greenbelt, MD 20771, USA

the implications for the negative ion formation process. We present data from a recent set of encounters where the latitude was varied between encounters, with other parameters fixed. Models are beginning to explain the low mass negative ions,[7] but the formation process for the higher mass ions is still not understood. It is possible that the structures may be chains, rings or even fullerenes.[8]

Negative ions, mainly water clusters in this case, were seen during Cassini's recent close flybys of Enceladus.[9] We present mass spectra from the Enceladus plume, showing water clusters and additional species. As at Titan, the negative ions indicate chemical complexities which were unknown before the Cassini encounters, and are indicative of a complex balance between neutrals and positively and negatively charged ions.

The Cassini mission to Saturn has already revealed a wealth of detailed information about Saturn and its moons.[10,11] It has recently been announced (February 2010) that the mission will continue to explore Saturn's system until 2017, through Saturn's solstice, to allow the first studies of seasonal dependences in the system and gather more detailed data on this complex system.

Of the moons, Titan is unique in the solar system as it is the only moon with a dense, gravitationally bound atmosphere. Orbiting at 20 Saturn radii, it is usually located inside Saturn's magnetosphere though occasionally at or outside the magnetopause.[12] Following Voyager there has been considerable controversy over the role of Titan as a possible source of plasma for the inner magnetosphere. Cassini data have been used to resolve the issue: although important for many reasons in its own right, Titan turns out to be a very minor plasma source for the magnetosphere.

Due to their restriction to the visible wavelength range, the imagers on Voyager were unable to see below Titan's orange, smog-like haze in the atmosphere. The view from Cassini, in particular in the infrared and ultraviolet, has allowed much better remote sensing measurements of the surface through the haze. The haze itself clearly plays an important role in the atmospheric chemistry. *In situ* measurements, including those reported here, are revealing that plasma interactions with Titan's environment may provide the source of the haze.[1,2,4]

In situ measurements by Cassini at Titan revealed a wealth of chemical complexity in the neutral[13] and positively charged[14] populations. In addition was the significant and unexpected discovery of heavy (up to 13 800 amu q^{-1}) negative ions in Titan's ionosphere.[1,2] These ions were suggested[1,2,4] to be precursors of the aerosols in the atmosphere which had hidden Titan's surface from the Voyager and Earth-based imagers.

In addition to establishing that they are present at all suitable encounters,[2] further analysis[8] showed that the largest ions ($\sim$13 800 amu q^{-1}) occur at the lowest observed altitudes (950 km). Some dependence on latitude and solar zenith angle is also observed; high masses are seen preferentially at high Titan latitudes and near the terminator. This structure provides some information on the production and destruction processes. The formation of heavy ions is apparently most efficient, or the dissociation processes are least efficient, when sunlight is attenuated or absent. However, it is not yet clear from these observations whether the heavy ions build up in size with decreasing altitude, or whether larger ions, or neutrals which then ionize, are brought upwards from below and then dissociate.

Recently, the first chemical model of Titan's upper ionosphere which included negative ions was provided.[7] Several production processes were considered and corresponding rates estimated, with the conclusion that the main production process is dissociative electron attachment. The main loss processes were suggested to be associative detachment with some photodetachment. The conclusion was that the main low mass species were CN^-, C_3N^- and C_5N^-.

In order to investigate these processes further, we present data from recent scans in latitude taken during Titan encounters in 2009.

Cassini has also dramatically changed our view of the small icy moon Enceladus orbiting at 4 Saturn radii. It was discovered that strong plumes emanate from warm 'tiger stripe' features on the surface.[15,16] These plumes of water vapour and ice grains are thought to be the long-suspected source of particles making up Saturn's E-ring, and also the dominant source for plasma in Saturn's magnetosphere. *In situ* observations here revealed primarily water vapour and trace amounts of hydrocarbon-based neutral gas,[17] as well as water-group -rich positive ions that slow, divert and even stagnate the magnetospheric flow.[18] Directly over the plume sources, charged nanograin populations have been observed that are related to the tiger stripes but dispersed in their motion by Saturn's magnetic field.[19] Negative water group ions, possibly with additional species consistent with hydrocarbons, are also seen.[9]

Here we further investigate the negative ion signatures at Enceladus, showing that multiple peaks are visible up to a multiple of ~100 times the mass of a water molecule.

2. Instrumentation

The CAPS electron spectrometer (ELS[20,21]) is an electrostatic 'top-hat' analyser with a field of view 160° × 5° (divided into 20° 'anode' segments) and an energy resolution ($\Delta E/E$) of 16.7%. It is mounted on top of the CAPS ion mass spectrometer (IMS) which has a similar field of view. Together with the ion beam spectrometer (IBS), the complement of three sensors is mounted on a rotatable actuator which partially compensates for the fact that Cassini is a 3-axis stabilized spacecraft. Operation of the actuator sweeps the fan-shaped fields of view around the spacecraft z-axis to increase the angular coverage at the expense of time resolution. During some of the intervals reported at Titan, and all of those at Enceladus, the actuator position was kept fixed near the spacecraft ram direction to improve time resolution and focus on cold particles.

Although designed to measure electrons, the ELS can act as a mass spectrometer for cold populations of negative ions with thermal speeds much less than the spacecraft ram velocity. Within the Titan ionosphere, the spacecraft potential is $V_{sc} \sim -0.5$ V,[22,23] decelerating the negative ions entering the ELS. During Titan encounters, the relative speed of the spacecraft is ~6 km s^{-1} (see Table 1), thus the observed energy per charge E_{eV} (eV q^{-1}) can be converted to mass per charge m_{amu} (amu q^{-1}) using $m_{amu} = c\,(E_{eV} - V_{sc})$ with $c = 5.32$ amu eV^{-1} and $V_{sc} = -0.5$ V. The charge q on the ions is assumed to be 1 elementary charge (1.6×10^{-19} C) in this paper, though the charge on the larger ions may be larger than 1 as discussed earlier[2] which would make the mass even higher. Note that for positive ion measurements the conversion is similar except that the negative spacecraft potential must be added to the observed energy reducing the energy from the observed value, as the positive ions are accelerated by the negative potential. For the Enceladus encounters E3 and E5 the speeds were higher and the values of c were 0.92 and 0.63 amu eV^{-1} respectively, with $V_{sc} \sim -2$ V there.[9]

The CAPS ion beam spectrometer (IBS[20]) is a crossed fan electrostatic analyser measuring positive ions with an energy resolution ($\Delta E/E$) of 1.7% and a field of view for each fan of 150° × 1.4°. It is not designed as a mass spectrometer, but as with ELS cold ion energy spectra such as those in Titan's ionosphere may be converted to mass spectra using $m_{amu} = 5.32 \cdot (E_{eV} + V_{sc})$ in this case. The energy range used during the Titan encounters shown here is 3–207 eV, and we use data from one of the fans. At Enceladus, the ion populations are more fully covered by IMS.[18]

CAPS includes a third sensor, the Ion Mass Spectrometer (IMS). This sensitive mass spectrometer was not designed for the intense cold ionospheric plasma at Titan; it therefore saturates near closest approach to Titan and data are not shown in this paper. Enceladus data from IMS are shown elsewhere.[18]

3. Titan

At Titan, negative ions are always seen in CAPS electron spectrometer data during flybys below ~1400 km whenever CAPS ELS is orientated in the ram direction.[2,6]

Table 1 Encounter parameters for the flybys shown in this paper

	Date DOY	UT	Minimum altitude/km	Local time Saturn (hh : mm)	Local time Titan (hh : mm)	Latitude (°N)	Relative velocity/km s^{-1}	Solar zenith angle SZA (°)	Cassini in Titan shadow (night)
T16	22/7/06 203	00 : 25 : 26	949.9	17 : 06	02 : 27	85.38	5.97	105.32	no
T55	21/5/09 141	21 : 26 : 41	965.7	21 : 53	21 : 57	−21.87	5.99	141.52	yes
T56	06/6/09 157	20 : 00 : 00	967.7	21 : 50	21 : 55	−31.93	5.99	135.06	yes
T57	22/6/09 173	18 : 32 : 35	955.0	21 : 46	21 : 52	−42.00	5.99	127.83	no
T58	08/7/09 189	17 : 04 : 03	965.8	21 : 42	21 : 50	−52.06	5.99	120.16	no
T59	24/7/09 205	15 : 34 : 03	956.2	21 : 36	21 : 47	−62.11	5.99	112.16	no

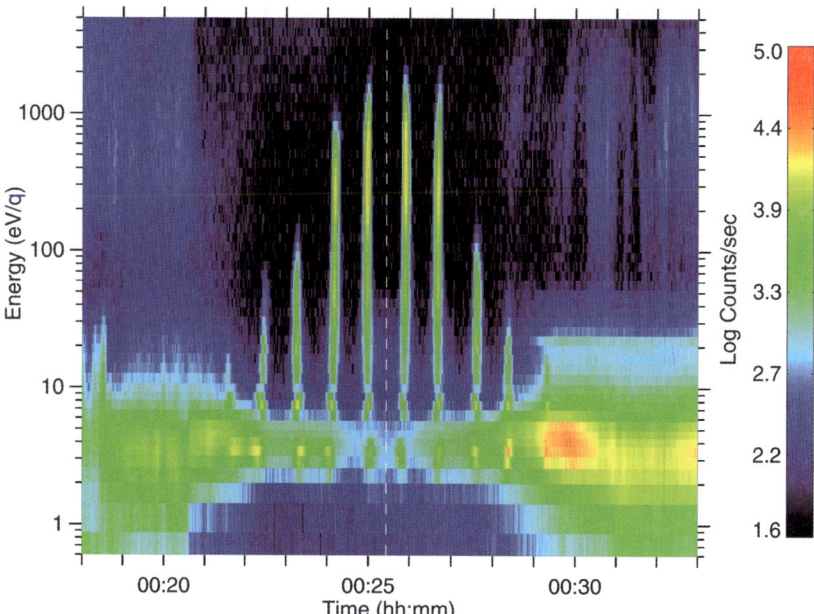

Fig. 1 CAPS ELS spectrogram for 25 min around closest approach during the T16 encounter. Count rate averaged over all anodes is colour scaled as a function of energy per charge (vertical axis) and time (horizontal axis). The vertical 'spikes' and associated peaks are identified as negative ions seen as the CAPS actuator sweeps the ELS field of view through the ram direction. Ionospheric electrons are seen below ~30 eV and magnetospheric electrons above this energy. During this encounter the highest mass ions so far (13) 800 amu q^{-1}, corresponding to ~2500 eV q^{-1}) were seen. Closest approach is indicated by a white dashed line.

The altitude and latitude dependence of the high-mass negative ions was recently examined,[6] and it was found that the maximum negative ion mass was higher at low altitude and at high latitudes. In addition, a weaker dependence of the maximum mass on solar zenith angle was found, though clearly illumination conditions were found to be important. The maximum ion mass (13 800 amu q^{-1}) was observed during Cassini's T16 Titan encounter, the lowest closest approach altitude to date. The ELS results from this encounter are shown in Fig. 1. Count rate (proportional to differential energy flux) is colour-coded in this energy spectrogram. The plot covers 25 min centred on the 950 km closest approach, and the altitude at the beginning and end of the plot is ~1800 km inbound and outbound in this case. A succession of negative ion 'spikes' is seen as the CAPS actuator sweeps the field of view through the ram direction, sampling the cold negative ion population. Each spike contains a number of discrete peaks in energy, which we interpret as different mass groups of negative ions. The maximum energy, and thus the maximum observed mass, increases towards closest approach. Note that during this particular encounter, the measured energies in IBS were unfortunately too low to determine the maximum positive ion mass, though lower mass spectra were observed, thus the IBS data are not shown here.

Further encounters were analysed,[6] and variations of the maximum negative ion mass with altitude, latitude and solar zenith angle were examined. It was suggested that the formation of the heaviest negative ions is most efficient, and/or the dissociation processes, including photodissociation and associative detachment, are less efficient, during times when sunlight is attenuated or even absent. However, as chemical growth is more usually associated with increased energy from sunlight,

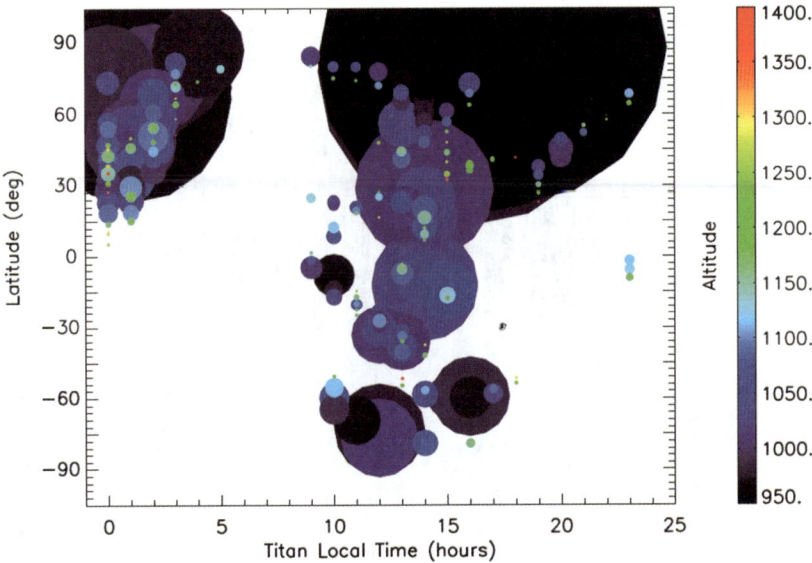

Fig. 2 Negative ion observations during 23 Titan encounters between TA and T48.[6] The circles represent a negative ion observation (radius proportional to mass). The circles are colour coded with altitude and plotted with respect to Titan latitude and local time.

another possibility is growth of particles on the day side and transport to the night side. Also, the production process for the heavy negative ions is more efficient at low altitudes, where the ambient electron density and total ion and neutral densities are high (at least two out of these three conditions are required). This, and the trend in mass with decreasing altitude, suggests that even heavier ions may exist at even lower altitudes and that the ions we observe may be the precursors of the aerosols, perhaps tholins (organic aerosols), observed in the atmosphere during UV occultations.[24] The heavy negative ions may thus ultimately be the source of particles forming Titan's haze layers at several hundred km. A summary of the altitude and latitude data is shown in Fig. 2 (from[6]).

New results from Titan are shown in Fig. 3. These data are from a recent sequence of encounters where many of the parameters of the encounters remained similar, but a range of Titan latitudes was covered (see Table 1). However the illumination conditions do vary during these encounters, with the solar zenith angle (SZA) varying between 142° and 112° at the different closest approaches; during each individual encounter the illumination conditions also vary substantially. In addition, the actuator angle was kept fixed in the ram direction to provide better altitude resolution.

For each latitude, Fig. 3 shows a pair of plots covering 20 min centred on closest approach while CAPS was measuring the ionosphere of Titan. In each case the minimum altitude is 955–975 km (see Table 1) and the plots start at 2150–2400 km altitude and end at 2350–2650 km. The upper panel shows the ELS energy spectra from 0.6–2,000 eV. The colour scale shows the count rate (proportional to differential energy flux) measured by ELS. In these plots, the population between ~20–1000 eV at the beginning and end of each plot corresponds to magnetospheric electrons. In each of the plots this population diminishes towards closest approach, corresponding to absorption of the magnetospheric electrons by the ionosphere of Titan. The population below ~10 eV corresponds to the ionospheric electrons. In each case, centred around closest approach, a succession of peaks is seen with energy, corresponding to the negative ion population. As mentioned above, the

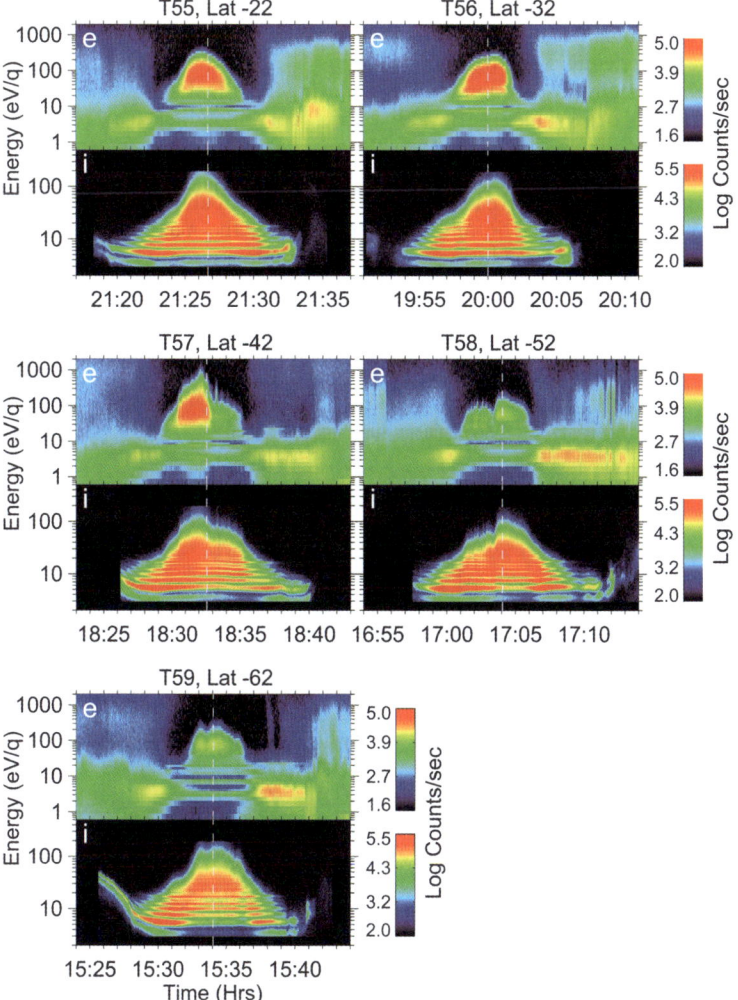

Fig. 3 CAPS ELS and IBS data for T55-59, representing a scan in Titan latitude (see text). Each panel shows ELS data averaged over all anodes (upper panel) and IBS data from fan 2 (lower panel) for 20 min around closest approach, which is indicated as a white dashed line in each case.

ram velocity of the spacecraft, combined with the energy analysis provided by ELS, acts as a mass spectrometer, and the observed energy of the spectral features can be converted to mass per charge using $m_{amu} = 5.32 \cdot (E_{eV} - V_{sc})$. In each case the maximum mass of the negative ions increases towards closest approach, which corresponds to the deepest penetration into the ionosphere.

The lower panel in each case shows the positive ion data from the IBS from 20 eV to 500 eV (upper energy of the instrument is 207 eV as used here). A series of peaks is seen, with an increase in the maximum energy seen towards closest approach. The energy peaks are interpreted as positive ions, with $m_{amu} = 5.32 \cdot (E_{eV} + V_{sc})$ in this case. Thus the maximum mass of the positive ions also increases towards closest approach reaching beyond 350 amu q^{-1} as previously observed[1,23] reaching >1000 amu q^{-1} in several of these cases.

Comparing the plots at different latitudes reveals several notable features:

1. The maximum negative ion mass appears during T57 at a latitude of −42° and where the solar zenith angle (SZA) is ∼128°.

2. The maximum positive ion mass was during T55; features are seen up to the highest energies observable by IBS (207 eV q^{-1}, corresponding to ∼1100 amu q^{-1}), during this encounter. For T55, the latitude is −22°, the solar zenith angle is ∼142° and the ionosphere is in shadow.

3. The positive and negative ion peak intensities are not always symmetrical about closest approach: during T56 and T58 for example, the maximum appears after closest approach, whereas during T57 the maximum intensity appeared before closest approach. T55 and T59 appear more symmetrical. This is probably due to the different illumination conditions before and after closest approach.

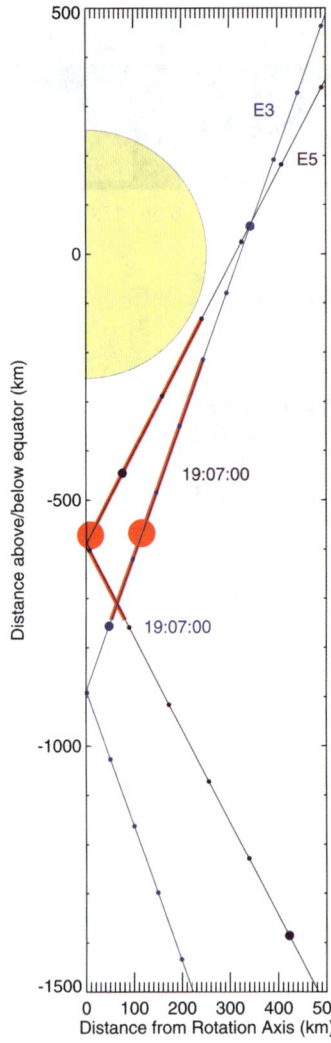

Fig. 4 Geometry of the spacecraft trajectory during the E3 and E5 Enceladus encounters. The overlaid red bars show the times during which negative ions were seen, and the red filled circles indicate the locations of the spectra in Fig. 5 and 6 respectively.

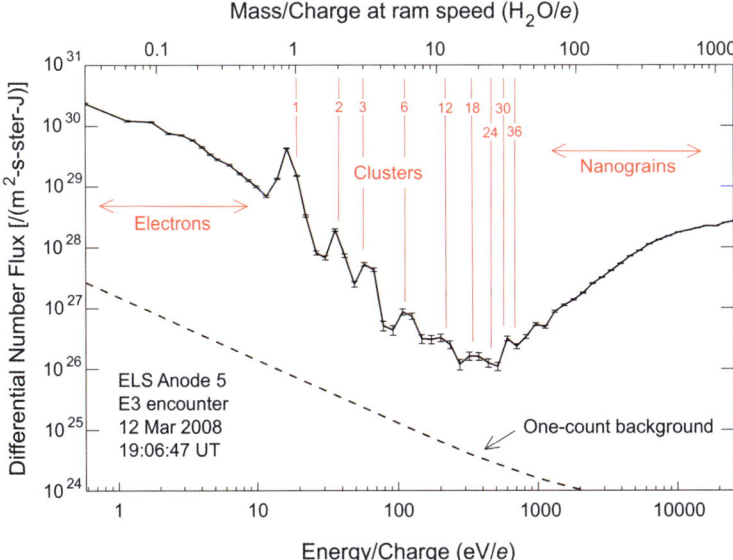

Fig. 5 CAPS ELS spectrum at 19 : 06 : 47 UT during the E3 encounter, plotted as differential number flux. The negative ion peaks are superimposed on electron and charged nanograin populations. The mass is also shown (details of conversion in the text) and multiples of 18 amu q^{-1} are indicated.

4. During some of the encounters, *e.g.*, T57 (outbound) and T58 (inbound), there are significant time variations in the negative ion intensity, usually correlated with positive ion intensity.

5. At the beginning of the T55 and T59 IBS plots, and at the end of all of the IBS plots, the low mass peaks appear to increase in energy, perhaps due to an increase in the negative spacecraft potential at these times.

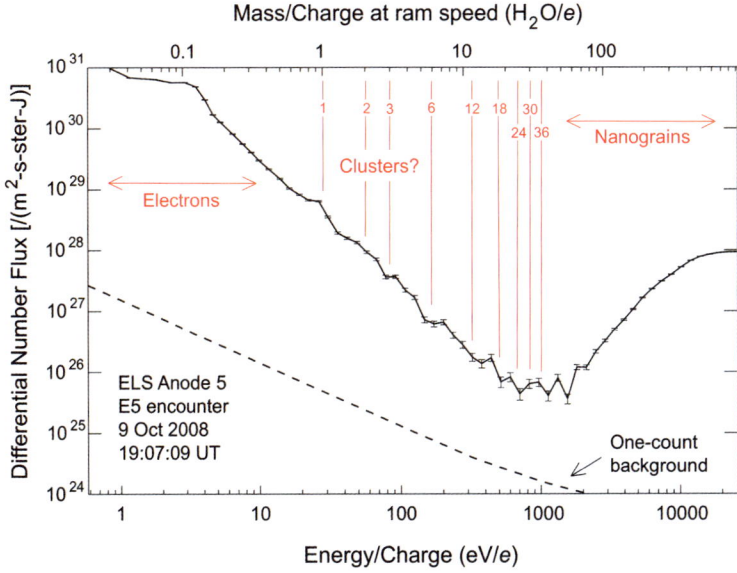

Fig. 6 As Fig. 5 but for the E5 encounter at 19 : 07 : 09 UT.

While it is difficult from these plots alone to reveal additional systematic trends beyond those observed already,[5] it is clear that there are significant differences between the different encounters. Moreover, the role of illumination, altitude and latitude are seen to be critical to the observation of high mass negative ions, confirming the earlier results.[6] Individual analyses of the encounters are however beyond the scope of the current paper.

4. Enceladus

Analysis of negative ion spectra during the E3 encounter[9] showed negative ion peaks in the spectra that plausibly correspond to water cluster ions. Here, we examine individual spectra from the E3 and E5 encounters, when the ELS was oriented in the ram direction, at the times of the maximum intensity of the negative ion peaks.

Fig. 4 illustrates the geometry with respect to Enceladus, indicating in particular the locations where negative ions were seen (red bars along the trajectories). The negative ions were observed well within the plume in both cases, and the times of the presented spectra are indicated by a filled red circle.

Fig. 5 shows an ELS spectrum from the E3 encounter. In each case, a negative ion spectrum is superimposed on the ambient electron population (at low energies, $cf.$[9,18]) and a negatively charged dust population at high energies.[19] The negative ion population indicates peaks at multiples of the mass of water, though due to the energy resolution ($\Delta E/E = 16.7\%$) of the ELS, the ions could also be OH$^-$ and multiples thereof. What is particularly notable is that not all multiples are present, 1,2,3 and 6 at low masses, and then multiples of 6 up to 30 and perhaps higher up to $\sim$100.

In Fig. 6 we present a spectrum from the E5 encounter. This also shows evidence for narrow peaks, at multiples of the mass of a water molecule, similar to the E3 encounter. Some of the peaks seen on E3 are absent in the E5 data, and the low mass peaks in particular are not as prominent in this case. The peaks which were observed are, however, statistically significant beyond the electron and charged dust nanograin spectra.

In both Fig. 5 and 6, not all of the observed peaks correspond with water (or OH) multiples, indicating that other ions could be present.

5. Discussion

Chemical modeling[7] successfully predicts the presence of low mass species of negative ions at Titan, consistent with our measurements. However the larger ions are at present unexplained and the following open questions for discussion emerge:
• What are the formation mechanisms for the large negative ions?
• What are the destruction mechanisms for the large negative ions?
• What is the lifetime of the ions?
• Sunlight clearly plays a key role in determining the maximum negative ion mass, but does it control production, or loss, or the balance between the two?
• Do the ions formed in Titan's high ionosphere subsequently fall towards the surface, or are they formed at lower altitudes and somehow transported upwards?
• Are the large ions the precursors for material that ultimately reaches the surface and modifies the surface features?
• Are the large ions chains, rings or even fullerenes?[8]
• Are the formation and loss processes for the low mass ions sufficiently explained?[7]
• What is the role of the negative ion processes in the coupled scheme between neutrals and ions in Titan's upper atmosphere?
• Are the negative ions picked up by the co-rotating plasma in a process analogous to that for positive ion pickup?
• Are there negative ions in the magnetosphere?

The water cluster ion observations at Enceladus also led to some questions, including
- What is the dominant formation process?
- What is the dominant loss process?
- What is the lifetime of these ions in the plume?
- What is the source of the ions – ice grains and gas in the plume, or the surface or sub-surface of Enceladus, or both?
- Why are some multiples of negative water (or OH) ions absent?
- What is the relation between the observed negative ions and neutral and charged ice grains?
- Are there other ions present in addition to water? Can they be identified?
- What are the coupling processes between neutrals, positive ions and negative ions in the plume of Enceladus?

Some of the Enceladus topics are being addressed by current work in the CAPS team (*e.g.* the relationship between the negative ion and charged nanograin population, Hill *et al.*, in preparation), but additional discussion would also be welcomed.

6. Summary and conclusions

In summary, negative ions are always observed at Titan below 1400 km. The ions exhibit resolved low mass peaks and a broader mass group structure at higher masses. The variation of their maximum mass with altitude, latitude and solar zenith angle were studied.[5] Currently, studies are underway to determine the dependence of density on similar parameters, though this is currently limited by the lack of knowledge of the detector efficiency for high mass ions, since the ELS was designed and calibrated for electrons.

In addition, negative ions are seen in the Enceladus plume during the E3 and E5 encounters. The ions are observed at multiples of the mass of water molecules (or hydroxyl radicals). Their location is well within the plume, indicating a short lifetime.

Negative ions at Titan and Enceladus represent two of the unexpected discoveries of the Cassini mission. While some work reported here has been done to characterize the populations and to explain them, there is a need for additional chemical modeling and process identification, particularly for the higher mass ions at both Titan and Enceladus.

Acknowledgements

We thank the CAPS and ELS operation teams, the STFC in the UK and NASA/JPL contract 1243218, for financial support of the CAPS investigation.

References

1. J. H. Waite, Jr., D. T. Young, T. E. Cravens, A. J. Coates, F. J. Crary, B. Magee and J. Westlake, The Process of Tholin Formation in Titan's Upper Atmosphere, *Science*, 2007, **316**, 870.
2. A. J. Coates, F. J. Crary, G. R. Lewis, D. T. Young, J. H. Waite, Jr. and E. C. Sittler, Jr., Discovery of heavy negative ions in Titan's ionosphere, *Geophys. Res. Lett.*, 2007, **34**, L22103.
3. A. J. Coates, Interaction of Titan's ionosphere with Saturn's magnetosphere, *Philos. Trans. R. Soc. London, Ser. A*, 2009, **367**, 773–788.
4. J. H. Waite, Jr., D. T. Young, A. J. Coates, F. J. Crary, B. A. Magee, K. E. Mandt and J. H. Westlake, The Source of Heavy Organics and Aerosols in Titan's Atmosphere, *Proc. Int. Astron. Union*, 2008, **4**, 321–326. Published online by Cambridge University Press.
5. C. Sagan, B. N. Khare, W. R. Thompson, G. D. McDonald, M. R. Wing, J. L. Bada, T. Vo-Dinh and E. T. Arakawa, Polycyclic aromatic hydrocarbons in the atmospheres of Titan and Jupiter, *Astrophys. J.*, 1993, **414**, 399–405 (ISSN0004-637X).

6 A. J. Coates, A. Wellbrock, G. R. Lewis, G. H. Jones, D. T. Young, F. J. Crary and J. H. Waite Jr., Heavy negative ions in Titan's ionosphere: altitude and latitude dependence, *Planet. Space Sci.*, 2009, **57**, 1866–1871.
7 V. Vuitton, P. Lavvas, R. V. Yelle, M. Galand, A. Wellbrock, G. R. Lewis, A. J. Coates and J.-E. Wahlund, Negative ion chemistry in Titan's upper atmosphere, *Planet. Space Sci.*, 2009, **57**, 1558–1572.
8 E. C. Sittler, Jr., A. Ali, J. F. Cooper, R. E. Hartle, R. E. Johnson, A. J. Coates and D. T. Young, Heavy Ion Formation in Titan's Ionosphere: Magnetospheric Introduction of Free Oxygen and a Source of Titan's Aerosols?, *Planet. Space Sci.*, 2009, **57**, 1547–1557.
9 A. J. Coates, G. H. Jones, G. R. Lewis, A. Wellbrock, D. T. Young, F. J. Crary, R. E. Johnson, T. A. Cassidy and T. W. Hill, Negative Ions in the Enceladus Plume, *Icarus*, 2010, **206**, 618–622.
10 *Titan from Cassini–Huygens*, ed. Robert H. Brown, Jean-Pierre Lebreton and J. Hunter Waite, Springer, Dordrecht, ISBN 978-1-4020-9214-5, 2009.
11 *Saturn from Cassini–Huygens*, ed. Michele K. Dougherty, Larry W. Esposito and Stamatios M. Krimigis, Springer, Dordrecht, ISBN 978-1-4020-9214-5, 2009.
12 C. Bertucci, N. Achilleos, M. K. Dougherty, R. Modolo, A. J. Coates, K. Szego, A. Masters, Y. Ma, F. M. Neubauer, P. Garnier, J.-E. Wahlund and D. T. Young, The magnetic memory of Titan's ionized atmosphere, *Science*, 2008, **321**, 1475–1478.
13 J. H. Waite, H. Niemann, R. V. Yelle, W. T. Kasprzak, T. E. Cravens, J. G. Luhmann, R. L. McNutt, W.-H. Ip, D. Gell, V. De La Haye, I. Müller-Wordag, B. Magee, N. Borggren, S. Ledvina, G. Fletcher, E. Walter, R. Miller, S. Scherer, R. Thorpe, J. Xu, B. Block and K. Arnett, Ion Neutral Mass Spectrometer Results from the First Flyby of Titan, *Science*, 2005, **308**, 982–986.
14 T. E. Cravens, I. P. Robertson, J. H. Waite Jr., R. V. Yelle, W. T. Kasprzak, C. N. Keller, S. A. Ledvina, H. B. Niemann, J. G. Luhmann, R. L. McNutt, W.-H. Ip, V. De La Haye, I. Mueller-Wodarg, J.-E. Wahlund, V. G. Anicich and V. Vuitton, Composition of Titan's ionosphere, *Geophys. Res. Lett.*, 2006, **33**, L07105.
15 M. K. Dougherty, K. K. Khurana, F. M. Neubauer, C. T. Russell, J. Saur, J. S. Leisner and M. E. Burton, Identification of a dynamic atmosphere of Enceladus with the Cassini magnetometer, *Science*, 2006, **311**, 1406–1409.
16 C. C. Porco, P. Helfenstein, P. C. Thomas, A. P. Ingersoll, J. Wisdom, R. West, G. Neukum, T. Denk, R. Wagner, T. Roatsch, S. Kieffer, E. Turtle, A. McEwen, T. V. Johnson, J. Rathbun, J. Veverka, D. Wilson, J. Perry, J. Spitale, A. Brahic, J. A. Burns, A. D. DelGenio, L. Dones, C. D. Murray and S. Squyres, Cassini observes the active South pole of Enceladus, *Science*, 2006, **311**, 1393–1401.
17 J. H. Waite, Jr, W. S. Lewis, B. A. Magee, J. I. Lunine, W. B. McKinnon, C. R. Glein, O. Mousis, D. T. Young, T. Brockwell, J. Westlake, M.-J. Nguyen, B. D. Teolis, H. B. Niemann, R. L. McNutt Jr., M. Perry and W.-H. Ip, Liquid water on Enceladus from observations of ammonia and ^{40}Ar in the plume, *Nature*, 2009, **460**, 487–490.
18 R. L. Tokar, R. E. Johnson, M. F. Thomsen, R. J. Wilson, D. T. Young, F. J. Crary, A. J. Coates, G. H. Jones and C. S. Paty, Cassini Detection of Enceladus' Cold Water-Group Plume Ionosphere, *Geophys. Res. Lett.*, 2009, **36**, L13203.
19 G. H. Jones, C. S. Arridge, A. J. Coates, G. R. Lewis, S. Kanani, A. Wellbrock, D. T. Young, F. J. Crary, R. L. Tokar, R. J. Wilson, T. W. Hill, R. E. Johnson, D. G. Mitchell, J. Schmidt, S. Kempf, U. Beckmann, C. T. Russell, Y. D. Jia, M. K. Dougherty, J. H. Waite Jr. and B. Magee, Fine jet structure of electrically-charged grains in Enceladus' plume, *Geophys. Res. Lett.*, 2009, **36**, L16204.
20 D. T. Young, J.-J. Berthelier, M. Blanc, J. L. Burch, A. J. Coates, R. Goldstein, M. Grande, T. W. Hill, R. E. Johnson, V. Kelha, D. J. McComas, E. C. Sittler, K. R. Svenes, K. Szegv, P. Tanskanen, K. Ahola, D. Anderson, S. Bakshi, R. A. Baragiola, B. L. Barraclough, R. Black, S. Bolton, T. Booker, R. Bowman, P. Casey, G. Dirks, N. Eaker, J. T. Gosling, H. Hannula, C. Holmlund, H. Huomo, J.-M. Illiano, P. Jensen, M. A. Johnson, D. Linder, T. Luntama, S. Maurice, K. McCabe, B. T. Narheim, J. E. Nordholt, A. Preece, J. Rutzki, A. Ruitberg, K. Smith, S. Szalai, M. F. Thomsen, K. Viherkanto, T. Vollmer, T. E. Wahl, M. Wuest, T. Ylikorpi and C. Zinsmeyer, Cassini Plasma Spectrometer Investigation, *Space Sci. Rev.*, 2004, **114**, 1–112.
21 D. R. Linder, A. J. Coates, R. D. Woodliffe, C. Alsop, A. D. Johnstone, M. Grande, A. Preece, B. Narheim, K. Svenes and D. T. Young, The Cassini CAPS electron spectrometer, in *Measurement Techniques in Space Plasmas: Particles*, AGU geophysical monograph **102**, ed. R. E. Pfaff, J. E. Borovsky and D. T. Young, 1998, 257–262.
22 J.-E. Wahlund, R. Boström, G. Gustafsson, D. A. Gurnett, W. S. Kurth, A. Pedersen, T. F. Averkamp, G. B. Hospodarsky, A. M. Persoon, P. Canu, F. M. Neubauer, M. K. Dougherty, A. I. Eriksson, M. W. Morooka, R. Gill, M. André, L. Eliasson and

I. Müller-Wodarg, Cassini Measurements of Cold Plasma in the Ionosphere of Titan, *Science*, 2005, **308**, 986–989.
23 F. J. Crary, B. A. Magee, K. Mandt, J. H. Waite Jr. and J. Westlake, Heavy ions, temperatures and winds in Titan's ionosphere: Combined Cassini CAPS and INMS observations, *Planet. Space Sci.*, 2009, **57**, 1847–1856.
24 M.-C. Liang, Y. L. Yung and D. E. Shemansky, Photolytically generated aerosols in the mesosphere and thermosphere of Titan, *Astrophys. J.*, 2007, **661**, L199–L202.

Chemical origins of the Mars ultraviolet dayglow

David L. Huestis,[*a] Tom G. Slanger,[a] Brian D. Sharpee[a] and Jane L. Fox[b]

Received 23rd February 2010, Accepted 31st March 2010
DOI: 10.1039/c003456h

Airglow optical emissions from planetary atmospheres provide remotely observable signatures of atmospheric composition, energy deposition processes, and the resulting chemical reactions. We may one day be able to detect airglow emissions from extrasolar planets. Reliable interpretation requires quantitative understanding of the energy sources and chemical mechanisms that produce them. The ultraviolet dayglow observations by the *Mariner 6* and *7* (1969) and *Mariner 9* (1971–72) motivated numerous modeling studies and laboratory experiments. The most obvious source reaction is photodissociation and photoionization of ambient CO_2, which is known in the laboratory to produce the four strong dayglow emitting states:

$$h\nu + CO_2 \rightarrow O(^1S), CO(a^3\Pi), CO_2^+(A^2\Pi_u \ \& \ B^2\Sigma_u^+) \qquad (1)$$

If this simplest of models were sufficient, then the high altitude dayglow emissions would all share the same scale height, which would be that of CO_2. The few *Mariner* dayglow observations provide weak statistics. Addition of 4 months of *Mars Express* dayglow data, and including radio occultation and mass spectrometry data from other missions, have made the analyses and conclusions more robust. The $CO(a^3\Pi)$ and $CO_2^+(B^2\Sigma_u^+)$ dayglow altitude profiles are consistent with Reaction (1). In contrast, the $O(^1S)$ dayglow scale heights are much larger and are consistent with source Reaction (2):

$$O_2^+ + e^- \rightarrow O(^1S) \qquad (2)$$

Both sets of scale heights change with respect to solar activity roughly as suggested by modeling studies.

Introduction

Introduction to the Mars atmosphere

The *Mariner 4* (1965), *Mariner 6* and *7* (1969), *Mariner 9* (1971–72), and *Viking 1* and *2* (1976–80) missions provided the first quantitative information about the structure, energetics, and dynamics of the Mars atmosphere.[1–14] These limited data sets provided the first observational foundation and constraints for semiempirical reference atmospheric models,[15–25] photochemical and diffusion models,[26–37] and general/global circulation (GCM) models.[38–42] Not until more than 20 years later did new generations of capable landers and orbiters revisit the planet.[41,43–56]

These various missions provided diverse measurement capabilities. For many missions the ionospheric plasma density was investigated by analyzing occultation and refraction of radio transmissions back to Earth. Landers and aerobraking

[a]*SRI International, Menlo Park, CA, USA. E-mail: david.huestis@sri.com; Fax: +650 859 6196; Tel: +650 859 3464*
[b]*Wright State University, Dayton, OH, USA*

orbiters recorded deceleration. Two landers (*Viking 1* and *2*) used mass spectrometers to record altitude profiles of the densities of neutral and ionized atoms and molecules during descent. Two flyby missions (*Mariner 6* and *7*) and two orbiters (*Mariner 9* and *Mars Express*) carried ultraviolet spectrographs that recorded limb spectra of dayglow emissions *versus* tangent ray height. The primary focus of the work reported here is analysis, interpretation, and modeling of these dayglow emissions, with the objectives of (1) determining the mechanistic origin of these emissions and (2) providing experimental guidance and constraints on modeling the vertical chemical and thermal structure of the Mars atmosphere and ionosphere and their responses to variations in solar irradiation.

The upper graph in Fig. 1 shows the spectrum of the Mars dayglow as recorded during the *Mariner 6* and *7* flyby missions. Publication of this spectrum motivated a surge of laboratory measurements in the 1970s[57–74] of the wavelength-dependent yields of excited states in the "obvious" source reaction

$$h\nu + CO_2 \rightarrow O(^1S), CO(a^3\Pi \ \& \ A^1\Pi), CO_2^+(A^2\Pi_u \ \& \ B^2\Sigma_u^+), CO^+(B^2\Sigma^+) \quad (1)$$

given that CO_2 was known to be the primary chemical component of the Mars atmosphere. If this simplest of models were sufficient, then the scale heights would be that of CO_2 for all the high-altitude dayglow emissions.

As we will describe and discuss below, the few published *Mariner* dayglow observations[3–7] provided weak statistics.[14] Addition of 4 months of *Mars Express* dayglow data[75] and including radio occultation and mass spectrometry data from other missions have made the analyses and conclusions more robust. The $CO(a^3\Pi)$ and

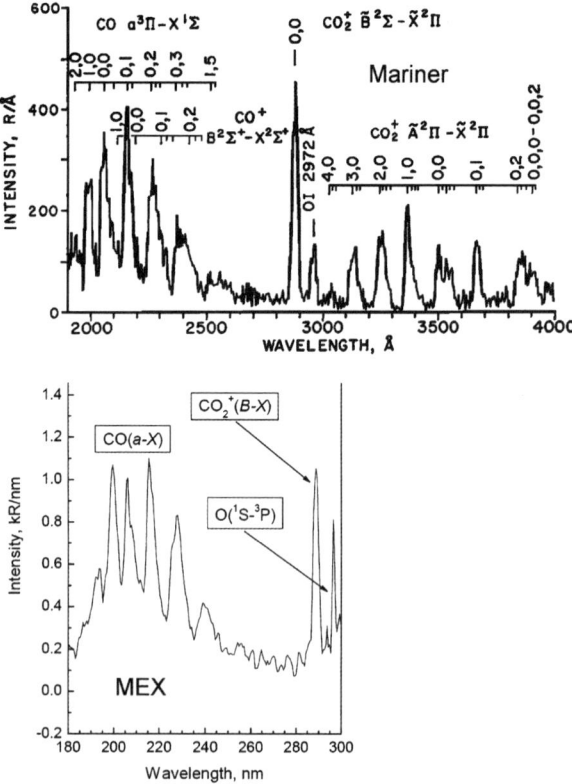

Fig. 1 Mars dayglow spectra from the *Mariner 4* and *Mars Express* (MEX) missions.

$CO_2^+(B^2\Sigma_u^+)$ altitude profiles are consistent with Reaction (1). In contrast, the $O(^1S)$ scale heights are much larger and are consistent with source Reaction (2):

$$O_2^+ + e^- \rightarrow O(^1S) \qquad (2)$$

which is known to be important in the Earth's ionosphere.[76–80] Both sets of Mars dayglow scale heights change with respect to solar activity roughly as suggested by modeling studies.[34,39]

Historical development of models of the Mars atmosphere

By the late 1960s and early 1970s, observationally validated, semiempirical yet quantitatively useful reference models of the chemical, thermal, and ionization structure of the Earth's atmosphere had been developed.[76,81–83] A key element of these models was adoption of an easy-to-measure proxy for the spectral distribution of hard ultraviolet light emitted by the sun.

Since the early development of radio and discovery of the ionosphere, it was observed that the occurrence of sunspots was correlated with ionospheric activity.[84] Sunspots themselves are hard to see and difficult to measure quantitatively. Eventually it was found that the sun emits radio waves at 10.7 cm wavelength whose intensity correlates well with sunspots, solar flares, and ionospheric activity.[85–89] Daily values of the 10.7 cm intensity, called $F_{10.7}$, have been recorded and tabulated since 1947 (ftp://ftp.ngdc.noaa.gov/STP/SOLAR_DATA/SOLAR_RADIO/FLUX/) in units of 10^{-22} W m^{-2} Hz^{-1}, with values ranging roughly from 30 to 300, and a value of about 70 corresponding to an average quiet sun.

In the Jacchia model,[82,83] the structure of the Earth's neutral upper atmosphere is parameterized by a single variable, the exospheric temperature, T_∞, which in turn is parameterized by $F_{10.7}$, latitude, and diurnal oscillation. $F_{10.7}$ is by far the dominant factor. For modeling atmospheric and ionospheric chemistry, the $F_{10.7}$ proxy has been incorporated into the EUVAC model[90–94] of the solar spectrum that provides a convenient means of calculating the solar-cycle variation of photoabsorption, photodissociation, and photoionization reactions. Other parameterizations of the solar spectrum are also available.[95–98]

The $F_{10.7}$ proxy is also useful in modeling the variability of the upper atmospheres of Mars and Venus.[99–107] Fig. 2 illustrates appropriate $F_{10.7}$ proxy values for some Mars missions. Table 3 in Hanson et al.[11] and Fig. 2 in Bougher et al.[40] also illustrate the differing solar conditions for Mars missions. See below for guidance about applying $F_{10.7}$ proxy values to planetary atmospheres.

Refinement of models of the Mars atmosphere began during the *Mariner* missions[15,16,26] and resumed after the *Viking* missions.[17,18,29,30,108,109] The un-published 1987 report by Stewart[18] was especially influential because it became the initial basis for development of the Mars-GRAM sequence of Mars Global Reference Atmospheric Models.[19–25] In spite of all this effort, the current understanding of the Mars atmosphere and ionosphere is still far from satisfactory. To quote Mazarico et al.,[47] "*Data measurements that can be used as upper boundary conditions for GCMs are critical but sparse below 200 km and almost nonexistent above.*" They found that the Stewart model[18] and Mars-GRAM 2001[22] gave quite different predictions, and neither provided an adequate representation of the exospheric densities (at 400 km) over two Mars years as inferred from satellite drag measurements from the Mars Odyssey spacecraft.[47] Quoting again, "*The density from the Stewart model is enhanced by a factor of $\approx$ 60 between solar minimum and solar maximum, whereas Mars-GRAM 2001 only shows a maximum fourfold increase. On the other hand, for a fixed $F_{10.7}$, the seasonal density variations are much larger in the Mars-GRAM model. Our density measurements are in general bounded by the predictions of the two models.*"

Of particular importance for the work discussed here, the only sources of observational information about the Mars exospheric temperature available to Stewart in

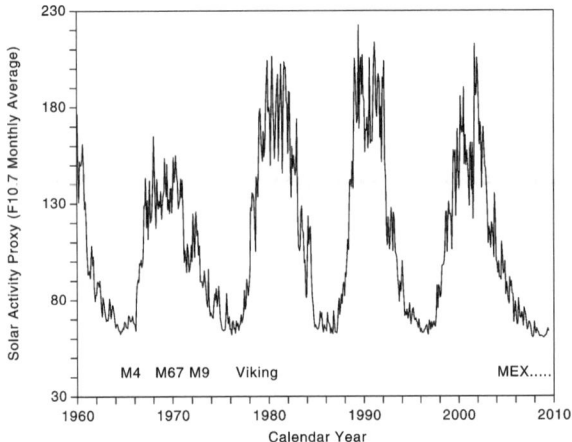

Fig. 2 Time correlation between Mars atmospheric science missions and solar activity. The time durations of the *Mariner 4* (M4) and *Mariner 6* and *7* (M67) flyby missions were much smaller than the widths of the symbols on the graph. The observing periods of the *Mariner 9* (M9), *Viking 1* and *2* (*Viking*), and *Mars Express* (MEX...) orbiters roughly match the width of the plotted symbols. The *Viking* landers touched down during the time indicated by the width of the letter "V".

1987 were the *Mariner* dayglow scale height measurements, *Mariner* and *Viking* plasma scale height measurements, and *Viking* mass spectrometry measurements. Interpretation of all of these measurements is challenging. The dayglow production mechanisms need to be confirmed. The plasma scale heights reflect the ion and electron temperatures, rather than that of neutral gas. Finally, above 120 km the *Viking 1* and *Viking 2* lander mass densities are not consistent with each other, both are noisy (or oscillatory),[9] and together or separately provide no obvious extrapolation toward a well-constrained exospheric temperature. Just as important, the 11 estimates of the exospheric temperature in Stewart's Table 2[18] are too few to separately constrain the dependence on solar activity ($F_{10.7}$) and heliocentric distance (indirectly L_S).

Applying the $F_{10.7}$ solar activity proxy to planetary atmospheres

Some planetary science researchers[103–105] have incorrectly supposed that the 10.7 cm solar radio emission intensity ($F_{10.7}$) is proportional to the *UV flux arriving at Earth* and thus should be adjusted by the square of the distance from the sun relative to that of Earth. This is a significant misinterpretation. The $F_{10.7}$ numerical value is a proxy for the numbers of sunspots and flares, and thus for the *wavelength dependence of the UV spectrum emitted by the sun*. The number of sunspots and the spectrum of solar radiation are the same for all objects in the solar system. We should not divide $F_{10.7}$ by the square of the distance from the sun (about a factor of 3 for Mars). The concept of "$F_{10.7}$ at Mars" is at best misleading. Rather, use $F_{10.7}$ in EUVAC or some other spectral model to calculate the solar ultraviolet spectrum and then divide the solar flux at all wavelengths. A small date-shift adjustment of the tabulated $F_{10.7}$ values is appropriate because Mars and Earth are not normally exposed simultaneously to the same emitting region of the rotating sun.[102,106]

Analysis of *Mars Express* SPICAM dayglow spectra

Overview of *Mars Express* SPICAM dayglow spectra

The *Mars Express* (MEX) SPICAM-UV instrument recorded numerous dayglow spectra from the Mars atmosphere at wavelengths between 100 and 300 nm.[52,55]

In the region between 190 and 300 nm these spectra are quite similar to those recorded by *Mariner 6* and *7*, as illustrated in Fig. 1. The strong $CO(a^3\Pi-X^1\Sigma^+)$ Cameron bands between 190 and 250 nm are clearly resolved. Also conspicuous are the $CO_2^+(B^2\Sigma_u^+-X^2\Pi_g)$ doublet at 289.0 nm and the $O(^1S-^3P)$ line at 297.2 nm. The *Mariner* dayglow spectra also include the $CO_2^+(A^2\Pi_u-X^2\Pi_g)$ emissions at longer wavelengths. SRI International, received from Francois Leblanc (Service d'Aéronomie, IPSL/CNRS), SPICAM UV dayglow spectra taken on 23 orbits between 0947 and 1457 (14 November 2004 through 6 March 2005).[75] This data set had already been reduced and intensity calibrated. Analyses of the altitude profiles for 20 of these orbits are reported here.

Subtracting scattered sunlight at low altitudes

Fig. 3 illustrates that at wavelengths longer than 250 nm, solar radiation features increasingly contaminate the dayglow spectrum as the tangent ray height is reduced below 130 km. Stellar occultation measurements have shown that there is haze opacity up to 100 km.[110] Depending on the details of the baffling of the spectrometer inlet, the recorded spectrum may include sunlight scattered by dust or clouds at much lower altitudes. The problem is especially severe on Venus, where clouds that are nominally around 60 km altitude interfere with long-wavelength dayglow observations until the tangent ray heights are above 150 km.[111] Fig. 3 (right frame) illustrates that there is valuable information about the $CO_2^+(B-X)$ doublet at 289.0 nm and the $O(^1S-^3P)$ atomic line at 297.2 nm if we can subtract the solar contribution. To do so we made a model of the solar spectrum in this region and subtracted its contribution to the dayglow spectrum, allowing us to track the 289.0 and 297.2 nm emissions down to below 80 km. These lower altitude data will be included in our discussion of dayglow altitude profiles below.

Graphical modeling and inferring chemical mechanisms

Our approach to analyzing observations of planetary atmospheres might be called graphical or "inverse" modeling, or letting the data themselves teach us as much as they can about how they should be grouped, categorized, and graphed without

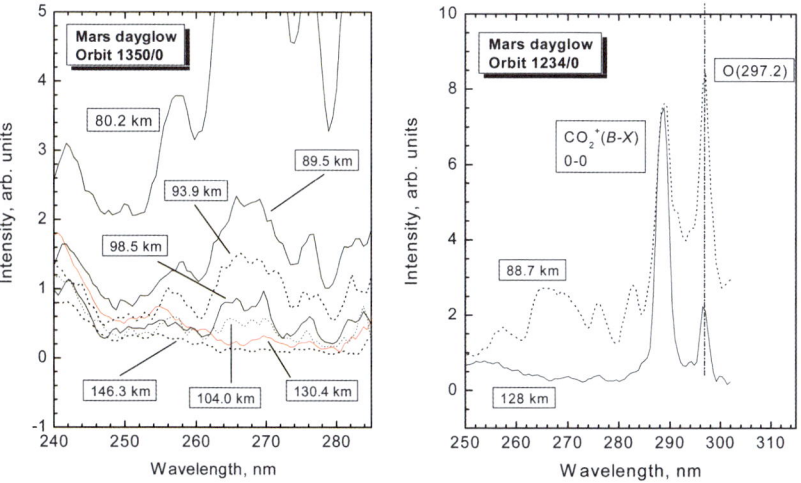

Fig. 3 SPICAM spectra of Mars dayglow. The left frame shows growth of spectral features from the solar spectrum as the tangent ray height decreases below 130.4 km. The right frame shows that the intensity ratio between the CO_2^+ 289.0 nm doublet and the $O(^1S)$ 297.2 nm line changes rapidly in the same altitude region where dust-scattered sunlight is significant.

our making too many assumptions about chemical mechanisms. Another aspect is that we are experienced physical chemists, a discipline that might be defined as "finding the type of graph paper on which the data produce a straight line," the slope of which is likely to be quantitatively and mechanistically meaningful. In addition, we have often found new insights from planetary observations by temporarily ignoring published interpretations. These ideas work well when the data are sufficiently numerous and recorded under diverse observational conditions.

During the *ancient Mariner/Viking* era in the 1960s and 1970s, the data were few in number with sparse sampling of the relevant input variables, such as planetary geometry and solar inputs. During the *modern* era in the 1990s and 2000s we have obtained large data sets from multiple orbiters and landers, spanning wide ranges of planetary and solar variables over multiple Mars years and solar cycles. With such rich data resources we have a choice of whether to concentrate on time-sequence analysis of atmospheric dynamics or time-averaged analysis of chemical reaction kinetics. The latter is the emphasis adopted here.

The simplest model of the vertical structure of a planetary atmosphere[112] begins with assumptions of a constant gravitational acceleration g, a constant molecular weight M, and a constant temperature T, and thus a constant mass scale height $H = kT/Mg$, resulting in a mass density or molecular number-density altitude profile of the form

$$\rho(z) = \rho(z_0) \exp[-(z - z_0)/H]$$

The message here is not about whether the constancy assumptions are accurate, but rather that without other information the default assumption should be that molecular densities decrease in altitude with characteristic scale heights H. The quantitative validity of this approach depends on the individual altitude scans being roughly exponential and that the collections of altitude scans being analyzed together are consistent with shared scale heights. Fig. 4 shows Mars dayglow altitude profiles for a single orbit, along with the exponential dependencies indicated in Fig. 5 and Tables 1 and 2 (see below). The exponential or constant scale height assumption is supported, except for the $CO_2^+(B^2\Sigma_u^+ \rightarrow X^2\Pi_g)$ and $CO(a^3\Pi \rightarrow X^1\Sigma^+)$ emissions at altitudes below 130 km, where the exciting species, *i.e.*, photons, has been depleted by absorption at higher altitudes. In addition, for limb and occultation

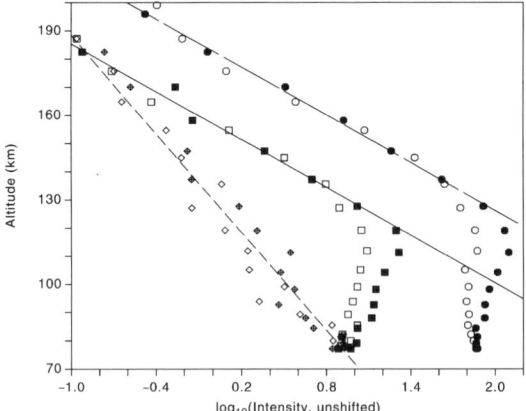

Fig. 4 Dayglow emissions on orbit 0947. Solid and open points indicate downward and upward scans, respectively. Circles and long dashed line indicates $CO(a^3\Pi \rightarrow X^1\Sigma^+)$ emission. Squares and solid line indicates $CO_2^+(B^2\Sigma_u^+ \rightarrow X^2\Pi_g)$ emission. Diamonds and short dashed line indicates $O(^1S-^3P)$ emission.

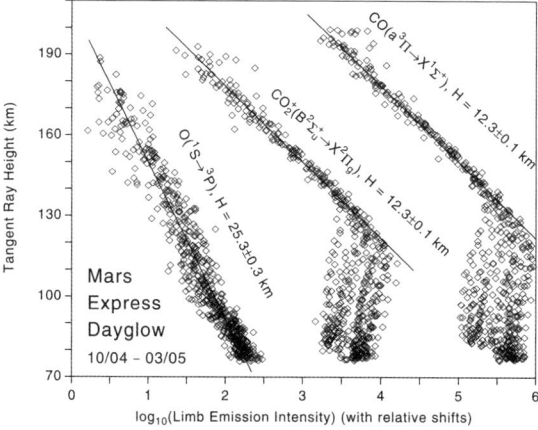

Fig. 5 Simultaneous fitting of *Mars Express* dayglow altitude profiles.

Table 1 CO(a–X) and CO_2^+(B–X) Mars dayglow emission scale heights for altitudes from 135 to 185 km and those for the inferred source reaction: $h\nu + CO_2 \rightarrow CO(a)$ and $CO_2^+(B)$.[a]

Mission	H/km	Method	Dates	F10.7	Ref.
Solar maximum					
M6,7,9	17.4 ± 0.4	Dayglow	1969, 1971	155, 110	3,4,6,115
MEX	12.3 ± 0.1	Dayglow	10/04–3/05	105–95	Present
Model	15.8	H(CO$_2$)		High	39
Model	13.8	H(CO$_2$)		Max	34
Solar minimum					
VL1,2	8.9 ± 0.2	H(mass)	1976	72, 74	9,10,116
Model	11.3, 11.6	Dayglow		low	30
Model	9.6	H(CO$_2$)		low	30
Model	11.1	H(CO$_2$)		low	39
Model	7.1	H(CO$_2$)		min	34

[a] M6,7,9 indicates the *Mariner 6, 7,* and *9* missions. MEX indicates the *Mars Express* mission. VL1,2 indicates the *Viking* landers. $F_{10.7}$ is the radio frequency proxy for solar UV activity, averaged over 81 prior days at Earth (see Fig. 2 for correspondence with observation dates). The upper half of the table is for high solar activity, while the lower half is for low solar activity.

measurements the effective integration path length is many times the local vertical scale height,[113,114] which produces significant averaging, thus reducing the influence of atmospheric dynamics over small spatial scales and resulting in more nearly-exponential observed altitude profiles.

Finally, by emphasizing higher altitudes and unattenuated external energy sources, the rates of chemical processes scale in altitude as do their chemical precursors. If the latter vary exponentially in altitude, then the former will do so as well (at least approximately when fit over several scale heights). In addition, if the altitude dependence is observed to be exponential, then an Abel inversion is not required.[113]

The exponential scale heights of the reaction products themselves reflect the balance between energy deposition, chemical reactions, and diffusion. As discussed above, the scale heights (and temperatures) for the Mars dayglow and ionosphere

Table 2 O(^{1}S-^{3}P) Mars dayglow emission scale heights for altitudes from 77 to 200 km and those for the primary inferred source reaction: e + O_2^+ → O(^{1}S).[a]

Mission	H/km	Method	Dates	F10.7	Refs
Solar maximum					
M6,7	25.6 ± 1.4	Dayglow	1969	155	3,4
M6,7	21.4, 23.4	½H(e$^-$)	1969	155	5,12
M9	18.9, 19.4	½H(e$^-$)	1971–72	110	8,12
MGS	24.7	½H(e$^-$)	2/01	162	MGS-M-RSS-5-EDS-V1.0
MEX	25.3 ± 0.3	Dayglow	10/04–3/05	105–95	Present
Model	23.5	½H(O_2^+)		High	39
Model	18	½H(O_2^+)		Max	34
Solar minimum					
M4	14.6	½H(e$^-$)	1965	79	1,2,12
VL1	14.5	½H(O_2^+)	1976	72	11
VO1,2	12.4,14.3, 17.9	½H(e$^-$)	1977–78	75, 87, 140	12
MGS	13.8	½H(e$^-$)	12/98	133	MGS-M-RSS-5-EDS-V1.0
MGS	16.1	½H(e$^-$)	6/05	78	MGS-M-RSS-5-EDS-V1.0
Model	11.8	Dayglow		Low	30
Model	13.4	½H(O_2^+)		Low	30
Model	9.6	H(CO_2)		Low	30
Model	14.3	½H(O_2^+)		Low	39
Model	11.4	½H(O_2^+)		Min	34

[a] M4 indicates the *Mariner 4* mission. M6,7 indicates the *Mariner 6* and *7* missions. M9 indicates the *Mariner 9* mission. MGS indicates the *Mars Global Surveyor* mission. MEX indicates the *Mars Express* mission. VL1 indicates the *Viking 1* lander. VO1,2 indicates the *Viking 1* and *2* orbiters. $F_{10.7}$ is the radio frequency proxy for solar UV activity, averaged over 81 prior days at Earth (see Fig. 2 for correspondence with observation dates). The upper half of the Table is for high solar activity, while the lower half is for low solar activity.

appear strongly correlated with solar activity (or $F_{10.7}$). For Mars, the second most important environmental variable is expected to be the heliocentric distance (indirectly, L_S), which varies with solar season much more than for most other solar system objects. Between perihelion and aphelion, the overall solar insolation increases by 45%. An increase in the $F_{10.7}$ proxy from 80 to 120 or 160 increases the ultraviolet solar radiation component and the resulting rate of CO_2 ionization by 46% or 92%, respectively [see Fig. 2 above]. At lower altitudes the solar zenith angle (SZA) becomes important.[35] Correlating the atmospheric observables with these environmental variables is a primary goal of future analysis of additional data sets.

Altitude dependence of *Mars Express* dayglow emissions

Fig. 5 shows the results of simultaneous fitting of Mars dayglow altitude profiles from 20 *Mars Express* orbits. For each orbit there is a descending (diamonds) and ascending (circles) scan, with a spectrum assigned to each corresponding tangent ray height. The different scans have been shifted left or right in intensity to obtain a best fit. The sloping solid lines are shared between all scans and all orbits. For the $CO_2^+(B^2\Sigma_u^+ \rightarrow X^2\Pi_g)$ and CO($a^3\Pi \rightarrow X^1\Sigma^+$) emissions, some outlying points above about 185 km were excluded from the fits, as were points below about 135 km. The O(^{1}S → ^{3}P) data are fit separately while the $CO_2^+(B^2\Sigma_u^+ \rightarrow X^2\Pi_g)$ and CO($a^3\Pi \rightarrow X^1\Sigma^+$) data are fit together with a common scale height. Nearly the same scale height values are obtained when the latter two emissions are fit separately. The rms fitting errors were ±30% (555 points) for O(^{1}S → ^{3}P) and ±18% (323 points) for $CO_2^+(B^2\Sigma_u^+ \rightarrow X^2\Pi_g)$ plus $CO(a^3\Pi \rightarrow X^1\Sigma^+)$. Note that for the latter, Simon et al.[37] indicate that "*Instrumental uncertainties remain under 15%.*"

Because the altitude profiles of the $CO_2^+(B^2\Sigma_u^+ \rightarrow X^2\Pi_g)$ and $CO(a^3\Pi \rightarrow X^1\Sigma^+)$ emissions are so well defined and so similar to each other, we are justified in suggesting that they share the same production mechanism, for which the "obvious" choice is photodissociation of CO_2. The reductions in emissions below about 135 km are suggestive of attenuation of the exciting vacuum ultraviolet (VUV) solar radiation for larger solar zenith angles.

Figures 9 and 11 from Simon et al.[37] also suggest that $CO_2^+(B^2\Sigma_u^+ \rightarrow X^2\Pi_g)$ and $CO(a^3\Pi \rightarrow X^1\Sigma^+)$ emissions share the same high-altitude exponential scale height. Note that the right frame in their Figure 11 indicates a fictitious increase in $CO_2^+(B^2\Sigma_u^+ \rightarrow X^2\Pi_g)$ emissions below 90 km resulting from their failure to subtract the contribution from dust-scattered solar radiation. In contrast, the Monte Carlo model of Shematovich et al.[36] does a plausible job of calculating the altitudes of the peak CO(a–X) and CO_2^+(B–X) emissions, but is very poor at calculating the scale heights in the high-altitude exponentially decreasing regions.

Our analyses of the scale heights for the $CO_2^+(B^2\Sigma_u^+ \rightarrow X^2\Pi_g)$ and $CO(a^3\Pi \rightarrow X^1\Sigma^+)$ Mars dayglow emissions are summarized in Table 1. Both the *Mariner* and early *Mars Express* dayglow observations occurred during periods of relatively high solar activity (see Fig. 2). The high-altitude fitted scale heights are consistent with modeling predictions of the CO_2 scale heights under similar conditions.[34,39] The corresponding agreement between the *Viking* lander mass spectra and low solar activity models suggests that we are approaching a quantitative characterization of the altitude–temperature–density profile of Mars' neutral atmosphere and that analyses of additional dayglow observations have much to offer.

Leblanc et al.[55] concluded from a different MEX data set that the MEX $CO(a^3\Pi \rightarrow X^1\Sigma^+)$ and $CO_2^+(B^2\Sigma_u^+ \rightarrow X^2\Pi_g)$ dayglow emissions were roughly exponential in altitude from 140 to 180 km, sharing roughly the same scale heights, with numerical values compatible with those reported here. They also concluded that measurement of these scale heights should be useful for determination of the Mars exospheric temperature and its variations with respect to solar illumination.

The almost purely exponential altitude profile, from 77 to 190 km, for the O(^{1}S → ^{3}P) dayglow emissions is a surprising result. The proposed excitation mechanism, $e^- + O_2^+ \rightarrow O(^1S)$, should produce maximum slant-height emission near the altitude of peak plasma density, which will depend on the solar zenith angle. The *Viking 1* dayside measurements suggest a peak around 130 km,[11] while the MGS (*Mars Global Surveyor*) occultation measurements suggest variable and higher altitudes, presumably due in part to solar zenith angles near 90°. Perhaps some of the low-altitude O(^{1}S → ^{3}P) dayglow emission comes from Lyman-α (121.6 nm) photodissociation of CO_2, where the absorption cross section is low, and the intense solar radiation penetrates to lower altitudes, as suggested by Fox and Dalgarno,[30] although we have found no unambiguous experimental evidence for the double-peaked altitude dependence that they predict. Curiously, Simon et al.[37] present no *Mars Express* O(^{1}S → ^{3}P) 297.2 nm observational data, yet assert consistency with their model and with the Fox and Dalgarno double-peak predictions The modeled curves are all they show. Our work described above suggests that finding a secondary low-altitude maximum in the limb dayglow data would demand great care in subtracting scattered sunlight.

Our analyses of the scale heights for the O(^{1}S-^{3}P) Mars dayglow emissions are summarized in Table 2. As mentioned above, both the *Mariner* and early *Mars Express* dayglow observations analyzed occurred during periods of relatively high solar activity. The corresponding high-altitude predictions from the observed and modeled scale heights of the electron density and the O_2^+ density agree well with the fitted dayglow scale heights *versus* solar activity.

In contrast, the initial modeling calculations by Fox and Dalgarno,[30] which were performed with *Viking* or low-solar-activity atmospheric parameters, suggested that the most important source of $O(^1S)$ is photodissociation of CO_2, as in Reaction (1) above. When we examine their Figure 12 we see that, at altitudes between 120 and 240 km, dissociative recombination of electrons with O_2^+, as in Reaction (2) above, is always the second most important $O(^1S)$ production mechanism (and growing in importance at higher altitudes). The newer Fox *et al.* model[39] predicts that at solar maximum the O_2^+ and dissociative recombination scale height has increased by 64%, while the CO_2 and photodissociation scale height has increased by only about 42%. This change could make dissociative recombination the dominant source mechanism from 120 km to above 240 km under solar maximum conditions. Unfortunately, we have not yet analyzed any Mars dayglow observations under solar minimum conditions, but these should become available.[37]

Interpretation and modeling

Modeling $O(^1S)$ production mechanisms and $O(^1S-^3P)$ Mars dayglow emissions

Optical emissions from the $O(^1S)$ and $O(^1D)$ metastable excited states of atomic oxygen provide valuable remotely observable signatures of energy deposition and chemical reaction processes in the atmospheres and ionospheres of Venus, Earth, and Mars. On Earth, the $O(^1S \rightarrow {}^1D)$ green line at 557.7 nm and the $O(^1D \rightarrow {}^3P)$ red line at 630.0 nm are prominent features in nightglow and auroral spectra. From space the $O(^1S \rightarrow {}^3P)$ uv line at 297.2 nm is also observed on Earth (dayglow and nightglow) and Mars (dayglow). Only recently has the $O(^1S \rightarrow {}^1D)$ green line been observed in telescope observations of the nightglow of Venus.[117,118]

Plausible production mechanisms for these emissions include:

$$e^- + O_2^+(v) \rightarrow O(^1S,^1D,^3P) + O(^1S,^1D,^3P) \tag{A}$$

$$O + O + M \rightarrow O_2^*; \text{ followed by } O_2^* + O \rightarrow O(^1S) \tag{B}$$

$$e^- + CO_2^+ \rightarrow O(^1S,^1D,^3P) + CO \tag{C}$$

$$e^- + CO_2 \rightarrow O(^1S,^1D,^3P) + CO + e^- \tag{D}$$

$$h\nu + CO_2 \rightarrow O(^1S,^1D,^3P) + CO \tag{E}$$

Mechanisms (A) and (B) are known to predominate in the Earth's ionosphere and mesosphere, respectively.[76–80,119] We do not yet know the altitude dependence of the Venus nightglow green line emission and thus cannot establish whether mechanism (A) or (B) is dominant.[120]

As indicated above, in the early modeling studies of Fox and Dalgarno,[30] mechanisms (A) and (E) were the leading candidates for the Mars $O(^1S \rightarrow {}^3P)$ dayglow. Our quantitative knowledge of the yield of $O(^1S)$ following photodissociation of CO_2 by solar radiation has improved little since then.[121–123] Fig. 6 summarizes the available quantum yield measurements, all of which are more than 30 years old. The CO_2 absorption cross section (illustrated in the upper frame) is highly structured at all wavelengths, except in the range from about 83 to 90 nm or shorter than about 65 nm. The structure is even more obvious when plotted on a logarithmic scale.[123,124] The experimental $O(^1S)$ quantum yields measurements (illustrated in the lower frame) were all performed using ultraviolet radiation from discharge sources selected with a monochromator with a bandpass between 0.1 and 0.4 nm (if specified). Only one study[73] reported the effects of varying the bandpass, which confirmed the very sharp minimum near 109 nm. More recent measurements by Stark *et al.*[124] found

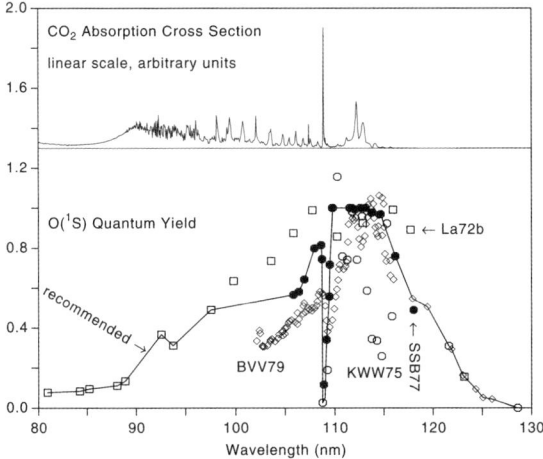

Fig. 6 CO_2 absorption cross section[123] and $O(^1S)$ photodissociation quantum yield. La72b = ref. 70, KWW75 = ref. 71, SSB77 = ref. 73, BVV79 = ref. 74.

that a spectral resolution of 0.005 nm or less is needed to resolve the sharp structures in the CO_2 absorption spectrum and to produce quantitatively reliable cross sections. There is strong evidence that the $O(^1S)$ photodissociation yield is high between 110 and 120 nm. The evidence is a little weaker between 100 and 105 nm. At shorter wavelengths (80 to 100 nm) there is only one data set, with too few data points to have resolved the structure in the absorption spectrum. It could easily be that overall yield of $O(^1S)$ from solar photodissociation of CO_2 is much less than previously believed.[30,121] A detailed review of all available information is recommended, as are additional experiments.[125] Of particular interest is the contribution of H-atom Lyman-α 121.6 nm radiation, which is expected to be the dominant source of $O(^1S)$ at low altitudes.[30] The CO_2 absorption cross section at 121.6 nm is so small that the abundant solar radiation at that wavelength can penetrate below 110 km.

In contrast, the quantitative characterization of mechanism (A) has improved significantly in recent years.[126–133] That $O_2^+(v)$ is the dominant ion in the Mars ionosphere was initially surprising. It is not produced directly but rather by ion-molecule reactions:[127,134]

$$O^+ + CO_2 \rightarrow O_2^+(v) + CO$$

$$O + CO_2^+ \rightarrow O_2^+(v) + CO$$

It is important that the $O_2^+(v)$ product ion from these reactions is observed to contain significant vibrational excitation because, in mechanism (A), excited vibrational levels produce much higher yields of $O(^1S)$, as indicated in Table 3. The vibrationally resolved dissociative recombination rate coefficients and $O(^1S,^1D,^3P)$ yields also

Table 3 Partial cross sections and quantum yields for $e^- + O_2^+(X^2\Pi_g, v = 0,1,2)$ dissociative recombination extrapolated to zero electron energy.[132,133]

v	σ_v	$O(^1S)$	$O(^1D)$	$O(^3P)$	$O(^1S)/O(^1D)$
0	1	0.06	0.94	1.00	0.062
1	0.31 ± 0.13	0.14	1.44	0.52	0.097
2	0.52 ± 0.16	0.21	1.02	0.76	0.207

depend on the electron energy. A more reliable model of the $O(^1S-^3P)$ Mars dayglow would result from including the latest laboratory information.

Modeling $CO_2^+(B^2\Sigma_u^+)$ and $CO(a^3\Pi)$ production mechanisms and $CO_2^+(B^2\Sigma_u^+ \rightarrow X^2\Pi_g)$ and $CO(a^3\Pi \rightarrow X^1\Sigma^+)$ Mars dayglow emissions

There are several plausible production mechanisms for these emissions:

$$e^- + CO_2^+(v) \rightarrow CO(a^3\Pi) + O \tag{F}$$

$$e^- + CO_2 \rightarrow CO(a^3\Pi) + O \tag{G}$$

$$h\nu + CO_2 \rightarrow CO(a^3\Pi) + O \tag{H}$$

$$e^- + CO_2 \rightarrow CO_2^+(B^2\Sigma_u^+) \tag{I}$$

$$h\nu + CO_2 \rightarrow CO_2^+(B^2\Sigma_u^+) \tag{J}$$

$$h\nu + CO_2^+(X^2\Pi_g) \rightarrow CO_2^+(B^2\Sigma_u^+) \tag{K}$$

In the early work by Fox and Dalgarno,[30] above 140 km mechanism (H) was the leading source of $CO(a^3\Pi)$, followed by mechanisms (F) and (G), each about a factor of 5 weaker. Between 120 and 200 km, mechanism (J) was the dominant source of $CO_2^+(B^2\Sigma_u^+)$, with mechanism (K) in second place below 160 km and mechanism (I) above. Although it is important to review the latest laboratory information on all the plausible mechanisms, the suggested focus is on verifying the high-altitude mechanism and evaluating what can be learned by modeling the low-altitude fall-off of the $CO_2^+(B^2\Sigma_u^+ \rightarrow X^2\Pi_g)$ and $CO(a^3\Pi \rightarrow X^1\Sigma^+)$ dayglow emissions below 130 km, especially *versus* the solar zenith angle.

Conclusions

Our analyses of the altitude dependencies of *Mariner* and *Mars Express* mission observations of the $CO(a^3\Pi \rightarrow X^1\Sigma^+)$, $CO_2^+(B^2\Sigma_u^+ \rightarrow X^2\Pi_g)$, and $O(^1S \rightarrow ^3P)$ ultraviolet dayglow emissions from the Mars atmosphere between 130 and 190 km and low solar zenith angles are consistent with, and best explained and modeled by the source reactions

$$h\nu + CO_2 \rightarrow CO(a^3\Pi) + O$$

$$h\nu + CO_2 \rightarrow CO_2^+(B^2\Sigma_u^+) + e^-$$

$$e^- + O_2^+(v) \rightarrow O(^1S) + O$$

Thus the scale heights of the tangent-altitude dependencies of the $CO(a^3\Pi \rightarrow X^1\Sigma^+)$ and $CO_2^+(B^2\Sigma_u^+ \rightarrow X^2\Pi_g)$ emission intensities reflect those of their ambient CO_2 precursor.

In contrast to what has been assumed previously, the altitude dependence of the $O(^1S \rightarrow ^3P)$ emission at 297.2 in the Mars dayglow is *NOT* a monitor of the Mars neutral atmosphere, but instead is a monitor of the Mars ionosphere. Thus, the exospheric temperature of the Mars atmosphere cannot be derived directly from the $O(^1S)$ dayglow emissions. This conclusion overturns an assumption that has been a central feature in Mars atmospheric models for almost four decades.

Acknowledgements

We would like to thank Dr Francois Leblanc (Service d'Aéronomie, IPSL/CNRS) for providing *Mars Express* SPICAM UV dayglow spectra. Dr Philip C. Cosby (SRI International) assisted in preliminary analysis of the MEX data sets. A copy of the Stewart 1987 report, JPL D-41612.pdf, was received from the Jet Propulsion Laboratory (JPL), California Institute of Technology, Pasadena, CA, following Freedom of Information Act (FOIA) Request JPL-08-046. This work was supported under grant NNX06AE04G from the NASA Mars Data Analysis program.

References

1 A. J. Kloire, D. L. Cain, G. S. Levy, V. R. Eshleman, G. Fjeldbo and F. D. Drake, *Science*, 1965, **149**, 1243–1248.
2 G. Fjeldbo and V. R. Eshleman, *Planet. Space Sci.*, 1968, **16**, 1035–1059.
3 C. A. Barth, W. G. Fastie, C. W. Hord, J. B. Pearce, K. K. Kelly, A. I. Stewart, G. E. Thomas, G. P. Anderson and O. F. Raper, *Science*, 1969, **165**, 1004–1005.
4 C. A. Barth, C. W. Hord, J. B. Pearce, K. K. Kelly, G. P. Anderson and A. I. Stewart, *J. Geophys. Res.*, 1971, **76**, 2213–2227.
5 G. Fjeldbo, A. Kliore and B. Seidel, *Radio Sci.*, 1970, **5**, 381–386.
6 C. A. Barth, C. W. Hord, A. I. Stewart and A. L. Lane, *Science*, 1972, **175**, 309–312.
7 C. A. Barth, A. I. Stewart, C. W. Hord and A. L. Lane, *Icarus*, 1972, **17**, 457–468.
8 A. J. Kloire, G. Fjeldbo, B. L. Seidel, M. J. Sykes and P. M. Woiceshyn, *J. Geophys. Res.*, 1973, **78**, 4331–4351.
9 A. O. Nier and M. B. McElroy, *J. Geophys. Res.*, 1977, **82**, 4341–4349.
10 A. Seiff and D. B. Kirk, *J. Geophys. Res.*, 1977, **82**, 4364–4378.
11 W. B. Hanson, S. Sanatani and D. R. Zuccaro, *J. Geophys. Res.*, 1977, **82**, 4351–4363 and Table 3.
12 G. F. Lindal, H. B. Hotz, D. N. Sweetnam, Z. Shippony, J. P. Brenkle, G. V. Hartsell and R. T. Spear, *J. Geophys. Res.*, 1979, **84**, 8443–8456.
13 P. Withers, R. D. Lorenz and G. A. Neumann, *Icarus*, 2002, **159**, 259–261.
14 D. L. Huestis, T. G. Slanger, B. D. Sharpee, P. C. Cosby, and J. L. Fox, presented at the *3rd International Workshop on The Mars Atmosphere: Modeling and Observations*, Williamsburg, VA, 2008.
15 G. E. Thomas, *J. Atmos. Sci.*, 1971, **28**, 859–868.
16 A. I. Stewart, *J. Geophys. Res.*, 1972, **77**, 54–68.
17 R. D. Culp and A. I. Stewart, *J. Astronaut. Sci.*, 1984, **32**, 329–341.
18 A. I. F. Stewart, Revised Time Dependent Model of the Sustenance Studies, Final Report JPL PO #NQ-802429, 1987 [an electronic copy is available upon request: david.huestis@sri.com].
19 C. G. Justus and G. Chimonas, Development of an Engineering Model Atmosphere for Mars, Georgia Inst. Tech. Semi-Annual Report, NASA Grant NAG8–078, 1988.
20 C. G. Justus, *J. Spacecraft*, 1991, **28**, 216–221.
21 C. G. Justus, B. G. James and D. L. Johnson, *Adv. Space Res.*, 1997, **19**, 1223–1231.
22 C. G. Justus and D. L. Johnson, Mars Global Reference Atmospheric Model 2001 Version (Mars-GRAM 2001): Users Guide, 2001, [http://trs.nis.nasa.gov/archive/00000549].
23 C. J. Justus, B. F. James, S. W. Bougher, A. F. C. Bridger, R. M. Haberle, J. R. Murphy and S. Engel, *Adv. Space Res.*, 2002, **29**, 193–202.
24 C. G. Justus, A. Duvall and D. L. Johnson, *Adv. Space Res.*, 2004, **34**, 1673–1676.
25 C. G. Justus, Mars Global Reference Atmospheric Model 2005, Version 1.2, readme.txt on the Mars-GRAM 2005 CD, 2007, [http://see.msfc.nasa.gov/tte/model_Marsgram.htm].
26 A. Dalgarno, T. C. Degges and A. I. Stewart, *Science*, 1970, **167**, 1490–1491.
27 J. C. McConnell and M. B. McElroy, *J. Geophys. Res.*, 1970, **75**, 7290–7293.
28 C. A. Barth, *Z. Naturforsch.*, 1974, **29**, 185–188.
29 M. N. Izakov, *Icarus*, 1978, **36**, 189–197.
30 J. L. Fox and A. Dalgarno, *J. Geophys. Res.*, 1979, **84**, 7315–7333.
31 R. Rodrigo, E. Garcia-Alvarez, M. J. Lopez-Gonzalez and J. J. Lopez-Moreno, *J. Geophys. Res.*, 1990, **95**, 14795–14810.
32 H. Nair, M. Allen, A. D. Anbar and Y. L. Yung, *Icarus*, 1994, **111**, 124–130.

33 J. L. Fox, in *Atomic, Molecular, and Optical Physics Handbook*, ed. G. W. F. Drake, AIP Press, Woodbury, NY, 1996, pp. 940–968.
34 V. A. Krasnopolsky, *J. Geophys. Res.*, 2002, **107**, DOI: 10.1029/2001E001809.
35 J. L. Fox and K. E. Yeager, *J. Geophys. Res.*, 2006, **111**, A10309.
36 V. I. Shematovich, D. V. Bisikalo, J.-C. Gerard, C. Cox, S. W. Bougher and F. Leblanc, *J. Geophys. Res.*, 2008, **113**, E02011.
37 C. Simon, O. Witasse, F. Leblanc, G. Gronoff and J.-L. Bertaux, *Planet. Space Sci.*, 2009, **57**, 1008–1021.
38 S. W. Bougher, R. G. Roble, E. C. Ridley and R. E. Ridley, *J. Geophys. Res.*, 1990, **95**, 14811–14827.
39 J. L. Fox, P. Zhou and S. W. Bougher, *Adv. Space Res.*, 1996, **17**, 203–216.
40 S. W. Bougher, S. Engel, R. G. Roble and B. Foster, *J. Geophys. Res.*, 2000, **105**, 17669–17692 and Fig. 2.
41 S. W. Bougher, S. Engel, D. P. Hinson and J. R. Murphy, *J. Geophys. Res.*, 2004, **109**, E03010.
42 S. W. Bougher, J. M. Bell, J. R. Murphy, and R. Lillis, *Proceedings of the 2nd Workshop on Mars Atmosphere Modeling and Observations*, Granada, Spain, 2006.
43 J. A. Magalhães, J. T. Schofield and A. Seiff, *J. Geophys. Res.*, 1999, **104**, 8943.
44 G. M. Keating, S. W. Bougher, R. W. Zurek, R. H. Tolson, G. J. Cancro, S. N. Noll, J. S. Parker, T. J. Schellenberg, R. W. Shane, B. L. Wilkerson, J. R. Murphy, J. L. Hollingsworth, R. M. Haberle, M. Joshi, J. C. Pearl, B. J. Conrath, M. D. Smith, R. T. Clancy, R. C. Blanchard, R. G. Wilmoth, D. F. Rault, T. Z. Martin, D. T. Lyons, P. B. Esposito, M. D. Johnston, C. W. Whetzel, C. G. Justus and J. M. Babicke, *Science*, 1998, **279**, 1672–1676.
45 P. W. Tracadas, M. T. Zuber, D. E. Smith and F. G. Lemoine, *J. Geophys. Res.*, 2001, **106**, 23349–23357.
46 P. Withers, S. W. Bougher and G. M. Keating, *Icarus*, 2003, **163**, 14–32.
47 E. Mazarico, M. T. Zuber, F. G. Lemoine and D. E. Smith, *J. Geophys. Res.*, 2007, **112**, E05014.
48 R. H. Tolson, A. M. Dwyer, J. L. Hanna, G. M. Keating, B. E. George, P. E. Escalera and M. R. Werner, *J. Spacecr. Rockets*, 2005, **42**, 435–443.
49 P. Withers and M. D. Smith, *Icarus*, 2006, **185**, 133–142.
50 J.-L. Bertaux, F. Leblanc, S. Perrier, E. Quemerais, O. Korablev, E. Dimarellis, A. Reberac, F. Forget, P. C. Simon, S. A. Stern and B. Sandel, the SPICAM Team, *Science*, 2005, **307**, 566–569.
51 J.-L. Bertaux, F. Leblanc, O. Witasse, E. Quemerais, J. Lilensten, S. A. Stern, B. Sandel and O. Korablev, *Nature*, 2005, **435**, 790–794.
52 J.-L. Bertaux, O. Korablev, S. Perrier, E. Quemerais, M. Montmessin, F. Leblanc, S. Lebonnois, F. Lefevre, F. Forget, A. Fedorova, E. Dimarellis, A. Reberac, D. Fonteyn, J. Y. Chaufray and S. Guibert, *J. Geophys. Res.*, 2006, **111**, E10S90.
53 E. Quemerais, J.-L. Bertaux, O. Korablev, S. Perrier, E. Dimarellis, C. Cot, B. R. Sandel and D. Fussen, *J. Geophys. Res.*, 2006, **111**, E09S04.
54 F. Leblanc, J. Y. Chaufray and J.-L. Bertaux, *Geophys. Res. Lett.*, 2007, **34**, L02206.
55 F. Leblanc, J. Y. Chaufray, J. Lilensten, O. Witasse and J.-L. Bertaux, *J. Geophys. Res.*, 2006, **111**, E09S11.
56 C. Cox, J.-C. Gerard, B. Hubert, J.-L. Bertaux and S. W. Bougher, *J. Geophys. Res.*, 2010, **115**, E04010.
57 C. E. Brion and K. H. Tan, *Chem. Phys.*, 1978, **34**, 141–151.
58 R. W. Carson, D. L. Judge and M. Ogawa, *J. Geophys. Res.*, 1973, **78**, 3194–3196.
59 T. Gustafsson, E. W. Plummer, D. E. Eastman and W. Gudat, *Phys. Rev. A: At., Mol., Opt. Phys.*, 1978, **17**, 175–181.
60 E. P. Gentieu and J. E. Mentall, *J. Chem. Phys.*, 1976, **64**, 1376–1380.
61 L. C. Lee, R. W. Carlson and D. L. Judge, *J. Phys. B: At. Mol. Phys.*, 1976, **9**, 855–863.
62 S. Leach, M. Devoret and J. H. D. Eland, *Chem. Phys.*, 1978, **33**, 113–121.
63 L. C. Lee and D. L. Judge, *J. Chem. Phys.*, 1972, **57**, 4443–4445.
64 S. Leach, P. R. Standard and W. E. Gelbart, *Mol. Phys.*, 1978, **36**, 1119–1132.
65 J. A. R. Samson and J. L. Gardner, *J. Chem. Phys.*, 1973, **58**, 3771–3774.
66 J. A. R. Samson and J. L. Gardner, *J. Geophys. Res.*, 1973, **78**, 3663–3667.
67 J. A. R. Samson, J. L. Gardner and J. E. Mentall, *J. Geophys. Res.*, 1972, **77**, 5560–5566.
68 T. S. Wauchop and H. P. Broida, *J. Geophys. Res.*, 1971, **76**, 21–26.
69 G. M. Lawrence, *J. Chem. Phys.*, 1972, **56**, 3435–3442.
70 G. M. Lawrence, *J. Chem. Phys.*, 1972, **57**, 5616–5617.
71 I. Koyano, T. S. Wauchop and K. H. Welge, *J. Chem. Phys.*, 1975, **63**, 110–112.
72 B. A. Ridley, R. Atkinson and K. H. Welge, *J. Chem. Phys.*, 1973, **58**, 3878–3880.
73 T. G. Slanger, R. L. Sharpless and G. Black, *J. Chem. Phys.*, 1977, **67**, 5317–5323.

74 N. K. Bibinov, F. I. Vilesov, I. P. Vinogradov, L. D. Mineev and A. M. Pravilov, *Sov. J. Quantum Electron.*, 1979, **9**, 838–844.
75 F. Leblanc, personal communication, 2005.
76 P. M. Banks and G. Kocharts, *Aeronomy Parts A and B*, Academic Press, New York, 1973.
77 D. R. Bates, *Planet. Space Sci.*, 1978, **26**, 897–912.
78 D. R. Bates, *Planet. Space Sci.*, 1990, **38**, 889–902.
79 J. H. A. Sobral, H. Takahashi, M. A. Abdu, P. Muralikrishna, Y. Sahai and C. J. Zamlutti, *Planet. Space Sci.*, 1992, **40**, 607–619.
80 J. W. Chamberlain, *Physics of the Aurora and Airglow*, Am. Geophys. Union Reprint Edition, Washington, DC, 1995.
81 L. G. Jacchia, *Revised Static Models of the Thermosphere and Exosphere with Empirical Temperature Profiles*, Smithsonian Astrophysical Observatory Special Report 332, 1971.
82 L. G. Jacchia, *Thermospheric Temperature, Density, and Composition: New Models*, Smithsonian Institute Astrophysical Observatory Special Report 375, 1977. [nssdcftp.gsfc.nasa.gov/models/atmospheric/jacchia/jacchia-77/j77sri.for].
83 D. L. Huestis, Jacchia 1977 Atmospheric Model, 2002. [nssdcftp.gsfc.nasa.gov/models/atmospheric/jacchia/jacchia-77/j77sri.for], 2002.
84 S. S. Kirby, T. R. Gilliland, E. B. Judson and N. Smith, *Phys. Rev.*, 1935, **48**, 849.
85 T. Kakinuma and G. Swarup, *Astrophys. J.*, 1962, **136**, 975–994.
86 O. Hachenberg, in *Solar System Radio Astronomy*, ed. J. Aarons, Plenum Press, NY, 1965, pp. 95–108.
87 L. G. Jacchia and J. W. Slowey, in *Space Research XIII*, ed. M. J. Rycroft and S. K. Runcorn, Akademie-Verlag, Berlin, 1973, pp. 343–348.
88 L. G. Jacchia, J. W. Slowey and I. G. Campbell, *Planet. Space Sci.*, 1973, **21**, 1835–1842.
89 J. E. Titheridge, *J. Atmos. Sol.–Terr. Phys.*, 2003, **65**, 1035–1052.
90 M. R. Torr, D. G. Torr, R. A. Ong and H. E. Hinteregger, *Geophys. Res. Lett.*, 1979, **6**, 771–774.
91 P. G. Richards, J. A. Fennelly and D. G. Torr, *J. Geophys. Res.*, 1994, **99**, 8981.
92 R. W. Schunk and A. F. Nagy, *Ionospheres – Physics, Plasma Physics, and Chemistry*, Cambridge University Press, Cambridge, 2000.
93 R. W. Schunk and A. F. Nagy, *Ionospheres – Physics, Plasma Physics, and Chemistry*, Cambridge University Press, Cambridge, 2nd edn, 2009.
94 D. L. Huestis, *EUVAC Solar EUV Flux Model*, 2009. [http://www-mpl.sri.com/software/euvac.html].
95 M. J. Buonsanto, P. G. Richards, W. K. Tobiska, S. C. Solomon, Y.-K. Tung and J. A. Fennely, *J. Geophys. Res.*, 1995, **100**, 14569–14580.
96 W. K. Tobiska, T. Woods, F. Eparvier, R. Viereck, L. Floyd, D. Bouwer, G. Rottman and O. R. White, *J. Atmos. Sol.–Terr. Phys.*, 2000, **62**, 1233–1250.
97 P. G. Richards, T. N. Woods and W. K. Peterson, *Adv. Space Res.*, 2006, **37**, 315–322.
98 T. Woods, *Tom Woods Rocket Result Page*, 2007. [http://lasp.colorado.edu/rocket/rocket_results.html].
99 M. H. Hantsch and S. J. Bauer, *Planet. Space Sci.*, 1990, **38**, 539–542.
100 C. E. Martinis, J. K. Wilson and M. J. Mendillo, *J. Geophys. Res.*, 2003, **108**(A10), 1383.
101 J. L. Fox, *J. Geophys. Res.*, 2004, **109**, A11310.
102 P. Withers and M. Mendillo, *Planet. Space Sci.*, 2005, **53**, 1401–1418.
103 J. M. Forbes, S. Bruinsma and F. G. Lemoine, *Science*, 2006, **312**, 1366–1368.
104 J. M. Forbes, F. G. Lemoine, S. L. Bruinsma, M. D. Smith and X. Zhang, *Geophys. Res. Lett.*, 2008, **35**, L01201.
105 S. W. Bougher, T. M. McDunn, K. A. Zoldak and J. M. Forbes, *Geophys. Res. Lett.*, 2009, **36**, L05201.
106 J. L. Fox and K. E. Yeager, *Icarus*, 2009, **200**, 468–479.
107 F. Gonzalez-Galindo, F. Forget, M. A. Lopez-Valverde, M. Angelats i Coll and E. Millour, *J. Geophys. Res.*, 2009, **114**, E04001.
108 M. B. McElroy, T. Y. Kong and Y. L. Yung, *J. Geophys. Res.*, 1977, **82**, 4379–4388.
109 A. J. Stewart and W. B. Hanson, *Adv. Space Res.*, 1982, **2**, 87–101.
110 F. Montmessin, E. Quemerais, J. L. Bertaux, O. Korablev, P. Rannou and S. Lebonnois, *J. Geophys. Res.*, 2006, **111**, E09S09.
111 M. A. LeCompte, L. J. Paxton and A. I. F. Stewart, *J. Geophys. Res.*, 1989, **94**, 208–216.
112 J. W. Chamberlain and D. M. Hunten, *Theory of Planetary Atmospheres*, Academic Press, San Diego, 2nd edn, 1987, p. 3.
113 D. L. Huestis, *J. Quant. Spectrosc. Radiat. Transf.*, 2001, 69, 709–721. [http://en.wikipedia.org/wiki/Chapman_function].
114 D. L. Huestis, *J. Quant. Spectrosc. Radiat. Transf.*, submitted, 2007.
115 A. I. Stewart, C. A. Barth, C. W. Hord and A. L. Lane, *Icarus*, 1972, **17**, 469–474.
116 A. O. Nier and M. B. McElroy, *Science*, 1976, **82**, 1298–1300.

117 T. G. Slanger, P. C. Cosby, D. L. Huestis and T. A. Bida, *Science*, 2001, **291**, 463–465.
118 T. G. Slanger, D. L. Huestis, P. C. Cosby, N. J. Chanover and T. A. Bida, *Icarus*, 2006, **182**, 1–9.
119 T. G. Slanger and G. Black, *Planet. Space Sci.*, 1977, **25**, 79–88.
120 T. G. Slanger and J. L. Fox, Proceedings of the 41st Meeting of the AAS Division for Planetary Sciences, Farjardo, Puerto Rico, 2009, Bull. AAS 41((3)), p. 1127.
121 D. L. Huestis and T. G. Slanger, Proceedings of the 38th Annual Meeting of the AAS Division for Planetary Sciences, Pasadena, CA, 2006, Bull AAS 38, p. 609.
122 D. L. Huestis and J. Berkowitz, *Proceedings of the Asia Oceania Geosciences Society (AOGS) 7th Annual Meeting*, Hyderabad, 2010.
123 D. L. Huestis, S. W. Bougher, J. L. Fox, M. Galand, R. E. Johnson, J. I. Moses and J. C. Pickering, *Space Sci. Rev.*, 2008, **139**, 63–105.
124 G. Stark, K. Yoshino, P. L. Smith and K. Ito, *J. Quant. Spectrosc. Radiat. Transfer*, 2007, **103**, 67–73.
125 K. S. Kalogerakis, C. Romanescu, D. L. Huestis, T. G. Slanger, L. C. Lee, M. Ahmed, and K. R. Wilson, *Proceedings of the Asia Oceania Geosciences Society (AOGS) 6th Annual Meeting*, Singapore, 2009.
126 J. L. Queffelec, B. R. Rowe, F. Vallee, J. C. Gomet and M. Morlais, *J. Chem. Phys.*, 1989, **91**, 5335–5342.
127 C. W. Walter, P. C. Cosby and J. R. Peterson, *J. Chem. Phys.*, 1993, **98**, 2860–2871.
128 R. Peverall, S. Rosen, M. Larsson, J. R. Peterson, R. Bobbenkamp, S. L. Guberman, H. Danared, M. Ugglas, A. Al-Khalili, A. N. Maurellis and W. J. van der Zande, *Geophys. Res. Lett.*, 2000, **27**, 481–484.
129 R. Peverall, S. Rosen, J. R. Peterson, M. Larsson, A. Al-Khalili, L. Vikor, J. Semaniak, R. Bobenkamp, A. Le Padellec and W. J. van der Zande, *J. Chem. Phys.*, 2001, **114**, 6679–6689.
130 P. C. Cosby, J. R. Peterson, and D. L. Huestis, in *Dissociative Recombination of Molecular Ions with Electrons*, ed. S. L. Guberman, Kluwer/Plenum, New York, 2003, pp. 101–108.
131 A. Petrignani, F. Hellberg, R. D. Thomas, M. Larsson, P. C. Cosby and W. J. van der Zande, *J. Chem. Phys.*, 2005, **122**, 234311.
132 A. Petrignani, W. J. van der Zande, P. C. Cosby, F. Hellberg, R. D. Thomas and M. Larsson, *J. Chem. Phys.*, 2005, **122**, 014302.
133 P. C. Cosby and D. L. Huestis, submitted as Final Report on NASA grant NAG5-11173, 2009.
134 J. L. Fox, *Adv. Space Res.*, 1985, **5**, 165.

PAPER

Laboratory chemistry relevant to understanding and modeling the ionosphere of Titan

Nigel G. Adams,* L. Dalila Mathews and David Osborne, Jr

Received 23rd February 2010, Accepted 12th April 2010
DOI: 10.1039/c003233f

Laboratory data have a dual and critical role in interpreting information obtained from the Cassini spacecraft in its passes through the Titan ionosphere. Firstly, *in situ* mass spectra are obtained by Cassini and their conversion into atmospheric molecular composition requires chemical modeling to create agreement between the observed mass spectra and those determined from the models. Secondly, once agreement is obtained, then the chemical model can be considered to represent the evolution of the Titan atmosphere. As a contribution to these endeavors in the past, laboratory measurements have been made in the Selected Ion Flow Tube (SIFT) of the reactions of a series of ring molecules with the important ionospheric ion CH_3^+. These reactions showed that a dominant reaction channel is association. In the present study, this work has been extended to reactions of another important Titan ion $C_3H_3^+$. These ion–molecule reactions have also been studied at room temperature using a SIFT. Reactions have been studied in detail with benzene, toluene and pyridine and show again that association is very important. The loss of ionization in the ionosphere is then controlled by electron–ion dissociative recombination of the association ions and their progeny. The recombination reactions have been studied as a function of temperature (300 to 550 K) using a flowing afterglow. These combined data have been used to develop a subset of the chemistry and test its viability. They have indicated that association of the important Titan ions with the abundant nitrogen, followed by switching of the nitrogen for the ring compounds, can build up larger species, perhaps resulting in multi-rings. Recombination of such species can affect the ionization balance and provide species which can contribute to the parallel neutral chemistry. Species are suggested that should be looked for in the *in situ* mass spectra.

Introduction

The eminently successful Cassini spacecraft, as part of its mission, has probed the atmosphere of Titan in a series of flybys (>62).[1-5] Data from the Ion Neutral Mass Spectrometer (INMS) have shown a rich chemistry with ions being detected at almost every mass in the range 1 to 99 amu with better than one unit mass resolution.[6-8] Based on an earlier flyby of this atmosphere by Voyager in 1980, radar occultation showed that the atmosphere is mainly N_2 with some hydrocarbons and this was supported by chemical modeling.[9] The data obtained by the Cassini INMS also supports this conclusion. Additional mass spectral information was obtained by the CAPS (Cassini Plasma Spectrometer) Ion Beam Spectrometer (IBS) which showed, at a lower mass resolution, that there were positive ions with masses from 100 to 350 amu. Also, the CAPS Electron Spectrometer (ELS) detected negative ions having masses between 50 and 8000 amu.[8] The former observation is

Department of Chemistry, University of Georgia, Athens, GA, 30602, USA

consistent with a hydrocarbon chemistry up to higher masses[10,11] and the latter indicate that PAHs and assorted nitriles are present, perhaps leading to tholins.[8] Even with all of this information, the molecular compounds are not directly identified. Rather the atmosphere is modeled with all the known hydrocarbon chemistry and a predicted mass spectrum generated to compare with the actual flyby spectra. Then the chemical model is adjusted to get agreement with the actual spectra and when this is achieved the atmospheric molecular composition is considered to be known. This of course relies on all the relevant chemistry being included in the chemical model. It also requires that all of the rate constants and ion and neutral product distributions be known. In addition, the indication that PAHs are present means that smaller rings are probably also present. Indeed, the chemical modeling of the INMS mass spectra has shown that the mass 79 amu ion is either $C_6H_7^+$ (protonated benzene) or ionized pyridine ($C_5H_5N^+$); in addition mass 80 and 93 have been tentatively identified as protonated pyridine ($C_5H_5NH^+$) and protonated toluene ($C_7H_9^+$), respectively.[10,12,13] This suggests that other single rings are present, both ion and neutral. Of these, the possible neutral rings will react with the abundant ions in the atmosphere which include CH_3^+, $C_2H_5^+$, $C_3H_3^+$, etc.[14,15] Thus, to add to the body of kinetic data, the reactions of CH_3^+ with the single ring compounds benzene (C_6H_6), cyclohexane (C_6H_{12}), pyridine (C_5H_5N), pyrimidine ($C_4H_4N_2$), piperidine ($C_5H_{11}N$), 1,4-dioxane ($C_4H_8O_2$), furan (C_4H_4O), pyrrole (C_4H_5N) and pyrrolidine (C_4H_9N) were previously investigated.[16] These reactions all proceed at close to the gas kinetic rate. What is surprising though is that the ion product distributions for aromatic species with π electrons in the ring are very different from the non-aromatic rings without π electrons. For aromatics, there is an association channel competitive with charge transfer (both dissociative and non-dissociative) and hydride ion (H^-) abstraction. Proton transfer, which is normally facile, hardly competes. Also, in contrast, association is not observed for non-aromatic rings.

Critical to the chemistry is electron–ion dissociative recombination (DR), which together with ionization of the major neutral species by solar photons and Saturn's magnetospheric electrons determines the ionization balance. Recent studies of the DR of the protonated cyanides $HCNH^+$, CH_3CNH^+ and $C_2H_5CNH^+$ have shown that these species recombine very rapidly.[17,18] In addition, under laboratory conditions, association

$$RCNH^+ + RCN + He \rightarrow (RCN)_2H^+ + He \quad (1)$$

(where R is H, CH_3 or C_2H_5) is competitive with $RCNH^+$ DR, and may also be so under Titan ionospheric conditions, allowing larger ion species to be created before recombination occurs.

The previous laboratory studies discussed above, were carried out primarily to contribute to an understanding of Titan ionospheric chemistry. The CH_3^+ reactions have shown that there is considerable association for the aromatic rings, which will be enhanced further at the low temperatures of the Titan ionosphere. Thus, this will result in the growth of unsaturated compounds as compared with those that are fully saturated. For the other ions that are important in the Titan ionosphere a series of reactions of $C_3H_3^+$ with rings have been investigated in the present study. In addition, much of the Titan chemistry is thought to center around proton transfer and we have suggested that a significant component could involve protonated ring compounds. If this were the case, DR of such species would be an important ionization loss process. Because of this, we have also studied such recombination reactions and present these data below.

Ion–molecule reactions

Rate constants and ion product distributions for ion–molecule reactions were determined in a Selected Ion Flow Tube (SIFT)[19] for reactions with $C_3H_3^+$. The SIFT consists of a stainless steel tube (~1 meter long, 8 cm in diameter) along which helium is flowed. Ions are injected into the upstream end of the tube from a remote ion source after mass selection in a quadrupole mass filter. A low pressure electron impact ion source containing propyne, C_3H_4 was used to produce the ions of interest. The ions flow along the tube and are sampled at the downstream end through an orifice in a nose cone. They are detected with a second quadrupole mass filter and counted by an electron multiplier. Gases, with which the ions react, are introduced at various positions along the flow tube giving a series of reaction distances and thus reaction times. The rate constants of the ion–molecule reactions are determined from the variation of the detected primary ion signal with reactant gas flow. Percentage ion product distributions can be obtained from the variation of the percentages of the various product ions with reactant gas flow. Extrapolation to zero reactant gas flow removes the effects of secondary reactions. Rate constants and ion product distributions are then calculated in the usual way.[20,21] Rate constants are accurate to within ±20 to 30% dependent on the degree of sticky nature of the reactant gas. Product ion distributions are accurate to ±5 in the percentage.

$C_3H_3^+$ reactions have been studied with the single hydrocarbon rings, benzene (C_6H_6) and toluene (C_7H_8), and with the nitrogen containing ring pyridine (C_5H_5N) to obtain rate constants and full ion product distributions. These reactions were studied in detail since the reactant neutrals are important observed species in the Titan atmosphere. Other ring compounds (the N-hetero rings, piperidine $C_5H_{11}N$, pyrimidine $C_4H_4N_2$, pyrrolidine C_4H_9N, and pyrrole C_4H_5N and the O-hetero rings tetrahydrofuran C_4H_8O, furan C_4H_4O, and 1,4-dioxane $C_4H_8O_2$) were studied to quantitatively determine the magnitude of the association channels.[22] An example of these data for reaction with pyridine is presented in Fig. 1. Here the $C_3H_3^+$ signal as a function of pyridine flow shows a linear decay on a semi-log plot with the concurrent increase of the products due to proton transfer (20%) and association (80%). Notice that a linear decay of $C_3H_3^+$ is not observed in all cases as is shown in Fig. 2 for the decay of the $C_3H_3^+$ signal with toluene flow where the decay is bi-exponential; compare this with the variation with pyridine flow. This behavior is due to the presence of the linear (propargyl) and cyclic

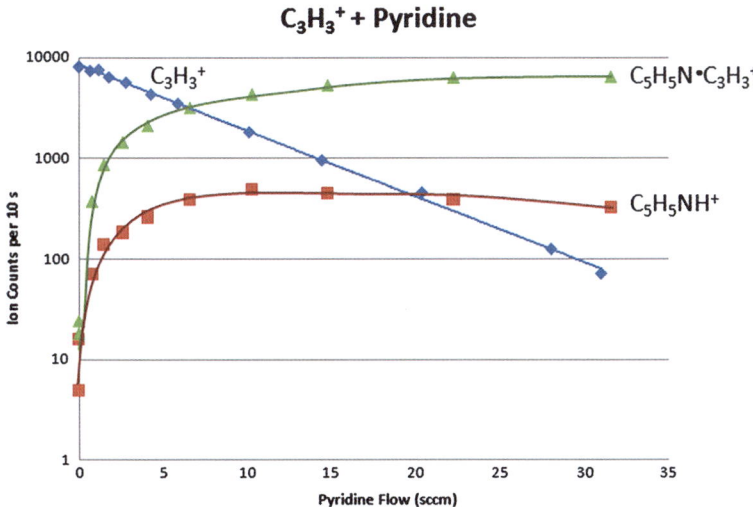

Fig. 1 Reactant and product ion counts for the reaction of $C_3H_3^+$ with pyridine (C_5H_5N) at 298 K. Note that the major product is association (80%) which dominates over the usually rapid proton transfer (20%).

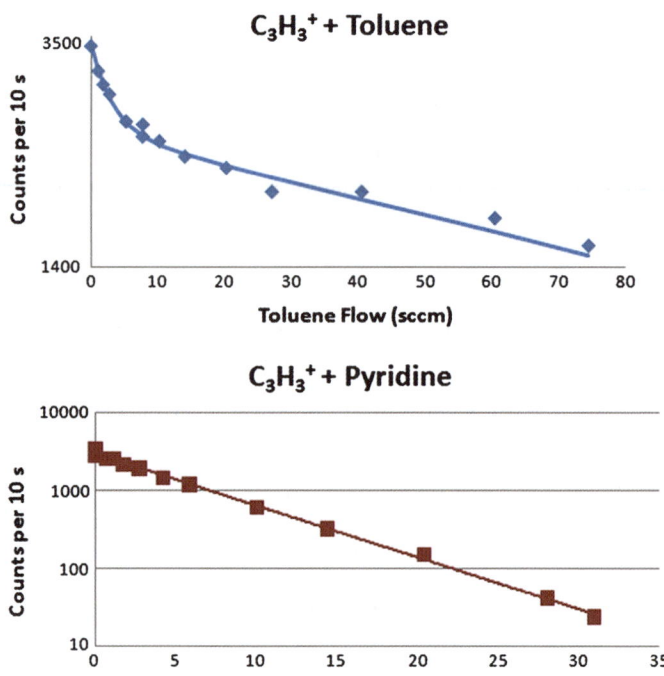

Fig. 2 Reactant ion counts for the reactions of $C_3H_3^+$ with toluene and pyridine. Notice the differences in the ordinate scales. The toluene data show the presence of linear (propargyl; 30%) and cyclic (cyclopropenium; 70%) isomers reacting differently. The pyridine data show the two isomers reacting at the same rate. Since the rate constant is almost gas kinetic, the linearity cannot be due to quenching of the higher energy isomer before reaction.

(cyclopropenium) $C_3H_3^+$ isomers as has been shown previously from reactions with benzene.[23] Rate constants for the individual isomers were obtained by fitting the bi-exponential decay. This also gave the proportions as ~30% l-$C_3H_3^+$ and ~70% c-$C_3H_3^+$ with c-$C_3H_3^+$ being less reactive. Thus, when a linear decay was observed, it implied that both isomers reacted at approximately the same rate and thus that the ion product distributions applied to a 30/70% mix. Note that with a bi-exponential decay, the product distribution deduced is closer to that for the more rapidly reacting species. The kinetic data for these ring reactions are given in Tables 1 and 2. In all cases, the reactions are gas kinetic for l-$C_3H_3^+$ and generally much slower for

Table 1 Rate constants at 298 K for the reactions of linear and cyclic $C_3H_3^+$ isomer ions with the ring compounds (C_6H_6, C_7H_8 and C_5H_5N) indicated. The efficiencies quoted are for the more reactive l-$C_3H_3^+$. Comparisons to literature values are given.[a]

Neutral molecule	l-$C_3H_3^+$ k_{exp}/cm^3 s^{-1}	c-$C_3H_3^+$ k_{exp}/cm^3 s^{-1}	k_{Theory}/cm^3 s^{-1}	Efficiency
Benzene	1.7(−9)	≤ 5(−13)	1.5(−9)	1.1
	1.40(−9)[45]			
Toluene	1.9(−9)	3.8(−11)	1.7(−9)	1.1
	1.75(−9)[45]	1.70(−11)[45]		
Pyridine	2.2(−9)	2.2(−9)	2.6(−9)	0.85

[a] k_{Theory} is calculated using combined variational transition state theory and classical trajectory theory.[46]

Table 2 Percentage ion product distributions (at 298 K) for the $C_3H_3^+$ isomer reactions with the rings (C_6H_6, C_7H_8 and C_5H_5N) as indicated. Since c-$C_3H_3^+$ is unreactive with benzene and toluene (see Table 1) the product distributions for these neutral reactants apply to l-$C_3H_3^+$. That for pyridine applies to the isomer mixture.[a],[c]

Neutral reactant	Ion products	Product channel	Yield
Benzene[b]	$C_6H_6{\cdot}C_3H_3^+$	A	84%
C_6H_6	$C_6H_7^+$	PT	16%
750.4 kJ mol^{-1}			
9.24 eV			
Toluene	$C_7H_8{\cdot}C_3H_3^+$	A	43%
C_7H_8	$C_9H_7^+$	AF	28%
784 kJ mol^{-1}	$C_{10}H_9^+$	AF	17%
8.83 eV	$C_{10}H_8^+$	AF	12%
Pyridine	$C_5H_5N{\cdot}C_3H_3^+$	A	80%
C_5H_5N	$C_5H_5NH^+$	PT	20%
930 kJ mol^{-1}			
9.26 eV			

[a] A is association; AF is association with fragmentation; PT is proton transfer. [b] A previous ICR study gave the products of this reaction as $C_7H_7^+$ and $C_9H_9^+$.[23] This was at a much lower pressure than in the SIFT and would not have shown an association channel. [c] The energies under the reactant neutrals refer to the proton affinity (kJ mol^{-1}) and ionization energy (eV).

c-$C_3H_3^+$ except for the reaction with pyridine. Indeed, for all of the N-hetero rings, the two isomers react at the same gas kinetic rate.[22] For the product distributions there are only two channels with proton transfer dominated by association; this is greater than in the CH_3^+ reactions as can be seen from Table 3. The dominance of association is consistent through all the $C_3H_3^+$ reactions studied whether with pure hydrocarbon or N-hetero rings. Other channels are charge transfer, proton

Table 3 Percentage association channels for the reactions of CH_3^+ and $C_3H_3^+$ with the series of single ring compounds indicated (determined at 298 K). Product ion structures are suggested where possible. In many cases this identification is not possible since there are many isomers (often 20 or 30). This information is obtained from the NIST Webbook.[47]

	CH_3^+[16]		$C_3H_3^+$	
Reactant neutral	% Association	Possible product identification	% Association	Possible product identification
Piperidine, $C_5H_{11}N$	0%		[a]$C_8H_{14}N^+$ 76%	n-Butylpyrrole H$^+$
Pyrrolidine, C_4H_9N	0%		[a]$C_7H_{12}N^+$ 69%	1-Azabicyclo[2.2.2]oct-2-ene H+
Benzene, C_6H_6	$C_7H_9^+$ 47%	Toluene H$^+$	$C_9H_9^+$ 84%	Indene H+
Toluene, C_7H_8	—		$C_{10}H_{11}^+$ 100%[b]	Naphthalene-1,2-dihydro H$^+$
Pyridine, C_5H_5N	$C_6H_8N^+$ 50%	Aniline H$^+$, methyl pyridine H$^+$	$C_8H_8N^+$ 80%	Indolizine H$^+$
Pyrimidine, $C_4H_4N_2$	$C_5H_7N_2^+$ 24%		[a]$C_7H_7N_2^+$ 89%	1H-Indazole H+
Pyrrole, C_4H_5N	$C_5H_8N^+$ 8%		[a]$C_7H_8N^+$ 84%	Pyridine, 3-ethenyl H+

[a] Details of these data subject of a separate study[22] [b] The percentage quoted includes the association channel plus channels where there is fragmentation after association.

transfer and hydride ion abstraction which are all small. The recombination energies for these ions are 8.67 eV (l-$C_3H_3^+$)[24,25] and 6.6 eV (c-$C_3H_3^+$),[26] respectively.

Reactions of the ring compounds with other important Titan ions, $C_2H_5^+$, $HCNH^+$, *etc.* need to be studied to more clearly establish the trends in reactivity. It should be noted that recently a detailed experimental and theoretical investigation of the $C_2H_5^+$ + C_6H_6 reaction was made.[27] Under single collision conditions, it was shown that the reaction proceeds *via* a long lived association complex implying that stabilized association will occur under the higher pressure conditions in the Titan lower atmosphere. Thus, those data seem to be consistent with the data presented here. Further investigations of the reactions of these ions with other ring compounds are obviously required.

Electron–ion dissociative recombination (DR)

DR is the dominant ionization loss process in most molecular plasmas including the Titan ionosphere. Studies of hydrocarbon ion chemistry have shown that N-hetero compounds could be important in rapidly generating large neutral species by association and recombination. To better understand this latter process and provide kinetic data, DR has been studied for a variety of ring compounds in particular, protonated benzene, pyridine, and pyrimidine. This has been accomplished in a temperature variable flowing afterglow using the Langmuir probe technique.

The flowing afterglow, like the SIFT, consists of a tube ~1 meter long and 8 cm in diameter down which helium is flowing. However, here unlike the SIFT, ionization is created directly in the upstream end of the flow tube with a microwave discharge in helium. The composition of the flowing electron–ion plasma is controlled by adding gases at various positions along the flow tube. Typical gas additions were the sequential addition of Ar and H_2 creating an Ar^+/electron plasma followed by an H_3^+/electron plasma. Introduction of the reactant gas then produces the recombining protonated reactant gas, $H^+ \cdot$ Ring, by proton transfer from H_3^+ and this recombines with the electrons as determined by the rate constant, α_e. The rate constants are

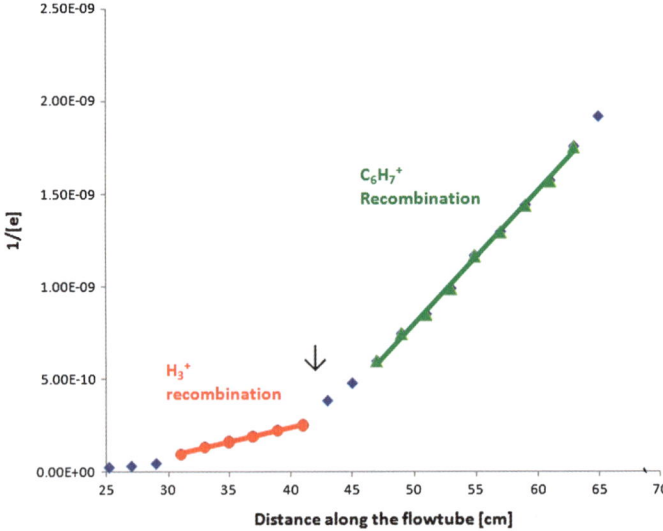

Fig. 3 Reciprocal electron density, 1/[e] (cm^3), as a function of distance along the flow tube for DR of protonated benzene, $C_6H_7^+$. Upstream of the C_6H_6 addition point (indicated by the arrow) is the H_3^+ DR and this converts to $C_6H_7^+$ by proton transfer. The slope of the line gives the recombination rate constant, α_e.

determined using the Langmuir probe to measure the electron density as a function of distance along the flow tube. Fig. 3 shows the variation of the reciprocal of the electron density with distance, the slope of which gives the recombination rate constant. Vibrational relaxation of the recombining ions is achieved by the resonant proton transfer

$$H^+ \cdot Ring^* + Ring \rightarrow H^+ \cdot Ring + Ring^* \qquad (2)$$

where * represents vibrational excitation. α_e was then measured at a series of reactant gas concentrations to determine the concentration where the ion of interest was the dominant recombining ion. In this way, the α_e for the protonated monomer ions ($H^+ \cdot Ring$) could be obtained. At higher flows of reactant gas, association becomes competitive with $H^+ \cdot Ring$ recombination allowing the proton bound dimer $H^+ \cdot (Ring)_2$ to be produced and its recombination studied.

$$H^+ \cdot Ring + Ring \xrightarrow{He} H^+ \cdot (Ring)_2 \xrightarrow{e} \text{neutral products} \qquad (3)$$

Recombination rate constants are accurate to ±15% for the monomers and ±20% for the dimers. Kinetic data can be obtained as a function of temperature from 80 to 650 K by cooling the flow tube with liquid nitrogen or using resistive heaters. With the recombining ions in this study, the full temperature range could not be used because of condensation of the reactant gas in the flow tube at the lower temperatures. This limited the lowest temperature to greater than 180 K.

In this study, the temperature dependencies of α_e have been determined for the protonated (Fig. 4) and proton bound dimer (Fig. 5) ions of benzene (C_6H_6), pyridine, (C_5H_5N) and pyrimidine ($C_4H_4N_2$). The data for the protonated ions show a mild reduction in α_e with increasing temperature, with the rate constant increasing with the sequential addition of N atoms to the ring. For the proton bound dimers, the rate constants are generally somewhat larger having an increase with temperature (the opposite of the dependence for protonated rings). Note that although the thermalized temperature is varied in the flowing afterglow, it is thought to be the electron temperature that controls the recombination rate. It should also be noted that some recombination measurements have been made previously for some PAHs ($C_{10}H_8^+$, $C_{12}H_{10}^+$, $C_{16}H_{10}^+$) in various isomeric forms.[28] These data have shown that the recombination rate constants vary significantly with isomeric form by as much as a factor of

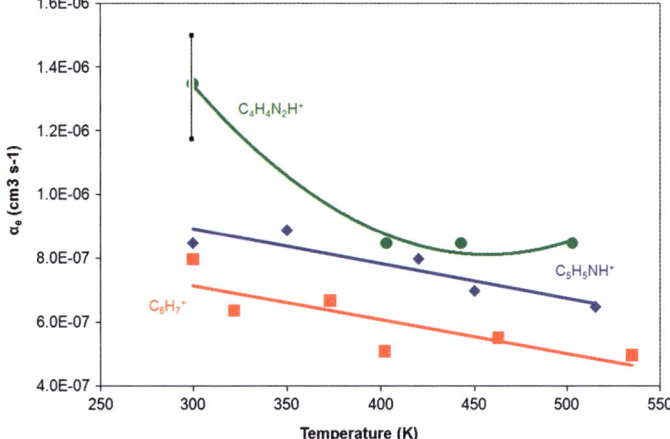

Fig. 4 Temperature dependencies of the recombination rate constants, α_e, for the DR of protonated benzene, $C_6H_7^+$; pyridine, $C_5H_5NH^+$; pyrimidine, $C_4H_4N_2H^+$.

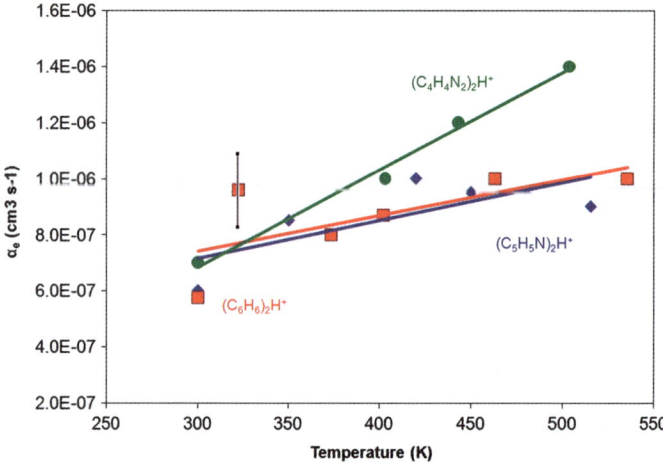

Fig. 5 Temperature dependencies of the recombination rate constants, α_e, for the DR of the proton bound dimers of benzene (C_6H_6), pyridine, (C_5H_5N) and pyrimidine ($C_4H_4N_2$). The error bar is included to illustrate the magnitude of the deviations that can occur at the lowest temperatures where condensation of the reactant is most likely.

five. Sufficient data, however, are not available to establish trends in the DR of protonated ring compounds. What is certain though is that the variations in these data will profoundly affect the ionization density in the Titan atmosphere. Also, they will influence the, as yet unknown, products of DR. Plans are underway to determine these neutral product distributions,[29] although for such polyatomic species and with the stability of the rings, it is likely that the ejection of an H-atom will occur maintaining the integrity of the rings and thus building up to larger ring species. However at the moment, little information is available for the neutral products of such polyatomic recombination. This is a very difficult task, but such data are urgently needed. The data that are available have been obtained mainly using the storage ring technique,[30,31] however it has not been possible to study ring recombination since, with these more massive ions, the laboratory kinetic energies required are beyond the limit imposed by the need to maintain the magnetic rigidity of the storage ring. At a stage earlier in the chemistry, recombination rate constants are also required for many ions and at the Titan ionospheric temperatures of both the ions and electrons. This is also not an easy task because of condensation of the reactant gases in flowing afterglow laboratory experiments at the lower temperatures. The problem is avoided in the measurements of recombination cross-sections using storage rings (SR).[32] However, here care has to be taken to internally relax the ions by storing them in the ring to allow them to internally cool radiatively. In addition, the temperature dependencies are not for truly thermalized species since the ion internal (vibrational and rotational) temperature is not varied as the ion–electron energy is changed. Only at room temperature and under its equivalent electron–ion relative kinetic energy is the situation close to thermalized.

Significance of ring compounds to Titan ionospheric chemistry

A large body of data is establishing that ion–molecule association is important in the chemistry of the Titan atmosphere. This type of channel is expected to increase in importance with decreasing temperature from the laboratory experiments at 298 K down to the ion temperature of the Titan atmosphere (170 K). With sequences of such reactions, large hydrocarbons with nitrogen incorporated could perhaps generate multi-ring species such as Polycyclic Aromatic Hydrocarbons

(PAHs and PAHNs, nitrogen containing PAHs) of the type thought to exist in the upper reaches of the atmosphere generating aerosols and maybe tholins.[8] Table 3 shows the association complexes that have been detected in laboratory studies and in some cases possible structures have been suggested. Particularly interesting are the products of the reactions with benzene and toluene. In the reaction of CH_3^+ with benzene, the association could produce protonated toluene. Theory has shown that the ion approaches the benzene ring perpendicularly due to an interaction with the π electrons in the ring.[33–36] On closer approach, the CH_3^+ moves into the plane of the ring with σ bonding and adopts the structure of protonated toluene. No calculations are available for the other relevant association reactions. DR of this ion is expected to be rapid, based on the recombination studies described above and, for species as complex as this, may dissociate by ejecting an H-atom and leaving the toluene intact. No product data are available for ion species of this complexity, but it appears that there are ample modes to distribute the excess energy thus minimizing fragmentation. This process could be repeated for the association of toluene with CH_3^+ producing protonated dimethyl, or ethyl, benzene. Subsequent recombination and association could produce ethyl methyl benzene and diethyl benzene, perhaps closing the second ring. Alternatively, toluene could react with $C_3H_3^+$ where the association complex could have the structure of protonated naphthalene 1,2-dihydro. This type of process could then provide a route for ring growth in converting single ring compounds to di-rings and beyond. Such reactions may occur in the other associations listed in Table 3 and may provide routes to PAHs thought to be present in the Titan ionosphere.[8]

Although recombination rate constants are available for many species, often they are not at temperatures appropriate for the Titan atmosphere. In the recombination studies above, rate constant data are presented for the temperature range 300 to 540 K. Literature data also do not cover an extensive range.[31] However on Titan at an altitude of 1000 km the electron temperature is of the order of 1200 K.[37] Thus, this would be a considerable extrapolation of the laboratory data and values would be uncertain. Therefore detailed models will have to wait for more extensive measurements. As mentioned above, storage ring studies can measure up to electron kinetic energies equivalent to 1200 K, but not for species as large as ring compounds. However some insights can be gained from the data already available. In the recombination studies above, data on temperature dependencies of rate constants are discussed. Further information can be obtained from several recent reviews.[30–32]

Association and recombination have been treated sequentially in the above, however, they also have to be considered to be in competition in the chemical models of the Titan ionosphere. Detailed modeling is beyond the scope of this paper, but a great deal of insight can be gained by just considering a subset of this chemistry. Following the solar photoionization and ionization by Saturn magnetosphere electrons, a hydrocarbon chemistry proceeds generating ions such as CH_3^+, $C_2H_5^+$, $C_3H_3^+$, etc. generically designated by I^+. These ions can then react with neutral species such as the ring neutrals benzene, toluene, pyridine, pyrimidine, and pyrrole represented generically by concentrations [Ring] and causing association which is stabilized by the very dominant N_2 viz

$$I^+ + \text{Ring} + N_2 \xrightarrow[k_d]{k_a} I^+ \cdot \text{Ring} + N_2 \quad (4)$$

where k_a is the forward association rate constant and k_d is the reverse collisional dissociation. Note that oxygen containing rings have been excluded from this modeling since oxygen is not very prevalent in the Titan atmosphere. During reaction (4), the association will compete with DR of I^+

$$I^+ + e \xrightarrow{\alpha_e} \text{products} \quad (5)$$

where α_e is the electron recombination rate constant. For reactions in which association occurs, the product ion, $I^+ \cdot Ring$, can then recombine as

$$I^+ \cdot Ring + e \xrightarrow{\alpha_a} products \qquad (6)$$

where α_a is the recombination rate constant for the association complex. Association will be more rapid when it involves dominant constituents as much as possible, viz

$$I^+ + 2N_2 \underset{k'_d}{\xrightarrow{k'_a}} I^+N_2 + N_2 \qquad (7)$$

with forward k_a' and reverse k_d' rate constants followed by the exchange of the N_2 molecule for the ring

$$I^+ \cdot N_2 + Ring \xrightarrow{k_s} I^+ \cdot Ring + N_2 \qquad (8)$$

where k_s is the rate constant for this switching reaction. Since $[N_2]$ is included, the sequential reactions (7) and (8) are a much more rapid alternative than the direct association reaction (4). Kinetic equations can be written for each of these reactions, for example, reaction (4) gives

$$\frac{d[I^+]}{dt} = -k_a[Ring][N_2][I^+] \qquad (9)$$

To see the competition between the various reactions, the equations can be written in the form of normalized rates.

$$\frac{1}{[I^+]} \frac{d[I^+]}{dt} = -k_a[Ring][N_2] \qquad (10)$$

Kinetic data for these reactions are estimated as average values determined from our previous measurements and the data given above. k_a is estimated as 2×10^{-26} cm^6 s^{-1} from the modeling of our data for the HCN system[17] while there is no value for available for k_d. From detailed modeling, it has been shown that smaller I^+ species are most abundant.[12] For these species, storage ring data have been reviewed[31] and α_e for the recombination of I^+ is estimated from those for small hydrocarbon ions as 1×10^{-6} cm^3 s^{-1}.[30,31,38] However, there are no data available for the recombination rate constants of the association complexes likely formed between the abundant Titan ions and ring compounds. In our studies, we were able to measure the recombination rates of protonated rings and proton bound ring dimers. The proton bound dimers are likely to compare most closely to the structure of the association complex. Therefore, in reaction (6) α_a is estimated to be 2.5×10^{-6} cm^3 s^{-1} using, from our data, an average α_e for dimers and extrapolated to 1200 K. k_a' is obtained as k_a reduced by a factor of 2 to account for the less efficient association of I^+ with the more abundant N_2. The switching reaction, with k_s rapid, makes reactions (7) and (8) equivalent to reaction (4) and we assume that k_s is gas kinetic at $\sim 10^{-9}$ cm^3 s^{-1}.

To determine the rapidity of the various reaction processes, the concentrations have been estimated from Cassini observations.[39] Concentrations estimated at an altitude of 1000 km are $[N_2] = 1 \times 10^{10}$ cm^{-3} determined from the pressure, $[e] = 10^2$ to 10^3 cm^{-3} and $[Ring] = 10^4$ cm^{-3} from the N_2 density and the mixing fraction of C_6H_6 assuming $[C_6H_6] = [Ring]$. Using these concentrations and the kinetic data, the normalized rates, e.g. $R = \frac{1}{[I^+]} \frac{d[I^+]}{dt}$, for the various reactions have been determined and are listed in Table 4. This information shows that the association reaction (4) is not competitive with recombination reaction (5), however, reaction sequence (7) and (8) are much more competitive (by about 6 orders of magnitude) with reaction (7) being the rate limiting step. If the $[N_2]$ and $[Ring]$ concentrations are

Table 4 Approximate normalized rates, e.g. $\frac{1}{[I^+]}\frac{E[I^+]}{dt}$, for reactions (4) to (8) at the conditions for 1000 km altitude in the Titan atmosphere. It can be seen that for the higher values of [N_2] and [Ring], association (reaction (7)) is becoming competitive with recombination (reaction (5)).

Reaction from text	Normalized rate/s^{-1}	Normalized rates at 1000 km/s^{-1}	Rates for [N_2] and [Ring], increased by 10
(4)	$\frac{1}{[I^+]}\frac{d[I^+]}{dt} = -k_a[\text{Ring}][N_2]$	2×10^{-12}	2×10^{-10}
(4)^a	$\frac{1}{[I^+\cdot\text{Ring}]}\frac{d[I^+\cdot\text{Ring}]}{dt} = -k_d[N_2]$	—	—
(5)	$\frac{1}{[I^+]}\frac{d[I^+]}{dt} = -\alpha_e[e]$	10^{-4} to 10^{-3}	10^{-4} to 10^{-3}
(6)	$\frac{1}{[I^+\cdot\text{Ring}]}\frac{d[I^+\cdot\text{Ring}]}{dt} = -\alpha_e[e]$	2.5×10^{-4} to 2.5×10^{-3}	2.5×10^{-4} to 2.5×10^{-3}
(7)	$\frac{1}{[I^+]}\frac{d[I^+]}{dt} = -k'_a[N_2]^2$	1×10^{-6}	1×10^{-4}
(7)^a	$\frac{1}{[I^+\cdot N_2]}\frac{d[I^+\cdot N_2]}{dt} = -k'_d[N_2]$	—	—
(8)	$\frac{1}{[I^+\cdot N_2]}\frac{d[I^+\cdot N_2]}{dt} = -k_s[\text{Ring}]$	1×10^{-5}	1×10^{-4}

^{a} Indicates the reverse reaction of reactions (4) and (7).

considered to be an order of magnitude larger, then reaction (4) is still not competitive, but reaction sequence (7) and (8) becomes comparable with the recombination steps, reactions (5) and (6). Thus, this reaction sequence will become more dominant with decreasing altitude and the loss of ionization will stem from the recombination of the larger species in eqn (6) (rather than reaction (5)) where there would be expected to be less fragmentation. Obviously, these conclusions are only semi-quantitative and the kinetic data need to be included in the detailed chemical models. However, they do give insights into directions that the models should explore.

Conclusions

The chemistry of the Titan ionosphere is very complex with many large species being produced. Laboratory studies of the various processes are essential to convert

Cassini mass spectra into molecular ionospheric abundances. The laboratory data will also guide the chemical models and show which reaction processes are critical to the chemistry. The present analysis has concentrated on reactions involving ring compounds and has shown how three body association followed by electron–ion dissociative recombination, can build up larger ring compounds and eventually lead to PAHs, aerosols and tholins, which have been suggested to be present.[8] Note that the association reactions used have been three body and no account has been taken of their temperature dependencies or the possibility of saturation. Temperature variation of these rate constants can be as much as $T^{\frac{-l}{2}+\delta}$, where l is equal to the number of rotational degrees of freedom in the separated reactants and δ accounts for the collisional stabilization step.[40–44] Thus, for polyatomic reactants, the temperature dependence can be as much as T^{-3} and, at the lower temperatures, the rate constant may even reach a temperature independent value where every intermediate complex is stabilized. This could give a saturated three body reaction which proceeds at the two body gas kinetic limit of the association step. Much more laboratory data will be required, especially as a function of temperature, before the evolution of the Titan ionosphere is understood.

Acknowledgements

This research was supported by the NASA Planetary Atmospheres program under grant number NNX07AF57.

References

1 R. H. Brown, J.-P. Lebreton and J. H. Waite, ed., *Titan from Cassini–Huygens*, Springer, Dordrecht, 2009.
2 J. H. Waite, H. Nieman, R. V. Yelle, W. T. Kasprzak, T. E. Cravens, J. G. Luhmann, R. L. McNutt, W.-H. Ip, D. Gell, V. de la Haye, I. Muller-Wordag, B. Magee, N. Borggren, S. Ledvina, G. Fletcher, E. Walter, R. Miller, S. Scherer, R. Thorpe, J. Xu, B. Block and K. Arnett, *Science*, 2005, **308**, 982–986.
3 J.-E. Wahlund, M. Galand, I. Mueller-Wodarg, J. Cui, R. V. Yelle, F. J. Crary, K. Mandt, B. Magee, J. H. Waite Jr., D. T. Young, A. J. Coates, P. Garnier, K. Aagren, M. Andre, A. I. Eriksson, T. E. Cravens, V. Vuitton, D. A. Gurnett and W. S. Kurth, *Planet. Space Sci.*, 2009, **57**, 1857–1865.
4 A. J. Coates, F. J. Crary, G. R. Lewis, G. T. Young, J. H. Waite and E. C. Sittler, *Geophys. Res. Lett.*, 2007, **34**, L22103.
5 A. Coates, A. Wellbrock, G. R. Lewis, G. H. Jones, D. T. Young, F. J. Crary and J. H. Waite, *Planet. Space Sci.*, 2009, **57**, 1866–1871.
6 J. H. Waite, Jr., W. S. Lewis, W. T. Kasprzak, V. G. Anicich, B. P. Block, T. E. Cravens, G. G. Fletcher, W.-H. Ip, J. G. Luhmann, R. L. McNutt, H. B. Niemann, J. K. Parejko, J. E. Richards, R. L. Thorpe, E. M. Walker and R. V. Yelle, *Space Sci. Rev.*, 2004, **114**, 113–231.
7 B. A. Magee, J. H. Waite, K. E. Mandt, J. Westlake, J. Bell and D. A. Gell, *Planet. Space Sci.*, 2009, **57**, 1895–1916.
8 J. H. Waite, Jr., D. T. Young, T. E. Cravens, A. J. Coates, F. J. Crary, B. Magee and J. Westlake, *Science*, 2007, **316**, 870–875.
9 C. N. Keller, T. E. Cravens and L. Gan, *J. Geophys. Res.*, 1992, **97**, 12117.
10 T. E. Cravens, I. P. Robertson, J. H. Waite, R. V. Yelle, W. T. Kasprzak, C. N. Keller, S. A. Ledvina, H. B. Nieman, J. G. Luhmann, R. L. McNutt, W.-H. Ip, V. de la Haye, I. Mueller-Wodarg, J.-E. Wahlund, V. G. Anicich and V. Vuitton, *Geophys. Res. Lett.*, 2006, **33**, L07105.
11 T. E. Cravens, I. P. Robertson, J. H. Waite, R. V. Yelle, V. Vuitton, A. J. Coates, J.-E. Wahlund, K. Agren, M. S. Richard, V. De la Haye, A. Wellbrock and F. M. Neubauer, *Icarus*, 2009, **199**, 174–188.
12 V. Vuitton, R. V. Yelle and M. J. McEwan, *Icarus*, 2007, **191**, 722–742.
13 V. Vuitton, R. V. Yelle and J. Cui, *J. Geophys. Res.*, 2008, **113**, E05007.
14 C. N. Keller, V. Anicich and T. E. Cravens, *Planet. Space Sci.*, 1998, **46**, 1157–1174.

15 V. Vuitton, P. Lavvas, R. V. Yelle, M. Galand, A. Wellbrock, G. R. Lewis, A. J. Coates and J.-E. Wahlund, *Planet. Space Sci.*, 2009, **57**, 1558–1572.
16 L. D. Fondren, N. G. Adams and L. Stavish, *J. Phys. Chem. A*, 2009, **113**, 592–598.
17 J. L. McLain, C. D. Molek, D. Osbourne Jr. and N. G. Adams, *Int. J. Mass Spectrom.*, 2009, **282**, 85–90.
18 J. McLain and N. G. Adams, *Planet. Space Sci.*, 2009, **57**, 1642–1647.
19 N. G. Adams and D. Smith, in *Techniques for the Study of Ion–Molecule Reactions*, ed. J. M. Farrar and J. W. H. Saunders, Wiley Interscience, New York, 1988, pp. 165–220.
20 N. G. Adams and D. Smith, *Int. J. Mass Spectrom. Ion Phys.*, 1976, **21**, 349–359.
21 N. G. Adams and D. Smith, *J. Phys. B: At. Mol. Phys.*, 1976, **9**, 1439–1451.
22 L. D. Fondren and N. G. Adams, *Int. J. Mass Spectrom.*, 2010, submitted.
23 K. C. Smyth, E. S. G. Lias and P. Ausloos, *Combust. Sci. Technol.*, 1982, **28**, 147–154.
24 D. W. Minsek and P. Chen, *J. Phys. Chem.*, 1990, **94**, 8399–8401.
25 F. P. Lossing, *Can. J. Chem.*, 1972, **50**, 3973–3981.
26 S. G. Lias, J. E. Bartmess, J. F. Liebman, R. D. Levin and W. G. Mallard, *J. Phys. Chem. Ref. Data Suppl.*, 1988, **17**, 1–861.
27 J. Žabka, M. Polášek, D. Ascenki, P. Tosi, J. Roithová and D. Schröder, *J. Phys. Chem. A*, 2009, **113**, 11153–11160.
28 L. Biennier, M. Alsayed-Ali, A. Foutel-Richard, O. Novotny, S. Carles, C. Rebrion-Rowe and B. R. Rowe, *Faraday Discuss.*, 2006, **133**, 289–301.
29 C. D. Molek, V. Poterya, N. G. Adams and J. McLain, *Int. J. Mass Spectrom.*, 2009, **285**, 1–11.
30 N. G. Adams, V. Poterya and L. M. Babcock, *Mass Spectrom. Rev.*, 2006, **25**, 798–828.
31 A. I. Florescu-Mitchell and J. B. A. Mitchell, *Phys. Rep.*, 2006, **430**, 277–374.
32 M. Larsson and A. E. Orel, *Dissociative Recombination of Molecular Ions*, Cambridge University Press, Cambridge, UK, 2008.
33 G. Raos, L. Astorri, M. Raimondi, D. L. Cooper, J. Gerratt and P. B. Karadakov, *J. Phys. Chem. A*, 1997, **101**, 2886–2892.
34 P. C. Miklis, R. Ditchfield and T. A. Spencer, *J. Am. Chem. Soc.*, 1998, **120**, 10482–10489.
35 Y. Ishikawa, H. Yilmaz, T. Yanai, T. Nakajima and K. Hirao, *Chem. Phys. Lett.*, 2004, **396**, 16–20.
36 F. Zheng, R. Sa, J. Cheng, H. Jiang and J. Shen, *Chem. Phys. Lett.*, 2007, **435**, 24–28.
37 J.-E. Wahlund, R. Bostrom, G. Gustafsson, D. A. Gurnett, W. S. Kurth, A. Pedersen, T. F. Averkamp, G. B. Hospodarsky, A. M. Persoon, P. Canu, F. M. Neubauer, M. K. Dougherty, A. I. Eriksson, M. W. Morooka, R. Gill, M. Andre, L. Eliasson and I. Muller-Wodarg, *Science*, 2005, **308**, 986–988.
38 J. L. McLain, V. Poterya, C. D. Molek, L. M. Babcock and N. G. Adams, *J. Phys. Chem. A*, 2004, **108**, 6704–6708.
39 T. E. Cravens, R. V. Yelle, J.-E. Wahlund, D. E. Shemansky and A. F. Nagy, in *Titan from Cassini–Huygens*, ed. R. H. Brown, J.-P. Lebreton and J. H. Waite, Springer, Dordrecht, 2009, pp. 259–295.
40 D. R. Bates, *J. Phys. B: At. Mol. Phys.*, 1979, **12**, 4135–4146.
41 E. Herbst, *J. Chem. Phys.*, 1979, **70**, 2201–2204.
42 E. Herbst, *J. Chem. Phys.*, 1980, **72**, 5284–5285.
43 D. R. Bates, *J. Chem. Phys.*, 1980, **73**, 1000–1001.
44 N. G. Adams and D. Smith, in *Swarms of Ions and Electrons in Gases*, ed. W. Lindinger, T. D. Mark and F. F. Howorka, Springer-Verlag, Vienna, 1984, pp. 194–217.
45 V. Anicich, *An Index of the Literature for Bimolecular Gas Phase Cation–Molecule Reaction Kinetics: JPL Publication 03–19*, Jet Propulsion Laboratory, Pasadena, 2003.
46 T. Su and W. J. Chesnavich, *J. Chem. Phys.*, 1982, **76**, 5183–5185.
47 NIST Standard Database 69, 2008.

Fast ion–molecule reactions in planetary atmospheres: a semiempirical capture approach

Alexandre Faure,[*a] Véronique Vuitton,[b] Roland Thissen,[b] Laurent Wiesenfeld[a] and Odile Dutuit[b]

Received 4th March 2010, Accepted 30th March 2010
DOI: 10.1039/c003908j

The description of planetary and interstellar chemistry relies strongly on ion–molecule reaction rate data collected at room temperature or above. However, the temperature in the ionospheres of planets and in the interstellar medium can decrease down to 100 K and 10 K, respectively. We present here a simple semiempirical method to extend available measurements towards those temperatures. Our approach is based on the long-range capture theory combined with room temperature data. Results are presented for cation–molecule and anion–molecule reactions. An overall good agreement is observed between our model and various experimental data in the temperature range 20–295 K. Deviations larger than a factor of 2 are found, however, with ion trap measurements below ~50 K. Predictions are also made for reactions of carbon chain and hydrocarbon ions with atomic hydrogen, of particular importance in Titan's atmosphere and in interstellar clouds.

Introduction

The exotic world of Titan, Saturn's largest moon, has emerged as a natural laboratory on a planetary scale to shed light on chemical evolution. Titan is the only Solar System solid body besides Earth and Venus with a thick atmosphere (1.5 bar at the surface). The temperature ranges from about 200 K in the upper atmosphere to 70 K at the tropopause.[1] The atmospheric composition consists of molecular nitrogen (N_2, 98%), methane (CH_4, 2%), as well as their photochemical progeny, which culminates in organic, brownish haze layers.[2,3] A substantial ionosphere constituted of both positive and negative ions is present at the top of the atmosphere and a complex ion chemistry is operating there.[4,5]

In order to interpret the observational data as well as understand the chemical mechanisms associated with the formation of the ions, numerous chemical models have been developed.[6,7] In these models, ion–molecule reaction rates are assumed equal to the rates at 300 K. This approach is justified by the fact that ion–molecule reaction rates are usually strongly exothermic and fast at room temperature and consequently do not vary much when the temperature decreases. However, a survey of the positive ion–molecule reaction rates in a typical Titan model shows that only about 70% of the rates are above a limit of 4×10^{-10} cm^3 s^{-1} and can be considered as being fast.[8] Although the remaining 30% are expected to exhibit a significant variation with temperature, only a few have been measured below room

[a]*Laboratoire d'Astrophysique, Observatoire de Grenoble, Université Joseph Fourier, CNRS UMR5571, B.P. 53, 38041 Grenoble Cedex 09, France. E-mail: afaure@obs.ujf-grenoble.fr*
[b]*Laboratoire de Planétologie, Observatoire de Grenoble, Université Joseph Fourier, CNRS UMR5109, B.P. 53, 38041 Grenoble Cedex 09, France*

temperature. Carrasco et al.[9,10] have investigated the sensitivity of a Titan ionospheric chemistry model to uncertainties on rate constants, taking into account only typical experimental uncertainties. They have studied, in particular, the impact of the evolution of reaction product distributions with temperature/collision energy on predicted ion densities. They were able to identify a small set of multiple pathway reactions as responsible for the major part of the uncertainty. However, their study did not take into account the uncertainty on the total reaction rates when extrapolated to low temperature.

In a recent paper, Faure et al.[11] have shown that the rate constants for neutral–neutral reactions measured at room temperature can be extended down to very low temperature ($T \sim 10$ K) by using a simple semiempirical model derived from the capture theory. This model was checked against a large sample (28 reactions) of low temperature CRESU (Cinétique de Réaction en Ecoulement Supersonique Uniforme) rate data involving the radicals $C(^3P)$, $CN(X^2\Sigma^+)$, $CH(X^2\Pi)$ and $C_2H(X^2\Sigma^+)$. A good agreement (to within a factor of 2 above 25 K) was observed for 26 of the 28 reactions. Here, we report on the extension of the model of Faure et al.[11] on neutral–neutral reactions to the case of ion–molecule reactions. Results are first presented for a number of cation–molecule and anion–molecule reactions for which a negative temperature dependence of the rates has been observed experimentally. Predictions are then computed for reactions between ions and atomic hydrogen, of particular importance in Titan's atmosphere and in the interstellar medium. Our method is described in Section 2 and the results are presented and discussed in Section 3. Our conclusions are drawn in Section 4.

Model

For many exothermic chemical reactions of ions with molecules, the rate constants can be calculated accurately using simple capture theories.[12] All of these theories assume that the rate constant is dominated by long-range effects and that the reaction occurs with unit probability if the ion–molecule pair surmount the centrifugal barrier and approach each other within a critical, so-called capture, distance. When the neutral molecule has no dipole, the rate can be estimated by the "Langevin" model[13] which considers only the (isotropic) ion-induced dipole interaction long-range term and predicts temperature independent rates of the order of 10^{-9} cm^3 s^{-1}. This prediction has been observed experimentally from room temperature and above down to temperatures of a few kelvin, see e.g. Rowe et al.[14] There are however many exceptions to the Langevin model. A well-known example is the reaction $NH_3^+ + H \rightarrow NH_4^+ + H$ which is slow at room temperature, $k(300 \text{ K}) \sim 5 \times 10^{-13}$ cm^3 s^{-1}, and becomes slower as the temperature decreases down to 50–100 K to finally accelerate at still lower temperature, $k(10 \text{ K}) \sim 1 \times 10^{-12}$ cm^3 s^{-1} (see Herbst et al.[15] and references therein). A similar temperature dependence has been predicted for the reaction $C_2H_2^+ + H \rightarrow C_2H_3^+ + H$ and, in both cases, it was attributed to tunneling through a small potential barrier coupled with a shallow entrance channel complex.[16] Another interesting example is the rate of the reaction $O_2^+ + CH \rightarrow CH_3O_2^+ + H$ which was observed to have a minimum near 300 K, $k(290 \text{ K}) = 5.4 \times 10^{-12}$ cm^3 s^{-1}, and to increase sharply at low temperatures ($k \propto T^{-1.8}$) with a value at 20 K that is close to the limiting capture rate.[17] This strong negative temperature dependence was attributed to the presence of an intermediate ($CH_4O_2^+$) complex whose mean lifetime increases with decreasing temperature. Similar results were observed more recently by Gerlich and coworkers for reactions of hydrocarbon ions with hydrogen molecules,[18,19] as shown below. From a dynamical point of view, negative temperature dependences can be produced by the presence of a small "submerged" barrier in the potential energy surface (see, e.g., Faure et al.[20]).

When the neutral molecule has a dipole†, the long-range (anisotropic) ion–dipole interaction term must be added to the ion-induced dipole term in the capture potential energy surface. A number of capture treatments have been developed for these reactions, with a high level of refinement, by many authors (see the review by Clary[12] and references therein). An interesting prediction of all these theories is the sharp negative temperature dependence of the rate constants. This has been also observed experimentally, e.g. for the (charge transfer) reaction $He^+ + HCl \rightarrow He + HCl^+$ down to 27 K.[21] As a result, the rate constants at low temperatures are much greater than the "canonical" value of 10^{-9} cm^3 s^{-1} and can be as large as 10^{-7} cm^3 s^{-1} for neutrals with large dipoles.[22] It is important for the following to note that the simple "locked dipole" approximation, which assumes that the dipole aligns with the incoming charge, provides good results in the very low temperature regime. In contrast, the popular average-dipole-orientation (ADO) theory has been shown to significantly underestimate the rate constants at low temperatures (e.g. Clary et al.[23]).

Additional interaction terms can also affect the magnitude and temperature dependence of the rates. Thus, for strongly polar ions reacting with polar neutral molecules, Clary[24] has shown that the effect of dipole–dipole, dipole-induced dipole and dispersion terms in the potential energy surface can produce a small enhancement (about 20%) of the rate. This effect was however found to decrease as the temperature is decreased. A similar enhancement of the rate for highly polarizable anions is also produced by the induced dipole-induced dipole interaction term, as shown by Eichelberger et al.[25] The effect of the ion-quadrupole term was also investigated experimentally in the case of neutrals with large quadrupole moments. In this case, rates were found to be temperature independent (within the experimental uncertainties), suggesting no influence of the quadrupole moment on the reaction efficiency.[26] Thus, the inclusion of long-range terms in addition to the ion-dipole and ion-induced dipole terms has a significant but only moderate effect on the rate constants, both in magnitude and temperature dependence.

In the case of open-shell species, spin–orbit and multi-potential energy surface effects can also play a role in the determination of the rate constants. Clary et al.[27] have shown that the rate constants for the open-shell ions C^+ and N^+ are significantly overestimated by capture theories simply because only a fraction of the potential energy surfaces are attractive. As a result, the rate constants were found to be bound above by the capture rate, k_c, and below by fk_c where f is the fraction of attractive surfaces. The description of the non-adiabatic transitions between the surfaces would be therefore necessary to determine the actual rate.

Thus, it appears that there are many important exothermic ion–molecule reactions which are not just determined by the long-range ion-induced dipole interaction term. These reactions do not obey the simple Langevin model and, in particular, can show strong negative temperature dependences ($k \propto T^{-n}$). In this case, the reaction rates are also found to approach the capture limit at very low temperature ($T \sim 10$ K), provided the rate constant is large enough at room temperature. From a survey of literature data, we have found that an ion–molecule reaction is expected to remain fast at low temperature if its rate is larger than $\sim 10^{-11}$ cm^3 s^{-1} at room temperature. We note that a similar conclusion was obtained by Smith[28] in the case of neutral–neutral reactions (with the additional requirement that the activation energy is zero or negative at 300 K). Therefore, in analogy with the model of Faure et al.[11] for neutral–neutral reactions, we suggest the following semiempirical model to extend the rate constant of a fast ion–molecule reaction from room temperature down to 10 K:

† This is particularly important in Titan's atmosphere where a lot of heteroatom molecules are expected.

• At room temperature ($T_2 \sim 295$ K), the (experimental or theoretical) rate constant must be greater than 10^{-11} cm^3 s^{-1}. (Otherwise the reaction is likely to possess an activation barrier and to be negligible at lower temperature.)
• Only cation reactions for which the charge transfer channel is endothermic are considered, in order to avoid processes which would not be governed by capture.
• At $T_1 = 10$ K, the rate constant is calculated from the capture theory by considering only the (isotropic) ion-induced dipole and ion-"locked in" dipole interaction terms. The corresponding expression can be found for example in Su and Bowers:[29]

$$k(T) = 2\pi q(\alpha/\mu)^{1/2} + 2\pi q \mu_D (2/\pi k_B \mu T)^{1/2} \quad (1)$$

where q is the charge of the ion, α is the polarizability of the neutral, μ is the reduced mass, μ_D is the dipole moment of the neutral molecule and k_B is the Boltzmann constant. It should be noted that that the first term in eqn (1) is temperature independent while the second is proportional to $T^{-1/2}$.
• The room temperature rate (k_2) and the capture rate (k_1) are analytically interpolated by the following 2-parameters power law (which can be easily implemented in any chemical models):

$$k(T) = \alpha(T/300)^\beta \quad (2)$$

It should be noted that for molecules with no dipole moment, eqn (1) reduces to the usual Langevin expression, $k_{\text{ion-ind}} = 2\pi q(\alpha/\mu)^{1/2}$. The ion-locked dipole term in the interaction potential thus simply provides an additive contribution to the rate constant, that one can express by the relationship:

$$k_{\text{ion-ind+ion-lock}} = k_{\text{ion-ind}} + k_{\text{ion-lock}} \quad (3)$$

Such additivity rule does not apply for other interaction terms, as noted for example by Clary.[24] In the following, polarizabilities and dipole moments were taken from Lide.[30]

Results and discussion

Comparisons to experimental results

A first application of the above model is presented in Fig. 1 for the reaction O_2^+ + $CH_4 \rightarrow CH_3O_2^+ + H$. This reaction was measured by Rowe et al.[14] using a CRESU apparatus and a selection ion flow tube (SIFT) apparatus in the temperature range 20–560 K. The rate constant at room temperature ($\sim 5 \times 10^{-12}$ cm^3 s^{-1}) is slightly below our threshold value of 10^{-11} cm^3 s^{-1} but we have selected this reaction as the prototype of strongly temperature dependent cation-(non polar) molecule reactions. Indeed, as shown in Fig. 1, the rate constant falls off by two orders of magnitude between 20 and 290 K! Moreover, Rowe et al.[17] noticed that the extrapolation of their data to even lower temperatures was close to the Langevin rate, $k = 1.2 \times 10^{-9}$ cm^3 s^{-1}. Our model is thus found to provide a very good description of the O_2^+ + CH_4 rate constant over the temperature range 20–290 K: the experimental data are reproduced within the experimental uncertainties. We note that below 10 K, the rate constant is expected to "saturate" at the Langevin value, as observed for other systems (see below).

Further comparisons between our semiempirical model and low temperature measurements are illustrated in Fig. 2 for hydrocarbon cations $C_nH_m^+$ reacting with H_2. In general, only the less hydrogenated ions ($m \leq 1$) shows a significant reactivity with H_2 molecules. These reactions also proceed mainly by H-atom abstraction, producing ions with an additional H atom, as discussed for example in Giles et al.[31] and Scott et al.[32] The following measurements are analyzed: $CH_4^+ + H_2$;[18]

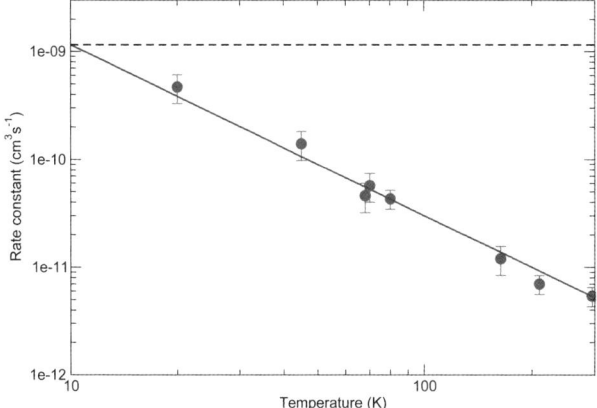

Fig. 1 The rate constant of the reaction $O_2^+ + CH_4$ as a function of temperature. The capture prediction, eqn (1) (here with $\mu_D = 0$), is given by the dashed line. The experimental data is represented by the filled circles. The present model is denoted by the solid line and corresponds to eqn (2) with α and β fixed by the experimental value at 295 K and by the capture value at 10 K. See text for details and references.

$C_3^+ + H_2$ and $C_3H^+ + H_2$;[19] $C_4^+ + H_2$, $C_4H^+ + H_2$ and $C_5^+ + H_2$.[31] For CH_4^+, C_3^+ and C_3H^+, reactions with H_2 have been studied in a variable temperature 22-pole ion trap from room temperature down to 10 K. For C_4^+, C_4H^+ and C_5^+, measurements have been performed in a SIFT apparatus at 80 K and 300 K. It should be noted that rate constants for association reactions (such as $C_6^+ + H_2$) and abstraction reaction with rate constants smaller than 10^{-11} cm^3 s^{-1} at room temperature (such as $C_6H^+ + H_2$) are not reported here. Again, the agreement between our model and

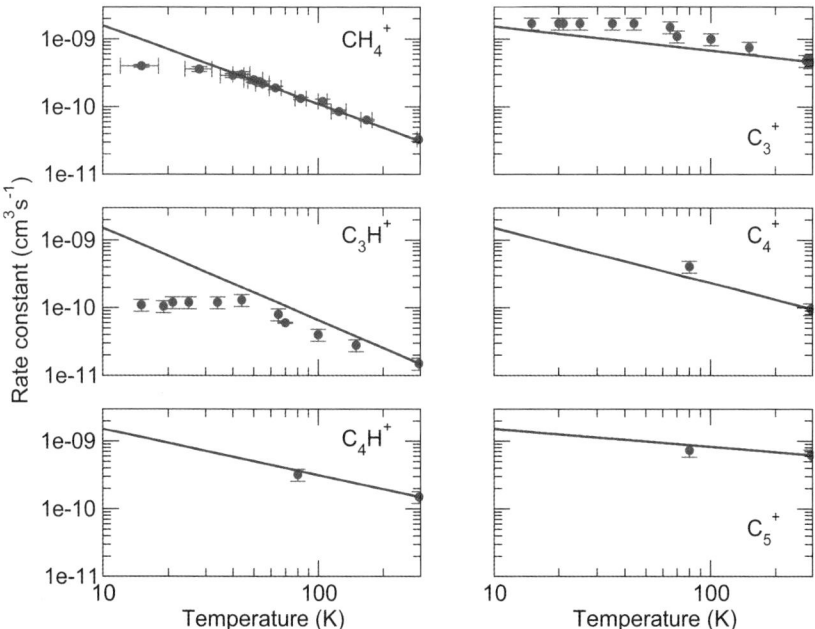

Fig. 2 Same as Fig. 1 for hydrocarbon ions, $C_nH_m^+$, reacting with H_2. See text for details and references.

Table 1 Fitting coefficients α (cm^3 s^{-1}) and β of eqn (2), to the rate constants of various cations reacting with hydrogen molecules. Fits are only valid in the temperature range 10–295 K. Powers of ten are given in parentheses

Reaction	α	β
$O_2^+ + CH_4$	5.26(−12)	−1.59
$CH_4^+ + H_2$	3.09(−11)	−1.16
$C_3^+ + H_2$	4.57(−10)	−0.354
$C_3H^+ + H_2$	1.47(−11)	−1.37
$C_4^+ + H_2$	9.47(−11)	−0.815
$C_4H^+ + H_2$	1.48(−10)	−0.683
$C_5^+ + H_2$	6.17(−10)	−0.263

the experimental data is generally good (within 50% for all reactions above 40 K). In the case of CH_4^+, the model even reproduces the data within error bars down to 30 K. For C_3^+, our model is found to underestimate the experimental rate in the whole temperature range but the agreement is within 50% down to 15 K. The experimental rate is actually found to saturate at the Langevin value (within experimental uncertainties) but for quite a large temperature, $T \sim 50$ K. For C_3H^+, the experimental rate is found to saturate at about the same temperature but with only a small fraction of the Langevin value ($\sim$1/10). In both cases, radiative association reactions were found to compete with the abstraction reactions at low temperature. In particular, in the case of C_3H^+, Savic and Gerlich[19] have shown that the rate for radiative association is larger than the rate for abstraction when $T \leq 20$ K. The sum of both rates is however still only a small fraction (about 1/5 at 20 K) of the Langevin rate. This behaviour is also observed, to a lesser extent, for CH_4^+ below 30 K. It is not clear why the low temperature rate constants for CH_4^+ and C_3H^+ do not approach the Langevin value more closely. Possible explanations include the influence of the rotation of H_2[19] but also uncertainties in the experimental temperature of the ions.[33] For C_4^+, C_4H^+ and C_5^+, the agreement between our model and the SIFT data at 80 K and 300 K is very good (within 30%). It is clear, however, that experimental data at lower temperatures is crucial to test our predictions. The α and β fitting coefficients of the above reaction rates are given in Table 1.

In the case of anions, there are very few (theoretical or experimental) studies that have investigated the temperature dependence of reaction rates. To our knowledge, there are actually only seven anion–molecule reactions that have been measured at low temperatures (below 200 K): $Cl^- + CH_3Br$;[34] $NH_2^- + H_2$;[35] $O^- + NO$, $O^- + C_2H_2$, $O^- + C_2H_4$, $S^- + CO$ and $S^- + O_2$.[36] Some of these anions are not particularly relevant for planetary or interstellar chemistry but they provide good benchmarks for our model.

The rate constant for the (S_N2) reaction $Cl^- + CH_3Br \rightarrow CH_3Cl + Br^-$ was measured in a CRESU apparatus. As shown in Fig. 3, the experimental rate constant was found to increase by over two orders of magnitude when the temperature is decreased from room temperature to 23 K, similarly to $O_2^+ + CH_4$, but with an even stronger negative temperature dependence. Again, the overall agreement with our model is good, although the calculated values are slightly higher than the CRESU data. It should be noted that CH_3Br has a relatively large dipole of 1.82 D and the ion-locked dipole rate was found to dominate over the ion-induced dipole rate in the whole plotted temperature range. Approximate quantum dynamical calculations have also been performed for this system for which a full dimensional potential energy surface is available.[37] Le Garrec et al.[34] have employed the rotating bond approximation (RBA) of Clary. They were able to reproduce a negative temperature dependence for the rate constant which they attributed to a submerged barrier in the potential surface. Very recently, Hennig and Schmatz[38]

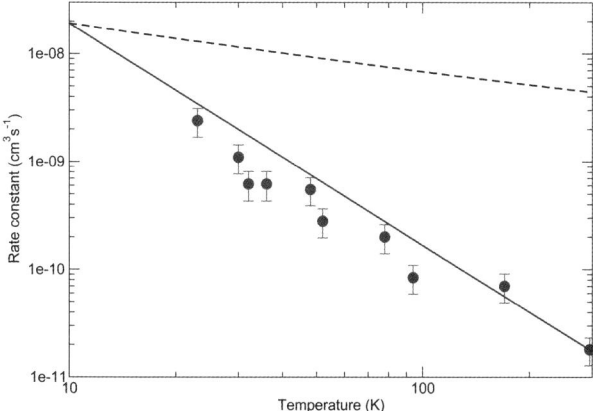

Fig. 3 Same as Fig. 1 for Cl⁻ reacting with CH₃Br. The capture prediction, eqn (1) (here with $\mu_D = 1.82$ D), is given by the dashed line. See text for details and references.

have shown that the initial rotational motion of CH₃Br also decreases the reaction probability. We note that similar effects were also observed for neutral–neutral reactions (see Faure et al.[11] and references therein).

The reaction $NH_2^- + H_2 \rightarrow NH_3 + H^-$ was measured recently from room temperature down to 8 K using a 22-pole radiofrequency ion trap. As shown in Fig. 4 a level-off of the experimental rate is observed below ~30 K at about 1/10 of the Langevin value. A slight decrease of the rate is also observed below 20 K. The agreement with our model is good (better than a factor of 2) down to ~60 K. However, below this temperature, a large deviation is observed, similar to that observed in

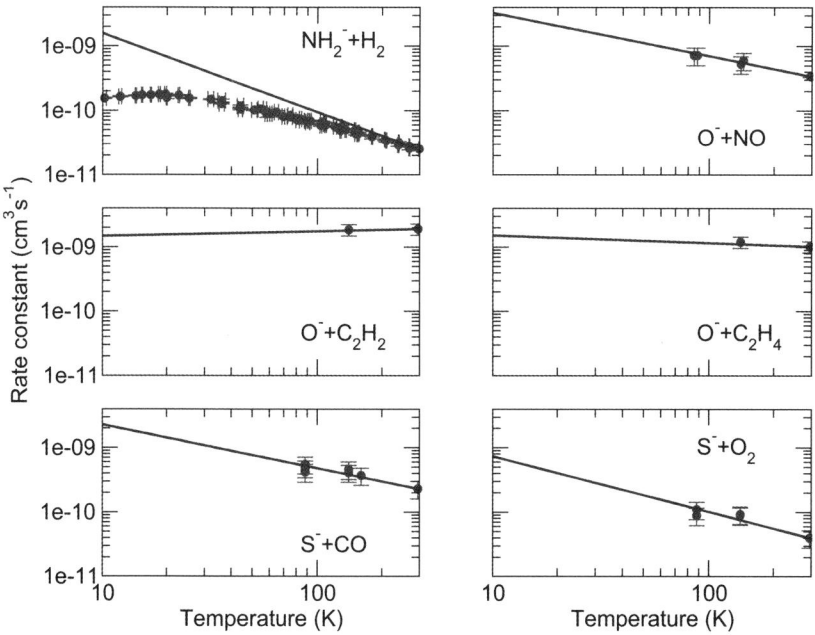

Fig. 4 Same as Fig. 1 for the anions NH_2^-, O^- and S^- reacting with neutral molecules. See text for details and references.

Table 2 Fitting coefficients α (cm^3 s^{-1}) and β of eqn (2), to the rate constants of various anions reacting with hydrogen molecules. Fits are only valid in the temperature range 10–295 K. Powers of ten are given in parentheses

Reaction	α	β
Cl$^-$ + CH$_3$Br	1.74(−11)	−2.06
NH$_2^-$ + H$_2$	2.46(−11)	1.22
O$^-$ + NO	3.36(−10)	−0.670
O$^-$ + C$_2$H$_2$	1.90(−9)	0.075
O$^-$ + C$_2$H$_4$	1.02(−9)	−0.117
S$^-$ + CO	2.27(−10)	−0.677
S$^-$ + O$_2$	3.94(−11)	−0.861

CH$_4^+$ + H$_2$ and C$_3$H$^+$ + H$_2$ (see Fig. 2). Again different explanations are possible, as discussed above. Measurements with *para*-H$_2$ could be particularly interesting to assess differences with normal H$_2$.

Reactions involving O$^-$ and S$^-$ were measured in a SIFT apparatus in the temperature range 88–468 K. Three of these are associative electron detachment reactions, O$^-$ + NO → NO$_2$ + e$^-$, S$^-$ + CO → COS + e$^-$ and S$^-$ + O$_2$ → SO$_2$ + e$^-$, while the other two, O$^-$ + C$_2$H$_2$ and O$^-$ + C$_2$H$_4$, have multiple pathways (including associative detachment). These five reactions were selected by Viggiano and Paulson[36] as having, possibly, a significant temperature dependence in either the rate constant or branching ratios. As can be noticed in Fig. 4, our model agrees with the SIFT data within the experimental error bars for all the five reactions. It should be noted that as the dipoles of NO and CO are small (0.15 D and 0.11 D, respectively), the ion-locked dipole rate was found to dominate over the Langevin rate only at low temperature ($T < 60$ K). We also note that for O$^-$ + C$_2$H$_2$ and O$^-$ + C$_2$H$_4$ we have reported the total rate constants, that is including all reaction pathways. These total rates remain essentially constant between 294 K and 140 K and agree with the Langevin value of $\sim$1.5 $\times$ 10^{-9} cm^3 s^{-1} within the experimental uncertainties. Viggiano and Paulson[36] have shown, however, that the dominant channels (C$_2$H$_2$O+e$^-$ and C$_2$H$^-$ + OH for C$_2$H$_2$; C$_2$H$_4$O+e$^-$ and C$_2$H$_2^-$ + H$_2$O for C$_2$H$_4$) exhibit a marked temperature dependence. We emphasize that our model, as any theoretical treatment based on the capture approximation, only considers the entrance channel to the reaction and can therefore not predict branching ratios (except in special cases such as isotopic reactions[39]). The prediction of reliable branching ratios requires the description of the potential energy surfaces all the way from reactants to products and is obviously beyond the scope of the present paper. The α and β fitting coefficients of the above reaction rates are given in Table 2.

Predictions for carbon chain and hydrocarbon ions reacting with H atoms

Although hydrogen atoms have not been formally detected in Titan's upper atmosphere, photochemical models predict their presence with a mole fraction of about 5 $\times$ 10^{-4} at 1000 km, making them the fifth most abundant species.[40,41] As a consequence, it is important to properly characterize the reactions involving hydrogen atoms. Notably, loss of negative ions occurs mostly through associative detachment with radicals, amongst which hydrogen atoms are the most abundant, along with methyls.[7] Atomic hydrogen is also very abundant in the interstellar medium, especially in the most diffuse molecular regions where the H to H$_2$ ratio is large (see, *e.g.*, the recent review by Snow and Bierbaum[42]).

Hydrogen atom reactions with hydrocarbon cations C$_n$H$_m^+$ behave very differently from the corresponding reactions with H$_2$: H-atom transfer usually occurs for the

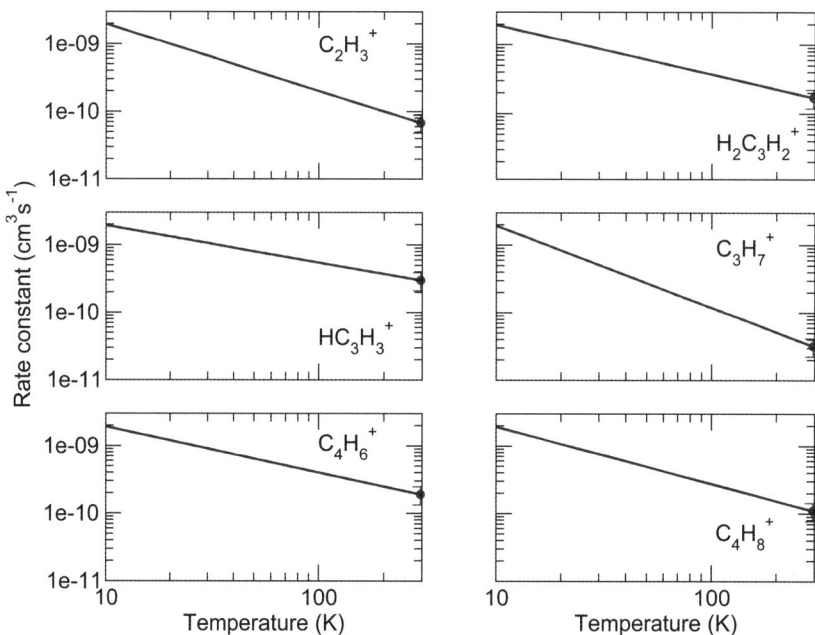

Fig. 5 Same as Fig. 1 for hydrocarbon cations, $C_nH_m^+$, reacting with atomic hydrogen. See text for details and references.

smaller species, provided that the pathway is exothermic, while larger unsaturated species ($n > 3$) undergo association reactions.[32] Thus, in contrast to H_2 reactions which result in an increase of hydrogenation, H-atom transfer results in a decrease of hydrogenation. From the experimental point of view, atomic hydrogen is a very difficult species to monitor in the laboratory. There are therefore very few laboratory investigations of ions reacting with H atom, in particular at low temperature. We note, however, the recent advances in the development of the 22-pole ion trap instrument to measure ion–H atom reactions at low temperature.[43]

We have selected a number of H-atom reactions with carbon chain and hydrocarbon ions for which the experimental rates are larger than 10^{-11} cm^3 s^{-1} at room temperature. Results for hydrocarbon cations are presented in Fig. 5. The rates at 298 K have been taken from the SIFT measurements of Scott et al.[32] At this temperature, all of the selected reactions proceed by H-atom transfer, except $C_4H_6^+ + H$ for which the dominant channel is $C_2H_5^+ + C_2H_2$.[32] Strong negative temperature dependences are observed, as expected, especially for $C_3H_7^+$ whose rate increases by almost

Table 3 Fitting coefficients α (cm^3 s^{-1}) and β of eqn (2), to the rate constants of various cations reacting with hydrogen atoms. Fits are only valid in the temperature range 10–295 K. Powers of ten are given in parentheses

Reaction	α	β
$C_2H_3^+ + H$	6.69(−11)	−0.995
$H_2C_3H_2^+ + H$	1.68(−10)	−0.719
$HC_3H_3^+ + H$	2.97(−10)	−0.551
$C_3H_7^+ + H$	3.14(−11)	−1.21
$C_4H_6^+ + H$	1.88(−10)	−0.685
$C_4H_8^+ + H$	1.08(−10)	−0.846

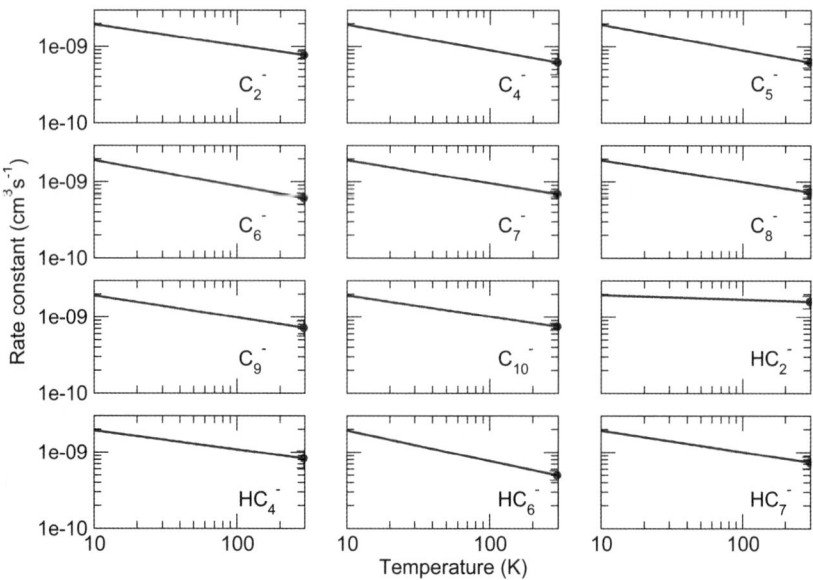

Fig. 6 Same as Fig. 1 for carbon chain anions reacting with atomic hydrogen. See text for details and references.

two orders of magnitude from room temperature down to 10 K. However, for some of these reactions, radiative association could play a major role at low temperatures ($T \leq 50$ K), resulting in overestimated predictions. The α and β fitting coefficients are given in Table 3.

Results of predictions for carbon chain and hydrocarbon anions are plotted in Fig. 6. The rates at 298 K have been taken from the flowing afterglow SIFT measurements of Barckholtz et al.[44] (see also Eichelberger et al.[45]). At room temperature, the smaller C_n^- ($n \leq 6$) proceed exclusively by associative detachment while association was found to dominate for larger species. The total rates, that is including associative detachment and association channels, are reported in Fig. 6. The only observed channel for the HC_n^- + H reactions is associative detachment. It is interesting to note

Table 4 Fitting coefficients α (cm³ s⁻¹) and β of eqn (2), to the rate constants of various anions reacting with hydrogen atoms. Fits are only valid in the temperature range 10–295 K. Powers of ten are given in parentheses

Reaction	α	β
C_2^- + H	7.66(−10)	−0.275
C_4^- + H	6.17(−10)	−0.336
C_5^- + H	6.17(−10)	−0.335
C_6^- + H	6.07(−10)	−0.340
C_7^- + H	6.86(−10)	−0.303
C_8^- + H	7.26(−10)	−0.286
C_9^- + H	7.16(−10)	−0.290
C_{10}^- + H	7.47(−10)	−0.278
HC_2^- + H	1.60(−9)	−0.059
HC_4^- + H	8.27(−10)	−0.250
HC_6^- + H	4.97(−10)	−0.398
HC_7^- + H	7.36(−10)	−0.282

that the rate for HC_2^- at 298 K is significantly larger than the rates measured for the larger HC_n^- anions. We also note that the temperature dependences are much less marked than for the hydrocarbon cations. Indeed, the room temperature rates for all of the reactions plotted in Fig. 6 are larger than 5×10^{-10} cm^3 s^{-1}, which is a significant fraction ($\sim$1/4) of the Langevin rate. As a result, weak temperature dependences are predicted for these reactions, as listed in Table 4.

Room temperature rates of H-atom reactions with carbon chain anions are thus significantly larger than those with hydrocarbon cations. However a direct comparison is not possible, as the type of reactions involved in both cases are not the same, *i.e.* associative detachment for anions and H-atom transfer for cations. Low temperature measurements are crucial to test our predictions and to assess the role of dynamical effects on the reaction efficiency.

Conclusions

We have reported on the extension of the model of Faure *et al.*[11] on neutral–neutral reactions to the case of ion–molecule reactions. This simple, semiempirical, model has been employed to extend the rate constants of fast cation–molecule and anion–molecule reactions from room temperature down to 10 K. Predicting rate constants down to these very low temperatures is crucial for Titan's and interstellar chemistry as they can increase by more than an order of magnitude when the temperature decreases from 300 K to 10 K. Sensitivity analysis of the Titan's and interstellar chemistry models to such changes should be undertaken.

The present model has been shown to apply to ion–molecule reactions for which the rate at room temperature is greater than 10^{-11}cm^3 s^{-1}: a good agreement with various experimental data (in the temperature range 10–295 K) was indeed observed for a sample of 14 reactions involving O_2^+, hydrocarbon cations, Cl$^-$, NH_2^-, O$^-$ and S$^-$. Deviations larger than a factor of 2 were found, however, with ion trap measurements below $\sim$50 K. Our approach is based on the long-range capture theory, combined with room temperature data. Only the long-range ion-induced dipole and ion-"locked" dipole interaction terms are considered in the capture calculations and, as a result, the present model simply requires the polarizabilities and dipole moments of the neutral reactants. It is therefore easily applicable to any fast reaction involving large polyatomic molecules, provided the reaction obeys the above criterion.

Predictions have been made for a number of H-atom reactions with carbon chain and hydrocarbon ions of planetary and interstellar interest. Obviously, our approach has to be further tested against the full amount of low temperature rate data available for charged species. This will be done in future works. We believe, still, that the present model should provide an accuracy better than a factor of 2 in many favorable cases. Such a precision is clearly valuable at the current level of astrochemical modeling. Finally, it should be noted that laboratory measurements of rate constants and reaction products for a number of important ion–molecule reactions do not exist. For these our model is not applicable at present.

Acknowledgements

A. F. thanks Oscar Asvany and Roland Wester for providing the rate data of CH_2^+ + H_2 and NH_2^- + H_2, respectively. Support by the CNRS "Programme National de Planétologie" is acknowledged.

References

1 M. Fulchignoni *et al.*, *Nature*, 2005, **438**, 785–791.
2 J. Cui, R. V. Yelle, V. Vuitton, J. H. Waite, W. T. Kasprzak, D. A. Gell, H. B. Niemann, I. C. F. Müller-Wodarg, N. Borggren, G. G. Fletcher, E. L. Patrick, E. Raaen and B. A. Magee, *Icarus*, 2009, **200**, 581–615.

3   S. Vinatier, B. Bézard, T. Fouchet, N. A. Teanby, R. de Kok, P. G. J. Irwin, B. J. Conrath, C. A. Nixon, P. N. Romani, F. M. Flasar and A. Coustenis, *Icarus*, 2007, **188**, 120–138.
4   A. J. Coates, F. J. Crary, G. R. Lewis, D. T. Young, J. H. Waite and E. C. Sittler, *Geophys. Res. Lett.*, 2007, **34**, L22103.
5   T. E. Cravens *et al.*, *Geophys. Res. Lett.*, 2006, **33**, L07105.
6   C. N. Keller, V. G. Anicich and T. E. Cravens, *Planet. Space Sci.*, 1998, **46**, 1157–1174.
7   V. Vuitton, P. Lavvas, R. V. Yelle, M. Galand, A. Wellbrock, G. R. Lewis, A. J. Coates and J. Wahlund, *Planet. Space Sci.*, 2009, **57**, 1558–1572.
8   I. W. M. Smith, A. M. Sage, N. M. Donahue, E. Herbst and D. Quan, *Faraday Discuss.*, 2006, **133**, 137.
9   N. Carrasco, O. Dutuit, R. Thissen, M. Banaszkiewicz and P. Pernot, *Planet. Space Sci.*, 2007, **55**, 141–157.
10 N. Carrasco, C. Alcaraz, O. Dutuit, S. Plessis, R. Thissen, V. Vuitton, R. Yelle and P. Pernot, *Planet. Space Sci.*, 2008, **56**, 1644–1657.
11 A. Faure, V. Vuitton, R. Thissen and L. Wiesenfeld, *J. Phys. Chem. A*, 2009, **113**, 13694–13699.
12 D. C. Clary, *Annu. Rev. Phys. Chem.*, 1990, **41**, 61.
13 P. Langevin, *Ann. Chem. Phys.*, 1905, **5**, 245.
14 B. R. Rowe, J. B. Marquette, G. Dupeyrat and E. E. Ferguson, *Chem. Phys. Lett.*, 1985, **113**, 403–406.
15 E. Herbst, D. J. Defrees, D. Talbi, F. Pauzat and W. Koch, *J. Chem. Phys.*, 1991, **94**, 7842–7849.
16 K. Yamashita and E. Herbst, *J. Chem. Phys.*, 1992, **96**, 5801–5807.
17 B. R. Rowe, G. Dupeyrat, J. B. Marquette, D. Smith, N. G. Adams and E. E. Ferguson, *J. Chem. Phys.*, 1984, **80**, 241–245.
18 O. Asvany, I. Savić, S. Schlemmer and D. Gerlich, *Chem. Phys.*, 2004, **298**, 97–105.
19 I. Savić and D. Gerlich, *Phys. Chem. Chem. Phys.*, 2005, **7**, 1026.
20 A. Faure, L. Wiesenfeld and P. Valiron, *Chem. Phys.*, 2000, **254**, 49–67.
21 C. Rebrion, J. B. Marquette, B. R. Rowe and D. C. Clary, *Chem. Phys. Lett.*, 1988, **143**, 130–134.
22 N. G. Adams, D. Smith and D. C. Clary, *Astrophys. J.*, 1985, **296**, L31–L34.
23 D. C. Clary, D. Smith and N. G. Adams, *Chem. Phys. Lett.*, 1985, **119**, 320–326.
24 D. C. Clary, *Chem. Phys. Lett.*, 1995, **232**, 267–272.
25 B. R. Eichelberger, T. P. Snow and V. M. Bierbaum, *J. Am. Soc. Mass Spectrom.*, 2003, **14**, 501–505.
26 C. Rebrion, J. B. Marquette, B. R. Rowe, N. G. Adams and D. Smith, *Chem. Phys. Lett.*, 1987, **136**, 495–500.
27 D. C. Clary, C. E. Dateo and D. Smith, *Chem. Phys. Lett.*, 1990, **167**, 1–6.
28 I. W. M. Smith, *Angew. Chem., Int. Ed.*, 2006, **45**, 2842.
29 T. Su and M. T. Bowers, *J. Chem. Phys.*, 1973, **58**, 3027–3037.
30 D. R. Lide *CRC Handbook of Chemistry and Physics*, 84th edn, CRC Press, 2003.
31 K. Giles, N. G. Adams and D. Smith, *Int. J. Mass Spectrom. Ion Processes*, 1989, **89**, 303.
32 G. B. I. Scott, D. A. Fairly, C. G. Freeman, M. J. McEwan, N. G. Adams and L. M. Babcock, *J. Phys. Chem. A*, 1997, **101**, 4973.
33 O. Asvany and S. Schlemmer, *Int. J. Mass Spectrom.*, 2009, **279**, 147–155.
34 J. Le Garrec, B. R. Rowe, J. L. Queffelec, J. B. A. Mitchell and D. C. Clary, *J. Chem. Phys.*, 1997, **107**, 1021–1024.
35 R. Otto, J. Mikosch, S. Trippel, M. Weidemüller and R. Wester, *Phys. Rev. Lett.*, 2008, **101**, 063201.
36 A. A. Viggiano and J. F. Paulson, *J. Chem. Phys.*, 1983, **79**, 2241–2245.
37 L. Wang, L. Zhu and W. L. Haze, *J. Phys. Chem.*, 1994, **98**, 1608.
38 C. Hennig and S. Schmatz, *J. Chem. Phys.*, 2009, **131**, 224303.
39 C. E. Dateo and D. C. Clary, *J. Chem. Soc., Faraday Trans. 2*, 1989, **85**, 1685.
40 P. P. Lavvas, A. Coustenis and I. M. Vardavas, *Planet. Space Sci.*, 2008, **56**, 67–99.
41 E. H. Wilson and S. K. Atreya, *J. Geophys. Res. (Planets)*, 2004, **109**, 6002.
42 T. P. Snow and V. M. Bierbaum, *Annu. Rev. Anal. Chem.*, 2008, **1**, 229–259.
43 D. Gerlich and M. Smith, *Phys. Scr.*, 2006, **73**, C25–C31.
44 C. Barckholtz, T. P. Snow and V. M. Bierbaum, *Astrophys. J.*, 2001, **547**, L171–L174.
45 B. Eichelberger, T. P. Snow, C. Barckholtz and V. M. Bierbaum, *Astrophys. J.*, 2007, **667**, 1283–1289.

Meteoric ion layers in the Martian atmosphere

Charlotte L. Whalley and John M. C. Plane*

Received 2nd March 2010, Accepted 6th April 2010
DOI: 10.1039/c003726e

Low-lying plasma layers have been observed sporadically in the Martian atmosphere by radio occultation measurements from spacecraft such as the Mars Express Orbiter and the Mars Global Surveyor. These layers are just a few km wide, and tend to occur around 90 km. It has been proposed that the layers consist of metallic ions, for two reasons: they occur in the aerobraking region of the planet where meteoroids ablate; and they resemble sporadic E layers in the terrestrial atmosphere which are known to be composed principally of Fe^+ and Mg^+ ions. This paper addresses the problem of how metallic ions can persist in a CO_2-rich atmosphere, where the ions should be neutralized rapidly by formation of metal–CO_2 cluster ions followed by dissociative electron recombination. Laboratory studies using the pulsed laser photolysis/laser induced fluorescence and flow tube/mass spectrometer techniques were used to measure the following rate coefficients: k ($Mg^+ + CO_2$ (+ CO_2) → $Mg^+.CO_2$, 190–403 K) = (5.3 ± 0.7) × 10^{-29} (T/300 K)$^{(-1.86 \pm 0.03)}$ cm^6 molecule^{-2} s^{-1}; $k(Mg^+.CO_2 + O_2 \rightarrow MgO_2^+ + CO_2$, 297 K) = (2.2 ± 0.8) × 10^{-11} cm^3 molecule^{-1} s^{-1}; $k(MgO_2^+ + O \rightarrow MgO^+ + O_2$, 297 K) = (6.5 ± 1.8) × 10^{-10} cm^3 molecule^{-1} s^{-1}; and $k(MgO^+ + O \rightarrow Mg^+ + O_2$, 297 K) = (5.9 ± 2.4) × 10^{-10} cm^3 molecule^{-1} s^{-1}. A model of magnesium and iron chemistry in the Martian atmosphere was then constructed, which includes meteoric differential ablation rates calculated with the Leeds CABMOD model, photo-ionization, and gas-phase ion–molecule and neutral chemistry. The model shows that nearly all the metallic ions between 70 and 110 km should be Mg^+, because the reactions of MgO_2^+ and MgO^+ with atomic O are fast enough to prevent these molecular ions undergoing dissociative electron recombination (unlike the analogous Fe species). There are enough Mg^+ ions to form sporadic layers of the observed plasma density, and the layers can have a lifetime against neutralization in excess of 20 h.

Introduction

The atmospheres of planetary bodies are subject to a continuous bombardment of interplanetary dust, most of which is of cometary origin.[1] In the case of the terrestrial atmosphere, the daily meteoric input over the whole planet is estimated to be around 30 t d^{-1}, although this is probably uncertain within a factor of 3.[2] Most of the incoming mass contribution is from particles with masses between 1 and 10 μg. Collisions with air molecules cause these small particles to undergo flash heating to over 1800 K, at which point they melt and their constituent elements evaporate.[3] Meteoric ablation tends to occur in the planetary aerobraking region, where the atmospheric pressure is about 1 μbar (around 95 km on Earth and 80 km on Mars). Since most meteoroids are stony chondrites with the bulk composition of olivine, the major meteoric constituents are Fe, Mg and Si—according to a recent model of chemical ablation, these elements should be injected into the atmosphere

---

School of Chemistry, University of Leeds, Woodhouse Lane, Leeds, LS2 9JT, UK. E-mail: cmclw@leeds.ac.uk; j.m.c.plane@leeds.ac.uk

in roughly equal quantities.[3] If the incoming meteoroid is travelling faster than about 25 km s^{-1}, the freshly evaporated metal atoms will tend to ionize through hyperthermal collisions with air molecules.[3] Metallic ions are also produced by photoionization, and charge transfer with the major lower ionospheric ions: NO^+ and O_2^+ on Earth,[2] and O_2^+ on Mars.[4]

Sporadic E layers are thin layers of concentrated plasma between 1 and 3 km wide, which occur intermittently at altitudes between 90 and 120 km in the terrestrial atmosphere.[5] Measurements with rocket-borne mass spectrometry have shown that the major constituents of sporadic E layers are Mg^+ and Fe^+.[6,7] Recently, low-lying plasma layers have been observed in the Martian atmosphere by using radio occultation from spacecraft such as the Mars Express Orbiter[8] and the Mars Global Surveyor.[9] These layers are just a few km wide and occur between 70 and 110 km. The majority of the layers are around 90 km with peak electron densities in the range $(1-2) \times 10^4$ cm^{-3}. Fig. 1 is an electron density profile retrieved from the Mars Global Surveyor Radio Science experiment,[10,11] when such a layer is present. The Martian ionosphere is characterised by the main layer around 140 km, a secondary layer at 120 km, and then the so-called *third* layer at 90 km.[12] Because these third layers are sporadic and localized, appearing in only about 10% of measured profiles,[9] it has been postulated that the layers consist of metallic ions, by analogy with terrestrial sporadic E layers.[8,9]

Further evidence for the third layer being composed of metallic ions is that the layer occurs within the height range where Fe and Mg should ablate from meteoroids in the Martian atmosphere. Fig. 2 illustrates the atomic Fe injection profile (Mg is almost identical) for Mars, calculated using our Chemical Ablation Model (CABMOD—further details of this model are provided in the modelling section).[3] The peak injection occurs around 80 km, about 10 km lower than on Earth. It is interesting to note that similar layers have also been observed in the aerobraking regions of other planetary bodies, including Venus[13] and probably Titan.[14]

The purpose of this paper is to investigate whether Mg^+ and Fe^+ can actually survive long enough in a CO_2-rich atmosphere to form sporadic layers. Indeed, one Mars Express observation detected a third layer in the same local area and

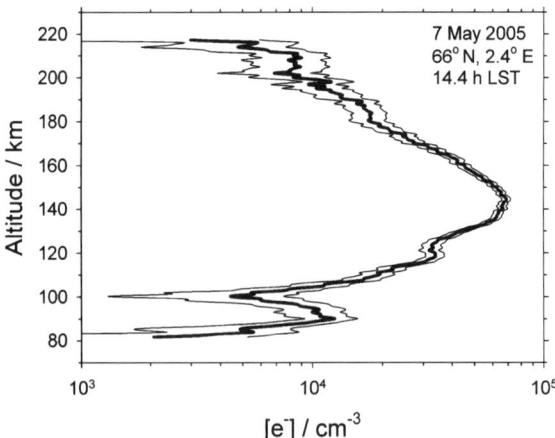

Fig. 1 Profile of electron density (thick black line) measured by the Mars Global Surveyor Radio Science experiment. The thin black lines indicate the 1σ uncertainty in the retrieved density. The ionosphere contains a main layer around 140 km, and a secondary ion layer at 120 km. This profile also exhibits the third layer, which occurs sporadically around 90 km and has been attributed to a layer of metallic ions. The data were provided by P. Withers (Boston University) from the NASA Planetary Data System.

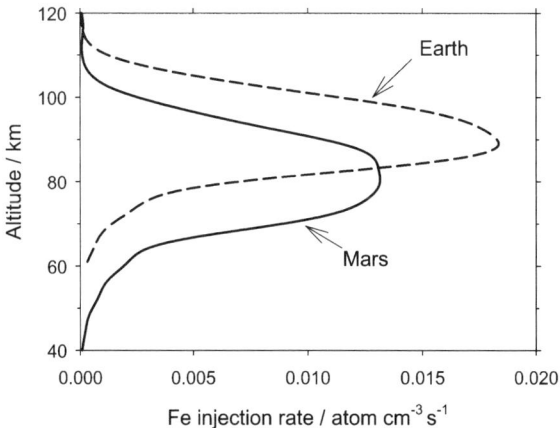

Fig. 2 Height profiles for Mars and Earth of the injection rates of Fe (including Fe$^+$) from the ablation of interplanetary dust, calculated using the CABMOD model.[3]

altitude range on two consecutive days, indicating that it had survived for about 20 h.[8] A major potential difficulty is that the ion–molecule chemistry of these metallic ions should lead to very rapid neutralization, because in the Martian atmosphere metallic ions such as Mg$^+$ should recombine rapidly with CO_2:

$$Mg^+ + CO_2 (+ M) \rightarrow Mg^+ \cdot CO_2 \text{ (M = third body, } CO_2 \text{ on Mars)} \quad (1)$$

The resulting cluster ion, or some other molecular ion produced from it by ligand-switching, will then undergo dissociative electron recombination ($Mg^+ \cdot X + e^- \rightarrow Mg + X$), a process which is extremely rapid (typical rate coefficient 3×10^{-7} cm^3 molecule^{-1} s^{-1} [15]). Two previous models of metal chemistry in the Martian atmosphere have assumed a rate coefficient for reaction 1 of $k_1 = 1 \times 10^{-30}$ cm^6 molecule^{-2} s^{-1}.[4,16] However, this would appear to be a serious underestimate. For instance, the analogous reaction Fe$^+$ + CO_2 + He has a rate coefficient of 1.01×10^{-29} (T/300 K)$^{-2.31}$ cm^6 molecule^{-2} s^{-1}.[17] Extrapolated to a typical temperature of 140 K in the Martian atmosphere at 90 km (see Fig. 3, bottom panel), the rate coefficient increases to 5.9×10^{-29} cm^6 molecule^{-2} s^{-1}, *even* before considering the enhanced efficiency of CO_2 over He as the third body. The lifetime of the metallic ions would therefore potentially be less than 2 min at 90 km. It should be noted that dielectric recombination of these atomic ions is unimportant: *e.g.*, the rate coefficient k(Fe$^+$ + e$^- \rightarrow$ Fe + $h\nu$) = 7×10^{-12} cm^3 molecule^{-1} s^{-1},[18] so a layer with [e$^-$] = 2×10^4 cm^{-3} would survive dielectric recombination for 80 days.

In the terrestrial ionosphere, reaction with O_3 is the most significant source of metallic molecular ions, for example:

$$Mg^+ + O_3 \rightarrow MgO^+ + O_2 \quad (2)$$

These reactions proceed close to their Langevin limits (Table 1) and so once again metallic ions such as Mg$^+$ and Fe$^+$ should not survive long enough to form sporadic E layers.[19] However, we have shown that metal oxide ions such as FeO$^+$ and FeO$_2^+$ are recycled back to Fe$^+$ by reaction with atomic O, so that sporadic E layers can exist for many hours above 95 km where the atomic O is greater than 10^{11} cm^{-3}.[19] More recently we have found that the reactions CaO$^+$ + O and CaO$_2^+$ + O proceed at similar rates to their Fe analogues,[20] so that the following reactions of magnesium oxide ions would be expected to control the lifetime of Mg$^+$ in a sporadic layer:

$$MgO_2^+ + O \rightarrow MgO^+ + O_2 \quad (3)$$

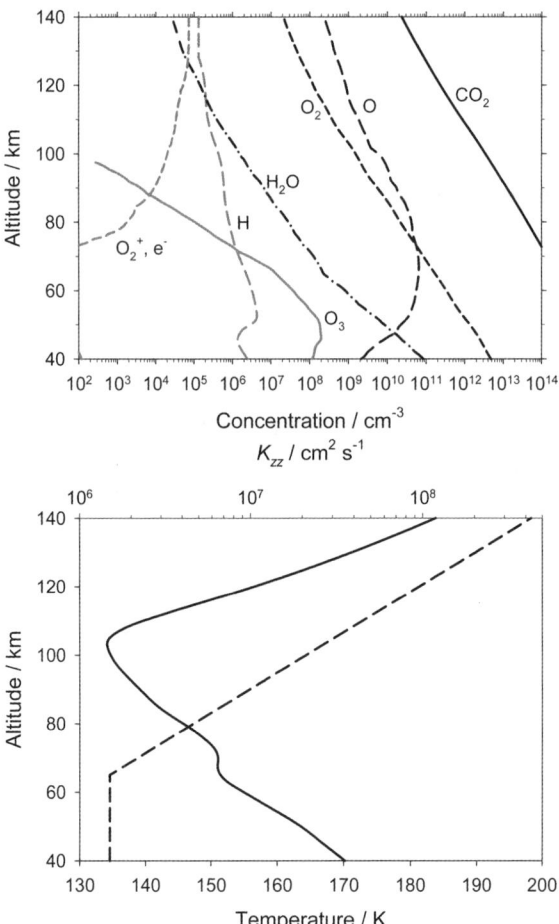

Fig. 3 Top panel: height profiles of constituents of the Martian atmosphere which play a role in the chemistry of meteoric metals. Bottom panel: height profiles of the temperature (solid line) and vertical eddy diffusion coefficient, K_{zz} (broken line). This data is a global daytime average—see the text for further details.

$$MgO^+ + O \rightarrow Mg^+ + O_2 \qquad (4)$$

Fig. 3 (top panel) shows height profiles of the atmospheric constituents of the Martian atmosphere which are relevant for the chemistry of metallic ions (the source of this data is discussed in the modelling section below). Note that atomic O is more abundant that O_2 above 75 km and becomes the major "trace" species. Nevertheless, its concentration at 90 km is only about 2×10^{10} cm^{-3}, more than an order of magnitude lower than the concentration of O in the terrestrial atmosphere at heights around 100 km where long-lived sporadic E layers occur. This makes the contribution of metallic ions to the Martian third ion layer seem even less likely. However, reactions 3 and 4 have not been studied previously (apart from a rough estimate[21] of k_4), and it is worth noting that the analogous reactions of the iron and calcium oxides with O proceed about 20 times slower than their Langevin rates.[19,20]

In this paper we will describe a laboratory study of reaction 1, carried out by first measuring the rate coefficient k_1 in He as a function of temperature, and then

Table 1 - Rate coefficients for ion–molecule reactions that are important in the Martian atmosphere

	Magnesium	Iron
Photo-ionization (at 90 km)	10^{-8} s^{-1}	
Mt + hv → Mt$^+$ + e$^-$	3.2^c	73^c
Bimolecular reactions	10^{-10} cm^3 molecule^{-1} s^{-1}	
Mt + O$_2^+$ → Mt$^+$ + O$_2$	12^d	11^e
Mt$^+$ + O$_3$ → MtO$^+$ + O$_2$	11.7 ± 1.9^{b,f}	1.3 ± 0.3^{b,g}
MtO$^+$ + O$_3$ → Mg$^+$ + 2O$_2$	5.1 ± 2.5^f	—
→ MgO$_2^+$ + O$_2$	3.4 ± 1.7^f	
MtO$^+$ + O → Mt$^+$ + O	5.9 ± 2.4^c	0.32 ± 0.15^h
MtO$_2^+$ + O → MtO$^+$ + O$_2$	6.5 ± 1.8^c	0.63 ± 0.27^h
Mt$^+$·CO$_2$ + O$_2$ → MtO$_2^+$ + CO$_2$	0.24 ± 0.11^c	21.6 ± 3.5^i
Mt.X$^+$ + e$^-$ → Mt + X (X = O, O$_2$ or CO$_2$)	3000 (200 K/T)$^{1/2}$ ^{aj}	3000 (200 K/T)$^{1/2}$ ^{aj}
Recombination reactions	10^{-28} cm^6 molecule^{-2} s^{-1}	
Mt$^+$ + CO$_2$ + CO$_2$ → Mt$^+$·CO$_2$ + CO$_2$	2.2 ± 0.6^{a,b,c}	4.5 ± 1.0^{a,b,g}
Mt$^+$ + O$_2$ + CO$_2$ → Mt$^+$·O$_2$ + CO$_2$	1.9 ± 0.4^{a,b,c}	1.4 ± 0.3^{a,b,g}

^{a} A recombination reaction where the rate coefficient measured in He has been multiplied by a factor of 7.7 (see text) to correct for CO$_2$ as third body. ^{b} Extrapolated to 140 K using the experimental temperature dependence. ^{c} This study. ^{d} Rutherford et al. (1971).[35] ^{e} Rutherford and Vroom (1972).[36] ^{f} Unpublished work, C. L. Whalley and J. M. C. Plane, University of Leeds. ^{g} Rollason and Plane (1998).[60] ^{h} Woodcock et al. (2006).[19] ^{i} Vondrak et al. (2006).[17] ^{aj} Estimate based on review of dissociative electron recombination by Florescu-Mitchell and Mitchell (2006)[15]

determining the relative efficiency of CO$_2$ to He as the third body. We will then report a study of the switching reaction

$$Mg^+·CO_2 + O_2 \rightarrow MgO_2^+ + CO_2 \quad (5)$$

This reaction is exothermic, $\Delta H_0(5) = -23$ kJ mol^{-1},[22] and could thus be rapid in the Martian atmosphere where the abundance of O$_2$ is about 2 orders of magnitude less than CO$_2$ (Fig. 3, top panel). The final part of the experimental section will describe a study of reactions 3 and 4 involving atomic O. We will then use the experimental rate coefficients from the present study, together with our previous studies of Fe ion–molecule chemistry,[17,19] to construct a model of Mg$^+$ and Fe$^+$ ion chemistry in the Martian atmosphere which will be used to determine whether the observed sporadic third layer can actually be produced by metallic ions.

Experimental

Pulsed laser photolysis/laser induced fluorescence (PLP-LIF)

The apparatus used to study the reaction between Mg$^+$ and CO$_2$ has been described in detail previously.[23] Mg$^+$ ions were produced by pulsed multi-photon photolysis of magnesium acetyl acetonate vapour (Mg(C$_5$H$_7$O$_2$)$_2$, or MgAcAc) in the central chamber of a stainless steel reactor in an excess of the He bath gas and CO$_2$ reactant. Powdered MgAcAc was located in a tantalum boat, located in a heat pipe connected to the central chamber, and heated to between 433 and 463 K, maintained to within ±2 K during a particular experiment. The resulting vapour was then entrained in the He bath gas and carried to the central chamber where it mixed with a larger flow of He and CO$_2$. The photolysis of the MgAcAc vapour by an ArF excimer laser (typical pulse energy = 50 mJ, repetition rate = 5 Hz) at 193 nm produced Mg^+ ions which were probed at 279.6 nm ($Mg^+(3^2P_1-3^2S_0)$) using a Nd:YAG-pumped dye laser (laser dye Rhodamine 590, pulse energy 10 μJ) frequency doubled with a BBO crystal. The excimer and dye laser beams were aligned collinearly and counter-propagated through the centre of the chamber, with a dichroic mirror to protect the dye laser from the excimer beam. The LIF signal was measured using a photomultiplier tube (Electron Tubes, Model 9816QB) after passing through an interference filter centred at 280 nm (fwhm = 15 nm), and captured with a gated integrator. Materials: He (99.9999%, BOC Gases) and CO_2 (99.995%, BOC Gases) were used without further purification. MgAcAc (Alfa, 98.0%) was purified by being pumped on in the heat pipe at between 433 and 463 K for several minutes prior to the experiments.

Fast flow tube-mass spectrometer (FT-MS)

This apparatus has been used previously to study the ion–molecule reactions of Fe^+ and Ca^+.[17,24] The flow tube, constructed from stainless steel with an internal diameter of 37.5 mm, has a total length of 1285 mm from the carrier gas entry point to the mass spectrometer skimmer cone at the downstream detection point. Laser ablation of a piece of MgO (Alfa Aesar, 99.95%) with a Nd:YAG laser at 532 nm (repetition rate = 10 Hz, pulse energy ~10 mJ), loosely focused onto the target using a quartz lens (focal length = 150 mm), was used to produce Mg^+ ions. The ablation target was mounted on a rotary feedthrough powered by a DC motor; rotation of the target ensured that a fresh surface was presented to each laser pulse in order to maintain a uniform Mg^+ signal. The pulses of Mg^+ ions were entrained in a flow of He which entered upstream of the ablation target. Typically, a total gas flow rate of 3000 sccm was used at a pressure of 1.2 Torr, so that the plug flow velocity was 50.0 m s^{-1}. The Reynolds number was therefore always <80, ensuring laminar flow within the tube.

Ions were detected using a differentially pumped 2-stage quadrupole mass spectrometer (VG Quadrupoles, Model SXP Elite) with a skimmer cone (0.1 mm diameter orifice) located 1160 mm downstream of the ablation target. To sample thermal ions from the flow tube this cone was biased by −11 V; to reduce the residence time of the Mg^+ ions in the skimmer zone, where otherwise cluster ions might form in the cold expansion, the exit skimmer cone (2 mm diameter orifice) to the second stage containing the quadrupole was biased by −94 V. A multi-channel scaler, synchronised with the laser pulses, was used to record the ion signals accumulated from ~ 10^3 laser shots.

Atomic O was produced by the microwave discharge of N_2 to generate N atoms, which were then titrated with NO (N + NO → O + N_2). The microwave cavity was mounted on a side-arm consisting of a 150 mm quartz tube which protruded into the flow tube, so that O atoms entered the flow 740 mm downstream of the ablation target. [O] was determined by titration with NO_2;[25] chemiluminescence from excited NO_2 generated from the NO + O reaction was measured through a photomultiplier with an interference filter at 590 ± 5 nm, situated just upstream of the mass spectrometer. The first-order loss of O to the flow tube walls was determined by measuring [O] at different flow velocities at constant pressure (and thus at different flow times in the flow tube). The change in relative [O] was monitored by the addition of NO just upstream of the photomultiplier tube and measuring the relative intensity of the resulting NO_2^* chemiluminescence. MgO^+ and MgO_2^+ ions were produced by the addition of O_3 through a side-arm located 125 mm downstream of the microwave discharge. Materials: He (99.995%, BOC Gases) was purified by passing through a molecular sieve held at 77 K. N_2 (99.998%, BOC Gases) and O_2 (99.999%, BOC Gases) were used without further purification. O_3 was produced by flowing O_2 through a commercial ozoniser. The resulting 5–8% O_3 in O_2 mixture was collected on silica gel held at 156 K by an ethanol slush bath and the O_2 pumped off at the same temperature to produce a 60–75% O_3/O_2 mixture.

Experimental results

Reaction 1 in the PLP-LIF system

In the study of reaction 1, the Mg^+ ions were in a large excess of CO_2 (by a factor $>10^9$), so that the loss of Mg^+ ions is characterised by the pseudo first-order decay coefficient k', where

$$k' = k_{diff} + k_{MgAcAc}[MgAcAc] + k_1^{He}[CO_2][He] + k_1^{CO2}[CO_2][CO_2] \quad (I)$$

where k_{diff} describes the diffusion of the Mg^+ ions out of the volume defined by the intersection of the laser beams within the field of view of the photomultiplier tube; k_{MgAcAc} is the rate coefficient for the reaction between Mg^+ and the organometallic precursor (MgAcAc); and k_1^{He} and $k_1^{CO_2}$ are the rate coefficients for reaction 1 where M = He and CO_2, respectively. The LIF signal decays obtained during the experiments were well fitted to the single exponential form $A.\exp(-k't)$. The value of $(k_{diff} + k_{MgAcAc})$ was determined when $[CO_2] = 0$, and varied from 3000–27000 s^{-1} depending upon the conditions and heat pipe temperature for a particular experimental run.

k_1^{He} was measured by working at low $[CO_2]$ ($[CO_2]/[He] < 0.01$), so that the contribution of the last term in eqn (I) to k' was not significant. k_{rec}, determined from the slopes of plots of k' versus $[CO_2]$, should then be equal to $k_1^{He}[He]$. Fig. 4 illustrates plots of k_{rec} against [He] at three different pressures. The linearity of these plots establishes that reaction 1 is close to the low pressure limit, and the slopes of the plots yield k_1^{He}. Fig. 5 is a plot of k_1^{He} versus temperature, from which the temperature dependence of reaction 1 can be determined:

$$k_1^{He}(190–403\ K) = (7.05 \pm 1.14) \times 10^{-30}\ (T/300\ K)^{(-1.86\pm0.03)}\ cm^6\ molecule^{-2}\ s^{-1}$$

where the uncertainties are obtained from the standard errors in the kinetic plots (7.9%) and the small systematic errors in the mass flow rates (0.8%) and total pressure (2%). For modelling purposes at temperatures below 300 K, a global uncertainty of 17% in k_1^{He} is appropriate.

Reaction 1 was then studied at higher $[CO_2]$, so that the last term in eqn (I) became significant. Fig. 6 illustrates the resulting upward curvature as CO_2 plays an increasing role as the third body. The broken line in the plot illustrates how k' would have changed if $k_1^{CO_2} = k_1^{He}$. The solid line is a second-order polynomial fit through

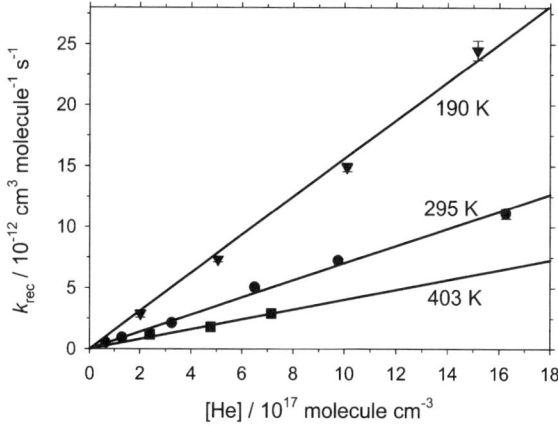

Fig. 4 Plots illustrating the pressure and temperature dependence of k_{rec} for the reaction Mg^+ + CO_2 + He. The solid lines are linear regression fits to the data at each temperature.

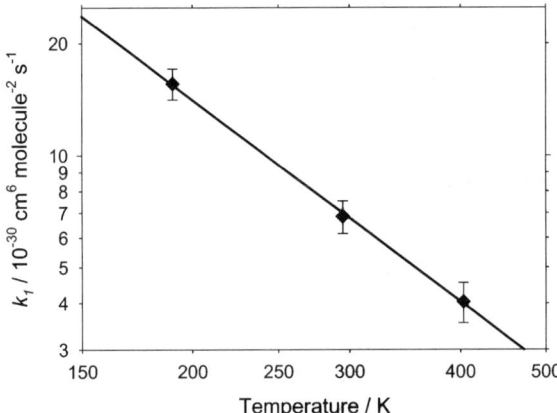

Fig. 5 Plot of $k_1(Mg^+ + CO_2 + He)$ versus T. The solid line is a weighted linear regression fit through the data.

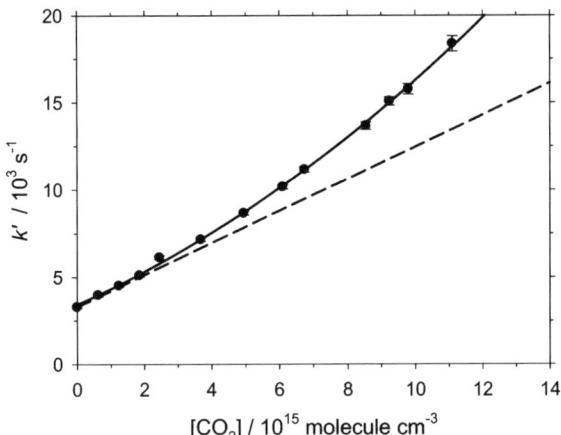

Fig. 6 Plot of k' versus $[CO_2]$ for the recombination of Mg^+ with CO_2 in the presence of He, at 295 K and 4 Torr. The solid line is a second-order polynomial fit through the data. The dashed line indicates how k' would vary if CO_2 had the same third-body efficiency as He.

the data, where the first term is the intercept ($k_{diff} + k_{MgAcAc}[MgAcAc]$), the second term corresponds to $k_1^{He}[He]$, and the third term to $k_1^{CO_2}$. Thus there is only one variable parameter in the fit. Three sets of experiments at total pressures in the reactor of 4, 6 and 10 torr yielded a consistent result of $k_1^{CO_2} = (5.3 \pm 0.7) \times 10^{-29}$ cm^6 molecule^{-2} s^{-1} at 295 K. That is, the relative efficiency of CO_2 to He is 7.7 ± 1.6.

Reactions 3, 4 and 5 in the FT-MS system

$$MgO^+ + O \text{ and } MgO_2^+ + O$$

MgO^+ ions were produced by the addition of O_3 downstream of the O atom injection point. Previously we have used N_2O to make FeO^+ and CaO^+ from their atomic ions.[19,20] However, the $Mg^+ + N_2O$ reaction has a large barrier (47 kJ mol^{-1} [26]), so that this oxidant is unsuitable. Reaction 2 is fast (Table 1), but the resulting MgO$^+$ itself reacts with O$_3$,

$$\text{MgO}^+ + \text{O}_3 \rightarrow \text{MgO}_2^+ + \text{O}_2 \quad (6a)$$

$$\rightarrow \text{Mg}^+ + 2\text{O}_2 \quad (6b)$$

so that it is not possible to produce MgO$^+$ cleanly without MgO$_2^+$.[26] Nevertheless, we actually need to measure k_3 for understanding the atmospheric chemistry of magnesium, and since we have recently measured k_2, k_{6a} and k_{6b} [26] (see Table 1), it is possible to construct a model from which k_3 and k_4 can be retrieved simultaneously. The branching ratio for reaction 6b is 0.35 ± 0.21 *i.e.* when MgO$^+$ reacts with O$_3$, 35% recycles back to Mg$^+$.

Reactions 3 and 4 were studied by measuring the relative [Mg$^+$] in the presence and absence of a fixed [O], as function of varying [O$_3$]. Fig. 7 illustrates an example of such an experiment. The [O$_3$] typically varied from 9×10^{10} to 2.7×10^{12} molecule cm^{-3} in these experiments; therefore the *absolute* [O$_3$] was too small to measure directly in the flow tube, and so was determined from the decay of Mg$^+$ as a function of O$_3$ in the absence of O, using the value for k_2 determined in a separate PLP-LIF experiment.[26] The rate coefficients k_3 and k_4 were then obtained with a model containing the following coupled differential equations:

$$\frac{d[\text{Mg}^+]}{dt} = -k_{\text{diff}}^{\text{Mg}^+}[\text{Mg}^+] - k_2[\text{Mg}^+][\text{O}_3] + k_4[\text{MgO}^+][\text{O}] + k_{6b}[\text{MgO}^+][\text{O}_3] \quad (II)$$

$$\frac{d[\text{MgO}^+]}{dt} = -k_{\text{diff}}^{\text{MgO}^+}[\text{MgO}^+] - k_4[\text{MgO}^+][\text{O}] - k_{6b}[\text{MgO}^+][\text{O}_3] \quad (III)$$
$$+ k_3[\text{MgO}_2^+][\text{O}] + k_1[\text{Mg}^+][\text{O}_3]$$

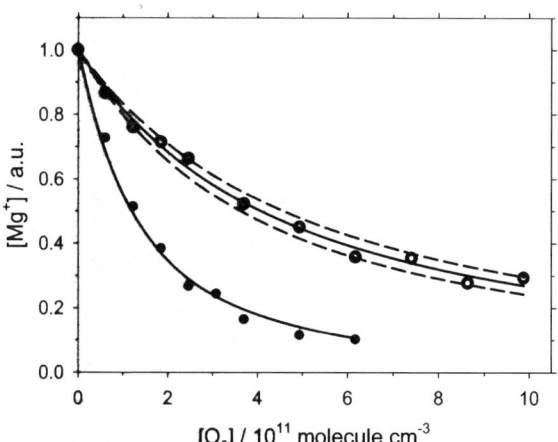

Fig. 7 Reaction of MgO$^+$ + O at 297 K, measured using the Mg$^+$ mass channel. The open circles and solid line are the experimental data and model fit in the presence of atomic O. The filled circles and solid line are the experimental data and model fit in the absence of O. The dashed lines show the model prediction when k_4 is varied within its uncertainty. Flow velocity = 52.9 m s^{-1}; [He] = 3.9×10^{16} molecule cm^{-3}; [O] = 9.7×10^{12} molecule cm^{-3} injected 420 mm upstream; O$_3$ injected 295 cm upstream of the mass spectrometer orifice.

$$\frac{d[MgO_2{}^+]}{dt} = -k_{diff}^{MgO2+}[MgO_2] - k_3[MgO_2][O] + k_{6a}[MgO^+][O_3] \qquad (IV)$$

$$\frac{d[O]}{dt} = -k_{diff}^{O}[O] - k_{O+O_3}[O][O_3] \qquad (V)$$

where k_{diff}^{X} refers to the first-order loss of species X by diffusion and uptake on the flow tube wall, and k_{O+O_3} is the rate coefficient for the reaction between O and O_3. k_{diff}^{Mg+} was measured by observing the change in [Mg$^+$] as a function of flight time at constant pressure,[20] yielding values which ranged from 233 to 264 s^{-1}. In the model, the same wall loss rate was assumed for MgO$^+$ and MgO$_2{}^+$ (the model is not sensitive to this because of the large turnover rates in the gas phase when O and O_3 are present). k_{diff}^{O} describes the loss of O atoms to the flow tube wall and was measured prior to each experiment. The average k_{diff}^{O} was 378 ± 12 s^{-1}.

The coupled differential equations were then solved using a fourth-order Runge–Kutta integrator, and the sum-of-squares difference with the experimental data points yielded χ^2. The optimal values for k_3 and k_4 were then determined from the global minimum on the 2-dimensional χ^2 surface.[19,20] The goodness-of-fit of the model is illustrated in Fig. 7. The whole experiment was repeated four times: k_3 ranged from 5.7 × 10^{-10} to 6.9 × 10^{-10} cm^3 molecule^{-1} s^{-1}, and k_4 from 4.9 × 10^{-10} to 6.8 × 10^{-10} cm^3 molecule^{-1} s^{-1}. The [O] at the injection point was also varied from (4.1 to 9.9) × 10^{12} molecule cm^{-3}, achieved by changing the power of the microwave discharge (from 70 to 110 W) and the flow of N$_2$ through the quartz injector (from 100–200 sccm). Varying [O] did not produce a systematic change in k_3 and k_4.

Averaging these results yields k_3 = (6.5 ± 1.8) × 10^{-10} cm^3 molecule^{-1} s^{-1} for MgO$_2{}^+$ + O and k_4 = (5.9 ± 2.4) × 10^{-10} cm^3 molecule^{-1} s^{-1} for MgO$^+$ + O. These overall uncertainties were obtained with a Monte Carlo procedure which combined the statistical standard deviations from the four measurements (5.0 × 10^{-11} and 8.4 × 10^{-11} cm^3 molecule^{-1} s^{-1} for k_3 and k_4, respectively) with the uncertainties of the following parameters: [O] at the point of injection (3.3%); k_{diff}^{O} (4.3–10.1%, depending on experiment); [O$_3$] (10%); and the uncertainties in k_2, k_{6a} and k_{6b} listed in Table 1. As the O$_3$ was injected downstream of the O atom injection point there is an uncertainty in the O$_3$ mixing length *i.e.* the distance over which O$_3$ mixes into the main flow. This was estimated to be 15% of the total reaction length, based on previous work with this flow tube for a study of CaO$^+$ + O.[20] Therefore this uncertainty was included in the Monte Carlo procedure, which allowed all these parameters to be varied simultaneously within their 1σ uncertainties. Systematic errors arising from the flow and pressure calibration were negligible: the flow speed calculated from the mass flow and pressure (with a correction for laminar flow[17]) was checked by timing the arrival of the Mg$^+$ pulse at the mass spectrometer (the flow times agreed within 2.3%).

One concern was the possible removal of O$_3$ by species produced in the microwave discharge, which would have led to an overestimation of k_3 and k_4. We have previously investigated the trace species produced in an N$_2$ discharge followed by downstream titration with NO.[19] These species include NO$^+$, NO$_2$, and N, depending on the proximity to the NO titration endpoint. Using the model described above, if O$_3$ were destroyed in the flow tube with a first-order rate of 95 s^{-1} this would decrease the retrieved k_3 and k_4 by a factor of 2. However, the rate coefficients for reaction of O$_3$ with NO$^+$, NO$_2$, and N are between 3.5 × 10^{-17} and 1 × 10^{-14} cm^3 molecule^{-1} s^{-1}, so that concentrations of at least 9 × 10^{15} molecule cm^{-3} would be needed to remove O$_3$ at this rate. This is quite unrealistic: in fact if *all* the N produced in the discharge and the NO added in the titration were available to react with O$_3$ (rather than reacting with each other to form O), then O$_3$ would be removed at only 1.2 s^{-1}.

$$Mg^+ \cdot CO_2 + O_2$$

Reaction 5 was studied at a pressure of 1.2 Torr and at 295 K. To produce $Mg^+ \cdot CO_2$, CO_2 was added 140 mm downstream of the ablation target, with O_2 added a further 200 mm downstream. The reaction was measured by monitoring the change in $Mg \cdot CO_2^+$ signal intensity as $[O_2]$ was varied. The kinetic behaviour of this reaction is complicated by the continued formation of $Mg^+ \cdot CO_2$ from the remaining Mg^+ downstream of the O_2 addition port, as well as the removal of $Mg^+ \cdot CO_2$ by recombination with a second CO_2:

$$Mg^+ \cdot CO_2 + CO_2 + He \rightarrow Mg^+(CO_2)_2 + He \qquad (7)$$

Production of MgO_2^+ *via* direct reaction between O_2 and Mg^+ is negligible due to the small concentration of O_2 used during experiments and the rate coefficient being an order of magnitude smaller than $Mg^+ + CO_2$.[26] To obtain k_5 and k_7 a model was constructed with the following coupled differential equations:

$$\frac{d[Mg^+]}{dt} = -k_{diff}^{Mg^+}[Mg^+] - k_1[Mg^+][CO_2][He] \qquad (VI)$$

$$\frac{d[Mg^+ \cdot CO_2]}{dt} = -k_{diff}^{Mg^+ \cdot CO_2}[Mg^+ \cdot CO_2] - k_5[Mg^+ \cdot CO_2][O_2]$$
$$- k_7[Mg^+ \cdot CO_2][CO_2][He] + k_1[Mg^+][CO_2][He] \qquad (VII)$$

k_7 was first determined in the absence of O_2, by monitoring the change in $Mg^+ \cdot CO_2$ as a function of $[CO_2]$: $k_7 = (1.2 \pm 0.4) \times 10^{-29}$ cm^6 molecule^{-2} s^{-1}. k_5 was then determined by a global fit of the model (eqn (VI) and VII) to the experimental points. This produced the satisfactory fit shown in Fig. 8, with $k_5 = (2.2 \pm 0.8) \times 10^{-11}$ cm^3 molecule^{-1} s^{-1}. The uncertainties were again estimated by combining the statistical errors from the model fit with various systematic errors, using the Monte Carlo procedure described above.

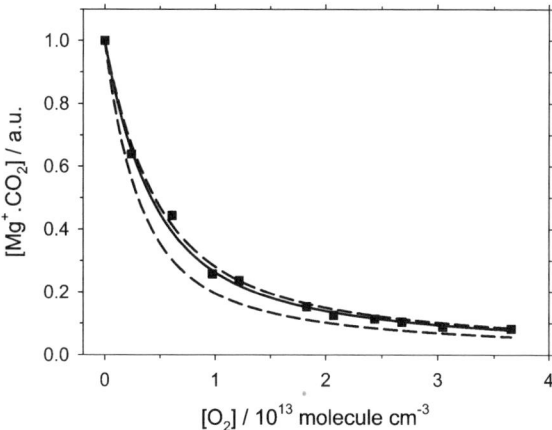

Fig. 8 Ligand-switching reaction of $Mg^+ \cdot CO_2 + O_2$ at 295 K, measured using the $Mg^+ \cdot CO_2$ mass channel. The solid line is a model fit to the experimental data (solid points). he dashed lines show the model prediction when k_6 is varied by the experimental uncertainty. Flow velocity = 50.0 m s^{-1}; [He] = 3.8 × 10^{16} molecule cm^{-3}; [CO$_2$] = 1.4 × 10^{14} molecule cm^{-3} injected 1020 mm upstream of the mass spectrometer orifice; O$_2$ is injected 770 mm upstream.

Discussion

Experimental results

In this study we have shown that CO_2 is significantly more efficient than He as the third body in the recombination of Mg^+ with CO_2 (reaction 1), by a factor of 7.7 ± 1.6. In neutral recombination reactions CO_2 is generally ~4 times more efficient.[27] There are at least two reasons for the greater efficiency in the ion–molecule reaction. First, the collision frequency of energised $Mg^+ \cdot CO_2^*$ with CO_2 is actually faster than with He (7.7×10^{-10} compared with 5.4×10^{-10} cm^3 molecule^{-1} s^{-1}), because the long-range ion-induced dipole interaction is much greater for CO_2 with a polarizability that is 15 times larger than that of He.[28] Second, the CO_2 binds relatively strongly to $Mg^+ \cdot CO_2$ ($D_0(CO_2 \cdot Mg^+ - CO_2) = 42$ kJ mol^{-1} [22]), which will encourage more efficient transfer of energy.

Reaction 3 ($MgO_2^+ + O$) and reaction 4 ($MgO^+ + O$) both have rate coefficients which are close to their collision frequencies. For, $k_4 = (5.9 \pm 2.4) \times 10^{-10}$ cm^3 molecule^{-1} s^{-1} compared with a collision frequency of 6.2×10^{-10} cm^3 molecule^{-1} s^{-1} if only the ion-induced dipole interaction is considered (*i.e.*, the Langevin model[29]). If the ion–quadrupole interaction is included (taking the electric quadrupole moments of O(^{3}P) to be $\Theta_{xx} = -4.3 \times 10^{-40}$ C m^2 and $\Theta_{zz} = 2.2 \times 10^{-40}$ C m^2 [30]), then the collision frequency increases to 8.1×10^{-10} cm^3 molecule^{-1} s^{-1}, with a negligible temperature dependence ($T^{-0.06}$). The only previous measurement with which to compare our results is a flowing after-glow study of reaction 4. Ferguson and Fehsenfeld[21] reported a rough estimate of $k_4 \sim 1 \times 10^{-10}$ cm^3 molecule^{-1} s^{-1}, but provided very few experimental details of their measurement so comparison with the present result is difficult.

Reactions 3 and 4 are strikingly fast compared with the analogous reactions of FeO_2^+ and FeO^+,[19] and CaO^+ [20] with O, which are all approximately 1 order of magnitude slower. One possible reason for this is that the bond energies of the magnesium oxide ions are significantly smaller: *cf.* $D_0(Mg^+\text{-O}) = 216$ [220 ± 15] kJ mol^{-1},[22] $D_0(Fe^+ - O) = 330$ [340 ± 6] kJ mol^{-1},[19] $D_0(Ca^+ - O) = 317$ [$318 - 344$] kJ mol^{-1},[31] and $D_0(Mg^+ - O_2) = 90$ [≤ 110] kJ mol^{-1},[22] $D_0(Fe^+ - O_2) = 114$ [105 ± 25] kJ mol^{-1},[17] $D_0(Ca^+ - O_2) = 227$ kJ mol^{-1} [24] (bond energies from electronic structure calculations are followed in square parentheses by experimental determinations, where available).

Magnesium and iron chemistry in the Martian atmosphere

Fig. 9 illustrates schematically the ion–molecule chemistry of magnesium which is likely to control the meteoric Mg^+ layer in the Martian atmosphere. The chemistry

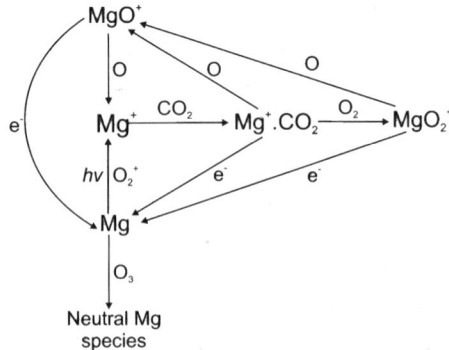

Fig. 9 Schematic diagram of the ion–molecule chemistry of magnesium in the Martian atmosphere.

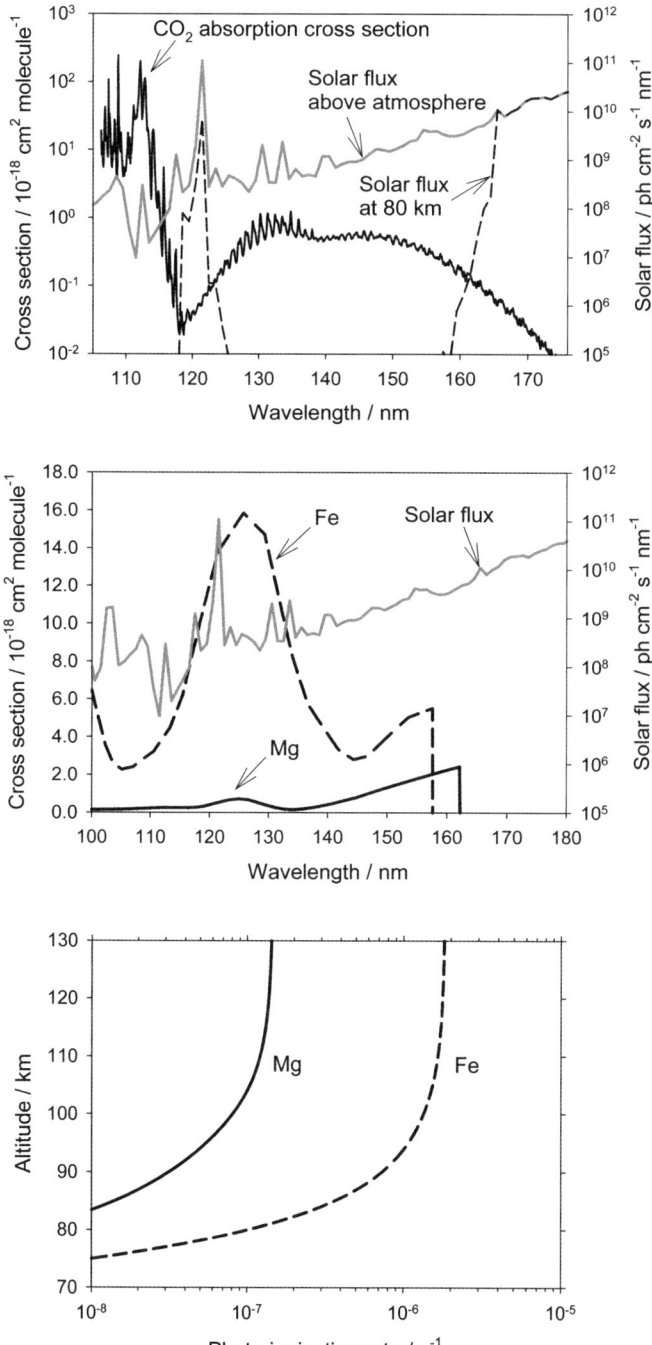

Fig. 10 Top panel: CO_2 absorption cross section between 106 and 176 nm, and the solar actinic flux incident above the Martian atmosphere, and at 80 km. Middle panel: photo-ionization cross sections of Fe and Mg superimposed on the incident solar flux. Bottom panel: photo-ionization rates between 70 and 130 km.

of the Fe^+ layer should be very similar. Table 1 lists the rate coefficients for the ion–molecule reactions of magnesium and iron which are used in the atmospheric model described below, extrapolated to 140 K where a measured T dependence is available. We have applied the CO_2/He efficiency factor from the present study to all the recombination reactions, and assume that the temperature dependence measured in He is the same when CO_2 is the third body. Note that all the relevant reactions for magnesium in Fig. 9, apart from dissociative recombination of the molecular ions with electrons, have now been studied in the laboratory (the same situation applies for iron). An atmospheric model of these metals can therefore be constructed with a fair degree of confidence.

First we consider the rates of photo-ionization of the metal atoms. The top panel of Fig. 10 illustrates: the solar actinic flux incident on the top of the atmosphere (taken to be the solar cycle averaged terrestrial flux scaled by a factor of 0.43); the absorption spectrum of CO_2 (at 195 K) from 106–176 nm;[32,33] and the resulting actinic flux at 80 km assuming a solar zenith angle of 45°. Note that because the CO_2 cross section is relatively small between 118 and 125 nm, the solar Lyman-α line at 121.6 nm penetrates strongly down to even as low as 80 km. The middle panel of Fig. 10 illustrates the photo-ionization cross sections of Fe and Mg, calculated from the astrophysical Opacity Project.[34] The particularly strong overlap of the Fe cross section with solar Lyman-α results in the photo-ionization rate for Fe being more than 1 order of magnitude larger than Mg between 70 and 130 km in the Martian atmosphere, as shown in the bottom panel of Fig. 10. Nevertheless, the rate for Fe is still only 7×10^{-7} s^{-1} at 90 km, corresponding to a lifetime against photo-ionization of 16 days. Hence, we conclude that photo-ionization is not an important source of metallic ions. It should be noted that the Fe photo-ionization rate profile in Fig. 2 is in good agreement with a previous calculation by Molino-Cuberos et al.,[4] while that for Mg is about a factor of 3 smaller. Instead of photo-ionization, charge transfer reactions of Mg and Fe with ambient O_2^+ ions (Fig. 2) must be the dominant source of the metallic ions, since these reactions proceed at their Langevin rates.[35,36] This conclusion was also reached by Molino-Cuberos et al.,[4] in contrast to an earlier study by Pesnell and Grebowsky.[16]

Once the ions form, recombination with CO_2 will be faster by at least 3 orders of magnitude compared to reaction with O_3 (the dominant reaction in the terrestrial atmosphere), and recombination with O_2 will be even slower. The resulting CO_2 cluster ions are then likely to undergo rapid switching with O_2 to form MgO_2^+ and FeO_2^+. At this point an important difference emerges between the metals. MgO_2^+ and MgO^+ both react with atomic O at least 1 order of magnitude faster than the analogous reactions of FeO_2^+ and FeO^+.[19] The result is that the Fe-cluster ions are much more likely to be destroyed by dissociative recombination with electrons, while the Mg-cluster ions are more likely to be reduced by atomic O back to Mg^+. Although the dissociative electron recombination reactions of these metallic molecular ions have not been studied, a recent major review[15] concludes that dissociative recombination reactions involving small molecular ions have pretty similar rate coefficients (within a factor of 3), and so all the reactions here are assigned a representative coefficient of 3×10^{-7} cm^3 $molecule^{-1}$ s^{-1}.

A model of metal chemistry in the Martian atmosphere

We now use the rate coefficients measured in this study to examine whether the sporadic third layers observed in the Martian atmosphere really can be produced by long-lived metallic ions. For this purpose we use a 1-dimensional time-resolved model, previously developed for modelling the terrestrial metal layers.[37] The model extends from 30 to 150 km with a resolution of 0.5 km. The chemistry scheme contains the ion–molecule chemistry in Table 1 (Fig. 9), and the neutral chemistry is taken from our previous models of the terrestrial Fe[38] and Mg[39] layers, with additional MgO reactions from Rollason and Plane.[40] A sink reaction is included where species like MgOH and MgCO$_3$ (and their Fe analogues) polymerize to form embryonic meteoric smoke particles—this process is assigned a rate coefficient of 9×10^{-10} cm^3 molecule^{-1} s^{-1}, reflecting the large electric dipole moments of these metal-containing molecules.[41] The eddy diffusion coefficient (K_{zz}) profile (see Fig. 2, bottom panel) is taken from Rosenqvist and Chassefiere.[42] The value of 1.5×10^6 cm^2 s^{-1} below 65 km was estimated from Phobos 2 solar occultation measurements of dust, ozone and clouds at low latitudes. The large values (>10^7 cm^2 s^{-1}) above 90 km are consistent with the K_{zz} required to model measurements made by the Viking 1 and 2 spacecraft,[43] and the values that have been used in other models.[16,43,44]

Here we will only investigate daytime conditions, since all the sporadic layers reported to date have been observed on the Martian dayside.[8,9] The height profiles of the relevant neutral species which play a role in the metal chemistry (see Fig. 2, top panel) are taken from the following sources: Molina-Cuberos et al. [4] (MCO2); Rodrigo et al.[44] (R90); and a Mars Atmosphere Modelling and Observations Workshop (MA03) held in Granada in 2003 which provides evaluated daytime profiles of many constituents (http://www.atm.ox.ac.uk/user/newmanc/workshop/intercomp.html). CO_2, N_2 and O_2 are from MA03; H, O_3 and H_2O from R90; O and O_2^+ (= electrons) from MC02. The O profile was multiplied by a factor of 3 to bring the [O]/[CO$_2$] ratio up to 1.2% at 135 km, which is consistent with the new atomic O density derived from observations of 130.4 nm triplet emission with the SPICAM-UV instrument on Mars Express.[45] Note that the important neutral parameters for controlling the metal ion lifetimes are CO_2, O_2 and O (Fig. 9). The temperature profile (Fig. 2, bottom panel) was taken from MA03—the precise profile is not important because the ion–molecule reactions have small T-dependences (Table 1).

The injection rate of Mg and Fe atoms was calculated using CABMOD,[3] which describes the differential ablation of individual elements from a meteoroid of specified mass, density and entry velocity. The differential ablation predicted by CABMOD has recently been tested successfully using meteor head echo data from the large aperture radar at Arecibo.[46] The processes included in the model are: sputtering by inelastic collisions with air molecules before the meteoroid melts; evaporation of atoms and oxides from the molten particle; a detailed thermodynamic treatment of the melt and the vapour around the particle; diffusion-controlled migration of the volatile constituents (Na and K) through the molten particle; Langmuir evaporation of each element; and impact ionization of the ablated fragments by hyperthermal collisions with air molecules.[3] In order to adapt CABMOD for Mars, the meteoroid mass distribution is assumed to be the same as for Earth, but the meteoroid velocity distribution is shifted to lower velocities using the procedure described by Molina-Cuberos et al.,[47] because the mean orbital velocity of Mars is 24.1 km s^{-1} and its escape velocity is 5.0 km s^{-1}, compared with 29.8 km s^{-1} and 11.2 km s^{-1} for Earth. Integrating over the mass and velocity distributions yields the total injection rate, which is illustrated for Fe in Fig. 2 (the profile for Mg is very similar[3]). The peak height and absolute injection rate for Mars is in good agreement with the previous model of Molina-Cuberos et al.[47]

We now use the model to answer two questions. The first is whether sufficient metallic ions occur between 80 and 100 km so that, if they are collected by some process into a layer a few km wide at 90 km, the resulting concentration would exceed 1×10^4 cm^{-3} as observed in typical sporadic third layers (e.g., Fig. 1). The second question is whether the lifetime of the metallic ions in such a layer would exceed 20 h, as indicated by measurements from Mars Express on sequential orbits.[8] Fig. 11 illustrates vertical profiles of the magnesium and iron species predicted by the 1D model. For magnesium, the Mg$^+$ layer peaks around 83 km and the Mg layer around 70 km. Below 60 km the dominant gas-phase species are MgOH and MgCO$_3$, which are largely converted into smoke particles below 50 km. In the case of iron, the Fe$^+$ layer peaks around 102 km and the Fe layer at 74 km.

Inspection of Fig. 11 reveals a very striking difference between the predicted quantities of metal ions: the Mg$^+$ column abundance is 4.5×10^9 ion cm^{-2}, whereas the

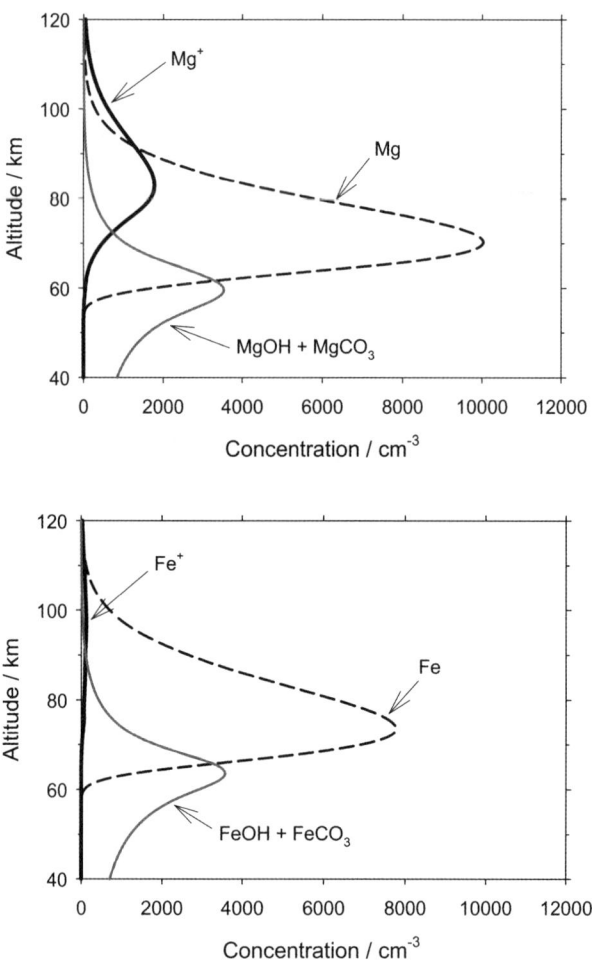

Fig. 11 Top panel: modelled height profiles of Mg^+, Mg and the neutral magnesium reservoirs, for the globally-averaged Martian atmosphere during daytime. Bottom panel: height profiles of the corresponding iron species.

Fe^+ abundance is only 4.6×10^8 ion cm^{-2} *i.e.* 1 order of magnitude lower. This major difference results from the much larger rate coefficients for the reactions of MgO^+ and MgO_2^+ with atomic O, compared with FeO^+ and FeO_2^+ (Table 1). Previous models[4,16] of Mg^+ and Fe^+ in the Martian atmosphere predicted layers with daytime peak concentrations around 10^4 cm^{-3}. However, these models assumed rate coefficients for the recombination of these ions with CO_2 that were more than 200 times smaller than the results from the present study (Table 1).

If the Mg^+ layer in Fig. 11 were concentrated into a 3 km wide layer, the resulting Mg^+ concentration would peak at $\sim 2 \times 10^4$ cm^{-3}, sufficient to form an observable sporadic third layer. By analogy with known mechanisms which cause sporadic *E* layers in the terrestrial atmosphere, we speculate briefly on how the concentration of Mg^+ ions into such a layer might occur on Mars. The classical mechanism for sporadic *E* formation at low- and mid-latitudes is through ion convergence caused by neutral wind shear in the Earth's magnetic field,[48] often driven by atmospheric tides and gravity waves.[5] At high latitudes the main process appears to be ion convergence in electric fields near auroral activity,[49] where charged particle precipitation produces excess ionization in the lower *E* region.[50]

In contrast to the Earth, Mars does not have a permanent magnetic field. However, there is crustal remanent magnetization, with intense magnetization mainly confined to the ancient, heavily cratered highlands in the south.[51] Thus a wind shear mechanism may operate in some locations. Reconnection with the interplanetary magnetic field can also take place in localized regions, permitting energetic particles to precipitate into the neutral atmosphere.[52] Indeed, auroral emissions have been observed very recently by the SPICAM UV instrument on Mars Express.[53] Enhanced particle precipitation events, which occur after solar coronal

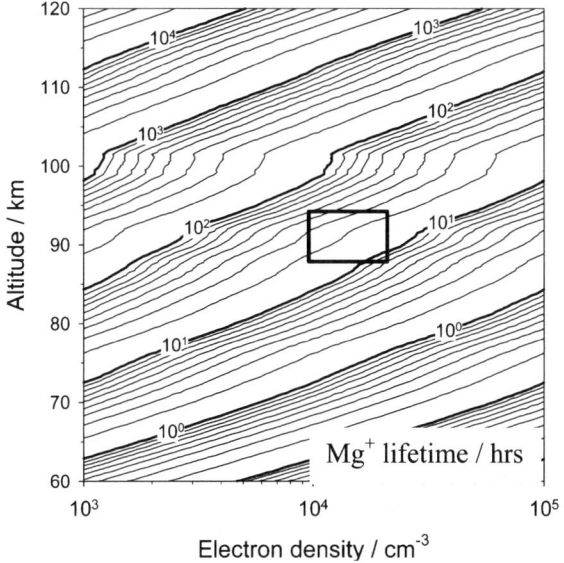

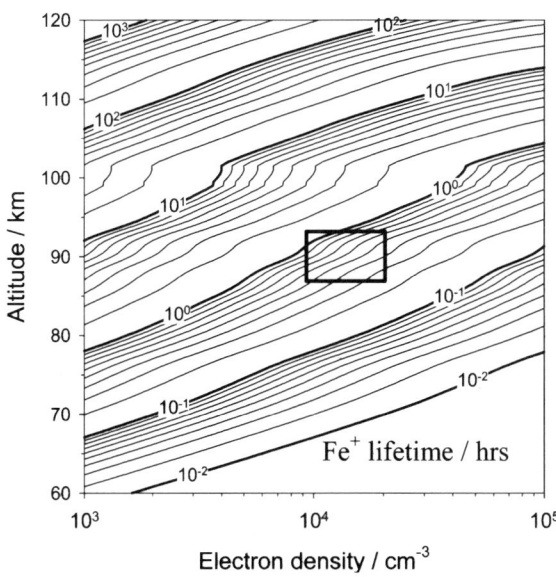

Fig. 12 Top panel: lifetime (in hours) for Mg^+ ions during daytime in the Martian atmosphere, as a function of height and total electron density. Bottom panel: analogous plot for Fe^+. The black-edged box in each plot brackets the height range and electron densities typical of the observed sporadic third layers.

mass ejections, will increase the general level of ionization in the lower ionosphere, converting more neutral metal atoms into ions through charge exchange. These events could therefore trigger the formation of sporadic third layers. Mars is also characterised by pronounced thermal tides which extend from the surface to the exosphere,[54] and very strong gravity waves have been measured in the thermosphere,[55] most likely driven by the topography.[56] Thus most of the mechanisms which form sporadic E in the terrestrial atmosphere should also operate on Mars.

Fig. 12 illustrates the lifetime of Mg^+ and Fe^+ against neutralization, as a function of altitude and electron density. The black box in each panel brackets the height range (85–95 km) and typical electron densities ($(1-2) \times 10^4$ cm^{-3}) of the observed third layers. Note that the Mg^+ lifetime within the box is generally between 10 and 40 h (bearing in mind the *caveat* that this is a daytime model). These long lifetimes would therefore account for the Mars Express observations that the layers can persist for 20 h.[8] In contrast, the Fe^+ lifetime within the box is mostly less than 1 h.

A final point to consider is whether the model predictions can be tested by future observations of Mg^+ and Mg. These species are observed in the terrestrial atmosphere from satellites during daytime, using their strong resonance lines at 280 and 285 nm, respectively.[57,58] The challenge with making similar measurements in the Martian atmosphere is that there is very strong emission in the dayglow between 280 and 300 nm from CO_2^+ ($B^2\Sigma - X^2\Pi$) in the lower ionosphere, which will therefore overlap with the emission lines from Mg^+ and Mg. Likewise, observations of Fe^+ and Fe at 260 and 248 nm, respectively, will be obscured by emissions from the strong Cameron bands of CO ($a^3\Pi - X^1\Sigma$) between 185 and 270 nm. Indeed, recent observations with the SPICAM UV instrument show that both these molecular bands have intensities of at least 1 kR nm^{-1} in the Martian dayglow.[59] Thus, retrieving the metallic species will require careful limb-scanning against a strong variable background.

Conclusions

The laboratory measurements described in this paper have enabled us to construct a model of the ion–molecule chemistry of magnesium and iron in the Martian atmosphere, where all the important reactions (apart from dissociative recombination with electrons) have been studied experimentally in the laboratory. The model indicates that Mg^+ plays a much more important role in the lower ionosphere of Mars than Fe^+, because atomic O reacts very rapidly with MgO^+ and MgO_2^+. The background layer of Mg^+ should be sufficient to form sporadic layers with a plasma density exceeding 10^4 cm^{-3}. Moreover, the chemical lifetime of the layer should easily exceed 10 h at 90 km. These results strongly support the conjecture[8,9] that the sporadic third layer in the Martian atmosphere is formed by metallic ions produced by meteoric ablation.

Acknowledgements

We acknowledge the European Office of Aerospace Research and Development for financial support (award no. FA8655-09-1-3015), and the Engineering and Physical Sciences Research Council for a research studentship (for C. L. W.). We thank Dr Tomas Vondrak (formerly of the University of Leeds) for performing the CABMOD runs for Mars, and assembling the library of minor constituents. We also acknowledge stimulating discussions with Dr Paul Withers (Boston University) on the subject of this paper.

References

1 I. P. Williams, in *Meteors in the Earth's Atmosphere*, ed. E. Murad and I. P. Williams, Cambridge University Press, Cambridge, 2002.

2 J. M. C. Plane, *Chem. Rev.*, 2003, **103**, 4963–4984.
3 T. Vondrak, J. M. C. Plane, S. Broadley and D. Janches, *Atmos. Chem. Phys.*, 2008, **8**, 7015–7031.
4 G. J. Molina-Cuberos, H. Lichtenegger, K. Schwingenschuh, J. J. Lopez-Moreno and R. Rodrigo, *J. Geophys. Res.*, 2002, **107**, 5027.
5 J. D. Mathews, *J. Atmos. Sol.–Terr. Phys.*, 1998, **60**, 413–435.
6 E. Kopp, *J. Geophys. Res.*, 1997, **102**, 9667–9674.
7 J. M. Grebowsky and A. C. Aikin, in *Meteors in the Earth's Atmosphere*, ed. E. Murad and I. P. Williams, Cambridge University Press, Cambridge, 2002.
8 M. Pätzold, S. Tellman, B. Häusler, D. Hinson, R. Schaa and G. L. Tyler, *Science*, 2005, **310**, 837–839.
9 P. Withers, M. Mendillo, D. P. Hinson and K. Cahoy, *J. Geophys. Res.*, 2008, **113**, A12314.
10 D. P. Hinson, in *Mars Global Surveyor Radio Occultation Profiles of the Ionosphere – Reorganized*, ed. R. A. Simpson, NASA Planetary Data System, 2008.
11 D. P. Hinson, R. A. Simpson, J. D. Twicken, G. L. Tyler and F. M. Flasar, *J. Geophys. Res.*, 1999, **104**, 26997–27012.
12 P. Withers, *Adv. Space Res.*, 2009, **44**, 277–307.
13 M. Pätzold, S. Tellmann, B. Häusler, M. K. Bird, G. L. Tyler, A. A. Christou and P. Withers, *Geophys. Res. Lett.*, 2009, **36**, L05203.
14 J. G. Molina-Cuberos, J. J. Lopez-Moreno and F. Arnold, *Space Sci. Rev.*, 2008, **137**, 175–191.
15 A. I. Florescu-Mitchell and J. B. A. Mitchell, *Phys. Lett.*, 2006, **430**, 277–374.
16 W. D. Pesnell and J. Grebowsky, *J. Geophys. Res.*, 2000, **105**, 1695–1707.
17 T. Vondrak, K. R. I. Woodcock and J. M. C. Plane, *Phys. Chem. Chem. Phys.*, 2006, **8**, 503–512.
18 S. N. Nahar, M. A. Bautista and A. K. Pradhan, *Astrophys. J.*, 1997, **479**, 497–503.
19 K. R. S. Woodcock, T. Vondrak, S. R. Meech and J. M. C. Plane, *Phys. Chem. Chem. Phys.*, 2006, **8**, 1812–1821.
20 S. Broadley, T. Vondrak, T. G. Wright and J. M. C. Plane, *Phys. Chem. Chem. Phys.*, 2008, **10**, 5287–5298.
21 E. E. Ferguson and F. C. Fehsenfeld, *J. Geophys. Res.*, 1968, **73**, 6215.
22 R. J. Plowright, T. J. McDonnell, T. G. Wright and J. M. C. Plane, *J. Phys. Chem. A*, 2009, **113**, 9354–9364.
23 J. M. C. Plane, T. Vondrak, S. Broadley, B. Cosic, A. Ermoline and A. Fontijn, *J. Phys. Chem. A*, 2006, **110**, 7874–7881.
24 S. L. Broadley, T. Vondrak and J. M. C. Plane, *Phys. Chem. Chem. Phys.*, 2007, **9**, 4357–4369.
25 F. Kaufman, *Proc. R. Soc. London, Ser. A*, 1958, **247**, 123–139.
26 C. L. Whalley and J. M. C. Plane, manuscript in preparation.
27 R. G. Gilbert and S. C. Smith, *Theory of Unimolecular and Recombination Reactions*, Blackwell, Oxford, 1990.
28 D. R. Lide, *Handbook of Physics and Chemistry*, CRC Press, Boca Raton, FL, 2006.
29 I. W. M. Smith, *Kinetics and Dynamics of Elementary Gas Reactions*, Butterworths, London, 1980.
30 G. L. Gutsev, P. Jena and R. J. Bartlett, *Chem. Phys. Lett.*, 1998, **291**, 547–552.
31 R. J. Plowright, T. G. Wright and J. M. C. Plane, *J. Phys. Chem. A*, 2008, **112**, 6550–6557.
32 G. Stark, K. Yoshino, P. L. Smith and K. Ito, *J. Quant. Spectrosc. Radiat. Transfer*, 2007, **103**, 67–73.
33 K. Yoshino, J. R. Esmond, Y. Sun, W. H. Parkinson, K. Ito and T. Matsui, *J. Quant. Spectrosc. Radiat. Transfer*, 1996, **55**, 53–60.
34 M. A. Bautista, P. Romano and A. K. Pradhan, *Astrophys. J. Suppl.*, 1998, **118**, 259–265.
35 J. A. Rutherford, R. F. Mathis, B. R. Turner and D. A. Vroom, *J. Chem. Phys.*, 1971, **55**, 3785.
36 J. A. Rutherford and D. A. Vroom, *J. Chem. Phys.*, 1972, **57**, 3091.
37 J. M. C. Plane, *Atmos. Chem. Phys.*, 2004, **4**, 627–638.
38 J. M. C. Plane, D. E. Self, T. Vondrak and K. R. I. Woodcock, *Adv. Space Res.*, 2003, **32**, 699–708.
39 J. M. C. Plane and M. Helmer, *Faraday Discuss.*, 1995, **100**, 411–430.
40 R. J. Rollason and J. M. C. Plane, *Phys. Chem. Chem. Phys.*, 2001, **3**, 4733–4740.
41 R. W. Saunders and J. M. C. Plane, *J. Atmos. Sol.–Terr. Phys.*, 2006, **68**, 2182–2202.
42 J. Rosenqvist and E. Chassefiere, *J. Geophys. Res.*, 1995, **100**, 5541–5551.
43 M. N. Izakov, *Icarus*, 1978, **36**, 189–197.
44 R. Rodrigo, E. Garcia-Alvarez, M. J. Lopez-Gonzalez and J. J. Lopez-Moreno, *J. Geophys. Res.*, 1990, **95**, 14795–14810.

45 J. Y. Chaufray, F. Leblanc, E. Quemerais and J. L. Bertaux, *J. Geophys. Res.*, 2009, **114**, E02006.
46 D. Janches, L. P. Dyrud, S. L. Broadley and J. M. C. Plane, *Geophys. Res. Lett.*, 2009, **36**, L06101.
47 G. J. Molina-Cuberos, O. Witasse, J. P. Lebreton, R. Rodrigo and J. J. Lopez-Moreno, *Planet. Space Sci.*, 2003, **51**, 239–249.
48 J. D. Whitehead, *J. Atmos. Terr. Phys.*, 1989, **51**, 401–424.
49 J. K. Olesen, F. Primdahl, F. Spangslev and N. D'Angelo, *J. Geophys. Res.*, 1975, **80**, 696.
50 J. W. MacDougall, J. M. C. Plane and P. T. Jayachandran, *J. Atmos. Sol.–Terr. Phys.*, 2000, **62**, 1169–1176.
51 J. E. P. Connerney, M. H. Acuna, P. J. Wasilewski, G. Kletetschka, N. F. Ness, H. Reme, R. P. Lin and D. L. Mitchell, *Geophys. Res. Lett.*, 2001, **28**, 4015–4018.
52 A. M. Krymskii, T. K. Breus, N. F. Ness, M. H. Acuna, J. E. P. Connerney, D. H. Crider, D. L. Mitchell and S. J. Bauer, *J. Geophys. Res.*, 2002, **107**, 1245.
53 F. Leblanc, O. Witasse, J. Lilensten, R. A. Frahm, A. Safaenili, D. A. Brain, J. Mouginot, H. Nilsson, Y. Futaana, J. Halekas, M. Holmstrom, J. L. Bertaux, J. D. Winningham, W. Kofman and R. Lundin, *J. Geophys. Res.*, 2008, **113**, A08311.
54 J. M. Forbes, S. L. Bruinsma, X. L. Zhang and J. Oberheide, *Geophys. Res. Lett.*, 2009, **36**, L15812.
55 J. E. Creasey, J. M. Forbes and G. M. Keating, *Geophys. Res. Lett.*, 2006, **33**, L22814.
56 Y. Moudden and J. M. Forbes, *J. Geophys. Res.*, 2008, **113**, E11009.
57 M. Scharringhausen, A. C. Aikin, J. P. Burrows and M. Sinnhuber, *J. Geophys. Res.*, 2008, **113**, D13303.
58 J. Correira, A. C. Aikin, J. M. Grebowsky, W. D. Pesnell and J. P. Burrows, *Geophys. Res. Lett.*, 2008, **35**, L06103.
59 C. Simon, O. Witasse, F. Leblanc, G. Gronoff and J. L. Bertaux, *Planet. Space Sci.*, 2009, **57**, 1008–1021.
60 R. J. Rollason and J. M. C. Plane, *J. Chem. Soc., Faraday Trans.*, 1998, **94**, 3067–3075.

PAPER | www.rsc.org/faraday_d | Faraday Discussions

Exploring extrasolar worlds: from gas giants to terrestrial habitable planets

Giovanna Tinetti,[*a] Caitlin A. Griffith,[b] Mark R. Swain,[c] Pieter Deroo,[c] Jean Philippe Beaulieu,[ad] Gautam Vasisht,[c] David Kipping,[ae] Ingo Waldmann,[a] Jonathan Tennyson,[a] Robert J. Barber,[a] Jeroen Bouwman,[f] Nicole Allard[d] and Linda R. Brown[c]

Received 14th April 2010, Accepted 27th April 2010
DOI: 10.1039/c005126h

Almost 500 extrasolar planets have been found since the discovery of 51 Peg b by Mayor and Queloz in 1995. The traditional field of planetology has thus expanded its frontiers to include planetary environments not represented in our Solar System. We expect that in the next five years space missions (Corot, Kepler and GAIA) or ground-based detection techniques will both increase exponentially the number of new planets discovered and lower the present limit of a $\sim$1.9 Earth-mass object [*e.g.* Mayor *et al.*, *Astron. Astrophys.*, 2009, **507**, 487]. While the search for an Earth-twin orbiting a Sun-twin has been one of the major goals pursued by the exoplanet community in the past years, the possibility of sounding the atmospheric composition and structure of an increasing sample of exoplanets with current telescopes has opened new opportunities, unthinkable just a few years ago. As a result, it is possible now not only to determine the orbital characteristics of the new bodies, but moreover to study the exotic environments that lie tens of parsecs away from us. The analysis of the starlight not intercepted by the thin atmospheric limb of its planetary companion (transit spectroscopy), or of the light emitted/reflected by the exoplanet itself, will guide our understanding of the atmospheres and the surfaces of these extrasolar worlds in the next few years. Preliminary results obtained by interpreting current atmospheric observations of transiting gas giants and Neptunes are presented. While the full characterisation of an Earth-twin might requires a technological leap, our understanding of large terrestrial planets (so called super-Earths) orbiting bright, later-type stars is within reach by current space and ground telescopes.

1 Introduction

Half a century ago, the space age began with the launch of Sputnik. Now at the completion of a fairly detailed study of the planets of our own solar system, we are at the dawn of the age of exoplanets. Almost 500 exoplanets, *i.e.* planets orbiting a star different from our Sun, are now known thanks to indirect detection

---

[a] *Department of Physics and Astronomy, University College London, Gower Street, London, UK WC1 E6BT. E-mail: g.tinetti@ucl.ac.uk*
[b] *LPL, University of Arizona, 1629 E. University Blvd, Tucson, AZ, 85721, USA*
[c] *Jet Propulsion Laboratory, 4800 Oak Grove Drive, Pasadena, CA, 91109, USA*
[d] *Institut d'Astrophysique de Paris, 98bis Boulevard Arago, Paris, France*
[e] *Harvard-Smithsonian Center for Astrophysics (CfA), 60 Garden Street, Cambridge, MA, 02138, USA*
[f] *Max-Planck-Institut fuer Astronomie, Koenigstuhl 17, 69117 Heidelberg, Germany*

techniques.[1] In the first decade after their initial discovery in 1995,[2] the task was to find more and more of these astronomical bodies: the biggest, the smallest; the hottest, the coolest; the system with the most planets in it. In recent years, attention has switched from finding planets to characterising them. Among the variety of exoplanets discovered so far, special attention has been devoted to those planets which transit their parent star, whose presence can be detected by the reduction in the brightness—the extinction—of the central star as the planet passes in front of it. More than 106 currently identified exoplanets are transiting planets, and for these objects planetary and orbital parameters such as radius, eccentricity, inclination, mass (given by radial velocity combined measurements), are known, allowing first order characterisation on the bulk composition and temperature (Gas giant? Neptune type? Terrestrial?). But it is clear that there is great variety even amongst the family of transiting exoplanets. The smallest, Corot-7b[3] and GJ1214b,[4] have masses of just 0.0151 and 0.0179 M_J. They orbit their parent star at distances of <0.02 AU. At the other end of the transiting planet distance scale, HD 80606 b orbits its G5 star with a period of more than 100 days and an eccentricity of 0.93. And at the high mass end, WASP-18b and XO-3b have a mass of 10.43 and 11.79 M_J and their atmospheric temperature is likely to be very hot.

Most importantly recent results have been able to demonstrate that for transiting exoplanets orbiting stars brighter than 12 Mag, it is possible to identify the main chemical components in the planet's atmosphere. A stellar occultation (called primary transit) occurs when the light from a star is partially blocked by an intervening body, such as a planet. With this method, we can indirectly observe the thin atmospheric ring surrounding the optically thick disk of the planet while the planet is transiting in front of its parent star.[5,6] In the secondary transit technique, we firstly observe the combined spectrum of the star and the planet. Then, we take a second measurement of the star alone when the planet disappears behind it: the difference between the two measurements consists of the planet's own spectrum.[7,8]

In the past 3 years, key observations with the Spitzer and Hubble Space Telescopes have, for the first time, given us real insights into the composition of some of the most unusual exoplanets so far discovered—the class known as hot-Jupiters. More specifically, infrared transmission and emission spectroscopy have revealed the presence of the primary carbon and oxygen species such as CH_4, CO_2, CO, and H_2O,[9–16] and provided constraints for the temperature profiles,[17–20] which are coupled to the composition. Today, broad-band or low-resolution spectroscopy from ground and space based observatories allow us to:

• determine planetary and orbital parameters
• constrain the albedo
• detect the main molecular species in the hot transiting planets' atmosphere
• constrain the horizontal and vertical thermal gradients in the hot exoplanets' atmospheres
• constrain the boundary conditions in the upper atmosphere
• detect the presence of clouds or hazes

Relatively high resolution spectroscopy data were recently obtained with ground-based telescopes in the optical[21,22] and NIR,[23,24a] confirming that alkali metals are present in hot-Jupiter atmospheres and showing non thermal emission processes. These achievements open up enormous possibilities in terms of atmospheric characterisation for the short term future prior to the launch of the next generation of space telescopes (the James Webb Space Telescope, launch 2014) or a dedicated mission (*e.g.* EChO, the Exoplanet Characterization Observatory, http://echo-spacemission.eu).

With current instruments we can already study the atmospheres of more than ten transiting hot-Jupiters and approach the case of hot Neptunes and warm super-Earths transiting later type stars, *e.g.* GJ436 b and GJ1214 b. Surveys aimed at detecting extrasolar planets are focusing on searches for ever smaller worlds and rocky planets in the habitable zones. While Corot and Kepler will increase the statistics of such objects with the ultimate goal of detecting earth-like planets around G-type stars, transit and radial velocity surveys from the ground (HARPS, MEarth, WFCAM Transit Survey) will actually provide the optimal targets for atmospheric characterisation, in particular super-Earths transiting bright M-dwarfs down to the habitable zone. Feasibility studies show that those objects will be easily studied by JWST and EChO-like missions.[24b]

2 Retrieving atmospheric parameters from exoplanet spectra

Within the past year, efforts to determine the abundances of atmospheric constituents found instead a range of degenerate temperature and composition solutions from the spectra,[16,20,26] see Fig. 1. Using an iterative forward model approach for spectral retrieval, we evaluated a variety of temperatures (T) as a function of pressure (P) together with the molecular absorption effects. Combining near-infrared spectra with mid-infrared measurements, we find that absorption due to H_2O, CH_4 and CO_2 explains most of the features present in the observed hot-Jupiter spectra (see Fig. 1). The additional contribution of CO is more than plausible and in few cases it even refines our fit, but we cannot discard the possibility that improved data lists for methane and/or CO_2 would provide the missing opacity. The radiative transfer calculations assume local thermal equilibrium (LTE) conditions—as expected for pressures exceeding 10^{-3} bar that are probed by the infrared spectra—and constant mixing ratios for the molecules.

For emission spectra, we obtain a family of plausible solutions for the molecular abundances and detailed temperature profiles for most of the hot-Jupiters observed. In Fig. 1 we show the example of HD 209458 b,[20] for which both NIR and MIR spectroscopical and photometrical data are available. Additional observational constraints on the atmospheric temperature structure and composition require either improved wavelength coverage/spectral resolution for the dayside spectrum or a transmission spectrum. The degeneracy is even higher when only a handful of photometrical observations are available (Fig. 2), calling for caution against premature theoretical classifications, such as the idea that hot-Jupiters may be divided in

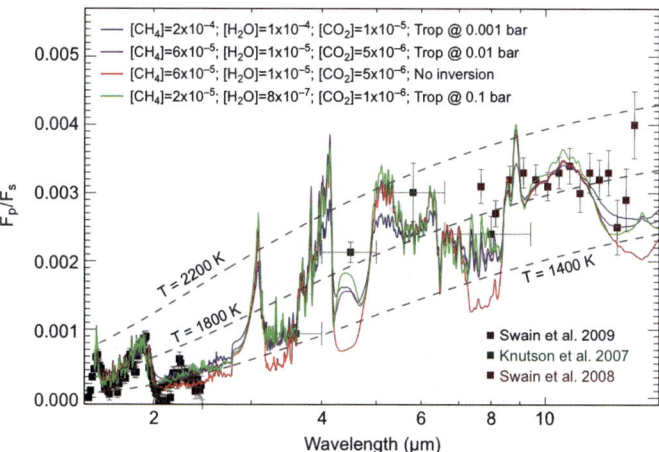

Fig. 1 Emission photometry and spectroscopy data for HD 209458 b.[20] The near-infrared and mid-infrared observations compared to synthetic spectra for four models that illustrate the range of temperature/composition possibilities consistent with the data. For each model case, the molecular abundance of CH_4, H_2O, & CO_2 and the location of the tropopause is given, these serve to illustrate how the combination of molecular opacities and the temperature structure cause significant departures from a purely single-temperature thermal emission spectrum.

two classes, where the presence or absence of a stratosphere is caused by the presence or absence of TiO/VO.[27]

Transmission spectra are less sensitive to the atmospheric temperatures, yet the derived composition at the terminator depends sensitively on the assumed radius. In particular the atmospheric temperature may play an important role in the overall scale height, and hence in the amplitude of the spectral signatures, as well as in the molecular absorption coefficients. For most cases a thermochemical equilibrium H_2O abundance of 4.5×10^{-4} relative to H_2[28–30] provide an excellent match to the data, see *e.g.* the case of XO-2b, Fig. 3. However, a ~1% difference in the estimate of the planetary radius at the ~1 bar pressure level, would result in a variation of the H_2O abundances by a factor of 10. Transit data at multiple wavelengths are needed to constrain the H_2O abundance. The mixing ratios determined for CH_4, CO and CO_2 depend on the data lists used and on the H_2O mixing ratio.

In Fig. 4 we consider the planet HD 189733 b. While most of the photometric and spectroscopic data are explainable with the presence of water vapour[31] and methane,[11] the recent observation of the photometric point at 4.5 μm[32] suggests

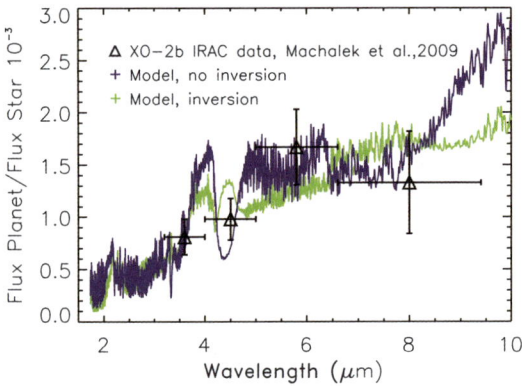

Fig. 2 Photometry secondary transit data obtained with the Spitzer IRAC instrument for the hot-Jupiters XO-2b.[25] In color are overplotted the atmospheric models for XO-2b containing water, methane, CO and CO_2. Blue line: simulated spectrum of XO-2b obtained using a T–P profile with no temperature inversion. Green line: simulated spectrum of XO-2b obtained using a T–P profile with temperature inversion.

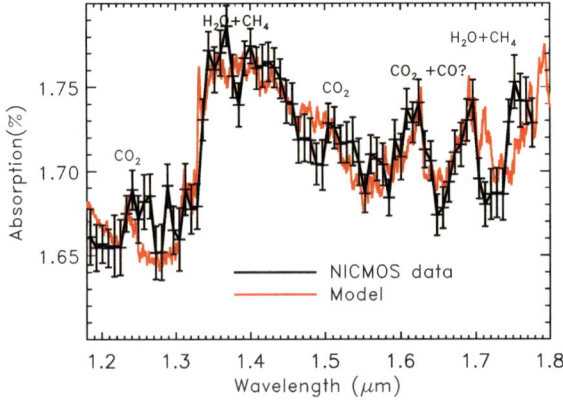

Fig. 3 Transmission spectrum of the hot-Jupiter XO-1b,[16] the fit was obtained with H_2O, CH_4, CO_2 and CO.

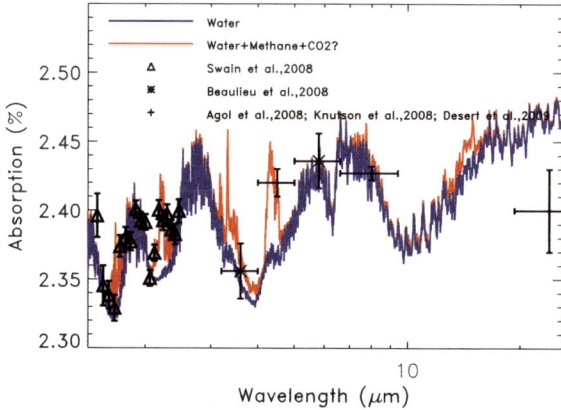

Fig. 4 Primary transit photometry and spectroscopy data of HD 189733 b recorded by multiple instruments and different teams. While most of the features can be explained by a combination of water vapour and methane, the IRAC band at 4.5 μm seems to indicate the additional presence of CO_2 and/or CO.

the additional presence of CO_2 and possibly CO in that planetary atmosphere. The addition of a small quantity of CO_2 and CO does not affect the fit at shorter wavelengths, and has the advantage of being consistent with the "day-side" composition observed with NICMOS.[13] We note that some of the temperature profiles/molecular mixing ratios consistent with the observations raise the question of whether the dayside atmosphere is in radiative and thermochemical equilibrium. Although advection of heat and/or photochemistry could support departures from radiative and thermochemical equilibrium (Fig. 5), our present lack of knowledge of molecular opacities at high temperatures for species such as CH_4, H_2S and C_2H_6 limits our ability to determine decisively whether this condition is met or not; thus there is an urgent need for further laboratory studies to obtain molecular databases for determining high temperature opacities of the most common molecules expected in hot-Jupiter/hot-Neptune atmospheres.

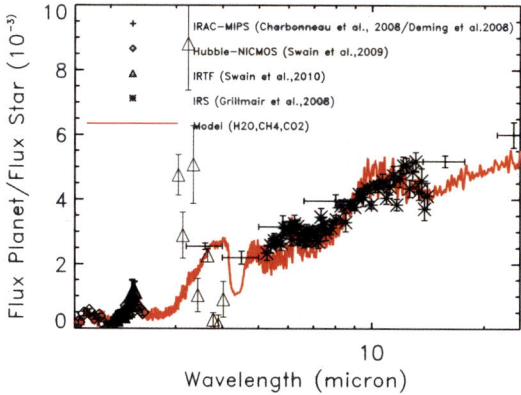

Fig. 5 Emission photometry and spectroscopy data for HD 189733 b. A radiative transfer model (red) assuming LTE conditions and consistent with the measurements made with Spitzer and Hubble fails to describe the emission structure at 3.1–4.1 μm,[23] and we find no plausible combination of atmospheric parameters that provides a good model of the observations under LTE conditions. The brightness temperature of the 3.25 μm emission feature indicates the likely presence of a non-LTE emission mechanism.

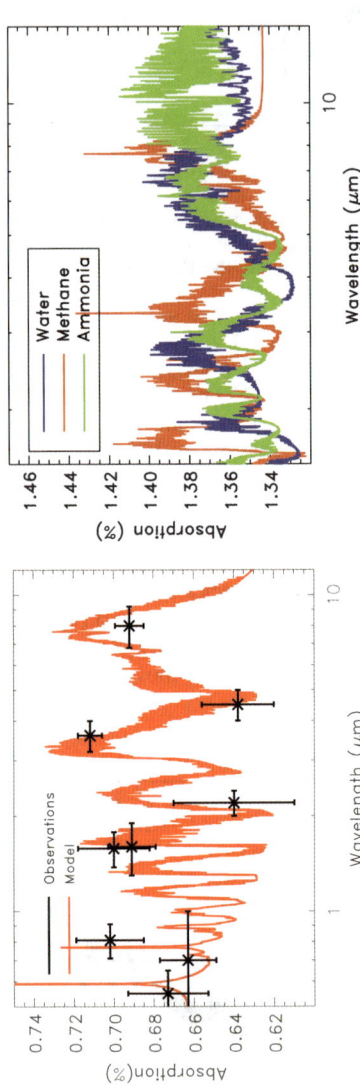

Fig. 6 Simulated transmission spectrum for the hot-Neptune GJ 436 b and the warm super-earth GJ1214 b. The models contain molecular hydrogen, methane, water vapour and ammonia. GJ1214 b atmosphere is here supposed to contain mainly molecular hydrogen and that is why the molecular features are very prominent. If the main component of its atmosphere is a heavier molecule, then the spectral features will be far less detectable. Observations will soon be available to confirm or reject the theoretical predictions.

We show in (Fig. 6) simulated transmission spectra for the hot-Neptune GJ436 b and the warm super-Earth GJ1214 b in the case its very extended atmosphere is mainly composed of molecular hydrogen. The spectral features in both cases are measurable with current space and ground-based observatories and more observations will probably become available in the next months.

3 The models

We model the transmission and emission spectra of transiting exoplanets using line-by-line radiative transfer models which account for the effects of molecular opacities[9,31,33] and hazes.[34,35] In our simulations we include H_2–H_2, H_2O, CH_4, CO, CO_2, NH_3, HCN *etc*. While the BT2 line list for water[36] can be calculated at the appropriate temperatures, the available data lists for methane at high temperature are inadequate to probe the modulations of the atmospheric thermal profile. To cover the spectral range from the visible to the Mid-IR, we have to use multiple data lists for methane, HITRAN 2008, PNNL, and hot-temperature measurements at 800, 1000 and 1273 K.[37] The Nassar and Bernath[37] data provide a much better fit to our observations in the region where they overlap with HITRAN 2008.[38] Compared to the results obtained with the Nassar and Bernath[37] line lists, mixing ratios 10–50 times larger are needed for methane if we use PNNL[39] or HITRAN 2008. The HITRAN 2008 data bank has the advantage of covering the entire spectral range measured by Hubble and Spitzer, with the downside (shared also by the PNNL list) that it results from measurements at room temperature, and therefore is quite inadequate to estimate the mixing ratio of methane at the temperatures of interest for hot planets. For CO_2 we use HITEMP[40] and CDSD-1000,[41] for CO we also use HITEMP. The contribution of H_2–H_2 at high temperatures was taken from ref. 42. The opacity was interpolated to the temperature of each atmospheric layer. As collision induced absorption scales with the square of the pressure, the H_2–H_2 contribution becomes important for pressures higher than 1 bar. The line shapes of alkali metals are calculated at different temperatures and interpolated for intermediate values. Their spectral contribution becomes important in the visible-NIR wavelength range.[43] An accurate line list for ammonia at has recently been calculated[44] and its extension to high temperatures achieved.[45]

4 Conclusions

An aspect of exoplanetary science that is both high-impact and cutting-edge is the study of extrasolar planet atmospheres. The ultimate goal is to obtain a high-resolution spectrum of an Earth-like planet, and although such a goal remains lofty, the key intermediate steps towards this end are already being taken with current technology for planets which are more massive and/or warmer that our own Earth. The characterisation of exoplanet atmospheres with current telescopes can be tackled with two main approaches: low resolution spectroscopy, from space using SPITZER and HST or the ground (*e.g.* NASA-IRTF), and high resolution spectroscopy from the ground (for example, VLT CRIRES). We can already probe the atmospheric constituents of several giant exoplanets, which orbit very close to their parent star, using transit techniques. The observations can be explained mainly with the combined presence of H_2O, CH_4, CO and CO_2 in the atmosphere of the planet. The photometric and spectroscopic emission data observed are consistent with the above composition but a variety of T–P profiles and mixing ratios are compatible with the data. Additional observations of transiting hot-Jupiters, especially spectroscopic data, will allow a more thorough classification of this type of planets unknown in our Solar System.

With current telescopes we can also approach the case of hot Neptunes and large terrestrial planets (super-Earths) transiting bright later type stars, *e.g.* GJ 436b or GJ 1214 b. Thanks to Corot, Kepler, ground-based transit surveys and the improvements in radial velocity measurements, many rocky planets and, possibly, few exomoons, are expected to be discovered in the next months/years. Further into the future, the James Webb Space Telescope will be the next generation of space telescopes to be online (launch 2014) and a dedicated mission to characterise transiting exoplanet atmospheres has been recently being proposed to NASA and ESA (EChO). Those observatories will guarantee high spectral resolution from space and the characterisation of smaller/colder targets, allowing us to expand the variety of characterisable extrasolar planets down to terrestrial planets and/or habitable zone of stars cooler than the Sun.

Acknowledgements

G. T. is supported by a Royal Society University Research Fellowship. Part of the research at the Jet Propulsion Laboratory (JPL), California Institute of Technology, was performed under contracts and grants with National Aeronautics and Space Administration.

References

1 J. Schneider, http://www.exoplanet.eu, 2010.
2 M. Mayor and D. Queloz, *Nature*, 1995, **378**, 355.
3 A. Leger, D. Rouan, J. Schneider, P. Barge, M. Fridlund, B. Samuel, M. Ollivier, E. Guenther, M. Deleuil, H. Deeg, M. Auvergne, R. Alonso, S. Aigrain, A. Alapini, J. Almenara, A. Baglin, M. Barbieri, H. Bruntt, P. Borde, F. Bouchy, J. Cabrera, C. Catala, L. Carone, S. Carpano, S. Csizmadia, R. Dvorak, A. Erikson, S. Ferraz-Mello, B. Foing, F. Fressin, D. Gandolfi, M. Gillon, P. Gondoin, O. Grasset, T. Guillot, A. Hatzes, G. Hebrard, L. Jorda, H. Lammer, A. Llebaria, B. Loeillet, M. Mayor, T. Mazeh, C. Moutou, M. Paetzold, F. Pont, D. Queloz, H. Rauer, S. Renner, R. Samadi, A. Shporer, C. Sotin, B. Tingley and G. Wuchterl, *Astron. Astrophys.*, 2009, **506**, 287.
4 D. Charbonneau, Z. K. Berta, J. Irwin, C. J. Burke, P. Nutzman, L. A. Buchhave, C. Lovis, X. Bonfils, D. W. Latham, S. Udry, R. A. Murray-Clay, M. J. Holman, E. E. Falco, J. N. Winn, D. Queloz, F. Pepe, M. Mayor, X. Delfosse and T. Forveille, *Nature*, 2009, **462**, 891–894.
5 T. M. Brown, *Astrophys. J.*, 2001, **553**, 1006.
6 D. Charbonneau, T. M. Brown, R. W. Noyes and R. L. Gilliland, *Astrophys. J.*, 2002, **568**, 377.
7 D. Deming, S. Seager, L. J. Richardson and J. Harrington, *Nature*, 2005, **434**, 740.
8 D. Charbonneau, L. E. Allen, S. T. Megeath, G. Torres, R. Alonso, T. M. Brown, R. L. Gilliland, D. W. Latham, G. Mandushev and F. T. O. nad Alessandro Sozzetti, *Astrophys. J.*, 2005, **626**, 523.
9 G. Tinetti, M. C. Liang, A. Vidal-Madjar, D. Ehrenreich, A. L. des Etangs and Y. Yung, *Astrophys. J.*, 2007, **654**, L99.
10 T. Barman, *Astrophys. J.*, 2008, **676**, L61.
11 M. Swain, G. Vasisht and G. Tinetti, *Nature*, 2008, **452**, 329.
12 C. J. Grillmair, A. Burrows, D. Charbonneau, L. Armus, J. Stauffer, V. Meadows, J. van Cleve, K. von Braun and D. Levine, *Nature*, 2008, **456**, 767.
13 M. Swain, G. Vasisht, G. Tinetti, J. Bouwman, P. Chen, Y. Yung, D. Deming and P. Deroo, *Astrophys. J.*, 2009, **690**, L114.
14 J. P. Beaulieu, S. Carey, I. Ribas and G. Tinetti, *Astrophys. J.*, 2008, **677**, 1343.
15 J. Beaulieu, D. Kipping, V. Batista, G. Tinetti, I. Ribas, S. Carey, J. A. Noriega-Crespo, C. A. Griffith, G. Campanella, S. Dong, J. Tennyson, R. Barber, P. Deroo, S. Fossey, D. Liang, M. R. Swain, Y. Yung and N. Allard, *Mon. Not. R. Astron. Soc.*, 2010 DOI: astroph 0909.0185.
16 G. Tinetti, P. Deroo, M. R. Swain, C. A. Griffith, G. Vasisht, L. R. Brown, C. Burke and P. McCullough, *Astrophys. J.*, 2010, **712**, L139.
17 J. Harrington, B. M. Hansen, S. H. Luszcz, S. Seager, D. Deming, K. Menou, J. Cho and L. J. Richardson, *Science*, 2006, **314**, 623.
18 H. A. Knutson, D. Charbonneau, L. E. Allen, J. J. Fortney, E. Agol, N. B. Cowan, A. P. Showman, C. S. Cooper and S. T. Megeath, *Nature*, 2007, **447**, 183.
19 A. Burrows, J. Budaj and I. Hubeny, *Astrophys. J.*, 2007, **668**, L171.

20 M. Swain, G. Tinetti, G. Vasisht, P. Deroo, C. Griffith, J. Bouwman, P. Chen, Y. Yung, A. Burrows, L. Brown, J. Matthews, J. Roe, R. Kuschnig and D. Angerhausen, *Astrophys. J.*, 2009, **704**, 1616.
21 S. Redfield, M. Endl, W. Cochran and L. Koesterke, *Astrophys. J.*, 2008, **673**, L87.
22 I. Snellen, S. Albrecht, E. de Mooij and R. L. Poole, *Astron. Astrophys.*, 2008, **487**, 357.
23 M. R. Swain, P. Deroo, C. A. Griffith, G. Tinetti, A. Thatte, G. V. P. Chen, J. Bouwman, I. J. Crossfield, D. Angerhausen, C. Afonso and T. Henning, *Nature*, 2010, **463**, 637.
24 (a) I. Snellen, R. de Kok, E. de Mooij and S. Albrecht, The orbital motion, absolute mass and high-altitude winds of HD 209458b, *Nature*, 2010, **465**, 1049; (b) Tessenyi *et al.*, submitted.
25 P. Machalek, P. R. McCullough, A. Burrows, C. J. Burke, J. L. Hora and C. M. Johns-Krull, *Astrophys. J.*, 2009, **701**, 514.
26 N. Madhusudhan and S. Seager, *Astrophys. J.*, 2009, **707**, 24.
27 J. J. Fortney, K. Lodders, M. S. Marley and R. S. Freedman, *Astrophys. J.*, 2008, **678**, 1419.
28 M.-C. Liang, C. D. Parkinson, A. Y. Lee, Y. L. Yung and S. Seager, *Astrophys. J.*, 2003, **596**, L247.
29 M.-C. Liang, S. Seager, C. D. Parkinson, A. Y. Lee and Y. L. Yung, *Astrophys. J.*, 2004, **605**, L61.
30 K. Zahnle, M. Marley, R. Freedman, K. Lodders and J. Fortney, *Astrophys. J.*, 2009, **701**, L20.
31 G. Tinetti, A. Vidal-Madjar, M.-C. Liang, J.-P. Beaulieu, Y. Yung, S. Carey, R. J. Barber, J. Tennyson, I. Ribas, N. Allard, G. E. Ballester, D. K. Sing and F. Selsis, *Nature*, 2007, **448**, 169.
32 J.-M. Desert, A. L. des Etangs, G. Hebrard, D. K. Sing, D. Ehrenreich, R. Ferlet and A. Vidal-Madjar, *Astrophys. J.*, 2009, **699**, 478.
33 G. Tinetti, V. S. Meadows, D. Crisp, W. Fong, T. Velusamy and H. Snively, *Astrobiology*, 2005, **5**(4), 461.
34 C. A. Griffith, R. V. Yelle and M. S. Marley, *Science*, 1998, **282**, 2063.
35 C. A. Griffith, T. Owen and R. Wagener, *Icarus*, 1991, **93**, 362.
36 R. J. Barber, J. Tennyson, G. J. Harris and R. N. Tolchenov, *Mon. Not. R. Astron. Soc.*, 2006, **368**, 1087.
37 R. Nassar and P. Bernath, *J. Quant. Spectrosc. Radiat. Transfer*, 2003, **82**, 279.
38 L. Rothman, I. Gordon, A. Barbe, D. Benner, P. Bernath, M. Birk, V. Boudon, L. Brown, A. Campargue, J. Champion, K. Chance, L. Coudert, V. Dana, V. Devi, S. Fally, J. Flaud, R. Gamache, A. Goldman, D. Jacquemart, I. Kleiner, N. Lacome, W. Lafferty, J. Mandin, S. Massie, S. Mikhailenko, C. Miller, N. Moazzen-Ahmadi, O. Naumenko, A. Nikitin, J. Orphal, V. Perevalov, A. Perrin, A. Predoi-Cross, C. Rinsland, M. Rotger, M. Simeckova, M. Smith, K. Sung, S. Tashkun, J. Tennyson, R. Toth, A. Vandaele and J. V. Auwera, *J. Quant. Spectrosc. Radiat. Transfer*, 2009, **110**, 533.
39 PNNL, http://www.pnl.gov/.
40 L. S. Rothman, I. E. Gordon, R. J. Barber, H. Dothe, R. R. Gamache, A. Goldman, V. I. Perevalov, S. A. Tashkun and J. Tennyson, *J. Quant. Spectrosc. Radiat. Transfer*, 2010, **111**, 2139–2150.
41 S. A. Tashkun, V. Perevalov, J. Teffo, A. D. Bykov and N. N. Lavrentieva, *J. Quant. Spectrosc. Radiat. Transfer*, 2003, **82**, 165.
42 A. Borysow, U. G. Jorgensen and Y. Fu, *J. Quant. Spectrosc. Radiat. Transfer*, 2001, **68**, 235.
43 N. F. Allard, F. Allard, P. H. Hauschildt, J. F. Kielkopf and L. Machin, *Astron. Astrophys.*, 2003, **411**, L473.
44 S. N. Yurchenko, R. J. Barber, A. Yachmenev, W. Theil, P. Jensen and J. Tennyson, *J. Phys. Chem. A*, 2009, **113**, 11845–11855.
45 S. N. Yurchenko, R. J. Barber and J. Tennyson, *Mon. Not. R. Astron. Soc.*, 2010, submitted.

General Discussion

Dr Faure opened the discussion of the paper by Professor Miller: In your non-LTE calculations, did you include the contribution of electron-impact excitation?

Professor Miller responded: Kokoouline *et al.*[1] have calculated the vibrational excitation rate of H_3^+ by electrons at $\sim 10^{-13}$ m^3 s^{-1}, comparable with the rate for dissociative recombination. This value is roughly two orders of magnitude greater than the vibrational excitation rate for collisions with molecular hydrogen, which Oka and Epp (2004) set to the Langevin rate of 2×10^{-15} m^3 s^{-1}. But the electron density would be 1% of the H_2 density in only the most highly ionised astrophysical environments. Indeed, under conditions that would produce such high electron densities, H_2 would almost certainly have been dissociated. So it is unlikely that electron impact excitation will ever be competitive with H_2 excitation.

More usual upper end values of [e$^-$]/[H$_2$] are between 10^{-7} to 10^{-5}, which would mean that electron excitation of H_3^+ would be less than 0.1% of the excitation by H_2. Given the other simplifications in our technique for calculating non-LTE effects, it is probably not worth putting in electron impact excitation if it is not going to contribute more than (an optimistic) 0.1% of the total effect.

1 V. Kokoouline, A. Faure, J. Tennyson and C. H. Greene, Calculation of rate constants for vibrational and rotational excitation of the H_3^+ ion by electron impact, *Mon. Not. R. Astron. Soc.*, in press.

Dr Yelle addressed Dr Miller: Considering their importance for the cooling rate from H_3^+ in the non-LTE regime, would you comment on the accuracy of the VT rates used in your calculations?

Professor Miller answered: In our original contribution to *Faraday Discussions* volume 147, we concentrated on corrections to the H_3^+ partition function, and what effect they might have on calculating the total emission of this ion in various situations. Dr Yelle is perfectly correct to highlight the importance of the (ro-)vibration-translation (VT) rate for collisional excitation and de-excitation between H_3^+ and H_2 for determining to what extent H_3^+ emission is in local thermodynamical equilibrium (LTE), however, and therefore to what extent this ion can act as a coolant in the atmospheres of giant planets, and under what conditions. But it is far from being the only important factor. So it is worthwhile outlining the steps we take in our H3Pcool programme to calculate the total cooling due to H_3^+ (which has been used, *mutans mutandis*, in ref. 22, 29 and 35, cited in our paper).

We have used the technique proposed by Oka and Epp (ref. 36 in our original paper) to do this. This approach was originally applied to rotational transitions in H_3^+ in order to understand some of the spectral features observed in the interstellar medium. The Oka–Epp approach provides an elegant and symmetric way of carrying out a detailed balance calculation under conditions in which LTE no longer pertains. Central to their calculations, rather than their overall approach, is that they assume that the collisional population and depopulation of H_3^+ ro-vibrational states is due to the proton hopping reaction:

$$H_3^+(rv = n, t = x) + H_2(rv = m, t = y) \rightarrow H_2(rv = n', t = x') + H_3^+(rv = m', t = y') \quad (1)$$

with the obvious proviso that the energy in the various $rv = RV$, $t = T$ vibration–translation states is conserved. This reaction is assumed to occur without an energy barrier at the Langevin rate of 2×10^{-15} m^3 s^{-1}. Clearly this is an assumption. In the absence of a better assumption or of detailed VT rate measurements, it is not an improbable working assumption, however. If collisional excitation and de-excitation of the relevant H_3^+ vibrational states is faster than this rate, the H_3^+ cooling efficiency per molecule is greater as the gas density decreases, and *vice versa*.

We now come to a second approximation to our calculations, that Oka and Epp do not have to make, since they are interested in the population of rotational sub-levels within the vibrational ground state: we assume that one can decouple the vibrational states of H_3^+ from the rotational sub-states. The decoupling between vibrational levels and their rotational sub-levels is not, strictly speaking, allowed, as the only good quantum numbers for H_3^+ are the angular momentum, J, the parity of the level, p, the nuclear spin state—*ortho* or *para*—and the level number in the J,p set. Indeed, one of the major advances in the calculation of the ro-vibrational levels of H_3^+ by the first-principles method was that it did not require any *a priori* allocation of vibrational "quantum numbers", but instead deduced them *a posteriori* (ref. 27). (Note the translational states of the ion do not enter further into our calculations.)

The resulting rotation-only radiative transitions have very small Einstein A_{if} coefficients, since the molecule has no permanent dipole moment, compared with the ro-vibrational transitions (see ref. 33). So we can assume that the rotational sub-levels within a vibrational manifold remain in LTE. Our calculations, then, concentrate on the extent to which the vibrational level populations depart from LTE. To do this, we carry out a detailed balance calculation making use of the first 17 vibrational levels and their associated vibration-only A_{if} values given in ref. 26 of our original paper.

Regions of a planetary atmosphere for which H_3^+ cooling is in any way significant have >90% of the gas as H_2; we assume that this is the only gas that takes part in collisional (de- and) excitation of H_3^+. The H_2 gas density, $N(H_2)$ is an input to our calculation. In our calculation of level populations, we start with an LTE distribution at a given temperature, T, which is also an input to the calculation. The method of Oka and Epp enables us to calculate the rate of change of a (fractional) level population—$df(v)/dt$—as a function of timestep. We have found that a timestep of $\Delta t = 10^{-10} \times (10^{20}/N(H_2))^{1/2}$ s ensures that the calculation is not subject to numerical instability, and that 5×10^5 timesteps ensure convergence.

In Table 1 we have given the outputs for three quite typical $N(H_2)/T$ values – 10^{12} m^{-3}/ 400 K, 10^{18} m^{-3}/700 K, 10^{20} m^{-3}/1600 K – that demonstrate the results we get. After printing out $N(H_2)$, T and the timestep, the programme gives five columns—the level number, $df(v)/dt$, the fractional population of the levels, $f(v)$, calculated after 5×10^5 timesteps, the value of $f(v)$ assuming LTE at the given temperature, and $r(v)$ given by $f(v)/f_{LTE}(v)$. (Note that level 1 is the ground vibrational state.) It is clear the calculation is well converged. For example, if we take level 2 for the 10^{12} m^{-3}/400 K calculation, $df(v)/dt$ is just -3.12×10^{-8} which, when multiplied by $\Delta t = 1.0 \times 10^{-6}$ s, produces a change of just -3.12×10^{-14} in $f(2)$, a mere -1.46×10^{-7} of its final value of just 2.13×10^{-6}. And so on.

In Melin's work (ref. 22, 34 and 35 of our original paper), we found it was sufficiently accurate to use $r(2)$ to calculate the scaling required on the H_3^+ LTE cooling rate to account for sub-thermal populations of all the vibrational levels. This is because, for temperatures of interest for that work, almost all of the cooling came from ro-vibrational transitions in the ν_2 band. At higher temperatures, however, of significance in hot(tish) Jupiters, such as that modeled by Koskinen and coworkers (ref. 29), we now take the scaling factor calculated from an LTE-weighted average of the level populations of levels 2, 4, 5 and 6 given by:

$$R = [r(2) \times f_{LTE}(2) + r(4) \times f_{LTE}(4) + r(5) \times f_{LTE}(5) + r(6) \times f_{LTE}(6)] / [f_{LTE}(2) + f_{LTE}(4) + f_{LTE}(5) + f_{LTE}(6)] \quad (2)$$

In Tables 2 and 3 we give the values of $r(2)$ and the resulting H_3^+ cooling, $E(H_3^+)$ for various values of $N(H_2)$ and T.

Table 1 Example output from H3Pcool programme

H3Pcooling for *Faraday Discussions* 147

Program H3Pcool, version of January 29, 2010; this program is part of Europlanet RI JRA3; please credit Europlanet when using

H_2 density = 0.100E+13; Temperature = 400.00; Time step = 0.100E−05

1	0.47222E−05	0.99999E+00	0.99987E+00	0.10001E+01
2	−0.31244E−07	0.21305E−06	0.11574E−03	0.18407E−02
3	−0.46909E−05	0.75499E−05	0.10905E−04	0.69234E+00
4	0.31480E−14	0.12661E−08	0.34805E−07	0.36378E−01
5	0.72388E−15	0.27141E−09	0.15759E−07	0.17223E−01
6	0.57373E−15	0.18900E−09	0.21330E−08	0.88608E−01
7	0.50134E−15	0.15430E−09	0.16717E−09	0.92296E+00
8	0.71256E−17	0.20863E−11	0.11629E−10	0.17940E+00
9	0.22884E−17	0.66441E−12	0.42644E−11	0.15580E+00
10	0.70834E−18	0.20471E−12	0.20037E−11	0.10216E+00
11	0.72054E−18	0.20667E−12	0.74147E−12	0.27874E+00
12	0.88646E−18	0.25339E−12	0.51844E−12	0.48876E+00
13	0.13333E−18	0.37832E−13	0.56148E−13	0.67378E+00
14	0.15212E−19	0.43066E−14	0.89869E−14	0.47921E+00
15	0.10743E−19	0.30396E−14	0.60305E−14	0.50404E+00
16	0.12664E−19	0.35767E−14	0.35944E−14	0.99509E+00
17	0.12303E−20	0.34750E−15	0.85629E−15	0.40582E+00

Emission/molecule = 0.21522E−24 W mol^{-1} sr^{-1}

H_2 density = 0.100E+19; Temperature = 700.00; Time step = 0.100E−08

1	0.10509E−04	0.99289E+00	0.99286E+00	0.10000E+01
2	−0.10510E−04	0.55594E−02	0.55910E−02	0.99435E+00
3	0.40173E−09	0.14497E−02	0.14497E−02	0.10000E+01
4	0.19841E−09	0.54257E−04	0.54325E−04	0.99875E+00
5	0.13281E−09	0.34437E−04	0.34543E−04	0.99692E+00
6	0.47115E−10	0.11009E−04	0.11017E−04	0.99925E+00
7	0.12026E−10	0.25714E−05	0.25713E−05	0.10000E+01
8	0.27910E−11	0.56024E−06	0.56060E−06	0.99936E+00
9	0.16019E−11	0.31573E−06	0.31601E−06	0.99913E+00
10	0.10497E−11	0.20492E−06	0.20523E−06	0.99846E+00
11	0.60312E−12	0.11623E−06	0.11629E−06	0.99951E+00
12	0.49478E−12	0.94767E−07	0.94786E−07	0.99980E+00
13	0.14231E−12	0.26611E−07	0.26614E−07	0.99989E+00
14	0.50457E−13	0.93383E−08	0.93413E−08	0.99968E+00
15	0.40402E−13	0.74349E−08	0.74371E−08	0.99970E+00
16	0.30191E−13	0.55334E−08	0.55333E−08	0.10000E+01
17	0.13441E−13	0.24364E−08	0.24377E−08	0.99944E+00

Emission/molecule = 0.46926E−20 W mol^{-1} sr^{-1}

H_2 density = 0.100E+21; Temperature = 1600.00; Time step = 0.100E−09

1	0.87526E−06	0.83046E+00	0.83043E+00	0.10000E+01
2	−0.76210E−06	0.86118E−01	0.86138E−01	0.99977E+00
3	−0.84053E−07	0.47724E−01	0.47723E−01	0.10000E+01
4	−0.86699E−08	0.11340E−01	0.11343E−01	0.99976E+00
5	−0.86701E−08	0.92998E−02	0.93047E−02	0.99947E+00
6	−0.43341E−08	0.56428E−02	0.56438E−02	0.99983E+00
7	−0.21664E−08	0.29862E−02	0.29862E−02	0.10000E+01
8	−0.10828E−08	0.15331E−02	0.15336E−02	0.99971E+00
9	−0.10832E−08	0.11928E−02	0.11934E−02	0.99953E+00
10	−0.10831E−08	0.98716E−03	0.98805E−03	0.99910E+00
11	−0.54007E−09	0.77036E−03	0.77062E−03	0.99966E+00
12	−0.53946E−09	0.70457E−03	0.70468E−03	0.99984E+00

Table 1 (*Contd.*)

H3Pcooling for *Faraday Discussions* 147

13	−0.26687E−09	0.40421E−03	0.40425E−03	0.99988E+00
14	−0.26677E−09	0.25561E−03	0.25569E−03	0.99967E+00
15	−0.13130E−09	0.23135E−03	0.23142E−03	0.99967E+00
16	−0.13139E−09	0.20334E−03	0.20334E−03	0.99998E+00
17	−0.13175E−09	0.14196E−03	0.14206E−03	0.99932E+00

H_2 local density = 0.10000E+21 m^{-3}; Temperature = 0.16000E+04 K
Emission/molecule = 0.12644E−18 W mol^{-1} sr^{-1}

Dr Sittler commented: Andrew Coates briefly touched on the subject of heavy negative ions and fullerenes. Up till now most of the discussions have been focused on the aromatics and PAHs, while fullerenes are unsaturated (no hydrogen) versions forming from long carbon chains and polycyclic rings. The fullerene subject has very exciting possibilities when it comes to pre-biological chemistry at Titan. There is an interesting synergy between Enceladus, also discussed by Andrew, which is dumping large quantities of water vapor into Saturn's magnetosphere where it gets ionized, picked up by Saturn's rotating magnetosphere to energies ~50–100 eV, charge transfers with the neutral water products and then gets ejected to Saturn's outer magnetosphere as fast neutrals to Titan's orbit. Some will leave the system while some will be on highly elliptical orbits and become ionized again, picking up with keV energies out by Titan's orbit. The water group ions (H_2O^+, OH^+, O^+) were initially observed during Cassini's TA encounter where they can bombard Titan's upper atmosphere as originally reported by Hartle *et al.*[1,2] The heavy negative ions as reported by Coates *et al.*[3] and shown in Fig. 1 are observed above 950 km and as shown in Sittler *et al.*,[4] could also be fullerenes which are hollow shells of 60 or 70 carbon atoms as shown in Fig. 2. Experiments have shown that noble gases colliding with fullerenes at center-of-mass energies ~100 eV can become trapped within fullerenes[5–7] and same was done for nitrogen trapping within fullerenes.[8,9] The keV oxygen ions can penetrate below 950 km so they can reach the fullerenes and become trapped within them (see Cravens *et al.*[10]). The fullerenes, which protect the oxygen from Titan's reducing

Table 2 Sub-LTE parameter of v_2 band

Log10 [N(H_2)]/m^{-3} T/K	12	14	16	18	20
400	0.18407E−02	0.12383E+00	0.93315E+00	0.99928E+00	0.99999E+00
500	0.12250E−02	0.54293E−01	0.84722E+00	0.99816E+00	0.99998E+00
600	0.12047E−02	0.31093E−01	0.76258E+00	0.99649E+00	0.99996E+00
700	0.13257E−02	0.21188E−01	0.70223E+00	0.99435E+00	0.99994E+00
800	0.14879E−02	0.16248E−01	0.66432E+00	0.99185E+00	0.99992E+00
900	0.16621E−02	0.13549E−01	0.64214E+00	0.98917E+00	0.99989E+00
1000	0.18394E−02	0.13549E−01	0.63066E+00	0.98655E+00	0.99986E+00
1100	0.20172E−02	0.11137E−01	0.62666E+00	0.98426E+00	0.99983E+00
1200	0.21952E−02	0.10676E−01	0.62813E+00	0.98248E+00	0.99981E+00
1300	0.23739E−02	0.10487E−01	0.63374E+00	0.98133E+00	0.99979E+00
1400	0.25538E−02	0.10486E−01	0.64260E+00	0.98084E+00	0.99978E+00
1500	0.27356E−02	0.10622E−01	0.65407E+00	0.98098E+00	0.99977E+00
1600	0.29195E−02	0.10860E−01	0.66768E+00	0.98169E+00	0.99977E+00

Table 3 H_3^+ emission $E(H_3^+)/W$ molecule^{-1} sr^{-1}

Log10 [N(H$_2$)]/m^{-3} T/K	12	14	16	18	20
400	0.21522E¬24	0.12628E¬22	0.93847E¬22	0.10050E¬21	0.10057E¬21
500	0.79612E¬24	0.36517E¬22	0.52246E¬21	0.61527E¬21	0.61639E¬21
600	0.24429E¬23	0.71270E¬22	0.15331E¬20	0.19992E¬20	0.20061E¬20
700	0.62550E¬23	0.11358E¬21	0.33290E¬20	0.46926E¬20	0.47187E¬20
800	0.13487E¬22	0.16362E¬21	0.60723E¬20	0.90177E¬20	0.90897E¬20
900	0.25461E¬22	0.22441E¬21	0.99059E¬20	0.15209E¬19	0.15370E¬19
1000	0.43620E¬22	0.30057E¬21	0.15004E¬19	0.23489E¬19	0.23798E¬19
1100	0.69285E¬22	0.39584E¬21	0.21487E¬19	0.33946E¬19	0.34473E¬19
1200	0.10390E¬21	0.51513E¬21	0.29514E¬19	0.46696E¬19	0.47513E¬19
1300	0.14933E¬21	0.66517E¬21	0.39343E¬19	0.61985E¬19	0.63167E¬19
1400	0.20713E¬21	0.85090E¬21	0.51123E¬19	0.79874E¬19	0.81485E¬19
1500	0.27905E¬21	0.10781E¬20	0.65041E¬19	0.10046E¬18	0.10255E¬18
1600	0.36685E¬21	0.13526E¬20	0.81274E¬19	0.12383E¬18	0.12644E¬18

atmosphere (which otherwise would form CO and CO_2), can then condense into larger clusters[11] and/or combine with PAHs and those with nitrogen additions to form aerosols that now have oxygen trapped within them. As discussed in ref. 4 it can take 100–1000 years for these aerosols to reach the surface and during this time, including the time on surface, GCRs and their secondaries are irradiating the aerosols and it is known from laboratory measurements such irradiated mixtures can form amino acids which are the building blocks of life as we know it (Reggie Hudson, private communication). Since aerosols are globally distributed around Titan, these processes must be global on a planetary scale including the surface (veneer of tholins covering the surface). More recently, fullerenes have been identified as the carbon constituent found in carbonaceous chondritic meteorites that traps such noble gases as neon, argon, krypton and xenon.[12] So, these processes must be going on.

In Table 4 we show reaction rates of interest that should be studied in the laboratory at temperature $T \sim 100$–200 K and pressures on the order of nanobars that may play an important role in both the formation of PAHs and long carbon chains from which fullerenes may form. The first reaction shows dissociative electron attachment which can make CN^-. This ion radical and others such as C_3N^- and C_5N^- are discussed in ref. 13. Here, we note in Fig. 1, that ions C_2H^-, C_4H^- and C_6H^- could also

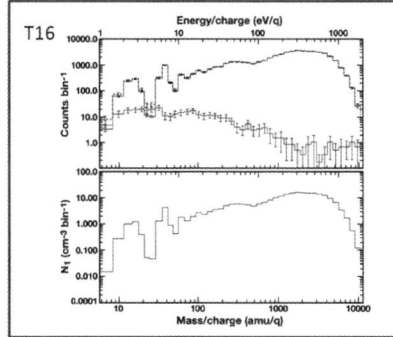
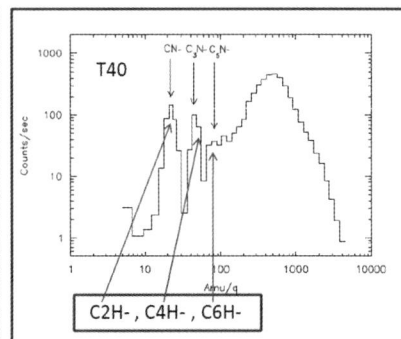

Fig. 1 Detection of negative ions at Titan.

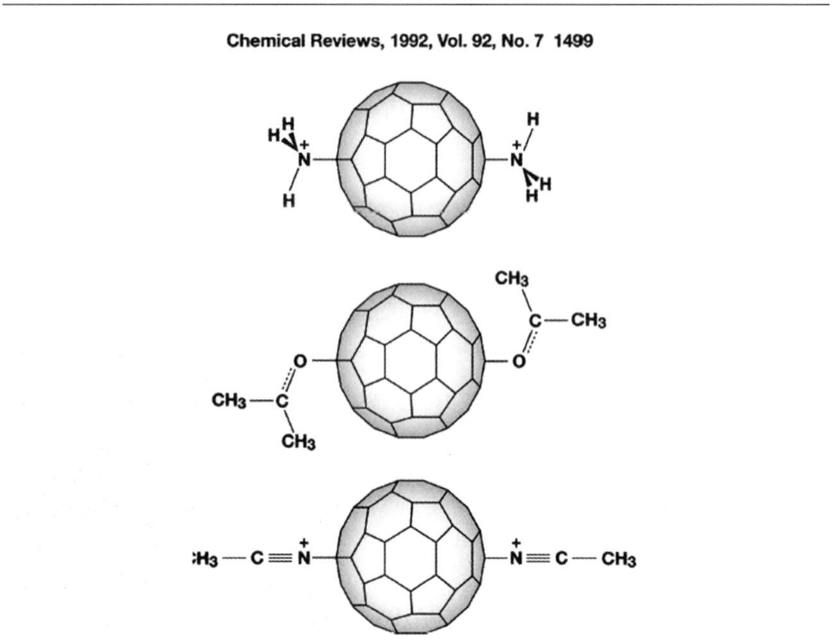

Fig. 2 Fullerenes.

apply. So, the formation of these ions, as shown in Table 4 for radiative electron attachment rates can lead to the formation of long carbon chains and thus fullerenes. We also show possible proton transfer reaction between ion CN^- and $HC_{2n+1}N$ to

Table 4 Reactions relevant to formation of intermediates initiating growth of larger negative ions

Dissociative electron attachment process
$e + HCN \rightarrow H + CN^-$

Proton transfer acid (AH) and base (B-) type fast reactions
$CN^- + HC_3N \rightarrow HCN + C_3N^-$ $CN^- + HC_5N \rightarrow HCN + C_5N^-$ $CN^- + C_2H_2 \rightarrow HCN + C_2N^-$ $CN^- + C_4H_2 \rightarrow HCN + C_4N^-$ $CN^- + C_6H_2 \rightarrow HCN + C_6N^-$ Note that O^- would also act as strong base similar to CN^- and would lead to fast acid–base type reactions with polyynes and cyanopolyynes, as shown above, by forming OH radicals. The measured rate coefficients using CRESU apparatus at RENNES FRANCE for the reaction of C_2H with C_2H_2 at 150 K were $>1.08 \times 10^{10}$ cm^3 s^{-1}

Radiative electron attachment to radicals (a few examples)
$C_{2n}H + e^- \rightarrow (C_{2n}H^-)^* \rightarrow C_{2n}H^- + h\nu$ ($n = 1,2,3,$ etc.)

Growth of PAHs *via* sequential neutral–neutral reactions
$C_2H + C_6H_6 \rightarrow PAH$ $C_2H_2 + C_6H_5 \rightarrow PAH$

Growth of long carbon chains from which fullerenes can form
$C_2H + HC_{2n}H \rightarrow HC_{2n} + 2H + H$ ($n = 1,2,3,$ etc.) $\rightarrow C_{60}$

make $C_{2n+1}N^-$ ions and CN^- interaction with acetylene to make $C_{2n}H^-$ ions. Since, the electron affinity of C_2H is less than CN the reaction for C_2H^- is not likely to be important but the others could occur. The ions C_4H^-, C_6H^- and C_8H^- have been detected in molecular clouds,[14–17] so must be going on. The same can be said for the radiative electron attachment reactions. Finally, the last set of reactions show radical neutral–neutral collisions that can make PAHs from the aromatics phenyl and benzene and carbon chains from C_2H and acetylene polymers $HC_{2n}H$. Many of the reactions in ref. 13 are based on calculations and have not been measured at the temperatures of Titan's upper atmosphere or that for molecular clouds. An example is the radiative attachment rate for C_2H^- is $\sim 10^{-15}$ cm^3 s^{-1} for T_e $\sim$1000 K, while C_4H^- $\sim 10^{-8}$ cm^3 s^{-1} or 7 orders of magnitude larger and both are based on calculations. So, we feel all these reactions should be measured, although some are more important than others. These measurements involving smaller molecules and ions are the building blocks for constructing the large polymers and will allow for more progress in this area of research. Making tholins in the laboratory is important, but it is very difficult to understand what is happening in detail due to their complexity and large number of atomic species, but by measuring the more basic reactions then one can better understand aerosol formation.

In Fig. 3 we show the CRESU experiment at the University of Rennes which has the ability to measure many of these reactions at temperatures more characteristic of

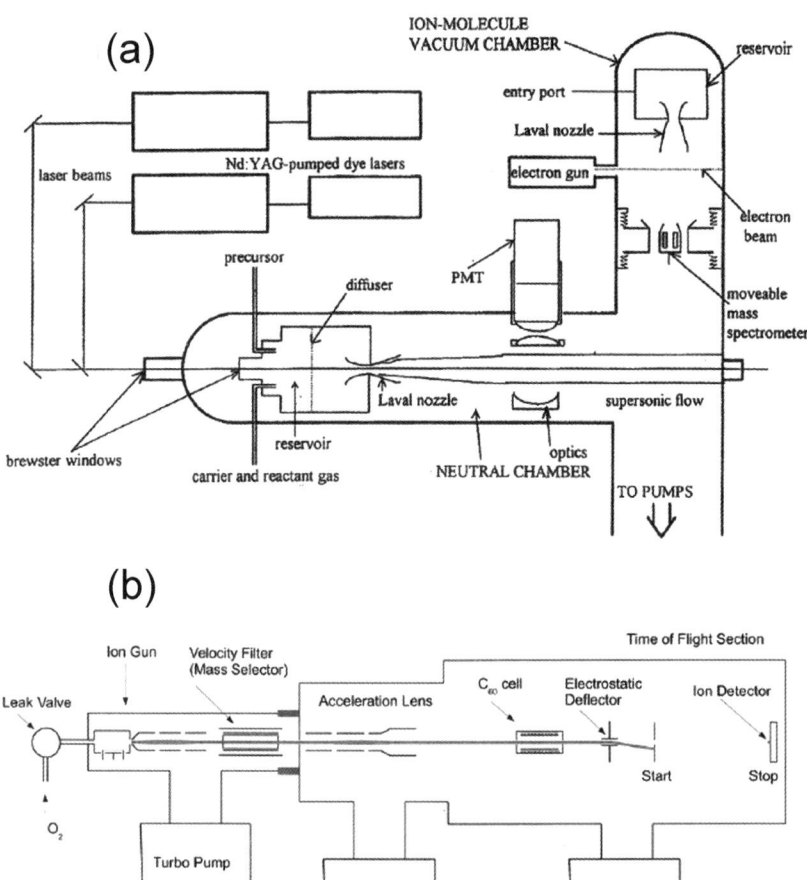

Fig. 3 (a) CRESU apparatus at Rennes University; (b) Goddard Ion Beam apparatus.

Titan's atmosphere, although at higher pressures so corrections will be required to make estimates more applicable to Titan's upper atmosphere. In the lower part of Fig. 3 we show the experimental setup that we would use at the Goddard Space Flight Center using an already-available facility to study high energy collisions between fullerenes and oxygen ions. To the best of our knowledge this has not yet been done. The experiments of nitrogen trapping do indicate that oxygen trapping should also occur and will also be investigated by the Goddard facility. The basic idea is that one has a calibrated O^+ ion source of known energy ~400 eV which then passes through a fullerene collision chamber. This is then followed by a gated time-of-flight (TOF) section. If O^+ ions strike fullerene and are trapped, the $(C_{60} + O)^+$ product ion has roughly same energy as the incident O^+ ion and thus TOF gives us the mass of the ion product.

1 R. E. Hartle et al., Preliminary interpretation of Titan plasma interaction as observed by the Cassini plasma spectrometer: comparisons with Voyager 1, Geophys. Res. Lett., 2006, **33**, L08201.
2 E. C. Hartle et al., Initial interpretation of Titan plasma interaction as observed by the Cassini plasma spectrometer: comparisons with Voyager 1, Planet. Space Sci., 2006, **54**, 1211.
3 A. J. Coates et al., Negative ions at Titan and Enceladus: recent results, Faraday Discuss., 2010, **147**, DOI: 10.1038/C004700G.
4 E. Sittler et al., Heavy ion formation in Titan's ionosphere: Magnetospheric introduction of free oxygen and a source of Titan's aerosols?, Planet. Space Sci., 2009, **57**, 1547–1557.
5 T. Weiske et al., Injection of helium atoms into doubly and triply charged C_{60} cations, J. Phys. Chem., 1991, **95**, 8451–8452.
6 K. A. Caldwell et al., Endohedral complexes of fullerene radial cations, J. Am. Chem. Soc., 1991, **113**, 8519–8521.
7 K. A. Caldwell et al., High energy collisions of fullerene radical cations with noble gases: capture of the target gas and charge stripping of C_{60}^+, C_{70}^+ and C_{84}^+, J. Am. Chem. Soc., 1992, **114**, 3743–3756.
8 B. Pietzak et al., Buckminsterfullerene C_{60}: a chemical Faraday cage for atomic nitrogen, Chem. Phys. Lett., 1997, **21**, 259–263.
9 A. Weidinger et al., Atomic nitrogen in C_{60}:N@C_{60}, Appl. Phys. A, 1998, **66**, 287–292.
10 T. E. Cravens et al., Energetic ion precipitation at Titan, Geophys. Res. Lett., 2008, **35**, 03103, DOI: 10.1029/2007GL032451.
11 T. P. Martin et al., Clusters of fullerenes molecules and metal atoms, Phys. Scr., 1996, **T66**, 38–47.
12 L. Becker et al., Fullerenes in meteorites and the nature of planetary atmospheres, in Natural Fullerenes and Related Structures of Elemental Carbon, ed. F. J. M. Rietmeijer, Springer, 2006, pp. 95–121.
13 V. Vuitton et al., Negative ion chemistry in Titan's upper atmosphere, Planet. Space Sci., 2009, **57**, 1558–1572.
14 M. C. McCarthy et al., Laboratory and astronomical identification of the negative molecular ion C_6H^-, Astrophys. J., 2006, **652**, L141–L144.
15 S. Brünken et al., Detection of the carbon chain negative ion C_8H^- in TMC-1, Astrophys. J., 2007, **664**, L43–L46.
16 A. J. Remijan et al., Detection of C_8H^- and comparison with C_8H toward IRC +10216, Astrophys. J., 2007, **664**, L47–L50.
17 N. Sakai et al., Tentative detection of C_4H^- toward the low-mass protostar IRAS 04368 +2557 IN 11527, Astrophys. J., 2008, **673**, L71–L74.

Professor Mitchell asked: It is interesting to learn of the role of H_3^+ in cooling the atmospheres of planets. In the new ITER reactor, cooling by formation of molecular ions is essential to prevent destruction of surfaces with which the hot plasma will come into contact as in the divertor region. Recombination of the ions in the volume will avoid the recombination energy being released in the material of the surface and also molecular species can easily lose energy by radiation. Indeed heavier species are deliberately injected into divertor regions of Tokamaks in order to favor energy loss by radiation.

Professor Miller replied: This is an interesting comment. As we say in our paper, there is already a line list for H_2D^+ and an accompanying cooling function. It looks as if there is a case for the UCL Group putting together a line list and cooling function for T_3^+.

Dr Sittler responded: I agree that the formation of C_2H^- by radiative electron attachment is a process not likely to occur and the auto-detachment rate of C_2H^- is exceedingly fast from the transient complex of C_2H^-. Therefore the first peak of CAPS-ELS cannot be ascribed to C_2H^- as first indicated. But, C_2H can be present. However, C_4H^-, C_6H^-, most likely would be formed. See Table 5 of Vuitton et al. in the *Planet. Space Sci.* paper; the negative ion/radical density ratio. The experimental values of radiative attachment rate coefficients are not yet measured. In Lavvas *et al.* (2008, paper I) on page 51 they discuss the importance of polyyne radicals C_2H, C_4H and C_6H, but from reading the text clearly biased against unsaturated polymer formation and larger aerosols. They also point out that reaction rates for higher order polyyne radicals have not been measured at low temperatures. Furthermore, one does not expect radicals to accumulate since they can react very fast but rather one must compare the rates of production and loss for such radicals to determine their importance. Coming back to the measurement issue, for example Table 6, most rates are estimates or measured at high temperatures $T > 300$ K. Looking at Table 5 of Lavvas *et al.* (2008. paper I) many of the reaction rates are either estimated or measured at $T > 300$ K. The best solution to this debate is to measure the electron attachment reaction rates for the polymer precursors such as the ethynyl radicals C_2H and CN radicals as prime examples. Finally, if molecular clouds, where polyyne radical negative ions have been observed, are analogs of Titan's upper atmosphere then one may expect the same at Titan.

Professor Mason noted: I would like to draw to the author's attention a recent paper[1] in which we operated a coronal discharge in a chemical mixture that simulated Titan's upper atmosphere (also presented in a poster at this meeting) and detected the anions produced. We only observed nitrile ions that were CN^--based; we did not see any anions that were CH based (*e.g.*, no C_nH^-).

Electron-induced anion formation (at least for light anions) is most likely by the process of dissociative electron attachment, for which the database is poor, with few absolute cross sections (rate constants) being known but the most common process for a hydrocarbon containing CN (M) is the lowest of the H with resultant anion (M less H), *e.g.* HCCCN forms $CCCN^-$. The cross sections will, however, be very strongly dependent on the internal (vibrational) energy of the target molecule and if the electron interaction occurs within as cluster (aerosol) new chemistry can occur that makes it difficult to predict the final anions. see ref. 2. However there is an active research community in the EU able to study these processes, and methods for making absolute cross sections have now been developed. So if you could indicate what you would like measured, it can be arranged (at least in the gaseous phase)!

1 G. Horvath, Y Aranda-Gonzalvo, N J Mason, M Zahoran and S Matejcik, Negative ions formed in $N_2/CH_4/Ar$ discharge—A simulation of Titan's atmospheric chemistry, *Eur. Phys. J. Appl. Phys.*, 2009, **49**, 13105.
2 N. J. Mason, Electron induced chemistry: a forward look, *Int. J. Mass Spectrom.*, 2008, **277**, 31–34.

Professor Coates responded: Thank you, we will consider this.

Professor Plane commented: Are you able to give even rough estimates of the number densities of these ions?

Professor Coates responded: We presented density per bin in the *Geophys. Res. Lett.* (2007) paper. Conversion of course depends on our best guess for assumptions on efficiency. My student Anne Wellbrock presented the Figure below at the meeting.

Miss Wellbrock continued: We are currently studying the effect of solar radiation on the negative ion densities in order to help constrain the chemical formation and destruction processes. We started by looking at an altitude range of 950–1050 km. At this range, solar radiation can reach the ions at solar zenith angles (SZA) of up to 135°, as can be seen in Fig. 4. However, at such high angles the amount of solar radiation may be very small. We use two sets of data. The first (left hand panel in Fig. 4) is from 22 Titan encounters where negative ions were observed. The second data set (right hand panel) is from one encounter only (T57) which took place at a SZA of ~135°. The actuator was fixed during this flyby which means that the data obtained is a continuous set of measurements. We study two different mass groups: light ions (10–30 amu) and heavier ions (110–200 amu). The density measurements shown have large uncertainties and should currently only be treated as relative densities.

The results show that the densities of the light ions are highest at low SZAs (*i.e.* large amount of solar radiation), whereas the densities of the heavier ions are highest at high SZAs (*i.e.* no solar radiation). Therefore, nightside reactions seem to yield the highest densities for higher masses and photochemical reactions yield the highest densities for the lower mass negative ions. Both mass groups have a density minimum in the terminator region where only a small amount of solar radiation reaches the ions. In addition, the data from T57 show that there is a sudden decrease in density from being in complete darkness (marked as 'in shadow', SZA just over 135°) to being exposed to a small amount of solar radiation (marked as 'not in shadow', SZA just under 135°), especially for the heavier mass group.

Professor Plane asked: I am struck by your observation of large negative ions close to 1000 km altitude. A comparison with the Earth's atmosphere is instructive. Meteoric ablation gives rise to meteoric smoke particles in the aerobraking region around 90 km. If these particles, which are mostly negatively-charged through electron attachment, grow larger than about 5 nm in radius, they sediment rapidly out of the upper atmosphere. Even the smaller particles around 1 nm, which can be observed by high-power large aperture radars such as the Arecibo radar or by particle detectors on rockets, do not get much above 100 km.

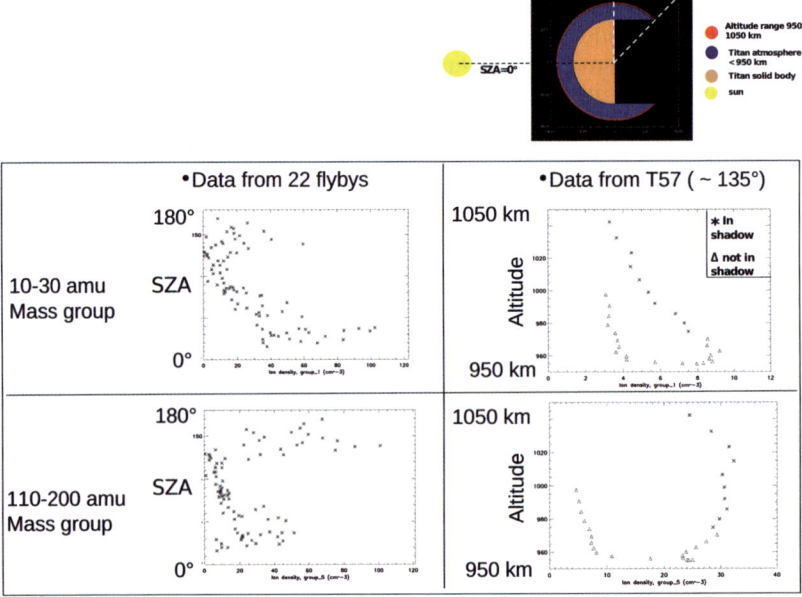

Fig. 4 Negative ions measured by Cassini CAPS electron spectrometer: Effects of solar zenith angle (SZA) on relative densities (horizontal axes).

The equivalent aerobraking region on Titan is around 500 km. Even though sedimentation on Titan is much slower because of the small gravitational force, the negative ions reported in this paper above 900 km must form *in situ* and then sediment to lower altitudes.

Professor Coates answered: Thanks for this useful comment which adds an interesting argument to our assumptions in the *Planet. Space. Sci.* (2009) paper.

Professor Schram asked: Regarding the negative ions differences in daylight and in shadow, could the following mechanism not be operationable: In sunlight the electron and positive ion densities become larger; if negative ions are produced they lead to a further enhancement of ion densities, as the transport is slower (negative ions tend to concentrate in the positive potential regions). Then, probably, mutual recombination could be a fast process therewith counteracting agglomoration to larger ions. In the dark part charges are less abundant (and electrons smaller in density?): and is there more time to grow in size?

Professor Coates responded: Thank you, we will consider this. There is some indication in the Wellbrock *et al.* poster (and Fig. 4, above, in reply to another comment) that this may be mass-dependent.

Professor Schram then opened the discussion of the paper by Dr Huestis, asking: Are there signs of non-equilibrium excitation in the observed spectra? This could be electronic (N ^{2}D *versus* N 4S: the ratio is far above equilibrium) but also in vibration and rotation. Regarding rotational excitation: does this not reflect a dissociating or associating reaction to excite rotational non-equilibrium? And if so, does such a non-equilibrium influence the kinetics (*e.g.* by enhancing the rate of dissociative attachment)?

Dr Huestis replied: We have no direct evidence for vibrational or rotational excitation in CO_2. The manuscript describes the role of vibrationally excited $O_2^+(v)$ dissociative recombination, the high altitude source of $O(^1S)$. However, the CO(A–X) dayglow emission bands suggest a rotational temperature of roughly 1600 K for low-*J* levels and 10 000 K for high-*J* levels, as a result of energy released in the UV photodissociation of CO_2 [Conway, *J. Geophys. Res.*, 1981, **86**, 4767].

Professor Mitchell replied: Dissociative recombination rates and dissociative electron attachment rates are often strongly affected by the internal energy of the molecular ions or molecules. In one experiment, (Mostefaoui *et al.*, *J. Phys. B*, 1999, **32**, 5247) where we were able to study the recombination of NO^+ ions with 14 vibrational states populated, we measured a rate that was a factor of three lower than rates measured for NO^+ ions in the ground vibrational state.

Professor Schram concurred: I fully agree, that the internal excitation of molecular ion is an interesting aspect also in the context of plasma. In addition to vibration, rotational excitation may be important and may be stable enough to be important. Also in the dissociative recombination rotationally excited fragments can result as is measured *e.g.* for OH in emission A–X in an atmospheric water plasma (*cf.* P. Bruggeman *et al.*, *Plasma Process. Polym.*, 2009, **6**, 751–762).

Professor Miller opened the first general discussion of the papers by noting: There have been several suggestions that H_3^+ is formed with over-populations of excited ~(ro-)vibrational states in planetary atmospheres. Extensive studies of Jupiter, however, have failed to find any spectroscopic evidence to support this. On the contrary, where hot band transitions have been found, they are generally less intense than one might expect from LTE calculations—a sign that non-LTE conditions are prevailing on those regions of the atmosphere where the higher vibrational levels are being populated.

Professor Schram continued: A second line of research (besides flames), which may be of interest to the planetary chemistry is that of low pressure plasma chemistry.[1] The motivation in this field is to understand the processes active in plasma deposition, *e.g.* for solar cells. Fast deposition and thin layers of good quality are needed and there is a massive search for new materials and methods. Among others, carbon and silicon based materials have been analyzed thoroughly, also in view of other applications. This field is thus presently more-and-more directed to understanding volume and surface chemistry. The densities are so low that three-particle formation reactions are absent and the plasma medium is primarily dissociating. Formation of molecules takes place at the surface, where also deposition and etching occur. Despite the low pressure cluster, particle growth is observed and has been analyzed; dust may add new surfaces and will thus enhance molecule formation and deposition.[2] The surface is modified by a strong absorption of radicals, as the conditions ensure that the radical flow is large and the LH desorption is low, because of the low temperature of the substrate.[3] When the radical fluxes are large, excited molecules come from the surface, also with (besides vibrational or electronic) rotational excitation. For a recombining plasma hydrogen molecule formation at the surface H + H → $H_2(r,v)$ has been measured in detail showing high non-equilibrium rotational (and vibrational) excitation up to the dissociation limit, see for recent results on $H_2(r,v)$, HD(r,v) and $D_2(r,v)$.[4] Other evidence for excited molecule formation is at the surface is observed for N_2^* and NO_2^*.[5] Excited molecules may facilitate negative ion formation and thus open a new route to association of molecules and cluster formation.

There is more-and-more evidence that significant molecule formation takes place at the surface by association of adsorbed fragments, leading to preferred new molecules as H_2, CO, H_2O, HCN, C_nH_m, NH_3, *etc.*, depending on conditions.[6] It proved that the resulting molecular content is quite similar, irrespective of the primary molecules used, as long as the atomic abundances in the dissociated molecules is the same.

It would be very interesting to investigate further interaction of the plasma chemistry with planetary science and to test models.

1 M. A. Lieberman and A. J. Lichtenberg, *Principles of Plasma Discharges and Materials Processing*, 1994, Wiley-Interscience, Chicester.
2 A. Bouchoule, *Dusty Plasmas: Physics, Chemistry and Technological Impacts in Plasma Processing*, 1999, Wiley, New York.
3 D. C. Schram, *Plasma Sources Sci. Technol.*, 2009, **18**, 014003.
4 O. Gabriel *et al.*, *J. Chem. Phys.*, 2010, **132**, 104305 and references therein.
5 J. van Helden *et al.*, *IEEE Trans. Plasma Sci.*, 2005, **33**, 390.
6 R. Zijlmans *et al.*, Molecule synthesis in a $CH_4\ N_2\ O_2$ plasma, *Plasma Sources Sci. Technol.*, 2006, **15**, 564

Professor Miller asked: The charge exchange reaction $H^+ + H_2(v \geq 4) \rightarrow H + H_2^+$ is thought to be extremely important in planetary atmospheres. We do not know this rate experimentally, however, and it would be very useful to have it. Are there any prospects of getting such measurements?

Professor Schram answered: I can quote a paper [*Astrophys. J.*, 2004, **606**, L167–L170] by Savin *et al.* with relevant references and names: Janev, from the fusion side, and Kristc. A serious discussion point refers to the possibility of vibrational excitation and even more important of rotational excitation of the $H_2(r,v)$ molecule. In experiments on recombining plasmas in Eindhoven and in Nieuwegein (a fusion lab) it appears that ions from the source, which are mainly H^+, are very efficiently lost by what is called molecular assisted recombination. This process requires that first the atomic ions are converted to molecular ions by charge transfer with molecules. This has been first described by de Graaf *et al.* in 1993 (*Phys. Rev. E*, **48**, 1993, 2098–2102): this was so efficient that an overall rate of in the order of 10^{-15} m^3 s^{-1} was needed to explain the ion loss. Later it was found that at least partly this may be due to strong rotational (and vibrational) excitation; the rotational excitation is more important as it involves a much larger statistical weight as, *e.g.*, described in a recent paper by Gabriel *et al.* (*J. Chem. Phys.*, 2010, **132**, 104305) The work in Nieuwegein also needs high rates to explain the observed ion loss in the expansion from the source (see, *e.g.*, Shumack *et al.*, *Phys. Rev. E*, 2008, **78**(4), 046405). There is still discussion about the rates: recent experiments in Nieuwegein seem to require higher rates than quoted and the matter is under investigation. My personal guess is that it is probably due to a difference in rotational excitation: in highly recycling plasmas as these expanding plasmas are with molecules formed at surfaces in an excited state there is a strong non thermal r,v excitation of molecules. In this view one also needs information on the state of excitation of hydrogen molecules. Is there information on the r,v state of hydrogen molecules in planetary conditions?

Professor Ashfold continued: Pursuing the question about possible sources, and consequences, of rotational excitation in atmospheric species it is worth noting that long wavelength photolysis of NH$_3$ (considered to be the major route for forming H atoms in the atmosphere of Titan) is known to produce NH$_2$ fragments with high levels of rotational excitation, largely concentrated about the *a*-inertial axis.[1] Similarly, and of possible relevance when considering the atmosphere of Enceladus, some of the OH fragments formed by Lyman-α photolysis of H$_2$O carry extremely high levels of rotational ($E_{\text{rot}} > 4$ eV) excitation, while others are formed in highly vibrationally excited ($v \leq 9$) levels.[2,3]

1 D. H. Mordaunt, M. N. R. Ashfold and R. N. Dixon, *J. Chem. Phys.*, 1996, **104**, 6460.
2 D. H. Mordaunt, M. N. R. Ashfold and R. N. Dixon, *J. Chem. Phys.*, 1994, **100**, 7360.
3 S. A. Harich, D. W. H. Hwang, X. F. Yang, J. J. Lin, X. M. Yang and X. M. Yang, *J. Chem. Phys.*, 2000, **113**, 10073.

Professor Suits spoke: These remarkable observations of negative ions bring to mind several questions. Of course photodetachment and its possible wavelength dependence is one issue. What role do you see for photodetachment as a loss process? Another is the question of multiply charged species. It is worthwhile to note that a species going from a higher charge state to a lower one (through photodetachment or loss of a small charged fragment) would exhibit an apparent increase in mass. Have you ever seen trends that might be consistent with this?

Professor Coates responded: According to the Vuittton *et al.* paper (*Planet. Space Sci.*, 2009), photodetachment is not dominant. We will have a look back in the light of the second part of your comment.

Dr Sittler addressed Dr Yelle: Roger, what about the formation of C$_2$H$^-$ from radiative attachment?

Dr Yelle replied: We did include these processes in the Vuitton *et al.* (2009) paper and calculated that the nitrile species (CN$^-$, C$_3$N$^-$, C$_5$N$^-$) had significantly higher densities than the C$_{2n}$H$^-$ species near the ionospheric peak near 1100 km. However, if you look at Fig. 9 in Vuitton *et al.* (2009) you will see that C$_2$H$^-$ is very nearly the most abundant negative ion near 800 km (equal to C$_3$N$^-$ and the second most abundant ion, after CN$^-$, at 1200 km. This is undoubtedly for the reason that you point out, which is included in our calculations. It is certainly true that many of our rates are based on estimates rather than measurements, and we need to bear this in mind, but at the present time there does not appear to be any reason to favor C$_{2n}$H$^-$ species over C$_{2n+1}$N$^-$ species near the ionospheric peak. I would also point out that the new information presented at this conference by Coates et al. on the variations of negative ion densities with latitude and local time suggest that we should examine a wider range of conditions than considered in Vuitton et al. (2009), and perhaps the composition could vary with location.

Dr Vuitton said: Radiative attachment to $C_{2n}H$ ($n = 1,2,3$) is considered in the Vuitton et al. (2009) model. However, the process is slow for C_2H according to calculations from Herbst and Osamura (2008). Although the calculations indicate that the rate is faster for C_4H and C_6H, the mole fraction of these radicals is small according to photochemical models (see for example Lavvas et al. 2008). As a consequence, radiative attachment is inefficient to produce $C_{2n}H^-$ ions.

1 E. Herbst and Y. Osamura, *Astrophys. J.*, 2008, **679**, 1670–1679.
2 P. Lavvas, A. Coustenis and I. M. Vardavas, *Planet. Space Sci.*, 2008, **56**, 67–99.
3 V. Vuitton, P. Lavvas, R. V. Yelle, M. Galand, A. Wellbrock, G. R. Lewis, A. J. Coates and J.-E. Wahlund, *Planet. Space Sci.*, 2009, **57**, 1558–1572.

Dr Sittler replied: You state that the C_2H forms in the lower atmosphere and is known to be of low abundance in Titan's upper atmosphere, but that is based on a model calculation? What about the dissociative attachment for C_2H_2 to make C_2H^-? I would also like to respond to Dr Vuitton's earlier comment: you state that the radiative reaction rate for C_2H^- is very slow, but it is relatively fast for C_4H^- and C_6H^-. All these are based on estimations and not measurements. Yes, the reaction rates for these radicals are expected to be fast, which will tend to make them less abundant, but they may also be forming rapidly too, from referring to Table 2 in Vuitton et al. (2009) paper.

Professor Mitchell stated: With regard to the dissociative attachment of electrons to molecules, size is not an important factor. What is critical is the position of the curve crossing between the initial and attaching negative ion states. For example, the difference in the attachment rate for 2 eV electrons to molecular hydrogen in the $v = 0$ and $v = 6$ vibrational states is six orders of magnitude (Wadehra, *Phys. Rev. A*, 1984, **29**, 106) because the curve crossing occurs at this latter level (Bardsley and Wadehra, *Phys. Rev. A*, 1979, **2**, 1398). For C_2H, therefore, the rate will be determined by the ro-vibrational state and the position of the curve crossing.

Dr Sittler asked: What about the dissociative attachment for C_2H_2 to make C_2H^- ($C_2H_2 + e^- \rightarrow C_2H^- + H$)?

Dr Yelle replied: This is included in the calculations and is responsible for the relatively large abundance of C_2H^-.

Professor Mitchell returned to the discussion of the paper by Professor Coates: In your presentation, you mentioned that you had observed ions with 13 800 amu q^{-1} so the units are in Thomsons rather than Daltons. The question is what is q? In flames, soot particles are predominantly electrically charged, either positively by thermionic emission of electrons or negatively by electron attachment. These are large structures and it is known that they can carry many charges. This is an intriguing question, I believe, in the present context.

Professor Coates responded: We had a brief look at this in the *Geophys. Res. Lett.* paper (2007) in order to examine charging in the ionospheric plasma environment, and made a simple calculation which limits the possible number of charges to a few (up to 5). If the charge were that large it would indeed affect our mass results making the molecules/aerosols even bigger. So I agree charge should certainly be taken into account and measured in the next mission—but for now we must use the data we have.

Professor Ashfold returned to the discussion of the paper by Dr Huestis: Huestis and colleagues present convincing evidence that, contrary to previous assumptions, solar photolysis of CO_2 is not the principal source of the $O(^1S-^3P)$ Mars day glow emission. This begs the following questions: Is it possible to put an upper limit on the possible contribution that CO_2 photolysis could make to the day glow emission (relative to that from the $e^- + O_2(v)$ reaction) without revealing itself in plots like that shown in Fig. 4? If so, does this upper limit encourage a reduction in the $O(^1S)$ quantum yield from CO_2 photolysis at 110–115 nm (which has traditionally been assumed to be unity)?

Dr Huestis replied: The shortness of the oral presentation prevented detailed discussion and interpretation of multiple possible production mechanisms of $O(^1S)$. Pages 10–12 of the manuscript (Interpretation and Modeling) cover this topic, including a Figure illustrating the current incomplete knowledge of the wavelength dependence of the quantum yield for $O(^1S)$ from photodissociation of CO_2. While we still believe that the yield is high, between 110 and 120 nm, we think that the overall yield integrated over the solar spectrum and the CO_2 absorption spectrum, may be substantially lower than previously estimated, largely because the absorption cross section is so highly structured. Our overall conclusion is that vibrationally excited $O_2^+(v)$ dissociative recombination is the primary source of $O(^1S)$ above 120 km, while 121.6 nm (Lyman α) photodissociation of CO_2 dominates between 80 and 120 nm. If one connects these two mechanisms in Fig. 12 from Fox and Dalgarno [*J. Geophys. Res.*, 1979, **84**, 7315], one obtains a reasonably straight line on log paper from 80 to 180 km, as in the data shown in the Fig. 5 in the manuscript. See also Fox and Hac [*Icarus*, 2010, **208**, 176] and Fig. 1 in Fox and Yeager [*Icarus*, 2009, **200**, 468].

Professor Mitchell asked: Could you comment on the role of dissociative recombination on isotope loss in planetary atmospheres?

Dr Huestis answered: Fox and Hac have recently reported a comprehensive investigation titled 'Isotope fractionation in photochemical escape of O from Mars' [*Icarus*, 2010, **208**, 176], following dissociative recombination. They model the altitude dependence of vibrational excitation in normal and isotopic $O_2^+(v)$. The lighter atomic isotope, ^{16}O, escapes preferentially, consistent with its lower mass.

Mr Thissen addressed Professor Coates: The z parameter of the m/z ratio is very difficult to evaluate in the heavy ions detected in the m/z range around 1000. As these objects might be close to the macroscopic scale, some signals might have been recorded by the CDA instrument on board Cassini. This detector is based on the ballistic impact of dusts and subsequent ionization by the resulting plasma. It is most probably sensitive to the mass, and not the m/z of the particles. If no signal was detected, even in the altitude range where CAPS detects heavy ions, this might put an upper limit to the possible mass of the negatively charged ions, and therefore also an upper limit on the z parameter.

Professor Coates answered: That is an interesting comment, many thanks—we will check with the CDA team.

Professor Plane opened the discussion of the paper by Dr Adams: Another source of negative ions in the upper atmospheres of solar system bodies is meteoric ablation, which produces iron, magnesium and silicon vapours in roughly equal quantities [T. Vondrak, J. M. C. Plane, S. L. Broadley and D. Janches, A chemical model of meteoric ablation, *Atmos. Chem. Phys.*, 2008, **8**, 7015–7031]. These elements should undergo oxidation and eventually form metal silicates, $MgSiO_3$ and $FeSiO_3$. I have used *ab initio* quantum calculations to show that these molecules have very large dipole moments—12.2 and 9.5 Debye, respectively—and electron affinities of 2.6 eV. The metal carbonates have similar properties. So these metal-containing molecules should polymerise readily into larger particles (known in the terrestrial atmosphere as "meteoric smoke") and attach electrons to form quite stable heavy negative ions.

Dr Sittler asked: All your association and proton transfer reaction rates using the SIFT apparatus were measured at 298 K and at high pressures so three-body collisions dominated. You note that condensation in the flow tube limits your measurements to $T > 180$ K. As shown by Morales *et al.* (this volume) using the CRESU techniques for temperature range between 23 K to 300 K one cannot reliably extrapolate measurements at 300 K to $T \sim 150$ K for Titan's upper atmosphere and ionosphere. All the Cassini INMS ($m/q \leq 100$ amu for neutral and positive ions), CAPS-IBS ($m/q < 350$ amu for positive ions) and CAPS-ELS ($m/q < 13\,000$ amu) for negative ions are all measured for heights ≥ 950 km where ion/neutral temperatures $T \sim 150$ K, electron temperatures $T_e \sim 1000$ K and pressures are very low (on the order of nanobars) so two-body collisions dominate. Even the CRESU observations are made under high pressure conditions. Will it ever be possible to make such measurements in the future that are more characteristic of Titan's upper atmosphere where all this complex chemistry is occurring or can one make reliable corrections to such measurements? You also note that the recombination reactions are more likely dominated by the electron temperature, and in this case one has the inverse condition where your electron temperatures are <550 K (collisional equilibrium with Helium carrier gas), while electron temperatures within Titan's upper atmosphere $T_e \sim 1000$ K. Like the question for association reactions, will it be possible to increase your measurements to $T \sim 1000$ K or make reliable corrections without using simple extrapolations?

Dr Adams answered: One of the main points of this work has been to show that association can compete with facile processes such as proton and charge transfer. Thus, large species can be rapidly produced and this warrants further study. Pressure dependence studies will show if association is pressure saturated. If so then association can be important at low pressures, especially with the longer lifetimes of the intermediate complex at the lower temperatures appropriate to the Titan atmosphere. On the second point, dissociative electron–ion recombination is difficult to study especially with the differing ion and electron temperatures. Both theoretical studies and, I agree, higher temperature measurements need to be made to understand this process and permit meaningful extrapolations to Titan conditions.

Dr Vuitton asked: Your experiments show that several reaction channels are association. Can you tell if these are collisionally stabilized or radiatively stabilized? This is a fundamental information for modelers. In the former case, these channels can be ignored in Titan's models (in the upper atmosphere, the pressure is too low for 3-body reactions to occur) but in the later case, they have to be included.

Dr Adams answered: The association channels are competitive with processes such as proton and charge transfer. That the latter processes usually occur at the gas kinetic rate, implies that association is occurring with high efficiency and thus is a saturated collisionally stabilized reaction. To confirm this, would require a study as a function of third body pressure; a saturated reaction would have a pressure independent binary rate constant. As the pressure is reduced, the association would eventually become an unsaturated ternary reaction. If the temperature is reduced, then the lifetime of the intermediate complex would increase and the reaction would remain saturated down to lower pressures. To be certain of the temperatures and pressures at which the changeover from saturated to unsaturated kinetics occurs would require detailed experimental studies. The main purpose of the present study is to make the community aware of these effects on their chemical models.

Professor Sims noted: The size of the ion–molecule reactive systems studied by Adams et al.[1] means that radiative association could be significant pathway to form association products, at least under the relatively low pressure conditions prevailing in Titan's ionosphere. Ryzhov et al.[2] show for example that the rate coefficient for formation of protonated acetone (a relatively small system in comparison to some of those studied here) by radiative association reaches a value of 8.4×10^{-11} cm^3 molecule^{-1} s^{-1} at 245 K, with a strong negative temperature dependence. Radiative association should therefore be taken into account as a pathway for the growth of larger ionic species in models of planetary ionospheres.

1 N. G. Adams, L. D. Mathews and D. Osborne, Jr, *Faraday Discuss.*, 2010, **147**, DOI: 10.1039/C003233F.
2 V. Ryzhov, Y. C. Yang, S. J. Klippenstein and R. C. Dunbar, *J. Phys. Chem. A*, 1998, **102**, 8865.

Dr Sittler asked: So, one can correct the high pressure measurements by considering their radiative relaxation to lower energy states after a 2-body collision, and that these radiative time scales are shorter than mean time between collisions? With three-body collision is some of the excess energy carried away by the none reactive component? Is that what you're trying to say?

Professor Sims replied: In general, radiative association is very difficult to observe experimentally, in particular for neutral–neutral associations, as it is hard to attain conditions where the time for stabilisation of the initial excited adduct by a third-body collision (which indeed does carry away some of the excess energy, by analogy with the Lindemann mechanism[1] for unimolecular reactions) is longer than the radiative lifetime. We have in the past used measurements on collisionally-assisted association at low temperatures coupled with simple calculations on deactivation by infrared emission to provide an estimate of the radiative association rate coefficient for the $CH + H_2 \rightleftharpoons CH_3$ reaction.

1 F. A Lindemann, *Trans. Faraday Soc.*, 1922, **17**, 598.
2 R. A. Brownsword, I. R. Sims, I. W. M. Smith, D. W. A. Stewart, A. Canosa and B. R. Rowe, *Astrophys. J.* , 1997, **485**, 195.

Dr Ellinger asked: $C_3H_3^+$ is known to exist under two structures, one linear, one cyclic. Could the different products of the reaction of $C_3H_3^+$ with pyridine be related to the two different isomers ?

Dr Adams answered: The reaction of $C_3H_3^+$ with toluene (Fig. 2 in the paper) shows that two isomers, linear and cyclic, are indeed present and react at different rates. Thus, in the reaction with pyridine, where both isomers react at the gas kinetic rate, the product distribution will be a combination of those for the two isomer reactions.

Professor Mitchell commented: Your Fig. 5 is surprising, as you show a positive temperature dependence for the recombination rates. I have only seen this previously in the work of Alan Hayhurst on flame studies of H_3O^+. Is this dependence due to ro-vibrational excitation of the recombining ions at higher temperatures?

Dr Adams replied: Positive temperature dependencies are unusual and prior to this study, temperature dependencies for proton bound dimer species had not been made. Thus, there has been no previous data for direct comparison.

Dr Biennier said: Dr Wolf Geppert from Stockholm University presented recently the measurement of the dissociative recombination of protonated benzene $C_6H_7^+$ performed with the CRYRING magnetic storage ring. The rate constant he derived, about 2×10^{-6} cm^3 molecule^{-1} s^{-1} at room temperature, is much higher than your results obtained in a flow tube. How do you explain this discrepancy?

Dr Adams responded: There are several points to consider. Geppert's measurement in the Storage Ring (SR) was for $C_6D_7^+$ whereas we studied $C_6H_7^+$ recombination in the Flowing Afterglow (FA) and hydrogenated species are known to recombine somewhat more rapidly. In addition, the SR measures cross sections and rate constants are derived by integrating over a Maxwell–Boltzmann energy distribution. Thus, the internal energy of the ions is not changed and this SR value is not a truly thermal rate constant, as is obtained in the FA. The rate constant of this reaction has previously been measured in a FA (J. L. McLain, V. Poterya, C. D. Molek, L. M. Babcock and N. G. Adams, *J. Phys. Chem. A*, 2004, **108**, 6704) with a value at room temperature close to Geppert's but with a different temperature dependence. From a consideration of the present study, this previous value may have been contaminated by proton bound dimers, $(C_6H_6)_2H^+$

Professor Mitchell commented: Storage ring experiments measure cross sections as a function of electron energy. They are really collision experiments rather than experiments that measure reactions under chemical conditions. When one measures a cross section at 1 eV but the ion is in its vibrational ground state, one cannot speculate on the rate of reaction of that ion at 11 000 K even if at that temperature the ion would still exist! This is an exaggeration and in fact one measures cross sections over a wide energy range to determine the effect of the Maxwellian distribution of electron energies on the ion being studied, usually in a cold state. Rate coefficients measured in flames yield rates that are very different from those measured in cooler experiments for the ions are ro-vibrationally excited and this aspect is not taken into account in storage ring experiments. A beautiful example of this is shown in the work of LeGarrec *et al.* who studied the dramatic effects of ro-vibrational excitation on the complementary process of dissociative electron attachment to neutral molecules (LeGarrec *et al.*, *J. Chem. Phys.*, 1997, **107**, 54).

Professor Schram said: My question concerns the possibility of the formation of negative ions during the flow time by dissociative attachment to fragments resulting from dissociative recombination? Radical fragments from the latter process could be (*e.g.* rotational) excited, as there is substantial energy available in the dissociative recombination. Could such processes influence your results?

Dr Adams replied: In the flowing afterglow, fragmentation products from dissociative electron–ion recombination will have a number density less than or equal to the ionization number density of 4×10^{10} cm^{-3}. Thus, unless the fragments have an attachment rate constant at the gas kinetic limit (about 4×10^{-7} cm^3 s^{-1} as is obtained for SF_6 and CCl_4) there will be no significant electron attachment. In addition, the variation of electron density with distance along the flow tube will be very different for the two processes. Indeed to study electron attachment, the electron density has to be reduced to about 10^9 cm^{-3} to inhibit recombination; but it is an interesting question.

Dr Loison opened the discussion of the paper by Alexandre Faure: Why do you choose to fit the experimental data between 10 K and 300 K and not between 1 K and 300 K?

Dr Faure replied: The choice of $\sim$10 K is indeed rather arbitrary and it could be actually optimized to each system (when low temperature data are available) in the typical range 1–100 K. The current choice of ~10 K provides however a good compromise for all the available data presented in our paper.

Dr Pernot noted: You said that we should reach the 10 percent uncertainty limit to get good predictions. I hope that this is not the message that has been retained from my earlier presentation. Otherwise, we had better stop working right now! My message is rather: "If you know that this interpolation method provides rate coefficients within a factor of two, please make use of this uncertainty information in model predictions, by all means!"

Dr Faure responded: I fully agree with this comment. Our hope is that the inclusion of our results in chemical models will improve the current predictions, even with a factor of two of uncertainty.

Dr Pernot asked: In the conclusion of your paper, you state that "a sensitivity analysis should be done". Do you have any project in this direction, or is this a bottle in the sea?

Dr Faure responded: We have currently no project on sensitivity analysis but we are interested in a possible collaboration with your group.

Professor Plane asked: While agreement is generally good (within a factor of 2) between your semi-empirical approach and measurements at low temperatures, in some cases illustrated in Fig. 2 and 4, the agreement is very poor (more than a factor of 10 difference). Could you discuss why this might be the case?

Dr Faure replied: As recognized in our paper, deviations larger than a factor of 2 (up to a factor of 10) are indeed found with ion trap measurements below 50 K, both for cations (CH_4^+ and C_3H^+) and an anion (NH_2^-). It is not clear why the measured rates are only a small fraction of the Langevin value at very low temperature. Possible explanations include dynamical bottlenecks caused by the influence of the rotation of H_2 and/or nuclear-spin effects, the competition between the abstraction and the radiative association channels but also uncertainties in the experimental temperature of the ions. Some of these effects are discussed in Savic and Gerlich, *Phys. Chem. Chem. Phys.*, 2005, **7**, 1026.

Professor Mason asked: Do you have an explanation of why the rate constant for some ions appears to plateau (remain constant) for low temperatures?

Dr Faure replied: Rate constants for (fast and exothermic) ion–non polar molecule reactions are expected to "saturate" at about the Langevin value when the temperature is decreased. The Langevin value, or capture limit, is indeed the upper limit of the reaction rate, when all collisions lead to reaction. This saturation effect has been observed for numerous reactions in the past and it is clearly illustrated in Fig. 2 of our paper for $C_3^+ + H_2$ where the rate remains constant, at the Langevin value, below ~50 K. Now for $C_4^+ + H_2$, $C_3^+ + H_2$ (Fig. 2) and $NH_2^- + H_2$ (Fig. 4), the rates are found to plateau at about the same temperature but with only a small fraction of the Langevin value. As discussed by Savic and Gerlich (*Phys. Chem. Chem. Phys.*, 2005, **7**, 1026), this "low" saturation can be only partly explained by the radiative association channel which becomes dominant at very low temperature. Nuclear spin effects and rotational bottlenecks probably play an important role, as discussed in our paper. Thus, the appearance of a plateau is expected from simple capture arguments but the saturation value depends on the nuclear rearrangement dynamics which is driven by the whole potential energy surface and is ignored in the capture approach.

Dr Moses opened the discussion of the paper by Ms Whalley: The maximum Mg^+ density in your model is smaller than the electron densities observed for the low-altitude layers in the Martian radio-occultation profiles. Is that of any concern to you?

Ms Whalley replied: The sporadic third layers on Mars are formed by atmospheric forces acting to concentrate some of the background layer of ions into a small (1–2 km thick), highly concentrated layer. The model output shows the concentration of the Mg^+ background layer in the Martian upper atmosphere. The purpose of the model was to see if the background layer of Mg^+ had a sufficient concentration, so that if formed into a sporadic E layer, it could create the concentrations (of the sporadic layers) observed by the radio occultation measurements.

Professor Atreya asked: This is a neat result for the specific set of data chosen in your study, but can you say whether the derived rate constant would be valid for other sets of data? As you may know, there are wide variations in the magnitude and the altitude of sporadic E layer seen in the MGS and MEX radio occultations data.

Ms Whalley replied: I think the derived rate coefficient would be applicable to other sets of data. The major constraint is knowing the atomic O concentration at different altitudes in the Martian atmosphere.

Professor Mason enquired: Your experiments on MgO^+ with ozone are interesting but are they relevant to Mars, where ozone concentrations are so much lower than on Earth?

Professor Plane answered: No, the O_3 reactions with Mg^+ and MgO^+ are not important in the Martian atmosphere. Recombination of Mg^+ with CO_2 is the most rapid reaction below 100 km. The resulting $Mg^+ \cdot CO_2$ cluster ion then undergoes slow ligand-switching with O_2. The reactions of $Mg^+ \cdot CO_2$ and $Mg^+ \cdot O_2$ with atomic O, followed by reaction of MgO^+ with O, then recycle the molecular ions to the atomic Mg^+.

Mr Thissen commented: Do you plan to study Mg^{2+} ions and what chemistry do you expect from these?

Ms Whalley responded: I personally don't intend to study Mg^{2+} ions so I'm not sure what behaviour to expect.

Professor Mitchell: You refer to magnesium on Mars and it is interesting because people are studying the combustion of magnesium in carbon dioxide as this could be used as a solid rocket fuel there. Aluminium which is usually used for this purpose on Earth does not burn in carbon dioxide, but magnesium does. My question, however, concerns the emission of electrons from magnesium oxide particles. Experiments in flames have shown that the work function of magnesium oxide particles seems to be about 0.75 eV and thus they should release lots of electrons by thermionic emission or photoemission. Can these effects be incorporated into your model?

Ms Whalley answered: We have briefly looked at these effects, so with further study they could be incorporated.

Professor Plane continued: We do not include MgO particle formation in our model of magnesium in the Martian atmosphere. The reason is that MgO recombines very rapidly with CO_2 to form $MgCO_3$ [R. J. Rollason and J. M. C. Plane, *Phys. Chem. Chem. Phys.*, 2001, **3**, 4733–4740], so that the concentration of MgO in the Martian atmosphere will be extremely low. Thus, MgO particles will not form by the direct condensation of MgO, but could form indirectly from the polymerization and subsequent rearrangement of species such as $MgCO_3$.

Dr Sittler remarked: Borucki and Whitten[1] had modeled nanoparticles or embryo aerosols in Titan's lower atmosphere. Their modeling had shown that they actually reduced the ionization within Titan's lower atmosphere since they tended to acquire negative charges, thus reduce electron densities and thus reduce electron conductivities in agreement with Huygens probe measurements as reported by Fulchignoni et al.[2] and Hamelin et al.[3] So, the nanoparticles could reduce electron densities and not form ionization layers as suggested by Brian Mitchell.

1 W. J. Borucki and R. C. Whitten, Influence of high abundances of aerosols on the electrical conductivity of the Titan atmosphere, *Planet. Space Sci.*, 2008, **56**, 19–26 (please also note my correction in the discussion below).
2 M. Fulchignoni, et al., In situ measurements of the physical characteristics of Titan's environment, *Nature*, 2005, **438**, 785–791.
3 M. Hamelin, et al., Titan's atmospheric electricity from measurements of the PWA-HASI instrument on board the Huygens probe, Titan's atmospheric electricity, PWA-HASI Team-EUROPLANET, Berlin, 2006.

Professor Mason addressed the meeting: In discussing the question of the role of light producing (or removing) electrons from dust (and meteorite dust) I would draw attention to the work of Rapp in the University of Rostock, *e.g.* M. Rapp, Charging of mesospheric aerosol particles; the role of photodetachment and photoionization from meteoretic smoke and ice particles , *Geophysics*, 2009, **27**, 2417–2422.

Professor Schram remarked: The general experience for plasma processing plasmas is that clusters are negatively charged. The charge for a spherical cluster would be in the order of 1 charge per nanometer radius if the electron temperature T_e is around 1 eV (proportional to r in nm and T_e in eV). The negative charging is a consequence of the large mobility of electrons, leading to fast charging of any surface in a plasma and an ambipolar field, causing an acceleration of the ions and deceleration of the electrons. The initial negative charging will be determined by the arrival of the first electrons and will depend on electron density n_e and electron velocity and though considerably limited by orbital restraints, will be fast.

For small clusters in the nanometer range it is believed that the charge is not stable and changes from 1 positive to 1 negative and neutral. If, with strong photon fluxes, clusters become positively charged, then the attraction of electrons becomes even of Coulomb character, with very large cross sections in particular for low electron temperatures. At the other hand the plasma negative charge will linearly depend on T_e and if T_e is small (100 K = 0.01 eV) then one charge (on average) is expected for a 100 nm cluster. It is not clear to me whether in that situation compensated charges would occur on the relatively large clusters in particular if it is made up by isolating material. It is, however, evident that in the situation of low T_e (small charge), low n_e (long charging time) and large photon fluxes (photoelectron emission) the possibility of positive clusters will be substantially higher. Based on these considerations a positive charge would be expected when the cluster comes just out of the shadow (low T_e and n_e) in the light.

One aspect that can also influence this discussion is that, in a plasma, positively charged clusters can be more easily observed than negatively charged ones, because of the above mentioned ambipolar fields: if one applies a positive potential to extract negative ions then first the electrons will be attracted leaving the heavy negative ions behind. This changes only if all electrons get lost in charging clusters. It could be of interest to compare experience in plasma chemistry and planetary science. In ref. 2–3 also various ion mass spectra can be found from work in groups of Winter and of Hollenstein.

1 A. Bouchoule, *Dusty Plasmas: Physics, Chemistry and Technological impacts in Plasma Processing*, 1999, Wiley, New York.
2 J. Winter *et al.*, *Plasma Sources Sci. Technol.*, 2009, **18**, 034010.
3 C. Deschenaux *et al.*, *J. Phys. D: Appl. Phys.*, 1999, **32**, 1876.

Professor Mitchell continued: This is an important point, for the fact that particles are charged will greatly influence their aggregation. In combustion, the great proportion of soot particles are charged. They can become positively charged by thermionic emission and ion attachment and negatively charged by electron attachment. When electronegative materials such as halons are added to a flame, the amount of negative charging increases and it is observed that soot particles are much bigger. Positive charging by photoemission will undoubtedly be important as will the attachment processes mentioned by Professor Schram. It is intriguing to think of what might happen in the case of the hot atmospheres of some Jupiter-like exoplanets where the temperatures can reach 1500 K or more, similar to that of a flame. Thermionic charging of graphite-like particles if they exist, might then be important.

Another charging mechanism for exoplanets near hot stars would be particle charging by X-ray absorption which can lead to very intense electron emission [J. B. A. Mitchell *et al.*, *Astron. Astrophys.*, 2002, **386**, 743].

Professor Plane asked: Although aerosols (ice particles and meteoric smoke particles) are generally negatively charged—at least in the earth's upper mesosphere—rocket flights occasionally measure layers of positively-charged aerosols. One explanation for this is that the particles attach metallic atoms which lower their work function. These metallic atoms are produced by the ablation of meteoroids in the planetary aerobraking region. We showed that this is the case for Na atoms attached to amorphous ice, where the work function dropped to about 2.3 eV and the photo-ionization cross section increased enormously [T. Vondrak, J. M. C. Plane and S. R. Meech, Photoemission from sodium on ice: A mechanism for positive and negative charge coexistence in the mesosphere, *J. Phys. Chem. B*, 2006, **110**, 3860–3863].

Professor Schram answered: Please refer to my remarks above.

Dr Sittler asked: I must clarify my earlier comment so I don't quote Borucki and Whitten (2008) incorrectly. Nanoparticles tend to have low ionization potentials <7 eV and even with Titan's thick atmosphere and aerosols at higher altitudes, which tend to extinct the solar UV that produce photoemission from the nanoparticles, they still found photoemissions too high and thus electron conductivities too high. Therefore, they speculated that the aerosols were composed of a tholin core with a liquid methane–ethane outer layer. This was based on the Huygens ACP measurements reported by Israel *et al.*[1] which are considered somewhat contentious. With this model, methane and ethane in their gas phase for altitudes >150 km would absorb the UV that would otherwise produce photoemission from the nanoparticles or aerosols. In the case of Mars, where there is no extinction of the UV from higher altitudes, the nanoparticles should have significant photoemission rates and as the modeling of Borucki and Whitten[2] for Titan show, the nanoparticles should produce significant ionization layers within Mars atmosphere as reported here by Whalley *et al.* in this volume.

1 G. Israel, *et al.*, Complex organic matter in Titan's atmospheric aerosols from *in situ* pyrolysis and analysis, *Nature*, 2005, **438**, 796–799.
2 W. J. Borucki, and R. C. Whitten, Influence of high abundances of aerosols on the electrical conductivity of the Titan atmosphere, *Planet. Space Sci.*, 2008, 56, 19–26.

Professor Plane opened the second general discussion of the papers, stating: Recombination reactions involving ions often occur in the "fall-off" region where the kinetics are changing from third- to second-order (becoming independent of pressure). Extrapolation of the rate coefficient over a range of pressures then requires knowledge of the high-pressure limiting rate coefficient. Often this is set equal to the ion–neutral capture rate, which can be estimated using Langevin theory (with a modification if the neutral reactant has a significant dipole moment). However, recent work has shown that ion–molecule recombination reactions sometimes have high pressure-limiting rate coefficients which can be more than an order of magnitude smaller than the capture rate. Explanations for this include the effect of an anisotropic potential surface, and collision-induced dissociation. We recently carried out a study of Mg^+ recombining with H_2O, where this was the case [E. Martinez-Nunez, C. L. Whalley, D. Shalashilin and J. M. C. Plane, Dynamics of Mg^+ + H_2O + He: capture, collisional stabilization and collision-induced dissociation, *J. Phys. Chem. A*, 2010, **114**, 6472–6479].

Dr Pernot noted: Dissociative recombination (DR) is a key process in the ionospheric chemistry of Titan. Experimental measurements of branching ratios of DR have shown a formidable variety of fragmentation patterns, often with bond breaking between heavy atoms [Florescu-Mitchell and Mitchell, *Phys. Rep.*, 2006, **430**, 277–374]. It is therefore striking that most modern ionospheric chemistry models implement a very impoverished representation of DR, mostly through a single H-loss channel. The introduction of three channels for the DR of $CH_2NH_2^+$ by Yelle *et al.* in Paper 2 is a remarkable exception. As many important ions in Titan's ionosphere are protonated neutrals, the reduction of DR to "H-loss" can be expected to bias significantly the production of stable/heavy neutrals over the production of reactive/energetic radicals. We (S. Plessis, N. Carrasco and P. Pernot) recently submitted a paper entitled "Knowledge-based probabilistic representations of branching ratios in chemical networks: the case of dissociative recombinations" [*J. Chem. Phys.*, 2010, **133**, 134110] addressing specifically this issue (see also poster P05). Our analysis is that the systematic implementation of the H-loss scenario is due to the incapacity of modelers to implement available data in their models. The main problem is that the measured DR branching ratios often correspond to incompletely determined products: for $C_xH_yN_z^+$ ions of high relevance to Titan, the number of H atoms within each fragment is not experimentally resolved, leaving the chemical identity of the products unspecified. For instance, the DR of acrylonitrile, CH_2CHCNH^+, leads to about 50% of C_3NH_x + yH + wH$_2$, and 50% of C_2H_x + CNH_y + zH + wH$_2$, with the constraint $x + y + z + 2w = 4$ [Vigren *et al.*, *Astrophys. J.*, 2009, **695**, 317–324]. Considering all the exoergic products, one is left with 15 channels (6 in the first group and 7 in the second), if one ignores the possible HCN/HNC isomers. The probability distribution within each group is totally unknown, which prevents a deterministic modeling of this complex process. As a solution, we suggest a probabilistic representation of uncertain data. We use nested Dirichlet distributions to represent the probabilistic tree imposed by the structured uncertainty in the branching ratios data. This enables us to implement for the first time all available DR branching ratios data within a ionospheric chemistry model. We compared the production rates of neutral species and radicals obtained through our representation to the production rates of a "H-loss"-based scenario representative of existing models. We observe a spectacular enhancement of the production rates for reactive radicals, such as C_3H and NH_2 (a factor of 10 000), NH and C_3N (a factor of 1000), *etc.* At the opposite, the production rate of stable neutrals such as C_3H_4, HC_3N or CH_3CN is reduced by a factor of about 10.

This study confirms that the present ionospheric models based on the H-loss paradigm unduly boost the production of stable neutrals at the expense of reactive radicals, depriving the radical chemistry of a rich chemodiversity. It appears also that DR can be a significant source of radical chemistry during the night-time. Implementation of our full RD scheme in a coupled ion–neutral ionospheric model is in progress.

Professor Mitchell enquired: The products of recombination are released with large energies and this can go into either kinetic energy or internal energy (ro-vibrational or even sometimes electronic excitation). Is thought ever given to the effect of this excitation on the subsequent reactivity of these radicals?

Dr Pernot answered: I have not made an exhaustive bibliographic search, but I know of two papers related to this kind of issue: "Photochemical sources of non-thermal neutrals for the exosphere of Titan", by T. E. Cravens *et al.* [*Planet. Space Sci.*, 1997, **45**, 889–896] and the more recent "Titan's corona: The contribution of exothermic chemistry" by V. de la Haye *et al.* [Icarus, 2007, **191**, 236–250]. Both papers treat of the escape of molecules and of suprathermal neutrals, but do not consider the effect of internal energy on reactivity. In both papers, DR is treated with a single dissociation channel for each ions, generally the most exothermic one, with all excess energy released into kinetic energy. In one of our papers "Sensitivity of a Titan ionospheric model to the ion–molecule reaction parameters" [N. Carrasco *et al.*, *Planet. Space Sci.*, 2008, 56, 1644–1657], we have considered the effect of the different reactivity of excited $N^+(^1D)$ on the ionospheric composition, and observed only minor effects. We are clearly in need of more comprehensive studies, which will not be facilitated by the lack of accurate data.

Dr Adams continued: In some of the recombination product distributions which you showed, the products were not fully separated and your statistical analysis seems reasonable. When the product distribution is fully resolved, I assume that a statistical analysis is not required?

Dr Pernot replied: I would rather say that in *most* cases of interest for Titan's atmosphere the products are not fully separated and require a statistical treatment. In those few cases where all products are resolved, the statistical modeling is indeed much simpler, but we still account for measurement uncertainty (which is not always negligible) and use the Dirichlet representation to ensure a unit sum of all branching ratios.

Professor Coates remarked: We would be interested to find out more about the processes at work in Titan's upper atmosphere which result in an interesting chemical mix between neutrals, heavy positive ions and even heavier negative ions, as seen in our observations. It would be good to learn more about the possible processes from people working in lab experiments and theory (see paper 13 in this volume by Coates *et al.*).

In addition, we use microchannel plates in the Cassini instruments, and it would be useful to understand the microchannel plate efficiency for very heavy negative ions (up to 13 800 amu q^{-1}) in particular better than we do at present. Our electron spectrometer was designed for electrons and has a post-acceleration of only ~+200 V. Experience from people working with high mass ions in the lab would be appreciated. This would help us make better density estimations.

Professor Plane said: Earlier in the discussion I referred to the formation of metal silicate molecules in the aerobraking regions of solar system bodies, and the fact that these molecules have unusually high electron affinities and should therefore form stable negative ions. Meteoric ablation on Titan should be a comparatively gentle affair, because of the very large atmospheric scale height—incoming meteoroids will not hit a "brick wall" of air, as they do with the terrestrial planets. Nevertheless, Fe, Mg and Si should be able to form metal silicates. Is it possible that any of the negative ions measured by the CAPS instrument are anions such as $MgSiO_3^-$?

Professor Coates replied: As I understand it this may be an effect lower in the Titan atmosphere than where we see the particles, but we will have a look at it.

Dr Moses asked: Regarding the production of negative ions on Titan, I think that the plan for the end of the extended Cassini mission is to pass the spacecraft through Saturn's upper atmosphere a time or two before plunging the spacecraft into Saturn to end the mission. It will be interesting to see whether high-molecular-weight negative ions are present in Saturn's upper atmosphere as they are on Titan; their presence or absence on Saturn should provide clues to the formation mechanism(s).

Professor Coates answered: Agreed—we had assumed that this was related to the Titan mix of atmospheric neutrals, but we will certainly look at that time.

Professor Mitchell noted: A good source of information on the efficiency of microchannel plates for detecting heavy ions is the following: G. W. Fraser, The ion detection efficiency of microchannel plates, *Int. J. Mass Spectrom.*, 2002, **215**, 13. Fig. 19 in this paper gives efficiency curves for heavy molecular ions as a function of impact energy on the detector. (Note that the reference to the review of Frank from which this Figure is derived should have the page number 375, and not 275 as in the text.)

Professor Coates answered: Thanks, we will look at this again and at information on proteins mentioned at the meeting.

Titan and habitable planets around M-dwarfs

Jonathan I. Lunine[ab]

Received 30th March 2010, Accepted 12th April 2010
DOI: 10.1039/c004788k

The Cassini–Huygens mission discovered an active "hydrologic cycle" on Saturn's giant moon Titan, in which methane takes the place of water. Shrouded by a dense nitrogen–methane atmosphere, Titan's surface is blanketed in the equatorial regions by dunes composed of solid organics, sculpted by wind and fluvial erosion, and dotted at the poles with lakes and seas of liquid methane and ethane. The underlying crust is almost certainly water ice, possibly in the form of gas hydrates (clathrate hydrates) dominated by methane as the included species. The processes that work the surface of Titan resemble in their overall balance no other moon in the solar system; instead, they are most like that of the Earth. The presence of methane in place of water, however, means that in any particular planetary system, a body like Titan will always be outside the orbit of an Earth-type planet. Around M-dwarfs, planets with a Titan-like climate will sit at 1 AU – a far more stable environment than the $\sim$0.1 AU where Earth-like planets sit. However, an observable Titan-like exoplanet might have to be much larger than Titan itself to be observable, increasing the ratio of heat contributed to the surface atmosphere system from internal (geologic) processes *versus* photons from the parent star.

1. Introduction

Titan is Saturn's largest moon and the only natural satellite with an atmosphere dense enough to create a true greenhouse effect in the same way as does our atmosphere.[1] It is the second-largest moon in the solar system, being slightly smaller than Ganymede, and is remarkably similar in mass and density to Callisto (which is slightly smaller) and Ganymede. All three bodies are roughly 50–60% rock with the bulk of the remainder being water ice. This suggests that some process truncates accretion of rock-ice bodies at this size, which corresponds to a gravitational potential energy per unit mass comparable to the latent heat of vaporization of water ice.[2] In other words, for larger bodies, the water simply evaporates away. I will return to this point later.

While hints of Titan's atmosphere came in observations over a century ago,[3] to be confirmed by the discovery of methane in the 1940s,[4] ground-based observations are frustrated by a thick layer of haze which shrouds Titan's surface and lower atmosphere from view at optical wavelengths (Fig. 1). It was spacecraft exploration that revealed the complex and intriguing nature of Titan's surface and atmosphere environment, first from the Voyager 1 close flyby in 1980, then Hubble Space Telescope images of the surface from Earth orbit,[5] and subsequently the ongoing Cassini–Huygens mission which entered Saturn orbit in 2004.[6] However, the advent of sensitive near-infrared detectors on ground-based telescopes equipped with adaptive optics capability have permitted monitoring of Titan's atmosphere and surface,[7]

---

[a]*Dipartimento di Fisica, University of Rome "Tor Vergata", Rome, Italy 00133*
[b]*Istituto di Fisica dello Spazio Interplanetario, Istituto Nazionale di Astrofisica, Rome, Italy 00133. E-mail: jlunine@roma2.infn.it*

 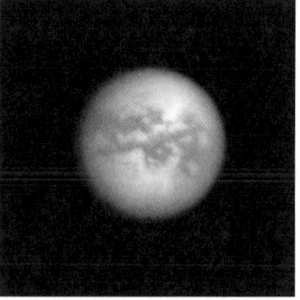

Fig. 1 (Left) Haze layers, hundreds of kilometres above the surface, are revealed in this Cassini image taken with the Sun behind Titan. The haze layers strongly obscure the surface at wavelengths below a micron, as shown in the Cassini image on the right at 0.94 microns wavelength. NASA/JPL/Space Science Institute image.

revealed transient cloud formation missed by the brief visits of flyby spacecraft (in which the Cassini Orbiter must be included, as its orbit about Saturn permits only flybys of Titan).

Titan's atmosphere is cold: 94 K at the equatorial surface, 70 K at the lowest altitude temperature minimum ("tropopause" in terrestrial parlance) at 44 km, increasing to 190 K at the top of the stratosphere. This cold is, of course, the result of the Saturn system's great distance from the Sun. The atmosphere is mostly nitrogen, with 5% mixing ratio (by number) of methane near the surface, which decreases to about 1.5% at the tropopause because of saturation in the lowermost atmosphere. This is consistent with the presence of clouds made of mostly methane in the troposphere down to eight km, though visible clouds are relatively scarce, and transient, compared with the Earth. In the stratosphere and above, sources of energy such as ultraviolet sunlight and charged particles from the solar wind and magnetic environment ("magnetosphere") around Saturn break methane into hydrogen and reactive fragments. Much of the hydrogen escapes thanks to Titan's low gravity (about 1/6 that of the Earth), and hence the radicals combine to form higher hydrocarbons and (with the participation of nitrogen) nitriles.[8] All of these are less volatile than methane itself, and hence form aerosols that precipitate to the surface. The atmosphere is thus a rich repository of organic species of complexity and type varying with altitude,[9] and it is this assemblage which is responsible for the layers of haze that obscure the surface. The dense atmosphere (1.4 bar pressure at the surface) screens out all but a limited flux of cosmic rays from the surface itself. This situation is unique in the solar system: active organic chemistry at high altitudes leading to deposition of condensed products on a surface largely protected from particle and UV radiation.

While Titan's atmosphere itself is noteworthy, it is the interaction of this atmosphere with the surface and possibly the interior that is striking. Cassini has discovered, over six years of Saturn system exploration at the time of this article, an active hydrologic cycle on Titan in which clouds form and dissipate, rain feeds broad valleys and narrow streams, and liquid resides in polar lakes and basins.[10] The working fluids are methane and its photochemical products from the stratosphere, ethane and propane. The ongoing cycling of methane and its organic products through vapor, liquid and solid phases, driven by solar energy, makes Titan only the second world in the solar system with such processes. It raises the issue of whether such a cycle, which is likely key to the long-term habitability of the Earth, might make possible a form of life or at least self-organizing chemistry in liquid methane on Titan.[11] The environment Cassini–Huygens has revealed on Titan some 9.5 astronomical units (AU) from the Sun is equivalent to that which a similar body would possess at about 1 AU from an M-dwarf, the smallest and least luminous of the stable hydrogen-burning stars.[12] However, it would be extremely difficult to observe a Titan-sized (10% the surface area of the Earth) or Titan-mass (1% that of the Earth) body around even a nearby M-dwarf. Consideration of larger planets in the same orbital position reveals some interesting physical and chemical differences, as I discuss below. The pace of exoplanet research is such that the opportunity to see such bodies will come soon, and so it is useful to think of Titan as a kind of avatar for bodies in what should be a stable and typical orbital configuration around the cosmos' most common kind of star.

The point of this paper is not to provide a general description of Titan, beyond what has been offered above. Recent comprehensive reviews are available.[6,13] Instead, I take as my point of departure the evidence for an active methane/ethane/propane cycle, consider the implications for surface organic chemistry, and then pursue the idea that the environment we see on Titan might in some respects be a very common one on planets around other stars.

2. Titan's methane cycle: evidence for active cycling

The so-called "methane cycle" on Titan is illustrated in Fig. 2. Although it must always be remembered that other hydrocarbons, in particular ethane and propane, are also involved, methane is the parent molecule and dominates on seasonal timescales. (One year on Titan – the Saturn orbital period around the Sun – is 29.5 Earth years). It differs from that of the Earth in three important ways: (a) more than one working fluid is involved in the phase changes; (b) one of these fluids (methane) is only weakly trapped in the troposphere by condensation, and hence is lost in the

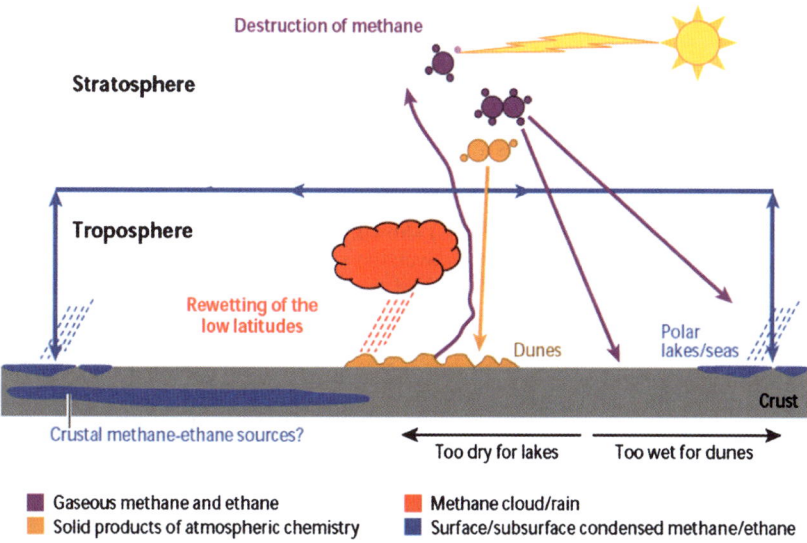

Fig. 2 The methane cycle on Titan, adapted from ref. 13. Molecules that are liquid on the surface (methane and ethane) are shown in the upper atmosphere in purple; they end up in the polar seas and lakes. Solids such as acetylene (yellow) form the particles in the equatorial dune fields. Methane is photolyzed by ultraviolet sunlight in the stratosphere and is converted to higher hydrocarbons and nitriles; in the troposphere it cycles seasonally. Some products of methane cycle from pole to pole on longer timescales associated with orbital variations. Erosional features seen at the equator suggest rainstorms despite the present dry state of the equatorial atmosphere. The long-term source of methane Titan's interior; methane may be stored in the crust as clathrate hydrate; ethane may be lost to the crust or deeper interior in the form of clathrate as well.

Fig. 3 Mosaic of strips of radar data, with a resolution between 300–1000 m, of the north polar region. The sea to the left of center, Ligeia Mare, is about 500 km in diameter. Above 60° north latitude, about 14% of the mapped area is covered by what are interpreted to be lakes or seas. NASA/JPL/USGS.

upper atmosphere by chemistry and direct escape; and (c) on Titan the main surface–atmosphere reservoir of the working fluid is the atmosphere, not a massive surface ocean as on the Earth. The evidence for methane and ethane being on the surface in exposed liquid basins at the poles and in the surface soil at the equator come directly from mass spectrometer measurements of vaporizing methane from underneath the landed Huygens probe,[14] from near-infrared Cassini Orbiter ("VIMS") spectra of one of the polar lakes (Ontario Lacus) showing ethane absorption features,[15] and changes in the size of Ontario Lacus over the summer season consistent with the presence of methane.[16] Propane is inferred from its presence in the stratosphere,[17] chemical models of lake solubility,[18] and a tentative identification in VIMS multi-band imagery around Ontario Lacus.[19] In addition, the emission at radio wavelengths of the seas and lakes *versus* surrounding terrains argues that they contain organic molecules,[20] and the morphology particularly of the larger seas is consistent with flooding of terrains that themselves have been eroded by liquid (Fig. 3).

Fig. 4 illustrates the different behaviors of these three constituents, along with three solid organic products of Titan chemistry tentatively identified to be present around Ontario Lacus. At plausible lake temperatures, derived from models[21] to be between 89–91 K, only methane has a sufficient vapor pressure to move a significant fraction of the liquid in the lakes annually from the summer to winter pole; ethane is marginal. The other components (excepting HCN) can be moved over a 50 000 Earth year cycle, or a 1700 Titan year cycle, equivalent to the Earth's Croll–Milankovitch cycles.[22] This cycle occurs because of the significant ellipticity of Saturn's orbit around the Sun and the perturbations of Jupiter on that orbit. Evidence that the longer term cycle might be operating on Titan comes from the dichotomy between the north and south polar regions; the former has much more surface area covered by lakes and seas than does the latter. Two analogs to Titan's multi-component volatile cycle are the CO_2–H_2O dual volatile cycles on Mars and

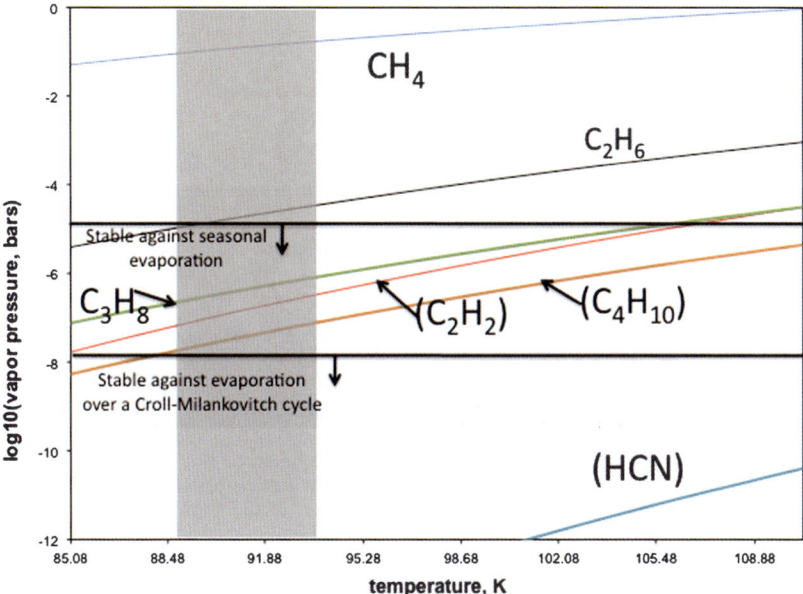

Fig. 4 Vapor pressure in bars *versus* temperature for methane and organic products of methane photolysis possibly present around the southern lake Ontario Lacus.[19] The shaded region shows plausible temperatures in the lake.[21] The two lines represent the limits for stability of the molecules against evaporative transport over a Titan year (upper) and the longer-term cycle associated with orbit perturbations.[22] The methane vapor pressure is so large that its transport is limited by the available energy flux from the Sun.[16]

the Earth. However, the former is essentially all solid–vapor exchange, and in the case of the latter, CO_2 by itself does not enter the condensed phase, but instead appears as a solute. In these respects, the operation of a methane–ethane, or methane–ethane–propane, liquid–vapor cycle on Titan, is unique, and presents a wealth of interesting problems coupling the physical chemistry with atmospheric and surface mass and heat transport.

Notwithstanding the impressive appearance of the northern polar region, lakes cover only a small fraction (2–3%) of the quarter of Titan's surface area imaged by the Cassini radar system.[23] The lack of an ocean means that the dominant known reservoir of methane is the atmosphere. The latent and sensible heat balances are quite different for the Earth and Titan, therefore. In place of the terrestrial ocean is an extensive equatorial belt of dune fields, which may stretch over as much as 20% of the surface of Titan.[24] Both radio emission observed by the Cassini Orbiter RADAR instrument[20] and spectra by VIMS[25] indicate that the principal material of the sand-sized dune particles is solid organics, presumably produced from the methane chemistry high in the atmosphere. Indeed, the dunes may represent the single largest surface reservoir of organic molecules,[26] which is somewhat puzzling in view of the prediction that ethane should be produced in the upper atmosphere in larger abundance than all the solid organic products combined.[27] It is not the size of the dune fields themselves, but rather the absence of even larger amounts of ethane exposed at Titan's surface that is the puzzle.

Assuming the dunes to have a thickness of the order of tens to hundreds of meters, they represent an amount of solid material roughly appropriate to continuous conversion of methane to higher hydrocarbons and nitriles over the age of the solar system (Fig. 5). However, the atmospheric inventory of methane – the largest reservoir known on the surface and in the atmosphere – is sufficient to sustain such

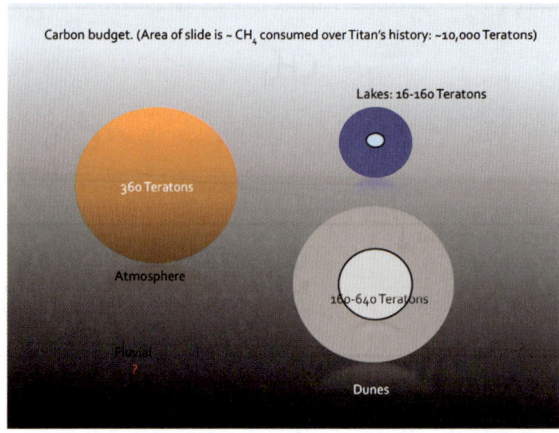

Fig. 5 Relative sizes of the different reservoirs on Titan of methane and its atmospheric chemical products are shown, proportional to the areas of the circles. Numbers are from ref. 26. Inner and outer circles represent lower and upper ranges of the estimates. The rectangular background represents, to roughly the same scale, the amount of methane converted to higher hydrocarbons assuming steady-state photochemistry over the age of the solar system.

chemistry for at most 200 million years, or 5% of the age of the solar system.[28] This may be an overestimate of the lifetime if direct escape of methane is competitive, as seems to be the case.[29,30] Titan's atmosphere cannot hold more than several times this amount of methane before it would be fully saturated and the methane rain out (this includes provision for a higher surface temperature associated with radiative forcing by the enhanced methane abundance). Indeed, the ethane abundance in the known lakes and seas – assuming the methane–ethane ratio in the seas to be that calculated by Cordier *et al.*[18] and mean depth estimated from the lack of radar return from much of the largest seas[31] – is coarsely consistent with what would be produced were photochemistry to have proceeded for only a few hundred million years. In other words, were one to postulate that Titan's atmosphere today is the result of a major outgassing event on the order of a few hundred million years ago, we would obtain the surface ethane abundance mapped out with the Cassini Orbiter.

Aside from the philosophical objection that such an hypothesis would make the current environment of Titan extremely unusual when measured against the whole history of the solar system, the abundance of solid *versus* liquid organics at the surface[26] is then grossly out of balance with what the photochemical models predict.[27] And it is more plausible to imagine that liquids could be sequestered out of the view of the surface as opposed to solids. Fractures and void spaces could accommodate liquids seeping from the surface[32] in what is likely to be a mostly water-ice crust[33] overlying a liquid water–ammonia layer some 60 km below the surface.[34] Ethane might also be sequestered during impacts, which earlier in Titan's history might have repeatedly punctured a thinner crust, releasing liquid water–ammonia to the surface and allowing the ethane to become trapped as clathrate hydrate.[35] This clathrate hydrate phase would have sunk to the bottom of the liquid water–ammonia layer. The crust itself could have been largely composed of methane clathrate hydrate, buoyant relative to the water–ammonia liquid mantle, up until 1/2–1 billion years ago when some thermal models predict thickening and formation of a pure water ice crust.[36] Methane might be periodically released from the clathrate in the crust by thermal plumes associated with convection, or by catastrophic events such as impacts.

The surface hydrologic connection from the equator to the polar lakes and seas is fluvial system poorly delineated because of the limited spatial resolution on Titan's

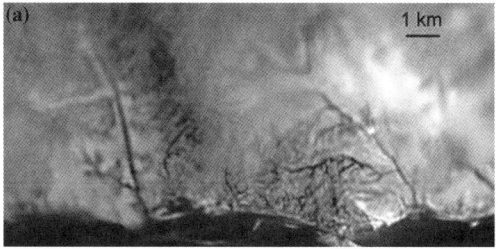

Fig. 6 (a) Two types of drainage features imaged by the Huygens probe as it floated toward its eventual landing site many kilometres to the south. The features on the right (east) are dendritic in nature, and probably caused by repeated episodes of methane rainfall.[42] ESA/NASA/JPL/ Univ. of Arizona image. (b) Larger scale fluvial features: valleys hundreds of meters wide cut into terrain that may be decreasing in elevation toward the southern pole. This Cassini RADAR image from ref. 43 covers several tens of thousands of square kilometres of area at high southern latitude.

surface available from the Cassini Orbiter, typically no better than 300–500 m. Only in the region of the Huygens probe landing site, where images were taken below the haze, could small-scale dendritic features like those ubiquitous on the Earth be detected. Indeed branching dendritic networks cutting steep sided walls were seen (Fig. 6), and these drain liquid, almost certainly methane, down to a flat plain.[37] The origin of the liquid, given the absence of an obvious source at the top of the hill (though fractures and what appear to be little ponds appear nearby), is heavy rainfall. The current atmospheric environment in equatorial latitudes like that of the Huygens site, coupled with the low solar flux, makes development of convective rainstorms difficult there, in contrast to the more nearly saturated polar region[38] where rainfall has been observed.[39] However, an outburst of low altitude cloud at near-equatorial latitude observed in 2008[7] indicates that mechanisms exist – perhaps even methane geysering – to generate convective clouds at all latitudes.

Titan's large scale topography[40] relative to a sphere suggests that liquids generally drain along fluvial and possibly subcrustal pathways from the equator to the pole; however most valleys imaged by Cassini seem follow local topography and a global network is not evident. Modeling of the channels suggest that centimetre or even larger-sized grains – the size of the dune particles – can be moved.[41] Because these particles include potentially reactive solids like acetylene, movement by fluvial transport may also permit exothermic chemical reactions converting (for example) acetylene to benzene to contribute to the surface energy budget. If hydrogen were available, conversion back to methane might occur. The presence of what is evidently an active hydrologic cycle, with rain, fluvial transport, dune formation, and changes in lake levels seasonally, in which the working materials are organic molecules, raises the question of how far toward a form of life might organic chemistry proceed within this environment.

3. Could Titan's methane cycle support life?

We know of only one form of life: that which we see on the Earth. All life on Earth is based on carbon as the key element, and requires liquid water to be viable. (In what follows I will restrict the discussion to life based on carbon because it is so abundant in Titan's surface–atmosphere system). The networks of organic reactions that sustain life are those appropriate to the temperature range for liquid water, narrowed by the constraint that proteins be stable. What is not understood is how critical are the properties of liquid water as the medium for life based on carbon. One can make a list of the "essential" properties of liquid water for life and they would seem to exclude other possibilities, but one cannot test the hypothesis that liquid *water* is essential without going to other environments in which the primary liquid is something other than water, and the other general requirements for life exist. These other general requirements can be expressed as (i) an environment not in or close to thermodynamic equilibrium, (ii) abundant carbon-containing atoms and heteroatoms, and (iii) a fluid environment.[44] Further development of the notion that a kind of life might exist in the surface methane/ethane/propane liquids on Titan includes observation that the absence of hydrogen bonding in the liquid hydrocarbons allows hydrogen bonding among the organic molecules of the biochemical system, generally precluded in liquid water.[11] However, both solubilities and temperatures are low in the Titan environment, slowing reactions, although the possibility that low-temperature reaction rates are much higher than given by extrapolation along an Arrhenius curve is well known.[45]

Titan has a number of other intriguing advantages for exotic biology at the surface. Atmospheric photochemistry and charged particle chemistry has pre-loaded the surface deposits with a variety of different molecules and polymers. Acetylene, which should be abundant as a product of methane chemistry, has the potential to undergo exothermic reactions at the surface – essentially serving to transport the energy of ultraviolet photons and energetic particles to a surface which is largely shielded from these energy sources. The lakes appear to shrink and expand on at least two timescales – seasonal and Croll–Milankovitch – serving to provide the equivalent in liquid methane of hydration and dehydration reactions. Convective rainstorms and possible geysering or other crustal thermal activity provides additional sources of mechanical and thermal energy. Finally, it appears that Titan may not be completely differentiated into a water ice mantle and a rocky core;[33] depending on whether the mixture of rock and ice is at the small-grain level or involves large boulders, there could be minerals present to catalyze reactions between organic molecules (there will always be some rocky debris near the surface anyway from meteoritic infall over time).

Curiously, acetylene is largely absent from regions where solid organic deposits exist on the surface; in contrast, benzene seems to be overabundant.[46] This indicates that further reactions are occurring on the surface, and chemical energy available from the acetylene is being released into the environment. It does not, however, imply biological activity (Fig. 7). A possible metabolic cycle on Titan would be conversion at the surface of acetylene with hydrogen back to methane,

$$C_2H_2 + 3H_2 \rightarrow 2CH_4$$

followed by transport of the methane into the upper atmosphere (convectively in the lower troposphere, diffusively thereafter), where it would either escape directly or be converted through the energy input of ultraviolet and particle radiation into higher hydrocarbons including acetylene.[47,48] The higher hydrocarbons would then condense and fall to the surface. The conversion of acetylene and hydrogen back to methane provides 334 kJ mol^{-1}, which is eight times the energy per mole required to power growth of terrestrial methanogens.[49] Analysis of the Cassini data on the hydrogen profile through the atmosphere indeed suggests a significant sink at the surface.[50]

While appealing as a conceptual metabolic cycle, this process does not provide a real mechanism for recycling methane and slowing the net loss over time because it requires a source of hydrogen, which is mostly lost to space during photolysis.[8] Hydrogen could be coming from the interior, perhaps associated with reactions between liquid water, organics, and rock trapped in the lower mantle, but is not much different from the hypothesis that methane itself is being outgassed.

Detection of a kind of biology, or (and this may be a question of semantics) a kind of self-organizing, auto-catalytic chemistry not accessible to laboratory spatial and

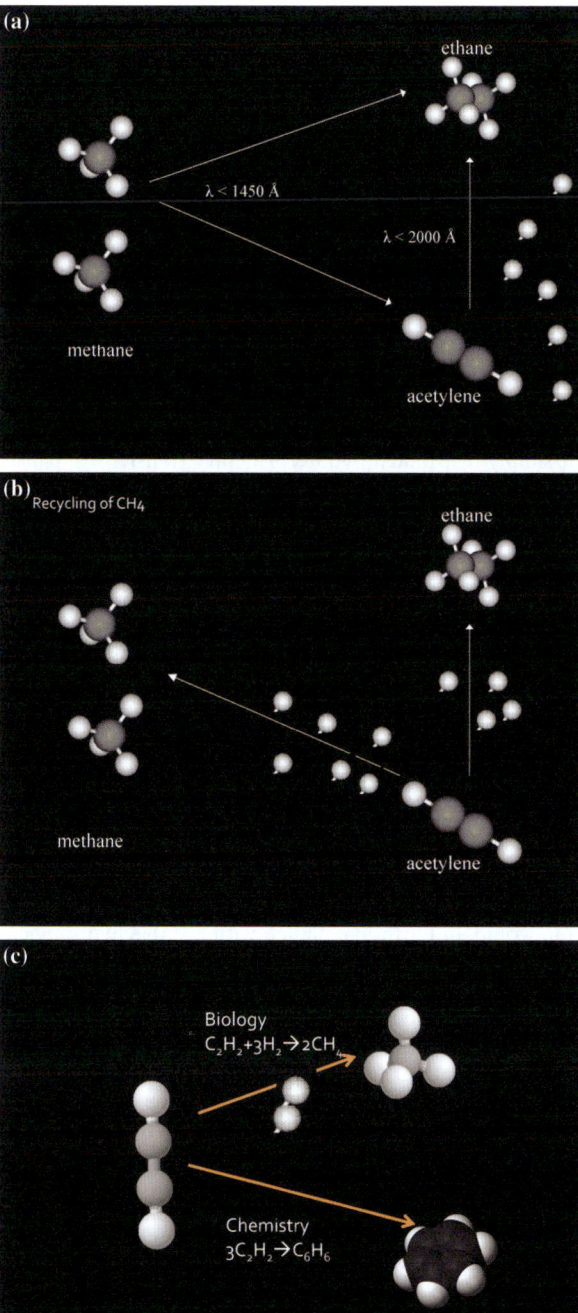

Fig. 7 (a). Photochemistry in Titan's upper atmosphere is the primary loss mechanism for methane, primarily generating acetylene and (directly or indirectly through acetylene) ethane, with loss of hydrogen to space. (b) Surface reactions may convert acetylene to ethane or, more likely, methane, if hydrogen were present. (c) The absence of acetylene and overabundance of benzene in the solid regions of Titan's surface suggest that abiotic pathways dominate; however the presence of chemical or biological cycles utilizing hydrogen to back convert acetylene to methane cannot be ruled out.

temporal scales would require landing in one of the large seas and doing chemical analyses to detect deviations in abundances from those expected from the atmosphere. The approach would require splashing something like the Huygens Probe, with advanced instrumentation, into the sea (ensuring flotation is not a major issue; the Huygens probe itself would have floated for a time had it landed in liquid). Computational or even laboratory work to elucidate putative metabolic pathways for chemical systems operating in liquid hydrocarbons beyond the conceptual cycles proposed so far would allow specific tests to be quantified. The accessibility of Titan's surface to landed vehicles, the low gravity and high atmospheric density, the abundance of organics and surface liquids, all make Titan an attractive target for future exploration and the search for life.[51] A bonus to searching for a kind of biology that works in liquid hydrocarbons is that it would be so different from terrestrial life – down to the monomers used by the biopolymers – that terrestrial contamination from the spacecraft or impact debris spalled off the Earth or Mars in the past could be immediately ruled out.[12] Therefore, Titan presents a very strict test of the hypothesis that life has formed elsewhere in our solar system than the Earth, and not simply distributed by interplanetary transfer or "panspermia".

4. Titan as an avatar for methane-cycle planets around M-dwarfs

Regardless of whether Titan hosts a form of exotic life in its methane/ethane/propane seas (or elsewhere on its surface), I argue that it may represent a more common type of world with a stable volatile cycle than does the Earth. Expressed another way, planets with stable methane cycles may be more common than planets with stable water hydrologic cycles. The argument rests on the relative abundances of main sequence stars – those undergoing hydrogen fusion leading to a stable production of photons and gradual evolution over time – as a function of mass, or equivalently, spectral "type". The abundance of main sequence stars goes roughly as the inverse square of the stellar mass, for masses near that of the Sun and below.[52] Smaller stars are less luminous and longer-lived because the fusion reactions in their interiors go more slowly. The smallest, coolest, longest-lived stars – the M-dwarfs – outnumber stars like the Sun by more than an order of magnitude. They are the most abundant type of star in the cosmos.

What position would an Earth-like planet, with a stable hydrologic cycle, occupy around an M-dwarf? This spectral class has a broad range of masses (from 0.1–0.4 times the mass of the Sun). For the typical M-dwarf – in the middle of its spectral class – the luminosity would be 100 times smaller than that of our Sun. To receive the same number of photons (leaving aside the fast that the distribution of photons is redder for an M-dwarf than for the Sun), we would require moving the Earth inward by a factor of 10 relative to its orbit around the Sun, because the photons received go as the square of the distance. Thus, a planet the size of the Earth around an M-dwarf, with a so-called effective temperature (defined by energy per area per time received and then thermalized) that of the Earth, would be at 0.1 AU. This is four times closer to the star than Mercury is to the Sun.

Much has been written about the stability of the environment of an Earth-type planet at 0.1 AU from an M-dwarf.[53] In a broad brush sense, such a planet would have a climate nothing like the Earth's, because tidal locking would force one face toward the parent star at all times. Freeze out of the atmosphere on the night side could be prevented by vigorous heat transport, but the resulting climate is one for which we can draw no parallel within our own solar system. Flares and ejection of high energy material from the stellar corona (coronal mass ejection) would impact the planet's atmosphere and surface more severely at 0.1 AU than at 1 AU. And, indeed, even though M-dwarfs are less luminous than the Sun, they tend to have more flare activity. A planet with a sufficiently large magnetic field might fend off some of the effects, but an Earth-type planet with a weak field could lose its atmosphere altogether. Arguments have been made that the delivery of water to such planets during formation might be more difficult because of their position deep in the gravity well of the star.[54]

Planets with persistent liquid water on their surfaces and hence a water hydrologic cycle may well exist around M dwarfs – indeed, the spectral class is so broad that the zone over which such planets would occur ranges from 0.02–0.2 AU.[55] However, the extent of the differences of the external and tidal conditions between such planets and the Earth is so great that our home planet's environment may be a very poor guide to their characteristics. Put another way, the most common type of planetary hydrologic cycle involving water is not that of the Earth. The most common type may or may not be habitable over lengths of time akin to that of our own biosphere, but we simply cannot assume this based on the Earth.

The situation with Titan is different: for a typical M-dwarf a planet with Titan's effective temperature would reside at around 1 AU, where tidal locking, coronal mass ejections, flares, and inefficiency in volatile delivery during formation are not of concern. Around the smallest M-dwarfs, this distance would shrink to 0.2 AU, but even were we to disregard these, the number of remaining M dwarfs vastly outnumber G dwarfs like the Sun. Because the 1 AU environment around M dwarfs is benign in the same sense as is that of our Sun, planets at that distance from an M-dwarf should have stable methane hydrologic cycles for which our own Titan is a good guide. And if Titan's hydrocarbon lakes and seas host an exotic form of life, such life may be the most abundant kind of life in the cosmos if Earth-like planets at 0.1 AU from M dwarfs turn out to have environments too unstable to be habitable.

Having made this claim, I must now proceed to address the myriad complications that are present. One is that our solar system's Titan is so small that it would not be observable with planned direct detection systems around even the nearest M-dwarf. And so observable methane-cycle planets would have to be larger. There is a problem with this, which I address in the next section. A second issue has to do with the origin of methane itself. On Titan, methane may have come from the original building blocks from which Titan formed, or been made deep in Titan's interior by reaction of carbon dioxide and water in the presence of rock, in a process called serpentinization.[56] Isotopic measurement of Titan and other bodies in the Saturn system suggest that the methane is primordial.[57] In the disk around the Sun, methane may have been grossly underabundant compared to more oxidized forms of carbon, so that the formation of Titan around a giant planet – with higher gas densities – may have been key to the abundance of methane.[58,59] More recent models suggest less of a distinction.[57] Nonetheless, a stand-alone planet forming within the disk around a parent star rather than around a giant planet might have a different, perhaps smaller, inventory of methane, so that production by serpentinization would be required in order to have a long-term methane cycle. In fact, more massive versions of Titan would possess just such a characteristic – the ability to process higher quantities of oxidized carbon into methane – in their interior.

5. Observability of Titan-class planets around M dwarfs

If Titan's methane-rich volatile cycle is the most common avatar for stable volatile cycles – and even habitable worlds – in the cosmos, what chance do we have of observing them? Some techniques of planet detection – for example the transit of a planet across its parent star – would require just about the same sensitivity to detect a Titan around an M-dwarf as an Earth around a star like the Sun. Spectra can even be constructed from transits.[60] However, for planets at 1 AU from their parent star, transits are rare, and the more likely approach is direct detection. For this approach it is the actual amount of light reflected by a planet that matters, as well as the ratio of the brightness of the planet to the star. The latter is comparable in the case of Titan and an M dwarf, but the former (the absolute brightness) is not. I assume conservatively in the exercise that follows that to obtain spectra of a Titan at 1 AU from an M dwarf with currently planned direct detection systems, one would need to increase the surface brightness of Titan to make it detectable. Increasing the size of our putative methane-cycle world has two effects: it raises the gravity and the heat flow from the interior. The former shrinks the scale of the atmosphere and shrinks the escape rate of both methane and the major atmospheric gas N_2; this acts to decrease the methane escape rate. (The UV flux received by a planet 1 AU from an M-dwarf is significantly less than that received by the Earth from the Sun, so this also slows the loss of methane from the hydrologic cycle).

With respect to the heat flow from Titan's interior, assuming an Earth-sized planet at 1 AU from an M-dwarf, the ratio of heat generated in Titan's interior relative to that thermalized by sunlight and available for convection goes from 0.007 to about 0.05 (Earth's heat flow today renormalized for Titan's rock-to-ice ratio of 60% by mass). If we demand instead that a detectable methane-cycle planet around an M dwarf have the same absolute brightness as that of the Earth around the Sun (ignoring the spectral differences), the planet would need a surface area roughly three times that of the Earth, or a mass five times that of the Earth, implying a ratio of internal heat *versus* thermalized sunlight between 0.2–0.3. All of these numbers are much larger than the ratio of the geothermal heat contribution to the Earth relative to the average received solar flux, of order 0.0001 today and several times larger in the Archean. The environment of our putative Earth-sized, or super-Earth-sized, methane-cycle planet might be vastly different from that of Titan thanks to the substantial heat flow from the crust. Indeed, a water ice crust under the most extreme of these cases might not be stable.

A more fundamental problem may exist with planets in which we imagine a Titan composition but a mass that of the Earth. As mentioned at the beginning of this piece, Titan, Ganymede and Callisto represent the largest ice-rock bodies in the solar system, and are all about the size at which the gravitational potential energy released during formation became comparable to the latent heat of vaporization of water ice.[2] A deep water vapor atmosphere would have formed, some water being lost to space and some raining back down on the growing planet which by this point would have had a deep liquid water ocean. The final product would likely have been a planet dominated by rock, with much smaller amounts of water than Ganymede, Callisto and Titan, though probably much more than the Earth. How much methane, or even carbon dioxide, could have been retained under such a scenario should be calculated, but the final volatile composition would have been much different – probably much less – than on Titan. There could still be a methane "hydrologic cycle", but one in which much less total methane is present and – thanks to the higher gravity and lower incident UV flux – much less is photolyzed. Ethane and propane might be relatively rare, and a single-volatile (methane) hydrologic cycle be dominant like on the Earth.

The lesson here is that we should not expect anything like the planetary environments of our own solar system, including Titan, to be type examples of what would be encountered around the most common stars in the cosmos. Indeed, one need only slightly twist Hamlet's words[61] and say that there are surely more kinds of Earths in the heavens than are dreamt of in our philosophy. Even the complexity of Titan's surface-atmosphere volatile cycle assures us so.

Acknowledgements

This work was financed within the scope of the Italian program "Incentivazione alla mobilità di studiosi straineri e italiani residenti all'estero", and by the NASA Astrobiology Institute.

References

1 C. P. McKay, J. B. Pollack and R. Courtin, *Science*, 1991, **253**, 1118.
2 D. J. Stevenson, A. W. Harris and J. I. Lunine, in *Satellites*, ed. J. A. Burns and M. S. Matthews, University of Arizona Press, Tucson, 1986, p. 39.
3 J. Comas Solà, *Astron. Nachr.*, 1908, **179**, 289.
4 G. P. Kuiper, *Astrophys. J.*, 1944, **100**, 378.
5 P. H. Smith, M. T. Lemmon, R. D. Lorenz, L. A. Sromovsky, J. J. Caldwell and M. D. Allison, *Icarus*, 1996, **119**, 336.
6 R. H. Brown, J.-P. Lebreton and J. H. Waite, *Titan from Cassini–Huygens*, Springer, Dordrecht, 2009, 535pp.
7 E. L. Schaller, H. G. Roe, T. Schneider and M. E. Brown, *Nature*, 2009, **460**, 873.
8 D. F. Strobel, *Icarus*, 1974, **21**, 466.
9 V. Vuitton, R. V. Yelle and M. J. McEwan, *Icarus*, 2007, **191**, 722.
10 J. I. Lunine and R. D. Lorenz, *Annu. Rev. Earth Planet. Sci.*, 2009, **37**, 299.
11 S. A. Benner, A. Ricardo and M. A. Carrigan, *Curr. Opin. Chem. Biol.*, 2004, **8**, 672.
12 J. I. Lunine, *Proc. Am. Philos. Soc.*, 2009, **153**, 403.
13 J. I. Lunine and S. K. Atreya, *Nat. Geosci.*, 2008, **1**, 159.
14 H. B. Niemann, *et al.*, *Nature*, 2005, **438**, 779.
15 R. H. Brown, L. A. Soderblom, J. M. Soderblom, R. N. Clark, R. Jaumann, J. W. Barns, C. Sotin, B. Buratti, K. H. Baines and P. D. Nicholson, *Nature*, 2008, **454**, 607.
16 J. I. Lunine, A. Hayes, O. Aharonson, G. Mitri, R. Lorenz, E. Stofan, S. Wall, C. Elachi and Cassini Radar Team, *Bull. Am. Astron Soc.*, 2009, **41**, 09-RC-164-AAS-DPS abstract.
17 A. Coustenis, *et al.*, *Icarus*, 2007, **189**, 35.
18 D. Cordier, O. Mousis, J. I. Lunine, P. Lavvas and V. Vuitton, *Astrophys. J.*, 2009, **707**, L128.
19 M. Moriconi, J. I. Lunine, A. Adriani, E. D'Aversa, A. Negrao, G. Filacchione and A. Coradini, *Icarus*, 2010, in press.
20 M. A. Janssen, *et al.*, *Icarus*, 2009, **200**, 222.
21 T. Tokano, *Astrobiology*, 2009, **9**, 147.
22 O. Aharonson, A. G. Hayes, J. I. Lunine, R. D. Lorenz, M. D. Allison and C. Elachi, *Nat. Geosci.*, 2009, **2**, 851.
23 A. Hayes, *et al.*, *Geophys. Res. Lett.*, 2008, **35**, L09204, DOI: 10.1029/2008GL033409.
24 J. Radebaugh, *et al.*, *Icarus*, 2008, **194**, 690.
25 J. W. Barnes, *et al.*, *Icarus*, 2008, **195**, 400.
26 R. D. Lorenz, *et al.*, *Geophys. Res. Lett.*, 2008, **35**, L02206, DOI: 10.1029/2007GL032118.
27 P. P. Lavvas, A. Coustenis and I. M. Vardavas, *Icarus*, 2008, **56**, 67.
28 K. E. Mandt, J. H. Waite, Jr., W. Lewis, B. Magee, J. Bell, J. Lunine, O. Mousis and D. Cordier, *Icarus*, 2009, **57**, 1917.
29 P. Paillou, K. Mitchell, S. Wall, G. Ruffiè, C. Wood, R. D. Lorenz, E. Stofan, J. Lunine, R. Lopes and P. Encrenaz, *Geophys. Res. Lett.*, 2008, **35**, L05202.
30 R. V. Yelle, J. Cui and I. C. F. Mueller-Wordag, *J. Geophys. Res.*, 2008, **113**, E10003.
31 D. F. Strobel, *Icarus*, 2008, **193**, 588.
32 D. J. Stevenson, in *Proc. Symp. Titan*, ESA, Paris, 1992, p. 29.
33 L. Iess, N. J. Rappaport, R. A. Jacobson, P. Racioppa, D. J. Stevenson, P. Tortora, J. W. Armstrong and S. W. Asmar, *Science*, 2010, **327**, 1367.
34 C. Bèghin, P. Canu, E. Karkoschka, C. Sotin, C. Bertucci, W. S. Kurth, J. J. Berthelier, R. Grard, M. Hamelin, K. Schwingenschuh and F. Simoes, *Planet. Space Sci.*, 2009, **57**, 1872.
35 J. Lunine, N. Artemieva and G. Tobie, *Lunar Planet. Sci.*, 2010, **41**, 1537.
36 G. Tobie, J. I. Lunine and C. Sotin, *Nature*, 2006, **440**, 61.
37 L. Soderblom, *et al.*, *Planet. Space Sci.*, 2007, **55**, 2015.
38 R. Hueso and A. Sanchez-Lavega, *Nature*, 2006, **442**, 428.
39 E. P. Turtle, J. E. Perry, A. S. McEwen, A. D. DelGenio, J. Barbara, R. A. West, D. D. Dawson and C. C. Porco, *J. Geophys. Res.*, 2009, **36**, L02204.
40 H. A. Zebker, B. Stiles, S. Hensley, R. Lorenz, R. L. Kirk and J. Lunine, *Science*, 2009, **324**, 921.
41 R. Jaumann, *et al.*, *Icarus*, 2008, **197**, 526.
42 J. T. Perron, M. P. Lamb, C. D. Koven, I. Y. Fung, E. Yager and M. Adamkovics, *J. Geophys. Res.*, 2006, **111**, E11001.
43 J. I. Lunine, *et al.*, *Icarus*, 2008, **195**, 415.
44 Committee on the Limits of Organic Life in Planetary Systems, Committee on the Origins and Evolution of Life, National Research Council, *The Limits of Organic Life in Planetary Systems*, 2007, National Academy Press, Washington, DC, 116 pp.
45 D. B. Atkinson, V. I. Jaramillo and M. A. Smith, *J. Phys. Chem. A*, 1997, **101**, 3356.

46 R. Clark, *et al.*, *J. Geophys. Res.*, 2010, DOI: 10.1029/2009JE003369.
47 O. Abbas and D. Schulze-Makuch, *Int. J. Astrobiol.*, 2002, **1**, 233.
48 D. Schulze-Makuch and D. H. Grinspoon, *Astrobiology*, 2005, **5**, 560.
49 C. P. McKay and H. D. Smith, *Icarus*, 2005, **178**, 274.
50 D. F. Strobel, *Icarus*, 2010, DOI: 10.1016/j.icarus.2010.03.003.
51 R. Shapiro and D. Schulze-Makuch, *Astrobiology*, 2009, **9**, 1.
52 A. N. Cox, ed., *Allen's Astrophysical Quantities*, 1999, AIP Press, New York.
53 H. Lammer, *Astrobiology*, 2007, **7**, 27.
54 J. J. Lissauer, *Astrophys. J.*, 2007, **660**, L149.
55 J. Tarter, *et al.*, *Astrobiology*, 2007, **7**, 30.
56 S. K. Atreya, E. Y. Adams, H. B. Niemann, J. E. Demick-Montelara, T. C. Owen, M. Fulchignoni, F. Ferri and E. H. Wilson, *Planet. Space Sci.*, 2006, **54**, 1177.
57 O. Mousis, J. I. Lunine, M. Pasek, D. Cordier, J. H. Waite, Jr., K. E. Mandt, W. S. Lewis and M.-J. Nguyen, *Icarus*, 2009, **204**, 749.
58 J. S. Lewis and R. G. Prinn, *Astrophys. J.*, 1980, **238**, 357.
59 R. G. Prinn and M. B. Fegley, Jr., *Astrophys. J.*, 1981, **249**, 308.
60 D. Deming and S. Seager, *Nature*, 2009, **462**, 301.
61 W. Shakespeare, Hamlet, Act 1 Scene 5.

PAPER

The fate of aerosols on the surface of Titan

S. I. Ramírez,[*a] P. Coll,[b] A. Buch,[c] C. Brassé,[b] O. Poch[b] and F. Raulin[b]

Received 9th March 2010, Accepted 21st April 2010
DOI: 10.1039/c003925j

A laboratory study based on the chemical transformation that Titan's aerosol analogues suffer when placed under putative surface conditions of the satellite was performed. In order to understand the role that aqueous ammonia may play on the chemical transformation of atmospheric aerosols once they reach the surface, we synthesized laboratory analogues of Titan's aerosols from an $N_2 : CH_4$ (98 : 2) mixture irradiated at low temperatures under a continuous flow regime by a cold plasma discharge of 180 W. The analogues were recovered, partitioned in several 10.0 mg samples and placed inside different ammonia concentrations during 10 weeks at temperatures as low as those reported for Titan's surface. After a derivatization process performed to the aerosols' refractory phase with MTBSTFA in DMF, the products were identified and quantified using a GC-MS system. We found derived residues related to amino acids as well as urea. The simplest amino acids aminoethanoic acid (glycine) and 2-aminopropanoic acid (alanine) as well as diaminomethanal (urea), are found regardless of the ammonia concentration and temperature value to which the aerosol analogues were exposed. Our results have important astrobiological implications to Titan's environment particularly if the existence of the suggested subsurface water–ammonia mixture and its deposition on the satellite's surface is validated.

Introduction

With the many data collected by the Cassini–Huygens mission, Titan the largest satellite of Saturn appears as a complex organics factory at planetary scale.[1–3] The chemistry starts in the high atmospheric layers of the satellite, with the formation of organic molecules and ions in the ionosphere[4,5] derived from the coupled chemistry of the two main constituents of Titan's atmosphere: molecular nitrogen (N_2) and methane (CH_4), induced by solar ultraviolet radiation and high energy electrons coming from Saturn's magnetosphere. The Cassini Plasma Spectrometer (CAPS) detected negative ions of up to 10 000 Dalton[5–7] meaning high molecular weights for these organics. They may contribute to form the haze layers in the high atmosphere. The particles from these layers diffuse to lower atmospheric levels and induce the condensation of lower molecular weight compounds, forming larger aerosols, which sediment down to the surface and accumulate. These aerosols thus very likely include macromolecular organic material, which represent the most complex organic products of Titan's atmospheric chemistry. With the help of the Gas Chromatograph and Mass Spectrometer (GC-MS) experiment,[8] and the Aerosol Collector

[a]Centro de Investigaciones Químicas, Universidad Autónoma del Estado de Morelos, México. E-mail: ramirez_sandra@uaem.mx; Fax: (52) 777 329 7998; Tel: (55) 777 329 7997
[b]Laboratoire Interuniversitaire des Systémes Atmosphériques, CNRS-IPSL, Universités Paris Est-Créteil et Denis-Diderot, France
[c]École Centrale de Paris, Chatenay-Malabry, France

and Pyrolyzer (ACP) experiment both carried by the Huygens probe,[9] some of the aerosols from the low stratosphere and mid troposphere of Titan were collected and analyzed by pyrolysis-GC-MS techniques. The results strongly suggest that the particles are made of refractory organics, containing carbon (C), nitrogen (N) and hydrogen (H) atoms and releasing, when pyrolyzed at 600 °C, hydrogen cyanide (HCN) and ammonia (NH_3).[9] However highly debated,[10] these results strongly suggest that the complex organic materials constituting the aerosol particles are similar to some of the Titan's tholins which have been produced by several laboratories, when simulating experimentally the chemical evolution of Titan's atmosphere, and which can be considered good analogues of Titan's atmospheric particles.

Many laboratory works have been carried out on analogues of Titan's aerosols to study their physical and chemical properties.[11] A variety of experimental conditions have been used. They include different radiation sources (UV photons or plasma produced from electrical discharges), gas composition, temperature, and pressure. And the characteristics of the analogues aerosol material, when obtained, vary widely as they are very related the experimental conditions. Overall, these laboratory experiments have produced two main classes of Titan tholins.[12] Tholins produced from UV photolysis, which have a low nitrogen content and include organic polymers, aliphatic unsaturated chains and low or no aromatic-hydrocarbon content. While, tholins produced by electrical-discharge plasma show a higher nitrogen content, they are made of organic oligomers, and include aromatic structures with N-heterocycles and aliphatic chains. Moreover, the plasma-produced tholins can release large amounts of HCN and NH_3 when pyrolyzed at 600 °C, contrary to the photochemical ones. They thus appear as better analogues of Titan's aerosols that can be used in the laboratory to study in more detail some of the properties of Titan's atmospheric particles.

The first images of Titan from the Cassini–Huygens mission revealed a highly diversified solid surface with features that suggest aeolian, tectonic, fluvial processes, and even an impact crater structure.[13] Since then, more detailed descriptions of dunes, channels, lakes, other smaller impact craters and cryovolcanic structures have been documented.[14] Moreover, the existence of an internal liquid water ocean, containing a few percent ammonia has been proposed.[14,15] This model of Titan's interior has recently been supported by Cassini observations of Titan rotation,[16] and even more strongly supported by Schumann resonance observations that suggest that the top of the subsurface ocean could be located at about 40 km below the icy surface.[17] It has also been proposed that ammonia–water mixtures can erupt from the putative subsurface ocean leading to cryovolcanism.[18] The Cassini Titan Radar Mapper obtained Synthetic Aperture Radar (SAR) images during 2004 and 2005 observations that revealed a highly complex geology occurring at Titan's surface, among which cryovolcanic features play a central role.[19] The composition of the cryomagma is mainly proposed to be a mixture of water ice and ammonia.[20] Recently, some changes were observed in the spectral behaviour of the surface, which seems to be consistent with some kind of exchanges of ammonia frost over a water ice substrate.[21] This might possibly be explained by episodic effusive events, which bring juvenile ammonia from the interior to the surface.

Titan's aerosols, once on the surface, may chemically evolve in spite of the low surface temperature (94–92 K) if they are in contact with water or water–ammonia mixtures. Big impacts on Titan's surface may episodically melt the ice crust and form liquid water oases that could stay in the liquid state for up to several thousands of years.[22] This could favour a chemical evolution of the organic components of atmospheric aerosols that settled down to the surface. Otherwise, these organics may also evolve at much slower chemical rates in the absence of violent-impact episodes. Could such events foster the hydrolysis of the macromolecular organics of the aerosols into compounds of biological interest, such as amino acids? Several studies have already been performed to bring answers to this question. The acidic hydrolysis of plasma-generated tholins produced a large variety of amino acids and other organics.[23,24] The production of amino acids is still observed, when hydrolysis is performed with liquid water at neutral pH.[24,25] Neish et al.[26] have studied the kinetic of incorporation of oxygen atoms into tholins, which is a necessary step toward the formation of biological molecules, and shown that it is a relatively fast process, even at 273 K. Somogyi et al.[27] exposed Titan tholins produced by an AC plasma to hot liquid water with dissolved ammonia. Their results indicate that tholins are reactive enough with the water/ammonia mixture to generate an oxidized solid material. Similar results were obtained recently by Neish et al.,[28] using water–ammonia mixtures at lower temperatures (down to 253 K), showing the production of complex organic molecules containing both oxygen and altered nitrogen functional groups, but without identification of the molecular species.

In this regard some questions still remain: Could laboratory synthesized tholins and by extrapolation Titan's aerosols, allow the production of biologically interesting compounds such as amino acids in the presence of low temperature water–ammonia mixtures? What could be the molecular nature of the products and their yields? What could be the chemical pathways that form these biologically interesting compounds? This paper tries to answer these questions and to understand the role that aqueous ammonia may play on the chemical transformation of atmospheric aerosols once they reach the surface of Titan. To do so, we synthesized laboratory analogues of Titan's aerosols from a N_2 : CH_4 (98 : 2) gas mixture irradiated in a low-temperature continuous-flow regime by a DC cold plasma discharge. The analogues were recovered, partitioned in several 10.0 mg samples and placed in aqueous ammonia solutions at low temperature for 10 weeks. After a chemical derivatization process performed on the refractory phase of the aerosol analogues, with MTBSTFA in DMF, the alkaline hydrolysis products were identified and quantified with the aid of a gas chromatography system coupled to a quadrupole mass spectrometer. We present and discuss here the results of this experimental study.

Experimental

The laboratory aerosol analogues were synthesized from a high-purity gas mixture of 2% methane (CH_4) and 98% nitrogen (N_2) (Praxair, France) that simulated Titan's atmosphere main composition. The gas mixture circulated inside a circuit formed by the gas-mixture cylinder, the plasma reactor, and a vacuum pump. The plasma reactor is made of Pyrex glass, has a "U" form and has an iron electrode attached to each arm. The electrodes have a glass protection to avoid their direct contact with the gas mixture. The plasma is generated along all the body of the reactor once the breakdown voltage is reached. Prior to the introduction of the simulated atmosphere and its irradiation with the cold plasma, all the circuit is evacuated for several minutes with the aid of the vacuum pump (Alcatel) to eliminate any contamination from terrestrial atmosphere, specially any traces of water vapor. The gas mixture used to simulate Titan's atmosphere was then allowed to continuously flow through the circuit during ~5 min. Then the cold plasma discharge was established between the two iron electrodes using a direct current power supply (Glassman, Inc.) that delivers a voltage of 1.8 kV and a current of 100.0 mA corresponding to a total power of 180 W. At this step, the body of the reactor is carefully placed inside a vessel that contains liquid nitrogen. In this way, the synthesis of the aerosol analogues occurs in a low-temperature regime. A gas flow controller was used to keep a constant flow of 845 ± 1 sccm of the simulated atmosphere inside the circuit. The system was allowed to operate in these conditions during some 8-hour periods. Liquid nitrogen was resupplied during this time to keep the low-temperature regime. After this time, the power supply was turned off, the gas mixture was allowed to flow for approximately two minutes and then was also stopped. The reactor was placed out of the liquid nitrogen vessel and allowed to reach room temperature. The light yellow aerosol analogues attached to the reactor

Table 1 Yields of amino acid derivatives produced by alkaline hydrolysis performed at different temperatures of aerosol analogues synthesized by irradiation of a mixture of CH_4 in N_2 (2 : 98). From left to right: T/K kept during the hydrolysis process, aqueous ammonia content (% NH_4OH), yield (μg) and transformation percentage (T (%)) for each of the identified products[a]

T/K	NH₄OH (%)	Alanine $C_3H_7NO_2$		Glycine $C_2H_5NO_2$		Valine $C_5H_{11}NO$		Leucine $C_6H_{13}NO$		Isoleucine $C_6H_{13}NO$		Proline $C_5H_9NO_2$		Serine $C_3H_7NO_3$		Urea CH_4N_2O	
		μg	T (%)	μg	T (%)	μg	T (%)	μg	T (%)	μg	T (%)	μg	T (%)	μg	T (%)	μg	T (%)
96	25.000	3.8	0.04	2.2	0.02											78.2	0.78
	12.500	16.2	0.16	6.3	0.06	2.1	0.02	3.3	0.03							59.6	0.60
	6.250	15.9	0.16	10.1	0.10	10.8	0.11	1.0	0.01	25.4	0.25	5.7	0.06	26.4	0.26	92.9	0.93
	3.125	57.6	0.58	5.2	0.05							50.3	0.50			49.8	0.50
253	25.000	5.8	0.06	2.2	0.02											62.3	0.62
	12.500 6.250	21.5 28.9	0.22	8.2	0.08	2.5	0.03							22.2	0.22	89.8	0.90
	0.29	6.9	0.07													129.9	1.30
	3.125	19.2	0.19	1.6	0.02	5.0	0.05									103.5	1.04
	25.000	31.7	0.32	41.8	0.42											71.8	0.72
277	12.500	75.6	0.76	69.6	0.70	2.4	0.02			10.1	0.10					9.0	0.09
	6.250	76.9	0.77	58.8	0.59	3.3	0.03									7.4	0.07
	3.125	91.0	0.91	58.3	0.58	1.7	0.02			8.0	0.08					7.2	0.07

[a] Molecular formulas are indicated below the common name. Systematic names recommended by the IUPAC: aminoethanoic acid = glycine; 2-aminopropanoic acid = alanine; 2-amino-3-methylbutanoic acid = valine; 2-amino-4-methylpentanoic acid = leucine; 2-amino-3-methylpentanoic acid = isoleucine; pyrrolidine-2-carboxylic acid = proline; 2-amino-3-hydroxypropanoic acid = serine; diaminomethanal = urea.

walls, were recovered by simple scratching and placed in a small glass container. There was no need of protection of the aerosol analogues from terrestrial atmosphere, since it has been demonstrated by our group, that once formed and kept at room temperature, they do not incorporate atoms from the gas components of terrestrial atmosphere. After approximately 33 h of tholins synthesis, 160.0 mg were recovered. The analogues were fractionated in 10.0 mg samples in order to expose them to basic hydrolysis conditions differentiated by the ammonia content and the temperature value as described in Table 1. As it is hard to estimate the concentration of ammonia on the transient liquid environments on Titan's surface, regardless of their origin, we have examined the chemistry that occurs when the aerosol analogues were placed in solutions whose percentage of ammonia was 25.000%, 12.500%, 6.250% and 3.125%. A sample of 10.0 mg of our aerosol analogues was placed into each of three glass-vials equipped with a screw hermetic cap and was added with an aliquot of 4.0 mL of the 25.000% ammonia solution. Then one vial was placed at 277 K, another at 253 K and the last at 96 K. The same procedure was followed for the three other ammonia solutions. A blank was prepared for each experimental condition to monitor an eventual contamination or artifacts originated by the analytical techniques.

All the samples were kept at the specified conditions during 10 weeks. Then they were prepared for their chemical characterization. Each vial contained a liquid and a refractory phase. The liquid phase was mainly aqueous ammonia while the refractory phase consisted of the aerosol analogues and their alkaline hydrolysis products. The refractory phase was dried up using a flow of nitrogen gas at 40 °C. On average, the time needed to evaporate the liquid phase from the vials was around 15 min. For the samples that were kept at 96 K, the phase change from ice to liquid was in fact very short, less than 30 s were enough. To allow the chemical analysis of the hydrolysis products, a derivatization process was performed using N-methyl-N-(tert-butyldimethylsilyl) trifluoroacetamide (MTBSTFA) in dimethylformamide (DMF). The analysis of refractory compounds by gas chromatography requires their transformation into volatile derivatives stable enough to be separated inside a chromatographic column kept at least at 90 °C. MTBSTFA was used as the chemical

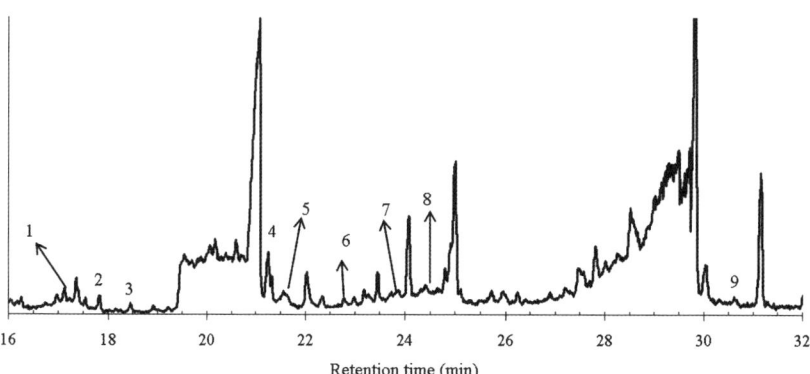

Fig. 1 Gas chromatogram reconstructed from mass spectra information corresponding to the separation of a mixture of amino acid derivatives produced by the alkaline hydrolysis at 96 K of aerosol analogues synthesized by the irradiation of a mixture of CH_4 in N_2 (2 : 98). Identifications: 1 = methyl laureate, 2 = alanine, 3 = glycine, 4 = urea, 5 = valine, 6 = leucine, 7 = isoleucine (not shown here), 8 = proline, 9 = serine (not shown here). Chromatographic conditions: column: 30 m × 0.25 mm × 0.25 μm TRx-1 (Restek); carrier flow (He) 1.0 mL min^{-1}; temperature programming: isothermal at 90 °C for 5 min., 4 °C min-1 up to 270 °C, isothermal at 270 °C for 5 min.; split: 1/20; detector: mass spectrometer using 70 eV EI. The corresponding retention time for peaks 7 and 9 is shown only for indicative purposes as in this specific chromatogram they are not displayed.

derivatization agent because it is sensitive to all compounds with labile hydrogen atoms. Those include primary and secondary amines, for example. Additionally it has the advantages of developing single-step derivatization reactions, it is less sensitive to hydrolysis conditions, and it does not require additional purification steps for the obtained derivatives prior to gas chromatography analysis.[29,30] A gas chromatography system coupled to a quadrupole mass spectrometer (ThermElectron DSQ II) was used for separation and molecular identification. Each separated compound was identified by its mass spectrum and its chromatographic parameters, such as its retention time. A standard solution of amino acids (Fluka) derivatized using the same analytical protocol, was used to calculate the quantity of each identified compound relative to the quantity of aerosol analogue placed under the alkaline hydrolysis conditions. Some amino acids as well as urea were found and their yield and transformation percentages are listed in Table 1. Fig. 1 shows a representative chromatogram for one of the samples.

Results and discussion

Seven amino acids were identified from the simplest ones, aminoethanoic acid or glycine up to the one with a 5-carbon chain, 2-amino-3-methylpentanoic acid or isoleucine. Methyl ramifications were found in some cases. Interestingly, we also found pyrrolidine-2-carboxylic acid, a heterocyclic amino acid, 2-amino-3-hydroxypropanoic acid, an amino acid containing an hydroxyl group, and diaminomethanal or urea. Table 1 includes the common and systematic names as well as the molecular formulas of the identified compounds. The yields were calculated in terms of the 10.0 mg aerosol analogue sample subjected to the hydrolysis and derivatization conditions. The standard amino acid solution was used to determine the retention time and the peak area related to the initial concentration of each amino acid derivative. Assuming a similar chromatographic response, we calculated the concentration of each synthesized compound relative to the peak area of alanine and its concentration in the standard solution. The yield of each of the synthesized compounds was then calculated from its concentration taking into consideration its molecular weight, the injection volume (33 μL), and the sample volume (4.0 mL). The transformation percentage (T (%) in Table 1) expresses the mass fraction of tholin sample (10.0 mg) that was transformed during the alkaline hydrolysis process into a given compound.

When analyzing the number of molecules formed of each identified compound as a function of the aqueous ammonia content, we noticed that almost all the identified amino acids report a positive slope, which means that as the content of ammonia increases, a higher amino acid yield is found, as expected. Surprisingly some yields show a negative slope, as is the case for valine at 96 K and 253 K, proline at 96 K and for alanine at the three tested temperatures. These three amino acids have their highest yield value at the lower ammonia content used for the hydrolysis. Then, for higher ammonia percentages, their formation rate rapidly decreases. This observation may have two explanations: either the higher ammonia percentages inhibit their formation or once these compounds are synthesized, they react with the ammonia solution and are transformed into another products. In either case, the dramatic decrease of their yield is something that requires more attention.

Urea is present for all the hydrolysis conditions, and shows the higher yields at 253 K, and 6.500% aqueous ammonia. When urea yields increase, those of alanine and glycine decrease. This may mean that urea, the smallest of all the synthesized compounds, has an important role in the interactions among the hydrolysis products. It is known that on Earth, urea is one of the main compounds found in metabolism processes such as the oxidation of amino acids or nitrogenated compounds. So our results may mean that urea is one of the by-products of amino acid further oxidation.

We can also see from Table 1, that the highest production yields for amino acids, particularly for the simplest ones, aminoethanoic acid (glycine) and 2-amonipropanoic acid (alanine) are observed at 277 K.

Conclusions and implication for Titan

According to Owen et al.[31] the interpretation of the low levels of primordial noble gases in Titan's atmosphere as a manifestation of a warm environment ($T \geq 75$ K) for the formation of constituent planetesimals leads to a requirement for nitrogen to be delivered as compounds, mainly ammonia (NH_3), while methane (CH_4) needed to be stored within the interior of the satellite. It is reasonable to assume that the icy planetesimals that formed Titan could easily supplied the amount of nitrogen that is now found in the atmosphere.[31] Interestingly, ammonia is a material not previously reported on Titan's surface but generally believed to be present in Titan's interior.[21] Several lines of evidence[15,16,26,28] suggest that Titan's present-day interior contains ammonia mixed with water, in the form of a liquid layer below a rigid water-ice crust that leads to the possibility that outgassing of NH_3 or NH_3-H_2O fluids could bring ammonia to the surface.[15,20,32] As part of the huge amount of information collected by the Cassini–Huygens mission, particularly from reflectance measurements, members of the Cassini's VIMS team[21] reported changes in spectral reflectance and size of a region located at the south-west side of Titan's globe. They propose these observations as strong evidence that ammonia frost or possibly fog is being deposited over a water ice substrate on a dynamic surface that is continually covered by precipitating atmospheric aerosols continually produced by photochemistry and other sources of energy from atmospheric methane. If this were the case, a chemical transformation of the aerosols on the postulated ammonia deposits, is expected. Nelson and colleagues[21] suggest that their observations are also evidence of slowly weathering, evaporation, and chemical processes occurring on Titan's surface with ammonia from Titan's interior as the chemical responsible agent.

Laboratory analogues of Titan's aerosols were exposed to a hydrolysis transformation under alkaline conditions using a wide range of aqueous ammonia concentrations and cold temperatures, conditions expected to be found on Titan's surface. We have found the formation of seven amino acids and urea that were identified and quantified by the coupled techniques of gas chromatography and mass spectrometry. Among these compounds, aminoethanoic acid (glycine), 2-aminopropanoic acid (alanine) and urea were identified in all the hydrolysis conditions. The fact that at 277 K the formation yields for these amino acid were the highest but those for urea were the lowest may mean that urea, being the chemically simplest compound formed, can be considered as the end-product from the more elaborate chemical transformations that lead to the formation of the observed amino acids. As the chemistry of the hydrolysis is affected by lower temperatures, chemical transformations are slower giving lower yields for amino acids regardless of the ammonia concentration, as can be seen in Table 1. It is interesting to notice that even when the hydrolysis rate is diminished at lower temperatures, we have identified the presence of an interesting diversity of amino acids, some of them with methyl ramifications, one including a heterocycle in its chemical structure, and one with a hydroxyl ramification. All of these signatures must have a close relationship with the initial chemical structure of Titan's aerosols.

These results mean that if the mixtures of ammonia and water postulated to exist on Titan's surface or subsurface are possible and detected, an interesting fate for the aerosols gravitationally deposited on that surface is expected. The mass flux of aerosols on Titan has been estimated[33] to be on the order of 2 to 3×10^{-14} g cm^{-2} s^{-1}. It will thus take about 1000 years to get one milligram of aerosols to be deposited over an area of 1 cm^2 on Titan's surface. Assuming a transformation rate into amino acids of 0.1% (from Table 1), the mass of produced amino acids would be of the order of the one microgram per cm^2, a significant quantity, *i.e.* detectable by *in situ* analysis, deposited on Titan's surface at its actual temperature.

A considerable number of laboratory experiments have demonstrated that the aerosols can be considered a complex mixture of hydrocarbons, nitriles, imines and amines molecules, or a mixture of oligomers including the mentioned chemical groups. A consensus about the polymeric, oligomeric or amorphous character of these aerosol analogues is still in debate, but what is true is the presence of these chemically active groups in the laboratory tholins and in Titan's aerosols.[11] So once these chemically active groups are placed in an environment that allows their chemical transformation, such as the one proposed in this study, interesting and biologically attractive products are detected. As our results have demonstrated, the freezing temperatures on Titan can significantly affect the formation rates of these compounds making necessary several years to get a significant quantity of amino acids formed on the surface of Titan, but as has been demonstrated earlier,[25] the conditions for these transformations are possible on Titan's surface.

Acknowledgements

The study was supported by the French Space agency (CNES) and by a CONACyT grant (105450). SIR acknowledges Laboratoire Interuniversitaire des Systémes Atmosphériques for the facilities provided for the development of experimental simulations.

References

1 S. Atreya, *Science*, 2007, **316**, 843–45.
2 F. Raulin, *Space Sci. Rev.*, 2008, **135**, 37–48.
3 F. Raulin, C. P. McKay, J. I. Lunine and T. Owen, in *Titan from Cassini–Huygens*, ed. R. Brown, J.-P Lebreton & H. Waite, Springer, 2009, ch. 9, pp. 215–233.
4 H. Waite Jr, H. Niemann, R. V. Yelle, W. T. Kasprzak, *et al.*, *Science*, 2005, **308**, 982–986.
5 H. Waite Jr, D. T. Young, T. E. Cravens, A. J. Coates, F. J. Crary, B. Magee and J. Westlake, *Science*, 2007, **316**, 870–875.
6 J.-E. Wahlund, M. Galand, I. Mueller-Wodarg, J. Cui, R. V. Yelle, *et al.*, *Planet. Space Sci.*, 2009, **57**, 1857–1865.
7 A. J. Coates, A. Wellbrock, G. R. Lewis, G. H. Jones, D. T. Young, F. J. Crary and J. H. Waite Jr., *Planet. Space Sci.*, 2009, **57**, 1866–1871.
8 H. B. Niemann, S. K. Atreya, S. J. Bauer, K. Biemann, *et al.*, *Space Sci. Rev.*, 2002, **104**, 551–590.
9 G. Israel, M. Cabane, J. F. Brun, H. Niemann, S. Way, W. Riedler, M. Steller, F. Raulin and D. Coscia, *Space Sci. Rev.*, 2002, **104**, 435–466.
10 K. Biemann, *Nature*, 2006, **444**, E6–E7.
11 See for instance: P. Coll, D. Coscia, N. Smith, M.-C. Gazeau, S. I. Ramírez, G. Cernogora, G. Israel and F. Raulin, *Planet. Space Sci.*, 1999, **47**, 1331–1340; S. I. Ramírez, P. Coll, A. da Silva, R. Navarro-González, J. Lafait and F. Raulin, *Icarus*, 2002, **156**, 515–539; H. Imanaka, B. N. Khare, E. L. O. Bakes, C. P. McKay, D. P. Cruikshank, S. Sujita, T. Matsui, R. N. Zare and J. E. Elsila, *Icarus*, 2004, **168**, 344–366; J.-M. Bernard, E. Quirico, O. Brissaud, G. Montagnac, B. Reynard, P. Mcmillan, P. Coll, M.-J. Nguyen, F. Raulin and B. Schmitt, *Icarus*, 2006, **185**, 301–307; M. J. Nguyen, F. Raulin, P. Coll, S. Derenne, C. Szopa, G. Cernogora, G. Israël and J. M. Bernard, *Planet. Space Sci.*, 2007, **55**, 2010–2014, and references therein.
12 T. Cassidy, P. Coll, F. Raulin, R. W. Carlson, R. E. Johnson, M. J. Loeffler, K. H. Hand and R. A. Baragiola, *Space Sci. Rev.*, 2010, DOI: 10.10007/s11214-009-9625-3.
13 C. Porco and the Cassini ISS team, *Nature*, 2005, **434**, 159–168.
14 *Titan from Cassini–Huygens*, ed. R. Brown, J.-P Lebreton and H. Waite, Springer, 2010.
15 A. D. Fortes, *Icarus*, 2000, **146**, 444–452; G. Tobie, O. Grasset, J. Lunine, A. Mocquet and C. Sotin, *Icarus*, 2005, **175**, 496–502.
16 R. D. Lorenz, B. W. Stiles, R. L. Kirk, M. D. Allison, P. P. del Marmo, L. Iess, J. I. Lunine, S. J. Ostro and S. Hensley, *Science*, 2008, **319**, 1649–1651.

17 C. Beghin, P. Canu, E. Karkoschka, C. Sotin, C. Bertucci, W. S. Kurth, J. J. Berthelier, R. Grard, M. Hamelin, K. Schwingenschuh and F. Simoes, *Planet. Space Sci.*, 2009, **57**, 1872–1888.
18 C. Sotin and the Cassini VIMS team, *Nature*, 2005, **435**, 786–789; G. Mitri, A. P. Showman, J. I. Lunine and R. M. C. Lopes, *Icarus*, 2008, **196**, 216–224.
19 R. M. C. Lopes and 43 colleagues, *Icarus*, 2007, **186**, 395–412.
20 J. S. Kargel, S. K. Croft, J. I. Lunine and J. S. Lewis, *Icarus*, 1991, **89**, 93–112; J. S. Kargel, in *Comparative Planetology with an Earth Perspective*, ed M. T. Chahine, M. F. A'Hearn and J. Rahe, Kluwer Academic, Boston, 1995, pp. 101–113; R. D. Lorenz, *Planet. Space Sci.*, 1996, **44**, 1021–1028.
21 R. M. Nelson, L. W. Kamp, D. L. Matson, P. G. J. Irwin, *et al.*, *Icarus*, 2009, **199**, 429–441.
22 D. P. O'Brien, R. D. Lorenz and J. I. Lunine, *Icarus*, 2005, **173**, 243–253; N. Artemevia and J. I. Lunine, *Icarus*, 2003, **164**, 471–480; N. Artemevia and J. I. Lunine, *Icarus*, 2005, **175**, 522–533.
23 B. N. Khare, C. Sagan, H. Ogino, B. Nagy, C. Er, K. H. Schram and E. T. Arakawa, *Icarus*, 1986, **68**, 176–184.
24 M. J. Nguyen PhD thesis, Université Paris 12, December 2007.
25 F. Raulin, M. J. Nguyen and P. Coll, Proc. SPIE 6694, *Instruments, Methods, and Missions for Astrobiology X*, ed. R. B. Hoover, G. V. Levin, A. Y. Rozanov and P. C. W. Davies, 2007, vol. 6694, p. L6940, DOI: 10.1117/12.732883.
26 C. D. Neish, A. Somogyi, H. Imanaka, J. I. Lunine and M. A. Smith, *Astrobiology*, 2008, **8**, 273–287.
27 A. Somogyi, C.-H. Oh, M. A. Smith and J. I. Lunine, *J. Am. Soc. Mass Spectrom.*, 2005, **16**, 850–859.
28 C. D. Neish, A. Somogyi, J. I. Lunine and M. A. Smith, *Icarus*, 2009, **201**, 412–421.
29 A. Buch, R. Sternberg, C. Szopa, C. Freissinet, C. Garnier, C. Rodier, R. Navarro-González, F. Raulin, M. Cabane, D. P. Glavin and P. R. Mahaffy, *Adv. Space Res.*, 2009, **43**, 143–151.
30 A. Buch, C. Marchetti, D. Meunier, R. Sternberg and F. Raulin, *J. Chromatogr., A*, 2003, **999**, 165–174.
31 T. C. Owen, H. Niemann, S. Atreya and M. Y. Zolotov, *Faraday Discuss.*, 2006, **133**, 387–391.
32 C. Sotin and G. Tobie, *C. R. Acad. Sci.*, 2004, **5**, 769–780.
33 M. G. Tomasko and R. A. West.In *Titan from Cassini–Huygens*, ed. R. Brown, J.-P Lebreton & H. Waite, Springer, 2009, ch. 12, pp. 297–321.

Untangling the chemical evolution of Titan's atmosphere and surface–from homogeneous to heterogeneous chemistry

Ralf I. Kaiser,*[a] Pavlo Maksyutenko,[a] Courtney Ennis,[a] Fangtong Zhang,[a] Xibin Gu,[a] Sergey P. Krishtal,[b] Alexander M. Mebel,*[b] Oleg Kostko[c] and Musahid Ahmed*[c]

Received 24th February 2010, Accepted 7th April 2010
DOI: 10.1039/c003599h

In this article, we first explored the chemical dynamics of simple diatomic radicals (dicarbon, methylidyne) utilizing the crossed molecular beams method. This versatile experimental technique can be applied to study reactions relevant to the atmospheres of planets and their moons as long as intense and stable supersonic beam sources of the reactant species exist. By focusing on reactions of dicarbon with hydrogen cyanide, we untangled the contribution of dicarbon in its singlet ground and first excited triplet states. These results were applied to understand and re-analyze the data of crossed beam reactions of the isoelectronic dicarbon plus acetylene reaction. Further, we investigated the interaction of ionizing radiation in form of energetic electrons with organic molecules ethane and propane sequestered on Titan's surface. These experiments presented compelling evidence that even at irradiation exposures equivalent to about 44 years on Titan's surface, aliphatic like organic residues can be produced on Titan's surface with thicknesses up to 1.5 m. Finally, we investigated how Titan's nascent chemical inventory can be altered by an *external* influx of matter as supplied by (micro)meteorites and possibly comets. For this, we simulated the ablation process in Titan's atmosphere, which can lead to ground and electronically excited atoms of, for instance, the principal constituents of silicates like iron, silicon, and magnesium, in laboratory experiments. By ablating silicon species and seeding the ablated species in acetylene carrier gas, which also acts as a reactant, we produced organo silicon species, which were then photoionized utilizing tunable VUV radiation from the Advanced Light Source. In combination with electronic structure calculations, the structures and ionization energies of distinct organo-silicon species were elucidated.

1. Introduction

The arrival of the Cassini–Huygens probe at Saturn's moon Titan – the only Solar System body besides Earth and Venus with a solid surface and a thick atmosphere with a pressure of 1.4 atm at surface level – in 2004 opened up a new chapter in the history of Solar System exploration. The mission revealed Titan as a world with striking Earth-like landscapes involving hydrocarbon lakes[1] and seas as well as sand dunes[2] and lava-like features[3] interspersed with craters and icy mountains of

---

[a]*Department of Chemistry, University of Hawaii at Manoa, Honolulu, HI, 96822, USA*
[b]*Department of Chemistry & Biochemistry, Florida International University, Miami, 33199, USA*
[c]*Chemical Sciences Division, Lawrence Berkeley National Laboratory, Berkeley, California, 94720, USA*

hitherto unknown chemical composition.[4] The discovery of a dynamic atmosphere and active weather system[5] illustrates further the similarities between Titan and Earth.[6] The aero-based haze layers,[7] which give Titan its orange-brownish color, are not only Titan's most prominent optically visible features,[8] but also play a crucial role in determining Titan's thermal structure and chemistry.[9] These smog-like haze layers are thought to be very similar to those that were present in Earth's atmosphere before life developed more than 3.8 billion years ago, absorbing the destructive ultraviolet radiation from the Sun, thus acting as 'prebiotic ozone' to preserve astrobiologically important molecules on Titan. Compared to Earth, Titan's low surface temperature of 94 K and the absence of liquid water preclude the evolution of biological chemistry as we know it. Exactly because of these low temperatures, Titan provides us with a unique prebiotic "atmospheric laboratory" yielding vital clues – at the frozen stage – on the likely chemical composition of the atmosphere of the primitive Earth. However, the underlying chemical processes, which initiate the haze formation from simple molecules, have been not understood well to date.

Titan's chemical inventory is the result of a rich, coupled photochemistry of two main atmospheric constituents: molecular nitrogen (N_2) and methane (CH_4).[10] Stratospheric trace constituents have been firmly identified *via* infrared spectroscopy on board the Voyager I and II spacecrafts[11] as well as in the framework of the Cassini–Huygens mission.[12] These are the hydrocarbon molecules acetylene (C_2H_2), ethylene (C_2H_4), ethane (C_2H_6), methylacetylene (CH_3CCH), propane (C_3H_8), diacetylene (C_4H_2), and benzene (C_6H_6), several nitriles like hydrogen cyanide (HCN), cyanoacetylene (HCCCN), cyanogen (C_2N_2), and dicyanodiacetylene (C_6N_2), as well as oxygen-bearing molecules carbon dioxide (CO_2), carbon monoxide (CO), and water (H_2O). The emission lines of these molecules can be utilized as valuable tracers to collect data on temperature profiles, on the potential existence of cold

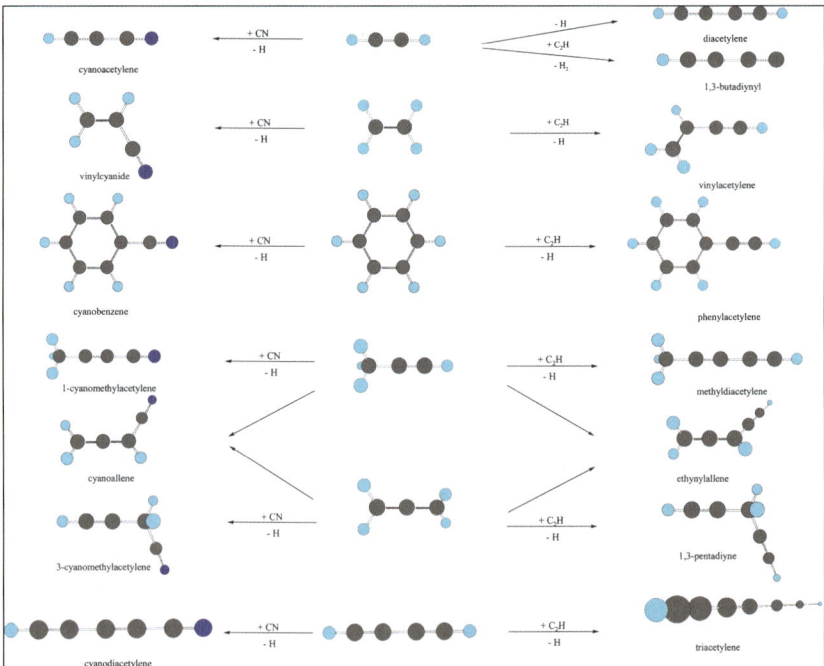

Fig. 1 Products formed in the bimolecular reactions of cyano ($CN(X^2\Sigma^+)$) and ethynyl radicals ($C_2H(X^2\Sigma^+)$) with unsaturated hydrocarbons.[109]

traps and freeze-out zones, and on molecular abundances in the stratosphere; they also help to understand the formation of the organic aerosol layers on Titan. Here, planetary chemists proposed that the formation of the aerosol layers is initiated by fast and barrierless reactions of small, carbon-bearing radicals such as simple diatomics (dicarbon, $C_2(X^1\Sigma_g^+/a^3\Pi_u)$;[13,14] methylidyne radicals, $CH(X^2\Pi_\Omega)$;[15] cyano radicals, CN $(X^2\Sigma^+)$[16]) and triatomic molecules like the ethynyl radical $(C_2H(X^2\Sigma^+))$[17,18] with unsaturated hydrocarbons *via* organic transient species. These considerations have led to the development of photochemical models of Titan[19] and also to extensive laboratory studies during the last decades.[8,20] However, the majority of these experiments were performed under bulk conditions.[21] Several limitations such as wall effects undermine their validity.[22] Also, the reaction products are often analyzed off-line and *ex situ*.[23] Hence, the detailed chemical dynamics of the reaction – the role of radicals and intermediates – cannot always be obtained, and reaction mechanisms can at best be inferred indirectly and qualitatively. Recently, a different experimental approach has been utilized. By conducting a series of detailed experiments on the elementary steps of cyano and ethynyl radical reactions and photodissociation studies of hydrocarbon molecules, a complete picture of the processes involved in the chemical processing of Titan's atmosphere is beginning to emerge. Since the macroscopic alteration of Titan's atmosphere consists of multiple elementary reactions that are a series of bimolecular encounters between radicals and molecules, this detailed understanding of the mechanisms involved at the microscopic level is crucial to unravel the chemical evolution and processing of low temperature environments in general. These are experiments under single collision conditions in which particles of one supersonic beam are made to 'collide' only with particles of a second beam (reactive collisions) or photons (photodissociation). Here, crossed beam experiments of cyano $(CN(X^2\Sigma^+))$[24] and ethynyl radicals $(C_2H(X^2\Sigma^+))$[17] with unsaturated hydrocarbons demonstrated that highly unsaturated nitriles – organic molecules carrying the cyano (CN) group – and hydrogen-deficient molecules, among them (substituted) polyynes up to triacetylene, can be formed (Fig. 1). Low temperature kinetic experiments of cyano and ethynyl radicals amplified the role of these neutral–neutral reactions in Titan's low temperature atmosphere as these studies depicted the barrierless nature of bimolecular encounters of ethynyl and cyano radicals with rate constants of a few 10^{-10} cm^3 s^{-1} being close to the gas kinetics limit.[14,25]

Among the diatomic species, closed shell, ground state dicarbon $C_2(X^1\Sigma_g^+)$ and its open shell first excited triplet state $C_2(a^3\Pi_u)$ counterpart have received considerable attention. In Titan's atmosphere, dicarbon can be formed as a transient species *via* photodissociation of the ethynyl radical, $C_2H(X^2\Sigma^+)$; the latter is the primary photodissociation product of acetylene (C_2H_2) at a wavelength less than 217 nm (5.7 eV).[26,27] Reactions of singlet and triplet dicarbon with unsaturated hydrocarbons such as acetylene and ethylene were shown to be rapid over the temperature range of 24 K to 300 K with reaction rates larger than 10^{-10} cm^3 s^{-1} for singlet dicarbon; reactions of triplet dicarbon were systematically slower than their singlet counterparts.[13,28] Nevertheless, since only the decay kinetics of the dicarbon reactants were followed, information on the reaction products were elusive. A series of crossed molecular beam experiments, in which reaction products can be identified under single collision conditions, unraveled a rich chemistry (Fig. 2) leading not only to hydrogen-terminated, polyyne-like carbon clusters (C_nH ($n = 4, 6$)), but also to resonantly stabilized free radicals (RSFR) of the generic formula C_nH_3 ($n = 4, 5$) – potential building blocks to form aromatic molecules in Titan's atmosphere. Since the electronic ground and first excited triplet states are very close in energy (718 cm^{-1}), they are both present in the crossed molecular beam studies and – due to the absence of any entrance barrier – both react with the hydrocarbon reactant. However, a recent crossed beam study of the reaction of dicarbon with hydrogen cyanide (HCN) provided evidence that at low collision energies, only singlet dicarbon reacted.[29] This system presents an unprecedented opportunity to discriminate

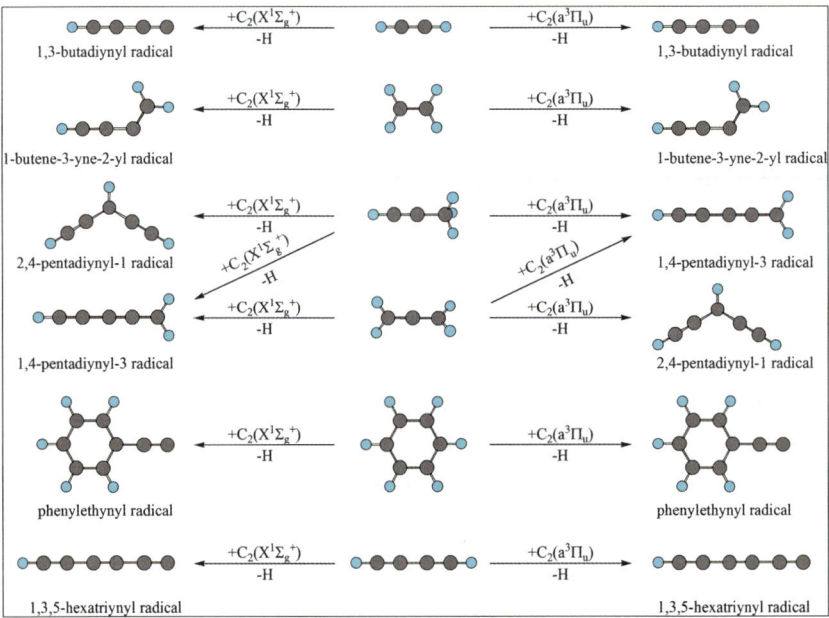

Fig. 2 Products formed in the bimolecular reactions of ground and excited states dicarbon $C_2(X^1\Sigma_g^+/a^3\Pi_u)$ with unsaturated hydrocarbons.[109]

the reaction dynamics of ground state (lower collision energy) and triplet dicarbon molecules, which reacts only at higher collision energies due to a barrier in the entrance channel of at least 29.3 kJ mol^{-1}. Therefore, we conducted the reaction of dicarbon molecules in both their singlet and triplet states with hydrogen cyanide at a collision energy of 42.4 kJ mol^{-1} and compare the findings with our previous study on the singlet surface. Note that our experimental collision energies are higher than the equivalent temperature conditions in Titan's atmosphere. Does this have any effect on the implications of these studies to Titan's chemistry? For this purpose, we have combined our experiments with electronic structure calculations (see below); the combined experimental and theoretical data suggest that only one reaction channel is open – at all temperatures and collision energies above the threshold. However, in case of multiple reaction pathways and product isomers, it is important to account for the collision-energy (and temperature) dependent change in the branching ratios of distinct product isomers formed. These findings are then applied to untangle the chemical dynamics of the isoelectronic dicarbon–acetylene system on the singlet and triplet surface. Recall that previous studies of this reaction not only failed to discriminate between the chemical dynamics on the singlet and triplet surface, but also led to contradictory results when comparing studies with continuous[30] and pulsed dicarbon beams.[31]

We also present preliminary data on the reactions of important methylidyne radicals (CH ($X^2\Pi_\Omega$)). Recall that in Titan's atmosphere, the methylidyne radical is expected to play a role in synthesizing higher hydrocarbon molecules. Since methane (CH$_4$) presents the most abundant hydrocarbon on Titan, the photolysis of methane is considered as the major source of methylidyne. Lacking an unsaturated bond, methane only absorbs light shorter than 145 nm; therefore, the photochemistry of methane mostly occurs in the stratosphere with a significant flux of Lyman-α photons at 121.6 nm (10.2 eV). Early laboratory works by McNesby et al.,[32] Laufer et al.,[33] Gorden et al.,[34] Rebbert et al.,[35] and Slanger et al.[36] suggested quantum yields (denoted ϕ) for methane photolysis at 121.6 nm as defined by eqn (1)–(4). Secondary processes were inferred to produce the methylidyne radical *via* reactions

(5)–(7).[37] In a recent theoretical study, Lodriguito et al.[38] investigated the photodissociation processes of methane theoretically on its lowest singlet potential surface at 122 nm. They found that non-adiabatic dynamics were important, and methyl ($CH_3(X^2A''_2)$) plus atomic hydrogen and the carbene ($CH_2(a^1A_1)$) plus molecular hydrogen were the major dissociation channels. The methyl radical was mostly formed via direct dissociation hopping to the ground state. On the other hand, carbene can either be formed by hopping to the ground state surface or through adiabatic dissociation involving carbene in its b^1B_1 state. They also suggested that the triple dissociation channel $CH + H_2 + H$ was less important; methylidyne is formed via a two step sequential mechanism, where either a molecular or atomic hydrogen elimination was followed by a atomic or molecular hydrogen loss. Later, Zhang et al.[39] reinvestigated this reaction utilizing the high resolution Rydberg tagging time-of-flight (TOF) technique after photo dissociating methane at 130 nm. Their results show an important single C–H bond fission channel from methane. A simulation of the TOF spectra indicated the formation of highly rotationally excited methyl radicals. These products are attributed to the conical intersection pathway between the excited state singlet (S_1) and ground state singlet (S_0) surface of methane.

$$CH_4 + h\nu \rightarrow {}^1CH_2 + H_2 \quad \phi_1 = 0.41 \qquad (1)$$

$$\rightarrow {}^3CH_2 + 2H \quad \phi_2 = 0.51 \qquad (2)$$

$$\rightarrow CH + H + H_2 \quad \phi_3 = 0.08 \qquad (3)$$

$$\rightarrow CH_3 + H \quad \phi_4 = 0.00 \qquad (4)$$

$${}^3CH_2 + h\nu \rightarrow CH + H \qquad (5)$$

$$CH_3 + h\nu \rightarrow CH + H_2 \qquad (6)$$

$${}^3CH_2 + H \rightarrow CH + H_2 \qquad (7)$$

However, due to the difficulty to probe quantitatively the hydrocarbon fragments, the branching ratios are debatable. Lee et al.,[40] for instance, reported excitation spectra of electronically excited carbene $CH_2(a^1B_1)$ showing that this channel is only a minor pathway in the methane photolysis. Utilizing the technique of hydrogen atom photofragment translational spectroscopy, Mordaunt et al.[41] found a simple carbon-hydrogen bond fission in methane to be the dominant primary process at 121.6 nm. The resulting methyl radical (CH_3) fragments are formed with sufficient internal energy that about 25% of them undergo secondary decomposition yielding predominantly methylidyne (CH) and molecular hydrogen (H_2). Brownsword et al.[42] measured the absolute hydrogen atom quantum yield as 0.47 for the methane Lyman-α photolysis in agreement with the conclusions of Mordaunt et al. Later, Mebel et al.'s[43] ab initio calculation and experimental works by Heck *et al.* and Wang *et al.* revealed that at 121.6 nm, the methyl radical channel can be considered as the major pathway.[27,44] Consequently, the methylidyne branching ratio should be higher than 0.08 as proposed originally. Kinetic studies in the range of 23 K to 295 K indicate that subsequent reactions of the methylidyne radical with hydrocarbon molecules are fast (few 10^{-10} cm^3 s^{-1}) and in the case of olefins, hold maximum rate constants at around 70 K.[15] Utilizing tunable vacuum ultraviolet (VUV) photoionization and time-resolved mass spectrometry, Leone and co-workers suggested that at 298 K, methylidyne reacts with ethylene to form 70 ± 8% allene (H_2CCCH_2), 30 ± 8% methylacetylene (CH_3CCH), and less than 10% cyclopropene (c-C_3H_4). Experiments with acetylene indicated the formation of mainly the cyclic C_3H_2 isomer with smaller fractions of triplet propargylene (HCCCH), in contrast to theoretical predictions. However, the authors emphasized that since the experiments were not conducted under single collision conditions, atomic hydrogen-triggered isomerization processes were likely responsible to change the nascent product distribution.[45] Therefore, experiments conducted under single collision conditions as provided in crossed molecular beams, in which the nascent reaction products can be probed, are clearly desired.

Secondly, since the gaseous molecules might also agglomerate to aerosol particles[46] or sequester to Titan's surface,[47] we also present new data on the interaction of potentially abundant organic solids (ethane, propane) with ionizing radiation in the form of energetic electrons, as generated in the track of high energy galactic cosmic ray particles. In a pioneering study, Sagan and Thomas outlined that energetic cosmic ray particles can penetrate deep into the lower atmospheric layers.[48] These energetic particles could incorporate part of their kinetic energy into chemical reaction and thus process simple organics in Titan's lower atmosphere. In a more recent study, Molina-Cubero *et al.* derived an energy deposition on Titan's surface of 4.5 × 10^9 eV cm^{-2} s^{-1}.[49] However, the radiation processing of these simple organics by energetic electrons is not well understood. To shed light on this matter and to gain a comprehensive picture of the hydrocarbon chemistry, not only in the gas phase as described above but also in the condensed phase, we present data on the interaction of ionizing radiation, in form of energetic electrons, with solid ices of ethane and propane – two of Titan's abundant saturated hydrocarbons. We will also investigate to what extent the radiation processing can lead to the formation of polymer-like macromolecules, which could present building blocks of Titan's organic aerosol layers, *via* heterogeneous chemistry.[50]

Finally, it has to be noted that Titan's nascent chemical inventory can be not only enriched, but also altered by an *external* influx of matter as supplied by (micro)meteorites and possibly comets.[51,52] Therefore, these processes must also be understood to gain a complete picture of the chemical inventory. Recent photochemical models suggest that Titan's oxygen-bearing species (carbon monoxide, carbon dioxide, and water) can be simultaneously reproduced using an oxygen flux consistent with the Cassini Plasma Spectrometer observations and a hydroxyl radical (OH) flux consistent with the predicted production from (micro)meteorite ablation.[52] The impact of (micro)meteorites with dense atmospheres also leads to an ablation of the nascent meteoritic material thus releasing ground and excited metal atoms (mostly iron (Fe), magnesium (Mg), and silicon (Si) from abundant silicates) and their ions.[53] As discussed by Petrie,[54] within the hydrocarbon-rich atmosphere of Titan, these species are expected to form unsaturated and polar organo-metallic molecules, which could provide effective nucleation sites for the condensation of polar molecules and highly unsaturated hydrocarbon molecules at high altitudes. This in turn could lead to metal- and silicon-doped tholin-like material. Although extensive work has been conducted on magnesium-bearing molecules,[53] an understanding of the organo-silicon chemistry is still in its infancy; only a single crossed molecular beam study on the reactions of ground state silicon atoms with acetylene has been conducted to date.[55] There have been a few theoretical studies on the ionization energy of Si(C_2H_2), and very early electron impact measurements for a few silicon carbides with large error uncertainties of a few tenths of an electron volt, but beyond this there is a paucity of information regarding organo-silicon compounds.[56] Due to this lack of data on silicon-carbon-bearing molecules, we report here on the *in situ* reaction of ablated silicon species – as simulated by laser ablation of neat silicon – with acetylene acting as a prototype unsaturated hydrocarbon molecule in Titan's atmosphere. These studies provide not only an inventory of potentially abundant neutral organo-silicon molecules in Titan's atmosphere, but also deliver accurate ionization energies of these molecules. The ionization energies are in turn crucial to predict to what extent newly formed organo-silicon molecules can be ionized in Titan's atmosphere by the harsh UV/VUV radiation field from the Sun possibly influencing the charge balance in Titan's upper atmosphere.

At this *Faraday Discussion*, we cover three fields of interest related to Titan's chemical evolution. First, we present the crossed molecular beams approach which can be applied to study bimolecular reactions in Titan's atmosphere. Data on the reactions of excited state dicarbon molecules with hydrogen cyanide (HCN) and of methylidyne radicals with methylacetylene (CH_3CCH) and diacetylene (C_4H_2) are discussed. Hereafter, we move to solid state chemistry and describe to what extent the interaction of ionizing radiation with closed shell hydrocarbons, ethane (C_2H_6) and propane (C_3H_8), present either in aerosol droplets and/or sequestered on Titan's surface can lead to the formation of an organic 'polymer' and hence influence the overall hydrocarbon budget. Finally, we present novel data on the formation and ionization energies of silicon-bearing organic molecules, which are of possible interest to understand the reactions of (micro)meteoritic ablated silicon atoms with acetylene in Titan's upper atmosphere.

2. Experimental approach

2.1. The crossed molecular beams setup

2.1.1. The crossed molecular beams approach.
Which experimental approach can be utilized to expose the chemical dynamics of reactions of diatomic molecules such as ground ($C_2(X^1\Sigma_g^+)$) and excited state dicarbon ($C_2(a^3\Pi_u)$) molecules as well as the methylidyne radical ($CH(X^2\Pi_\Omega)$) in Titan's hydrocarbon-rich atmosphere? Since the macroscopic alteration of atmospheres of planets and their moons involves multiple elementary reactions, that are a series of *bimolecular encounters*, a detailed understanding of the mechanisms involved at the most fundamental, microscopic level by eliminating any wall effects is desirable. These are experiments conducted under single collision conditions, in which particles of one supersonic beam – predominantly an unstable species such as dicarbon or methylidyne – are made to 'collide' only with particles of a second supersonic beam.[57,58] In strong contrast to bulk experiments, where reactants are mixed and where the product distribution might be influenced by wall effects of the reactant vessel, the crossed beam approach has the unique capability of generating the radicals in separate supersonic beams. In principle, both reactant beams can be prepared in well-defined quantum states before they cross at a specified energy under single collision conditions. This provides an unprecedented opportunity to observe the consequences of a single collision event, excluding secondary collisions and most importantly wall effects. In principle, the products can be detected *via* spectroscopic detection schemes such as laser induced fluorescence (LIF)[59] or Rydberg tagging,[60] ion imaging probes,[61,62] or *via* a quadrupole mass spectrometric detector (QMS) with universal electron impact ionization or photoionization. Crossed beam experiments can therefore help to untangle the chemical dynamics, to infer the intermediates, and to identify the nascent reaction products under single collision conditions; neither bulk nor kinetic experiments can supply this information. It should be mentioned that recent kinetics experiments pioneered an isomeric-specific detection of reaction products utilizing time-resolved multiplexed photoionization mass spectrometry *via* synchrotron radiation.[63] Under

those experimental conditions, the reaction intermediates may undergo up to a few thousand collisions with the bath molecules so that three-body encounters cannot be eliminated, and true single collision conditions are not provided. On the other hand, in 'real' atmospheres, stabilizations due to collisions are important if the collision times are shorter than the lifetime of the reaction intermediates, and they can be only probed in collisional environments. Therefore, crossed beams and kinetics studies must be regarded as highly complementary.

Over the past decades, the use of crossed molecular beams has led to an unprecedented advancement in our understanding of fundamental principles underlying chemical reactivity. Detailed experimental studies of simple three-atom reactions established experimental benchmarks such as the reactions of chlorine,[64] fluorine,[65] deuterium,[66] carbon,[67] nitrogen,[68] oxygen,[69] and sulfur atoms[70] with molecular hydrogen. This approach has been extended to tetra atomic systems like OH/CO[71] and OH/H$_2$ together with their isotopic variants,[72] and CN/H$_2$ (D$_2$).[73] These simple systems are prototypical reactions in bridging our theoretical understanding of reactive scattering, *via* dynamics calculations on chemically accurate potential energy surfaces, with experimental observations.[74] These dynamics calculations are needed to turn the *ab initio* results into quantities that can be compared with experiments. Quasi classical trajectory (QCT) calculations are of particular significance to investigate the effect of the collision energy on the chemical dynamics and to derive the experimental observables such as the collision energy dependence of the reaction cross section, the translational energy and angular distributions, and the differential cross section as a function of the center-of-mass angle and product center of mass velocity. Although interest in these light elementary reactions still continues, with the development of powerful theoretical models, attention has turned during the last years to more complex systems of significant practical interest such as in combustion processes,[62,75] catalysis,[76] atmospheric chemistry,[77] interstellar chemistry,[78] organo metallic chemistry,[79] and planetary chemistry as presented at the present *Faraday Discussion*.[80,81]

2.1.2. The crossed molecular beams machine. In case of reactions of dicarbon and methylidyne, the crossed molecular beam approach with universal mass-spectrometric detection presents the most versatile technique to study elementary reactions with reaction products of *unknown* spectroscopic properties. This helps to elucidate the chemical dynamics and – in the case of polyatomic reactions – the primary reaction products of bimolecular reactive encounters.[82] The crossed beams machine consists of two source chambers at a crossing angle of 90°, a stainless steel scattering chamber, and an ultra-high-vacuum tight, rotatable, differentially pumped quadrupole mass spectrometric (QMS) detector which can be pumped down to a vacuum in the high 10^{-13} torr range.[14] Multiple techniques can be applied to generate highly reactive beams. In the primary source, a *pulsed* beam of unstable (open shell) species are generated either by laser ablation (C, C$_2$, C$_3$),[83] laser ablation coupled with *in situ* reaction (CN, C$_2$D),[81] photolysis (C$_2$H, C$_2$H$_3$, C$_3$H$_3$),[84] or flash pyrolysis (C$_3$H$_5$, C$_6$H$_5$).[85] The pulsed primary beam is passed through a skimmer into the main chamber; a chopper wheel located after the skimmer and prior to the collision center selects a part of the pulse with well-defined velocity which reaches the interaction region. This section of the beam then intersects a *pulsed* reactant beam released by a second pulsed valve under well-defined collision energies. The incorporation of *pulsed beams* allows reactions with often expensive (partially) deuterated and/or highly toxic chemicals to be carried out. Also, pulsed sources with high beam densities allow the pumping speed and hence cost to be reduced drastically.

To analyze the product(s), our machine incorporates a triply differentially pumped, *universal* quadrupole mass spectrometric detector coupled to an electron impact ionizer. Here, any reactively scattered species from the collision center after a single collision event has taken place can be ionized in the electron impact ionizer, and – in principle – it is possible to determine the mass (and the gross formula) of all the products of a bimolecular reaction by varying the mass-to-charge ratio, m/z, in the mass filter. Since the detector is rotatable within the plane defined by both beams, this detector makes it possible to map out the angular (LAB) and velocity distributions of the scattered products. Measuring the time-of-flight (TOF) of the products, *i.e.* selecting a constant mass-to-charge value in the controller and measuring the flight time of the ionized species, from the interaction region over a finite flight distance at different laboratory angles allows extracting the product translational energy and angular distributions in the center-of-mass reference frame. This provides insight into the nature of the chemical reaction (direct *vs.* indirect), intermediates involved, the reaction product(s), their branching ratios, and in some cases the preferential rotational axis of the fragmenting complex(es) and the disposal of excess energy into the products' internal degrees of freedom as a function of scattering angle and collision energy. However, despite the triply differential pumping setup of the detector chambers, molecules desorbing from wall surfaces, which are on a straight line to the electron impact ionizer, cannot be avoided. Their mean free path is of the order of 10^3 m compared to maximum dimensions of the detector chamber of about 1 m. To reduce this background, a copper plate attached to a two-stage closed cycle helium refrigerator is placed right before the collision center and cooled down to 4 K. In this way, the ionizer views a cooled surface which traps all species with the exception of hydrogen and helium.

What information can we obtain from these measurements? The experimental observables contain some basic information. Every species can be ionized at the typical electron energy used in the ionizer and, therefore, it is possible to determine the mass and the gross formula of all the possible species produced from the reactions by simply selecting different m/z in the quadrupole mass spectrometer. Even though some problems such as dissociative ionization and background noise limit the method, the advantages with respect to spectroscopic techniques are obvious, since the applicability of the latter needs the knowledge of the optical properties of the products. In our setup, we are operating the ionizer during the scattering experiments at electron energies of 80 eV, *i.e.* at an energy at which the ionization cross section of the organic molecules is at their maxima.[56] Based on signal calculations, the crossed beam reactions of dicarbon and methylidyne radicals are very challenging and we need all the intensity (here: detectable ion counts) we can get. Note that our ionizer can be also operated *via soft* electron impact ionization as pioneered by Casavecchia *et al.*[58] This approach utilizes electrons with low, tunable energy (8–30 eV) to reduce strongly or even eliminate the problem of dissociative ionization from interfering species. However, soft ionization has – in case of the present experiments – one disadvantage: at electron energies of 8–30 eV, the ionization cross sections of the newly formed molecules are *at least* a factor of 20 lower than the electron impact ionization cross sections with 80 eV electrons.[86] Therefore, the low cross sections, the expected signal-to-noise, and the inherent data accumulation times make the application of soft ionization impractical for the present experiments utilizing *pulsed* beams. However, *soft electron impact ionization* and laser induced fluorescence (LIF) can be utilized to characterize the reactant beams on axis and *in situ* as described below. Another important aspect is that, by measuring the product velocity distributions, one can immediately derive the amount of the total energy available to the products and, therefore, the enthalpy of reaction of the reactive collision. This is of great help when different structural isomers with different enthalpies of formation can be produced. For a more detailed physical interpretation of the reaction mechanism it is necessary to transform the laboratory (LAB) data into the center-of-mass (CM) system using a forward-convolution routine.[87] This approach initially assumes an angular distribution $T(\theta)$ and a translational energy distribution $P(E_T)$ in the center-of-mass reference frame (CM). TOF spectra and the laboratory angular distribution are then calculated from these center-of-mass functions. The essential output of this process is the generation of a product flux contour map, $I(\theta,u) = P(u) \times T(\theta)$. This function reports the flux of the reactively scattered products (I) as a function of the center-of-mass scattering angle (θ) and product velocity (u) and is called the reactive *differential cross section*. This map can be seen as the *image* of the chemical reaction and contains *all* the information on the scattering process.

2.1.3. The supersonic beam sources.

2.1.3.1. The ablation source: supersonic dicarbon beam. Pulsed supersonic beams of dicarbon molecules are generated *via* laser ablation of graphite utilizing a home-built ablation source.[88] Here, a 266 nm laser beam originating from a neodymium-doped yttrium aluminium garnet (Nd:YAG) laser is tightly focused by a 1.5 m lens with pulse energies of 5–10 mJ onto a graphite rod performing a helical motion. The ablated species are seeded into helium carrier gas (99.9999%; Gaspro) at a backing pressure of 4 atm. Then, the supersonic beam of *in situ* generated dicarbon molecules is chopped and crosses a pulsed hydrogen cyanide beam (5% hydrogen cyanide premix in 99.9999% helium; Matheson Gas) perpendicularly in the interaction region at a collision energy of 42.4 ± 1.7 kJ mol^{-1}. Peak velocities of the dicarbon beam are 2019 ± 64 ms^{-1} with a speed ratio of 4.2 ± 0.5. The hydrogen cyanide beam segment crossing the dicarbon pulse is characterized by a peak velocity of 1612 ± 6 ms^{-1} and speed ratio of 15 ± 3. Note that the ablation beam also contains ground state carbon atoms (C(^3P_j)). However, as demonstrated earlier in our group, atomic carbon does not react with hydrogen cyanide to form C_3N molecules plus atomic hydrogen at our collision energy.[89] Finally, the reaction of co-ablated tricarbon molecules with unsaturated bonds is hindered by entrance barriers larger than our collision energy.[16] Also, we did not see reactive scattering signal at masses higher than *m/z* = 50. Therefore, the only species in the primary beam reacting with hydrogen cyanide are dicarbon molecules.

It is important to discuss the electronic states of the dicarbon molecules. Due to the low energy gap between the ground and first excited triplet state (718 cm^{-1}), the supersonic ablation beams contain dicarbon in its $X^1\Sigma_g^+$ electronic ground state as well as in its first electronically excited $a^3\Pi_u$ state.[90] Crossed molecular beam reactions of dicarbon molecules with hydrogen sulfide (H_2S) and hydrogen cyanide (HCN) form the HCCS($X^2\Pi_\Omega$)[91] and CCCN($X^2\Sigma^+$)[29] products *only* on the singlet surfaces. The corresponding reactions of triplet dicarbon are either repulsive (hydrogen sulfide reaction) or hold a significant entrance barrier of about 30 kJ mol^{-1} (hydrogen cyanide). Therefore, those previous crossed beam reactions provided direct evidence that dicarbon molecules in their $X^1\Sigma_g^+$ electronic ground state exist in the ablation beams. In order to verify the presence of triplet C_2 in the ablation beam we are conducting first a comprehensive laser induced fluorescence (LIF) study probing the dicarbon molecules in the first electronically excited state *via* the Swan transition ($d^3\Pi_g$–$a^3\Pi_u$). Here, triplet dicarbon was excited by the fundamental output of a Lambda Physics Scanmate dye laser using Coumarin 503 dye at about 516.5 nm at laser power of 45 µJ per pulse. The dye laser itself was pumped by an internal neodymium–yttrium–aluminium–garnet (Nd:YAG) laser operating at 355 nm at 10 Hz with an output power of 50 mJ per pulse. The dye laser was fired around 200 µs after the pulse valve opening with a pulse energy of a few µJ to intercept the peak of the dicarbon beam.[90] The geometry of the LIF detection experiment of triplet dicarbon is shown in Fig. 3 with the pulse sequence compiled in Fig. 4. The main experimental challenge of the LIF measurement was to suppress the scattered laser light in a tight volume of the crossed beams setup. To achieve this, the detection laser beam was focused to a 2–3 mm diameter spot at the crossing point by a 1.5 m lens. An antireflective coated lens has been employed to reduce multiple reflections in the lens that produce divergent beam components. A baffle tube containing eight irises of 4 mm and 5 mm diameter trapped divergent components in the detection laser beam. The skimmer of the secondary molecular beam has been removed to allow an unobstructed exit of the laser light. An ultra-low-backscatter laser beam trap (Thorlabs BT510, 6 × 10^{-6} fraction for integral

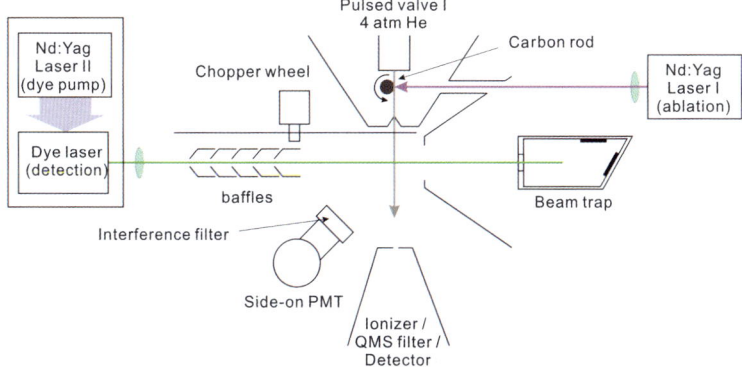

Fig. 3 Schematic geometry of the laser induced fluorescence (LIF) detection setup of triplet dicarbon as incorporated in the crossed beams machine.

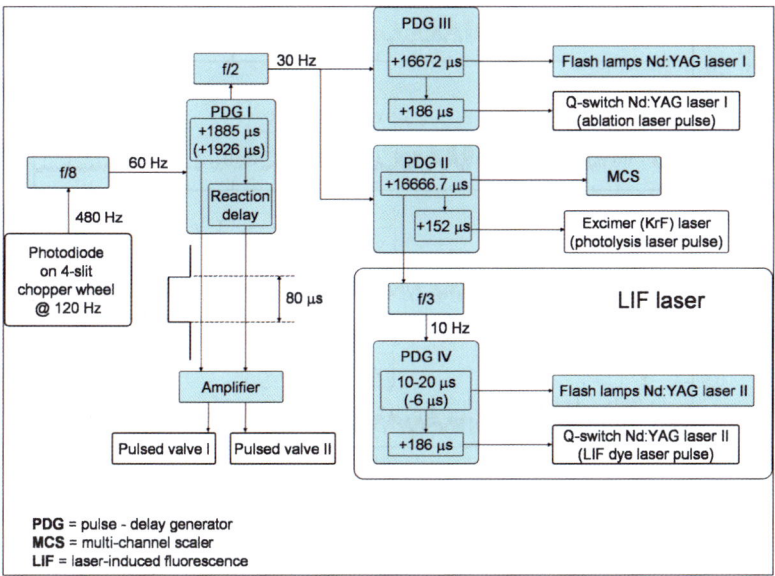

Fig. 4 Pulse sequence for the crossed beams experiments and LIF detection. Delay times are shown for distinct dicarbon beams produced in the primary chamber. For the methylidyne source, delay times different from the dicarbon experiment are shown in parentheses.

backscatter intensity) has been placed 50 cm behind the front wall of the secondary source to trap the detection beam. The fluorescence was detected by a Hamamatsu R955 photomultiplier tube (PMT) placed between the baffle tube and the detector chamber at about 10 cm from the intersection point. A band pass interference filter of 10 nm bandwidth centered at 562 nm (Andover) was placed in front of the PMT to block the scattered laser light and to pass the fluorescence to the first vibrationally excited ground electronic states of dicarbon. No spatial filtering of collected light was introduced. The signal was amplified by a built in preamplifier of the Hamamatsu C7247 PMT socket assembly prior to feeding into a digital oscilloscope and a computer for data collecting and processing. The LIF spectra were then analyzed utilizing the diatomic spectral simulation program by Tan[92] with spectroscopic constants from Bernath.[93] The corresponding LIF spectra of dicarbon seeded

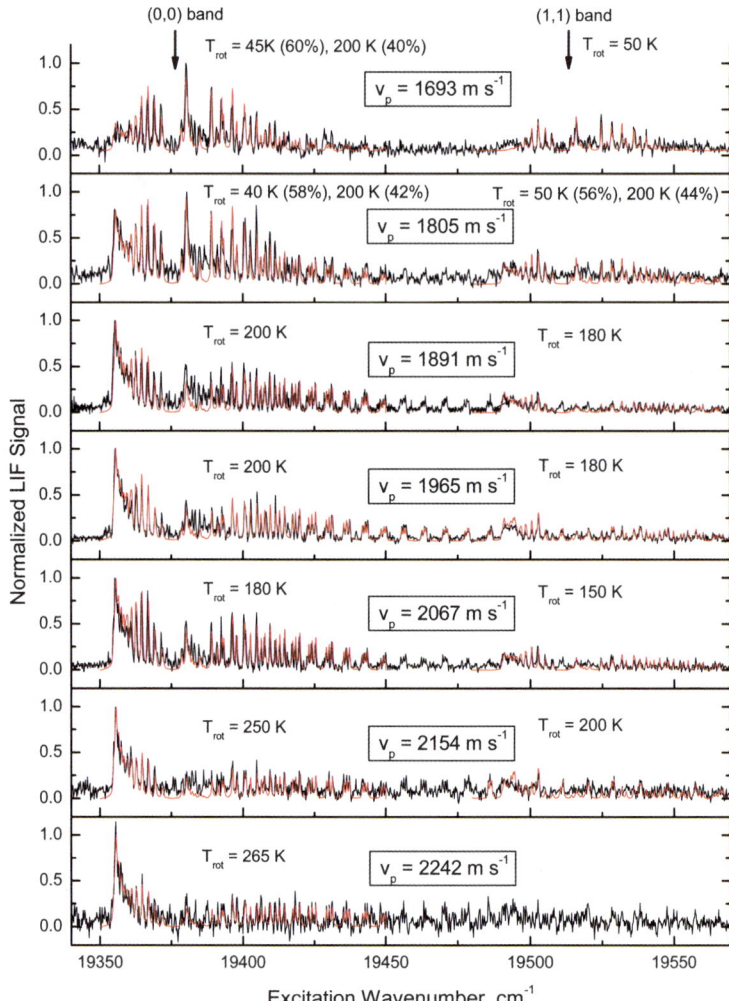

Fig. 5 LIF excitation spectra of dicarbon seeded in helium carrier gas for different velocities of the beam. Temperatures of the best-fit for rotational energy distributions are listed separately for the (0,0) and (1,1) bands. Some distributions have distinct non-equilibrium character; they are fit with two temperature components.

in helium and neon carrier gases are shown in Fig. 5 and 6, respectively. Different peak velocities (v_p) of the beam were chosen by selecting distinct delay times between the pulsed valve and the chopper wheel. The peaks in the spectra correspond to the excitation from different rotational states of the ground vibrational state of triplet dicarbon ((0,0) band) and of the first excited vibrational state ((1,1) band). Vibrational (0,0) and (1,1) bands are separated on the energy scale. Their integral intensity ratio is determined by the relative populations of the $\nu = 0$ and $\nu = 1$ states. We did not observe transitions from vibrational states higher than $\nu = 1$.

Fig. 7 addresses the dependence of the vibrational and rotational temperatures as well as speed ratios on the peak velocities of distinct parts of the chopped dicarbon beam. The vibrational temperature is expressed in practical terms of the fraction of triplet dicarbon in $\nu = 1$. The majority of the LIF spectra can be fitted well if we treat the vibrational and rotational temperatures for each vibrational state separately. In some spectra, however, we have to introduce two temperatures and a bimodal

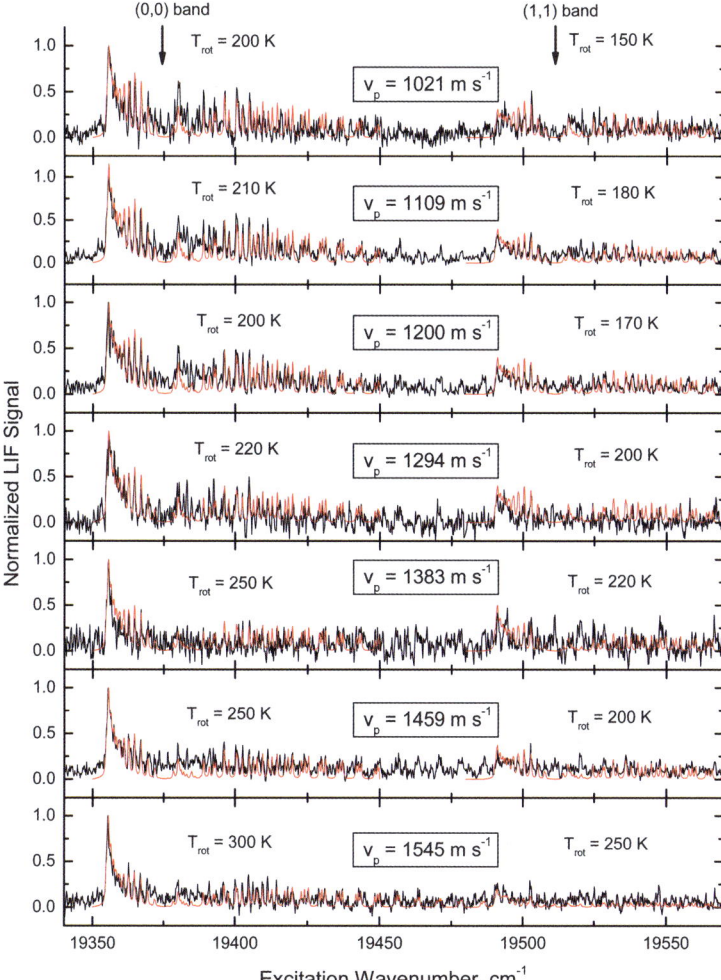

Fig. 6 LIF excitation spectra of dicarbon seeded in neon carrier gas for different velocities of the beam. Temperatures of the best-fit for rotational energy distributions are listed separately for the (0,0) and (1,1) bands.

distribution to describe the distribution of rotational state populations within one vibrational state. This probably indicates highly non-equilibrium energy distribution in the dicarbon molecules following graphite laser ablation events. The general trend for the speed ratio and the rotational temperature is that lower velocities provide cooler molecules; rotational temperatures as low as 50 K can be achieved. This is consistent with the fact that the ablated species seeded in the fast (front) parts of the helium and neon pulses have less collisions with the noble gas causing lower cooling efficiency. Rotational temperatures for $v = 1$ are systematically lower than for $v = 0$ suggesting a better rotational cooling of vibrationally excited species. There are some irregularities in the velocity dependence of the beam characteristics; these can be partially attributed to variations in the independently adjusted delay times of the pulsed valve and ablation laser as well as slight variations in the ablation laser power leading to distinct seeding conditions. Also, the vibrational population dependence on velocity does not exhibit a pronounced trend. Vibrational relaxation is about equally inefficient in the whole range of experimental conditions.

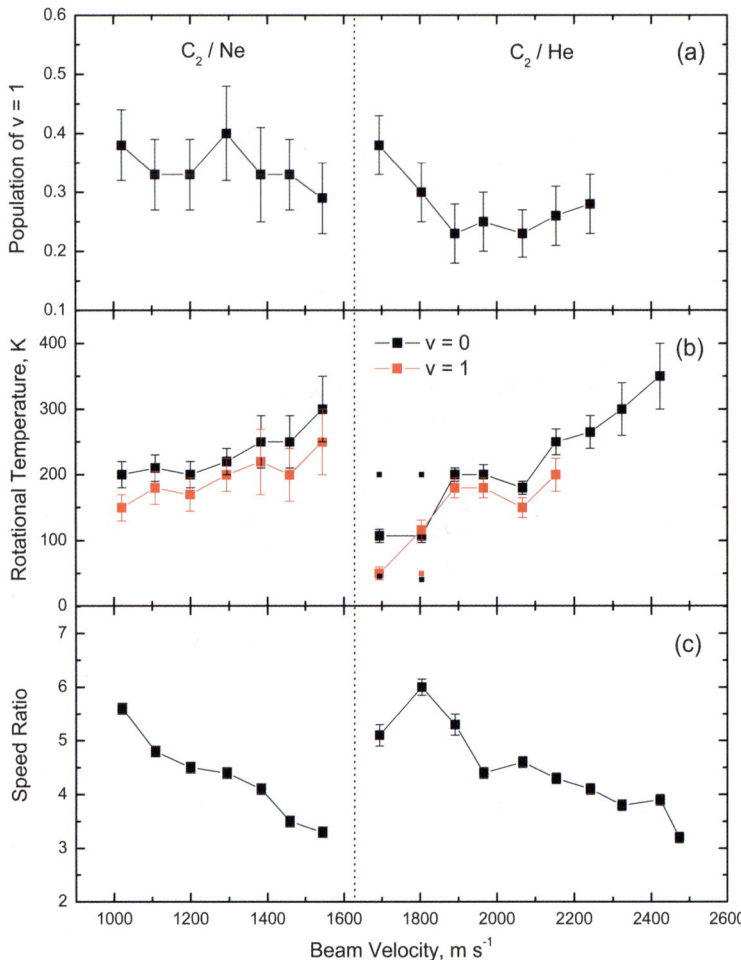

Fig. 7 Supersonic vibrational (a), rotational (b), and speed ratio (c) dependence on the dicarbon beam peak velocity. The dicarbon molecules are seeded in neon (left part) and helium (right part) carrier gases. The cooling efficiency is characterized in terms of the population of the first vibrationally excited state (a), rotational temperature (b), and speed ratio of the beam (c). Black and red curves in (b) represent rotational temperatures of $v = 0$ and $v = 1$, respectively. The smaller squares plot temperature components of a non-equilibrium energy distribution.

As compiled in Fig. 4, the 17 cm diameter four-slit chopper wheel with 0.76 mm slits operates at 120 Hz between the skimmer of the primary source and the interaction region. An infrared diode attached to the top of the chopper unit detects the slit passage thus providing the time zero of the experiment. By selecting the time delay between the diode pulse and the pulsed valve, distinct parts of the ablation pulse can be selected. An SRS DG535 delay/pulse generator (PDG I) is triggered at 60 Hz by the frequency divided output of the diode. The two outputs, AB and CD, of the pulse generator (50 Ω, +3.5 V, 80 μs pulse width) lead to a homemade pulse shaper, which in turn is connected to the Physik Instumente P-286.23 high voltage pulse amplifier. The output of the amplifier drives both piezoelectric Proch–Trickl valves at repetition rates of 60 Hz, opening times of 80 μs, and pulse amplitude of minus 400 V to 500 V. The delay time between the two valves provides for simultaneous arrival of the most intensive parts of the two reactant beams to the interaction region. The A output of the PDGI pulse generator (TTL, high impedance) passes a frequency divider (division by two, 50 Ω output) fed as an external trigger to a second delay/pulse generator (PDG II). The time delayed A output (A = T_0 + 16 666.66 μs, high impedance TTL) is fed into the SRS 430 multichannel scaler (MCS) utilizing trigger and discriminator levels of +0.5 and +0.2 V. A second output of the frequency divider serves as an external trigger of PDG III. This unit controls the time sequence of Spectra Physics Nd:YAG laser (30 Hz, 120 mJ per pulse at 266 nm). Channel AB (A = T_0 + 16,672 μs, B = A + 5 μs, 50 Ω, TTL) triggers the flash lamps and CD the Q-switch (C = A + 186 μs, D = C + 5 μs, 50 Ω, TTL). The time sequence for Nd:YAG pumped dye laser (10 Hz repetition rate) used for LIF detection is set by PDG IV triggered by the frequency divided A output of PDG II. Channel AB (A = T_0 + (10–20) μs, B = A+ 5 μs, 50 Ω, TTL) triggers the flash lamps and CD the Q-switch (C = A + 186 μs, D = C + 5 μs, 50 Ω, TTL) of the internal Nd:YAG laser. Delays for the primary pulsed valve and ablation laser were varied slightly depending on desired velocity of the dicarbon beam.

2.1.3.2. *The photolytic source: supersonic methylidyne beam.* We generated a pulsed supersonic beam of methylidyne radicals *via* photolysis of helium-seeded bromoform ($CHBr_3$) at seeding fractions of 0.12% at 248 nm at 30 Hz by bubbling helium gas (99.9999%; Gaspro) at a pressure of 2.2 atm through a stainless steel bubbler which houses the bromoform at a temperature of 283 K and feeding this gas mixture into a pulsed piezoelectric valve. The latter is operated at a repetition rate of 60 Hz, pulse widths of 80 μs, and a voltage of minus 400 V to 450 V. Here, by focusing 60 mJ per pulse output of excimer laser (KrF) with a 1 meter focus lens downstream of the nozzle to an area of about 4 mm by 0.7 mm, a few 10^{12} radicals cm^{-3} can be formed in the interaction region of the scattering chamber. The timing of the experiment is shown in Fig. 4 alongside with the timing for the dicarbon experiment. If the delay times differ, the values for the methylidyne experiment are shown in parentheses.

Methylidyne radicals are only produced in the $^2\Pi$ ground state. The A and B states have lifetimes of 440 ± 20 ns and 470 ± 20 ns and relax to the ground state before they reach the skimmer.[94] The photodissociation of bromoform to $CH(X^2\Pi)$ is a multiphoton process initiated by the cleavage of the C–Br bond to yield $CHBr_2$ + Br[95] (σ(248 nm) = 1.9 × 10^{-18} cm^2).[96] Utilizing photoionization photofragment translational spectroscopy, North *et al.* observed also CHBr, CBr, HBr, and Br_2 fragments which were attributed to higher-order photodissociation processes of $CHBr_2$ and CHBr. Mebel computed the photodissociation cross sections of $CHBr_2$ and CHBr at 248 nm to be 1.6 ± 0.4 × 10^{-18} and 2.0 ± 0.3 × 10^{-18} cm^2, respectively.[97] Although it is not feasible to eliminate $CHBr_2$, CHBr, and CBr in the supersonic beam, these molecules have – due to the heavy bromine atom – distinct center-of-mass angles when reacting with the hydrocarbon molecules. Therefore, the dynamics can be distinguished from those of the CH reactions based on the distinct mass-to-charge ratios and due to different center-of-mass angles and hence scattering ranges of the products. This presents a unique advantage of the present experimental setup. The velocity and speed ratio of the radical beam can be determined on-axis in the TOF mode. Since signal at $m/z = 13$ (CH^+) also originates from dissociative ionization of, for instance, non-photolyzed bromoform, operating the electron impact ionizer in the *soft ionization* mode, here at 34 eV, for the beam characterization is important. This translates into a narrow range of peak velocities of 1680–1780 ms^{-1} with speed ratios of 12–18.

We also utilized laser induced fluorescence to characterize the rotational and vibrational modes of the methylidyne radical, $CH(X^2\Pi)$, in the interaction region of the scattering chamber. Methylidyne radicals are detected using $A^2\Delta-X^2\Pi$ transitions: (0,0) vibrational band for excitation near 431 nm and (0,1) band for detection near 490 nm. The interference filter in front of the photomultiplier tube (PMT; Andover Corp.) is centered at 490 nm with a 10 nm bandwidth; this discriminates against scattered laser light. Two major modifications have been made to LIF detection setup compared to dicarbon experiment (Fig. 3). Firstly, the skimmer of the

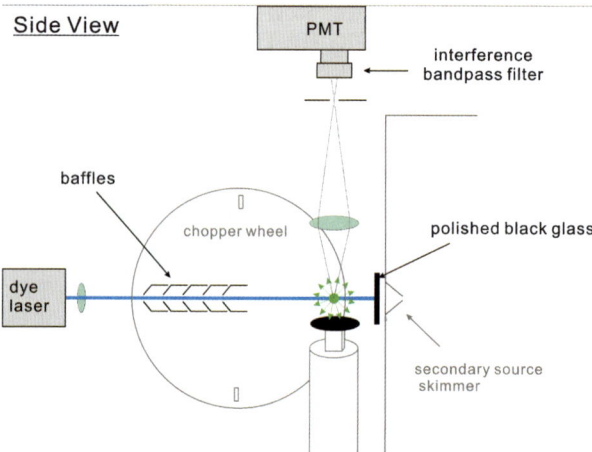

Fig. 8 Schematic geometry of the laser induced fluorescence detection setup of the methylidyne radical as incorporated in the crossed beams machine.

secondary source does not have to be removed in the new configuration (Fig. 8). The incoming detection laser beam is mainly absorbed by a piece of polished black glass (ThorLabs; neutral density filter; 40–20 surface quality); the reflected part travels back into the baffle tube. Secondly, spatial filtering of fluorescence signal is introduced. The fluorescence spot in the interaction region is projected by a 35 mm focus lens onto the center of the iris in front of the PMT, which is mounted on the lid of the machine. This vertical orientation of the detector allows also us to minimize the collection of Raleigh scattered light of the vertically polarized laser on the atoms and molecules in the beam. Another piece of polished black glass is placed under the interaction region to eliminate the propagation of scattered laser light in the light

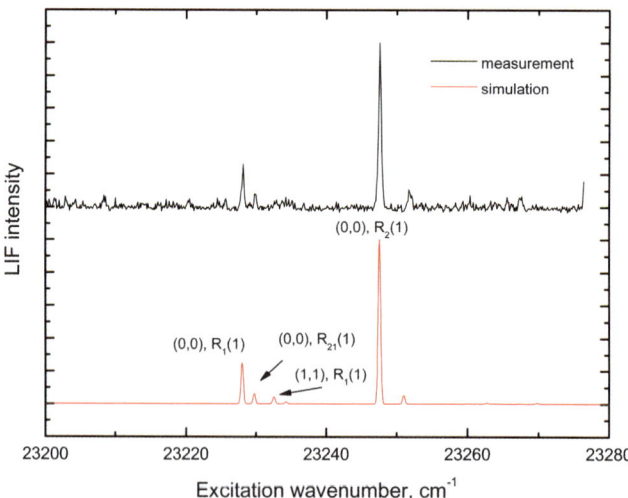

Fig. 9 LIF spectrum of helium-seeded methylidyne radicals (top) together with the simulation. Parameters of the best fit simulation suggest a rotational temperature of 14 K and relative populations of $v = 1$ level of less than 6% based on the (1,1), $R_1(1)$ peak. Note that we cannot distinguish between different spin–orbit states of methylidyne radical ($\Omega = 1/2$ vs. $\Omega = 3/2$) because for the observed transitions, the largest spectroscopic splitting (0.11 cm^{-1} for $R_2(1)$ transition) would be still smaller than the line width of the detection laser of 0.15 cm^{-1}.

collection cone. Fig. 9 shows the LIF spectrum of the methylidyne radical beam with a peak velocity of 1700 ms^{-1} as characterized with the TOF technique. Sixteen detection laser shots were averaged for each point. The spectrum was analyzed utilizing a LIFBASE database and spectral simulation for diatomic molecules by Jorge Luque.[98] The best fit simulation suggests a rotational temperature of 14 ± 1 K in the vibrational ground state; less than 6% of the radicals are in the first vibrationally excited state population. We conducted two test reactions of this beam by crossing it with argon-seeded diacetylene (C_4H_2) and neat methylacetylene (C_3H_4). The diacetylene beam has a peak velocity of 600 ± 15 ms^{-1} at a speed ratio of 8; the neat methylacetylene beam was characterized by a velocity of 840 ± 10 ms^{-1} and a speed ratio of 9.

2.2. The surface scattering machine

A surface scattering machine, in which an ultrahigh vacuum in the low 10^{-11} torr range can be achieved (Fig. 10),[99] was utilized to simulate the interaction of energetic electrons with organics as present on Titan in form of aerosol 'droplets' and solids on Titan's surface. Ethane (C_2H_6) and propane (C_3H_8) ices were prepared in separate experiments by passing ethane (99.999%; Gaspro) and propane (99.9%; Specialty Gas Group) for 3 min through a glass capillary array at hydrocarbon pressures of 1.5 × 10^{-8} Torr onto a highly polished silver mono-crystal. The silver substrate is attached to the freely rotating arm of a closed cycle helium refrigerator (CTI-Cryogenics CP-1020) and was held at 11.3 ± 0.5 K to allow for the rapid condensation of the gases. The temperature of the ice samples was measured by a silicon diode connected to a Lakeshore 331 temperature controller. Recall that Titan's surface temperature of 94 K is well above the 11.3 K of the cold head. Therefore, the ices were heated to 50 K (C_2H_6) and 65 K (C_3H_8) – just before the sublimation temperature under UHV conditions – and then irradiated with 5 keV electrons supplied by an electron gun (SPECS EQ 22) for 3 h at a beam current of 500 nA. The manufacturer states an electron extraction efficiency of 78.8%; this resulted in an exposure of 1.5 × 10^{16} electrons per cm^2 over the irradiation time with an irradiation area of 1.8 ± 0.3 cm^2. The analysis of the irradiated ice samples was performed on line and *in situ* with a Nicolet 6700 FTIR spectrometer operating in absorption–reflection–absorption mode (reflection angle $\alpha = 75°$) to the surface. Spectra were recorded at a resolution of 2 cm^{-1} over the mid-IR ranges

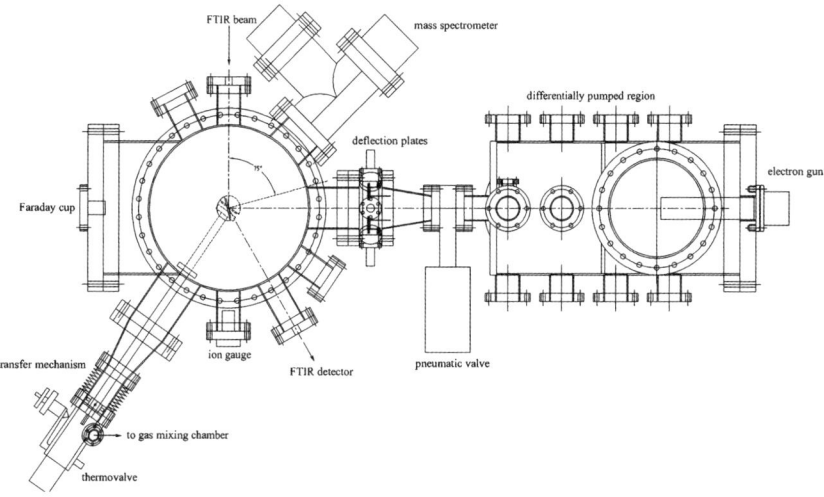

Fig. 10 Schematic top view of the surface scattering machine.[99]

(4000–500 cm^{-1}) utilizing a liquid nitrogen cooled mercury–cadmium–telluride detector. Gaseous species were monitored by a Balzer QMG 420 QMS operating in residual gas analyzer mode with an electron impact ionization energy of 100 eV and a mass range of up to 200 amu. Upon completion of the irradiation period, the ice samples were then heated to 300 K by a controlled heating program at a rate of 0.5 K per minute to allow for the analysis of volatile products as they sublimed from the target.

The analysis of the infrared bands of pristine ethane (C_2H_6) and propane (C_3H_8) allowed for the calculation of the thickness of the deposited ice layers using the integral absorption coefficients for the methyl group (CH_3) deformation bands provided by Bohn *et al.*[100] and the densities of ethane (0.713 ± 0.002 g cm^{-3}) and propane (0.763 ± 0.004 g cm^{-3}) at 77 K as given in Stewart and La Rock.[101] The thicknesses were calculated to be 200 ± 50 nm for ethane (C_2H_6) and 180 ± 30 nm for propane (C_3H_8). To gain a quantitative handle on the energy absorbed by the hydrocarbon molecules, the electron trajectories in the ice samples and the energy transfer were then simulated by the CASINO code.[102] This code calculates average transmission energies of the energetic electrons of 4.56 ± 0.01 keV and 4.59 ± 0.01 keV for ethane and propane ices, respectively. Therefore, each electron deposits on average 440 ± 10 eV into the ethane sample (linear energy transfer LET = 3.7 ± 0.2 keV μm^{-1}) and 410 ± 10 eV into the propane ice (LET = 3.7 ± 0.4 keV μm^{-1}). From here, the *average* amount of energy absorbed per target molecule, the dose, is calculated to be 36 ± 2 and 53 ± 5 eV per molecule for the ethane and propane samples.

2.3. The laser ablation apparatus

The experiments were performed with a laser ablation apparatus coupled to a 3 meter monochromator of the Chemical Dynamics Beamline at the Advanced Light Source.[103] The apparatus was described previously.[104] Compared to the original design, the ablation source was modified to incorporate an external motor assembly as described below (Fig. 11). Briefly, the ablation source consists of an aluminium block and a pulsed piezoelectric valve. The 6.35 mm diameter silicon rod (ESPI) rotates and simultaneously translates vertically inside of the aluminium ablation block, which acts as a guide for the ablation rod. The latter is connected to an in-vacuum translation-rotation stage, which is driven by a computer controlled stepper motor (RMS technologies) connected through a speed reducing gear box. A frequency doubled (532 nm) Nd:YAG laser operating at 50 Hz and output powers of about 0.8 mJ per pulse ablates the silicon rod. The ablated species are entrained in acetylene (C_2H_2; Airgas) carrier gas released by a Proch–Trickl piezo valve. Acetone, which acts as a stabilizer in the pressurized acetylene cylinder, is filtered out before the gas enters the piezo valve at a stagnation pressure of 1.5 atm. The ablated silicon-bearing species travel together with the gas pulse inside a 4 mm diameter 30 mm long extension channel before exiting the ablation block. The acetylene acts both as a carrier gas and as a reagent media; therefore, the organo-silicon molecules are formed *in situ* in the supersonic beam *via* reaction of acetylene with the ablated silicon species. This technique has been applied recently to produce and to derive the ionization energies of highly reactive organic transient radicals like linear and cyclic C_3H radicals.[105] A pair of deflection plates, producing an electrical field of about 660 V cm^{-1}, are located between the ablation block and skimmer assembly thus removing charged species generated by the ablation process; this allows only neutral particles to pass through the differential pumping wall, which is equipped with a 2 mm skimmer hole; this section separates the source chamber from the main photoionization chamber.

The neutral, supersonic beam is interrogated in the ionization region of a commercially available reflectron time-of-flight mass spectrometer (R. M. Jordan) by tunable monochromatic synchrotron radiation in the vacuum ultraviolet (VUV) region of the electromagnetic spectrum. Typically 10^{13} photons s^{-1} are available at

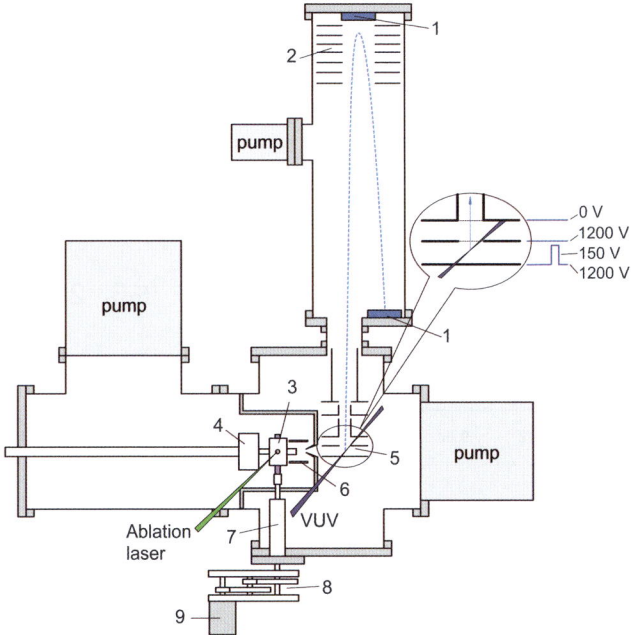

Fig. 11 Scheme of the laser ablation setup machine: 1) micro-channel plates, 2) reflectron, 3) ablation block with a silicon rod, 4) pulsed piezo valve, 5) Wiley-McLaren ion optics, 6) ion deflection plates, 7) translation-rotation stage, 8) gear box, 9) step motor.

this terminal. The photoionization region is situated 12 cm downstream from the ablation region. As the synchrotron light is quasi-continuous (500 MHz), a start pulse for the time-of-flight (TOF) ion packet is provided by pulsing the repeller plate (the lowest electrode in the Wiley-McLaren ion optics) of the time-of-flight ion optics. The pulsing sequence with voltages are shown in the inset of Fig. 11. The ions hit a microchannel plate (MCP) detector; the signal from these ions are collected with a multichannel-scalar card (FAST Comtec 7886) triggered by the repeller plate pulse. Time-of flight spectra, *i.e.* the flight time of the ion *versus* the intensity of the ion counts, are recorded for the photoionization energy range between 8.0 eV and 10.5 eV. The typical step size used for these experiments is 50 meV; the signal was collected for $5-7 \times 10^3$ laser shots. The detected signal was optimized using LabView (National Instruments) routines by changing the delay times (Nd:YAG lamp to Q-switch to attenuate laser pulse intensity; laser pulse to trigger pulse to piezo valve; laser pulse to repeller plate pulse) and voltages of the TOF ion optics.[104] Previously, we estimated that typically 10^8–10^9 molecules cm^{-3} are being ionized in the 1 mm^3 interaction region.

The photoionization efficiency (PIE) curves of a well-defined ion of a mass-to-charge ratio (*m/z*) can be obtained by plotting the integrated ion signal at the mass-to-charge *versus* the photoionization energy between 8.0 eV and 10.5 eV, normalized by the photon flux and the number of laser shots. The synchrotron VUV photon flux is measured by a Si photodiode (IRD, SXUV-100). These PIE curves can be exploited to extract the adiabatic ionization energies of the newly formed silicon bearing species. To calibrate the photon energy, auto ionization peaks of xenon and a resonance feature in the PIE curve of atomic silicon are used. Thus, to measure the energy resolution of the VUV light, scans of a resonance in the PIE spectrum of atomic silicon due to its atomic transitions were undertaken. The resulting PIE curves were measured in the photon energy range from 9.82 eV to 9.93 eV for three different sizes of the 3 meter monochromator exit slits; these are

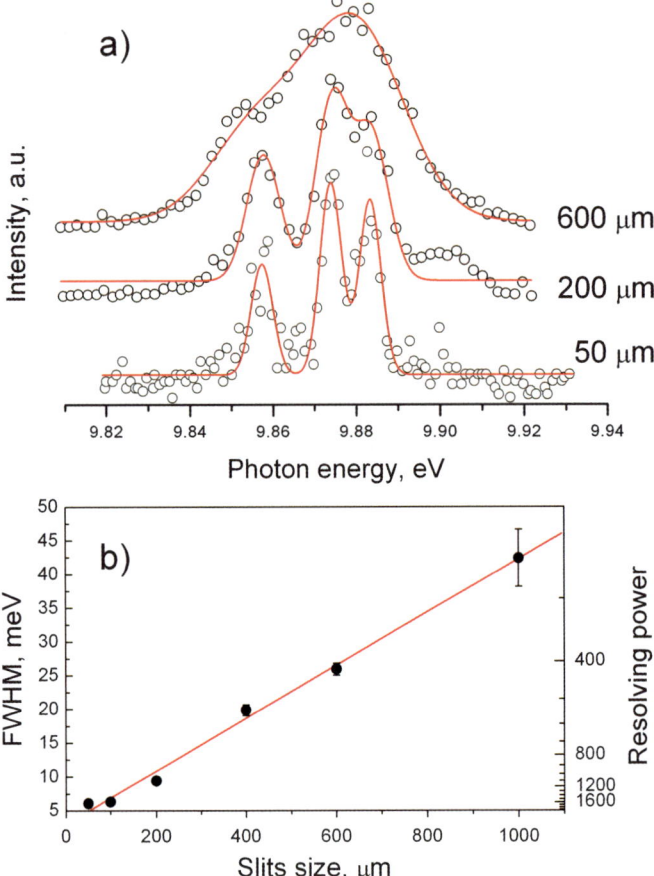

Fig. 12 a) Resonance features in the photoionization efficiency curves of atomic silicon for 50 μm, 200 μm, and 600 μm monochromator exit slits; b) the dependence of the full width at half maximum (FWHM) of the synchrotron VUV radiation and its resolving power on the monochromator exit slit size.

shown in Fig. 12 and are fitted with a Gaussian function. It is observed that the energy resolving power ($E/\Delta E$) is 1650 for 50 μm slits; this degrades to 250 for a slit width of 1 mm.

3. Theoretical approach

Molecular geometries and vibrational frequencies of reactants, intermediates, and transition states on the potential energy surface of the reaction of hydrogen cyanide (HCN) with triplet dicarbon ($C_2(^3\Pi_u)$) were calculated at the hybrid density functional B3LYP/6-311G** level of theory[106] using the GAUSSIAN 98 program package.[107] Relative energies were refined utilizing the coupled cluster CCSD(T) method as implemented in the MOLPRO package[108] with extrapolation to the complete basis set (CBS) limit. To achieve this, we computed CCSD(T) total energies for each stationary point with Dunning's correlation-consistent cc-pVDZ, cc-pVTZ, cc-pVQZ, and cc-pV5Z basis sets and projected them to CCSD(T)/CBS total energies using eqn (8):

$$E_{\text{tot}}(x) = E_{\text{tot}}(\infty) + Be^{-Cx} \qquad (8)$$

where x is the cardinal number of the basis set (2, 3, 4, and 5) and $E_{tot}(\infty)$ is the CCSD(T)/CBS total energy. With respect to calculations of the organo silicon neutral molecules and ions, we used the same B3LYP/6-311G** approach for the geometry optimization and evaluation of vibrational frequencies. Relative energies of various isomers as well as vertical and adiabatic ionization energies were then refined by single-point energy calculations at the CCSD(T)/cc-pVQZ level of theory.

4. Results & discussion

4.1. Crossed beam reactions

In our experiments of dicarbon with hydrogen cyanide, we recorded time-of-flight (TOF) spectra at various laboratory angles at mass to charge ratios of $m/z = 50$ (C_3N^+) and higher. Similar to the reactions of ground state singlet dicarbon conducted previously in our group at collision energies of 22.4 kJ mol^{-1} and of 25.8 kJ mol^{-1}[29] we detected signal at mass to charge ratios $m/z = 50$ (C_3N^+) (Fig. 13). Also, we could confirm that even at a collision energy of 42.4 kJ mol^{-1}, no products at higher masses were monitored. It is important to stress that all TOF spectra were fit with a single channel leading to the synthesis of a molecule of the gross formula C_3N formed through a dicarbon molecule *versus* hydrogen atom exchange; the formation of the thermodynamically less stable isocyano radical (CCNC) is endoergic by 102 kJ mol^{-1} and hence can be ruled out considering our collision energy of only 42.4 kJ mol^{-1}. Consequently, we propose that the cyanoethynyl radical (CCCN) is also the reaction product at $m/z = 50$. Note that the corresponding laboratory angular distribution is very narrow and spread only 30° in the scattering plane as defined by the dicarbon and hydrogen cyanide beams. Therefore, we can conclude that both in the previous reactions of singlet dicarbon and now with singlet and triplet dicarbon, the cyanoethynyl radical can be formed under single collision conditions.

Having analyzed the laboratory data, we now focus our attention to the derived center-of-mass functions. As outlined before, an acceptable fit could be achieved with a single reaction channel. Here, a close look at the center-of-mass translational

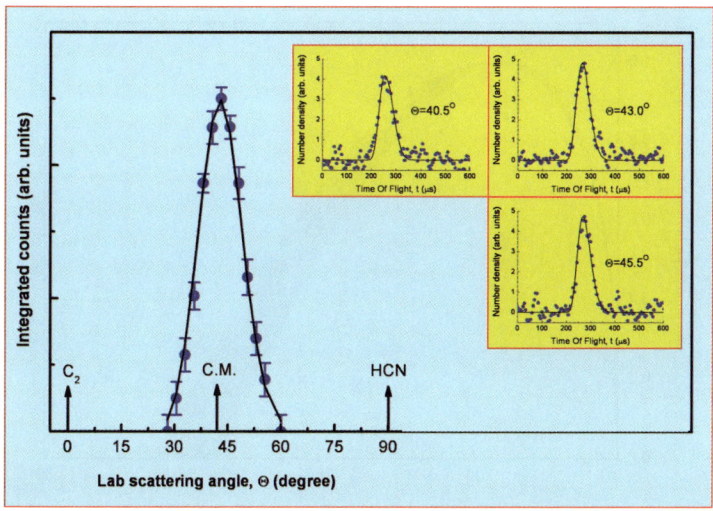

Fig. 13 Laboratory angular distribution and selected time-of-flight (TOF) spectra of the C_3N radical, which was formed in the reaction of dicarbon with hydrogen cyanide, recorded at $m/z = 50$ at a collision energy of 42.4 kJ mol^{-1} utilizing a single channel fit with the center-of-mass functions as shown in Fig. 14.

energy distribution (Fig. 14) assists to compare the experimentally derived reaction energy with theoretically predicted value. By subtracting the collision energy (E_C) of 42.4 kJ mol^{-1} from the maximum translational energy released into the translational degrees of freedom of the reaction products, here 60 ± 5 kJ mol^{-1}, the reaction was found to be exoergic by 18 ± 10 kJ mol^{-1}. This data is in good agreement with our *ab initio* data of 21 ± 5 kJ mol^{-1} for the triplet dicarbon reaction. Secondly, the translational energy distribution shows a plateau ranging from about 12 to 20 kJ mol^{-1}. As found from previous reactions of singlet and triplet dicarbon with acetylene and ethylene, these patterns could suggest the existence of two reaction pathways on the singlet and on the triplet surface. Recall that the translational energy distribution obtained at lower collision energies of 22.4 kJ mol^{-1} and of 25.8 kJ mol^{-1}, *i.e.* only from the singlet dicarbon reaction, peaked at zero translational energy, suggesting a barrierless decomposition of an intermediate – in this case singlet

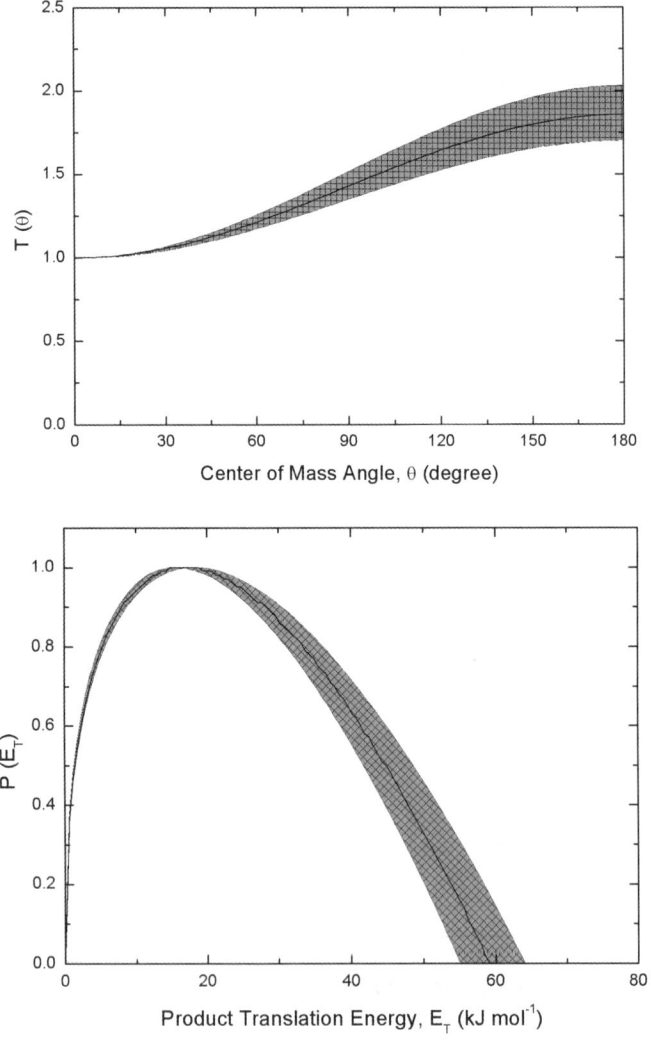

Fig. 14 Center-of-mass angular (top) and translational energy distributions (bottom) of the C$_3$N radical plus atomic hydrogen channel open in the reaction of dicarbon with hydrogen cyanide; the collision energy of the experiment was 42.4 kJ mol^{-1}.

cyanoacetylene – to form atomic hydrogen and the CCCN radical. However, in case of Fig. 14 and the involvement from channels on the singlet and triplet surface, we can therefore propose that the triplet surface should involve a reaction intermediate which decomposes *via* a rather tight exit transition state. Further, the center-of-mass angular distribution (Fig. 14) depicts intensity over the complete scattering range from $\theta = 0°$ to $\theta = 180°$. This pattern is indicative of the involvement of reaction intermediates. Most intriguing, at both lower collision energies of 22.4 kJ mol^{-1} and of 25.8 kJ mol^{-1}, in which only singlet dicarbon reacted, the authors observed a forward-scattered distribution with the intensity in the forward hemisphere increasing as the collision energy rose. Therefore, if only the singlet channel reacted at an even higher collision energy of 42.4 kJ mol^{-1} as in the present experiment, we would predict an even greater forward-peaking of the center-of-mass angular distribution (typical osculating complex model if only one reaction channel and one decomposing intermediate, here cyanoacetylene HC$_3$N, is involved). However, this is clearly not observed. Therefore, the backward scattering is likely attributable to the involvement of the reactions of triplet dicarbon.

Having unraveled that both the singlet and the triplet dicarbon molecules lead to distinct scattering dynamics of an osculating complex (singlet surface) and a backward-scattering (triplet surface), we are now refitting the laboratory data and superimpose a two channel fit for the singlet and triplet channel separately in an attempt to untangle the contribution of singlet *versus* triplet dicarbon to the reactive scattering signal. Again, it must be stressed that we could also fit the data of $m/z = 50$ with a single channel; the separation of the single center of mass function into two is carried out because we have explicit evidence of the reaction on the singlet and triplet surfaces. The results of this procedure are shown in Fig. 15 and 16. Here, the TOF spectra and the laboratory angular distribution can be fit with two reaction channels accounting for singlet and triplet dicarbon with relative weighting factors of about one to six. Assuming a similar reaction cross section, this demonstrates that dicarbon in its first excited triplet state is much more abundant in the ablation beam at a collision energy of 42.4 kJ mol^{-1} compared to singlet dicarbon. On the singlet surface, the center-of-mass translational energy distribution is

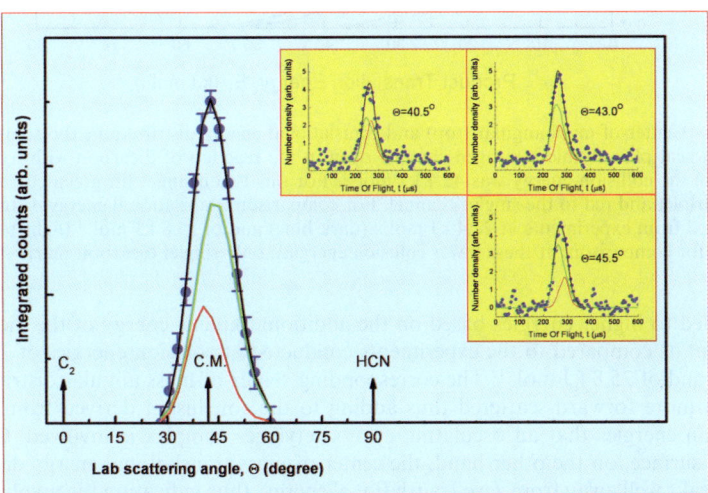

Fig. 15 Laboratory angular distribution and selected time-of-flight (TOF) spectra of the C$_3$N radical, which was formed in the reaction of dicarbon with hydrogen cyanide, recorded at $m/z = 50$ at a collision energy of 42.4 kJ mol^{-1} utilizing a two-channel fit with the center-of-mass functions as shown in Fig. 16; red: singlet channel; green: triplet channel.

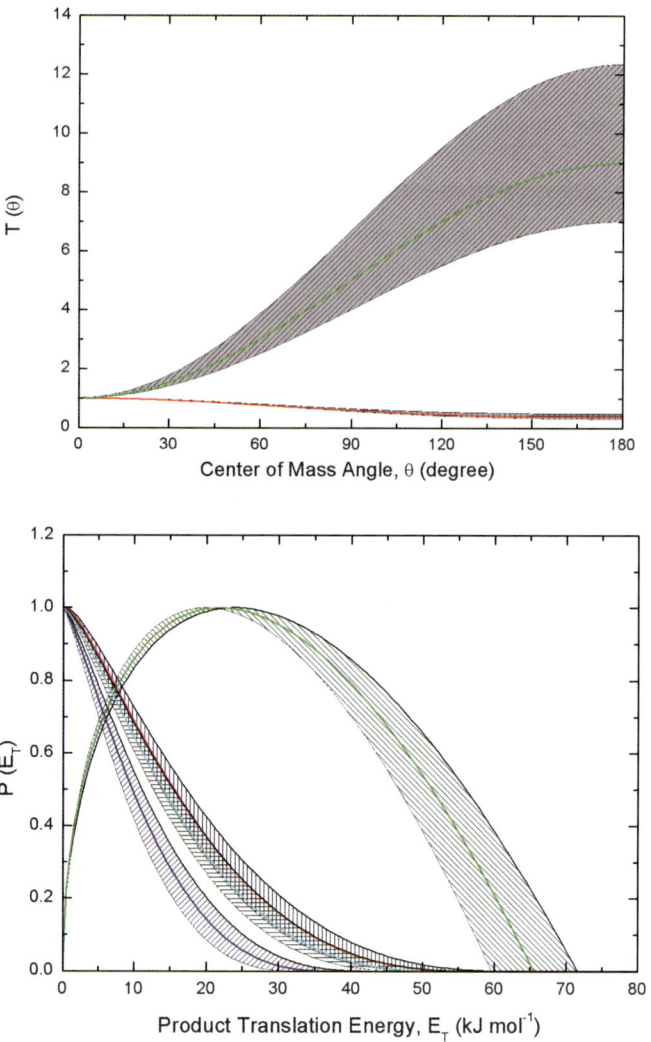

Fig. 16 Center-of-mass angular (top) and translational energy distributions (bottom) of the C_3N radical plus atomic hydrogen channel open in the reaction of dicarbon with hydrogen cyanide; the collision energy was 42.4 kJ mol^{-1}. For this two channel fit, green corresponds to the triplet and red to the singlet channel. For comparison, translational energy distribution extracted from experiments at 22.4 kJ mol^{-1} (dark blue) and of 25.8 kJ mol^{-1} (light blue) are shown for comparison; at these lower collision energies, only singlet dicarbon reacts.

extended to higher energies based on the additional kinetic energy of the dicarbon reactant as compared to the experiments conducted at collision energies of 22.4 kJ mol^{-1} and of 25.8 kJ mol^{-1}. The corresponding center-of-mass angular distribution is also more forward-scattered thus adding to the conclusion derived from lower collision energies that an osculating cyanoacetylenes complex is involved. On the triplet surface, on the other hand, the center-of-mass translational energy distribution peaks well away from zero translational energy, thus indicating the involvement of a tight exit transition state. The pronounced backward scattered center-of-mass angular distribution presents a distinct and unique feature on the triplet surface and is in strong contrast to the osculation complex patterns as found on the singlet manifold.

What reaction mechanism can be proposed on the triplet surface to account for these findings? A comparison of the experimental data with the triplet potential energy surface (Fig. 17) can shed some light on this issue. Here, the computations identified nine reaction intermediates **t1** to **t9**. Intermediate **t8** can be likely ruled out as the decomposing complex since a fragmentation of the latter has no exit barrier; therefore, we would expect a close to zero peaking of the center-of-mass

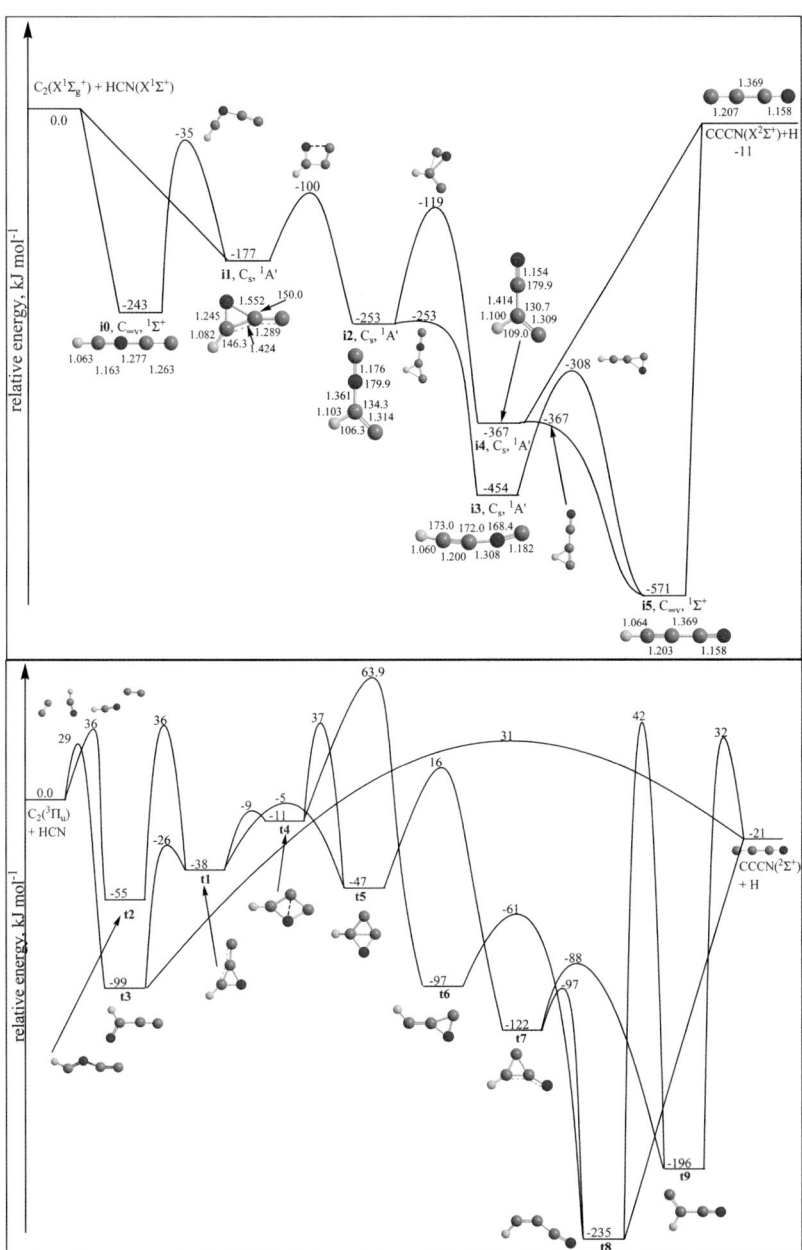

Fig. 17 Schematic triplet potential energy surface of the reaction of triplet dicarbon with hydrogen cyanide. The singlet surface is taken from ref. 29 and shown for comparison.

translational energy distribution. This was clearly not observed on the triplet surface. An alternative reaction pathway of **t8** is the isomerization to **t9** via a barrier located 42 kJ mol^{-1} above the separated products. Since the energetically more favorable pathway of a decomposition of **t8** to CCCN plus atomic hydrogen was not observed, we can conclude that a [1,3]-hydrogen shift in **t8** to **t9** is unlikely to proceed. Therefore, we can conclude that **t8** does not play a significant role in the scattering dynamics. This holds an important conclusion. According to the calculations, **t8** can be only formed from **t7** via a barrier located 98 kJ mol^{-1} below the separated reactants; a competing pathway of **t7** presents the isomerization to **t8**. Since this barrier is even higher than the calculated barrier of the non-observed rearrangement of **t7** to **t8**, we may conclude that **t9** cannot be accessed from **t7** either. Therefore, **t7** together with **t8** and **t9** are likely unimportant reaction intermediates on the triplet surface. Is this a reasonable conclusion? It is important to stress that **t7** can be formed only from isomerization of **t5**; the isomerization of **t6** to **t4** can be excluded since the barrier involved ranges at 64 kJ mol^{-1} with respect to the separated reactants; this is higher than our collision energy of 42.4 kJ mol^{-1}. Intermediate **t5** on the other hand is accessible via **t1** and through a two step sequence involving **t1** and **t4**. On the other hand, considering the locations of the barriers involved, **t1** isomerizes preferentially to **t3**. Therefore, we can conclude that the formation of **t5** is hindered due to the facile isomerization of **t1** to **t3**. Based on these arguments, we limited our discussion to intermediates **t1** to **t3**. The calculations identified two entrance channels, *i.e.* an addition of triplet dicarbon to the carbon and nitrogen atom of the hydrogen cyanide molecule leading to intermediates **t3** and **t2**. Both pathways have entrance barriers of about 29 and 36 kJ mol^{-1}. Recall that in previous experiments at lower collision energies of 22.4 kJ mol^{-1} and of 25.8 kJ mol^{-1}, these entrance channels were closed. Intermediate **t2** either decomposes back to the reactants or isomerizes to **t1**. Considering the location of the barriers, **t1** in turn isomerizes to intermediate **t3**. The latter can decompose to the observed CCCN isomer in an exoergic reaction (-21 kJ mol^{-1}) via a tight exit transition state located about 50 kJ mol^{-1} above the separated products. A tight exit transition state was predicted based on the off-peaking of the center-of-mass translational energy distribution of the triplet dicarbon channel (Fig. 16). Therefore, we propose that triplet dicarbon adds preferentially to the carbon atom of the hydrogen cyanide molecule forming intermediate **t3**; this pathway favors the lowest entrance barrier. The latter is relatively short lived; this intermediate can fragment by emitting the heavy CCCN molecule in the backward direction with respect to the triplet dicarbon beam, while the hydrogen atom leaves in the forward hemisphere. These findings amplify the difference in reaction dynamics of singlet *versus* triplet dicarbon with triplet dicarbon.

Having verified the distinct reactivities of singlet and triplet dicarbon with hydrogen cyanide, it is interesting to compare these findings with the isoelectronic dicarbon–acetylene system investigated earlier in our group.[31] Recall that the reaction of singlet and triplet dicarbon with acetylene was conducted at collision energies between 10.6 and 47.5 kJ mol^{-1}. Here, a singlet dicarbon was found to add without barrier to the acetylene molecule forming intermediate **s1** and/or **s2** (Fig. 22). These collision complexes isomerized yielding ultimately the linear diacetylene molecule (**s3**) which decomposed without exit barrier to the linear 1,3-butadiynyl radical plus atomic hydrogen.[109] Since the diacetylene molecule belongs to the $D_{\infty h}$ point group, the hydrogen atom could be emitted with equal probability from either carbon atom of the rotationally excited diacetylene molecule. Since diacetylene can only be excited to B-like rotations, this results in a forward–backward symmetric center-of-mass angular distribution at all collision energies. On the triplet surface, the rising collision energy resulted in a more pronounced backward scattering of the 1,3-butadiene product. However, the corresponding TOF spectra and laboratory angular distributions could be fit with a single reaction channel. Since the LIF studies provided explicit evidence on the presence of triplet dicarbon, we have re-analyzed the laboratory data for this system with a two-channel fit – one for the singlet and a second one for the triplet surface. On the singlet surface, we further imposed a forward–backward symmetric center-of-mass angular distribution due to the 'symmetric' diacetylene intermediate. The results are compiled in Fig. 18–21. Indeed, the two channel approach also leads to fits of an identical quality as the single channel approach. Further, on the singlet surface, the translational energy angular distributions all peaked at zero translational energy as expected for a barrierless decomposition of the diacetylene molecule. These are identical patterns as observed for the singlet dicarbon–hydrogen cyanide system. On the other hand, the center-of-mass functions for the triplet channel are quite distinct. First, the

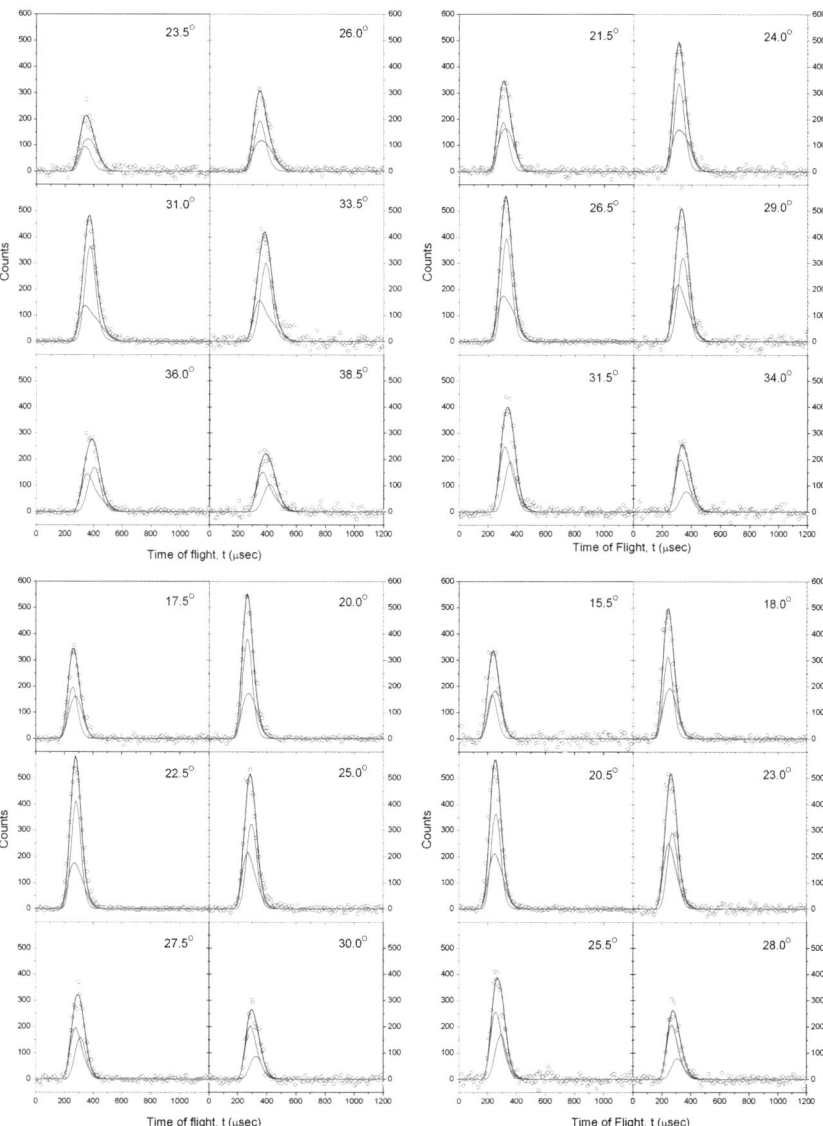

Fig. 18 Time-of-flight spectra recorded at mass-to-charge, m/z, of $m/z = 49$ (C_4H^+) utilizing a two channel fit of the reaction of acetylene with dicarbon leading to the 1,3-butadiynyl radical plus atomic hydrogen at four collision energies of 21.6 (upper left), 29.0 (upper right), 39.9 (lower left), and 47.5 kJ mol^{-1} (lower right). Red: singlet channel; blue: triplet channel.

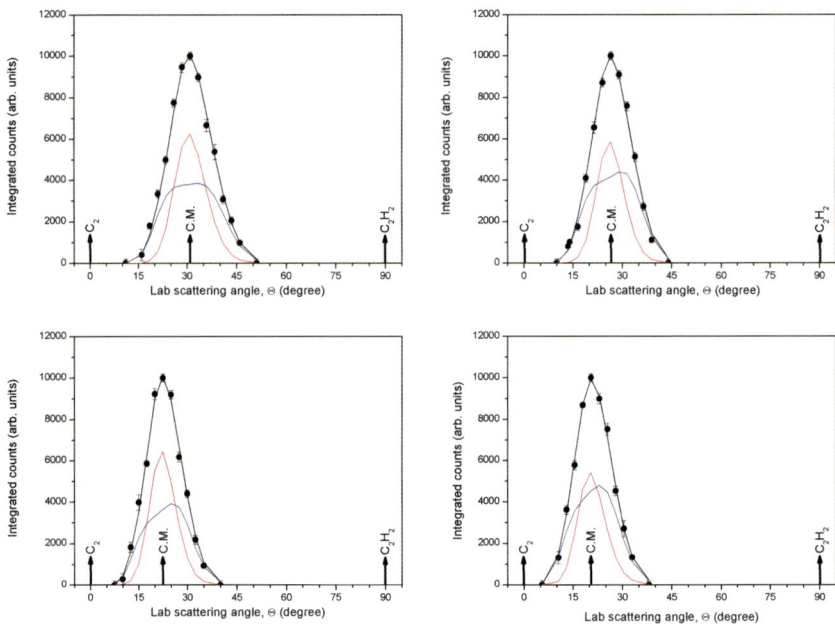

Fig. 19 Laboratory angular distributions at mass-to-charge, m/z, of $m/z = 49$ (C_4H^+) utilizing a two channel fit of the reaction of acetylene with dicarbon leading to the 1,3-butadiynyl radical plus atomic hydrogen at four collision energies of 21.6 (upper left), 29.0 (upper right), 39.9 (lower left), and 47.5 kJ mol^{-1} (lower right). Red: singlet channel; blue: triplet channel.

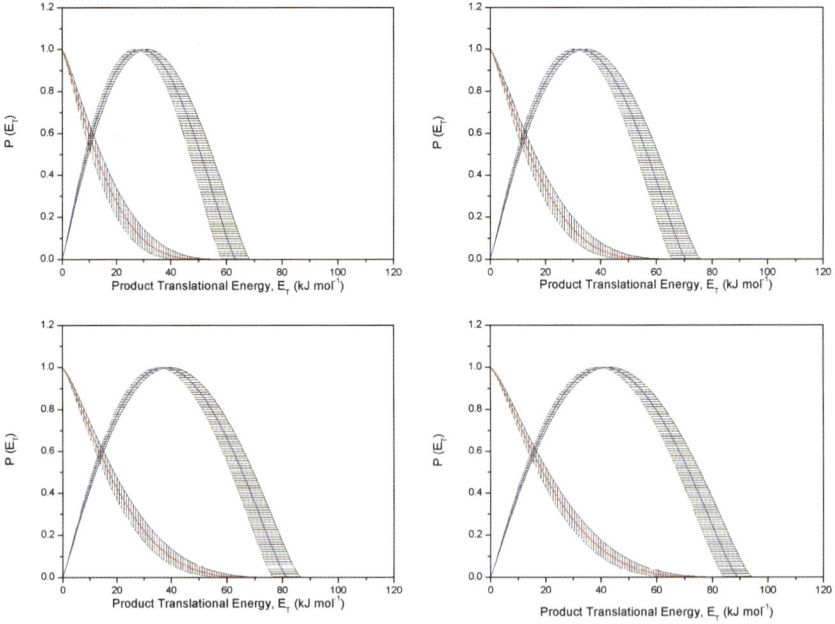

Fig. 20 Center-of-mass translational energy distributions utilizing a two channel fit of the reaction of acetylene with dicarbon leading to the 1,3-butadiynyl radical plus atomic hydrogen at four collision energies of 21.6 (upper left), 29.0 (upper right), 39.9 (lower left), and 47.5 kJ mol^{-1} (lower right). Red: singlet channel; blue: triplet channel.

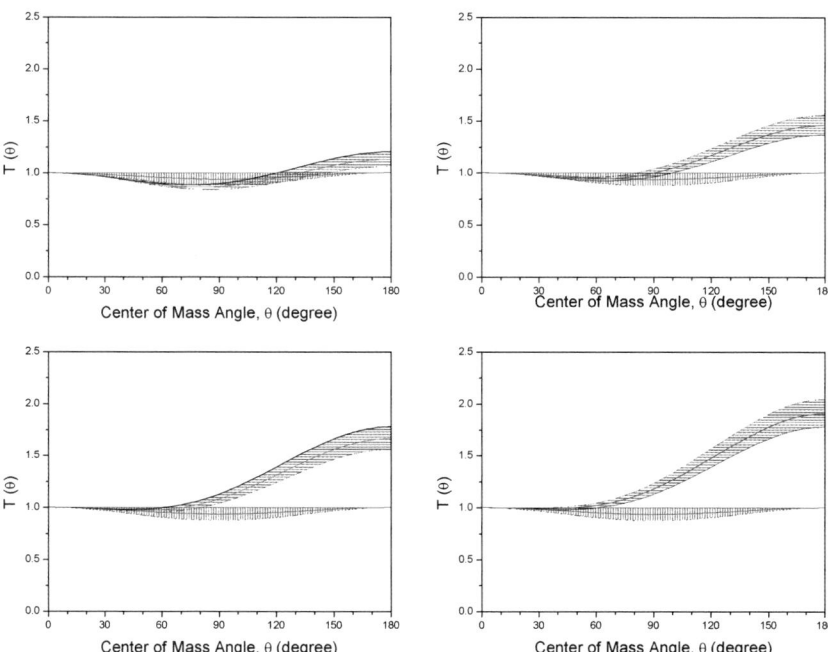

Fig. 21 Center-of-mass angular distributions utilizing a two channel fit of the reaction of acetylene with dicarbon leading to the 1,3-butadiynyl radical plus atomic hydrogen at four collision energies of 21.6 (upper left), 29.0 (upper right), 39.9 (lower left), and 47.5 kJ mol^{-1} (lower right). Red: singlet channel; blue: triplet channel.

translational energy distributions (Fig. 20) depict well-pronounced maxima as characteristic for a tight exit transition state. The existence of a tight exit transition state is verified by the triplet potential energy surface; here, intermediate **t3** decomposes to 1,3-butadienyl radicals plus atomic hydrogen (Fig. 22); **t3** is formed initially *via* isomerization of the initial collision complexes **t1** and **t2**. Secondly, the center-of-mass angular distribution are increasingly backward scattered as the collision energy rises (for all systems investigated here, we have almost constant triplet to singlet dicarbon ratios of about three to one). Therefore, this trend cannot be explained with an increase in the concentration of triplet dicarbon in the beam. However, as extracted from the LIF studies, as the collision energy rises from 21.6 to 47.5 kJ mol^{-1}, so does the rotational temperature increase from about 100 K to about 350 K. Therefore, these results suggest that the increased rotational temperature of the triplet dicarbon beam and the enhanced collision energy might be responsible for an enhancement in the backward-scattering. Recall that the reaction channel of triplet dicarbon with isoelectronic hydrogen cyanide also leads to a backward-scattered center-of-mass angular distribution (Fig. 16). There seems to be – at least in the case of acetylene and hydrogen cyanide – some consistency that the enhanced rotational temperature of dicarbon and/or the rising collision energy leads to an amplification of the scattering signal in the backward hemisphere.

Finally, we would like to comment briefly on preliminary results of the reaction of methylidyne radicals with diacetylene (C_4H_2) and methylacetylene (C_3H_4) (Fig. 23). Here, preliminary data suggest that in both systems, the methylidyne radical formally replaces at least a hydrogen atom leading to the products of the generic formulae C_5H_2 and C_4H_4, respectively. Those preliminary data demonstrate the feasibility of the reactions of methylidyne radicals with hydrocarbons of relevance

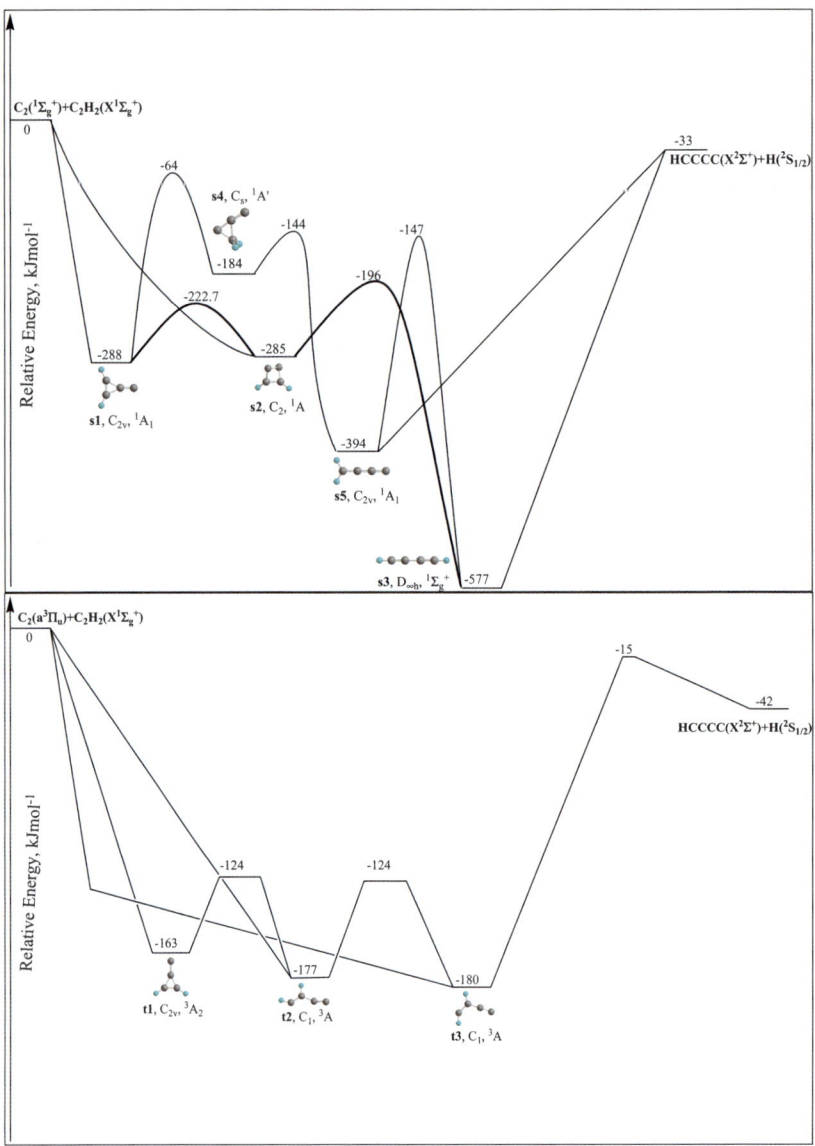

Fig. 22 Potential energy surfaces (PES) of the reactions of $C_2(X^1\Sigma_g^+)$ (upper) and $C_2(a^3\Pi_u)$ (lower) with acetylene, $C_2H_2(X^1\Sigma_g^+)$, adapted from ref. 31.

to Titan's atmosphere and depict the capability to conduct scattering experiments under single collision condition with this important diatomic radical species.

4.2. Surface scattering experiments

4.2.1. Ethane ices. Fig. 24(a) displays the mid-infrared spectrum of pristine ethane (C_2H_6) ice at 50 K. In the higher frequency region of the spectrum, the fundamental bands associated with the ν_{10} (2970 cm^{-1}) and the ν_5 (2878 cm^{-1}) CH$_3$ stretching vibration modes are clearly visible as the strongest absorption features. The deformation modes associated with the methyl (CH$_3$) group are located at lower frequencies, specifically the ν_{11} CH$_3$ deformation region (1468–1450 cm^{-1}), the ν_6

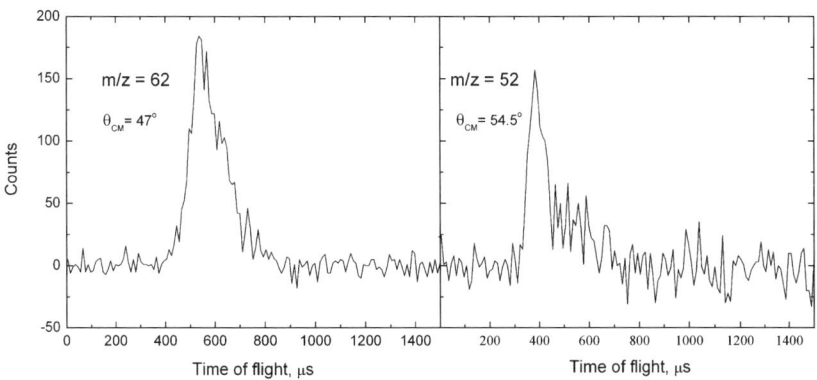

Fig. 23 Time-of-flight spectra of the C_5H_2 and C_4H_4 reaction products as detected *via* their molecular ions at $m/z = 62$ and 52 in the reaction of the methylidyne radical with diacetylene (left) and methylacetylene (right) at the corresponding center-of-mass angles.

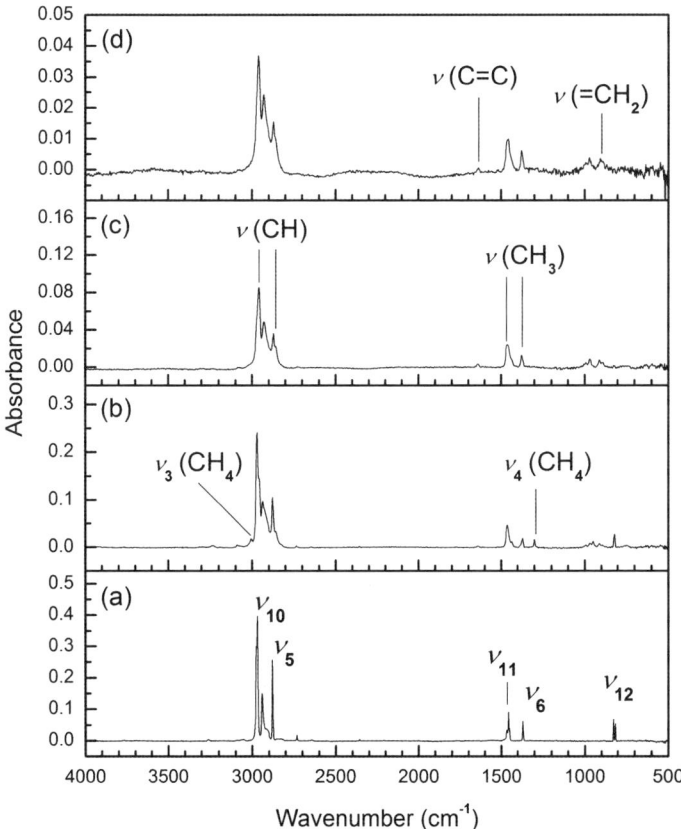

Fig. 24 Mid-infrared spectra of (a) pristine ethane ice at 50 K, (b) electron irradiated (500 nA, 3 h) ethane ice at 50 K, (c) electron irradiated ethane ice heated to 94 K, and (d) electron irradiated ethane ice heated to 300 K. Band assignments are compiled in Tables 1 and 2.

Table 1 Vibrational assignments of pristine ethane ice (C_2H_6) at 50 K

Pristine absorption/cm^{-1}	Literature value[a]/cm^{-1}	Assignment
2970	2972	ν_{10} (CH_3 stretch)
2941	2941	$\nu_8 + \nu_{11}$ (combination)
2878	2879	ν_5 (CH_3 stretch)
2733	2736	$\nu_2 + \nu_6$ (combination)
1468–1450	1464	ν_{11} (CH_3 deform)
1369	1370	ν_6 (CH_3 deform)
825–815	820	ν_{12} (CH_3 rock)

[a] Kim et al.[112]

CH_3 deformation band positioned at 1369 cm^{-1}, and the ν_6 CH_3 rocking doublet located at the 825 and 815 cm^{-1} positions. All peaks, which are summarized in Table 1, are in good agreement with literature values assigned to crystalline phase ethane ices.[110]

Having discussed the assignments in the pristine ethane sample, Fig. 24(b) depicts the mid-infrared spectrum of the sample after the 3 h exposure to 5 keV electron irradiation at a nominal current of 500 nA. The absorption features observed in the irradiated ethane sample are compiled in Table 2. Most absorption bands previously assigned to the CH stretching modes of the ethane molecule are observed to decrease in intensity and become broader in appearance. In the lower frequency region, a similar result is observed for the CH_3 deformation modes, where the ν_6 CH_3 rocking doublet originally located at 825 and 815 cm^{-1} for the neat C_2H_6 ice sample combines into the one broad peak located at 820 cm^{-1} in the irradiated sample spectrum. This indicates that the crystalline ethane ice converts to an amorphous phase over the irradiation process.[111] A number of new absorption features, not previously identified in the pristine ethane sample are also observed in the lower frequency region of Fig. 24(b) at 1642, 1435, 1299, 993, 969, 949 and 911 cm^{-1}; these vibrations

Table 2 Proposed vibrational assignments for residue formed from irradiated ethane ice (C_2H_6)

50 K/cm^{-1}	94 K/cm^{-1}	300 K/cm^{-1}	Assignment[a]	Literature[a]/cm^{-1}
3006	—	—	ν_3 (CH_4)[b]	3008[b]
2971	2958	2960	(Asym. CH_3 stretch)	2975–2950
2938	2929	2930	(Asym. CH_2 stretch)	2940–2915
2878	2871	2871	(Sym. CH_3 stretch)	2885–2865
2736	2724	—	Combination	—
1642	1642	1640	(C=C stretch)	1665–1630
1463	1462	1458	(CH_3 deform)	1465–1440
1435	—	—		
1371	1374	1375	(CH_3 deform)	1390–1370
1299	—	—	ν_4 (CH_4)[b]	1300[b]
993	993	—	(HC=CH wag)	1000–910
969	969	970	(CH_3 rock)	1060–900
949	—	—		
911	911	911	(=CH_2 wag)	980–810
—	891	—		
820	820	—	(CH_2 rock)	800–700

[a] Tejada et al.[110] [b] Kim et al.[112]

are assigned predominantly to saturated carriers and to a minor amount to olefinic functional groups. We then warmed up the sample at a rate of 0.5 K min^{-1}. Fig. 24(c) displays the mid-infrared spectrum of the irradiated ethane sample heated to 94 K with the observed absorption features compiled in Table 2. With the sublimation of ethane occurring in the 60–70 K temperature range, it is expected that absorption features associated with the original ethane sample would no longer be observed at 94 K. However, absorption features located in the high frequency region of the C–H stretching fundamentals are still clearly visible, displaying intensities ∼20% of the original ethane bands. These absorption bands are all shifted to the red in the order of ∼10 cm^{-1} from the original positions of the corresponding ethane peaks. This suggests that although the neat ethane sample sublimes at 60–70 K, the irradiated sample and the newly formed molecules can still retain ethane at elevated temperatures higher than the nominal sublimation temperature. In addition, the CH$_3$ deformation bands are also still present at 94 K at 1462 cm^{-1} and 1374 cm^{-1}; they do not display a significant shift from the positions assigned in non-irradiated ethane at 50 K. The irradiation induced 1435, 1299 and 949 cm^{-1} bands, previously identified in the 50 K irradiation spectrum, were observed to disappear after heating to 94 K. Finally, Fig. 24(d) displays the mid-infrared spectrum of remaining residues that were formed from the electron irradiation of ethane after heating the substrate to 300 K. The spectrum is similar in appearance to the spectrum recorded at 94 K in relation to peaks and their positions, except for the observation that the absorption intensities for all bands appear to be approximately 50% of those recorded at 94 K. These absorptions are indicative of a polymeric residue, containing mostly saturated, aliphatic groups and only to a minor amount olefinic units.

In an effort to correlate the observation of new absorption bands to species formed *via* electron irradiation of ethane ices, a mass spectroscopic analysis of the gas phase was performed during controlled heating of the sample from 50 K to 300 K. Fig. 25 displays the ion count profiles of methane (CH$_4$) and ethane (C$_2$H$_6$) generated during the warm-up phase of the pristine ethane ice (Fig. 25(a)) and after the electron irradiation (Fig. 25(b)). No new species were detected in the mass spectra of the gas phase during the actual irradiation period, but only during the sublimation phase of the irradiated ices. Signals for *m/z* = 30 (C$_2$H$_6$$^+$) developed in the range of 60 K to 70 K for the heated pristine sample, confirming the sublimation temperature for neat ethane under the present experimental conditions. However, after exposure to electron irradiation, the signal for the C$_2$H$_6$$^+$ species (*m/z* = 30) also started to increase at 60 K before reaching a 2 × 10^{-9} ion count maxima at 70 K. However, the signal for C$_2$H$_6$$^+$ in the irradiation experiment differed from the neat sample by the observation of an extended sublimation

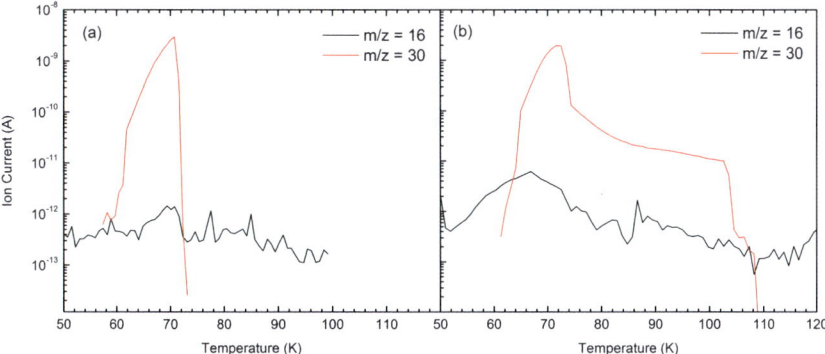

Fig. 25 Ion-count profiles of methane (CH$_4$) and ethane (C$_2$H$_6$) released during the warm-up phase after (a) 0 nA (b) 500 nA irradiation of ethane at 50 K.

temperature profile where ethane was only observed to stop out-gassing at 105 K. In addition, the signal for $m/z = 16$ (CH_4^+) increased between in the 52 K to 80 K temperature range after heating of the irradiated sample. This sublimation temperature range corresponds well with the sublimation profile of CH_4 previously confirmed by Kim et al.[122] using similar experimental conditions, allowing for the positive identification of methane as a degradation product of the irradiated ethane ices in the present study. However, it is important to note that ethane itself contributes to $m/z = 16$ due to electron impact ionization as shown in Fig. 24(a), meaning the blank contribution in the pristine ices must be subtracted to quantifying the methane signal in the irradiation experiments. It should be stressed that no higher molecular weight alkanes were observed to sublime in the irradiated sample.

The electron irradiation of the ethane ice at 50 K resulted in the observation of methane as a primary deposition product. This species is identified in the solid phase by the 3004 cm^{-1} ν_3 (CH_4) stretching fundamental band and the 1299 cm^{-1} ν_4 (CH_4) deformation fundamental band in the associated mid-IR spectrum (Fig. 24(b)). Further evidence of methane formation is provided by the mass spectrum recorded during the controlled heating of the irradiated ethane film. Here, the ion count profile for $m/z = 16$ (CH_4^+) increased between 50 K and 75 K. The complete sublimation of methane after heating the cold target to 94 K is confirmed by the mid-infrared spectrum recorded in Fig. 24(c), where the absorption bands previously attributed to the methane fundamentals disappeared.

As the mass spectrum recorded for the irradiated ethane sample confirms the total sublimation of the original ethane ice at 105 K, we can now shift our attention to the residue remaining on the silver substrate after heating to 300 K. The infrared spectrum presented in Fig. 24(d) clearly displays absorption features that can be attributed to a solid phase hydrocarbon formed as a product of the irradiation. High-frequency absorption bands in the 3000–2850 cm^{-1} region are clearly visible, which are typical of C–H stretching vibrations in aliphatic hydrocarbons. The observation of a distinct absorption band located at 2938 cm^{-1}, a region generally associated with the asymmetric stretching mode of C–H in alkane –CH_2– chain groups (2940–2915 cm^{-1}), suggests that ethane ices form extended aliphatic hydrocarbons upon irradiation.

Recent experiments conducted by Kim et al.[112] using 10 and 100 nA electron beam currents and lower irradiation time of only 60 min have confirmed the irradiation induced formation of n-butane. Although absorption bands could be tentatively assigned to the fundamental modes of the n-butane species in the present study, the absence of a ion count signal at $m/z = 58$ ($C_4H_{10}^+$) over its 90 K to 100 K sublimation

Table 3 Vibrational assignments of pristine propane ice (C_3H_8) at 65 K

Pristine absorption/cm^{-1}	Literature value[a]/cm^{-1}	Assignment
2964	2962	ν_2 (CH_3 stretch)
2935	—	ν_{23} (CH_2 stretch)
2893	2887	ν_{16} (CH_3 stretch)
2871	2887	ν_3 (CH_2 stretch)
2731	2752	$\nu_2 + \nu_6$ (combination)
1472 – 1461	1475–1460	$\nu_4, \nu_5, \nu_{17}, \nu_{24}$ (CH_3 deform)
1389 – 1368	1382–1368	ν_6, ν_{18} (CH_3 deform)
1186	1185	ν_{25} (CH_2 rock)
1155	1155	ν_7 (CH_2 rock)
1049	1050	ν_{20} (CC stretch)
868	869	ν_8 (CC stretch)
747	745	ν_{26} (CH_2 rock)

[a] Goodman et al.[113]

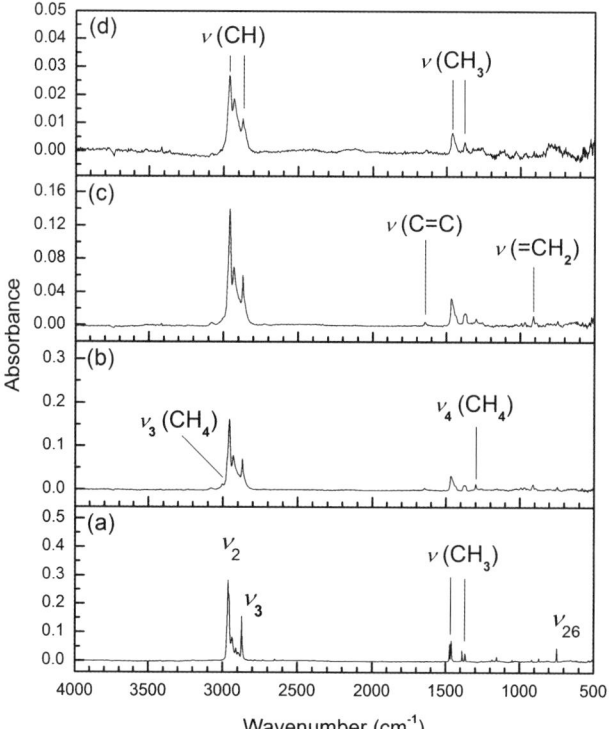

Fig. 26 Mid-infrared spectra of (a) pristine propane ice at 65 K, (b) electron irradiated (500 nA, 3 h) propane ice at 65 K, (c) electron irradiated propane ice heated to 94 K, and (d) electron irradiated propane ice heated to 300 K. Band assignments are compiled in Tables 3 and 4.

range in the corresponding mass spectra indicates that n-butane was not formed under the present irradiation conditions. After consideration of the absorbed energy dosages per ethane molecule, where the previous study calculated a total dosage of 1.4 ± 0.2 eV molecule^{-1} compared to 36 ± 2 eV per molecule calculated for the present experiments, it could be expected that the much higher irradiation dose would induce a continued 'polymerization' of any n-butane formed into higher order aliphatic alkanes.

It is observed through the absorption intensities assigned to the 2960 cm^{-1} asymmetric CH$_3$ stretching vibration and the 2960 cm^{-1} asymmetric CH$_2$ mode of the residue, that the methyl functional group is the stronger of the two absorbing modes. This indicates that the product residue may display a considerable amount of branching in its structure, as opposed to linear aliphatic conformations which would display a greater absorption intensity for the –CH$_2$– chain species. In addition, the 300 K mid-infrared spectrum of the residue provides evidence for unsaturated vinylene centers at 1640 cm^{-1} (C=C stretch) and terminal vinyl groups (=CH$_2$) at 911 cm^{-1}. This indicates also hydrogen loss under the irradiation conditions.

4.2.2. Propane ices. Fig. 26(a) displays the mid-infrared spectrum for pristine propane ice at 65 K. The associated peak positions and assignments are summarized in Table 3. In the C–H stretching region of the spectrum, the absorption bands associated with the terminal ν_2 (2964 cm^{-1}) and ν_{16} (2893 cm^{-1}) CH$_3$ stretching modes can be identified. This is in addition to the main chain ν_{23} (2935 cm^{-1}) and ν_3 (2871 cm^{-1}) CH$_2$ stretching fundamental vibrations that are also observed as strong absorption peaks. The CH$_3$ deformation vibrations are identified in their characteristic positions for aliphatic hydrocarbons, specifically in the regions located at (1472–1461 cm^{-1}) and (1389–1368 cm^{-1}) where a number of CH$_3$ deformation vibrations share similar vibrational frequencies (Table 3). The final low frequency absorption bands identified in the spectrum of pristine C$_3$H$_8$ ice can be assigned to the ν_{20} and ν_8 C–C stretching vibrations located at 1049 and 868 cm^{-1} respectively and the ν_{25}, ν_7 and ν_{26} CH$_3$ rocking vibrations in their respective 1186, 1155 and 747 cm^{-1} positions. All absorption bands observed in the current experimental spectra appear to correspond well with published values previously assigned to the crystalline ethane at 77 K.[113]

Fig. 26(b) depicts the mid-infrared spectrum of the sample after 3 h exposure to 5 keV electron irradiation at 500 nA at 65 K. Absorption bands and associated peak assignments for the irradiated propane ice spectrum are compiled in Table 4. There appears to be significant alteration to the original C$_3$H$_8$ band positions for the C–H stretching modes after irradiation, with most peaks experiencing a 4–6 cm^{-1} shift to lower frequency. This is in addition to displaying an approximate 40% decrease in absorption intensity and a noticeable peak broadening effect. A clear loss in fine structure is also displayed for the CH$_3$ deformation bands following irradiation, where the previously well-defined peaks in the 1472–1461 cm^{-1} and 1389–1368 cm^{-1} regions appear to combine into two separate diffuse features. In addition, novel absorption features that were not previously identified in the non-irradiated C$_3$H$_8$ ice spectrum, are also observed across the entire mid-infrared region of Fig. 26(b). These weak features can be observed at 3075, 3004, 1645, 1299, 993, 970 and 910 cm^{-1}. Again, these can be mainly attributed to hydrogen-rich aliphatic molecules.

Fig. 26(c) displays the mid-infrared spectrum of the irradiated propane sample at 94 K. The observed absorption features are compiled in Table 4. Sublimation of propane was observed to occur in the 70 K to 90 K temperature range from the controlled heating of the pristine propane ice sample; therefore it is expected that very little of the original propane sample would be left on the silver substrate to contribute to the absorption spectrum at 94 K. However, it can be observed in Fig. 26(c) that, similar to the delayed emission of the ethane ice, the high frequency C–H stretching bands are still clearly visible with only slightly lower absorption intensities than those recorded at 65 K after irradiation. Furthermore, the new

Table 4 Possible vibrational assignments for residue formed from irradiated propane ice (C$_3$H$_8$)

65 K/cm^{-1}	94 K/cm^{-1}	300 K/cm^{-1}	Assignment[a]	Literature[a]/cm^{-1}
3075	3072	3075	(= CH$_2$ stretch)	3150–3000
3004	—	—	ν_3 (CH$_4$)[b]	3008[b]
2957	2957	2959	(Asym. CH$_3$ stretch)	2975–2950
2930	2929	2930	(Asym. CH$_2$ stretch)	2940–2915
2870	2869	2871	(Sym. CH$_3$ stretch)	2885–2865
1645	1645	1635	(C=C stretch)	1665–1630
1467	1466	1460	(CH$_3$ deform)	1465–1440
1375	1373	1375	(CH$_3$ deform)	1390–1370
1299	1299	—	ν_4 (CH$_4$)[b]	1300[b]
993	991	—	(HC=CH wag)	1000–910
970	968	967	(CH$_3$ rock)	1060–900
910	910	910	(=CH$_2$ wag)	980–810
748	747	—	ν_{26} (C$_2$H$_6$)	745[c]

[a] Goodman et al.[113] [b] Kim et al.[112]

absorption peaks that were identified in the irradiated ice spectrum at 65 K could all be accounted for at 94 K.

Finally, Fig. 26(d) displays the mid-infrared spectrum of the irradiated propane ice after heating to 300 K. At 300 K any original propane should have long been sublimated from the cold target, leaving only a residue on the substrate. It is clear from the spectrum at 300 K that a residue is formed from the irradiation of the propane ice, with weakly absorbing C–H peaks clearly identified above the baseline in the 2960–2870 cm^{-1} region. Furthermore, absorption features corresponding to the CH_3 deformation modes at 1460 and 1375 cm^{-1} can be identified in the lower frequency regions, as well as the 1635, 967 and 910 cm^{-1} bands that remain after first being identified as irradiation induced features at 65 K.

Mass spectroscopy was employed for the gas phase analysis of sublimated propane during the controlled heating of the sample from 50 K to 300 K. Fig. 27(a) displays the ion count profiles of propane generated during the warm-up phase of the pristine propane ice, while Fig. 27(b) displays the ion count profile of propane during the controlled heating of the sample following electron irradiation. Surprisingly, no newly formed alkanes were observed in the mass spectra collected, neither during the actual irradiation period nor during the warm-up phase of the irradiated propane ices. Signals for $m/z = 44$ ($C_3H_8^+$) generated by the heated pristine sample were observed to increase in the range of 70–90 K (2×10^{-9} maximum ion count), correlating to the sublimation temperature for propane under the present experimental conditions. Following the irradiation of the propane sample, the ion current for the $C_3H_8^+$ species reached a maximum ion count of 5×10^{-10} at 86 K. The sublimation profile of the irradiated propane sample was observed to extend to 94 K, the only discernable difference between the shape of the pristine and irradiated propane ice gas phase sublimation profiles.

The mid-infrared spectrum obtained for irradiated propane ices after controlled heating to 300 K depicts the remaining hydrocarbon residue after sublimation of volatile species from the substrate surface. The general appearance of the residue spectrum is similar to the spectrum of the residue obtained from the irradiated ethane ices discussed earlier. Although the peaks observed in the propane spectrum are about 25% lower in absorption intensity than those observed in the irradiated ethane experiment, the basic patterns are generally the same. The distinctive C–H stretching modes identified at 2959 (asymmetric CH_3), 2930 (asymmetric CH_2) and 2871 cm^{-1} (symmetric CH_3); in addition to the positive identification of CH_3 group deformation modes at 1460 and 1375 cm^{-1}, all indicate that the residue consists of large hydrocarbon molecules with sublimation temperatures above 300 K. It can therefore be concluded that the irradiated propane molecules, each

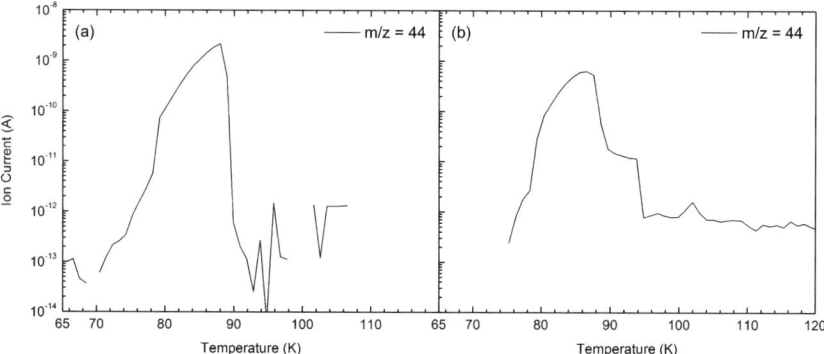

Fig. 27 Ion-count profiles of propane (C_3H_8) released during warm-up phase after (a) 0 nA (b) 500 nA irradiation of propane at 65 K.

exposed to 53 ± 5 eV per molecule, undergo extensive polymerization to form higher-order aliphatic hydrocarbons. In an effort to quantify the conversion of propane to the polymer, an integration of the $m/z = 44$ ($C_3H_8^+$) signals for both the 0 nA and 500 nA mass spectra was performed. A comparison of the signal of propane of the irradiation experiment (3.7×10^{-7} As) to the blank experiment (1.1×10^{-6} As) suggests that about 66% of the propane molecules underwent a irradiation induced alteration; this translates to a loss of $7.2 \pm 0.2 \times 10^{16}$ propane molecules. In our experiments, we irradiated the sample with 1.5×10^{16} electrons per cm^2. Since each electron deposits 410 ± 10 eV into the sample, this leads to a total energy deposition of 6.2×10^{18} eV cm^{-2}. Recall that Titan's energy deposition is of 4.5×10^{9} eV cm^{-2} s^{-1} on its surface. Therefore, the irradiation time in our laboratory corresponds to about 44 years exposure time on Titan. In these 44 'Titan' years, about 5×10^{-6} g of propane are transformed into a polymeric structure per cm^2. Over a typical time of 10^9 years, this results into an accumulation of about 116 g polymeric material per cm^2. Considering a typical density of 0.94 g cm^{-3} for a low density organic polymer, this results in a layer of about 1.2 meter polymer per cm^2. Therefore, our laboratory experiments show that in principle, propane can be converted to aliphatic-type polymers over geological timescales on Titan's surface *via* interaction with cosmic ray particles.

4.3. VUV photoionization studies

Fig. 28 shows a mass spectrum of the products formed from ablation of a silicon rod in a carrier gas of acetylene recorded at a photon energy of 10.5 eV. The inset in Fig. 28 depicts an expanded spectrum between 49–55 amu to demonstrate the resolution and quality of the mass spectrum. The spectrum is complex, but in conjunction with tunable VUV radiation allows for elucidation of the chemical composition *via* ionization energy determinations coupled with calculated results performed in this work and literature values. In Table 5, the most prominent masses are compiled along with their chemical identification. Also presented are calculated and measured ionization energies for a number of these species and previous literature results are

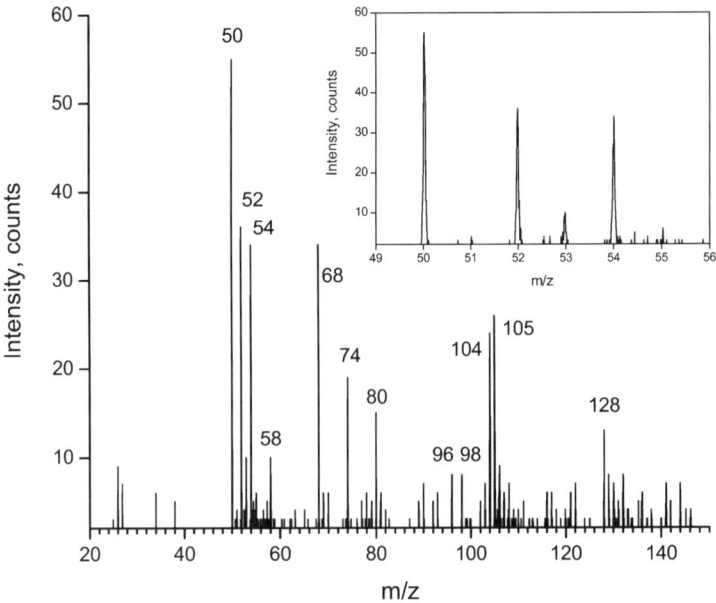

Fig. 28 Mass spectrum at photon energy of 10.5 eV.

Table 5 Mass, chemical species, measured and calculated ionization energies. All energies in eV.[a]

Mass	Species	IE$_{exp}$	IE$_{calc}$	Literature IE
52	SiC$_2$	9.75(±0.025)		10.2 ± 0.3[b], 10.321[d], 9.69(l)[f], 9.79(c)[f], 10.4 ± 0.3[h]
53	SiC$_2$H	9.1		
			7.26 (v) 7.26 (a)	7.06[e]
			8.10 (v) 6.96 (a)	
			9.54 (v) 9.02 (a)	
			9.26 (v)	
54	SiC$_2$H$_2$	9.3(±0.05)		
			9.50 (v) 9.31 (a)	8.97(a)[g], 9.83(v)[g], 9.00(c)[e]
			9.14 (v) 8.72 (a)	
			8.37 (v) 8.26 (a)	
68	Si$_2$C	9.2(±0.025)		9.2 ± 0.3[b], 9.06–9.33[c], 9.626[d], 9.18(l)[f], 9.19(c)[f], 9.5 ± 0.5[h]
80	Si$_2$C$_2$	8.95(±0.05)		7.731[i], 8.2 ± 0.3[b], 8.259[d], 8.24[j], 8.97(c)[f], 8.8 ± 0.5[h]
81	Si$_2$C$_2$H	8.2(±0.1)		
82	Si$_2$C$_2$H$_2$	9.4		

Table 5 (*Contd.*)

Mass	Species	IE$_{exp}$	IE$_{calc}$	Literature IE
92	Si$_2$C$_3$	8.7($\pm$0.1)		8.12(l)^f
96	Si$_3$C	8.4($\pm$0.1)		8.2 $\pm$ 0.3^b, 7.95–8.3^c, 8.679^d, 7.81(c)^f, 7.9 $\pm$ 0.5^h
104	Si$_2$C$_4$	8.45($\pm$0.05)		
105	Si$_2$C$_4$H	8.45($\pm$0.05)		
106	Si$_2$C$_4$H$_2$	8.5($\pm$0.05)		
108	Si$_3$C$_2$	8.5($\pm$0.1)		7.27(c)^f
116	Si$_2$C$_5$	<8.0		
128	Si$_2$C$_6$	8.5($\pm$0.1)		
129	Si$_2$C$_6$H	—		
130	Si$_2$C$_6$H$_2$	8.5		
136	Si$_4$C$_2$	8.5($\pm$0.1)		

^{a} V – vertical ionization energy, A – adiabatic ionization energy, (l) – linear, (c) – cyclic.
^{b} Reference, Drowart et al.[116] ^{c} Reference, Boldyrev et al.[121] ^{d} Reference, Pradhan et al.[120]
^{e} Reference, Ketvirtis et al.[115] ^{f} Reference, Yadav et al.[119] ^{g} Reference, Ikuta et al.[114]
^{h} Reference, Schumde et al.[117] ^{i} Reference Hou et al.[124] ^{j} Reference Parent et al.[122]

also reported in this Table. The following species are ionized in the molecular beam at 10.5 eV: polyynes (C$_4$H$_2$, C$_6$H$_2$, C$_8$H$_2$, C$_{10}$H$_2$), molecules belonging to the SiC$_2$H$_x$ family (x = 0,1,2), to the Si$_2$C$_2$H$_x$ family (x = 0,1,2), silicon–carbon clusters Si$_2$C$_x$ (x = 1–6), Si$_3$C, Si$_3$C$_2$, and Si$_4$C$_2$, as well as Si$_2$C$_4$H$_x$ (x = 1,2) and Si$_2$C$_6$H$_x$ (x = 1,2). A comprehensive investigation of the ionization energies of all these species is beyond the scope of this paper. To demonstrate the feasibility of our approach and to highlight the implications to Titan, we focus on the SiC$_2$H and SiC$_2$H$_2$ systems as case studies. Fig. 29 shows the PIE curves of m/z = 53 and 54 corresponding to the ions of SiC$_2$H and SiC$_2$H$_2$.

Considering SiC$_2$H$_2$, calculations at the CCSD(T)/cc-pVQZ//B3LYP/6-311G** level suggest that a cyclic structure holding C$_{2v}$ symmetry and in the ^{1}A$_1$ electronic ground state is the lowest lying state. A vertical ionization energy (VIE) of 9.5 eV and adiabatic ionization energy (AIE) of 9.3 eV is derived for ionization to a C$_{2v}$ symmetry and ^{2}A$_1$ state of the cation. The calculated adiabatic ionization energy agrees perfectly with our 9.3 eV experimental determination. An L shaped isomer in which silicon inserts between a carbon and hydrogen of the acetylene lies 78.3 kJ mol^{-1} above the ground state in the neutral channel with a calculated VIE and AIE of 9.1 and 8.7 eV respectively. There is no signal in our PIE spectrum at these energies and all the other possible isomers lie much higher in energy to be populated in our molecular beam. This suggests that we are exclusively producing the cyclic isomer of SiC$_2$H$_2$ in our molecular beam. Ikuta et al.[114] performed a comprehensive study on the ionization energies for various isomers of SiC$_2$H$_2$ at the similar CCSD(T)/cc-pVQZ level of theory. They suggest that the global minima for the neutral and cation, (C$_{2v}$, ^{1}A$_1$) and (C$_{2v}$, ^{2}B$_2$) are silacyclopropenylidine structures with VIE and AIE of 9.83 and 8.97 eV respectively. Ketvirtis et al.[115] calculated an AIE of 9.00 eV at the QCISD(T)-(full) level of theory. Our measured IE of 9.3 eV is discrepant with these calculations. The difference between our present calculated results and those by Ikuta et al.[114] can be explained by the discrepancy in the electronic states of the silacyclopropenylidine cation. Our calculations give the ^{2}A$_1$ state as the lowest cationic state at the vertical geometry with VIE of 9.50 eV, lower than 9.83 eV obtained by Ikuta et al. for the ^{2}B$_2$ state. Geometry relaxation in the ^{2}B$_2$ state of the cation is larger than in ^{2}A$_1$ resulting in the lower AIE for

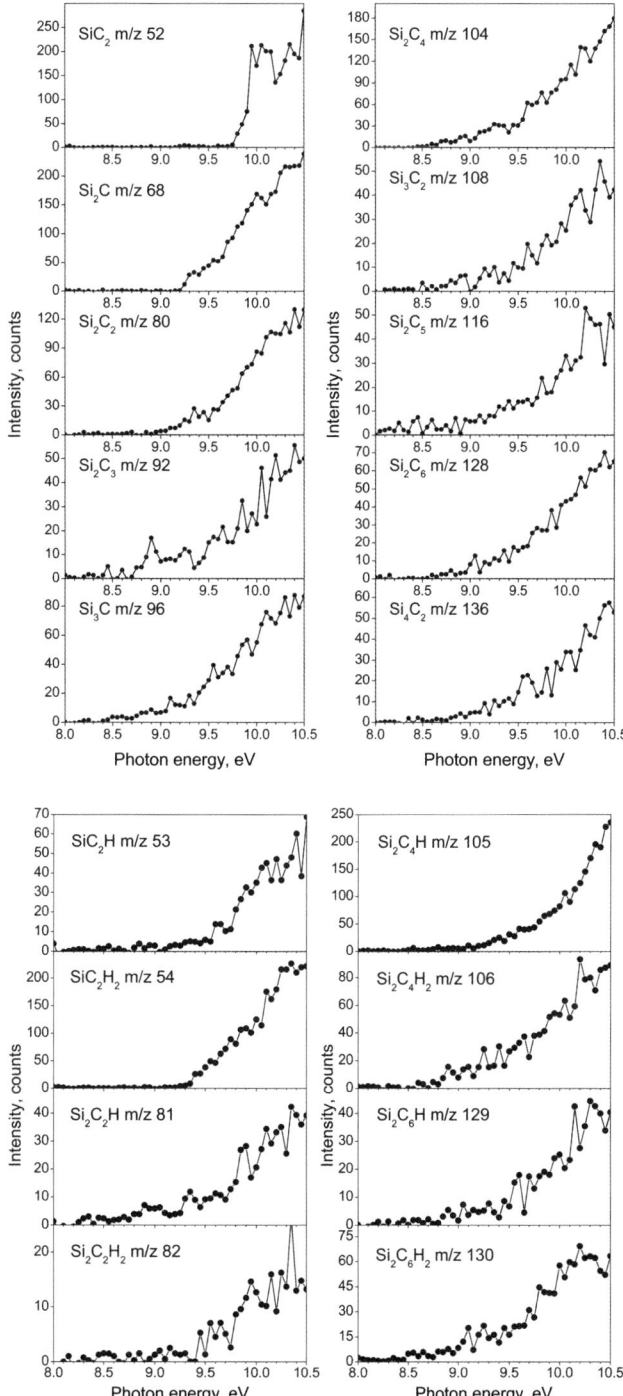

Fig. 29 top: PIE curves for SiC_2, Si_2C_x, ($x = 1–6$), Si_3C, Si_3C_2, Si_4C_2. bottom: PIE curves for SiC_2H, SiC_2H_2, Si_2C_2H, $Si_2C_2H_2$, Si_2C_4H, $Si_2C_4H_2$, Si_2C_6H, and $Si_2C_6H_2$.

$^2B_2$, 8.97 eV, as compared to that for $^2A_1$, 9.31 eV. This means that the $^2A_1$ cationic state has better Franck–Condon factors for ionization from the neutral silacyclopropenylidine SiC_2H_2 than the $^2B_2$ state and the comparison with the experimental PIE curve supports the hypothesis that the $^2A_1$ state of the ion was actually produced.

For SiC_2H, the story is not so straightforward. Calculations show the lowest lying isomer in the neutral channel is of the linear SiCCH radical form with $C_{\infty v}$ symmetry and $^2\Pi$ doublet state with VIE and AIE of 7.3 eV in both cases. Note that Ketvirtis et al.[115] calculated an AIE of 7.06 eV for this isomer at the QCISD(T)-(full) level. From our PIE it would appear that signal arises from around 8.0 eV above the base line and starts rising around 9.0 eV. A C_s, $^2A'$ cyclic isomer is calculated to lie 29.3 kJ mol^{-1} above the linear ground state. Calculations show that this state upon ionization rearranges to a linear SiCCH cation holding $C_{\infty v}$ symmetry with VIE and AIE of 8.10 eV and 6.96 eV, respectively. It is plausible that these states because of bad Franck–Condon factors do not show much intensity in the ionized cationic states. The isomers which have calculated ionization energies around 9.0 eV, include chain C_s-symmetric CCSiH and cyclic non-planar HSiCC, lying 192.5 and 175.8 kJ mol^{-1} above the lowest SiCCH structure. Although these high-energy isomers seem unlikely to be populated in the molecular beam, a plausible assignment of the observed PIE can be made based on the chain CCSiH isomer with calculated VIE and AIE of 9.54 and 9.02 eV, respectively. It should be noted that CCSiH can be produced directly from the L shaped isomer of SiC_2H_2 by the C–H bond cleavage. As the low-lying L shaped SiC_2H_2 isomer has not been observed in the PIE spectra, it is possible that it does not survive and dissociates to CCSiH, which in turn gives rise to the PIE curve measured from SiC_2H. On the other hand, according to our calculations, the cyclic H–SiCC isomer with a VIE of 9.26 eV decomposes when ionized indicating bad Franck–Condon factors for ionization.

There are very few measurements in the literature to which we can compare our measured appearance energies for the SiC systems. Drowart et al.[116] reported electron impact ionization results for SiC_2, Si_2C, Si_2C_2, and Si_3C to be 10.2, 9.2, 8.2, and 8.2 eV respectively with quoted errors of ±0.3 eV. Our measurements for SiC_2 and Si_2C_2 fall outside these error bars. Schmude and Gingerich[117] also reported electron impact appearance energies of SiC_2 (10.4 ± 0.3 eV), Si_2C (9.5 ± 0.5 eV), Si_2C_2 (8.8 ± 0.5 eV) and Si_3C (7.9 ± 0.5 eV). The value for SiC_2 falls outside the error limits compared to our measured value, and the relatively large errors in these electron impact measurements for the other species do not allow for reliable thermodynamic information to be extracted. Furthermore, the value for the silicon dimer (Si_2) measured in both electron impact measurements, 7.0 ± 0.5 eV and 7.4 ± 0.3 eV, fall well outside a recent VUV photoionization determination of 7.92 ± 0.05 eV reported by our group.[118] A recent theoretical calculation[119] at the DFT B3LYP/6-311G(3df) level report ionization energies of 9.69 eV (linear SiCC), 9.79 eV (triangular CSiC), 9.18 eV (linear SiCSi), 9.19 eV (triangular SiCSi), and 8.97 eV (rhombus Si_2C_2). These ionization energies agree well with our experimental results and, since the calculations are likely to be vertical in nature, would suggest good Franck–Condon factors for ionization. Their calculated value of 7.81 eV for rhomboidal Si_3C does not agree with our measured value of 8.4 eV and could likely arise from a different geometry being present in our molecular beam. The agreement with the other values does suggest that our measured ionization energies are more reliable than the early electron impact measurements. In terms of structure, it is widely believed that SiC_2 exists in the cyclic form and our results would indicate that this is indeed possible. A recent LDA-DFT study by Pradhan and Ray[120] report adiabatic ionization energies of 10.38–10.78 eV (SiC_2), 9.629 eV (Si_2C), 8.26 eV (Si_2C_2), and 8.68 eV (Si_3C) which are systematically above our measured energies suggesting an error in the local density approximation adopted by these authors. Our measured ionization energy of 9.2 eV for Si_2C agrees extremely well with the calculated adiabatic value of 9.2 ± 0.2 eV at various levels of theory for a transition to a linear $^2\Sigma_u^+$ state in the cation.[121] The same authors also calculated ionization energies for Si_3C using similar methods and suggest a vertical IE of 8.2 ± 0.2 eV for a C_{2v} $^1A_1$ state rhomboid structure transition to a C_{2v} $^2B_2$ state in the cation. They also suggest that the geometry of the neutral varies considerably from the cation leading to poor Franck–Condon factors and an adiabatic ionization energy of 7.8 ± 0.2 eV. In this particular case our appearance energy of 8.4 eV does not fall in with this prediction, however, the shape of the PIE curve is not steep but a gentle rise suggesting poor Franck–Condon factors for this species. For Si_2C_2, a charge transfer energy bracketing method[122] posits the ionization energy of 8.24 ± 0.2 eV which does not agree with our value of 8.95 eV. For the unrelaxed ion in the charge bracketing measurement, an ionization energy of 9.35 ± 0.1 eV was measured. It is possible that there was incomplete relaxation of the ion in the time frame of the charge bracketing experiment giving rise to a lower ionization energy. Ignatyev and Schaefer[122] report values of the ionization energies to the lowest energy state of the cation ($^2\Pi_g$) holding linear structure and the next highest state (^2A_g), a slightly distorted rhombic structure which is similar in structure to the neutral ground state. This rhombic transition could be what is observed in our photoionization experiment as it has been suggested by numerous theoretical calculations[123] that the linear Si–C–C–Si and a rhombic structure are almost degenerate in energy. It is also interesting to point out that Yadav et al.[119] calculate an IE of 8.97 eV for this transition which is in very good agreement with our measured appearance energy. For Si_2C_3, Si_2C_4, Si_2C_5, Si_2C_6, Si_3C_2, and Si_4C_2, there are no experimental or theoretical data on ionization energies in the literature to the best of our knowledge, and the results reported here are the first experimental determinations. There could be multiple isomers present in the molecular beam, coupled with different Franck–Condon factors for these isomers and these are probably reflected in the gentle rise and absence of sharp onsets in this set of PIE curves.

5. Conclusions

At this *Faraday Discussion*, we presented three research fields of interest to explore the chemical evolution of Titan's atmosphere and surface. First, we explored the chemical dynamics of simple diatomic radicals (dicarbon, methylidyne) utilizing the crossed molecular beams method. This versatile experimental technique can be applied to study reactions relevant to the atmospheres of planets and their moons as long as intense and stable supersonic beam sources of the reactant species exist. By focusing on reactions of dicarbon with hydrogen cyanide, we could untangle for the first time the contribution of dicarbon in its singlet ground and first excited triplet states. These results were applied to understand and re-analyze the data of crossed beam reactions of the isoelectronic dicarbon plus acetylene reaction. The reactions of dicarbon with hydrogen cyanide (HCN) and acetylene (C_2H_2) lead to doublet radical by hydrogen emission:

$$C_2(X^1\Sigma_g^+/a^3\Pi_u)) + HCN(X^1\Sigma^+) \rightarrow CCCN(X^2\Sigma^+) + H(^2S) \qquad (9)$$

$$C_2(X^1\Sigma_g^+/a^3\Pi_u)) + C_2H_2(X^1\Sigma_g^{\ +}) \rightarrow CCCCH(X^2\Sigma^+) + H(^2S) \qquad (10)$$

These radicals can react further with unsaturated hydrocarbons to form highly unsaturated hydrocarbons and nitriles, which could act as precursor molecules to the organic aerosol layers. Note that the kinetics of the 1,3-butadiynyl radical reactions with unsaturated hydrocarbons are very fast even at low temperatures relevant to Titan's atmosphere.[124] However, the reaction products are elusive. Crossed beam experiments are planned to untangle the dynamics of these reactions in a similar way as those of the important methylidyne reactions.

Second, we investigated the interaction of ionizing radiation in form of energetic electrons with organic molecules ethane and propane sequestered on Titan's surface. These experiments presented compelling evidence that even at irradiation exposures equivalent to about 44 years on Titan's surface, aliphatic-like organic residues can be produced. Scaling this to a geological time scale of, for instance, 10^9 years on Titan, this results in case of propane irradiation exposure into an accumulation of 1.5 meter organic polymer per cm^2. Therefore, our laboratory experiments show that in principle, propane can be converted to aliphatic-type solid polymers over geological timescales on Titan's surface *via* interaction with cosmic ray particles. It is important in future models of Titan to account for the radiation-induced modification of Titan's surface and to evaluate objectively its contribution compared to deposition from organics by 'sedimentation' from higher atmospheric layers. Nevertheless, a galactic cosmic ray driven chemistry will certainly influence Titan's overall hydrocarbon budget.

Finally, we investigated how Titan's nascent chemical inventory can be altered by an *external* influx of matter as supplied by (micro)meteorites and possibly comets. For this, we simulated the ablation process in Titan's atmosphere, which can lead to ground and electronically excited atoms of, for instance, the principal constituents of silicates like iron, silicon, and magnesium, in laboratory experiments. By ablating silicon species and seeding the ablated species in acetylene carrier gas, which also acts as a reactant, we produced organo silicon species, which were then photoionized utilizing tunable VUV radiation from the Advanced Light Source. In combination with electronic structure calculations, the structures and ionization energies of distinct organo-silicon species were elucidated. We plan to expand these studies also to iron and magnesium-bearing species in the near future.

In summary, laboratory experiments such as crossed molecular beams scattering experiments, surface scattering studies, and laser ablation studies coupled with electronic structure calculations and VUV photoionization provide a powerful tool to shed light on Titan's chemistry not only of the atmosphere, but also of Titan's surface. To understand Titan's chemistry comprehensively, surfaces, aerosols, and the gas phase cannot be treated separately, but must be incorporated into a single model. Naturally, this will present a nice challenge to the modeling community. However, by providing feedback between experimentalists, modelers, and observers, in principle these models can be refined iteratively until a satisfactory agreement between experiments, observations, and models is reached – not only with respect to Titan, but for any body in our Solar System.

Acknowledgements

The crossed beam (P. M., C. E., F. Z., X. G., R. I. K.) and electronic structure calculations (S. P. K., A. M. M.) were supported by the Chemistry Division of the US National Science Foundation within the framework of the Collaborative Research in Chemistry (CRC) Program (NSF-CRC CHE-0627854). The experiments at the Advanced Light Source (OK, MA) was supported by the Director, Office of Energy Research, Office of Basic Energy Sciences, and Chemical Sciences Division of the U.S. Department of Energy under contracts No. DE-AC02-05CH11231.

References

1 D. Cordier, O. Mousis, J. I. Lunine, P. Lavvas and V. Vuitton, *Astrophys. J.*, 2009, **707**, L128.
2 A. Coustenis and M. Hirtzig, *Res. Astron. Astrophys.*, 2009, **9**, 249; J. L. Mitchell, *J. Geophys. Res.*, 2008, **113**, E08015.
3 L. Le Corre, S. Le Mouelic, C. Sotin, J. P. Combe, S. Rodriguez, J. W. Barnes, R. H. Brown, B. J. Buratti, R. Jaumann, J. Soderblom, L. A. Soderblom, R. Clark, K. H. Baines and P. D. Nicholson, *Planet. Space Sci.*, 2009, **57**, 870.
4 C. Elachi, S. Wall, M. Allison, Y. Anderson, R. Boehmer, P. Callahan, P. Encrenaz, E. Flamini, G. Franceschetti, Y. Gim, G. Hamilton, S. Hensley, M. Janssen, W. Johnson, K. Kelleher, R. Kirk, R. Lopes, R. Lorenz, J. Lunine, D. Muhleman, S. Ostro, F. Paganelli, G. Picardi, F. Posa, L. Roth, R. Seu, S. Shaffer, L. Soderblom,

B. Stiles, E. Stofan, S. Vetrella, R. West, C. Wood, L. Wye and H. Zebker, *Science*, 2005, **308**, 970.
5 E. L. Schaller, H. G. Roe, T. Schneider and M. E. Brown, *Nature*, 2009, **460**, 873.
6 C. A. Griffith, *Nature*, 2006, **442**, 362.
7 P. Lavvas, R. V. Yelle and V. Vuitton, *Icarus*, 2009, **201**, 626.
8 Y. Sekine, H. Imanaka, T. Matsui, B. N. Khare, E. L. O. Bakes, C. P. McKay and S. Sugita, *Icarus*, 2008, **194**, 186.
9 M. G. Trainer, A. A. Pavlov, H. L. DeWitt, J. L. Jimenez, C. P. McKay, O. B. Toon and M. A. Tolbert, *Proc. Natl. Acad. Sci. U. S. A.*, 2006, **103**, 18035.
10 T. B. McCord, G. B. Hansen, B. J. Buratti, R. N. Clark, D. P. Cruikshank, E. D'Aversa, C. A. Griffith, E. K. H. Baines, R. H. Brown, C. M. Dalle Ore, G. Filacchione, V. Formisano, C. A. Hibbitts, R. Jaumann, J. I. Lunine, R. M. Nelson and C. Sotin, *Planet. Space Sci.*, 2006, **54**, 1524.
11 A. Coustenis, B. Bezard and D. Gautier, *Icarus*, 1989, **80**, 54; A. Coustenis, B. Bezard, D. Gautier, A. Marten and R. Samuelson, *Icarus*, 1991, **89**, 152.
12 A. Coustenis, R. K. Achterberg, B. J. Conrath, D. E. Jennings, A. Marten, D. Gautier, C. A. Nixon, F. M. Flasar, N. A. Teanby, B. Bezard, R. E. Samuelson, R. C. Carlson, E. Lellouch, G. L. Bjoraker, P. N. Romani, F. W. Taylor, P. G. J. Irwin, T. Fouchet, A. Hubert, G. S. Orton, V. G. Kunde, S. Vinatier, J. Mondellini, M. M. Abbas and R. Courtin, *Icarus*, 2007, **189**, 35.
13 A. Canosa, A. Paramo, S. D. Le Picard and I. R. Sims, *Icarus*, 2007, **187**, 558.
14 X. Gu, Y. Guo, F. Zhang, A. M. Mebel and R. I. Kaiser, *Faraday Discuss.*, 2006, **133**, 245.
15 A. Canosa, I. R. Sims, D. Travers, I. W. M. Smith and B. R. Rowe, *Astron. Astrophys.*, 1997, **323**, 644.
16 R. I. Kaiser and N. Balucani, *Acc. Chem. Res.*, 2001, **34**, 699.
17 X. Gu, Y. S. Kim, R. I. Kaiser, A. M. Mebel, M. C. Liang and Y. L. Yung, *Proc. Natl. Acad. Sci. U. S. A.*, 2009, **106**, 16078.
18 F. Zhang, Y. S. Kim, R. I. Kaiser, S. P. Krishtal and A. M. Mebel, *J. Phys. Chem.*, 2009, **113**, 11167.
19 X. Zhang, J. M. Ajello and Y. L. Yung, *Astrophys. J.*, **708**, L18; E. H. Wilson and S. K. Atreya, *J. Phys. Chem.*, 2009, **113**, 11221; E. Hebrard, M. Dobrijevic, P. Pernot, N. Carrasco, A. Bergeat, K. M. Hickson, A. Canosa, S. D. Le Picard and I. R. Sims, *J. Phys. Chem.*, 2009, **113**, 11227; J. Cui, R. V. Yelle, V. Vuitton, J. H. Waite, Jr., W. T. Kasprzak, D. A. Gell, H. B. Niemann, I. C. F. Muller-Wodarg, N. Borggren, G. G. Fletcher, E. L. Patrick, E. Raaen and B. A. Magee, *Icarus*, 2009, **200**, 581; V. A. Krasnopolsky, *Icarus*, 2009, **201**, 226.
20 D. Cordier, O. Mousis, J. I. Lunine, A. Moudens and V. Vuitton, *Astrophys. J.*, 2008, **689**, L61; J. I. Lunine and S. K. Atreya, *Nat. Geosci.*, 2008, **1**, 159; H. Imanaka and M. A. Smith, *J. Phys. Chem.*, 2009, **113**, 11187; V. Vuitton, J. F. Doussin, Y. Benilan, F. Raulin and M. C. Gazeau, *Icarus*, 2006, **185**, 287.
21 C. A. Arrington, C. Ramos, A. D. Robinson and T. S. Zwier, *J. Phys. Chem.*, 1999, **103**, 1294.
22 D. W. Clarke and J. P. Ferris, *Icarus*, 1997, **127**, 158.
23 A. Aflalaye, R. Sternberg, F. Raulin and C. Vidal-Madjar, *J. Chromatogr., A*, 1995, **708**, 283; R. Sternberg, C. Szopa, D. Coscia, S. Zubrzycki, F. Raulin, C. Vidal-Madjar, H. Niemann and G. Israel, *J. Chromatogr., A*, 1999, **846**, 307; A. Aflalaye, R. Sternberg, D. Coscia, F. Faulin and C. Vidal-Madjar, *J. Chromatogr., A*, 1997, **761**, 195.
24 F. Zhang, S. Kim, R. I. Kaiser, A. Jamal and A. M. Mebel, *J. Chem. Phys.*, 2009, **130**, 234308.
25 D. Chastaing, P. L. James, I. R. Sims and I. W. M. Smith, *Faraday Discuss.*, 1998, **109**, 165; D. Carty, V. Le Page, I. R. Sims and I. W. M. Smith, *Chem. Phys. Lett.*, 2001, **344**, 310.
26 W. M. Jackson and A. Scodinu, *Astrophys. Space Sci. Libr.*, 2004, **311**, 85; G. Apaydin, W. H. Fink and W. M. Jackson, *J. Chem. Phys.*, 2004, **121**, 9368; N. S. Smith and F. Raulin, *J. Geophys. Res.*, 1999, **104**, 1873; N. S. Smith, Y. Benilan and P. Bruston, *Planet. Space Sci.*, 1998, **46**, 1215; R. J. Cody, P. N. Romani, F. L. Nesbitt, M. A. Iannone, D. C. Tardy and L. J. Stief, *J. Geophys. Res.*, 2003, **108**, 5119; R. J. Cody, W. A. Payne, Jr., R. P. Thorn, Jr., F. L. Nesbitt, M. A. Iannone, D. C. Tardy and L. J. Stief, *J. Phys. Chem.*, 2002, **106**, 6060; G. P. Smith, *Chem. Phys. Lett.*, 2003, **376**, 381; M. J. Davis and S. J. Klippenstein, *J. Phys. Chem.*, 2002, **106**, 5860; K. Seki and H. Okabe, *J. Phys. Chem.*, 1993, **97**, 5284; B. A. Balko, J. Zhang and Y. T. Lee, *J. Chem. Phys.*, 1991, **94**, 7958; A. Lauter, K. S. Lee, K. H. Jung, R. K. Vatsa, J. P. Mittal and H. R. Volpp, *Chem. Phys. Lett.*, 2002, **358**, 314; A. M. Wodtke and Y. T. Lee, *J. Phys. Chem.*, 1985, **89**, 4744; J. Segall, Y. Wen,

R. Lavi, R. Singer and C. Wittig, *J. Phys. Chem.*, 1991, **95**, 8078; H. Okabe, *J. Chem. Phys.*, 1983, **78**, 1312; S.-H. Lee, Y. T. Lee and X. Yang, *J. Chem. Phys.*, 2004, **120**, 10983; E. F. Cromwell, A. Stolow, M. J. J. Vrakking and Y. T. Lee, *J. Chem. Phys.*, 1992, **97**, 4029; B. A. Balko, J. Zhang and Y. T. Lee, *J. Chem. Phys.*, 1992, **97**, 935; A. H. H. Chang, A. M. Mebel, X. M. Yang, S. H. Lin and Y. T. Lee, *Chem. Phys. Lett.*, 1998, **287**, 301; J. J. Lin, D. W. Hwang, Y. T. Lee and X. Yang, *J. Chem. Phys.*, 1998, **109**, 2979; A. Pena-Gallego, E. Martinez-Nunez and S. A. Vazquez, *Chem. Phys. Lett.*, 2002, **353**, 418.
27 J.-H. Wang, K. Liu, Z. Min, H. Su, R. Bersohn, J. Preses and J. Z. Larese, *J. Chem. Phys.*, 2000, **113**, 4146.
28 A. Paramo, A. Canosa, S. D. Le Picard and I. R. Sims, *J. Phys. Chem. A*, 2008, **112**, 9591.
29 X. Gu, R. I. Kaiser, A. M. Mebel, V. V. Kislov, S. J. Klippenstein, L. B. Harding, M. C. Liang and Y. L. Yung, *Astrophys. J.*, 2009, **701**, 1797.
30 N. Balucani, F. Leonori, R. Petrucci, K. M. Hickson and P. Casavecchia, *Phys. Scr.*, 2008, **78**, 058117.
31 X. Gu, Y. Guo, A. M. Mebel and R. I. Kaiser, *J. Phys. Chem.*, 2006, **110**, 11265.
32 J. R. McNesby and H. Okabe, *Adv. Photochem.*, 1964, 157.
33 A. H. Laufer and J. R. McNesby, *J. Chem. Phys.*, 1968, **49**, 2272.
34 R. Gorden, Jr. and P. Ausloos, *J. Chem. Phys.*, 1967, **46**, 4823.
35 R. E. Rebbert and P. Ausloos, *J. Photochem.*, 1972, **1**, 171.
36 T. G. Slanger and G. Black, *J. Chem. Phys.*, 1982, **77**, 2432.
37 Y. L. Yung and W. B. DeMore, *Photochemistry of Planetary Atmospheres*, Oxford University Press, New York, 1997.
38 M. D. Lodriguito, G. Lendvay and G. C. Schatz, *J. Chem. Phys.*, 2009, **131**, 224320.
39 Y. Zhang, K. Yuan, S. Yu and X. Yang, *J. Phys. Chem. Lett.*, 2010, **1**, 475.
40 L. C. Lee and C. C. Chiang, *J. Chem. Phys.*, 1983, **78**, 688.
41 D. H. Mordaunt, I. R. Lambert, G. P. Morley, M. N. R. Ashfold, R. N. Dixon, C. M. Western, L. Schnieder and K. H. Welge, *J. Chem. Phys.*, 1993, **98**, 2054.
42 R. A. Brownsword, M. Hillenkamp, T. Laurent, R. K. Vatsa, H. R. Volpp and J. Wolfrum, *Chem. Phys. Lett.*, 1997, **266**, 259.
43 A. M. Mebel, S.-H. Lin and C.-H. Chang, *J. Chem. Phys.*, 1997, **106**, 2612.
44 A. J. R. Heck, R. N. Zare and D. W. Chandler, *J. Chem. Phys.*, 1996, **104**, 4019; A. J. R. Heck, R. N. Zare and D. W. Chandler, *J. Chem. Phys.*, 1996, **104**, 3399; J.-H. Wang and K. Liu, *J. Chem. Phys.*, 1998, **109**, 7105.
45 F. Goulay, A. J. Trevitt, G. Meloni, T. M. Selby, D. L. Osborn, C. A. Taatjes, L. Vereecken and S. R. Leone, *J. Am. Chem. Soc.*, 2009, **131**, 993.
46 O. F. Sigurbjornsson and R. Signorell, *Phys. Chem. Chem. Phys.*, 2008, **10**, 6211.
47 M. G. Tomasko, L. R. Doose, L. E. Dafoe and C. See, *Icarus*, 2009, **204**, 271.
48 C. Sagan and W. R. Thompson, *Icarus*, 1984, **59**, 133.
49 G. J. Molina-Cuberos, J. J. Lopez-Moreno, R. Rodrigo, L. M. Lara and K. O'Brien, *Planet. Space Sci.*, 1999, **47**, 1347.
50 M.-C. Liang, Y. L. Yung and D. E. Shemansky, *Astrophys. J.*, 2007, **661**, L199.
51 S. H. Abbas and D. Schulze-Makuch, Abstracts of Papers, 234th ACS National Meeting, Boston, MA, United States, August 19–23, 2007, 2007, GEOC; J. Parnell, M. Baron and P. Lindgren, *J. Geochem. Explor.*, 2006, **89**, 322; N. Artemieva and J. Lunine, *Icarus*, 2003, **164**, 471; R. D. Lorenz, *Meteorit. Planet. Sci.*, 2004, **39**, 617; N. Artemieva and J. I. Lunine, *Icarus*, 2005, **175**, 522; M. E. Kress and C. P. McKay, *Icarus*, 2004, **168**, 475.
52 S. M. Horst, V. Vuitton and R. V. Yelle, *J. Geophys. Res.*, 2008, **113**, E10006.
53 S. Petrie, *Icarus*, 2004, **171**, 199; W. D. Pesnell and J. M. Grebowsky, *Adv. Space Res.*, 2001, **27**, 1807.
54 S. Petrie and R. C. Dunbar, *AIP Conf. Proc.*, 2006, **855**, 272.
55 R. I. Kaiser and X. Gu, *J. Chem. Phys.*, 2009, **131**, 104311.
56 P. J. Linstrom and W. G. Mallard, in *NIST Standard Reference Database Number 69*, National Institute of Standards and Technology, Gaithersburg, MD, June 2005.
57 K. Liu, *J. Chem. Phys.*, 2006, **125**, 132307; Y. T. Lee, *Atomic and Molecular Beam Methods*, Oxford University Press, Oxford, 1988; I. Kaiser Ralf, *Chem. Rev.*, 2002, **102**, 1309.
58 P. Casavecchia, G. Capozza and E. Segoloni, *Adv. Ser. Phys. Chem.*, 2004, **14**, 329.
59 M.-J. Nam, S.-E. Youn and J.-H. Choi, *J. Chem. Phys.*, 2006, **124**, 104307.
60 M. F. Witinski, M. Ortiz-Suarez and H. F. Davis, *J. Chem. Phys.*, 2006, **124**, 094307.
61 P. L. Houston, *J. Phys. Chem.*, 1996, **100**, 12757; M. Ahmed, D. S. Peterka and A. G. Suits, *Chem. Phys. Lett.*, 1999, **301**, 372; D. Townsend, W. Li, K. Lee Suk, L. Gross Richard and G. Suits Arthur, *J. Phys. Chem. A*, 2005, **109**, 8661; W. Li, C. Huang, M. Patel, D. Wilson and A. Suits, *J. Chem. Phys.*, 2006, **124**, 011102.
62 X. Liu, R. L. Gross and A. G. Suits, *J. Chem. Phys.*, 2002, **116**, 5341.

63 F. Goulay, D. L. Osborn, C. A. Taatjes, P. Zou, G. Meloni and S. R. Leone, *Phys. Chem. Chem. Phys.*, 2007, **9**, 4291; G. Meloni, T. M. Selby, F. Goulay, S. R. Leone, D. L. Osborn and C. A. Taatjes, *J. Am. Chem. Soc.*, 2007, **129**, 14019; D. L. Osborn, *Adv. Chem. Phys.*, 2008, **138**, 213; T. M. Selby, G. Meloni, F. Goulay, S. R. Leone, A. Fahr, C. A. Taatjes and D. L. Osborn, *J. Phys. Chem.*, 2008, **112**, 9366; C. A. Taatjes, N. Hansen, D. L. Osborn, K. Kohse-Hoeinghaus, T. A. Cool and P. R. Westmoreland, *Phys. Chem. Chem. Phys.*, 2008, **10**, 20; C. A. Taatjes, D. L. Osborn, T. A. Cool and K. Nakajima, *Chem. Phys. Lett.*, 2004, **394**, 19.
64 M. Alagia, N. Balucani, L. Cartechini, P. Casavecchia, E. H. van Kleef, G. G. Volpi, F. J. Aoiz, L. Banares and D. W. Schwenke, *et al.*, *Science*, 1996, **273**, 1519; D. Skouteris, H.-J. Werner, F. J. Aoiz, L. Banares, J. F. Castillo, M. Menendez, N. Balucani, L. Cartechini and P. Casavecchia, *J. Chem. Phys.*, 2001, **114**, 10662; N. Balucani, D. Skouteris, G. Capozza, E. Segoloni, P. Casavecchia, M. H. Alexander, G. Capecchi and H.-J. Werner, *Phys. Chem. Chem. Phys.*, 2004, **6**, 5007.
65 S.-H. Lee, F. Dong and K. Liu, *J. Chem. Phys.*, 2006, **125**, 133106.
66 B. D. Bean, J. D. Ayers, F. Fernandez-Alonso and R. N. Zare, *J. Chem. Phys.*, 2002, **116**, 6634.
67 N. Balucani, G. Capozza, E. Segoloni, A. Russo, R. Bobbenkamp, P. Casavecchia, T. Gonzalez-Lezana, E. J. Rackham, L. Banares and F. J. Aoiz, *J. Chem. Phys.*, 2005, **122**, 234309.
68 L. A. Pederson, G. C. Schatz, T.-S. Ho, T. Hollebeek, H. Rabitz, L. B. Harding and G. Lendvay, *J. Chem. Phys.*, 1999, **110**, 9091; N. Balucani, M. Alagia, L. Cartechini, P. Casavecchia, G. G. Volpi, L. A. Pederson and G. C. Schatz, *J. Phys. Chem.*, 2001, **105**, 2414; N. Balucani, L. Cartechini, G. Capozza, E. Segoloni, P. Casavecchia, G. G. Volpi, F. Javier Aoiz, L. Banares, P. Honvault and J.-M. Launay, *Phys. Rev. Lett.*, 2002, **89**, 013201; N. Balucani, P. Casavecchia, L. Banares, F. J. Aoiz, T. Gonzalez-Lezana, P. Honvault and J.-M. Launay, *J. Phys. Chem.*, 2006, **110**, 817.
69 N. Balucani, P. Casavecchia, F. J. Aoiz, L. Banares, J. F. Castillo and V. J. Herrero, *Mol. Phys.*, 2005, **103**, 1703; D. J. Garton, A. L. Brunsvold, T. K. Minton, D. Troya, B. Maiti and G. C. Schatz, *J. Phys. Chem.*, 2006, **110**, 1327; M. Alagia, N. Balucani, L. Cartechini, P. Casavecchia, E. H. van Kleef, G. G. Volpi, P. J. Kuntz and J. J. Sloan, *J. Chem. Phys.*, 1998, **108**, 6698; S. K. Gray, E. M. Goldfield, G. C. Schatz and G. G. Balint-Kurti, *Phys. Chem. Chem. Phys.*, 1999, **1**, 1141; F. J. Aoiz, L. Banares, J. F. Castillo, V. J. Herrero, B. Martinez-Haya, P. Honvault, J. M. Launay, X. Liu, J. J. Lin, S. A. Harich, C. C. Wang and X. Yang, *J. Chem. Phys.*, 2002, **116**, 10692.
70 S. H. Lee and K. Liu, *Appl. Phys. B: Lasers Opt.*, 2000, **71**, 627.
71 M. Alagia, N. Balucani, P. Casavecchia, D. Stranges and G. G. Volpi, *J. Chem. Phys.*, 1993, **98**, 8341.
72 B. R. Strazisar, C. Lin and H. F. Davis, *Science*, 2000, **290**, 958.
73 T. Takayanagi and G. C. Schatz, *J. Chem. Phys.*, 1997, **106**, 3227.
74 J. M. Bowman and G. C. Schatz, *Annu. Rev. Phys. Chem.*, 1995, **46**, 169.
75 B. Zhang, W. Shiu, J. J. Lin and K. Liu, *J. Chem. Phys.*, 2005, **122**, 131102; X. Liu and A. G. Suits, *Adv. Ser. Phys. Chem.*, 2004, 105; Q. Ran, C. H. Yang, Y. T. Lee, I. C. Lu, G. Shen, L. Wang and X. Yang, *J. Chem. Phys.*, 2005, **122**, 044307; G. Capozza, E. Segoloni, F. Leonori, G. G. Volpi and P. Casavecchia, *J. Chem. Phys.*, 2004, **120**, 4557; D. Troya, G. C. Schatz, D. J. Garton, A. L. Brunsvold and T. K. Minton, *J. Chem. Phys.*, 2004, **120**, 731; A. M. Mebel, R. I. Kaiser and Y. T. Lee, *J. Am. Chem. Soc.*, 2000, **122**, 1776; R. I. Kaiser, N. Balucani, O. Asvany and Y. T. Lee, *Symp. – Int. Astron. Union*, 2000, **197**, 251; N. Balucani, A. M. Mebel, Y. T. Lee and R. I. Kaiser, *J. Phys. Chem. A*, 2001, **105**, 9813; R. I. Kaiser, T. N. Le, T. L. Nguyen, A. M. Mebel, N. Balucani, Y. T. Lee, F. Stahl, P. v. R. Schleyer and H. F. Schaefer, III, *Faraday Discuss.*, 2001, **119**, 51; R. I. Kaiser, O. Asvany and Y. T. Lee, *Planet. Space Sci.*, 2000, **48**, 483; R. I. Kaiser, O. Asvany, Y. T. Lee, H. F. Bettinger, P. v. R. Schleyer and H. F. Schaefer, III, *J. Chem. Phys.*, 2000, **112**, 4994; L. Vereecken, J. Peeters, F. Bettinger Holger, I. Kaiser Ralf, R. v. Schleyer Paul and F. Schaefer Henry, 3rd, *J. Am. Chem. Soc.*, 2002, **124**, 2781; J. Zhou, B. Zhang, J. J. Lin and K. Liu, *Mol. Phys.*, 2005, **103**, 1757; W. Shiu, J. J. Lin, K. Liu, M. Wu and D. H. Parker, *J. Chem. Phys.*, 2004, **120**, 117; J. Zhou, W. Shiu, J. J. Lin and K. Liu, *J. Chem. Phys.*, 2004, **120**, 5863; W. Shiu, J. J. Lin and K. Liu, *Phys. Rev. Lett.*, 2004, **92**, 103201; R. I. Kaiser, A. M. Mebel, A. H. H. Chang, S. H. Lin and Y. T. Lee, *J. Chem. Phys.*, 1999, **110**, 10330; N. Balucani, H. Y. Lee, A. M. Mebel, Y. T. Lee and R. I. Kaiser, *J. Chem. Phys.*, 2001, **115**, 5107; H. F. Davis, J. Shu, S. Peterka Darcy and M. Ahmed, *J. Chem. Phys.*, 2004, **121**, 6254; R. L. Gross, X. Liu and A. G. Suits, *Chem. Phys. Lett.*, 2003, **376**, 710; X. Liu, R. L. Gross, G. E. Hall, J. T. Muckerman and A. G. Suits, *J. Chem. Phys.*, 2002, **117**, 7947; J. Shu, J. J. Lin, Y. T. Lee and X. Yang, *J. Chem. Phys.*, 2001, **115**, 849;

P. Casavecchia, G. Capozza, E. Segoloni, F. Leonori, N. Balucani and G. G. Volpi, *J. Phys. Chem.*, 2005, **109**, 3527; C. C. Wang, J. Shu, J. J. Lin, Y. T. Lee, X. Yang, T. L. Nguyen and A. M. Mebel, *J. Chem. Phys.*, 2002, **116**, 8292; Y.-C. Sun, I. T. Wang, T. L. Nguyen, H.-F. Lu, X. Yang and A. M. Mebel, *J. Phys. Chem.*, 2003, **107**, 6986; R. I. Kaiser, Y. T. Lee and A. G. Suits, *J. Chem. Phys.*, 1995, **103**, 10395; R. I. Kaiser, D. Stranges, Y. T. Lee and A. G. Suits, *J. Chem. Phys.*, 1996, **105**, 8721; R. I. Kaiser, C. Ochsenfeld, M. Head-Gordon, Y. T. Lee and A. G. Suits, *J. Chem. Phys.*, 1997, **106**, 1729; R. I. Kaiser, D. Stranges, Y. T. Lee and A. G. Suits, *Astrophys. J.*, 1997, **477**, 982; C. Ochsenfeld, R. I. Kaiser, Y. T. Lee, A. G. Suits and M. Head-Gordon, *J. Chem. Phys.*, 1997, **106**, 4141; R. I. Kaiser, C. Ochsenfeld, M. Head-Gordon and Y. T. Lee, *Astrophys. J.*, 1999, **510**, 784.
76 R. Z. Hinrichs, P. A. Willis, H. U. Stauffer, J. J. Schroden and H. F. Davis, *J. Chem. Phys.*, 2000, **112**, 4634; H. U. Stauffer, R. Z. Hinrichs, J. J. Schroden and H. F. Davis, *J. Phys. Chem.*, 2000, **104**, 1107; J. J. Schroden, C. C. Wang and H. F. Davis, *J. Phys. Chem.*, 2003, **107**, 9295; R. Z. Hinrichs, J. J. Schroden and H. F. Davis, *J. Am. Chem. Soc.*, 2003, **125**, 860; J. J. Schroden, M. Teo and H. F. Davis, *J. Chem. Phys.*, 2002, **117**, 9258; R. Z. Hinrichs, J. J. Schroden and H. F. Davis, *J. Phys. Chem.*, 2008, **112**, 3010; J. J. Schroden and H. F. Davis, *Adv. Ser. Phys. Chem.*, 2004, **14**, 215; H. U. Stauffer, R. Z. Hinrichs, J. J. Schroden and H. F. Davis, *J. Chem. Phys.*, 1999, **111**, 10758.
77 M. J. Perri, A. L. Van Wyngarden, J. J. Lin, Y. T. Lee and K. A. Boering, *J. Phys. Chem.*, 2004, **108**, 7995; A. L. Van Wyngarden, K. A. Mar, K. A. Boering, J. J. Lin, Y. T. Lee, S.-Y. Lin, H. Guo and G. Lendvay, *J. Am. Chem. Soc.*, 2007, **129**, 2866; Y.-J. Lu, T. Xie, J.-W. Fang, H.-C. Shao and J. J. Lin, *J. Phys. Chem.*, 2008, **128**, 184302.
78 R. I. Kaiser, W. Sun and A. G. Suits, *J. Chem. Phys.*, 1997, **106**, 5288; R. I. Kaiser, C. Ochsenfeld, M. Head-Gordon and Y. T. Lee, *Science*, 1998, **279**, 1181; R. I. Kaiser, C. Ochsenfeld, M. Head-Gordon and Y. T. Lee, *J. Chem. Phys.*, 1999, **110**, 2391; R. I. Kaiser, I. Hahndorf, L. C. L. Huang, Y. T. Lee, H. F. Bettinger, P. v. R. Schleyer, H. F. Schaefer, III and P. R. Schreiner, *J. Chem. Phys.*, 1999, **110**, 6091; H. F. Bettinger, P. v. R. Schleyer, H. F. Schaefer, III, P. R. Schreiner, R. I. Kaiser and Y. T. Lee, *J. Chem. Phys.*, 2000, **113**, 4250; I. Hahndorf, Y. T. Lee, R. I. Kaiser, L. Vereecken, J. Peeters, H. F. Bettinger, P. R. Schreiner, P. v. R. Schleyer, W. D. Allen and H. F. Schaefer, III, *J. Chem. Phys.*, 2002, **116**, 3248; R. I. Kaiser, Y. T. Lee and A. G. Suits, *J. Chem. Phys.*, 1996, **105**, 8705; R. I. Kaiser, D. Stranges, H. M. Bevsek, Y. T. Lee and A. G. Suits, *J. Chem. Phys.*, 1997, **106**, 4945; I. Hahndorf, H. Y. Lee, A. M. Mebel, S. H. Lin, Y. T. Lee and R. I. Kaiser, *J. Chem. Phys.*, 2000, **113**, 9622; L. C. L. Huang, H. Y. Lee, A. M. Mebel, S. H. Lin, Y. T. Lee and R. I. Kaiser, *J. Chem. Phys.*, 2000, **113**, 9637; T. N. Le, H.-y. Lee, A. M. Mebel and R. I. Kaiser, *J. Phys. Chem. A*, 2001, **105**, 1847; T. L. Nguyen, A. M. Mebel and R. I. Kaiser, *J. Phys. Chem. A*, 2001, **105**, 3284; R. I. Kaiser, T. L. Nguyen, T. N. Le and A. M. Mebel, *Astrophys. J.*, 2001, **561**, 858; W. D. Geppert, C. Naulin, M. Costes, G. Capozza, L. Cartechini, P. Casavecchia and G. Gualberto Volpi, *J. Chem. Phys.*, 2003, **119**, 10607.
79 N. Balucani, O. Asvany, Y. T. Lee, R. I. Kaiser, N. Galland and Y. Hannachi, *J. Am. Chem. Soc.*, 2000, **122**, 11234; N. Balucani, O. Asvany, Y. T. Lee, R. I. Kaiser, N. Galland, M. T. Rayez and Y. Hannachi, *J. Comput. Chem.*, 2001, **22**, 1359; W. D. Geppert, F. Goulay, C. Naulin, M. Costes, A. Canosa, S. D. Le Picard and B. R. Rowe, *Phys. Chem. Chem. Phys.*, 2004, **6**, 566.
80 N. Balucani, L. Cartechini, M. Alagia, P. Casavecchia and G. G. Volpi, *J. Phys. Chem.*, 2000, **104**, 5655; N. Balucani, O. Asvany, A. H. H. Chang, S. H. Lin, Y. T. Lee, R. I. Kaiser, H. F. Bettinger, P. v. R. Schleyer and H. F. Schaefer, III, *J. Chem. Phys.*, 1999, **111**, 7457; R. I. Kaiser, C. C. Chiong, O. Asvany, Y. T. Lee, F. Stahl, P. v. R. Schleyer and H. F. Schaefer, III, *J. Chem. Phys.*, 2001, **114**, 3488; F. Stahl, P. v. R. Schleyer, H. F. Bettinger, R. I. Kaiser, Y. T. Lee and H. F. Schaefer, III, *J. Chem. Phys.*, 2001, **114**, 3476; R. I. Kaiser, F. Stahl, P. v. R. Schleyer and H. F. Schaefer, III, *Phys. Chem. Chem. Phys.*, 2002, **4**, 2950; F. Stahl, P. v. R. Schleyer, H. F. Schaefer, III and R. I. Kaiser, *Planet. Space Sci.*, 2002, **50**, 685; L. C. L. Huang, Y. T. Lee and R. I. Kaiser, *J. Chem. Phys.*, 1999, **110**, 7119; L. C. L. Huang, N. Balucani, Y. T. Lee, R. I. Kaiser and Y. Osamura, *J. Chem. Phys.*, 1999, **111**, 2857; N. Balucani, O. Asvany, A. H. H. Chang, S. H. Lin, Y. T. Lee, R. I. Kaiser, H. F. Bettinger, P. v. R. Schleyer and H. F. Schaefer, III, *J. Chem. Phys.*, 1999, **111**, 7472; N. Balucani, O. Asvany, Y. Osamura, L. C. L. Huang, Y. T. Lee and R. I. Kaiser, *Planet. Space Sci.*, 2000, **48**, 447; N. Balucani, O. Asvany, A. H. H. Chang, S. H. Lin, Y. T. Lee, R. I. Kaiser and Y. Osamura, *J. Chem. Phys.*, 2000, **113**, 8643; N. Balucani, O. Asvany, L. C. L. Huang, Y. T. Lee, R. I. Kaiser, Y. Osamura and H. F. Bettinger, *Astrophys. J.*, 2000, **545**, 892; L. C. L. Huang,

O. Asvany, A. H. H. Chang, N. Balucani, S. H. Lin, Y. T. Lee, R. I. Kaiser and Y. Osamura, *J. Chem. Phys.*, 2000, **113**, 8656; R. I. Kaiser, *Eur. Space Agency, [Spec. Publ.]*, 2001, **SP-496**, 145; N. Balucani, O. Asvany, R. I. Kaiser and Y. Osamura, *J. Phys. Chem. A*, 2002, **106**, 4301.
81 R. I. Kaiser, J. W. Ting, L. C. L. Huang, N. Balucani, O. Asvany, Y. T. Lee, H. Chan, D. Stranges and D. Gee, *Rev. Sci. Instrum.*, 1999, **70**, 4185.
82 X. Gu and R. I. Kaiser, *Acc. Chem. Res.*, 2009, **42**, 290.
83 R. I. Kaiser and A. G. Suits, *Rev. Sci. Instrum.*, 1995, **66**, 5405.
84 F. Zhang, S. Kim and R. I. Kaiser, *Phys. Chem. Chem. Phys.*, 2009, **11**, 4707; R. I. Kaiser, C. Ochsenfeld, D. Stranges, M. Head-Gordon and Y. T. Lee, *Faraday Discuss.*, 1998, **109**, 183.
85 Y. Guo, A. M. Mebel, F. Zhang, X. Gu and R. I. Kaiser, *J. Phys. Chem. A*, 2007, **111**, 4914; F. Zhang, X. Gu, Y. Guo and R. I. Kaiser, *J. Org. Chem.*, 2007, **72**, 7597.
86 K. A. Berrington, K. L. Bell and Editors, Proceedings of the Second International Conference on Atomic and Molecular Data and Their Applications (ICAMDATA), held 26–30 March 2000, in Oxford England. [In: AIP Conf. Proc., 2000; 543] It is crucial to note that the ionization cross section drops almost linearly from its maximum toward lower electron energies (Wigner relation).
87 M. S. Weiss, University Of California, Berkeley, Berkeley, CA, 1986.
88 X. Gu, Y. Guo, E. Kawamura and R. I. Kaiser, *J. Vac. Sci. Technol., A*, 2006, **24**, 505.
89 A. M. Mebel and R. I. Kaiser, *Astrophys. J.*, 2002, **564**, 787.
90 F. Zhang, B. Jones, P. Maksyutenko, R. I. Kaiser, C. Chin, V. V. Kislov and A. M. Mebel, *J. Am. Chem. Soc.*, 2010, **132**, 2672.
91 M. Yamada, Y. Osamura and R. I. Kaiser, *Astron. Astrophys.*, 2002, **395**, 1031.
92 X. Tan, *CyberWit*, version 1.4, Santa Clara, 2004.
93 C. V. V. Prasad and P. F. Bernath, *Astrophys. J.*, 1994, **426**, 812; A. Tanabashi, T. Hirao, T. Amano and P. F. Bernath, *Astrophys. J. Suppl.*, 2007, **169**, 472.
94 M. Danielsson, P. Erman, A. Hishikawa, M. Larsson and E. Rachlew-Kaellne, *J. Chem. Phys.*, 1993, **98**, 9405.
95 P. Zou, J. Shu, T. J. Sears, G. E. Hall and S. W. North, *J. Phys. Chem. A*, 2004, **108**, 1482; W. S. McGivern, O. Sorkhabi, A. G. Suits, A. Derecskei-Kovacs and S. W. North, *J. Phys. Chem. A*, 2000, **104**, 10085.
96 R. Atkinson, D. L. Baulch, R. A. Cox, R. F. Hampson, Jr., J. A. Kerr and J. Troe, *J. Phys. Chem. Ref. Data*, 1992, **21**, 1125.
97 A. M. Mebel, Florida International University, unpublished results, 2009.
98 J. Luque, LIFBASE, see http://www.sri.com/psd/lifbase/.
99 C. J. Bennett, C. Jamieson, A. M. Mebel and R. I. Kaiser, *Phys. Chem. Chem. Phys.*, 2004, **6**, 735.
100 R. B. Bohn, S. A. Sandford, L. J. Allamandola and D. P. Cruikshank, *Icarus*, 1994, **111**, 151.
101 J. W. Stewart and R. I. La Rock, *J. Chem. Phys.*, 1958, **28**, 425.
102 D. Drouin, A. R. Couture, R. Gauvin, P. Hovington, P. Horny and H. Demers, *Monte Carlo Simulation of Electron Trajectory in Solids (CASINO ver 2.42)*, University of Sherbrook, Sherbrook, 2001.
103 P. A. Heimann, M. Koike, C. W. Hsu, D. Blank, X. M. Yang, A. G. Suits, Y. T. Lee, M. Evans and C. Y. Ng, *et al.*, *Rev. Sci. Instrum.*, 1997, **68**, 1945.
104 C. Nicolas, J. N. Shu, D. S. Peterka, M. Hochlaf, L. Poisson, S. R. Leone and M. Ahmed, *J. Am. Chem. Soc.*, 2006, **128**, 220–226.
105 R. I. Kaiser, L. Belau, S. R. Leone, M. Ahmed, Y. Wang, B. J. Braams and J. M. Bowman, *Chem. Phys. Chem.*, 2007, **8**, 1236.
106 A. D. Becke, *J. Chem. Phys.*, 1993, **98**, 5648; C. Lee, W. Yang and R. G. Parr, *Phys. Rev. B: Condens. Matter*, 1988, **37**, 785.
107 M. J. Frisch, G. W. Trucks, H. B. Schlegel, G. E. Scuseria, M. A. Robb, J. R. Cheeseman, V. G. Zakrzewski, J. A. Montgomery, Jr., R. E. Stratmann, J. C. Burant, S. Dapprich, J. M. Millam, A. D. Daniels, K. N. Kudin, M. C. Strain, O. Farkas, J. Tomasi, V. Barone, M. Cossi, R. Cammi, B. Mennucci, C. Pomelli, C. Adamo, S. Clifford, J. Ochterski, G. A. Petersson, P. Y. Ayala, Q. Cui, K. Morokuma, P. Salvador, J. J. Dannenberg, D. K. Malick, A. D. Rabuck, K. Raghavachari, J. B. Foresman, J. Cioslowski, J. V. Ortiz, A. G. Baboul, B. B. Stefanov, G. Liu, A. Liashenko, P. Piskorz, I. Komaromi, R. Gomperts, R. L. Martin, D. J. Fox, T. Keith, M. A. Al-Laham, C. Y. Peng, A. Nanayakkara, M. Challacombe, P. M. W. Gill, B. G. Johnson, W. Chen, M. W. Wong, J. L. Andres, C. Gonzalez, M. Head-Gordon, E. S. Replogle and J. A. Pople, *GAUSSIAN 98 (Revision A.11)*, Gaussian, Inc., Pittsburgh, PA, 2001.
108 R. D. Amos, A. Bernhardsson, A. Berning, P. Celani, D. L. Cooper, M. J. O. Deegan, A. J. Dobbyn, F. Eckert, C. Hampel, G. Hetzer, P. J. Knowles, T. Korona, R. Lindh,

A. W. Lloyd, S. J. McNicholas, F. R. Manby, W. Meyer, M. E. Mura, A. Nicklass, P. Palmieri, R. Pitzer, G. Rauhut, M. Schütz, U. Schumann, H. Stoll, A. J. Stone, R. Tarroni, T. Thorsteinsson and H.-J. Werner, *MOLPRO, a package of* ab initio *programs designed by H.-J. Werner and P. J. Knowles, Version 2002.6*, 2002.
109 R. I. Kaiser, in *Carbon-Centered Radicals: Structure, Dynamics and Reactivity*, ed. M. D. E. Fobes, Wiley, 2010, pp. 221.
110 S. B. Tejada and D. F. Eggers, Jr., *Spectrochim. Acta, Part A*, 1976, **32**, 1557.
111 R. L. Hudson, M. H. Moore and L. L. Raines, *Icarus*, 2009, **203**, 677.
112 Y. S. Kim, C. J. Bennett, L.-H. Chen, K. O'Brien and R. I. Kaiser, *Astrophys. J.*, 2010, **711**, 744.
113 M. A. Goodman, R. L. Sweany and R. L. Flurry, Jr., *J. Phys. Chem.*, 1983, **87**, 1753.
114 S. Ikuta, T. Saitoh and S. Wakamatsu, *J. Chem. Phys.*, 2004, **121**, 3478.
115 A. E. Ketvirtis, D. K. Bohme and A. C. Hopkinson, *J. Phys. Chem.*, 1995, **99**, 16121.
116 J. Drowart, G. de Maria and M. G. Inghram, *J. Chem. Phys.*, 1958, **29**, 1015.
117 R. W. Schmude, Jr. and K. A. Gingerich, *J. Phys. Chem.*, 1997, **101**, 2610.
118 O. Kostko, S. R. Leone, M. A. Duncan and M. Ahmed, *J. Phys. Chem. A*, 2010, **114**, 3176.
119 P. S. Yadav, R. K. Yadav, S. Agrawal and B. K. Agrawal, *J. Phys.: Condens. Matter*, 2006, **18**, 7085.
120 P. Pradhan and A. K. Ray, *Eur. Phys. J. D*, 2006, **37**, 393.
121 A. I. Boldyrev, J. Simons, V. G. Zakrzewski and W. von Niessen, *J. Phys. Chem.*, 1994, **98**, 1427.
122 D. C. Parent, *Int. J. Mass Spectrom. Ion Processes*, 1994, **138**, 307.
123 J. Hou and B. Song, *J. Chem. Phys.*, 2008, **128**, 154304.
124 C. Berteloite, S. D. Le Picard, P. Birza, M.-C. Gazeau, A. Canosa, Y. Benilan and I. R. Sims, *Icarus*, 2008, **194**, 746.

Mechanisms of formation of nitrogen-containing polycyclic aromatic compounds in low-temperature environments of planetary atmospheres: A theoretical study

Alexander Landera and Alexander M. Mebel

Received 23rd February 2010, Accepted 31st March 2010
DOI: 10.1039/c003475d

Polycyclic aromatic hydrocarbons (PAHs) are believed to be responsible for the formation of organic haze layers in Titan's atmosphere, but the nature of PAHs on Titan and their formation and growth mechanisms are not well understood. Considering the high abundance of nitrogen in Titan's atmosphere, it is likely that the haze layers hold not only pure hydrocarbon PAHs but also their nitrogenated analogs, N-containing polycyclic aromatic compounds (N-PACs) with 'hetero' N atoms in aromatic rings. Laboratory studies of Titan's tholins also support the hypothesis that, together with pure PAHs and their cations, N-PACs may be the fundamental building blocks of microphysical tholin particles. In the present work, we carried out *ab initio* quantum chemical calculations of potential energy surfaces for various reaction mechanisms of incorporation of nitrogen atoms into aromatic rings of polycyclic aromatic compounds, which may lead to the formation of N-PACs under the low-temperature and low-pressure conditions of Titan's atmosphere. This includes mechanisms analogous to the Ethynyl Addition Mechanism (EAM) recently proposed by us for the growth of PAH by sequential C_2H additions to benzene. We consider consecutive C_2H and CN additions to C_6H_6, C_6H_6 + CN → C_6H_5CN + H, C_6H_5CN + C_2H → $C_6H_4(CN)(C_2H)$ + H, C_6H_5CN + CN → $C_6H_4(CN)_2$ + H, $C_6H_4(CN)(C_2H)$ + C_2H → 2-aza-4-ethynyl-1-naphthyl/2-aza-1-ethynyl-4-naphthyl, $C_6H_4(CN)_2$ + C_2H → $C_6H_4(CN)(NCCCH)$, and $C_6H_4(CN)(NCCCH)$ + C_2H → 1,4-diethynylphthalazine. Although these reactions are found to be barrierless and exothermic and therefore feasible at low temperatures, the steps leading to the aza-ethynyl-naphthyl radicals, $C_6H_4(CN)(NCCCH)$, and 1,4-diethynylphthalazine can give N-PACs as final products only upon their collisional or radiational stabilization. Alternatively, an N-PAC can be synthesized *via* the reaction of 2-methyleneaminobenzonitrile with C_2H, producing 4-ethynyl-quinoline + H without an entrance barrier *via* a three-step sequence including C_2H addition to C of CN, ring closure, and H elimination. 2-Methyleneaminobenzonitrile itself can be formed in the reaction of methyleneaminobenzene with cyano radical, $C_6H_5(NCH_2)$ + CN → $C_6H_5(NCH_2)(CN)$ → $C_6H_4(NCH_2)(CN)$ + H, which also does not have any entrance barrier. Methyleneaminobenzene can be produced through recombination of phenyl and methylene-amidogen radicals followed by collisional stabilization of the product, *via* the barrierless C_6H_5 + CH_3N → $C_6H_5(NCH_3)$ → $C_6H_5(NCH_2)$ + H reaction, or in the reaction of phenyl with methyleneimine, C_6H_5 + CH_2NH → $C_6H_5(NHCH_3)$ → $C_6H_5(NCH_2)$ + H. The

---

Florida International University, Department of Chemistry & Biochemistry, Miami, 33199

latter would be slow at low-temperature conditions owing to the barriers of 4.5 and 2.8 kcal mol^{-1} relative to the initial reactants, but feasible if the reactants possess sufficient internal energy to overcome these barriers. We anticipate that the presented mechanisms are viable to form N-PACs in hydrocarbon and nitrogen rich, low temperature atmospheres of planets and their moons such as Titan.

Introduction

Polycyclic aromatic hydrocarbons (PAHs), present in the Earth environment in the form of atmospheric aerosols, soot, and volatile particles, are also believed to play an important role in interstellar chemistry[1-6] and in the atmospheric chemistry of planets and theirs moons in the Solar System.[7] PAHs have been identified in laboratory experiments simulating Jupiter's atmosphere and also the atmospheric chemistry of Saturn's moon Titan together with its colored organic haze layers.[8] The detection of the simplest aromatic hydrocarbon, benzene, which is considered as a critical precursor toward complex PAHs, in the low-temperature atmospheres of Jupiter and Saturn[9] and recently on Titan[10] strongly suggests that the PAH synthesis may proceed in these environments. It is widely believed that PAHs are responsible for the formation of organic haze layers in Titan's atmosphere,[7] but the nature of PAHs on Titan and their formation and growth mechanisms are not well understood. Considering the high abundance of nitrogen in Titan's atmosphere, it is probable that the haze layers contain not only pure hydrocarbon PAHs but also their nitrogenated analogs, *i.e.*, hetero-PAHs or PAHs including 'hetero' N atoms in their aromatic rings. Laboratory studies of a laboratory analogue of Titan's haze, called tholin, also support the hypothesis that, together with pure PAHs and their cations, nitrogenated heterocyclic PAHs may be the fundamental building blocks of microphysical particles in Titan's hazes.[11,12] For instance, nitrogen-containing heteroaromatics and PAHs were previously detected in tholins by Khare *et al.*,[13] Sagan *et al.*,[14] and Ehrenfreund *et al.*[15] In the last decade, Khare and co-workers synthesized a Titan tholin and measured the time variation of its infrared spectrum as a thin film developed.[11] They found that one of the earliest features to emerge in the IR spectra was a very strong absorption feature at $\sim$1351 cm^{-1}, which was ascribed to skeletal ring vibrations[16] between members of C and N rings, perhaps arranged in relatively dehydrogenated, positively charged heterocyclic ring structures reminiscent of graphitic structures. The IR spectra of the Titan tholin indicated an upper limit of 6% by mass in benzenes, heterocyclics, and PAHs with more than four rings.[11] Quirico *et al.*[17] generated powdered samples of carbon–nitrogen–hydrogen tholins that mimic Titan's atmosphere aerosols under levitation conditions in the laboratory with a dusty plasma and studied them using various spectroscopic techniques. The observed first-order UV Raman bands at 690 and 980 cm^{-1} suggested that C_3N_3 rings are present as a species inserted in the macromolecular network. A recent laboratory study of Titan's tholin IR spectra by Carrasco *et al.*[18] showed intense absorption bands peaking at $\sim$1560 and 1630–1650 cm^{-1}, which involve the contribution of C=N groups, as imine and/or in N-bearing heteroaromatic or heterocyclic groups. Imanaka *et al.*[12] produced tholins in an inductively coupled plasma from a nitrogen–methane gas mixture at various pressures ranging from 13 to 2300 Pa. A careful structural analysis by spectroscopic methods allowed them to conclude that tholins formed at low pressures include clusters of nitrogen-containing polycyclic aromatic compounds (N-PACs) in a matrix of carbon and nitrogen branched chain networks, which are connected tightly to each other with hydrogen bonding of N–H bonds. Nitrogen was found to be incorporated into tholin more efficiently at lower deposition pressures. Applying the experimental results to Titan, Imanaka *et al.* suggested[12] that the tholins formed at low pressures (13 and 26 Pa) may be better representations of Titan's haze than those formed at high pressures and thus, the N-PACs found in tholin formed at low pressure may be present in Titan's haze. These aromatic macromolecules might have a significant influence over the thermal structure and complex organic chemistry in Titan's atmosphere, because they are efficient absorbers of UV radiation and efficient charge exchange intermediaries.[12]

For understanding the origin and evolution of Titan's haze layers, it is important to find synthetic routes leading to the N-PACs from small molecules, abundant in Titan's atmosphere, and radicals, which can be produced by their photodissociation. To our best knowledge, the only theoretical study of the mechanism of nitrogen inclusion into an aromatic ring to form N-PACs has been reported by Ricca et al.[19] The authors used density functional B3LYP calculations to compute potential energy profiles for reaction sequences analogous to the Hydrogen-Abstraction–C_2H_2-Addition (HACA) mechanism of PAH growth,[20] but with HCN or cyanoacetylene HCCCN replacing C_2H_2 at certain reaction steps. They found that the presence of a nitrogen atom increases the barrier heights relative to the pure hydrocarbon species to about 15 kcal mol^{-1} and concluded that the calculated barriers are probably too high to allow reactions to occur in the atmosphere of Titan unless the reaction rates are enhanced by vibrational energy of the aromatic molecule, although the presence of a nitrogen atom in the aromatic ring promotes the formation of an additional ring. Similar to the HACA mechanism, we anticipate that the reactions suggested by Ricca et al.[19] would be more applicable to high-temperature conditions. As a practicable alternative to the HACA sequences to form PAH-like species under low temperature conditions in the interstellar medium and in hydrocarbon rich atmospheres of planets and their moons, we have recently established computationally a novel Ethynyl Addition Mechanism (EAM).[21] Initiated by an addition of the ethynyl radical (C_2H) to the *ortho* carbon atom of the phenylacetylene ($C_6H_5C_2H$) molecule, the reactive intermediate rapidly loses a hydrogen atom forming 1,2-diethynylbenzene. The latter can react with a second ethynyl molecule *via* addition to a carbon atom of one of the ethynyl side chains. A consecutive ring closure of the intermediate leads to an ethynyl-substituted naphthalene core. In the present work, we report *ab initio* quantum chemical calculations of potential energy surfaces (PESs) for various consecutive reactions of C_2H and CN addition to benzene combined with statistical calculations of reaction rate constants to extend the EAM to the formation and growth of N-PACs containing, for instance, pyridine and pyridazine aromatic rings. Similar to C_2H, which can be produced on Titan by the photolysis of acetylene,[22–25] the cyano radical, CN, can be formed by photodissociation of hydrogen cyanide (HCN) by short wavelength photons (λ < 231 nm) penetrating to the stratosphere.[26] As the presence of benzene in Titan's atmosphere has been firmly established[10] and both ethynyl and cyano radicals are integral parts of photochemical models describing chemical evolution of the stratosphere,[7,26–33] the reactions involving C_6H_6, C_2H, and CN may play an important role in the synthesis of hetero-polycyclic aromatic compounds. The goal of the present study is to elucidate this role and the conditions at which these reactions can take place. In addition, we investigate some other possible mechanisms of N-PAC formation involving different N-containing species (potentially) present on Titan, such as methyleneimine CH_2NH and its CH_3N isomer, methylene-amidogen radical CH_2N, and electronically excited N_2 in the lowest triplet state.

Computational methods

Geometries of all species on PESs considered in the present study were optimized using the hybrid density functional B3LYP method[34] with the 6-311G** basis set. The same method was used to obtain vibrational frequencies, zero-point energy (ZPE) corrections, and to characterize the stationary points as minima or first-order saddle points. To obtain more accurate energies, we applied the G3(MP2,CC)//B3LYP modification[35] of the original Gaussian 3 (G3) scheme[36] for high-level single-point energy calculations. The final energies at 0 K were obtained using the B3LYP optimized geometries and ZPE corrections according to the following formula $E_0[G3(MP2,CC)] = E[CCSD(T)/6-311G(d,p)] + \Delta E_{MP2} + E(ZPE)$, where $\Delta E_{MP2} = E[MP2/G3large] - E[MP2/6-311G(d,p)]$ is the basis set correction and E(ZPE) is the zero-point energy. ΔE(SO), a spin-orbit correction, and ΔE(HLC), a higher level correction, from the original G3 scheme were not included in our calculations, as they are not expected to make significant contributions into relative energies. We used the Gaussian 98[37] program package to carry out B3LYP and MP2 calculations, and the MOLPRO 2006[38] program package to perform calculations of spin-restricted coupled cluster RCCSD(T) energies.

Where necessary, rate constants of individual unimolecular reactions under single-collision conditions were evaluated within Rice–Ramsperger–Kassel–Marcus (RRKM) theory.[39] We calculated rate constants as functions of available internal energy of each intermediate or transition state; the internal energy was taken as the energy of chemical activation in each reaction, assuming zero collision energy. Only a single total-energy level was considered throughout, as for single-collision conditions (zero-pressure limit). The harmonic approximation was used in calculations of numbers and densities of state needed to evaluate the rate constants.

Results and discussion

Formation of 2-ethynylbenzonitrile and 1,2-dicyanobenzene

The high reactivity of benzene molecules with C_2H and CN radicals has been firmly established both experimentally and computationally. For instance, Goulay and Leone have measured thermal rate constants for the $C_6H_6 + C_2H$ reaction and found

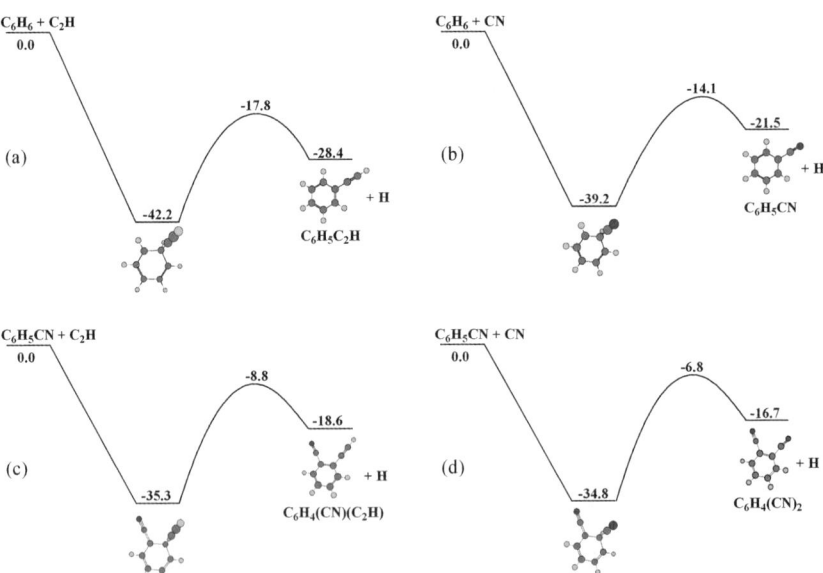

Fig. 1 Potential energy profiles of (a) $C_6H_6 + C_2H \rightarrow C_6H_5C_2H + H$ reaction calculated at the G3(MP2,CC)//B3LYP level (from ref. 42); (b) $C_6H_6 + CN \rightarrow C_6H_5CN + H$ reaction calculated at the B3LYP/6-31+G** level (from ref. 41); (c) $C_6H_5CN + C_2H \rightarrow C_6H_4(CN)(C_2H) + H$ reaction calculated at the G3(MP2,CC)//B3LYP level; and (d) $C_6H_5CN + CN \rightarrow C_6H_4(CN)_2 + H$ reaction calculated at the G3(MP2,CC)//B3LYP level. Relative energies of all species are shown in kcal mol^{-1}.

this reaction to be very fast even at low temperatures, with $k = (3.0–4.0) \times 10^{-10}$ cm^3 molecule^{-1} s^{-1} at 105–298 K.[40] Their results indicated that this reaction has no barrier and may play an important role in the formation of large molecules and aerosols at low temperatures. The results were consistent with the formation of a short-lifetime intermediate that decomposes to give the final products. Although the nature of these products was not determined experimentally, theoretical calculations of the reaction PES, rate constants, and product branching ratios consistently showed that the reaction should dominantly form C$_6$H$_5$C$_2$H (phenylacetylene/ethynylbenzene) + H (see Fig. 1(a)).[41,42] The C$_6$H$_6$ + CN reaction has been recently studied by Trevitt *et al.*,[43] who found the rate coefficient at 105, 165, and 295 K to be relatively constant over this temperature range, (3.9–4.9) $\times$ 10^{-10} cm^3 molecule^{-1} s^{-1}. These rapid kinetics, along with the observed negligible temperature dependence, were consistent with a barrierless reaction entrance channel. Trevitt *et al.* also probed the products of the C$_6$H$_6$ + CN reaction at room temperature using synchrotron VUV photoionization mass spectrometry and concluded that C$_6$H$_5$CN (cyanobenzene/benzonitrile) is the only product recorded. These experimental observations confirmed earlier theoretical calculations of the reaction PES (Fig. 1(b)) and relative product yields by Woon.[41] As both reactions of benzene with C$_2$H and CN have been found to be rapid at temperatures around 100 K, appearing to have barrierless entrance channels, it follows that they will proceed efficiently in Titan's atmosphere and are also likely to proceed at the temperature of interstellar clouds (10–20 K) to produce C$_6$H$_5$C$_2$H and C$_6$H$_5$CN, respectively.

Having established the feasibility of the production of ethynyl- and cyanobenzenes under low-temperature conditions, we now consider their reactions with CN. The calculated PESs for these reactions are illustrated in Fig. 1(c) and (d). One can see that the ethynyl radical can add to the *ortho* carbon atom of cyanobenzene without an entrance barrier to produce the C$_6$H$_5$(CN)(C$_2$H) adduct with exothermicity of 35.3 kcal mol^{-1}. Next, the adduct can lose the hydrogen atom from the C(H)(C$_2$H) group to form 2-ethynylbenzonitrile *via* a barrier of 26.5 kcal mol^{-1}. The H loss transition state lies 8.8 kcal mol^{-1} lower in energy than the initial reactants and the overall C$_6$H$_5$CN + C$_2$H → C$_6$H$_4$(CN)(C$_2$H) + H reaction is computed to be exothermic by 18.6 kcal mol^{-1}. In a similar way, the cyano radical can attack the *ortho* position in the benzene ring of cyanobenzene and form the C$_6$H$_5$(CN)$_2$ adduct without a barrier and with the energy gain of 34.8 kcal mol^{-1}. At the next step, C$_6$H$_5$(CN)$_2$ loses the hydrogen atom from C(H)(CN) *via* a transition state residing 6.8 kcal mol^{-1} below the reactants, producing 1,2-dicyanobenzene with the overall exothermicity of 16.7 kcal mol^{-1}. By the analogy with the C$_6$H$_6$ + C$_2$H/CN reactions, which have qualitatively similar potential energy profiles, we can expect that the C$_6$H$_5$CN + C$_2$H → C$_6$H$_5$(CN)(C$_2$H) → C$_6$H$_4$(CN)(C$_2$H) + H and C$_6$H$_5$CN + CN → C$_6$H$_5$(CN)$_2$ → C$_6$H$_4$(CN)$_2$ + H reactions should be fast at the low temperatures characteristic of Titan's atmosphere and to produce 2-ethynylbenzonitrile and 1,2-dicyanobenzene, respectively, as the dominant products, at least, at low pressures where collisional stabilization of the adducts formed at the initial step is not significant. Similarly, 2-ethynylbenzonitrile may be produced through CN addition to ethynylbenzene followed by an H loss: C$_6$H$_5$C$_2$H + CN → C$_6$H$_5$(C$_2$H)(CN) → C$_6$H$_4$(CN)(C$_2$H) + H. Below we consider possible reaction mechanisms for incorporation of one or two nitrogen atoms into an aromatic ring to form N-PACs, which use 2-ethynylbenzonitrile and 1,2-dicyanobenzene as starting reactants.

Incorporation of one nitrogen atom in the aromatic ring: Reactions of 2-ethynylbenzonitrile with ethynyl radical

To achieve the formation of a pyridine ring fused with the benzene ring present in 2-ethynylbenzonitrile, C$_2$H addition may proceed to either CCH or CN side chains. The potential energy diagram for the reaction channels involving the additions of

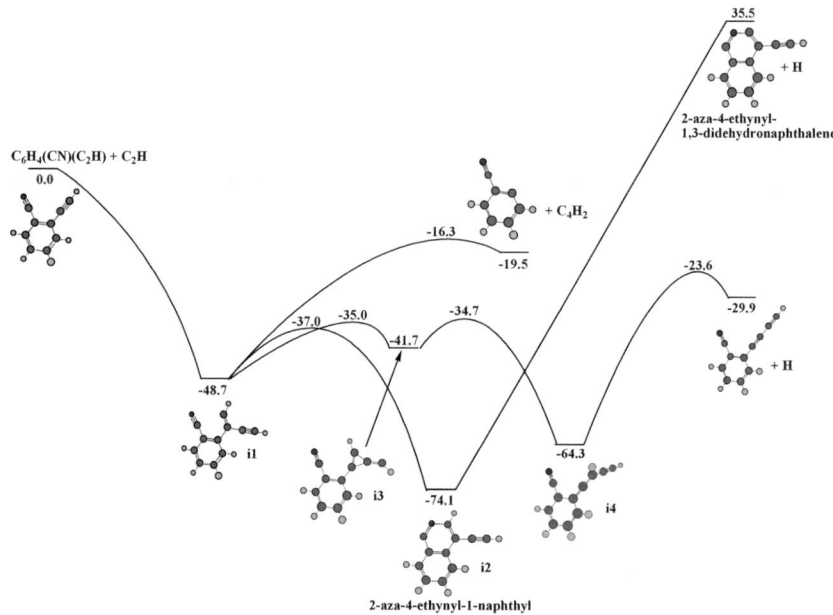

Fig. 2 Potential energy diagram of the $C_6H_4(CN)(C_2H) + C_2H$ reaction calculated at the G3(MP2,CC)//B3LYP level for the channels corresponding to ethynyl radical addition to the C_2H side chain. Relative energies of all species are shown in kcal mol^{-1}.

ethynyl radical to the CCH group is illustrated in Fig. 2. One can see that the reaction of 2-ethynylbenzonitrile with C_2H features two barrierless entrance channels leading to the formation of **i1**, with a relative energy of −48.7 kcal mol^{-1} with respect to the initial reactants (the addition to the β-C atom of CCH), or **i4** residing 64.3 kcal mol^{-1} below the reactants (α-C addition). The intermediate **i1** can isomerize to **i4** via a two-step C_2H migration to the adjacent carbon center α-C. The first step on this pathway exhibits an energy barrier of 13.7 kcal mol^{-1} relative to **i1** and leads to a transient bicyclic species **i3**, which can then further rearrange to **i4** through a relatively low ∼7 kcal mol^{-1} barrier. Alternatively, **i1** may undergo cyclization occurring with a modest 11.7 kcal mol^{-1} energy barrier producing **i2**, 2-aza-4-ethynyl-1-naphthyl radical, which lies 74.1 kcal mol^{-1} lower in energy than the $C_6H_4(CN)(C_2H) + C_2H$ reactants. Another possible pathway from **i1** involves elimination of diacetylene C_4H_2 leading to the formation of cyano-substituted phenyl radical, 19.5 kcal mol^{-1} below the reactants. However, the barrier for the C_4H_2 loss is relatively high, 32.4 kcal mol^{-1} with respect to **i1**, and hence this reaction channel is not expected to be competitive with either **i1** → **i3** → **i4** C_2H migration or the ring closure to form an N-PAC radical **i2**. Besides isomerizing to **i1**, intermediate **i4** can be subjected to an H loss from the side chain leading to the formation of 2-buta-1,3-diynyl-benzonitrile. The H-loss occurs via a relatively high barrier of 40.6 kcal mol^{-1} and the $C_6H_4(CN)(C_4H) + H$ products are 29.9 kcal mol^{-1} exothermic as compared with the $C_6H_4(CN)(C_2H) + C_2H$ reactants. Elimination of the hydrogen atom from the pyridine ring of 2-aza-4-ethynyl-1-naphthyl **i2** producing 2-aza-4-ethynyl-1,3-didehydronaphthalene is highly unfavorable, as the computed strength of the breaking C–H bond is 110.1 kcal mol^{-1} and the product resides 35.5 kcal mol^{-1} above the initial reactants. Elimination of hydrogen atoms from the benzene ring in **i2** is also expected to be highly endothermic as this would lead to a diradical product. A loss of the external C_2H group would result in the formation of a nitrogenated *p*-benzyne ring, in contrast to a nitrogenated *m*-benzyne ring present in the

2-aza-4-ethynyl-1,3-didehydronaphthalene product of H elimination. Since p-benzyne is known to be less stable than m-benzyne,[44] we do not consider the C_2H loss channel.

It is informative to compare the energetics of various intermediates, transition states, and products of the C_2H + 2-ethynylbenzonitrile reaction with those computed earlier[21] for C_2H + 1,2-diethynylbenzene, which leads to the production of the naphthalene core. The relative energy positions of the initial adducts $C_6H_4(CN)(C_2H)$-C_2H and $C_6H_4(C_2H)_2$-C_2H, the products, $C_6H_4(CN)(C_4H)$ + H and $C_6H_4(C_2H)(C_4H)$ + H, $C_6H_4(CN)$ + C_4H_2 and $C_6H_4(C_2H)$ + C_4H_2, as well as of majority of the transition states, relative to the respective reactants are rather similar. The exception is the relative energies of the polycyclic aromatic 2-aza-4-ethynyl-1-naphthyl and 4-ethynyl-1-naphthyl radicals, residing 74.1 and 113.6 kcal mol^{-1} below the reactants, respectively. This difference can be better understood if we compare the energy of ring formation in benzene from three acetylene molecules *versus* that of pyridine from $2C_2H_2$ + HCN. Using the standard enthalpies of formation of benzene, pyridine, C_2H_2, and HCN from the NIST Database,[45] we find that the $3C_2H_2 \rightarrow C_6H_6$ and $2C_2H_2$ + HCN $\rightarrow C_5NH_5$ are exothermic by 142.8 and 107.2 kcal mol^{-1}, respectively. Thus, the ring formation energy in pyridine is $\sim$36 kcal mol^{-1} lower than that in benzene and hence the nitrogen substitution in the aromatic ring has a destabilizing effect. Such a destabilizing trend continues as more nitrogen atoms are incorporated in the aromatic ring, as the ring formation energies decrease to 52.3 kcal mol^{-1} for C_2H_2 + 2HCN $\rightarrow C_4N_2H_4$ (pyridazine) and to 42.9 kcal mol^{-1} for 3HCN $\rightarrow C_3N_3H_3$ (s-triazine). Hence, we can expect that the formation of N-PACs becomes more and more difficult as the number of N-substitutions in the same aromatic ring increases.

Relative product yields in the reaction of C_2H with 2-ethynylbenzonitrile in case of ethynyl addition to the CCH side chain will depend on the reaction conditions. It is apparent from the calculated potential energy diagram in Fig. 2 that under single-collision conditions (as in crossed molecular beams experiments), the major reaction products are expected to be 2-buta-1,3-diynylbenzonitrile + H, with a minor contribution from C_6H_4CN (cyanophenyl radical) + C_4H_2. Meanwhile, RRKM calculations of rate constants for isomerization and dissociation of the energized initial adduct **i1** possessing an internal energy of 48.7 kcal mol^{-1}, corresponding to zero collision energy between C_2H and 2-ethynylbenzonitrile, show the ratio of 98.45 : 1.53 : 0.02 for the rate constants of the reaction steps leading to 2-aza-4-ethynyl-1-naphthyl **i2**, $C_6H_4(CN)(C_4H)$ + H, and $C_6H_4(CN)(C_2H)$ + C_2H, respectively. This result indicates that the **i1** $\rightarrow$ **i2** ring closure is clearly the preferable pathway. Under single-collision conditions, however, **i2** cannot be collisionally stabilized and will have to rearrange back to **i1** and then decompose either to 2-buta-1,3-diynylbenzonitrile + H *via* **i4** or directly to cyanophenyl + C_4H_2. On the other hand, if the pressure is not zero, the thermodynamically most favorable 2-aza-4-ethynyl-1-naphthyl N-PAC product can be stabilized by collisions. Sophisticated multichannel-multiwell RRKM/Master Equation (RRKM-ME) calculations of pressure and temperature dependent rate constants and product branching ratios will be required to verify this hypothesis. Nevertheless, some evidence in support of the likelihood of collisional stabilization of **i2** comes from the recent RRKM-ME study of the C_6H_6 + C_2H/CN reactions by Woon.[41] According to his calculations, the relative yield of the stabilized $C_6H_6C_2H$ adduct in the reaction of benzene with ethynyl radical exceeds 50% at pressures higher than 4.1, 11.5, 22.0, 39.9, 66.7, and 117.9 mbar at temperatures of 50, 100, 150, 200, 250, and 300 K, respectively. At the low pressures in Titan's haze-forming region and in the interstellar media (ISM), the yield of $C_6H_6C_2H$ is smaller than that of the $C_6H_5C_2H$ + H decomposition products, but a possibility of the adduct collisional stabilization in the range of 1–10% still exists at temperatures between 50 and 100 K at 1 mbar; the lower the temperature, the higher the yield of the stabilized radical adduct. Hence, the

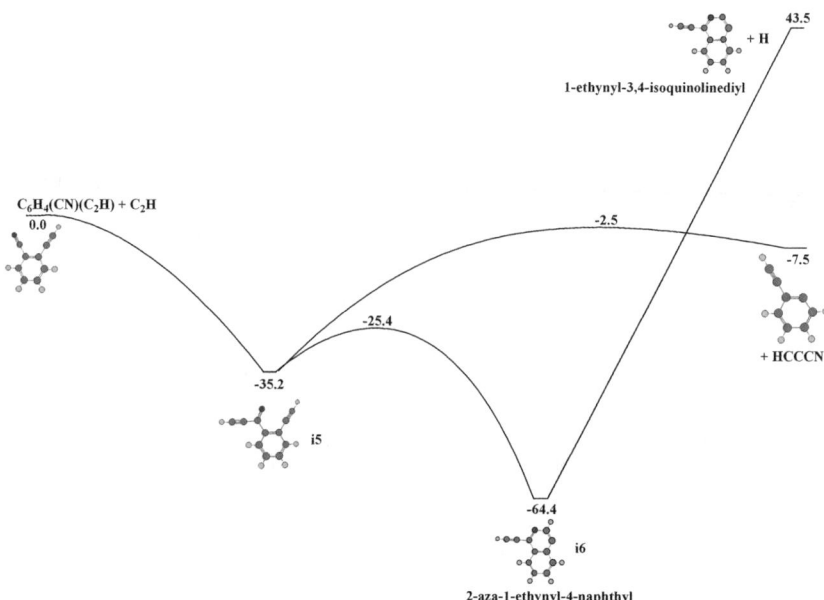

Fig. 3 Potential energy diagram of the $C_6H_4(CN)(C_2H) + C_2H$ reaction calculated at the G3(MP2,CC)//B3LYP level for the channels corresponding to ethynyl radical addition to the CN side chain. Relative energies of all species are shown in kcal mol^{-1}.

formation of a small amount of stabilized 2-aza-4-ethynyl-1-naphthyl radical in the C_2H + 2-ethynylbenzonitrile reaction may occur under Titan's stratospheric conditions.

As seen in Fig. 3, an alternative entrance channel of the C_2H + 2-ethynylbenzonitrile reaction is the addition of ethynyl radical to the CN group. This leads to the formation of intermediate **i5**, with a relative energy of −35.2 kcal mol^{-1} with respect to the initial reactants. By overcoming a modest ∼10 kcal mol^{-1} energy barrier, **i5** can cyclize to produce a polycyclic intermediate **i6**, 2-aza-1-ethynyl-4-naphthyl radical, which lies 64.4 kcal mol^{-1} below the reactants. Elimination of a hydrogen atom from **i6** would lead to the production of 1-ethynyl-3,4-isoquinolinediyl, but this channel is highly unfavorable because the product resides 43.5 kcal mol^{-1} above C_2H + 2-ethynylbenzonitrile. Besides the isomerization to **i6**, the adduct **i5** can decompose to ethynylphenyl + HCCCN by elimination of the cyanoacetylene fragment. This process is calculated to have a barrier of 32.7 kcal mol^{-1} relative to **i5**, with the transition state and products lying 2.5 and 7.5 kcal mol^{-1} lower in energy than the initial reactants, respectively. One can see that in the absence of collisional or radiational stabilization, ethynylphenyl + cyanoacetylene are expected to be the sole reaction products, as they represent the only exothermic exit channel, through which the energy of chemical activation due to the C_2H addition can be disposed into kinetic energy of the fragments. On the other hand, the RRKM rate constant for the **i5** → **i6** rearrangement calculated at zero collision energy is several orders of magnitude higher than that for **i5** → $C_6H_4C_2H$ + HCCCN. This means that the 2-aza-1-ethynyl-4-naphthyl radical **i6** can be easily produced in the reaction and the possibility of its collisional or radiational stabilization exists at non-zero pressures.

Incorporation of two nitrogen atoms in the aromatic ring

C_2H reactions with 1,2-dicyanobenzene (phthalonitrile). As seen in Fig. 4, an attack of phthalonitrile by C_2H to the C atom of one of CN groups can produce, *via*

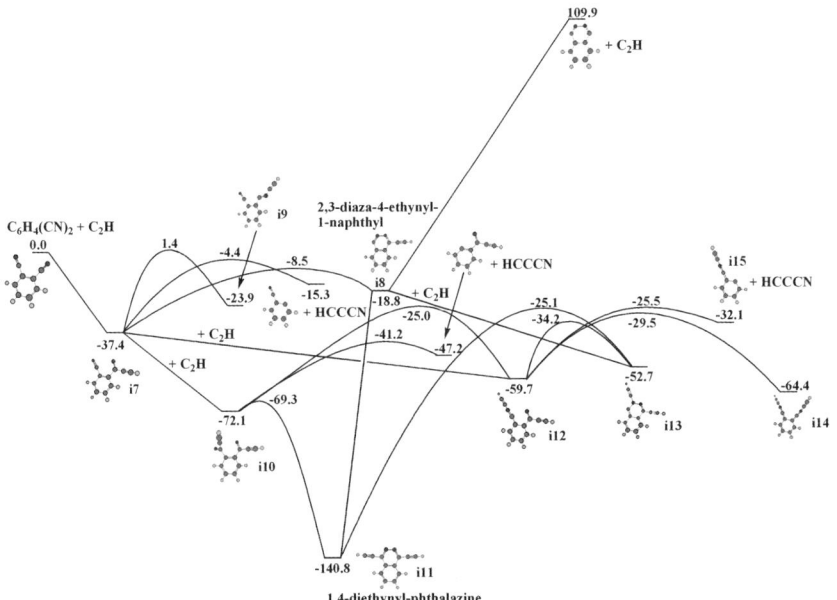

Fig. 4 Potential energy diagrams of the $C_6H_4(CN)_2 + C_2H$ and $C_6H_4(CN)(NCCCH) + C_2H$ reactions calculated at the G3(MP2,CC)//B3LYP level. Relative energies of all species are shown in kcal mol^{-1}.

a barrierless entrance channel, intermediate **i7**, residing 37.4 kcal mol^{-1} below the reactants. Next, **i7** may ring-close by overcoming a 29 kcal mol^{-1} barrier and produce a bicyclic structure **i8**, 2,3-diaza-4-ethynyl-1-naphthyl radical. This radical lies only 18.8 kcal mol^{-1} lower in energy than the $C_2H + C_6H_4(CN)_2$ reactants, which is a much shallower well on the potential energy surface as compared to those for 2-aza-4-ethynyl-1-naphthyl and 2-aza-1-ethynyl-4-naphthyl, −74.1 and −64.4 kcal mol^{-1} relative to $C_2H + C_6H_4(CN)C_2H$, respectively, and especially that for 4-ethynyl-1-naphthyl, −113.6 kcal mol^{-1} with respect to $C_2H + C_6H_4(C_2H)_2$. Thus, the destabilization trend of the aromatic ring with incorporation of nitrogen atoms is clearly observed. As an alternative to the **i7** → **i8** cyclization process, a loss of cyanoacetylene from **i7** may produce cyanophenyl radical via a 33 kcal mol^{-1} barrier. Otherwise, **i7** may pursue a C_2H migration route leading to **i9**, but this process proceeds via a 38.8 kcal mol^{-1} activation barrier with the transition state positioned 1.4 kcal mol^{-1} above the C_2H + phthalonitrile reactants and hence, this channel is unlikely at low-temperature conditions. The intermediate **i9**, which lies 23.9 kcal mol^{-1} below the reactants, can be formed directly from them by barrierless C_2H addition to one of the nitrogen atoms in phthalonitrile. **i9** represents a dead end on the PES because its isomerization to **i7** requires a higher barrier than the energy needed for the reverse **i9** → $C_2H + C_6H_4(CN)_2$ reaction and the alternative dissociation to cyanophenyl + isocyanoacetylene (28 kcal mol^{-1} less stable than cyanoacetylene) is also unfavorable as it is computed to be 36.6 kcal mol^{-1} endothermic.

Do some conditions exist at which the $C_2H + C_6H_4(CN)_2$ reaction can yield 2-aza-1-ethynyl-4-naphthyl **i8** as a product? At zero-pressure (single-collision) conditions this will not happen because cyanophenyl + C_4H_2 represent the only decomposition products of $C_6H_4(CN)_2$-C_2H exothermic with respect to the initial reactants. H loss from **i8** is highly endothermic and therefore forbidden. Although the computed ratio of RRKM rate constants for the transformation of the initial adduct **i7** to **i8** and cyanoethynyl + HCCCN, is 4 : 1, at collisionless conditions **i8** will isomerize back to **i7** and the latter will decompose to cyanoethynyl + HCCCN to dissipate the available internal energy, and thus the N-PAC intermediate **i8** will not survive. At non-zero pressure, the intermediate, which has a chance to be stabilized, is **i7** rather than **i8** because the former is both thermodynamically and kinetically more stable than the latter. Hence, 2-aza-1-ethynyl-4-naphthyl **i8** is not likely to be produced at any reaction conditions, but **i7** might be formed. Based on this, we consider further reactions of **i7** to see whether they could lead to an N-PAC synthesis.

As before, we take the ethynyl radical as a reaction facilitator and a source of chemical energy. C_2H addition to **i7** can occur at different sites. As the unpaired electron in **i7** is mostly localized at the nitrogen atom of the HCCN group, the most favorable ethynyl addition would proceed to that atom to form a closed-shell $C_6H_4(CN)(HC_2NCC_2H)$ molecule. However, since this route does not lead to the formation of a second aromatic ring, we did not follow this channel here. Alternatively, the C_2H addition to the C atom of the external cyano group in **i7** may lead to the formation of intermediate **i10** without an entrance barrier and with the energy gain of 34.6 kcal mol^{-1}. From the energized adduct **i10**, the reaction can proceed by cyclization, by elimination of one of the two external cyanoacetylene groups, or by migration of the added C_2H fragment from C to N of the cyano group. The cyclization occurs through a rather low activation barrier of 2.8 kcal mol^{-1} and leads to intermediate **i11**, a symmetric N-PAC 1,4-diethynylphthalazine (1,4-DPZ) molecule. 1,4-DPZ is 103.3 kcal mol^{-1} more stable than **i7** + C_2H. In the competing pathways, **i10** may lose cyanoacetylene via a 30.9 kcal mol^{-1} activation barrier or migrate the C_2H group from C to N via a higher barrier of 47 kcal mol^{-1} to produce intermediate **i12**. A comparison of the barrier heights for the considered transformations of the energized adduct **i10**, **i10** → **i11**, **i10** → C_6H_4(HCCN) + HCCN, and **i10** → **i12**, clearly shows that the cyclization pathway to 1,4-DPZ **i11** should dominate. An alternative C_2H addition to the N atom of the CN group in **i7** may lead, without a barrier, to the other initial adduct **i12** with exothermicity of 22.3 kcal mol^{-1}. Next, **i12** can ring-close to a bicyclic structure **i13** via a barrier of 25.4 kcal mol^{-1}, undergo a C to N C_2H shift to form **i14**, $C_6H_4(CNCCH)_2$, with a barrier of 30.2 kcal mol^{-1}, lose cyanoacetylene producing C_6H_4(CNCCH) **i15** through a 34.2 kcal mol^{-1} barrier, or isomerize back to **i10** overcoming a barrier of 34.6 kcal mol^{-1}. The bicyclic intermediate **i13**, which is ~7 kcal mol^{-1} less stable than **i12**, can in turn be subjected to C_2H migration from N to C in the nitrogen containing ring to produce the N-PAC 1,4-DPZ structure **i11** after overcoming a 27.6 kcal mol^{-1} barrier.

Under single-collision conditions, the **i7** + C_2H reaction is expected to produce either C_6H_4(NCCCH) + cyanoacetylene (ethynyl addition to C of the CN group) or C_6H_4(CNCCH) + HCCCN in case of C_2H attack to N of CN. C_2H loss from either **i11** or **i13** cannot occur in low-temperature environments because the products, **i8** (2-aza-1-ethynyl-4-naphthyl radical) + C_2H, reside 18.6 kcal mol^{-1} higher in energy than the **i7** + C_2H reactants. To qualitatively evaluate the possibility of the formation of the most stable intermediate 1,4-DPZ **i11** and **i14**, if they could be collisionally stabilized, we carried out RRKM calculations for the **i7** + C_2H reaction system. After computing energy-dependent rate constants for all feasible isomerization and dissociation pathways starting from the energized initial adducts **i10** and **i12**, we employed the steady-state approximation to calculate product branching ratios. In this consideration, we artificially supposed **i11** and **i14** to be final products assuming that their reverse reactions will not proceed due to their collisional stabilization. The results show that, if this were the case, the reactions of the energized **i10** would exclusively produce 1,4-DPZ, whereas the reactions of **i12** would yield 58.5% of 1,4-DPZ, 25.5% of **i14**, and 16% of C_6H_4(CNCCH) **i15** + HCCCN at zero collision energy between **i7** and C_2H. More quantitative RRKM-ME calculations of temperature and pressure dependent rate constants and product branching ratios in the **i7** + C_2H reaction are required to confirm the possibility of the formation of the N-PAC 1,4-DPZ species and to determine the reaction conditions under which this may actually occur. Nevertheless, our present results indicate

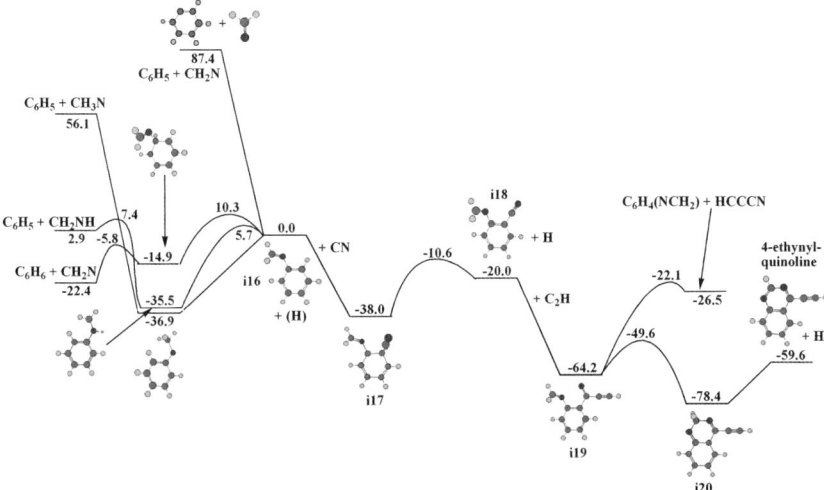

Fig. 5 Potential energy diagrams for various reactions leading to the formation of methyleneaminobenzene, C$_6$H$_5$(NCH$_2$), and the C$_6$H$_5$(NCH$_2$) + CN → C$_6$H$_4$(NCH$_2$)(CN) + H and C$_6$H$_4$(NCH$_2$)(CN) + C$_2$H → 4-ethynylquinoline + H reactions calculated at the G3(MP2,CC)//B3LYP level. Relative energies of all species are shown in kcal mol^{-1}.

that 1,4-diethynylphthalazine can be in principle synthesized through two consecutive C$_2$H additions to 1,2-dicyanobenzene (phthalonitrile) at non-zero pressure conditions where the association products can be collisionally stabilized.

Reactions involving 2-methyleneaminobenzonitrile. Another possible N-PAC precursor is 2-methyleneaminobenzonitrile **i18** (see Fig. 5). It can react with C$_2$H forming the initial adduct **i19** without a barrier and with the energy gain of 44.2 kcal mol^{-1}. The energized **i19** radical can undergo ring closure to the bicyclic structure **i20** via a relatively low barrier of 14.6 kcal mol^{-1}. This reaction step is exothermic by 14.2 kcal mol^{-1} and the resulting **i20** intermediate lies 58.4 kcal mol^{-1} lower in energy than the **i18** + C$_2$H reactants. The C–H bond in the CH$_2$ group of **i20** is weak and it takes only 18.8 kcal mol^{-1} to break it and to produce the polycyclic aromatic 4-ethynylquinoline product. Overall, the 2-methyleneaminobenzonitrile + C$_2$H → 4-ethynylquinoline + H reaction is calculated to be 39.6 kcal mol^{-1} exothermic. The only feasible competitive channel is elimination of cyanoacetylene from the initial adduct **i19** leading to the formation of the 2-methyleneaminophenyl radical. However, the barrier for HCCCN loss from **i19** is 42.1 kcal mol^{-1}, 27.5 kcal mol^{-1} higher than that for the ring closure. Also, the HCCCN elimination transition state resides 37.5 kcal mol^{-1} higher in energy than the 4-ethynylquinoline + H products. RRKM calculation of rate constants for the **i19** → **i20** and **i19** → 2-methyleneaminophenyl + H reaction steps confirms that the ring-closure process should dominate, and decomposition of **i20** to 4-ethynylquinoline + H is faster than its reverse isomerization to **i19**. These results indicate that 4-ethynylquinoline should be the major product of the 2-methyleneaminobenzonitrile + C$_2$H reaction under single-collision conditions. Since the C$_2$H addition to **i18** is barrierless and should be fast even at very low temperatures, the **i18** + C$_2$H reaction opens a synthetic route to the N-PAC 4-ethynylquinoline molecule, which is feasible under Titan's stratospheric conditions.

The question however remains how and at what conditions 2-methyleneaminobenzonitrile can be produced. The plausible precursor appears to be methyleneaminobenzene **i16**. Cyano radical can attack the *ortho* carbon atom in the benzene ring to form an adduct **i17** with the energy gain of 38.0 kcal mol^{-1} and without an entrance barrier. Next, **i17** can eliminate the hydrogen atom from the newly formed C(H)CN group and form 2-methyleneaminobenzonitrile. This process is computed to be 18.0 kcal mol^{-1} endothermic and to proceed *via* a barrier of 27.4 kcal mol^{-1}, but the H loss transition state resides 10.6 kcal mol^{-1} lower than the $C_6H_5(NCH_2)$ + CN reactants. Overall, the methyleneaminobenzene **i16** + CN → 2-methyleneaminobenzonitrile **i18** + H reaction is ~20 kcal mol^{-1} exothermic and, since it is barrierless in the entrance channel, the reaction should be fast at low-temperature conditions and should be able to produce 2-methyleneaminobenzonitrile at low pressure.

In turn, we need to consider potential synthetic pathways leading to methyleneaminobenzene. In principle, this molecule can be formed by recombination of phenyl and CH_2N (methylene-amidogen) radicals, which occurs without a barrier and is calculated to be exothermic by 87.4 kcal mol^{-1}. This process would be feasible only if both radicals are available and the atmospheric layers where they are abundant overlap. In addition, collisional stabilization of the energized methyleneaminobenzene produced in such radical recombination would be necessary. Vuitton *et al.* have recently inferred the existence of neutral CH_2NH (methyleneimine) in Titan's atmosphere, based on the detection of its protonated $CH_2NH_2^+$ counterpart by *Cassini* and estimated its mole fraction to be ~10 ppm, although with a large uncertainty.[46] According to crossed molecular beam experiments, CH_2NH is formed in the N(^{2}D) + CH_4 reaction.[47] If CH_2NH is present, the methylene-amidogen radical can be produced by its photodissociation.[48] In fact, Balucani *et al.*[47] suggested a synthetic route involving the N(^{2}D) + CH_4 reaction and its CH_2NH product as well CH_2N, leading to the formation of HCN, a relatively abundant molecule in Titan's atmosphere. Methylene-amidogen radical was also detected in laboratory studies as a major product of the N(^{4}S) + CH_3 reaction.[49] Since a reactive encounter between two radicals, like C_6H_5 and CH_2N, is a relatively rare event in a dense medium such as the atmosphere of Titan, we additionally considered reactions between a radical and a closed-shell molecule, which may produce methyleneaminobenzene **i16**. For instance, we computed PESs for the C_6H_5 + CH_2NH → **i16** + H and C_6H_6 + CH_2N → **i16** + H reactions, which are illustrated in Fig. 5. One can see that the C_6H_6 + CH_2N → $C_6H_6(NCH_2)$ → **i16** + H reaction is not expected to be important because it exhibits a large entrance barriers of 16.6 kcal mol^{-1}, is endothermic by 22.4 kcal mol^{-1}, and the H loss transition state for the second step lies as high as 32.7 kcal mol^{-1} above the reactants. On the other hand, the C_6H_5 + CH_2NH → $C_6H_5(NHCH_2)$ → **i16** + H reaction is more favorable, as it has entrance and H loss barriers of only 4.5 and 2.8 kcal mol^{-1} relative to the initial reactants and is 2.9 kcal mol^{-1} exothermic. Thus, the C_6H_5 + CH_2NH → $C_6H_5(NCH_2)$ + H reaction represents a potential synthetic route for the formation of methyleneaminobenzene under single-collision or low pressure conditions when the 4.5 kcal mol^{-1} barrier can be overcome. Although this reaction is likely to be too slow at low temperatures, it may be enhanced if the energized C_6H_5 + CH_2NH reactants are somehow generated. As recently established by Balucani *et al.*,[47] another significant product of the N(^{2}D) + CH_4 reaction is triplet CH_3N. The C_6H_5 + CH_3N recombination is highly exothermic (by 93.0 kcal mol^{-1}) and takes place without an entrance barrier. After the initial adduct $C_6H_5(NCH_3)$ is produced, the reaction proceeds by an H loss from the CH_3 group without an exit barrier and leads to **i16** + H. C_6H_5 + CH_3N → **i16** + H reaction is 56.1 kcal mol^{-1} kcal mol^{-1} exothermic, can proceed at zero- and low-pressure conditions and hence may serve as an efficient source of methyleneaminobenzene on Titan, where the phenyl and CH_3N radicals are available.

If methyleneaminobenzene can be produced, one can expect that its two consecutive reactions with C_2H can result in the formation of 1-aza-4-ethynyl-naphthalene. In the first reaction, methyleneaminobenzene + C_2H → $C_6H_5(NCH_2)(C_2H)$ → $C_6H_4(NCH_2)(C_2H)$ + H, 1-methyleneamino-2-ethynylbenzene can be formed, which is an analog of 2-methyleneaminobenzonitrile **i18**, with the C_2H group replacing

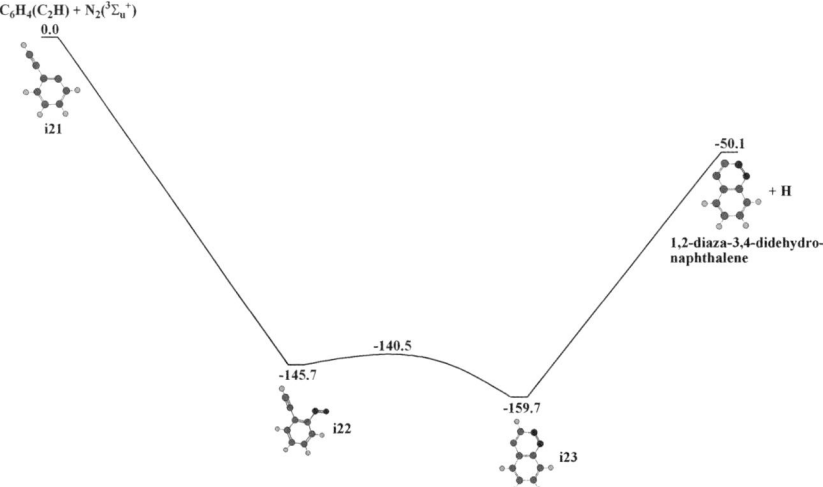

Fig. 6 Potential energy diagrams of the $C_6H_4(C_2H) + N_2$ ($^3\Sigma_u^+$) → 1,2-diaza-3,4-didehydronaphthalene + H reaction calculated at the G3(MP2,CC)//B3LYP level. Relative energies of all species are shown in kcal mol^{-1}.

CN. The second reaction would be similar to the **i18** + C_2H → **i19** → **i20** → 4-ethynylquinoline + H pathway: 1-methyleneamino-2-ethynylbenzene + C_2H → $C_6H_4(NCH_2)(C_4H_2)$ → ring-closure intermediate → 1-aza-4-ethynyl-naphthalene + H. We plan to carefully investigate the PES for this mechanism and its alternative channels in the future.

Reaction of 2-ethynylphenyl radical with triplet N_2. If molecular nitrogen is available in the excited triplet electronic state, it would be very reactive towards other radicals and can participate in the formation of N-PACs. For instance, the reaction of 2-ethynylphenyl radical **i21** with triplet N_2 is calculated to be highly exothermic (by as much as 145.7 kcal mol^{-1}) and to produce the $C_6H_4(C_2H)(N_2)$ radical **i22** without a barrier (see Fig. 6). Next, **i22** can undergo a facile ring closure to form 1,2-diaza-4-naphthyl radical **i23** overcoming a barrier of only 5.2 kcal mol^{-1}. The latter can dissipate its internal energy through dissociation taking place by a cleavage of the C–H bond in the diaza ring and producing 1,2-diaza-3,4-didehydronaphthalene. Although the breaking C–H bond is strong (109.6 kcal mol^{-1}), the 1,2-diaza-3,4-didehydronaphthalene + H products are calculated to be 50.1 kcal mol^{-1} exothermic relative to the 2-ethynylphenyl + triplet N_2 reactants and hence the reaction can easily take place even at very low temperatures and the zero- or low-pressure conditions of Titan's stratosphere. For this mechanism to be viable, the availability of electronically excited N_2 in the triplet state and the 2-ethynylphenyl radical is required. Although so far there is no experimental evidence of the existence of these species on Titan, one can imagine that triplet N_2 can be formed by collisions of atmospheric nitrogen molecules with energized electrons or by recombination of N_2^+ with an electron. On the other hand, the 2-ethynylphenyl radical may be produced by photodissociation of ethynylbenzene or *via* the reaction of *o*-benzyne with C_2H.

Conclusions

We have investigated several reaction mechanisms of incorporation of nitrogen atoms into aromatic rings of polycyclic aromatic compounds, which may lead to the formation of N-PACs under the low-temperature and low-pressure conditions of Titan's atmosphere. The most promising mechanism involves the reaction of 2-methyleneaminobenzonitrile with ethynyl radical, $C_6H_4(NCH_2)(CN) + C_2H$, producing 4-ethynylquinoline + H with an overall exothermicity of 39.6 kcal mol^{-1} and without an entrance barrier. 2-Methyleneaminobenzonitrile itself can be formed in the reaction of methyleneaminobenzene with cyano radical, $C_6H_5(NCH_2) + CN \rightarrow C_6H_5(NCH_2)(CN) \rightarrow C_6H_4(NCH_2)(CN) + H$, which is computed to be 20.0 kcal mol^{-1} exothermic and also does not have any entrance barrier. In turn, methyleneaminobenzene can be produced through recombination of phenyl and methylene-amidogen radicals followed by collisional stabilization of the product, *via* the barrierless $C_6H_5 + CH_3N \rightarrow C_6H_5(NCH_3) \rightarrow C_6H_5(NCH_2) + H$ reaction, exothermic by 56.1 kcal mol^{-1}, or by the reaction of phenyl with methyleneimine $C_6H_5 + CH_2NH \rightarrow C_6H_5(NHCH_3) \rightarrow C_6H_5(NCH_2) + H$. The latter would be slow at low-temperature conditions, as it exhibits barriers of 4.5 and 2.8 kcal mol^{-1} relative to the initial reactants, but feasible if the reactants possess sufficient internal to energy to overcome these barriers. The suggested mechanism awaits future experimental verification which can be achieved, for instance, *via* crossed molecular beam studies of the $C_6H_5 + CH_2NH$, methyleneaminobenzene + CN, and 2-methyleneaminobenzonitrile + C_2H reactions. Planetary spectroscopic observations of methyleneaminobenzene, 2-methyleneaminobenzonitrile, or 4-ethynylquinoline would be also helpful to support the proposed N-PACs' formation mechanism.

In principle, N-PACs containing one or two N atoms in an aromatic ring can be synthesized by consecutive C_2H and CN additions to benzene, $C_6H_6 + CN \rightarrow C_6H_5CN + H$, $C_6H_5CN + C_2H \rightarrow C_6H_4(CN)(C_2H) + H$, $C_6H_5CN + CN \rightarrow C_6H_4(CN)_2 + H$, $C_6H_4(CN)(C_2H) + C_2H \rightarrow$ 2-aza-4-ethynyl-1-naphthyl/2-aza-1-ethynyl-4-naphthyl, $C_6H_4(CN)_2 + C_2H \rightarrow C_6H_4(CN)(NCCCH)$, and $C_6H_4(CN)(NCCCH) + C_2H \rightarrow$ 1,4-diethynylphthalazine. However, although these reactions are barrierless and exothermic and therefore feasible at low temperatures, the steps leading to the azaethynylnaphthyl radicals, $C_6H_4(CN)(NCCCH)$, and 1,4-diethynylphthalazine can give these species as final products only upon their collisional or radiational stabilization. Whether such stabilization would be efficient enough under the conditions of Titan's stratosphere remains to be seen, and future RRKM-ME kinetic calculations of temperature and pressure dependent product branching ratios can address this question. Finally, the reaction of ethynylphenyl radical with electronically excited N_2 in the lowest $^3\Sigma_u^+$ triplet state can easily produce 1,2-diaza-3,4-didehydronapthalene + H *via* a barrierless and highly exothermic process. Although this reaction itself is facile, it requires an overlap of atmospheric regions where $C_6H_4C_2H$ and $N_2(^3\Sigma_u^+)$ can be generated and the lifetime of these radicals to be sufficient to 'find' each other.

Acknowledgements

This work was funded by the Collaborative Research in Chemistry (CRC) Program of the National Science Foundation (Award No. CHE-0627854).

References

1 D. J. Cook, S. Schlemmer, N. Balucani, D. R. Wagner, B. Steiner and R. J. Saykally, *Nature*, 1996, **380**, 227.
2 T. P. Snow, V. Le Page, Y. Keheyan and V. M. Bierbaum, *Nature*, 1998, **391**, 259.
3 A. Leger and L. D'Hendecourt, *Astron. Astrophys.*, 1985, **146**, 81.
4 M. K. Crawford, A. G. G. M. Tielens and L. J. Allamandola, *Astrophys. J.*, 1985, **293**, L45.
5 G. P. Van der Zwet and L. J. Allamandola, *Astron. Astrophys.*, 1985, **146**, 76.
6 (*a*) N. L. J. Cox and P. Ehrenfreund, *AIP Conf. Proc.*, 2006, **855**, 225, Astrochemistry; (*b*) P. Ehrenfreund and M. A. Sephton, *Faraday Discuss.*, 2006, **133**, 277, Chemical Evolution of the Universe; (*c*) P. Ehrenfreund, R. Ruiterkamp, Z. Peeters, B. Foing, F. Salama and

Z. Martins, *Planet. Space Sci.*, 2007, **55**, 383; (*d*) P. Ehrenfreund and M. Spaans, *ACS Symp. Ser.*, 2007, **981**, 232, Chemical Evolution across Space & Time.
7   E. H. Wilson and S. K. Atreya, *Planet. Space Sci.*, 2003, **51**, 1017.
8   C. Sagan, B. N. Khare, W. R. Thompson, G. D. McDonald, M. R. Wing and J. L. Bada, *Astrophys. J.*, 1993, **414**, 399.
9   B. Bézard, P. Drossart, T. Encrenaz and H. Feuchtgruber, *Icarus*, 2001, **154**, 492.
10 (*a*) A. Coustenis, A. Salama, B. Schulz, S. Ott, E. Lellouch, T. Encrenaz, D. Gautier and H. Feuchtgruber, *Icarus*, 2003, **161**, 383; (*b*) H. Niemann, S. K. Atreya, S. J. Bauer, G. R. Carignan, J. E. Demick, R. L. Frost, D. Gautier, J. A. Haberman, D. N. Harpold, D. M. Hunten, G. Israel, J. I. Lunine, W. T. Kasprzak, T. C. Owen, M. Paulkovich, F. Raulin, E. Raaen and S. H. Way, *Nature*, 2005, **438**, 779; (*c*) A. Coustenis, R. K. Achterberg, B. J. Conrath, D. E. Jennings, A. Marten, D. Gautier, C. A. Nixon, F. M. Flasar, N. A. Teanby, B. Bezard, R. E. Samuelson, R. C. Carlson, E. Lellouch, G. L. Bjoraker, P. N. Romani, F. W. Taylor, P. G. J. Irwin, T. Fouchet, A. Hubert, G. S. Orton, V. G. Kunde, S. Vinatier, J. Mondellini, M. M. Abbas and R. Courtin, *Icarus*, 2007, **189**, 35; (*d*) J. H. Waite, Jr., D. T. Young, T. E. Cravens, A. J. Coates, J. F. Crary, B. Magee and J. Westlake, *Science*, 2007, **316**, 870.
11 B. N. Khare, E. L. O. Bakes, H. Imanaka, C. P. McKay, D. P. Cruikshank and E. T. Arakawa, *Icarus*, 2002, **160**, 172.
12 H. Imanaka, B. N. Khare, J. E. Elsila, E. L. O. Bakes, C. P. McKay, D. P. Cruikshank, S. Sugita, T. Matsui and R. N. Zare, *Icarus*, 2004, **168**, 344.
13 B. N. Khare, C. Sagan, W. R. Thompson, E. T. Arakawa, F. Suits, T. A. Callcott, M. W. Williams, S. Shrader, H. Ogino, T. O. Willingham and B. Nagy, *Adv. Space Res.*, 1984, **4**, 59.
14 C. Sagan, B. N. Khare, W. R. Thompson, G. D. McDonald, M. R. Wing, J. L. Bada, T. Vo-Dinh and E. T. Arakawa, *Astrophys. J.*, 1993, **414**, 399.
15 P. Ehrenfreund, J. J. Boon, J. Commandeur, C. Sagan, W. R. Thompson and B. Khare, *Adv. Space Res.*, 1995, **15**, 335.
16 L. J. Allamandola, A. G. G. M. Tielens and J. R. Barker, *Astrophys. J. Suppl.*, 1989, **71**, 733.
17 E. Quirico, G. Montagnac, V. Lees, P. F. McMillan, C. Szopa, G. Cernogora, J.-N. Rouzaud, P. Simon, J.-M. Bernard, P. Coll, N. Fray, R. D. Minard, F. Raulin, B. Reynard and B. Schmitt, *Icarus*, 2008, **198**, 218.
18 N. Carrasco, I. Schmitz-Afonso, J.-Y. Bonnet, E. Quirico, R. Thissen, O. Dutuit, A. Bagap, O. Laprevote, A. Buch, A. Giulani, G. Adande, F. Ouni, E. Hadamcik, C. Szopa and G. Cernogora, *J. Phys. Chem. A*, 2009, **113**, 11195.
19 A. Ricca, C. W. Bauschlicher, Jr. and E. L. O. Bakes, *Icarus*, 2001, **154**, 516.
20 M. Frenklach, *Phys. Chem. Chem. Phys.*, 2002, **4**, 2028.
21 A. M. Mebel, V. V. Kislov and R. I. Kaiser, *J. Am. Chem. Soc.*, 2008, **130**, 13618.
22 W. M. Jackson and A. Scodinu, *Astrophys. Space Sci. Lib.*, 2004, **31**, 185.
23 K. Seki and H. Okabe, *J. Phys. Chem.*, 1993, **97**, 5284.
24 B. A. Balko, J. Zhang and Y. T. Lee, *J. Chem. Phys.*, 1991, **94**, 7958.
25 A. Läuter, K. S. Lee, K. H. Jung, R. K. Vatsa, J. P. Mittal and H.-R. Volpp, *Chem. Phys. Lett.*, 2002, **358**, 314.
26 (*a*) Y. L. Yung and W. B. DeMore, *Photochemistry of Planetary Atmospheres*, Oxford University Press, Oxford, UK, 1999; (*b*) M. C. Liang, A. N. Heays, B. R. Lewis, S. T. Gibson and Y. L. Yung, *Astrophys. J.*, 2007, **664**, L115.
27 E. H. Wilson and S. K. Atreya, *Bull. Amer. Astron. Soc.*, 2001, **33**, 1139.
28 A.-S. Wong, Y. L. Yung and A. J. Friedson, *Geophys. Res. Lett.*, 2003, **30**, 1447.
29 P. Rannou, F. Hourdin, C. P. McKay and D. Luz, *Icarus*, 2004, **170**, 443.
30 E. H. Wilson and S. K. Atreya, *J. Geophys. Res.*, 2004, **109**, E06002.
31 R. D. Lorenz, C. P. McKay and J. I. Lunine, *Science*, 1997, **275**, 642.
32 T. L. Roush and J. B. Dalton, *Icarus*, 2004, **168**, 158.
33 R. Lorenz, J. Mitton, *Lifting Titan's Veil*, Cambridge University Press, Cambridge, UK, 2002.
34 (*a*) A. D. Becke, *J. Chem. Phys.*, 1993, **98**, 5648; (*b*) C. Lee, W. Yang and R. G. Parr, *Phys. Rev. B: Condens. Matter*, 1988, **37**, 785.
35 (*a*) A. G. Baboul, L. A. Curtiss, P. C. Redfern and K. Raghavachari, *J. Chem. Phys.*, 1999, **110**, 7650; (*b*) L. A. Curtiss, K. Raghavachari, P. C. Redfern, A. G. Baboul and J. A. Pople, *Chem. Phys. Lett.*, 1999, **314**, 101.
36 L. A. Curtiss, K. Raghavachari, P. C. Redfern, V. Rassolov and J. A. Pople, *J. Chem. Phys.*, 1998, **109**, 7764.
37 M. J. Frisch, G. W. Trucks, H. B. Schlegel, G. E. Scuseria, M. A. Robb, J. R. Cheeseman, V. G. Zakrzewski, J. A. Montgomery, Jr., R. E. Stratmann, J. C. Burant, S. Dapprich, J. M. Millam, A. D. Daniels, K. N. Kudin, M. C. Strain, O. Farkas, J. Tomasi, V. Barone, M. Cossi, R. Cammi, B. Mennucci, C. Pomelli, C. Adamo, S. Clifford,

J. Ochterski, G. A. Petersson, P. Y. Ayala, Q. Cui, K. Morokuma, P. Salvador, J. J. Dannenberg, D. K. Malick, A. D. Rabuck, K. Raghavachari, J. B. Foresman, J. Cioslowski, J. V. Ortiz, A. G. Baboul, B. B. Stefanov, G. Liu, A. Liashenko, P. Piskorz, I. Komaromi, R. Gomperts, R. L. Martin, D. J. Fox, T. Keith, M. A. Al-Laham, C. Y. Peng, A. Nanayakkara, M. Challacombe, P. M. W. Gill, B. G. Johnson, W. Chen, M. W. Wong, J. L. Andres, C. Gonzalez, M. Head-Gordon, E. S. Replogle and J. A. Pople, *GAUSSIAN 98 (Revision A.11)*, Gaussian, Inc., Pittsburgh, PA, 2001.
38 R. D. Amos, A. Bernhardsson, A. Berning, P. Celani, D. L. Cooper, M. J. O. Deegan, A. J. Dobbyn, F. Eckert, C. Hampel, G. Hetzer, P. J. Knowles, T. Korona, R. Lindh, A. W. Lloyd, S. J. McNicholas, F. R. Manby, W. Meyer, M. E. Mura, A. Nicklaß, P. Palmieri, R. Pitzer, G. Rauhut, M. Schütz, U. Schumann, H. Stoll, A. J. Stone, R. Tarroni, T. Thorsteinsson and H.-J. Werner, *MOLPRO version 2002.6*, University of Birmingham, Birmingham, UK, 2003.
39 (*a*) P. J. Robinson and K. A. Holbrook, *Unimolecular Reactions*, Wiley, New York, 1972; (*b*) H. Eyring, S. H. Lin and S. M. Lin, *Basic Chemical Kinetics*; Wiley, New York, 1980; (*c*) J. Steinfield, J. Francisco and W. Hase, *Chemical Kinetics and Dynamics*; Prentice-Hall, Englewood Cliffs, NJ, 1989.
40 F. Goulay and S. R. Leone, *J. Phys. Chem. A*, 2006, **110**, 1875.
41 D. E. Woon, *Chem. Phys.*, 2006, **331**, 67.
42 A. Landera, A. M. Mebel and R. I. Kaiser, *Chem. Phys. Lett.*, 2008, **459**, 54.
43 A. J. Trevitt, F. Goulay, C. A. Taatjes, D. L. Osborn and S. R. Leone, *J. Phys. Chem. A*, 2010, **114**, 1749.
44 (*a*) R. Lindh, T. J. Lee, A. Bernhardsson, B. J. Persson and G. Karlström, *J. Am. Chem. Soc.*, 1995, **117**, 7186; (*b*) V. V. Kislov, T. L. Nguyen, A. M. Mebel, S. H. Lin and S. C. Smith, *J. Chem. Phys.*, 2004, **120**, 7008.
45 NIST Chemistry WebBook, NIST Standard Reference Database Number 69: http://webbook.nist.gov/chemistry/.
46 V. Vuitton, R. V. Yelle and V. G. Anicich, *Astrophys. J.*, 2006, **647**, L175.
47 N. Balucani, A. Bergeat, L. Cartechini, G. G. Volpi, P. Casavecchia, D. Skouteris and M. Rosi, *J. Phys. Chem. A*, 2009, **113**, 11138.
48 (*a*) M. T. Nguyen, D. Sengupta and T.-K. Ha, *J. Phys. Chem.*, 1996, **110**, 6499; (*b*) C. Larson, Y. Y. Ji, P. Samartzis, A. M. Wodtke, S. H. Lee, J. J. M. Lin, C. Chaudhuri and T. T. Ching, *J. Chem. Phys.*, 2006, **125**, 133302.
49 G. Marston, F. L. Nesbitt and L. J. Stief, *J. Chem. Phys.*, 1989, **91**, 3483.

Very high resolution mass spectrometry of HCN polymers and tholins

Véronique Vuitton,*[a] Jean-Yves Bonnet,[a] Maeliss Frisari,[a] Roland Thissen,[a] Eric Quirico,[a] Odile Dutuit,[a] Bernard Schmitt,[a] Léna Le Roy,[b] Nicolas Fray,[b] Hervé Cottin,[b] Ella Sciamma-O'Brien,[c] Nathalie Carrasco[c] and Cyril Szopa[c]

Received 2nd March 2010, Accepted 17th April 2010
DOI: 10.1039/c003758c

HCN polymers are complex organic solids resulting from the polymerization of hydrogen cyanide (HCN) molecules. They have been suspected to contribute to the refractory carbonaceous component of comets as well as the distributed CN sources in cometary atmospheres. Titan's tholins are also organic compounds produced in a laboratory setting but result from the complex chemistry between N_2 and CH_4 induced by UV radiation or electric discharges. Some of these compounds have optical properties in the visible range fairly similar to those of Titan's aerosols or those of the reddish surfaces of many icy satellites and small bodies. It has been proposed that HCN polymers are constituents of tholins but this statement has never received any clear demonstration. We report here on the comparative analysis of tholins and HCN polymers in order to definitely establish if the molecules identified in the HCN polymers are present in the tholins as well. First, we present a global comparison of HCN polymers with three kinds of tholins, using elemental analysis measurements, infrared spectroscopy and very high resolution mass spectrometry of their soluble fraction. We show that the chemical composition of the HCN polymers is definitely simpler than that of any of the tholins studied. Second, we focus on six ions representative of the composition of HCN polymers and using mass spectrometry (HRMS and MS/HRMS), we determine that these tholins contain at best a minor fraction of this kind of HCN polymers.

Introduction

Tholins and HCN polymers are complex polymeric organic solids mostly composed of H, C and N atoms. The term tholins designates solid products formed by organic chemistry in gas mixtures exposed to electrical discharges or ultraviolet radiation.[1,2] Since the pioneering experiments by Urey and Miller,[3] which were focused on the origin of life on Earth, numerous studies have been prompted by new fields of interest: formation of cometary refractory organics,[4,5] formation of Titan's aerosols[6] and composition of reddish icy surfaces of satellites in the outer solar system.[7] The so-called Titan's tholins are formed from N_2–CH_4 gas mixtures, in the general

---

[a]*LPG, Laboratoire de Planétologie de Grenoble, CNRS, Université J. Fourier, Grenoble, France. E-mail: veronique.vuitton@obs.ujf-grenoble.fr; Fax: +33 (0)4 76 51 41 46; Tel: +33 (0)4 76 63 52 78*
[b]*LISA, Laboratoire Interuniversitaire des Systèmes Atmosphériques, CNRS, Université Paris Est Créteil, Université Paris Diderot, Créteil, France*
[c]*LATMOS, Laboratoire Atmosphères, Milieux, Observations Spatiales, CNRS, Université Versailles St-Quentin, Université P. et M. Curie, Guyancourt, France*

framework of simulating the chemistry in Titan's atmosphere. They have received considerable attention for the last 30 years, as some of these compounds have optical properties in the visible range fairly similar to those of Titan's aerosols.[8] Titan's tholins have also been extensively used to simulate the reddish properties of the surfaces of many icy satellites and small bodies, like centaurs and trans-Neptunian objects.[7] Their global albedo and spectral variations in the visible range have been reproduced by mixtures of dark amorphous carbon and tholins.[9]

Titan's tholins (termed simply tholins below) are generally produced in cold plasma conditions, using either Direct Current (DC) or RadioFrequency (RF) discharges. Sophisticated experimental setups have been developed worldwide, allowing reproducible experiments and systematic studies. Tholins are recovered either as thin films deposited onto substrate windows or wall reactors,[6,10] or as fine-grained powders formed and grown in levitation (PAMPRE experiment[11]). Systematic studies have described the effect of pressure and gas composition on the composition of tholins, on the relative distribution of the major functional groups and on the optical properties. The control of the chemical composition and structure on optical properties has also been investigated, emphasizing in particular the role of nitrogen abundance on the absorption strength in the visible range.[10,12]

Tholins are complex combinations of C–N–H molecules, with m/z ranging from a few u up to macromolecules insoluble in organic solvents. A major issue lies in the elucidation of their chemical composition and structure, which are still poorly known. Tholins solubility in standard polar solvents ranges between ∼20–35%, depending on the N abundance.[13] The soluble fraction analyzed by High Resolution Mass Spectrometry (HRMS), appears as a highly complex mixture, with at least one peak at every m/z and m/z that extend up to 800 u.[13–15] The molecules detected can be classified along families, consisting of the addition of CH_2 functional groups on an initial structure with a fixed number of nitrogen atoms. Periodicity is observed between patterns formed by different families, pointing to a polymeric structure. In all molecules detected, unsaturation involves nitrogen atoms.

The tholins' insoluble macromolecular fraction is more complicated to investigate. Non-destructive spectroscopic techniques like infrared spectroscopy (IR) and Nuclear Magnetic Resonance (NMR) have revealed major chemical groups, mostly terminal ones: CH_2/CH_3, NH/NH_2, –CN and possibly –NC, and imines C=N, which are similar to those observed in the soluble fraction.[12,13,16] These techniques also exclude the presence of double bonded carbons as polyaromatic species and olefinic bonds. This observation suggests that the key role of nitrogen in the unsaturation of the soluble fraction also applies to the insoluble macromolecular fraction.

HCN polymers are also complex organic solids. They result from the polymerization of HCN molecules under various conditions, generally in the presence of a catalyst.[17] HCN polymers exhibit various solubilities, chemical structures and compositions. They form a broad family of compounds, which makes the term "HCN polymer" misleading. The control of the experimental conditions is definitely more complex than in the case of tholins, and a major issue lies in the reproducibility of the synthesis experiments.

HCN polymers are also, like tholins, very reactive compounds and are easily hydrolyzed, forming amino acids and other prebiotic molecules.[17–19] For this reason, they are considered as compounds of high exobiological interest. HCN polymers have also been suspected to contribute to the refractory carbonaceous component of comets. Their detection in cometary grains analyzed in the laboratory by secondary ion mass spectrometry has been claimed, but there is to date no chemical evidence of their presence.[20,21] They are also suspected to be a contributor to the distributed CN sources in cometary atmospheres.[22] Unlike tholins, HCN polymers have never been analyzed by HRMS. IR spectroscopy and NMR show that both kinds of samples exhibit similar chemical groups, but spectra appear simpler in the case of HCN polymers.[12,16,21] HCN polymers have been proposed as constituents of tholins,[23] however this statement has never received a clear demonstration.

In this publication, we present a comparison of the soluble fraction of tholins and of HCN polymers. The tholin samples were produced with the PAMPRE experiment at LATMOS (Guyancourt – France), as fine-grained powders grown in levitation. Previous studies have shown that tholins generated in the PAMPRE experiment are spherical in shape and 0.1 to 2.5 μm in diameter depending on the experimental conditions (CH_4 concentration, gas flow, RF power, plasma duration[24]). This size range is in agreement with the size of Titan's aerosols as deduced from the DISR observational data,[25,26] *i.e.* aggregates of thousands of 0.1 μm diameter monomers. The HCN polymers are one of a series synthesized at LISA (Créteil – France), containing a negligible oxygen amount and being almost completely soluble in methanol.

We report here on the comparative analysis of tholins and HCN polymers in order to establish if the molecules identified in the HCN polymers are present in tholins as well. First, we use elemental analysis, infrared spectroscopy and very high resolution mass spectrometry to present a global comparison of HCN polymers with three kinds of tholins. These tholins have been synthesized with a range of relative abundances of CH_4 in the initial gas mixture representative of those used for the simulation of Titan's atmosphere. Second, we focus on six ions representative of the composition of HCN polymers. Using HRMS, we determine if these ions are present in the tholins studied and with tandem mass spectrometry (MS/HRMS) techniques, we compare the structure of the ions present in both HCN polymers and tholins.

Experimental synthesis

HCN polymers

The HCN polymers are synthesized by hydrogen cyanide polymerization. As HCN is not a commercial product, this synthesis includes two steps: first HCN is synthesized and purified, and then it is polymerized.

HCN is produced in a vacuum manifold by reaction of NaCN (cyanide sodium, Aldrich, 98%) and $CH_3(CH_2)_{16}COOH$ (stearic acid, Merck, 97%) powders in an equimolar amount. The reagents are placed under vacuum ($<10^{-5}$ mbar) for a few hours in order to remove any water adsorbed on the powder mixture. They are subsequently heated to 350 K as the melting of the stearic acid drives the reaction. All gases produced during this reaction are trapped into a flask immersed into a liquid nitrogen bath. Then, the flask is very slowly heated at about 140 K to sublimate all the ices except the HCN that remains frozen in the flask. The purity of the remaining sample is checked by infrared spectroscopy and the main contaminant is found to be HCOOH. Any CO_2 produced from the thermal decomposition of stearic acid is eliminated during distillation.

Many methods have been proposed to synthesize HCN polymers.[19,27–33] The protocols, which propose a direct polymerization of HCN, differ by the nature of the catalyst. As NH_3 is detected in numerous astrophysical media and particularly in comets, we choose to catalyze the HCN polymerization with ammonia. We can note that Matthews and Moser[19,28] have already used ammonia to catalyze HCN polymerization.

HCN and NH_3 (Air Liquide, 99.995%) are mixed in an Erlenmeyer (10 : 1, HCN–NH_3). After the introduction of the reactants, the Erlenmeyer is closed with a glass cap and put in a water bath at room temperature in order to prevent high temperatures due to the exothermicity of the polymerization reaction. At room temperature, HCN is liquid while ammonia stays in the gaseous state in the Erlenmeyer. The liquid HCN darkens and some solids appear slowly. In a few hours, it turns to yellow, orange, brown and finally black. To reach the most complete polymerization as possible, we wait 51 days before stopping the reaction and collecting the resulting solids.

Tholins

The tholins used in this study were generated in the PAMPRE experiment. This experimental setup uses an RF capacitively coupled plasma (RF CCP) discharge as the energy source to induce chemistry in a N_2–CH_4 gas mixture. The RF CCP discharge is produced between two electrodes in the shape of a grid cage, thus confining the plasma.[11] The tholins generated are charged solid particles that are formed and grow in suspension in the plasma. They are maintained in levitation by the electrostatic force between the electrodes until they are large enough for other forces (ion and neutral drag, gravity) to compensate the latter. The tholins are then ejected from the plasma through the grid anode and deposited onto a glass crystallizer surrounding the plasma grid cage. After a production run, tholins are collected under atmospheric conditions and stored in plastic vials at room temperature for further *ex situ* analysis.

For the study presented here, tholins were produced using a 30 W, 13.56 MHz RF discharge at a pressure of 1 mbar, and at ambient temperature (neutral gas temperature $\sim 70~^\circ\text{C}$). Before producing tholins, the chamber was baked out at 110 °C for several hours, while being pumped down to 6×10^{-5} mbar, in order to remove H_2O adsorbed on the chamber walls and the electrodes. After this baking procedure, an additional cleaning of the electrodes was done by running an argon plasma discharge. Tholins were produced in three different N_2–CH_4 gas mixtures with 2% (SA98), 5% (SA95) and 10% (SA90) CH_4 concentrations, for 8 h each. These gas mixtures were chosen in order to compare tholins with known different chemical compositions. The production rates obtained for tholins produced at 2%, 5% and 10% were 32 mg h^{-1}, 29 mg h^{-1} and 3.3 mg h^{-1}, respectively.

Analytical techniques

Elemental analysis

An elemental composition analysis of the tholins produced in PAMPRE has been achieved at the Institut de Chimie des Substances Naturelles (ICSN, France) in order to measure the C, H, N, and O mass percentages in the tholins. For C, H and N, the analysis is done using a CHN 2400 Perkin Elmer analyzer which uses combustion under helium flow. A precisely weighed 1.5 mg sample is placed in the analyzer chamber which is first flushed with pure helium. The sample is then burnt at 940 °C with a 2-s long pure oxygen injection. The gases released, *i.e.* CO_2, H_2O and N_2, are mixed and analyzed by gas chromatography. The oxygen quantification is obtained *via* a Vario EL III Elementar analyzer which uses carbon pyrolysis at 1150 °C under helium flow. CO is then measured using a thermal conductivity detector.

Infrared spectroscopy

Fourier Transform Infrared measurements were performed using a Bruker Hyperion micro-spectrometer operating in the spectral range 4000–600 cm^{-1} and equipped with a MCT detector. The spectral resolution was 4 cm^{-1}. Samples were measured as crushed grains onto a ZnS window. The infrared spectra of the tholin samples were acquired under ultra-vacuum ($\sim 10^{-7}$ mbar) in order to minimize the spectral contribution of adsorbed water, while the spectrum of the HCN polymers was measured at ambient conditions.

Very high resolution mass spectrometry

While HCN polymers are almost completely soluble in methanol, tholins are not[13] and therefore a strict protocol of solubilisation of the samples was used. HCN polymers and tholins were dissolved in pure methanol (1 mg mL^{-1}), followed by 30 min sonication and 3 min centrifugation at 9000 g in order to collect the clear solution. Solubility of the tholins is estimated to be up to 30% in these conditions.[13] In order to check for possible hydrolysis of the sample before analysis, the same protocol of solubilisation was performed with acetonitrile instead of methanol. No significant difference could be observed in the spectra although the ionisation efficiency was systematically poorer.

An LTQ-Orbitrap hybrid mass spectrometer (Thermo Fisher Scientific, Bremen, Germany) equipped with the standard IonMax electrospray ionization system (ESI) was operated in the positive ionization mode with a spray voltage of 3.5 kV. Sheath gas was set to 6 arbitrary units and the ion transfer tube was maintained at 275 °C. Tube lens potential was kept at 50 V. Transfer capillary was changed and thoroughly washed between samples. Samples were introduced into the mass spectrometer by flow infusion at 3 µL min^{-1}. The Orbitrap mass analyzer was calibrated daily according to the manufacturer's directions using a mixture of caffeine, MRFA peptide, and Ultramark. No internal calibration was used in these measurements.

Full scan profile MS data (50–300 u window and 150–1000 u window) were acquired in the Orbitrap with resolving power setting of 100 000. The injection time was from 3 to 150 ms with a target of 100 000 ions in the Orbitrap adjusted by automatic gain control (AGC). All MS^n measurements were performed in the LTQ ion trap with a normalized collision energy of 30%. The activation time was set at 30 ms with the activation parameter $q = 0.25$. The isolation window was set to 1.2 u. The data presented here were all obtained with the FT-Orbitrap as the final mass analyser (MS/HRMS).

Data acquisition and processing were performed using the Xcalibur software (version 2.0.7; Thermo Fisher Scientific, San Jose, CA, USA). Elemental formula assignments were made by comparison of the observed *m/z* with calculated elemental formulas using H, C, N and O. We assumed that the elemental formula whose calculated *m/z* was closest to the observed *m/z* was the correct formula. In our tholins' spectra, low intensity peaks are often attributed to oxygen-bearing ions while this is seldom the case for HCN polymers. This observation is consistent with the lower C/O ratio in tholins than HCN polymers as determined from their elemental analysis (see next section).

Results and discussion

Global analysis of the samples

Elemental analysis. From an elemental analysis of the HCN polymers and tholins,[34] we calculated the C/H, C/N and C/O elemental ratios, as shown in Table 1. The oxygen content originates both from residual oxygen present during the synthesis and from oxidation/adsorption of water vapor during sample collection. Thermogravimetry experiments as well as IR spectra suggested that a non-trivial fraction of the tholins' H content originates from water adsorbed on the samples.[34] This fraction is probably similar in all tholins as the O content does not vary much from one sample to the other. The lower O fraction in HCN polymers indicates

Table 1 Elemental analysis of HCN polymers and tholins.[34] Relative errors are typically of 1%

14	C/H	C/N	C/O
SA90	0.74	2.4	6.9
SA95	0.84	1.7	5.8
SA98	0.95	1.2	5.9
HCN polymers	0.83	0.95	10

that the quantity of water adsorbed on HCN polymers is smaller than on tholins. These observations imply that the relative trend in the C/H ratio in tholins is real but that the C/H ratio of tholins cannot be directly compared to the C/H ratio of HCN polymers.

The results presented in Table 1 indicate that the tholins composition depends on the CH_4 content in the initial gas mixture. Tholins become more hydrogen rich and nitrogen poor when the amount of CH_4 increases. The C/N ratio is lower in HCN polymers than in any of the tholins, indicating higher nitrogen content. This shows that the global composition of the HCN polymers is different from the composition of all the tholin samples although some specific molecules constituting the HCN polymers might be present in tholins as well.

Infrared spectroscopy. The infrared spectra in Fig. 1 highlight the main chemical groups present in the samples. The broad feature in the spectral region 3500–3000 cm^{-1} shows evidence of the presence of primary and likely secondary amines. In SA98 and SA95, aliphatic groups CH_2 and CH_3 exhibit features around 2950, 1450 and 1380 cm^{-1}. In the four samples, cyanides and possibly isocyanides are identified by a group of features at $\sim$2200 cm^{-1}. Finally, the peak at $\sim$1650 cm^{-1} is attributed to the C=N bond (imine). Nevertheless, the spectral region 1650–1000 cm^{-1} is very congested pointing at a complex spectral mixture of different components. Beside the identification of these chemical groups, we observe that the spectra of tholins are smooth and composed of broad features, while the spectrum of HCN polymer is simpler, exhibiting narrow features. This suggests that the HCN polymer is chemically simpler than the tholins. A detailed assignment of the HCN polymer spectrum is beyond the scope of this paper. Note that the spectral contribution of adsorbed water, either in tholins and HCN polymers, does not significantly alter their spectral signatures.

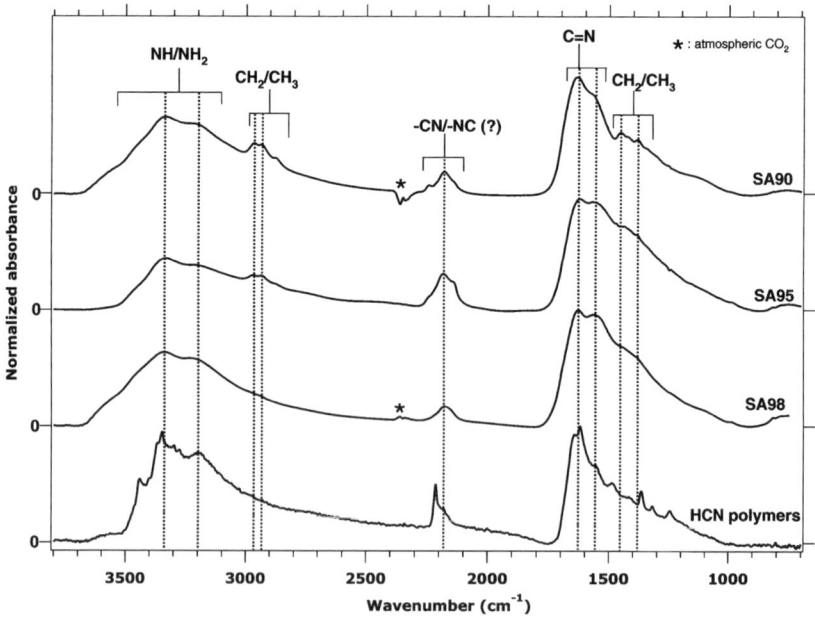

Fig. 1 Infrared spectra in the 3800–700 cm^{-1} region of the three tholin samples and of the HCN polymers.

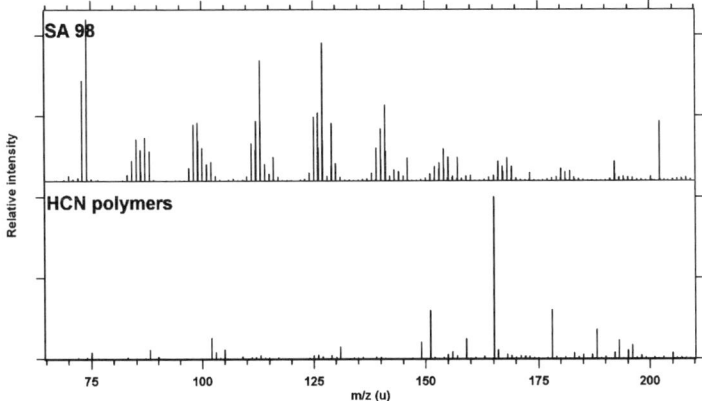

Fig. 2 Positive mode ESI Orbitrap spectrum in the m/z 65–210 region. (a) SA98, (b) HCN polymers.

Very high resolution mass spectrometry. ESI-HRMS allows the analysis of highly complex mixtures of organic compounds by direct infusion without prior separation, and therefore provides a snapshot of the thousands of molecules that can ionize under selected experimental conditions. ESI is considered to be a "cold" ionization method, which strips molecules from solution predominantly as protonated (or deprotonated) species into the gas phase. In general, no significant fragmentation of the protonated molecules occurs during the ESI process.

We compare ESI-Orbitrap spectra of HCN polymers and SA98 on Fig. 2. The global Orbitrap spectra of SA95 and SA90 are not reproduced here as their general shape and structure are similar to that of SA98. This is consistent with previous observations that every tholin spectrum is identical to the first order, not depending on the synthesis process.[14,15,35] They all show simple and regular patterns in the molecular weight distribution of the ions, *i.e.*, regularly spaced clusters separated by 13–14 u.[13] However, complex statistical representations based on Kendrick and van Krevelen diagrams reveal that the SA90 sample is richer in CH_2 than the SA98 sample.[15] This is consistent with their respective elemental analysis, showing a lower C/H ratio and a higher C/N ratio in SA90 than in SA98[34] (see Table 1). The general aspect of the HRMS spectrum of the HCN polymers is very different, with peaks extending to higher m/z and no obvious structure. However, several families of molecules could be identified, as discussed in the next section.

In order to better compare the general aspect of the mass spectra of HCN polymers and the three tholin samples, we focus in Fig. 3 on the m/z 150–185 region. This m/z region corresponds to three characteristic clusters of about 13 ions in the tholin samples while the HCN polymers spectrum is much simpler, with only a few ions in the same m/z range. Moreover, the representative close-up on m/z 165 in the inset shows that while a single ion is present at a given nominal m/z in the HCN polymers, at least three ions with the same nominal mass are observed in the tholins. Very high resolution and exact mass measurements in the Orbitrap allow the identification of these ions: $C_6H_9N_6^+$ (measured at +1.8 ppm of its exact m/z), $C_8H_{13}N_4^+$ (measured at +0.12 ppm of its exact m/z) and $C_{10}H_{17}N_2^+$ (measured at −0.24 ppm of its exact m/z). Although $C_6H_9N_6^+$, the ion identified in the HCN polymers, is also present in SA98, its relative abundance decreases in SA95 and it is almost absent from SA90. At the same time, the relative abundance of $C_8H_{13}N_4^+$ and $C_{10}H_{17}N_2^+$ increases from SA98 to SA90. This observation is consistent with the results of the elemental analysis, showing a lower C/N ratio in the HCN polymers than in any of the tholins and an increasing C/N ratio from SA98 to SA90.

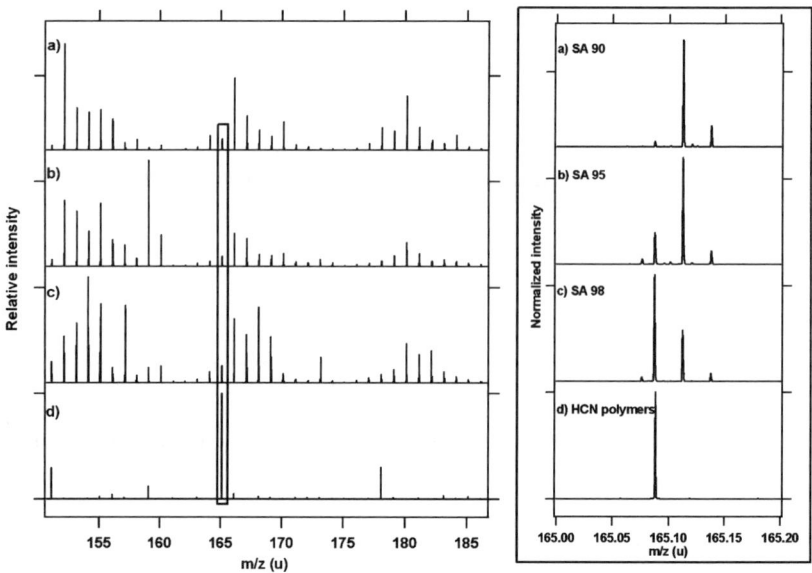

Fig. 3 Positive mode ESI Orbitrap spectrum in the 150–185 m/z region. (a) SA90, (b) SA95, (c) SA98, (d) HCN polymers. The inset gives a close-up on m/z 165 with intensities normalized for better visualization.

The elemental analysis results, infrared spectra and very high resolution mass spectra clearly indicate that the HCN polymers have a much simpler composition than the three tholins studied. At first, their composition appears closer to the composition of SA98 than of SA95 or SA90. In order to better determine if some of the molecules identified in the HCN polymers are present as well in the tholins, we next focus on a few specific ions representative of the molecules present in the HCN polymers. We first determine if ions with the same chemical composition are present in the tholin samples and then we perform MS/HRMS measurements in order to show if these ions have the same chemical structure.

Comparison of the HCN polymers and tholins

Fig. 4 presents a very high resolution mass spectrum of HCN polymers in the mass range m/z 150–300. From these data, we have identified several families of molecules in the HCN polymers. All molecules of a same family present an offset of m/z 27, the mass of HCN. For instance we have identified pure $(HCN)_n$ polymers from m/z 54 to m/z 891 (i.e. from $n = 2$ to $n = 33$). In order to compare tholins and HCN polymers, we have selected six molecules (m/z 109, 165, 178, 195, 244 and 261) from five distinct families. Three of these families are highlighted in Fig. 4. The m/z 109 ion (($HCN)_4$ molecule) was selected because of its importance in HCN polymerization processes evoked in the literature.[17,36] The m/z 244 ion (($HCN)_9$ molecule) was selected because it is the base peak in the mass spectrum of HCN polymers. The other molecules were selected because they are the most intense of their respective family, which are in turn the most intense families after the pure HCN one. The m/z 165 ion belongs to the family identified as $(HCN)_xH_2$ ($x = 6$), the m/z 178 ion to the $(HCN)_xNH$ ($x = 6$) family, the m/z 195 ion to the $(HCN)_xN_2H_4$ ($x = 6$) family, and finally the m/z 261 ion to the $(HCN)_xNH_3$ ($x = 9$) family. The different ions and their respective families are summarized in Table 2.

A summary of the (non)-detection of six selected HCN polymer ions in the three tholin samples is presented in Table 2. The Table shows that the trends presented

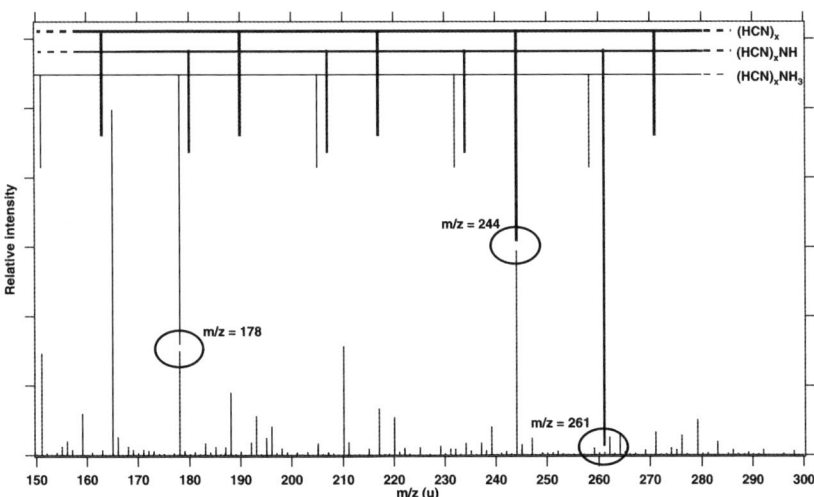

Fig. 4 Positive mode ESI Orbitrap spectrum of HCN polymers in the 150–300 m/z region. Three different families are highlighted along with the specific ions studied here (Table 2).

Table 2 Chemical formula, measured and theoretical m/z and Δppm as well as their HCN polymers family of the six ions studied here. The (non)-detection of these ions in the three tholin samples is also given. "++": major ion at its nominal m/z, "+": minor ion at its nominal m/z, "-": non-detected

HCN polymers family	Main ion in the HCN polymer sample				Tholins		
	Chemical formula	Measured m/z	Theoretical m/z	Δppm	SA98	SA95	SA90
$(HCN)_x$	$C_4H_5N_4^+$	109.0506	109.0509	−2.6	++	+	—
$(HCN)_xH_2$	$C_6H_9N_6^+$	165.0882	165.0883	−0.58	++	++	+
$(HCN)_xNH$	$C_6H_8N_7^+$	178.0833	178.0836	−1.2	++	+	—
$(HCN)_xN_2H_4$	$C_6H_{11}N_8^+$	195.1096	195.1101	−2.7	+	+	—
$(HCN)_x$	$C_9H_{10}N_9^+$	244.1055	244.1054	+0.35	++	+	—
$(HCN)_xNH_3$	$C_9H_{13}N_{10}^+$	261.1320	261.1319	+1.3	—	+	—

above for $C_6H_9N_6^+$ can be generalized to the other HCN polymer ions: they are detected in SA98 and to a lesser extent in SA95 but are mostly absent from SA90. The HCN polymer ions are systematically the ions containing the most N atoms at a given nominal m/z in the tholins. This is again consistent with the lower C/N ratio in HCN polymers than in any of the tholins and the increasing C/N ratio from SA98 to SA90 (Table 1). Only $C_9H_{13}N_{10}^+$ has a slightly different behavior: it is present in SA95 but is absent from both other samples.

Although HCN polymer ions are present in SA98, their relative intensity in a given ion cluster is fairly low, as illustrated in Fig. 3 for $C_6H_9N_6^+$ at m/z 162. However, the relative intensity of the peaks in the mass spectra is not directly correlated with the molecules concentration in the samples as their solubility and ionization efficiency may be different. Our HCN polymers are almost entirely soluble in methanol suggesting that those present in the tholins must have been extracted during the solubilization protocol. Ionization efficiency with an electrospray in the positive mode mostly depends on the proton affinity of the molecules, which is difficult to determine without knowing their exact structure. However, the ionization efficiency of the HCN

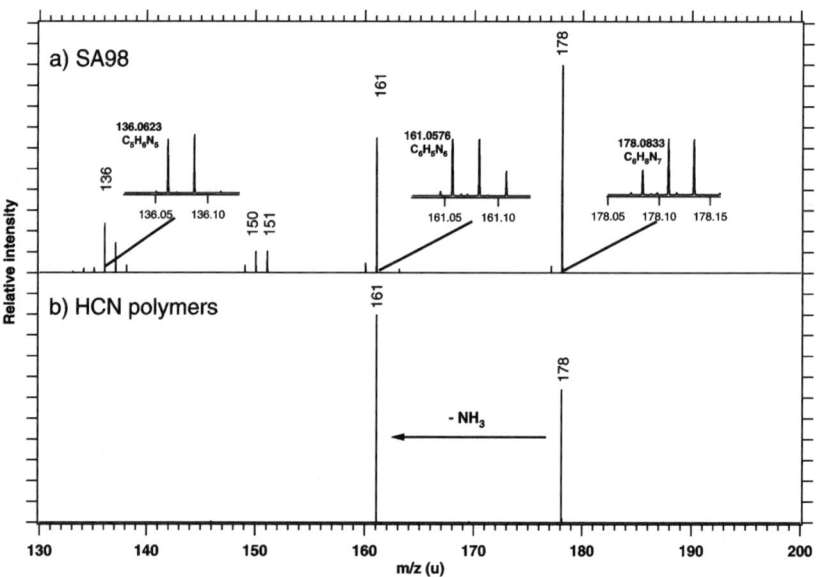

Fig. 5 Positive mode ESI Orbitrap MS² spectrum of m/z 178. (a) SA98, the inset gives a close-up on m/z 178, m/z 161 and m/z 136, (b) HCN polymers.

polymers was excellent and it is difficult to believe that their ionization efficiency is systematically lower than that of the tholins. To summarize, the low intensity of the HCN polymer ions in the tholins most likely reflects a real low abundance of these molecules in the tholin samples.

While the very high resolution mass spectra provide unambiguously the atomic composition of the molecules, they do not give much information on their structure as only general trends can be derived. In order to determine whether the same isomers are present in both HCN polymers and tholins, representative ions have been selected and used as precursor ions in MS/HRMS experiments. This technique has already been used on tholins and supported the presence of amino and nitrile functionalities in the samples studied.[35] Fig. 5 shows a characteristic MS² spectrum of m/z 178 for HCN polymers and SA98. The results of MS² experiments for the four major HCN polymer ions identified in SA98 (Table 2) are compared to the results obtained for the HCN polymer sample in Fig. 6. The low intensity of the HCN polymer ions in the tholin samples precludes from performing any further MS^n.

We first note that the loss of NH_3 is the main fragment for m/z 165, 178 and 244 in HCN polymers. This indicates that NH_2 represents a significant fraction of the HCN polymers, in agreement with the IR and NMR spectra. The two ions at m/z 109 and 244 are from the same family ($(HCN)_x$ with $x = 4$ and 9, respectively) and it came as a surprise that these two ions exhibit a very different fragmentation pattern. The ion at m/z 109 loses HCN exclusively, while the ion at m/z 244 loses mainly NH_3, a small fraction of HCN and NH_2CN. This implies a non-trivial structural evolution inside a same family, probably invalidating the process of linear polymerization.

While the analysis of the fragmentation patterns of HCN polymers is straightforward, those of the tholins must be interpreted with caution. First, for instrumental reasons, the selection window for the precursor ion cannot be narrower than 1 u and a mixture of ions with a different chemical composition is actually selected in the case of the tholin samples (Fig. 3). This implies that most of the fragment ions can originate from more than one parent ion. As shown for example in Fig. 5, for the MS² of m/z 178 in SA98, the fragment at m/z 136 $C_5H_6N_5{}^+$ can be attributed to a loss of NH_2CN from $C_6H_8N_7{}^+$ as well as a loss of C_3H_6 from $C_8H_{12}N_5{}^+$.

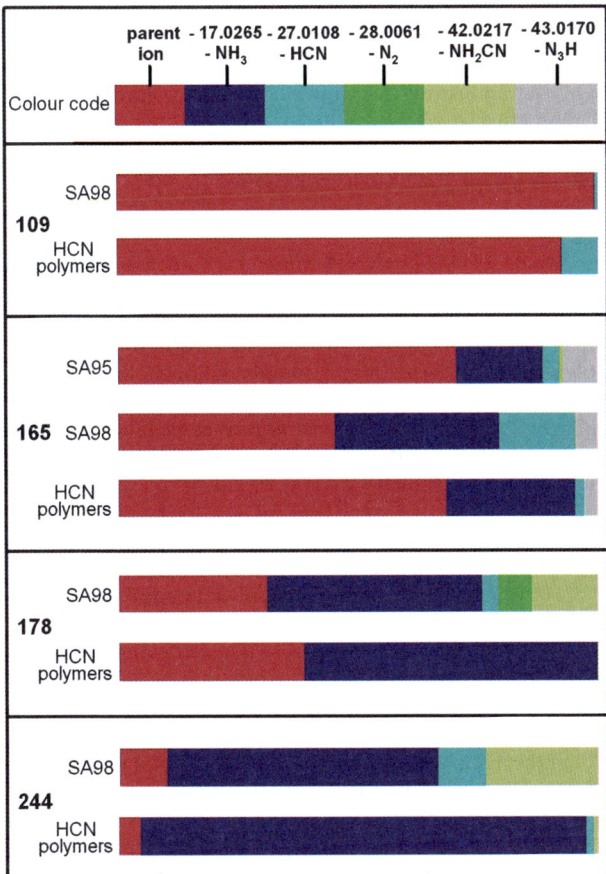

Fig. 6 Graphical representation of the MS² data for m/z 109, m/z 165, m/z 178 and m/z 244 from SA98 and HCN polymers. MS² data for m/z 165 from SA95 are included as well.

Only the loss of NH_3 and HCN can be systematically traced back to $C_6H_8N_7^+$ as the single parent ion. Next, tholins are obviously complex mixtures, and it may be possible that at a given same exact mass isomeric molecules are present, which will each exhibit their specific fragmentation signatures in the MS/HRMS. MS² experiments were obtained for m/z 165 in SA95 as well as in SA98 and in both cases major fragments include the loss of NH_3, HCN and N_3H, although the relative intensity of the loss of HCN and N_3H are reversed. This comparison illustrates the typical variability in the fragments intensity for two different samples with a same group of ions having different relative intensities.

Taking all these considerations into account, it is remarkable to see how similar are the tholins' and HCN polymers' dissociation patterns for the same nominal m/z. First, the percentage of fragmentation (intensity of the parent ion *vs.* sum of the intensity of the fragments) of the four parent ions is similar in SA98 and HCN polymers, with increasing fragmentation from m/z 109 to 165 to 178 to 244. Next, in both samples, the loss of NH_3 is by far the main fragment for the three heaviest ions, while for m/z 109 the only major fragment corresponds to the loss of HCN. Other major fragments in SA98 (loss of N_2 and NH_2CN for m/z 178, loss of NH_2CN for m/z 244) observed with very low intensities in HCN polymers may be explained by the further presence of other ions/isomers at this same nominal m/z in the tholin sample.

MS² fragments from SA98 are hence compatible with the fragments from HCN polymers, suggesting that the ions (or at least the major isomer) have a similar structure. This shows the very detailed description of complex mixtures that can be obtained with our LTQ-Orbitrap instrument. However, the interpretation of the data is rather difficult *a priori* and further work is needed in order to totally confirm our results. For example, further studies using pure compounds of known structure are necessary to gain understanding on the fragmentation pathways of molecules.

Conclusion

Mass spectrometry confirmed the relative molecular simplicity of HCN polymers compared to tholins already suggested by IR and NMR spectroscopy. Very high resolution and exact mass measurements coupled to MS/HRMS techniques revealed that HCN polymers are at best a minor component of tholins. This result indicates that the similar appearance of the IR spectrum of tholins and HCN polymers is only due to a similar distribution of functional groups rather than a detailed structural identity. An in-depth study of the composition of HCN polymers from their very high resolution mass spectrometry and MS^n will be presented in a forthcoming work.[37]

Titan's tholins are currently produced in a wide variety of experimental setups that employ different gas mixtures, energy sources, temperatures, and pressures.[1,2,6,10,11,35] However, the refractive index of only one kind of tholins has been characterized over a broad wavelength range (0.01–1000 μm).[1] As a consequence, these tholins have been used as a standard when retrieving the optical properties of Titan.[8] It appears that their albedo resembles that of Titan's actual aerosols in the visible range but that there are significant differences in the infrared.[26] For example, the VIMS results from occultation data show a lack of the 3 and 4.6 μm tholin features in the Titan aerosols.[38] The characterization of the optical properties of the tholins generated in the PAMPRE experiment is currently underway[39] and it will be interesting to compare the Titan albedos to these measurements.

MS/HRMS experiments on tholins proved to be difficult to interpret because of the high number of peaks at each nominal *m/z* making the assignment of a single parent to each fragment almost impossible. Therefore, it is challenging to retrieve some information on the structure of each ion individually. However, a broad picture of the tholins' composition in terms of chemical functionality could be obtained by performing systematic analyses of a large number of ions. For example, $C_2N_3^-$ has been identified as a major ion as well as an ubiquitous fragment in APPI-TOF spectra of SA98, suggesting that it is an important building brick in the growth of this tholins.[13] The most common fragments will be compared to a database of MS^n fragments of known standards in both positive and negative ionization mode. Another approach could be to simplify the spectra by running the sample through an HPLC column before injecting it into the mass spectrometer.

These results highlight the importance and necessity of very high mass resolution, accurate mass measurements and MS^n experiments for a more coherent and in-depth characterization of complex organic samples. Development of very high resolution ($m/\Delta m > 10^5$) mass spectrometers for spaceflight capable of *in situ* sampling of the atmosphere and the surface of Solar System bodies is mandatory.

Acknowledgements

This work was partly supported by ANR-07-BLAN-0123, Cible 2007 of the Rhone-Alpes Region and the CNRS interdisciplinary program Origines Planetes Vie (OPV). V. V. is grateful to the European Commission for the Marie Curie International Reintegration Grant No. 231013. R. T. thanks the CNRS Chemistry Institute for an ATIPE grant. L. Le Roy is the recipient of a CNES/Région Centre grant for her PhD work.

References

1 B. N. Khare, C. Sagan, E. T. Arakawa, F. Suits, A. Callcott and M. W. Williams, *Icarus*, 1984, **60**, 127–137.
2 B. N. Tran, J. C. Joseph, J. P. Ferris, P. D. Persans and J. J. Chera, *Icarus*, 2003, **165**, 379–390.
3 S. L. Miller, H. C. Urey and J. Oro, *J. Mol. Evol.*, 1976, **9**, 59–72.
4 H. Cottin, M.-C. Gazeau and F. Raulin, *Planet. Space Sci.*, 1999, **47**, 1141–1162.
5 W. Hagen, L. J. Allamandola and J. M. Greenberg, *Astrophys. Space Sci.*, 1979, **65**, 215–240.
6 P. Coll, D. Coscia, N. Smith, M.-C. Gazeau, S. I. Ramirez, G. Cernogora, G. Israel and F. Raulin, *Planet. Space Sci.*, 1999, **47**, 1331–1340.
7 D. P. Cruikshank, H. Imanaka and C. M. Dalle Ore, *Adv. Space Res.*, 2005, **36**, 178–183.
8 P. Rannou, F. Hourdin, C. P. McKay and D. Luz, *Icarus*, 2004, **170**, 443–462.
9 C. deBergh, B. Schmitt, L. V. Moroz, E. Quiricoand D. P. Cruikshank, in *The Solar System beyond Neptune*, ed. M. A. Barucci, H. Boehnhardt, D. P. Cruikshank and A. Morbidelli, University of Arizona Press, Tucson, 2008, pp. 483–506.
10 H. Imanaka, B. N. Khare, J. E. Elsila, E. L. O. Bakes, C. P. McKay, D. P. Cruikshank, S. Sugita, T. Matsui and R. N. Zare, *Icarus*, 2004, **168**, 344–366.
11 C. Szopa, G. Cernogora, L. Boufendi, J. J. Correia and P. Coll, *Planet. Space Sci.*, 2006, **54**, 394–404.
12 E. Quirico, G. Montagnac, V. Lees, P. F. McMillan, C. Szopa, G. Cernogora, J.-N. Rouzaud, P. Simon, J.-M. Bernard, P. Coll, N. Fray, R. D. Minard, F. Raulin, B. Reynard and B. Schmitt, *Icarus*, 2008, **198**, 218–231.
13 N. Carrasco, I. Schmitz-Afonso, J.-Y. Bonnet, E. Quirico, R. Thissen, O. Dutuit, A. Bagag, O. Laprévote, A. Buch, A. Giulani, G. Adandé, F. Ouni, E. Hadamcik, C. Szopa and G. Cernogora, *J. Phys. Chem. A*, 2009, **113**, 11195–11203.
14 N. Sarker, A. Somogyi, J. I. Lunine and M. A. Smith, *Astrobiology*, 2003, **3**, 719–726.
15 P. Pernot, N. Carrasco, R. Thissen and I. Schmitz-Afonso, *Anal. Chem.*, 2010, **82**, 1371–1380.
16 S. Derenne, E. Quirico, C. Szopa, G. Cernogora, B. Schmitt, V. Lees and P. F. McMillan, *Proceedings of the 39th Lunar and Planetary Science Conference*, League City, Texas, 2008.
17 C. N. Matthews and R. D. Minard, *Faraday Discuss.*, 2006, **133**, 393–401.
18 C. D. Neish, A. Somogyi, J. I. Lunine and M. A. Smith, *Icarus*, 2009, **201**, 412–421.
19 C. N. Matthews and R. E. Moser, *Nature*, 1967, **215**, 1230–1234.
20 J. Aleon, C. Engrand, F. Robert and M. Chaussidon, *Geochim. Cosmochim. Acta*, 2001, **65**, 4399–4412.
21 E. Quirico, C. Szopa, G. Cernogora, V. Lees, S. Derenne, P. F. McMillan, G. Montagnac, B. Reynard, J.-N. Rouzaud, N. Fray, P. Coll, F. Raulin, B. Schmitt and R. D. Minard, *Proceedings of the IAU Symposium 251: Organic Matter in Space*, Hong Kong, China, 2008.
22 N. Fray, Y. Bénilan, H. Cottin, M.-C. Gazeau, R. D. Minard and F. Raulin, *Meteorit. Planet. Sci.*, 2004, **39**, 581–587.
23 D. E. Budil, J. L. Roebber, S. A. Liebman and C. N. Matthews, *Astrobiology*, 2003, **3**, 323–329.
24 E. Hadamcik, J.-B. Renard, G. Alcouffe, G. Cernogora, A. C. Levasseur-Regourd and C. Szopa, *Planet. Space Sci.*, 2009, **57**, 1631–1641.
25 M. G. Tomasko, B. Archinal, T. Becker, B. Bézard, M. Bushroe, M. Combes, D. Cook, A. Coustenis, C. de Bergh, L. E. Dafoe, L. Doose, S. Douté, A. Eibl, S. Engel, F. Gliem, B. Grieger, K. Holso, E. Howington-Kraus, K. Karkoschka, H. U. Keller, R. Kirk, R. Kramm, M. Küppers, P. Lanagan, P. Lellouch, M. Lemmon, J. Lunine, E. McFarlane, J. Moores, G. M. Prout, B. Rizk, M. Rosiek, P. Rueffer, S. E. Schröder, B. Schmitt, C. See, P. Smith, L. Soderblom, N. Thomas and R. West, *Nature*, 2005, **438**, 765–778.
26 M. G. Tomasko, L. Doose, S. Engel, L. E. Dafoe, R. West, M. Lemmon, E. Karkoschka and C. See, *Planet. Space Sci.*, 2008, **56**, 669–707.
27 C. N. Matthews, *Planet. Space Sci.*, 1995, **43**, 1365–1370.
28 C. N. Matthews and R. E. Moser, *Proc. Natl. Acad. Sci. U. S. A.*, 1966, **56**, 1087–1094.
29 R. D. Minard, W. Yang, P. Varma, J. Nelson and C. N. Matthews, *Science*, 1975, **190**, 387–389.
30 R. E. Moser, A. R. Clagget and C. N. Matthews, *Tetrahedron Lett.*, 1968, **9**, 1599–1608.
31 R. E. Moser and C. N. Matthews, *Experientia*, 1968, **24**, 658–659.
32 Völker, *Angew. Chem.*, 1960, **72**, 379–384.
33 C. B. Warren, R. D. Minard and C. N. Matthews, *J. Org. Chem.*, 1974, **39**, 3375–3378.
34 E. Sciamma-O'Brien, N. Carrasco, C. Szopa, A. Buch and G. Cernogora, Icarus, 2010, DOI: 10.1016/j.icarus.2010.04.009, in press.

35 A. Somogyi, C.-H. Oh, M. A. Smith and J. I. Lunine, *J. Am. Soc. Mass Spectrom.*, 2005, **16**, 850–859.
36 J. P. Ferris, P. C. Joshi, E. H. Edelson and J. G. Lawless, *J. Mol. Evol.*, 1978, **11**, 293–311.
37 J.-Y. Bonnet, in preparation.
38 A. Bellucci, B. Sicardy, P. Drossart, P. Rannou, P. D. Nicholson, M. Hedman, K. H. Baines and B. Burrati, *Icarus*, 2009, **201**, 198–216.
39 E. Sciamma-O'Brien, in preparation.

Volatile inventories in clathrate hydrates formed in the primordial nebula

Olivier Mousis,[*ab] Jonathan I. Lunine,[c] Sylvain Picaud[d] and Daniel Cordier[ef]

Received 26th February 2010, Accepted 18th March 2010
DOI: 10.1039/c003658g

The examination of ambient thermodynamic conditions suggests that clathrate hydrates could exist in the Martian permafrost, on the surface and in the interior of Titan, as well as in other icy satellites. Clathrate hydrates are probably formed in a significant fraction of planetesimals in the solar system. Thus, these crystalline solids may have been accreted in comets, in the forming giant planets and in their surrounding satellite systems. In this work, we use a statistical thermodynamic model to investigate the composition of clathrate hydrates that may have formed in the primordial nebula. In our approach, we consider the formation sequence of the different ices occurring during the cooling of the nebula, a reasonable idealization of the process by which volatiles are trapped in planetesimals. We then determine the fractional occupancies of guests in each clathrate hydrate formed at a given temperature. The major ingredient of our model is the description of the guest–clathrate hydrate interaction by a spherically averaged Kihara potential with a nominal set of parameters, most of which are fitted to experimental equilibrium data. Our model allows us to find that Kr, Ar and N_2 can be efficiently encaged in clathrate hydrates formed at temperatures higher than $\sim$48.5 K in the primitive nebula, instead of forming pure condensates below 30 K. However, we find at the same time that the determination of the relative abundances of guest species incorporated in these clathrate hydrates strongly depends on the choice of the parameters of the Kihara potential and also on the adopted size of cages. Indeed, by testing different potential parameters, we have noted that even minor dispersions between the different existing sets can lead to non-negligible variations in the determination of the volatiles trapped in clathrate hydrates formed in the primordial nebula. However, these variations are not found to be strong enough to reverse the relative abundances between the different volatiles in the clathrate hydrates themselves. On the other hand, if contraction or expansion of the cages due to temperature variations are imposed in our model, the Ar and Kr mole fractions can be modified up to several orders of magnitude in clathrate hydrates. Moreover, mole fractions of other molecules such as N_2 or CO are also subject to strong changes with the variation of the size of the cages. Our results may affect the predictions of the composition of the planetesimals

[a] *Université de Franche-Comté, Institut UTINAM, CNRS/INSU, UMR 6213, France*
[b] *Université européenne de Bretagne, Université de Rennes 1, Institut de Physique de Rennes, CNRS, UMR 6251, France. E-mail: olivier.mousis@obs-besancon.fr*
[c] *Dipartimento di Fisica, Università degli Studi di Roma "Tor Vergata", Rome, Italy*
[d] *Université de Franche-Comté, Institut UTINAM, CNRS/INSU, UMR 6213, France*
[e] *Ecole Nationale Supérieure de Chimie de Rennes, CNRS, UMR 6226, France*
[f] *Université européenne de Bretagne, Université de Rennes 1, Institut de Physique de Rennes, CNRS, UMR 6251, France*

formed in the outer solar system. In particular, the volatile abundances calculated in the giant planets' atmospheres should be altered because these quantities are proportional to the mass of accreted and vaporized icy planetesimals. For similar reasons, the estimates of the volatile budgets accreted by icy satellites and comets may also be altered by our calculations. For instance, under some conditions, our calculations predict that the abundance of argon in the atmosphere of Titan should be higher than the value measured by Huygens. Moreover, the Ar abundance in comets could be higher than the value predicted by models invoking the incorporation of volatiles in the form of clathrate hydrates in these bodies.

1 Introduction

Clathrate hydrates (hereafter clathrates) were discovered in 1810 by Sir Humphrey Davy. Initially considered as laboratory curiosities, it is only from the 1930s that the study of their formation conditions became of significant interest because of the clogging pipelines during transportation of gas under cold conditions. Clathrates are crystalline solids which look like ice and form when water molecules constitute a cage-like structure around small "guest molecules". The most common guest molecules in terrestrial clathrates are methane (the most abundant), ethane, propane, butane, nitrogen, carbon dioxide and hydrogen sulfide. Water crystallizes in the cubic system in clathrates, rather than in the hexagonal structure of normal ice. Several different clathrate structures are known, the two most common ones being named "structure I" and "structure II". In structure I, the unit cell is formed of 46 water molecules and can incorporate up to 8 guest molecules. In structure II, the unit cell consists of 136 water molecules and can incorporate at most 24 guest molecules.

The thermodynamic conditions prevailing in many bodies of the solar system suggest that clathrates could also exist in the Martian permafrost,[1–3] on the surface and in the interior of Titan as well as in other icy satellites.[4–8] Moreover, it has been suggested that the activity observed in some cometary nuclei results from the dissociation of these crystalline structures.[9] Generally speaking, clathrates probably participated in the formation of planetesimals in the solar system. Indeed, formation scenarios of the protoplanetary nebula invoke two main reservoirs of ices that took part in the production of icy planetesimals. The first reservoir, located within 30 Astronomical Units (AU) of the Sun, contains ices (mostly water ice) originating from the Interstellar Medium (ISM) which, due to their proximity to the Sun, were initially vaporized.[10] With time, the decrease of temperature and pressure conditions allowed the water in this reservoir to condense at $\sim$150 K (for a total gas pressure of $\sim$10^{-7} bar) in the form of microscopic crystalline ice.[11] It is then considered that a substantial fraction of the volatile species were trapped as clathrates as long as free water ice was available within $\sim$30 AU in the outer solar nebula.[12] On the other hand, the remaining volatiles that have not been enclathrated due to the lack of available water ice have probably formed pure condensates at lower temperatures in this part of the nebula.[13,14] The other reservoir, located at larger heliocentric distances, is composed of ices originating from ISM that did not vaporize when entering into the disk. In this reservoir, water ice was essentially in the amorphous form and the other volatiles remained trapped in the amorphous matrix.[15,16] Consequently, icy planetesimals formed at heliocentric distances below 30 AU mainly agglomerated from clathrates while, in contrast, those produced at higher heliocentric distances (*i.e.* in the cold outer part of the solar nebula) are expected to be formed from primordial amorphous ice originating from ISM. Thus, clathrates may have been accreted in comets, in the forming giant planets and in their surrounding satellite systems.[4,12–14,17–24]

During the twentieth century, many theoretical and experimental studies allowed characterization of the crystalline structures of the most common clathrates. Meanwhile, a classification has been established to identify the nature of the clathrate and the form of occupation of the trapped molecules (single clathrate, multiple guest clathrates, *etc*). From the knowledge of the structure of clathrates, predictive rigorous methods have been developed to determine their thermodynamic properties. In particular, van der Waals & Platteeuw[25] laid the foundations of a statistical thermodynamics model to determine the properties of clathrates. This method is an excellent modern example of the use of statistical thermodynamics to predict macroscopic quantities such as temperature and pressure, using the microscopic properties like potential interactions. This approach, used today in industry and science, has saved substantial experimental effort for the determination of: (i) the equilibrium pressure of a clathrate formed from various mixtures; and (ii) the mole fraction of the different species trapped in the clathrate from a given fluid phase.

In the present work, we use a statistical thermodynamic model derived from the approach of van der Waals & Platteeuw in order to investigate the composition of clathrates that may have formed in the solar nebula. Indeed, many works published in the last decade and detailing the formation conditions of ices in the solar nebula have neglected the possibility of multiple guest trapping in clathrates.[14,17,18,20,24,26,27] In our approach, we consider the formation sequence of the different ices occurring during the cooling of the solar nebula and that is usually used to describe the process by which volatiles are trapped in planetesimals[14,17,18,20,24,26,27] (see Fig. 1). We then determine the fractional occupancies of guests in each clathrate formed at a given temperature. Similarly to papers following the approach of van der Waals & Platteeuw, our model is based on the use of intermolecular potentials, which themselves depend on parameters describing the interaction between the molecule and the cage, called "Kihara parameters". Because the models are extremely sensitive to the choice of these parameters, and because different sets of data exist in the literature,[28–34] we

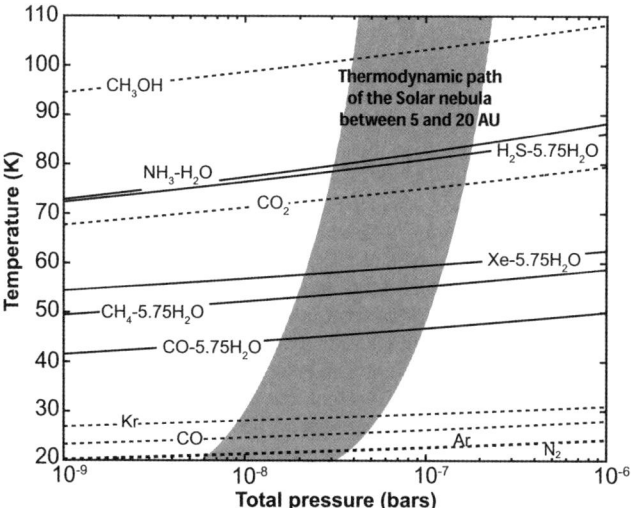

Fig. 1 Equilibrium curves of NH_3–H_2O hydrate, H_2S, Xe, CH_4 and CO clathrates (solid lines), CH_3OH, CO_2, Kr, CO, Ar and N_2 pure condensates (dotted lines), and the thermodynamic path followed by the solar nebula between 5 and 20 AU as a function of time, respectively, assuming a full efficiency of clathration. Abundances of various elements are solar, with CO : CO_2 : CH_3OH : CH_4 = 70 : 10 : 2 : 1, H_2S : H_2 = 0.5 × $(S/H_2)_\odot$, and N_2 : NH_3 = 10 : 1 in the gas phase of the disk. Species remain in the gas phase above the stability curves. Below, they are trapped as clathrates or simply condense.

examine the influence of the interaction potential parameters on our calculations. Moreover, it has been shown that the size of the clathrate cages depends on their formation temperature.[35–37] This also led us to investigate the influence of the size of cages on the resulting composition of clathrates formed in the low pressure and temperature conditions of the nebula.

The paper is organized as follows. In Section 2, we briefly depict the formation sequence of the different ices, including clathrates, in the outer solar nebula. In Section 3, we describe the statistical model based on the work of van der Waals and Platteeuw, and used to calculate the precise composition of clathrates. In Section 4, we define an optimal set of Kihara parameters used in our statistical model in order to calculate the nominal fractional occupancies of the different guests incorporated in clathrates formed in the solar nebula. Section 5 is devoted to the investigation of the influence of the potential parameters and structural characteristics (*i.e.* size of cages) of clathrates on our calculations of the fractional occupancies. In Section 6, we discuss the implications of our calculations for the composition of the outer solar system. In particular, we focus our attention on the volatile abundances in the atmosphere of Saturn, the issue of the measured argon deficiency in Titan and the prediction of the noble gas content in comets. Section 7 is devoted to the summary and discussion of our results.

2 Formation of clathrates in the primordial nebula

In the present work, we focus our attention on the formation sequence of the different ices produced in the giant-planets formation-zone in the primordial nebula. Once formed, these ices will add to the composition of the planetesimals accreted by the giant planets and their surrounding satellites during their growth. The composition of the initial gas phase of the disk is defined as follows: we assume that the abundances of all elements, including oxygen, are protosolar[38] and that O, C, and N exist only under the form of H_2O, CO, CO_2, CH_3OH, CH_4, N_2, and NH_3. The abundances of CO, CO_2, CH_3OH, CH_4, N_2 and NH_3 are then determined from the adopted CO : CO_2 : CH_3OH : CH_4 and N_2 : NH_3 gas phase molecular ratios. Once the abundances of these molecules are fixed, the remaining O gives the abundance of H_2O. Concerning the distribution of elements in the main volatile molecules, we set CO : CO_2 : CH_3OH : CH_4 = 70 : 10 : 2 : 1 in the gas phase of the disk, values that are consistent with the ISM measurements considering the contributions of both gas and solid phases in the lines of sight.[39–42] In addition, S is assumed to exist in the form of H_2S, with H_2S : H_2 = 0.5 × $(S/H_2)_\odot$,† and other refractory sulfide components.[43] We also consider N_2 : NH_3 = 10 : 1 in the nebula gas-phase, a value compatible with thermochemical models of the solar nebula[44] and with observations of cometary comae.[27] In the following, we adopt these mixing ratios as our nominal model of the solar nebula gas phase composition (see Table 1).

The process by which volatiles are trapped in icy planetesimals, illustrated in Fig. 1, is calculated using the stability curves of stochiometric hydrates, clathrates and pure condensates, and the thermodynamic path (hereafter cooling curve) detailing the evolution of temperature and pressure between 5 and 20 AU roughly corresponding to the formation locations of the giant planets in the solar nebula.‡ We refer the reader to the works of Papaloizou & Terquem[48] and Alibert *et al.*[49] for a full description of the turbulent model of accretion disk used here. The stability

---

† $(S/H_2)_\odot$ means the solar abundance of S relative to H_2. $\odot$ is the astronomical symbol for the Sun.

‡ Recent models of giant planet formation show that instead of forming at 20 and 30 AU, *i.e.* their current orbits, Uranus and Neptune underwent most of their growth among proto-Jupiter and proto-Saturn and were scattered outward when Jupiter acquired its massive gas envelope, and subsequently evolved toward their present orbits.[45–47]

Table 1 Gas phase abundances in the solar nebula.[a]

Species X	(X : H_2)	Species X	(X : H_2)
O	1.16×10^{-3}	N_2	7.62×10^{-5}
C	5.82×10^{-4}	NH_3	7.62×10^{-6}
N	1.60×10^{-4}	CO	4.91×10^{-4}
S	3.66×10^{-5}	CO_2	7.01×10^{-5}
Ar	8.43×10^{-6}	CH_3OH	1.40×10^{-5}
Kr	4.54×10^{-9}	CH_4	7.01×10^{-6}
Xe	4.44×10^{-10}	H_2S	1.83×10^{-5}
H_2O	5.15×10^{-4}		

[a] Elemental abundances derive from Lodders (2003). Molecular abundances result from the distribution of elements following the approach given in the text.

curves of hydrates and clathrates derive from Lunine & Stevenson's compilation[4] of published experimental work, in which data are available at relatively low temperatures and pressures. On the other hand, the stability curves of pure condensates used in our calculations derive from the compilation of laboratory data given in the CRC *Handbook of Chemistry and Physics*.[50] The cooling curve intercepts the stability curves of the different ices at particular temperatures and pressures. For each ice considered, the domain of stability is the region located below its corresponding stability curve. The clathration process stops when no more crystalline water ice is available to trap the volatile species. Note that, in the pressure conditions of the solar nebula, CO_2 is the only species that crystallizes at a higher temperature than its associated clathrate. We then assume that solid CO_2 is the only existing condensed form of CO_2 in this environment. In addition, we have considered only the formation of pure ice of CH_3OH in our calculations since, to the best of our knowledge, no experimental data concerning the stability curve of its associated clathrate have been reported in the literature.

In this study, we assume that the clathration efficiency is total, implying that guest molecules had the time to diffuse through porous water-ice solids before their growth into planetesimals and their accretion by proto-planets or proto-satellites. This statement remains plausible only if collisions between planetesimals have exposed essentially all the ice to the gas over time scales shorter or equal to planetesimals' lifetimes in the nebula.[4] In this case, NH_3, H_2S, Xe, CH_4 and $\sim$38.6% of CO form NH_3–H_2O hydrate and H_2S, Xe, CH_4 and CO clathrates with the available water in the outer nebula. The remaining CO, as well as N_2, Kr, and Ar, whose clathration normally occurs at lower temperatures, remain in the gas phase until the nebula cools enough to allow the formation of pure condensates. Note that, because we assume that the gas phase composition of the disk does not vary with the heliocentric distance, the calculated clathration conditions remain the same in the 5–20 AU range of the nebula, as shown recently by Marboeuf *et al.*[51] in the case of their study of the composition of ices produced in protoplanetary disks. Once crystallized, these ices will agglomerate and form planetesimals large enough to decouple from the nebular gas and will be accreted by the forming planets and satellites.

3 The statistical–thermodynamic model

To calculate the relative abundances of guest species incorporated in a clathrate from a coexisting gas of specified composition at given temperature and pressure, we follow the method described by Lunine & Stevenson[4] and Thomas *et al.*[3,7,52] which uses classical statistical mechanics to relate the macroscopic thermodynamic properties of clathrates to the molecular structure and interaction energies. It is based on the original ideas of van der Waals & Platteeuw for clathrate formation, which assume that trapping of guest molecules into cages corresponds to the three-dimensional generalization of ideal localized adsorption. This approach is based on four key assumptions:[4,31]

1. The host molecules contribution to the free energy is independent of the clathrate occupancy. This assumption implies in particular that the guest species do not distort the cages.

2. (a) The cages are singly occupied. (b) Guest molecules rotate freely within the cage.

3. Guest molecules do not interact with each other.

4. Classical statistics is valid, *i.e.*, quantum effects are negligible.

In this formalism, the fractional occupancy of a guest molecule K for a given type t (t = small or large) of cage can be written as

$$y_{K,t} = \frac{C_{K,t} P_K}{1 + \sum_J C_{J,t} P_J} \quad (1)$$

where the sum in the denominator includes all the species which are present in the initial gas phase. $C_{K,t}$ is the Langmuir constant of species K in the cage of type t, and P_K is the partial pressure of species K. This partial pressure is given by $P_K = x_K \times P$ (we assume that the sample behaves as an ideal gas), with x_K the mole fraction of species K in the initial gas phase given in Table 1, and P the total gas pressure, which is dominated by H_2.

The Langmuir constant depends on the strength of the interaction between each guest species and each type of cage, and can be determined by integrating the molecular potential within the cavity as

$$C_{K,t} = \frac{4\pi}{k_B T} \int_0^{R_c} \exp\left(-\frac{w_{K,t}(r)}{k_B T}\right) r^2 dr \quad (2)$$

where R_c represents the radius of the cavity assumed to be spherical, k_B the Boltzmann constant, and $w_{K,t}(r)$ is the spherically averaged Kihara potential representing the interactions between the guest molecules K and the H_2O molecules forming the surrounding cage t. This potential $w(r)$ can be written for a spherical guest molecule, as[53]

$$w(r) = 2z\varepsilon \left[\frac{\sigma^{12}}{R_c^{11} r}\left(\delta^{10}(r) + \frac{a}{R_c}\delta^{11}(r)\right) - \frac{\sigma^6}{R_c^5 r}\left(\delta^4(r) + \frac{a}{R_c}\delta^5(r)\right)\right] \quad (3)$$

with

$$\delta^N(r) = \frac{1}{N}\left[\left(1 - \frac{r}{R_c} - \frac{a}{R_c}\right)^{-N} - \left(1 + \frac{r}{R_c} - \frac{a}{R_c}\right)^{-N}\right] \quad (4)$$

In eqn (3), z is the coordination number of the cell. This parameter, which depends on the structure of the clathrate (I or II) and on the type of the cage (small or large), is given in Table 2. The Kihara parameters a, σ and ε for the molecule–water interactions employed in this work have been taken from Diaz Peña *et al.*[30] for CO and from Parrish & Prausnitz[28,29] or from Sloan & Koh[34] for all the other molecules of interest (see Table 3).

Finally, the mole fraction f_K of a guest molecule K in a clathrate can be calculated with respect to the whole set of species considered in the system as

$$f_K = \frac{b_s y_{K,s} + b_\ell y_{K,\ell}}{b_s \sum_J y_{J,s} + b_\ell \sum_J y_{J,\ell}} \quad (5)$$

where b_s and b_ℓ are the number of small and large cages per unit cell respectively, for the clathrate structure under consideration, and with $\sum_K f_K = 1$.

Table 2 Parameters for the cavities[28,29,a]

Clathrate structure	I		II	
Cavity type	Small	Large	Small	Large
R_c/Å	3.975	4.300	3.910	4.730
b	2	6	16	8
z	20	24	20	28

[a] R_c is the radius of the cavity. b represents the number of small (b_s) or large (b_ℓ) cages per unit cell for a given structure of clathrate (I or II), z is the coordination number in a cavity.

Table 3 Two different sets for the Kihara potential[a]

Set	Ref.	Molecule	σ/Å	(ϵ/k_B)/K	a/Å
(1)	PP72	H_2S	3.1558	205.85	0.36
	PP72 + SK08	CO_2	2.9681	169.09	0.6805
	PP72	CH_4	3.2398	153.17	0.300
	PP72	N_2	3.2199	127.95	0.350
	PP72	Xe	3.1906	201.34	0.280
	PP72	Ar	2.9434	170.50	0.184
	PP72	Kr	2.9739	198.34	0.230
	DP82	CO	3.101	134.95	0.284
(2)	SK08	H_2S	3.10000	212.047	0.3600
	SK08	CO_2	2.97638	175.405	0.6805
	SK08	CH_4	3.14393	155.593	0.3834
	SK08	N_2	3.13512	127.426	0.3526
	SK08	Xe	3.32968	193.708	0.2357
	PP72	Ar	2.9434	170.50	0.184
	PP72	Kr	2.9739	198.34	0.230
	DP82	CO	3.101	134.95	0.284

[a] σ is the Lennard-Jones diameter, ϵ is the depth of the potential well, and a is the radius of the impenetrable core. PP72, DP82 and SK08 correspond to the data taken from Parrish & Prausnitz,[28,29] Diaz Peña et al.[30] and Sloan & Koh,[34] respectively.

4 Predicted clathrate hydrates occupancies in the primordial nebula

We calculate here the relative abundances of guests that can be incorporated in H_2S, Xe, CH_4 and CO clathrates at the time of their formation in the solar nebula. As far as we know, the two most complete sets of Kihara parameters available in the literature concerning astrophysical molecules are those published by Parrish & Prausnitz and Diaz Peña et al. The parameters given by Parrish & Prausnitz have been obtained by comparing calculated chemical potentials based on the structural data of the clathrate cages with experimental results based on clathrate dissociation pressure data. These parameters have been used in recent work that aimed at investigating the composition of clathrates that may exist at the surfaces of Titan and Mars.[2,3,7,52] On the other hand, the parameters given by Diaz Peña et al. have been fitted to experimentally measured interaction virial coefficients for binary mixtures. These parameters have been used to quantify the trapping by clathrates of gases contained in volatiles observed in comets.[54] Thomas et al.[52] did a comprehensive comparison between these two sets of data and concluded that the parameters given by Parrish & Prausnitz are the most reliable because they have been self-consistently determined on experimentally measured clathrate properties and also because the results are similar to those obtained from other recent compilations (such as that of Sloan & Koh) whose molecules of astrophysical interest are listed in set (2) of Table 3). For these reasons, we have used the potential and structural parameters given by Parrish & Prausnitz for H_2S, CH_4, N_2, Xe, Ar and Kr in our nominal calculations. However, in the case of CO, we have used the Kirara parameters from Diaz Peña et al. because these data are absent from the compilation of Parrish & Prausnitz. Finally, Thomas et al.[3] found that, for common molecules, the parameters from Parrish and Prausnitz are very similar to those published in recent compilations also obtained from the fit of Langmuir constants to simple clathrate formation experimental data,[31–33] except for the value of the parameter a for CO_2 (the radius of the impenetrable core), which is almost twice larger in the recent sets of parameters than in Parrish and Prausnitz's parameters. Following the conclusions of Thomas et al.,[3] we have used the CO_2 Kirara parameters from Parrish & Prausnitz, except for the suspicious a_{CO_2} value, which has been replaced by the one given by Sloan & Koh. Set (1) of Table 3 enumerates the list of Kihara parameters for various molecules considered in our system and used in our nominal calculations.

In our calculations, any volatile already trapped or condensed at a higher temperature than the formation temperature of the clathrate under consideration is excluded from the coexisting gas phase composition. This implies that CO_2, Xe, CH_4, CO, Kr, Ar and N_2 are considered as possible guests in the case of H_2S clathrate. On the other hand, only N_2, Ar, Kr can become guests in CO clathrate. Fig. 2 represents the mole fraction f (eqn (5)) of volatiles encaged in structure I and structure II clathrates a priori dominated by H_2S, Xe, CH_4 and CO and formed in the primordial nebula. Interestingly enough, this Figure shows that, contrary to H_2S, Xe and CH_4, which remain the dominating guest species in the clathrates considered, CO becomes a minor compound in the clathrate that is expected to be dominated by this molecule. Indeed, whatever the structure considered, Kr and N_2 become more

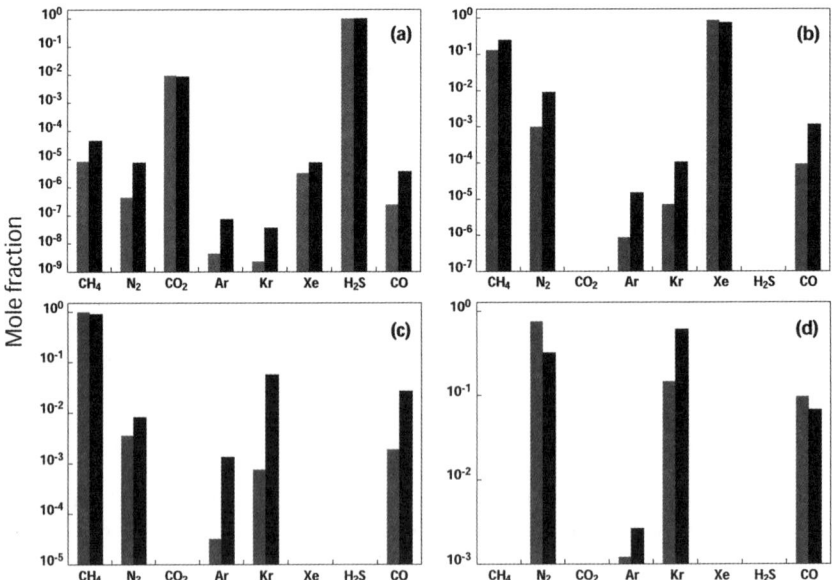

Fig. 2 Mole fraction of volatiles encaged in H_2S (a), Xe (b), CH_4 (c) and CO (d) clathrates. Grey and dark bars correspond to structure I and structure II clathrates, respectively.

abundant than CO in this clathrate, irrespective of their initial abundances in the gas phase of the nebula. This behavior results from the fact that the interaction potential between CO and the water molecules forming the surrounding cages is weaker than those involving the other species.

Table 4 gives the temperature and pressure formation conditions of these clathrates in the nebula and the relative abundance f_K^* defined as the ratio of K/X in X clathrate (where X = H_2S, Xe, CH_4 or CO). Fig. 3 represents the abundance ratio F_K^* which is defined as the ratio of K/X in X clathrate (*i.e.* f_K^*) to K/X in the initial nebula gas phase. A guest K which incorporates completely into a given clathrate displays a F_K^* value of 1 or greater, provided that there is enough water in the nebula available for clathration. This Figure shows that volatile species such as Kr, Ar or N_2 that are expected to form pure condensates below 30 K in the primitive nebula can be efficiently encaged in clathrates formed at higher temperature (~48.5 K and above). Indeed, irrespective of the particular structure, and depending on the amount of water available for clathration, these three volatiles can be efficiently trapped in CO clathrate at $T = 48.5$ K in the primitive nebula (panel (d) of Fig. 3). Kr can also be efficiently enclathrated at at $T = 55.5$ K in CH_4 clathrate (panel (c) of Fig. 3). Generally speaking, Fig. 3 illustrates the fact that some volatiles can be efficiently trapped at relatively high temperature in multiple guest clathrates compared to the temperatures at which they are expected to condense or to form single guest clathrates in the primordial nebula. It is important to note that the present results strongly depend on the choice of the initial gas phase conditions and on the amount of water ice available for clathration in the formation zone of planetesimals.

Table 4 Relative abundance f^* of volatiles encaged in structure I and structure II clathrates.[a]

Clathrate	Species	Structure I f_K^*	Structure II
H_2S	CO_2	9.67×10^{-3}	8.74×10^{-3}
$T = 82.3$ K	Xe	3.28×10^{-6}	7.58×10^{-6}
$P = 1.7 \times 10^{-7}$ bar	CH_4	8.46×10^{-6}	4.51×10^{-5}
	CO	2.47×10^{-7}	3.66×10^{-6}
	Kr	2.35×10^{-9}	3.63×10^{-8}
	Ar	4.48×10^{-9}	7.54×10^{-8}
	N_2	4.50×10^{-7}	7.62×10^{-6}
Xe	CH_4	1.52×10^{-1}	3.32×10^{-1}
$T = 59.8$ K	CO	1.07×10^{-4}	1.52×10^{-3}
$P = 1.2 \times 10^{-7}$ bar	Kr	8.30×10^{-6}	1.41×10^{-4}
	Ar	1.01×10^{-6}	2.02×10^{-5}
	N_2	1.13×10^{-3}	1.20×10^{-2}
CH_4	CO	1.91×10^{-3}	2.95×10^{-2}
$T = 55.3$ K	Kr	7.64×10^{-4}	6.27×10^{-2}
$P = 1.1 \times 10^{-7}$ bar	Ar	3.24×10^{-5}	1.48×10^{-3}
	N_2	3.62×10^{-3}	9.01×10^{-3}
CO	Kr	1.51	9.08
$T = 47.0$ K	Ar	1.25×10^{-2}	3.90×10^{-2}
$P = 1.0 \times 10^{-7}$ bar	N_2	7.84	4.78

[a] f_K^* is defined as the relative abundance of guest K to X in X clathrate (where X = H_2S, Xe, CH_4 or CO). Values of T and P correspond to the temperature and pressure of the H_2-dominated gas at which the cooling curve (here at 5 AU) intercepts the equilibrium curves of the considered clathrates (see Fig. 1) Only the species that are not yet condensed or trapped prior the epoch of clathrate formation are considered in our calculations.

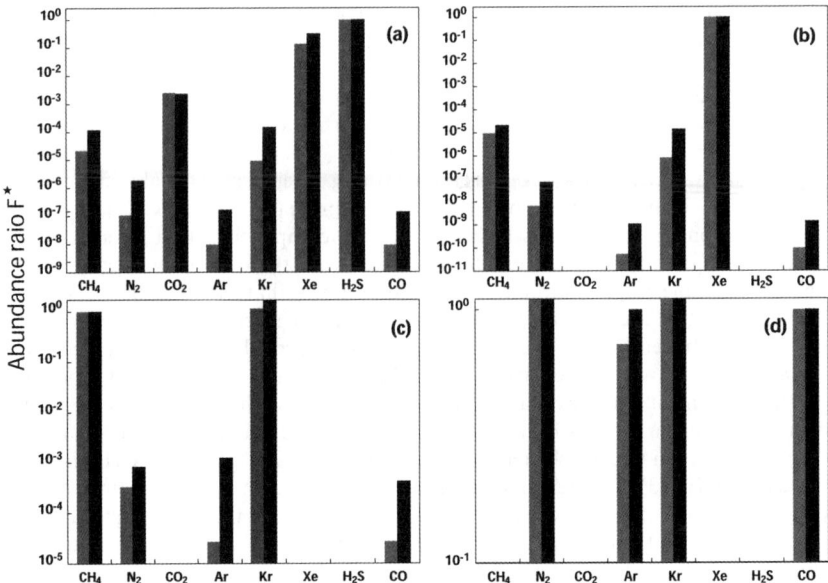

Fig. 3 Abundance ratio F_K^* is defined as the ratio of K/X in X clathrate (*i.e.* f_K^*) to K/X in the initial nebula gas phase (where X = H_2S (a), Xe (b), CH_4 (c) and CO (d)). Grey and dark bars correspond to structure I and structure II clathrates, respectively.

5 Sensitivity to parameters

The determination of the relative abundances of guest species incorporated in clathrates formed in the primitive nebula, and more generally in clathrates formed in any thermodynamic condition, strongly depends on their structural characteristics (*i.e.* size of the cages) and also on the parameters of the Kihara potential. Here, we investigate the influence of these structural characteristics and potential parameters on the fractional occupancies of guests in clathrates formed in the nebula.

5.1 Influence of the interaction potential parameters

It has been recently shown that, perturbing the σ and ϵ Kihara parameters taken from a given compilation in the 1–10% range, leads to strong variations of the values of the Langmuir constants and thus of the fractional occupancies of enclathrated molecules.[55] In the present case, in order to investigate the sensitivity of the composition of clathrates formed in the nebula to the variation of Kihara parameters, we have tested a second compilation of parameters close to our nominal set of values taken from Parrish & Prausnitz (set 1 of Table 3) but which derives from Sloan & Koh for H_2S, CO_2, CH_4, N_2 and Xe. Because we did not find any Kihara parameter for Ar, Kr and CO in the recent published compilations, we have adopted the same values as in our nominal set for these molecules (see set 2 of Table 3). As a result, the two sets of Kihara parameters used in our comparison are almost identical since several volatiles share the same data and the dispersion of data between other compounds is narrow. Two comparisons, represented in Fig. 4, have been made in the cases of Xe and CH_4 clathrates produced in the primordial nebula. In both cases, calculations have been performed for structure I and II clathrates. Despite the similarity of the potential parameters used, one can note dispersions up to more than one order of magnitude in the resulting mole fractions (eqn (5)) for a given volatile encaged in the same clathrate. We then conclude that even minor dispersions between the different existing sets of Kihara parameters can lead to large variations

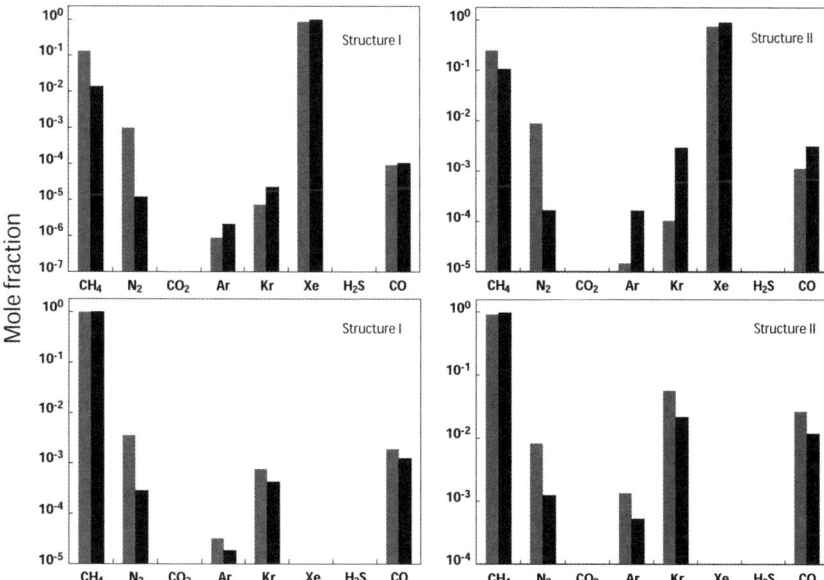

Fig. 4 Mole fraction of volatiles encaged in structure I and structure II clathrates dominated by Xe (top panels) and CH_4 (bottom panels). Grey and dark bars correspond to sets (1) and (2) of parameters of the Kihara potential given in Table 3, respectively.

into the determination of the volatiles trapping in clathrates formed in the primordial nebula. However, these variations were not found to be strong enough to modify the relative trapping efficiencies between the different volatiles in the clathrates considered here.

5.2 Influence of cage variations

In the present work, we have adopted the structural parameters of clathrates given by Parrish & Prausnitz and shown in Table 2. Up to now, we have assumed that the size of the cages R_c is unaffected by the trapping conditions of the different guests in the primordial nebula. However, laboratory measurements have shown that the size of the cages could increase with temperature and also with the size of the incorporated guest species.[35–37,56] We have thus investigated the influence of variations of the cage sizes on the mole fractions of guests encaged in clathrates by modifying by up to ±5% the values of R_c given in Table 2. This large variation is consistent with typical thermal expansion or contraction measured in the temperature range 90–270 K.[35–37,56] Note that because clathration of volatiles occurs at lower temperature in the nebula, slightly larger variations of the cage sizes may be expected.

The evolution of the mole fractions of all guests in structure I clathrates formed in the primitive nebula is given in Fig. 5 as a function of the size of the cages, at clathration temperatures and pressures given in Table 4. This Figure shows that the contraction or expansion of the cages clearly affects the mole fractions of some volatiles in clathrates and that the magnitude of these changes strongly depends on the interaction parameters between the guest species and the cages. Indeed, irrespective of the clathrate considered, Fig. 5 shows that the mole fractions of Ar and Kr can vary up to several orders of magnitude by changing the size of the cages in the range considered. Moreover, depending on the particular clathrate, mole fractions of other molecules such as N_2 or CO are also subject to strong changes with the variation of the size of the cages. Similar trends have been revealed by performing calculations for clathrates of structure II. Our calculations confirm that the trapping propensity

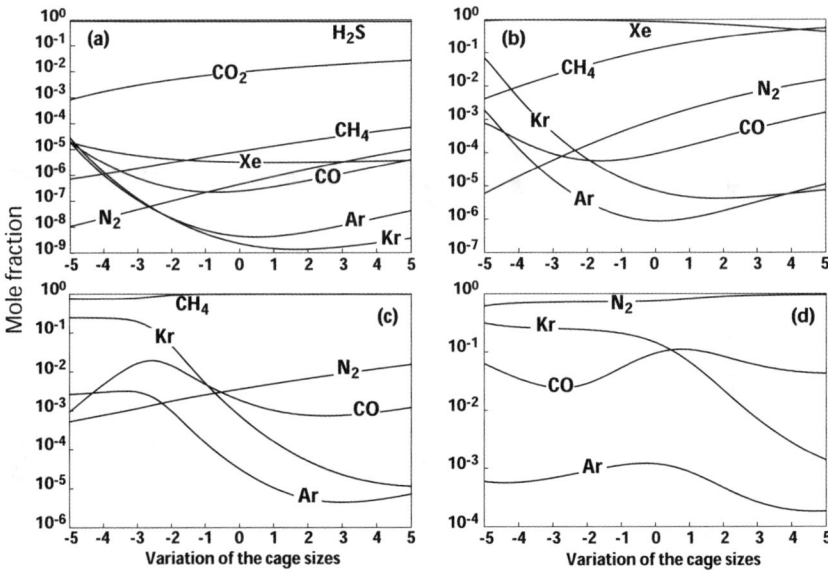

Fig. 5 Mole fraction of volatiles encaged in structure I clathrates dominated by H_2S (a), Xe (b), CH_4 (c) and CO (d) as a function of the cage sizes.

of a given molecule in clathrates is related rather sensitively to the shape of its intermolecular potential, which is determined from the adopted Kihara parameters and also from the adopted size of the cages (see eqn (3)). This is to be expected given that the Kihara parameters enter exponentially into the statistical partition function which determines the probability of occupancy. The tradeoff between the attractive part of the potential and the repulsive part is a function of the guest molecule shape and size, as well as the size of the cage, and given that clathrate structures feature two cage sizes, it is not easy to predict what happens to the relative incorporation of a molecule as cages shrink or expand thanks to temperature. The statistical mechanical model is therefore essential for predictive purposes.

Fig. 5 suggests that thermal variations of the cages need to be taken into account in particular if these variations are greater than a few percent. Similar conclusions have been obtained by Thomas et al.[52] in their study of clathrate formation and composition at the surface–atmosphere interface of Titan. As discussed in the following Section, these results may affect the predictions of the composition of the bodies formed in the outer solar system. On the other hand, variations with temperature are often not well constrained due to the small number of specific systems that have been studied, such as for example the pure methane clathrate for which the variations of the cages have been found to be small (∼0.3%) between 83 and 173 K(37). In the present study, several species are encaged in the same clathrate and the temperature range of interest is lower than those considered in experiments. For these reasons, it is difficult to infer variation laws describing the expansion/contraction of clathrates at temperatures relevant for the solar nebula.

6 Implications for the composition of the outer solar system

Our calculations of the multiple guest trapping in clathrates formed in the primordial nebula have implications for the formation and composition models of the giant planets, their surrounding satellites and also comets. Indeed, the fact that several compounds expected to be trapped or condensed at low temperature are incorporated at relatively higher temperature in clathrates in the nebula modifies the predictions of the composition of the icy planetesimals from which these bodies were presumably formed relative to what is obtained from simple condensation models. A few examples are discussed here.

6.1 Volatile enrichments in Saturn

Measurements by the mass spectrometer aboard the Galileo probe have shown that the abundances of C, N, S, Ar, Kr and Xe are all enriched by similar amounts with respect to their solar abundances in the atmosphere of Jupiter.[15,57,58] Similarly, recent Cassini CIRS observations have also confirmed what was inferred from previous measurements, that C is substantially enriched in the atmosphere of Saturn.[59,60] In order to interpret these volatile enrichments, it has been proposed that the main volatile compounds initially existing in the solar nebula gas phase were essentially trapped by crystalline water ice in the form of clathrates or hydrates in the feeding zones of Jupiter and Saturn.[14,17–19,24,26,27] These ices then agglomerated and formed planetesimals that were ultimately accreted by the forming Jupiter and Saturn. This is then the fraction of these icy planetesimals that vaporized when entering the envelopes of the two growing planets which engendered the observed volatile enrichments.

On the other hand, our statistical model allows us to infer that multiple guest clathrates were more likely formed than single clathrates in the solar nebula, implying substantial changes in the presumed composition of the planetesimals formed from these ices. As a result, the volatile enrichments calculated in the giant planets' atmospheres should also be altered because these quantities are proportional to the mass of accreted and vaporized icy planetesimals. Indeed, for instance, a recent interpretation of the carbon abundance in the atmosphere of Saturn is based on the hypothesis that the giant planet never formed at a disk temperature below 30 K,[27] implying that the planetesimals accreted by the giant planet were impoverished in Ar, Kr, CO and N_2 (*i.e.* volatiles whose condensation curves are located below 30 K in the nebula – see Fig. 1). However, our nominal calculations predict that Kr can be entirely trapped in CH_4-dominated clathrates at ~55 K in the nebula, provided that there is enough water available for clathration at this temperature in the feeding zone of Saturn. In this scenario, the trapping of Kr at higher temperature in the planetesimals accreted by Saturn implies that the atmospheric abundance of this noble gas should also be enhanced by an amount similar to that of carbon compared to solar (at least ~9 times solar[14,60]), instead of being predicted in solar abundance in Saturn's envelope.[27] However, this conclusion is valid only for the structural characteristics and the Kihara parameters adopted in our nominal calculations. As shown in Fig. 5, any expansion of the cage sizes during the formation of CH_4-dominated clathrate in the solar nebula could strongly reduce the trapping efficiency of Kr in this clathrate and thus decrease its resulting enhancement in the atmosphere of Saturn.

6.2 The argon deficiency in Titan

A puzzling feature of the atmosphere of Titan is that no primordial noble gases other than argon were detected by the Gas Chromatograph Mass Spectrometer (GCMS) aboard the Huygens probe during its descent to Titan's surface on January 14, 2005. The observed argon includes primordial ^{36}Ar, *i.e.* the main isotope, and the radiogenic isotope ^{40}Ar, which is a decay product of ^{40}K.[61] In any case, the $^{36}Ar/^{14}N$ is lower than the solar value by more than five orders of magnitude.[61] The other primordial noble gases Kr and Xe (and ^{38}Ar) were not detected by the GCMS instrument down to upper limits of 10 parts per billion relative to nitrogen.[61] The Kr and Xe deficiencies could simply be explained by the presence of clathrates on the surface of Titan that would have efficiently incorporated these noble gases.[7,52] On the other hand, in order to interpret the Ar deficiency in the atmosphere of Titan, it has been proposed that the satellite was formed from icy planetesimals initially produced in the solar nebula and that were partially devolatilized at a temperature not exceeding $\sim$50 K during their migration within Saturn's subnebula.[13] In this case, because Ar is poorly trapped in clathrates formed above $\sim$50 K in the nebula, only tiny amounts of this compound would have been incorporated in the building blocks of the forming Titan, in agreement with the observations. In particular, our nominal model predicts that this noble gas would remain essentially trapped in CH_4-dominated clathrate,§ and subsequently in the satellite, with $Ar:CH_4$ of 2.7×10^{-5} (see Fig. 3), in good match with the abundances of ^{36}Ar observed in Titan's atmosphere. However, this statement remains valid only if one considers our nominal calculations of multiple guest trapping in clathrates formed in the feeding zone of Saturn. Indeed, Fig. 5 shows that the trapping of Ar can strongly increase by a factor of more than 300 in CH_4-dominated clathrate formed in the solar nebula if the size of the cages decreases by a few percent. In this case, the amount of argon predicted by our calculations in the atmosphere of Titan would be higher than the value measured by Huygens or an alternative scenario must be invoked to explain its apparent depletion.

6.3 Noble gas content in comets

It has been proposed that the composition of volatiles observed in comets could be explained on the basis of their trapping in the form of clathrates in the primordial nebula.[54] In this model, the key parameter to explain the volatile content observed in comets is the amount of water ice available in the region of the nebula where the clathration took place. The mass of available water is then varied between the amount needed to trap the whole mass of volatiles present in the nebula (high-mass ice scenario) and that which is just required to enclathrate volatiles at temperatures above $\sim$50 K in the nebula (low-mass ice scenario). In this model, the case of argon is interesting because its relative abundance in comets is found to be very small ($Ar:H_2O \sim 10^{-8}$) when the low-mass ice scenario is considered. On the other hand, if enough water ice is present, $Ar:H_2O$ jumps to potentially detectable values ($Ar:H_2O \sim 10^{-4}$–10^{-3}). However, as in the case of Titan described above, taking into account the uncertainties of our calculations (in particular on the cage sizes), substantial amounts of Ar can be trapped in CH_4-dominated clathrates. In this case, we find $Ar:H_2O \sim 10^{-5}$–10^{-4} in the low-mass ice scenario, a value which is close to the one determined in the case of the high-mass ice scenario. Our estimate for the low-mass ice scenario implies then that it is difficult to consider the measurement of the abundance of argon in comets as a key test to constrain the mass of water ice that was available in the nebula for forming comets.

7 Summary and discussion

In this paper, we have calculated the relative abundances of guest species that can be incorporated in clathrates formed in the gas phase and thermodynamic conditions of the primordial nebula. We have assumed that the clathration efficiency is total in the primitive nebula, implying that guest molecules had the time to diffuse through porous water-ice solids before their growth into planetesimals and their accretion by proto-planets or proto-satellites. This statement remains plausible only if collisions between planetesimals have exposed essentially all the ice to the gas over time scales shorter or equal to planetesimals lifetimes in the nebula.[4] However, it is important to note that the efficiency of collisions between planetesimals to expose all the "fresh" ice over such a time scale still remains questionable and that we have no evidence that clathration was important in the primordial nebula.

---

§ The amount of Ar found in Xe and H_2S-dominated clathrates is negligible.

The results presented here derive from the usual statistical model based on the van der Waals and Platteeuw approach, generalized by Parrish & Prausnitz for the calculations of dissociation pressures of multiple guest clathrates. The major ingredient of our model is the description of the guest–clathrate interaction by a spherically averaged Kihara potential with a nominal set of parameters, most of which being fitted on experimental equilibrium data. Our model allows us to find that Kr, Ar and N_2 can be efficiently encaged in clathrates formed at temperatures higher than $\sim$48.5 K, the temperature of CO clathrate, in the primitive nebula, instead of forming pure condensates below 30 K. However, we find at the same time that the determination of the relative abundances of guest species incorporated in these clathrates strongly depends on the choice of the parameters of the Kihara potential and also on their adopted structural characteristics. Indeed, testing different potential parameters, we have noted that even minor dispersions between the different existing sets can lead to non-negligible variations in the determination of the volatiles trapped in clathrates formed in the primordial nebula. However, these variations are not found to be strong enough to reverse the relative abundances between the different volatiles in the considered clathrates. Moreover, we have found that the contraction or expansion of the cages, due to temperature variations, can alter the mole fractions of some volatile molecules up to several orders of magnitude in clathrates. On the other hand, due to the lack of laboratory experiments describing properly the variation of the size of cages as a function of temperature for molecules of astrophysical interest, it is actually difficult to quantify the influence of this parameter on the composition of planetesimals formed in the nebula. The determination of specific laws describing the variation with temperature of the size of cages is important because it could help to constrain the thermodynamic conditions encountered by planetesimals during their formation in the primordial nebula. Indeed, given a sufficient number of species whose abundances are determined in a particular object, our statistical mechanical model could be used to predict the composition and temperature (through the thermal expansion/contraction of the ice which affects cage size) of the planetesimals from which the object formed and hence the composition of bodies issued from the same family. Such a measurement will be achievable by the Rosetta spacecraft mission towards Comet 67P/Churyumov-Gerasimenko or by the next generation of missions designed to investigate the outer solar system.

Our results alter the predictions of the composition of the planetesimals formed in the outer solar system. In particular, the volatile abundances calculated in the giant planets' atmospheres should be recalculated because these quantities are proportional to the mass of accreted and vaporized icy planetesimals. For the same reasons, the estimates of the volatile budgets accreted by icy satellites and comets must be reconsidered based on our calculations. For instance, our calculations predict that the abundance of argon in the atmosphere of Titan might for some initial conditions be higher than the value measured by Huygens because substantial amounts of this volatile may be trapped in the satellite's building blocks during their formation. Similarly to the case of Titan, if comets were agglomerated from clathrates in the nebula, then the Ar abundance in these bodies should still remain potentially detectable, irrespective of their formation temperature ($\sim$50 K and below) because substantial amounts of this volatile are expected to be trapped in CH_4-dominated clathrate at $\sim$55.5 K.

It is important to note that the validity of the statistical model to determine the composition of clathrates in the nebula probably meets some limitations. In particular, the transferability of the Kihara parameters to temperatures and pressures beyond the range for which they have been fitted is uncertain.[62,63] Moreover, inconsistencies have been evidenced between Kihara parameters derived from different sets of experimental data.[64] In these conditions, supplementary measurements of the equilibrium pressure of clathrates at low temperature would be needed in order to obtain a set of Kihara parameters consistent with the low temperature and pressure conditions of the nebula. Moreover, it has been claimed in the literature that

Kihara potential may not accurately describe the interaction between guest and water molecules.[65] Recent works have thus been based on an atom–atom description of the intermolecular guest–clathrate interactions, in which effective parameters for these interactions have been fitted from results of *ab initio* quantum mechanical methods.[65–67] However, the accuracy of this atom–atom approach is strongly dependent on the ingredients of the fit, which can be the number of water molecules considered in the *ab initio* calculations, the number of sites chosen to represent the interacting molecules, the level of accuracy of the quantum methods used... It is then unfortunately very difficult to discriminate which approach (atom–atom or spherically averaged) is the most accurate.

References

1 B. K. Chastain and V. Chevrier, *Planet. Space Sci.*, 2007, **55**, 1246.
2 T. D. Swindle, C. Thomas, O. Mousis, J. I. Lunine and S. Picaud, *Icarus*, 2009, **203**, 66.
3 C. Thomas, O. Mousis, S. Picaud and V. Ballenegger, *Planet. Space Sci.*, 2009, **57**, 42.
4 J. I. Lunine and D. J. Stevenson, *Astrophys. J. Suppl.*, 1985, **58**, 493.
5 K. P. Hand, C. F. Chyba, R. W. Carlson and J. F. Cooper, *Astrobiology*, 2006, **6**, 463.
6 G. Tobie, J. I. Lunine and C. Sotin, *Nature*, 2006, **440**, 61.
7 C. Thomas, O. Mousis, V. Ballenegger and S. Picaud, *Astron. Astrophys.*, 2007, **474**, L17.
8 O. Mousis and B. Schmitt, *Astrophys. J.*, 2008, **677**, L67.
9 U. Marboeuf, O. Mousis, J.-M. Petit and B. Schmitt, *Astrophys. J.*, 2010, **708**, 812.
10 K. M. Chick and P. Cassen, *Astrophys. J.*, 1997, **477**, 398.
11 A. Kouchi, T. Yamamoto, T. Kozasa, T. Kuroda and J. M. Greenberg, *Astron. Astrophys.*, 1994, **290**, 1009.
12 O. Mousis, D. Gautier, D. Bockelée-Morvan, F. Robert, B. Dubrulle and A. Drouart, *Icarus*, 2000, **148**, 513.
13 O. Mousis, J. I. Lunine, C. Thomas, M. Pasek, U. Marboeuf, Y. Alibert, V. Ballenegger, D. Cordier, Y. Ellinger, F. Pauzat and S. Picaud, *Astrophys. J.*, 2009, **691**, 1780.
14 O. Mousis, U. Marboeuf, J. I. Lunine, Y. Alibert, L. N. Fletcher, G. S. Orton, F. Pauzat and Y. Ellinger, *Astrophys. J.*, 2009, **696**, 1348.
15 T. Owen, P. Mahaffy, H. B. Niemann, S. Atreya, T. Donahue, A. Bar-Nun and I. de Pater, *Nature*, 1999, **402**, 269.
16 G. Notesco and A. Bar-Nun, *Icarus*, 2005, **175**, 546.
17 D. Gautier, F. Hersant, O. Mousis and J. I. Lunine, *Astrophys. J.*, 2001, **550**, L227.
18 D. Gautier, F. Hersant, O. Mousis and J. I. Lunine, *Astrophys. J.*, 2001, **559**, L183.
19 Y. Alibert, O. Mousis and W. Benz, *Astrophys. J.*, 2005, **622**, L145.
20 Y. Alibert and O. Mousis, *Astron. Astrophys.*, 2007, **465**, 1051.
21 O. Mousis, *Astron. Astrophys.*, 2004, **413**, 373.
22 O. Mousis and D. Gautier, *Planet. Space Sci.*, 2004, **52**, 361.
23 O. Mousis and Y. Alibert, *Astron. Astrophys.*, 2006, **448**, 771.
24 O. Mousis, Y. Alibert and W. Benz, *Astron. Astrophys.*, 2006, **449**, 411.
25 J. H. van der Waals and J. C. Platteeuw, *Adv. Chem. Phys.*, 1959, **2**, 1.
26 F. Hersant, D. Gautier and J. I. Lunine, *Planet. Space Sci.*, 2004, **52**, 623.
27 F. Hersant, D. Gautier, G. Tobie and J. I. Lunine, *Planet. Space Sci.*, 2008, **56**, 1103.
28 W. R. Parrish and J. M. Prausnitz, *Ind. Eng. Chem. Process Des. Dev.*, 1972, **11**(1), 26.
29 W. R. Parrish and J. M. Prausnitz, *Ind. Eng. Chem. Process Des. Dev.*, 1972, **11**(3), 462.
30 M. Diaz Peña, C. Pando and J. A. R. Renuncio, *J. Chem. Phys.*, 1982, **76**, 325.
31 E. D. Sloan, *Clathrate Hydrates of Natural Gases*, 2nd ed.; Marcel Decker, Inc.: New York, 1998.
32 M. Jager, High pressure studies of hydrate phase inhibition using Raman spectroscopy. Ph.D. Thesis, 2001.
33 S. P. Kang, H. Lee, C. S. Lee and W. M. Sung, *Fluid Phase Equilib.*, 2001, **185**, 101.
34 E. D. Sloan and C. A. Koh, *Clathrate Hydrates of Natural Gases*, 3rd ed.; CRC Press, Taylor & Francis Group, Boca Raton, 2008.
35 V. P. Shpakov, J. S. Tse, C. A. Tulk, B. Kvamme and V. R. Belosludov, *Chem. Phys. Lett.*, 1998, **282**(2), 107.
36 V. R. Belosludov, T. M. Inerbaev, O. S. Subbotin, R. V. Belosludov, J. Kudoh and Y. Kawazoe, *J. Supramol. Chem.*, 2002, **2**(4–5), 453.
37 S. Takeya, M. Kida, H. Minami, H. Sakagami, A. Hachikubo, N. Takahashi and et al, *Chem. Eng. Sci.*, 2006, **61**(8), 2670.
38 K. Lodders, *Astrophys. J.*, 2003, **591**, 1220.
39 M. A. Frerking, W. D. Langer and R. W. Wilson, *Astrophys. J.*, 1982, **262**, 590.

40 M. Ohishi, W. M. Irvine and N. Kaifu, *Astrochem. Cosmic Phenom.*, 1992, **150**, 171.
41 P. Ehrenfreund and W. A. Schutte, *Adv. Space Res.*, 2000, **25**, 2177.
42 E. L. Gibb, *et al.*, *Astrophys. J.*, 2000, **536**, 347.
43 M. A. Pasek, J. A. Milsom, F. J. Ciesla, D. S. Lauretta, C. M. Sharp and J. I. Lunine, *Icarus*, 2005, **175**, 1.
44 J. S. Lewis and R. G. Prinn, *Astrophys. J.*, 1980, **238**, 357.
45 E. W. Thommes, M. J. Duncan and H. F. Levison, *Astron. J.*, 2002, **123**, 2862.
46 K. Tsiganis, R. Gomes, A. Morbidelli and H. F. Levison, *Nature*, 2005, **435**, 459.
47 P. S. Lykawka, J. Horner, B. W. Jones and T. Mukai, *Mon. Not. R. Astron. Soc.*, 2009, **398**, 1715.
48 J. C. B. Papaloizou and C. Terquem, *Astrophys. J.*, 1999, **521**, 823.
49 Y. Alibert, C. Mordasini, W. Benz and C. Winisdoerffer, *Astron. Astrophys.*, 2005, **434**, 343.
50 D. R. Lide, *CRC Handbook of Chemistry and Physics: A Ready-Reference Book of Chemical and Physical Data*, 83rd edn, Boca Raton, CRC Press, ISBN 0849304830, 2002.
51 U. Marboeuf, O. Mousis, D. Ehrenreich, Y. Alibert, A. Cassan, V. Wakelam and J.-P. Beaulieu, *Astrophys. J.*, 2008, **681**, 1624.
52 C. Thomas, S. Picaud, O. Mousis and V. Ballenegger, *Planet. Space Sci.*, 2008, **56**, 1607.
53 V. McKoy and O. Sinanoğlu, *J. Chem. Phys.*, 1963, **38**(12), 2946.
54 N. Iro, D. Gautier, F. Hersant, D. Bockelée-Morvan and J. I. Lunine, *Icarus*, 2003, **161**, 511.
55 N. I. Papadimitriou, I. N. Tsimpanogiannis, A. G. Yiotis, T. A. Steriotis and A. K. Stubos, in *Physics and Chemistry of Ice. Proceedings of the 11th International Conference on the Physics and Chemistry of Ice*, ed. W. Kuhs, 2007, vol. 311, p. 475.
56 K. C. Hester, Z. Huo, A. L. Ballard, C. A. Koh, K. T. Miller and E. D. Sloan, *J. Phys. Chem. B*, 2007, **111**, 8830.
57 P. R. Mahaffy, H. B. Niemann, A. Alpert, S. K. Atreya, J. Demick, T. M. Donahue, D. N. Harpold and T. C. Owen, *J. Geophys. Res.*, 2000, **105**, 15061.
58 M. H. Wong, P. R. Mahaffy, S. K. Atreya, H. B. Niemann and T. C. Owen, *Icarus*, 2004, **171**, 153.
59 F. M. Flasar, *et al.*, *Science*, 2005, **308**, 975.
60 L. N. Fletcher, G. S. Orton, N. A. Teanby, P. G. J. Irwin and G. L. Bjoraker, *Icarus*, 2009, **199**, 351.
61 H. B. Niemann and et al, *Nature*, 2005, **438**, 779.
62 A. L. Ballard and E. D. Sloan, *Chem. Eng. Sci.*, 2000, **55**, 5773.
63 M. A. Clarke and P. R. Bishnoi, *Fluid Phase Equilib.*, 2003, **211**, 51.
64 L. S. Tee, S. Gotoh and W. E. Stewart, *Ind. Eng. Chem. Fundam.*, 1966, **5**, 363.
65 R. Sun and Z. Duan, *Geochim. Cosmochim. Acta*, 2005, **69**, 4411.
66 J. B. Klauda and S. I. Sandler, *J. Phys. Chem. B*, 2002, **106**, 5722.
67 J. B. Klauda and S. I. Sandler, *Chem. Eng. Sci.*, 2003, **58**, 24.

General Discussion

Dr Sittler opened the discussion of the paper by Professor Lunine with a number of questions:
Question 1: You mention that an M-dwarf star may be just as active as our Sun, with flares and coronal mass ejections (CMEs), so I assume also have a solar wind?
Question 2: You mention an Earth-type planet with water based hydrological systems, would reside at 0.1 AU and could lose its entire atmosphere if it had a weak magnetic field (*i.e.*, solar wind scavenging of its upper atmosphere). You state, in the case of a Titan-type planet residing at 1 AU from the M-dwarf star, it would have a better chance of retaining its atmosphere where the solar wind is more benign. You then state, in order for it to be visible like an Earth-type planet, a Titan-type planet would need to be 3 times the size of the Earth and 5 times its mass. This would reduce the scale height of the atmosphere and make it more likely to retain its atmosphere. But, would the solar wind, flares and CMEs be greater for M-dwarf stars than our Sun, and thus increase atmospheric loss for a Titan-type planet at 1 AU?
Question 3: Since it is larger than the Earth, and Ganymede also has a magnetic field, would a Titan-type planet have a sufficiently large internal magnetic field to protect its atmosphere? Venus and Mars are examples of rocky planets larger and smaller than the Earth, respectively, that do not have magnetic fields. The former retains its atmosphere, while the latter has not. You say the UV would be less for M-dwarfs, but can one expect the same photochemistry at the top of a Titan-type planet as described above, like we have at Titan with rich organic and nitrile chemistry and formation of large positive and negative ions? As shown by Coates *et al.* (this volume) heavy negative ions are more prevalent in shadow than sunlight, which means photoemission less important. So, would one expect to see large negative ions at Titan-like planets where the solar UV is less intense (*i.e.*, photoemission less important and condensation to larger molecules increases)? Finally, since the Titan-type planet would not be gravitationally locked, like Titan is with Saturn (which has an effective rotation rate of 16 days relative to the Sun, due to its orbit around Saturn), what effect would a faster rotating Titan have on its upper atmosphere chemistry (*i.e.*, time in sunlight and darkness would be shorter)?

Professor Lunine replied: Most of the questions you ask require rather involved answers. I would strongly recommend two special issues of *Astrobiology*: the one on M dwarf habitability (2007, vol. 7, issue 1) and the more recent "Habitability Primer" (2010, vol. 10, issue 1). Beyond that, here are some brief and, therefore, perhaps inadequate answers to your questions:
Response to question 1: Yes, M dwarfs have winds.
Response to question 2: M dwarfs are active flare stars with XUV radiation up to 100 times greater than that of the present Sun, but variable, so that it is not easy to say for sure what the typical atmospheric loss would be for a planet at 1 AU from am M dwarf.
Response to question 3: This is difficult to predict in a general way without a detailed model, but let me point out that rotation periods of 16 days (like Titan) would still represent "fast rotation" relative to the orbital period at 1 AU from an M dwarf. There is nothing special about 24 h as a rotation period—it could be days or (like Venus) months and still not represent a tidally-locked configuration.

Professor Mitchell commented: What do you mean when you refer to acetylene as a "solid"?

Professor Lunine responded: The freezing point of acetylene is −82 °C. The surface temperature of Titan is, at the equator, −179 °C and at the poles about −182 °C. Therefore, condensed acetylene will be a solid on Titan's surface. There will also be some dissolved acetylene (~1%; Cordier *et al.*, Astrophys. J., 2009, **707**, L128) in the methane–ethane lakes and seas.

Dr Moses commented: In the first talk of this meeting, Sushil Atreya showed mass spectrometer data that indicate that C_2H_2 evaporated or was otherwise released from the surface during the landing of the Huygens probe, yet you mention that spectral signatures for acetylene at Titan's surface are not present in other Cassini remote-sensing data. Why do you think this is? Could the acetylene be coated by other materials that hide the spectral signatures? Do you have other ideas?

Professor Lunine responded: Indeed, the Huygens GCMS data have been reanalyzed in detail and in contrast to the original GCMS report (Niemann *et al.*, *Nature*, 2005, **438**, 779), in which CH_4, C_2H_6, CO_2, C_2N_2 (cyanogen) and C_6H_6 were reported in post-landing spectra, a new paper has been submitted reporting the presence of C_2H_2 in post-landing spectra. (The benzene detection is now labelled as tentative because of a large background in that mass range.) The introduction of all of these species into the GCMS was through an electrically heated inlet[1] in the base of the probe, and so it is generally thought that the species were resident in the "soil" but not necessarily exposed at the surface. For example, the surface is evidently solid from the Huygens DISR images and so the methane and ethane must be in the immediate subsurface and not exposed. Therefore, one can imagine that a near-IR spectrometer like Cassini VIMS, sampling only the exposed surface of Titan, might miss acetylene that is just underneath, or coated/buried by other materials.

1 R. Lorenz *et al.*, *Meteor. Planet. Sci.*, 2006, **41**, 1704.

Dr Sittler remarked: Yes, VIMS data[1] show an overabundance of benzene relative to acetylene on the surface. But in the earlier talk by Professor Atreya, the Huygens GCMS observed acetylene down near the surface, but that the mass 78 channel of the GCMS had high background, so detection of benzene could not be confirmed. Therefore, can one say that the Huygens GCMS data is still consistent with the VIMS observations, which must remove atmospheric effects from its observations of the surface?

1 Clark *et al.*, *J. Geophys. Res.*, 2010, in press.

Professor Atreya replied: The Huygens GCMS data detect acetylene *only* from surface evaporation, not in the air, not even near the surface. This is because acetylene lies in a very busy part of the mass spectrum where its presence cannot be distinguished from that of other competing or dominating molecules. Acetylene must be in fairly large concentration for it to be detected from the surface (CIRS gives on the order of ppm levels in the atmosphere). On the other hand, benzene, if present in the surface, must be at so low in concentration that it cannot be separated from the instrument background. Remember, also, concentration builds up tremendously in the surface deposits, which is the reason why acetylene and some other molecules could be detected from surface evaporation even though they were not seen during the descent. I cannot comment on the VIMS data without going through it carefully myself, for analysis of it has yet to be published. If everyone (VIMS and GCMS) turns out to be right, it might signal that not all locations on Titan are alike!

Professor Atreya then commented: The Huygens GCMS data from evaporation of the material at Titan's surface after landing cannot unambiguously confirm presence of benzene in the surface. If present, it's too low to distinguish from the benzene that was permanently present as background in the instrument throughout the probe's descent through the atmosphere.

Dr Bézard remarked: You mention the formation of benzene from acetylene ice as a possible explanation for the possible lack of acetylene and overabundance of benzene at the surface of Titan. Are there any laboratory measurements of the kinetics of this reaction in conditions prevailing at the surface of Titan? Is it possible to quantify the loss of C_2H_2 on Titan's surface from this process *vs.* its photochemical production?

Professor Lunine replied: During the meeting Professor Kaiser commented that his group has done some work on the cosmic ray mediated conversion of acetylene to benzene at the surface of Titan, using an energy deposition rate for cosmic rays calculated by Molina-Cuberosa *et al.*, *Planet. Space Sci.*, 1999, **47**, 1347.

Professor Kaiser asked: In your presentation you commented on the formation of the benzene molecule and that the formation of benzene *via* surface chemistry on Titan at 95 K is not feasible since there is a significant 'activation energy' involved.

It should be stressed that this is certainly true for thermal chemistry at a thermal equilibrium. Here, acetylene cannot form benzene. However, our recent study on the interaction of ionizing radiation with acetylene–carbon monoxide ices[1] and with pure acetylene ices in particular[2] showed that the ionizing radiation can lead no non-equilibrium chemistry involving metastable triplet reactants such as triplet carbon monoxide and/or triplet acetylene in the solid state. These reactions have—at most—only minor entrance barriers, which could be overcome if the triplet species are vibrationally exited. In the end, this can lead to the formation of benzene molecules as demonstrated recently in our group under conditions simulating Titan.

So the ultimate goal of chemical models would be to objectively address the key question to what extend galactic cosmic ray processing of Titan's aerosol particles and/or Titan's surface can lead to the formation of complex organics. Laboratory data on the processing of Titan's main hydrocarbon constituents such as methane, ethane, and acetylene[2] are well documented, but an incorporation of these data into Titan's chemical models has not been fully accomplished yet. This objective approach is crucial to judge what fraction of organics can be formed in the gas phase (*via* ion–molecule and/or neutral–neutral reaction) and also in the solid state (aerosol particles, surface).

1 L. Zhou, R. I. Kaiser, L. G. Gao, A. H. H. Chang, M. C. Liang, Y. Y. Yung, Pathways to oxygen-bearing molecules in the interstellar medium and in planetary atmospheres: cylopropenone (*c*-C_3H_2O) and propynal (HCCCHO), *Astrophys. J.*, 2008, **686**, 1493–1502.
2 L. Zhou, W. Zheng, R. I. Kaiser, A. Landera, A. M. Mebel, M. C. Liang, Y. L. Yung, Cosmic-ray mediated formation of benzene on the surface of Saturn's moon Titan, *Astrophys. J.*, 2010, **710**, 1243–1251.
3 C. J. Bennett, C. S. Jamieson, Y. Osamura, R. I. Kaiser, Laboratory studies on the irradiation of methane in interstellar, cometary, and solar system ices, *Astrophys. J.*, 2006, **653**, 792–811.
4 Y. S. Kim, C. J. Bennett, L. Sheng, K. O'Brian, R. I. Kaiser, Mechanistical studies on the radiation processing of crystalline and amorphous ethane ices and their implications to the chemistry of Titan and KBOs, *Astrophys. J.*, 2010, **711**, 744–756.

Professor Lunine responded: Indeed, having heard your discussion and read the Zhou *et al.* paper, I now agree that the surface conversion of acetylene to benzene is more favorable than I had originally thought. I can only echo the point of view that conferences, especially the *Faraday Discussions* (given their unique format), are crucial to giving all of us the opportunity to learn new things.

Professor Lunine addressed Professor Mebel: The work you cited on the facile cosmic-ray conversion of triplet acetylene to benzene is interesting and might help to explain the preponderance of benzene on the surface as seen in Cassini VIMS data. It is important, however, to put it in the context of potential sources of mechanical disturbance such as fluvial and aeolian transport, impacts, and (more speculatively) cryovolcanism. These may also act to convert acetylene into benzene, as well as do other chemistry on Titan's surface.

Professor Mebel responded: I agree with this comment and only would like to add that the calculated barrier for the $C_2H_2 + C_2H_2$ reaction in the ground electronic state is about 39 kcal mol^{-1} (to form cyclobutadiene), whereas a barrier of ~48 kcal mol^{-1} needs to be overcome to produce vinylacetelene.[1] On the other hand, the computed energy gap between the ground singlet and the first excited triplet electronic states of acetylene is 3.82 eV (103.9 kcal mol^{-1}).[2]

1 A. M. Mebel, V. V. Kislov and R. I. Kaiser, *J. Chem. Phys.*, 2006, **125**, 133113 (2006).
2 J. K. Lundberg, R. W. Field, C. D. Sherrill, E. T. Seidl, Y. Xie, H. F. Schaefer III, *J. Chem. Phys.*, 1993, **98**, 8384; Y. Yamaguchi, G. Vacek, H.F. Schaefer III, *Theor. Chim. Acta*, 1993, **86**, 97.

Professor Suits asked: Given that acetylene polymerization might proceed rapidly once it has begun, are there means of initiating this with lightning, static discharge or cosmic rays?

Professor Lunine responded: Initiation of acetylene polymerization with cosmic rays has been discussed by Ralf Kaiser and his group (at this meeting and in Zhou *et al.*, *Astrophys. J.*, 2010, DOI:10.1088/0004-637X/715/1/1). There is no detection of lightning on Titan, consistent not with the absence of storms necessarily but rather with nonpolar methane as the working meteorological fluid. Static discharge associated with dunes and smaller particles blown around by the winds is an intriguing additional possibility which has not been quantitatively investigated.

Professor Schram remarked: A question from my side is: how large could the electric fieldstrengths be?

Professor Lunine responded: Electric field measurements were made by the Huygens probe during descent and are described in Beghin *et al.*, *Planet. Space Sci.*, 2009, **57**, 1872. The field seems to peak at about 80–90 km at a value just above 5 mV m^{-1} Hz$^{-1/2}$. As to how large fields *could* be, one can only note that the air is mostly molecular nitrogen, with methane present, and very little in the way of polar species. Hence, it ought to be a good insulator, but there has been no detection of lightning.

Professor Plane asked: The "missing ethane" problem on Titan remains an unsolved problem. Would you care to comment on current theories to explain the absence of an ethane ocean covering the satellite?

Professor Lunine replied: If methane has been photolyzed without major interruptions over the age of the solar system, hundreds of meters of equivalent depth of ethane should have been produced during this time. Disposal of the ethane during volcanic and impact events might explain its relative absence on the surface. It is possible that less ethane actually survives to the surface from the stratosphere than is predicted by photochemical models, or it is incorporated somehow into other organic aerosols, or both. One must also not exclude the possibility that methane photochemistry is episodic, and a model to reconcile this with Cassini data is in preparation. The citations for the other possibilities may be found in J. I. Lunine and R. D. Lorenz, Rivers, lakes, dunes and rain: Crustal processes in Titan's methane cycle, *Annu. Rev. Earth Planet. Sci.*, 2009, **37**, 299–320.

Professor Atreya opened the discussion of the paper by Dr Tinetti: Do you find any evidence of PH_3 (or its products, P_2 and P_4)? I suspect PH_3 and its product could be present in the sensible atmosphere of hot Jupiters. In Jupiter's atmosphere, PH_3 is thermochemically stable at high temperature (1000–2000 K) in the planet's interior. Convection does pump certain quantities of PH_3 to the stratosphere where PH_3 is expected to form P_2 and P_4. Oxides of phosphorus could form also as the gas convects through the water cloud. I think these are the type of temperatures of the region of the EGP spectroscopic data.

Dr Tinetti replied: There is the suspicion that PH_3 is present in one of the hot-Jupiters' atmospheres (Bouwman *et al.*, in preparation), but the spectral resolution and S/N of the observations are not good enough to be conclusive.

Mr Hargreaves asked: It is outlined in the paper that there is an urgent need for laboratory studies to determine high temperature molecular opacities of species relevant to exoplanet atmospheres. We would like to report that we are recording high resolution (0.01 cm^{-1}) Fourier transform infrared (FT-IR) spectra of NH_3 and CH_4 at 100 °C intervals from 300–1400 °C. We will then produce detailed infrared line lists of these molecules to be used in the atmospheric modelling of exoplanets and brown dwarfs. In response to our work, which molecules are considered to have a large contribution to the opacity of exoplanet spectra but have little or no 'hot' data, thus hindering the accurate production of atmospheric models?

Dr Tinetti replied: We have currently good data for hot water and ammonia, CO and CO_2. The data for hot methane, hydrocarbons and H_2S are missing or incomplete.

Dr Manzanares commented: Considering the possibility of detection in the near infrared and visible regions, it is well known that between 5000 cm^{-1} and 14 000 cm^{-1}, the number of absorption bands decreases, reducing the superposition of the bands. The C–H, N–H, and O–H bonds have larger anharmonicities than other bonds allowing them to show stronger absorption bands in the near-IR and visible regions. These vibrational overtones are well separated in these two spectral regions. The only problem is that the bands are weak. They will only show at high pressures and/or with long optical paths.

Professor Miller noted: One reason for choosing the (near) infrared for observing exoplanets is that you get a better contrast between the parent star, which tends to emit most in the visible and the planet which, even at 1000 K, will have its maximum emission at 3 microns (and longer wavelengths if cooler).

Dr Tinetti replied: I agree with Professor Miller's response; the contrast in the optical is less favorable for emission spectroscopy (secondary transit technique). For transmission spectroscopy (primary transit), it is not necessarily the case though.

Dr Bézard asked: Photochemical models of hot Jupiters predict that CO is more abundant than CO_2; CO_2 being produced by H_2O photolysis followed by reaction of OH with CO. On the other hand, you mentioned that CO_2 has been detected but that spectroscopic evidence for CO is not so clear. Why is this? I would suppose that, if abundant enough, CO could be relatively easily detected through its strong (1–0) band at 4.7 μm.

Dr Tinetti replied: The problem is that the 4.7 μm absorption cannot be captured conclusively with Spitzer-IRAC photometric observations, and the low resolution observations with Hubble cover the 1.2–2.5 μm range, where CO absorption may overlap with one of methane or CO_2. The final, conclusive, evidence of CO in one of the hot-Jupiters (HD 209458b) was most recently provided by VLT-CRIRES measurements (Snellen *et al.*, *Nature*, 2010).

Dr Moses asked: What are your views on the 3-micron emission observed by Swain *et al.* (2010) for HD 189733b? Do you think it is an observational artifact, or is it real? If real, do you agree with the interpretation of fluorescence in the ν_3 methane band? Can you rule out thermospheric emission itself? For example, the parent star is likely variable, and one can imagine that the location of the base of the thermosphere might fluctuate in response to either variations of the star or from atmospheric dynamics of the planet (*e.g.*, waves). One can imagine scenarios in which the upper part of the molecular region becomes heated to over 2000 K, and the radiative cooling may not instantaneously drive the atmosphere back to pre-perturbation temperatures. Is that scenario possible?

Dr Tinetti replied: We have had additional observations since the publication of Swain *et al.* (2010). The five nights confirm the signature at 3 μm, and allow an improvement in the spectral resolution. The spectral signature at 3 μm now really looks like methane in emission. New papers are in preparation (Deroo *et al.*, Waldmann *et al.*). Pierre Drossard has proposed at DPS an interesting explanation of the spectral features observed; they could be due to non LTE methane emission.

Professor Miller commented:
Auroral methane emission is known from Jupiter. Normally, mesospheric emission absorbs radiation coming from the planet's interior and incident solar infrared radiation, but in the case of energetic auroral events, precipitating particles will penetrate sufficiently deep in the atmosphere to get below the homopause and excite CH_4 lines to show up in emission.

Dr Tinetti agreed with Professor Miller's comments.

Professor Mason concluded: I would like to draw delegates attention to the ESF Cost Action the Chemical Cosmos (CM0805). The Action supports collaborative visits within Europe (and non EU labs may join). All the research in this meeting would be covered by this Action so delegates may chose to look at and ask to join one or more of the Working groups, attend its meetings (the next one is in Grenoble, October 5–8) and use short visits to develop links established at this meeting. For details see http://cost.astrochemistry.eu/impressum

Ms Horst opened the discussion of the paper by Professor Raulin: Have you looked for amino acids in your tholins without hydrolyzing them? As we've shown in Horst *et al.*, 2010 (submitted), tholin samples produced in the PAMPRE experiment contain amino acids and nucleotide bases without any hydrolysis. The oxygen comes from contamination and also from oxygen that we put into the samples intentionally through the introduction of CO. Of course hydrolysis clearly increases the yield of the amino acids, but I think their presence in tholins without hydrolysis is potentially more interesting.

Professor Raulin answered: Although the gas mixture we used to produce Titan tholins (N_2 : CH_4 98 : 2) was free of oxygen, we did see oxygen contamination in our tholins and in their non-hydrolysed samples. Indeed, performing a direct derivatization of tholins, we did see some amino acids, but their yield was smaller than that observed after derivatization of hydrolysed tholins samples. For example the quantity of glycine is multiplied by more than a factor of three in tholins on which a 10-week cold-hydrolysis was performed, compared to our non-hydrolysed tholins. Urea is also present in non-hydrolysed tholins as well as in hydrolysed tholins.

The oxygen contamination may occur during the synthesis, the recovery and/or the storage of the tholins samples and also during the derivatization process.

Possible explanations for this production of urea and amino acids may be the presence of traces of dioxygen (O_2) in the reactor during the synthesis, and/or the reaction of highly reactive fresh synthesised tholins with water vapour from the air of the laboratory, performing a partial hydrolysis of the samples. It could be at least partly—but still not totally—avoided using a glove box filled with dinitrogen (N_2), as done in previous studies in our group (*cf.* Ramirez *et al.*, *Icarus*, 2002, **156**, 515–530, and references therein). Nevertheless, it must be pointed out that in these studies when using a glove box the oxygen contamination was very low, with C/O ~ 1500.

Another possibility could be the presence of water during the derivatization process. We have detected, in parallel experiments, a peak corresponding to the derivatization product of water which can be attributed to the hydroscopic character of our tholins, they adsorb atmospheric water as soon as in contact with the air.

Professor Coates remarked: I wondered if you saw charged particles in you experiments? We see positive and negative ions, as well as neutrals, in Titan's upper atmosphere.

Professor Raulin answered: Our experiment was essentially dedicated on the behaviour of Titan tholins upon ammonia aqueous solution hydrolysis. Thus we did not study the nature of the species formed in the reactor, such as charged species, in the present work.

Professor Schram remarked: Can emission of neutrals and neutral fragments not be caused by dissociative recombination? If so, do give intensity and ro-vibrational distribution of the emission information about process? Can that then be compared with observations in laboratory plasmas?

Professor Raulin replied: In previous works, we did use spectroscopy in the visible range to identify radicals produced in a plasma similar to the one produced for these experiments (J.-M. Bernard, PhD Thesis, University Paris 7, September 2004, Experimental simulation of Titan's stratosphere chemistry: monitoring of the produced species and comparison with a kinetic model). We were thus able to identify several radical species, and to interpret the chemical mechanism involved in the dissociation of methane and nitrogen. It clearly involves radical reaction, but probably also ion reactions.

Professor Atreya commented: How did you decide on the range of NH_3 in your lab experiments? I suggest trying to include H_2S also in future experiments, as it could be a component of cryovolcanic outgassing. Its vapor pressure even at Titan's 94 K surface temperature is appreciable. I realize it's challenging to work with H_2S.

Professor Raulin replied: Models of the internal structure of Titan suggest the presence of an ocean with up to about 25–30% NH_3. We thus used a maximum concentration of NH_3 in that range.

Besides using the maximum NH_3 concentration, we also tested lower concentrations having in mind that once these putative internal fluids emerge to the surface, they might be affected by dilution phenomena, caused for example by their contact with hydrocarbons fluids.

Concerning the comment on H_2S, we can make 2 main remarks: (1) it should be *a priori* much more efficient to hydrolyse tholins with concentrated solution of ammonia, instead of using slightly acidic solutions of low concentrations of H_2S (which has a low solubility in water, specially at low temperatures – and consequently does not provide a pH low enough for a good acid hydrolysis); (2) It is indeed more difficult to work with H_2S, particularly for safety reasons, but this is an idea to look at in the future.

Professor Mason remarked: Are you keeping any aerosol samples you make for longer than 10 weeks? Could you look at the heating effect, *e.g.* chemical changes when T is raised from 96 K? Do you think the chemistry is set early in the process or really changes steadily over time? Could you not analyse after 2, 5, 10 weeks and compare?

Professor Raulin responded: For the sets of experiments reported in this paper, we have chosen two variable parameters: ammonia concentration and temperature of hydrolysis, and have fixed the hydrolysis time, in order to be able to compare the results. This already required the use of an important amount of tholins and all the tholins synthesized under the described conditions were used for the reported experiment. However in a new series of experiments, more tholins are being synthesized and we are planning to look now at the influence of hydrolysis time, keeping samples at constant temperatures for different durations of hydrolysis.

Concerning the question about the heating effect, there was no specific monitoring of the effect of increasing temperature from 96 K to room temperature. However, as mentioned in the last version of the manuscript, for the samples that were kept at 96 K, the phase change from ice to liquid was in fact very short, less than 30 s were enough. Thus we think that this step is too short to have a noticeable effect on the hydrolysis.

Finally, concerning the question whether the chemistry is set early in the process or really changes steadily over time, with the present information and results it is hard to provide an answer. A comparison at different exposure times was not performed, as the goal of our experiments was to demonstrate that our tholins were sensitive to any chemical modification when placed in aqueous ammonia solution at low temperature. Now that we have a positive result we can plan more detailed studies aimed to better understand the hydrolysis process in the frame of Titan environment. As mentioned before, a new set of experiments are being conducted and we will be able to study specific aspects.

Professor Casavecchia (speaking on behalf of also N. Balucani, F. Leonori, R. Petrucci, K. M. Hickson, S. D. Le Picard and P. Foggi) opened the discussion of the paper by Professor Kaiser[1]: I would like to make some comments on the role of the internal state population of C_2 in the reaction $C_2 + C_2H_2$.

In a recent work, we have also reported on crossed molecular beam (CMB) experiments on the reaction $C_2(X^1\Sigma_g^+, a^3\Pi_u)$ with acetylene at two different collision energies (E_c) of 37.4 kJ mol^{-1} and 13.6 kJ mol^{-1}.[2] In our experiments we have produced supersonic beams of the C_2 transient species *via* a high-pressure, high-power radiofrequency (RF) discharge source for the production of supersonic beams of atomic and molecular radicals,[3] starting from dilute mixtures of CO and O_2 in He or Ne (we have used He and Ne as carrier gases to change the beam velocity and the experimental E_c). The mechanism of formation of C_2 in the plasma formed by the RF discharge is unknown, but presumably it involves the recombination of carbon atoms, which are, indeed, present in the beam in a large percentage.[4] Because of the small energy difference between the ground, $C_2(X^1\Sigma_g^+)$, and first excited, $C_2(a^3\Pi_u)$, electronic states (the electronically excited triplet state, $a^3\Pi_u$, lies only 7.3 kJ mol^{-1} above the ground state, $X^1\Sigma_g^+$) when compared to the temperature of the plasma (1200–1500 K), both states are formed in a significant amount. As a matter of fact, if we assume that the two electronic states are populated according to a Boltzmann distribution and if we consider the plasma temperature in the RF discharge (which is ~1200 K when the carrier gas is He and ~1500 K when the carrier gas is Ne), the population ratio $P(a^3\Pi_u)/P(X^1\Sigma_g^+)$ is 2.9 and 3.3 in the cases of He and Ne, respectively, that is roughly the same. This means that at the two E_cs of our experiments, the triplet-to-singlet ratio is practically constant. In addition to that, as C_2 is a diatomic species, the characterisation of the internal rovibrational levels of both electronic states becomes crucial for a correct interpretation of the scattering results. The limited number of relaxing collisions during the supersonic expansion,[4] in fact, does not allow the rovibrational distributions to be efficiently relaxed. Since the outcome of the reactive scattering experiments can be affected by the rovibrational populations of C_2, we have characterized the $C_2(X^1\Sigma_g^+, a^3\Pi_u)$ internal populations by using a new experimental apparatus consisting of a RF discharge source identical to that employed in CMB experiments and a vacuum chamber with a pulsed tunable laser (a frequency tripled Nd YAG laser at 355 nm and a dye laser) coupled to an optical detection system for radical detection by LIF (see ref. 2 and 5 for more details). We have first looked at the rovibrational spectra of the triplet state *via* the (0,0), (1,1) and (2,2) bands of the $d^3\Pi_g$–$a^3\Pi_u$ Swan system at ~516.5, 512.9 and 509.8 nm respectively.[2] The rotational temperatures were found to be ~250 K and 200 K for all bands of $C_2(a^3\Pi_u)$ with He and Ne as the carrier gases, respectively, while the vibrational temperature was determined to be 3500 K. Then, a similar characterization has been performed for the internal states of $C_2(X^1\Sigma_g^+)$ with He and Ne as carrier gases.[5] In this case we have looked at the (0,0), (1,1), (2,2) and higher transitions of the Mulliken ($D^1\Sigma_u^+$-$X^1\Sigma_g^+$) bands at approximately 231 nm. The rovibrational populations of $C_2(X^1\Sigma_g^+)$ are comparable to those obtained for the triplet state, but higher vibrational states ($v \geq 3$) are also partially populated. It should be noted that, even though a direct evaluation of the triplet-to-singlet density ratio in not possible by using LIF, we had a comparable loss of signal intensity of both singlet and triplet states in moving from He to Ne as the beam carrier gas. This observation is very important, because it implies that the singlet/triplet state ratio remains constant, thus confirming the reliability of the estimate based on the plasma temperatures.

If we try to use the information on the C_2 internal states in the interpretation of the observed scattering distributions, several interesting considerations can be made.[2,5] First of all, there was no need to consider the extra amount of energy supplied by the vibrational energy of C_2, even though the energy content of the vibrationally excited C_2 levels is quite sizable (for instance, for the $a^3\Pi_u$ state $\Delta v_{1-0} = 19.36$ kJ mol^{-1}, $\Delta v_{2-0} = 38.43$ kJ mol^{-1}, $\Delta v_{3-0} = 57.23$ kJ mol^{-1}). This is not surprising, as the vibrational excitation is expected to affect a reaction mostly when it involves a breaking bond. According to the *ab initio* calculations of the relevant triplet and singlet potential energy surfaces (PESs),[6] the C_2 moiety is actually incorporated in the C_4H product. Therefore, it is reasonable to expect that the vibrational excitation of the reactant will be retained as vibrational excitation of the molecular product and the extra amount of energy available to the system when the reaction involves C_2 in one of the excited vibrational levels is not converted into product translational energy (the reaction is vibrationally adiabatic).[2,5] The possible effect of the C_2 rotational excitation is more difficult to predict and quantify. Only detailed quantum dynamical calculations or state-to-state differential cross section measurements could fully elucidate this issue if and when they become amenable for a system of this complexity. We limit ourselves to saying that we have practically the same rotational distributions in both the He and Ne beams used for the two experiments, so that we expect a similar behaviour for the experiments at the two E_cs.

Since the two electronic states of C_2 are nearly isoenergetic, the difference in the energetics of the triplet and singlet reactions is not large enough to allow an easy disentanglement of the two contributions in the data analysis (as done for other systems, see for instance ref. 7), unless one of the states is not able to react (this appears to be the case for the reaction C_2 + HCN at low E_c investigated by Kaiser and coworkers[1]). In the cases of the C_2 reactions with unsaturated hydrocarbons, neither in ours[2] nor in Kaiser's experiments[6,8] has it been possible to discriminate without ambiguity the contributions of the two reactions to the observed signal. In fact, not are only the energetics of the processes very close, but also the mechanisms followed by the two reactions are of the same kind, that is, electrophilic addition of C_2 to the π bonds of unsaturated hydrocarbons with the formation of bound intermediates. Our laboratory data at the lower E_c were fit by using a backward–forward symmetric isotropic CM angular distribution and a product translational energy distribution with an average fraction of energy released into product translational energy, $f_T = E_T/E_{TOT}$, of about 0.4 (where E_{TOT} is the one associated to the more exoergic reaction of $C_2(a^3\Pi_u)$ in its ground ro-vibrational level). At the higher E_c of 37.4 kJ mol^{-1} the CM angular distribution became forward peaked, with an intensity ratio $T(0°)/T(180°) = 2$.

Our experimental findings have been interpreted in the light of *ab initio* calculations of the triplet and singlet state C_4H_2 PESs,[6] but in a different way with respect to the interpretation of Kaiser et al.[1] The favoured reaction approach in the singlet PES is C_2 addition to the π system of acetylene and leads to two different cyclic intermediates that easily rearrange to a very stable singlet HCCCCH(diacetylene) intermediate, lying about 580 kJ mol^{-1} below the reagent asymptote and about 550 kJ mol^{-1} below the product asymptote. Since HCCCCH resides in a very deep well with respect to both reactants and products, it is reasonable to expect that, at the low E_c of 13.6 kJ mol^{-1}, the singlet C_4H_2 intermediate would survive for several rotational periods before fragmenting into products. The formation of a long-lived complex is fully consistent with our finding of an isotropic CM angular distribution at this E_c. A similar reaction mechanism is also plausible for the reaction of $C_2(a^3\Pi_u)$; in this case, the $C_2(a^3\Pi_u)$ addition to the π system of acetylene can occur in three different ways leading to three different isomers that easily interconvert. Remarkably, the triplet C_4H_2 intermediates are much less stable than the singlet one, being about 180 kJ mol^{-1} below the reagent asymptote and 140 kJ mol^{-1} below the product asymptote. Our results indicate that at low E_c of 13.6 kJ mol^{-1} the triplet reaction also proceeds through the formation of a long-lived complex (*i.e.*, the triplet complex lifetime is at least 4–5 times longer than its rotational period) or, alternatively, an efficient inter-system crossing (ISC) between the triplet and singlet C_4H_2 PESs takes place. At the higher collision energy of 37.4 kJ mol^{-1}, the global CM angular distribution becomes forward peaked (with respect to the C_2 beam direction). This can be readily rationalized in terms of the osculating complex model for chemical reactions for the triplet reaction: the lifetime of the triplet complexes is expected to become significantly shorter than that at the lower E_c (because the wells associated with the triplet intermediates are much more shallow than the one associated with the very stable singlet C_4H_2). Alternatively, a less efficient ISC from the triplet to the singlet PES can still explain the forward bias. In any case, the forward scattering preference has been attributed to the triplet C_2 reaction.[2]

In summary, an important achievement of the spectroscopic characterization of our C_2 beams is that the $C_2(a^3\Pi_u)$ and $C_2(X^1\Sigma_g^+)$ internal populations in the beams used in our experiments are similar: the rovibrational distributions of the two electronic states are substantially the same in both Ne and He beams while the singlet/triplet state signal ratio remains approximately constant for experiments conducted with the two carrier gases. In this view, the change observed in our experimental distributions can be attributed exclusively to the change in the reaction mechanism of one or both the reactions of the two electronic states and not to a change in the beam composition or C_2 internal populations. In this scenario, the most plausible explanation is that the triplet reaction can be described by the osculating model of chemical reaction at the higher E_c investigated, or that at both E_cs the triplet complex lifetime is short, but at the lower E_c there is an efficient triplet-to-singlet ISC.

1 R. I. Kaiser, P. Maksyutenko, C. Ennis, F. Zhang, X. Gu, S. P. Krishtal, A. M. Mebel, O. Kostko and M. Ahmed, *Faraday Discuss.*, 2010, **147**, DOI: 10.1039/c003599h.
2 F. Leonori, R. Petrucci, K. M. Hickson, E. Segoloni, N. Balucani, S.D. Le Picard, P. Foggi and P. Casavecchia, Planet. Space Sci., 2008, **56**, 1658.
3 M. Alagia, V. Aquilanti, D. Ascenzi, N. Balucani, D. Cappelletti, L. Cartechini, P. Casavecchia, F. Pirani, G. Sanchini and G. G. Volpi, *Isr. J. Chem.*, 1997, **37**, 329.
4 F. Leonori, K. M. Hickson, S. D. Le Picard, X. Wang, R. Petrucci, P. Foggi, N. Balucani and P. Casavecchia, *Mol. Phys.*, 2010, **108**, 1097.
5 M. Costes, N. Daugey, C. Naulin, A. Bergeat, F. Leonori, E. Segoloni, R. Petrucci, N. Balucani and P. Casavecchia, *Faraday Discuss.*, 2006, **133**, 157; F. Leonori, R. Petrucci, E. Segoloni, A. Bergeat, K. M. Hickson, N. Balucani and P. Casavecchia, *J. Phys. Chem. A*, 2008, **112**, 1363.
6 R.I. Kaiser, N. Balucani, D. O. Charkin, and A. M. Mebel, *Chem. Phys. Lett.*, 2003, **382**, 112; X. B. Gu, Y. Guo, A. M. Mebel and R. I. Kaiser, *J. Phys. Chem. A*, 2006, **110** 1089.
7 N. Balucani, D. Stranges, P. Casavecchia and G. G. Volpi, *J. Chem. Phys.*, 2004, **120**, 9571.
8 N. Balucani, A. M. Mebel, Y. T. Lee and R. I. Kaiser, *J. Phys. Chem. A*, 2001, **105**, 9813; R. I. Kaiser, T. N. Le, T. L. Nguyen, A. M. Mebel, N. Balucani, Y. T. Lee, F. Stahl, P. V. Schleyer and H. F. Schaefer, *Faraday Discuss.*, 2001, **119**, 51; Y. Guo, X. B. Gu, N. Balucani and R. I. Kaiser, *J. Phys. Chem. A*, 2006, **110**, 6245; Y. Guo, X. B. Gu, F. T. Zhang, A. M. Mebel and R. I. Kaiser, *J. Phys. Chem. A*, 2006, **110**, 10699.

Professor Suits commented: I would just like to emphasize that Professors Kaiser and Casavecchia have different sources of dicarbon, with differing internal energy and different distributions of triplet and singlet. We can't directly compare the experiments at this stage.

Dr Moses asked: Can you do atomic H with cross-beam studies? For photochemical models of the giant planets, we have a particular need for quantitative information on the rate coefficients and products of atomic H with hydrocarbons, and especially for reactions of H with C_3H_x species.

Professor Kaiser responded: Yes, crossed beam experiments of H atom reactions can be done. It is important to note that in planetary atmospheres, reactions of suprathermal hydrogen atoms holding kinetic energies of a few eV are of particular importance (R. J. Morton, R. I. Kaiser, Kinetics of suprathermal hydrogen atom reactions with saturated hydrides in planetary and satellite atmospheres, *Planet. Space Sci.*, 2003, **51**, 365–373). Here, a treatment of thermal kinetic data *via* the inverse Laplace transformation shows an increase of rate constants of hydrogen abstraction reactions from simple hydrides of up to 40 orders of magnitude.

Professor Casavecchia added: Yes, we may be able, at least in principle, to do crossed-beam studies of atomic H reactions. This because we have very recently developed a continuous supersonic beam of H atoms. However, H atom reactions with hydrocarbons in the thermal energy range are most difficult to study with the crossed-beam technique, because they have usually (small) entrance energy barriers, which may be difficult to overcome because of the low relative translational energies that can be attained when working with light H atoms. Furthermore, reactions of H atoms have unfavourable kinematics, being the center of mass of the system, located very close to the heavy molecular reagent. Finally, reactions of H atoms with C_3H_x species are particularly challenging, because these are reactions between two radicals, and in addition to the above difficulties you face also intensity problems (radical beams are typically one or two order of magnitude less intense than molecular beams).

Professor Casavecchia opened the discussion of the paper by Proffessor Mebel: In your theoretical study of the mechanisms of formation of nitrogen-containing polycyclic aromatic compounds in low-temperature environments of planetary atmospheres, such as that of Titan, you did not consider any reaction of excited nitrogen atoms, N(^{2}D), with aromatic hydrocarbons, such as benzene. Is there a reason? Do you think that these reactions maybe be relevant? Are you planning to include N(^{2}D) reactions in the future in your theoretical studies?

Professor Mebel responded: We have not considered the reactions of N(^{2}D) so far. However, I agree that such reactions may be relevant to the formation of NPACs in Titan's atmosphere, especially if they proceed without or with a very low entrance barrier. The reaction of N(^{2}D) with benzene seems to be the first plausible candidate to consider and we plan to investigate the corresponding potential energy surface in the near future. I should also note that the reaction of the $C_6H_4(C_2H)$ radical with molecular nitrogen in its first excited triplet electronic state has been studied in our paper and has been shown to easily produce 1,2-diaza-3,4-didehydronaphthalene under single-collision conditions.

Professor Coates asked: Do you see heavy charged particles as well as neutrals in your work? Is there any process to produce those?

Professor Mebel answered: As our work is theoretical, we can compute potential energy surfaces for ions as well as for neutrals. Indeed, in some of our works we compute dissociation mechanisms of large cations in order to predict their fragmentation pattern. We can also investigate mechanisms of ion–molecular reactions through similar PES calculations and compute their product branching ratios under various conditions.

Dr Klippenstein said: At low temperatures the probability for stabilizing a chemically activated species is a strong function of the molecular size of the system. At some large-enough size one might find that essentially all complexes are stabilized, even when there are significant exothermic product channels. This increase in stabilization probability might have some affect on the applicability of your scheme to larger and larger PAH growth. Can you comment on this?

Also, in a number of places you discuss the branching ratios between various channels for zero-collision energy. It would be helpful if you also presented the absolute decomposition rate coefficients that entered into these branching ratio determinations. These absolute rate coefficients are of some use in making zeroth order estimates of the stabilization probabilities.

Professor Mebel answered: To address the first part of the question, we computed radiative stabilization rate constants k_{rad} for several NPAC intermediates, which can be produced in the reactions considered in our paper, including 2-aza-4-ethynyl-1-naphthyl **i2**, 2-aza-1-ethynyl-4-naphthyl **i6**, and 1,4-diethynyl-phthalazine **i11**. These calculations were carried out using the approach described by Klippenstein et al.[1] and utilizing B3LYP/6-311G** vibrational frequencies and IR intensities. The rate constants were computed for average temperatures corresponding to available internal energies of these intermediates, if they were produced in bimolecular reactions $C_6H_4(CN)(C_2H) + C_2H$ (for **i2** and **i6**, see Fig. 2 and 3 in the manuscript, respectively) or **i7** + C_2H (for **i11**, Fig. 4) at zero collision energy, assuming that all the energy of chemical activation is transformed into vibrational energy. The following k_{rad} values were obtained: 31, 24, and 57 s^{-1} for **i2**, **i6**, and **i11**, respectively. To judge on the importance of the radiative stabilization process, its rate constants need to be compared with those for unimolecular dissociation processes. Indeed, as the stabilization rates increase with the molecular size, the radiative processes are expected to become more important, and may eventually overtake dissociation. This would increase the probability of PAH and NPAC molecules to be produced *via* the mechanism suggested in our paper, because their stabilization would then be faster than dissociation.

Table 1 RRKM calculated rate constants (in s^{-1}) for individual unimolecular reaction steps of isomerization and dissociation of various intermediates in the C$_6$H$_4$(CN)(C$_2$H) + C$_2$H reaction (Fig. 2 and 3 in our paper) corresponding to different collision energies.

Collision energy/kcal mol^{-1}	0.0	1.0	2.0	3.0	4.0	5.0
Fig. 2						
i1 → i3	6.5 × 10^9	7.3 × 10^9	8.1 × 10^9	9.0 × 10^9	9.9 × 10^9	1.1 × 10^{10}
i3 → i1	4.7 × 10^{11}	5.1 × 10^{11}	5.4 × 10^{11}	5.7 × 10^{11}	6.1 × 10^{11}	6.4 × 10^{11}
i1 → i2	1.9 × 10^9	2.1 × 10^9	2.2 × 10^9	2.4 × 10^9	2.6 × 10^9	2.8 × 10^9
i2 → i1	9.2 × 10^3	1.3 × 10^4	1.8 × 10^4	2.4 × 10^4	3.2 × 10^4	4.3 × 10^4
i1 → C$_6$H$_4$CN + C$_4$H$_2$	4.2 × 10^5	6.5 × 10^5	9.8 × 10^5	1.4 × 10^6	2.1 × 10^6	3.0 × 10^6
i3 → i4	1.1 × 10^{12}	1.1 × 10^{12}	1.2 × 10^{12}	1.3 × 10^{12}	1.4 × 10^{12}	1.4 × 10^{12}
i4 → i3	9.2 × 10^6	1.1 × 10^7	1.3 × 10^7	1.6 × 10^7	1.9 × 10^7	2.2 × 10^7
i4 → C$_6$H$_4$(CN)(C$_4$H) + H	2.0 × 10^4	2.9 × 10^4	4.0 × 10^4	5.5 × 10^4	7.5 × 10^4	1.0 × 10^5
k_{diss}(–C$_4$H$_2$)	2.0 × 10^0	3.9 × 10^0	7.5 × 10^0	1.4 × 10^1	2.5 × 10^1	4.4 × 10^1
k_{diss}(–H)	1.5 × 10^2	2.5 × 10^2	4.1 × 10^2	6.7 × 10^2	1.1 × 10^3	1.6 × 10^3
Fig. 3						
i5 → i6	1.6 × 10^9	1.8 × 10^9	2.0 × 10^9	2.2 × 10^9	2.4 × 10^9	2.7 × 10^9
i6 → i5	2.9 × 10^5	3.9 × 10^5	5.2 × 10^5	6.9 × 10^5	9.0 × 10^5	1.2 × 10^6
i5 → C$_6$H$_4$(C$_2$H) + HCCCN	2.3 × 10^0	1.1 × 10^1	3.8 × 10^1	1.2 × 10^2	3.2 × 10^2	7.7 × 10^2
k_{diss}	4.1 × 10^{-4}	2.3 × 10^{-3}	1.0 × 10^{-2}	3.7 × 10^{-2}	1.2 × 10^{-1}	3.3 × 10^{-1}

To respond to the second part of the question, we choose the C$_6$H$_4$(CN)(C$_2$H) + C$_2$H reaction as an example; the RRKM calculated rate constants for unimolecular dissociation and isomerization reaction steps for all intermediates involved are collected in Table 1. Using these individual rate constants, we also computed overall decomposition rate constants for the **i2** and **i6** complexes within the steady-state approximation. These rate constants, k_{diss}(–C$_4$H$_2$) for **i2** → C$_6$H$_4$CN + C$_4$H$_2$, k_{diss}(–H) for **i2** → C$_6$H$_4$(CN)(C$_4$H) + H, and k_{diss} for **i6** → C$_6$H$_4$(C$_2$H) + HCCCN, are also shown in Table 1. One can see that at zero collision energy, k_{diss}(–H) ≈ 150 s^{-1}, is a factor of 5 higher than the rate of radiative stabilization of this molecule. Therefore, the C$_6$H$_4$(CN)(C$_4$H) + H decomposition channel is favored over stabilization of 2-aza-4-ethynyl-1-naphthyl **i2** produced in the C$_6$H$_4$(CN)(C$_2$H) + C$_2$H reaction. Alternatively, the decomposition rate constant for 2-aza-1-ethynyl-4-naphthyl **i6**, 4.1 × 10^{-4} s^{-1}, is significantly lower than its stabilization rate constant, 24 s^{-1}, indicating that this NPAC radical can be formed through the mechanism of C$_2$H addition to the CN group in C$_6$H$_4$(CN)(C$_2$H) shown in Fig. 3 in our paper. This is also true for the formation of 1,4-diethynyl-phthalazine **i11** in the **i7** + C$_2$H reaction, as the radiative stabilization rate constant of **i11** appears to be much higher than that for its decomposition. Thus, radiative stabilization appears to be a rather important process for larger molecules and should not be neglected.

1 S. J. Klippenstein, Y.-C. Yang, V. Ryzhov and R. C. Dunbar, *J. Chem. Phys.*, 1995, **104**, 4502.

Professor Suits commented: For experiments involving these large systems, it is helpful to know the lifetimes. The traditional crossed-beam approach is sensitive to lifetime: for intermediates that live dozens of microseconds or more, they may not decompose before arriving at the detector. In this case, they would only be seen at the center-of-mass angle. For systems that decompose on a timescale of less than a microsend, there is no issue. In between these timescales, the distribution one detects could be in error, as the intermediate could dissociate on the way to the detector.

Professor Zwier asked: Did you consider isocyano isomers of your cyano compounds? Do they compete with their cyano analogs and/or do they produce unusual end products worth considering further?

Professor Mebel answered: We have considered some isocyano isomers of our cyano compounds. Normally, the isocyano isomers are significantly higher in energy than the corresponding cyano compounds and therefore their formation is expected to be less competitive. However, the question whether the isocyano isomers can produce some unusual end products is important and merits further detailed calculations.

Professor Zwier noted: As a follow-up comment, I think it is interesting to note that mass 128, which we all tend to associate with naphthalene, is also the mass of dicyanobenzene. So one needs to be careful in associating a peak at *m/z* 128 in a mass spectrum with naphthalene, and perhaps even in associating it with a pure hydrocarbon.

Professor Raulin commented: You've studied the case of *ortho* dicyano benzene isomer. Did you look at other isomers (*meta*, *para*)? It would be interesting to check if the *ortho* is favored (or not) compared to the two others, since the *ortho* is more likely to induce a second ring formation.

Professor Mebel responded: At the moment, we have not looked at *meta* and *para* isomers of dicyano benzenes. We expect that the *meta* and *para* dicyano benzenes can be easily formed in the reactions of the CN radical with cyano benzene, just as is the *ortho* dicyano benzene. Most probably, the energy difference between the *ortho*, *meta*, and *para* isomers of dicyano benzene will be rather small, but this question indeed deserves further investigation. In the present paper, we focused at the *ortho* isomer because this isomer would be the most likely precursor for the closure of the second ring and the formation of an N-containing polycyclic aromatic molecule.

Dr Lavvas asked: Would other nitrogen containing radicals, such as C_3N or HCCN behave in a similar way as CN? Once a nitrogen containing aromatic is formed, does the presence of N in the ring structure affect its further growth to larger aromatic structures on reaction with C_2H or CN?

Professor Mebel answered: The answer to the first question depends on the existence of a barrier for the addition of these radicals to unsaturated molecules and on the height of the barrier if it exists. For instance, the C_3N radical is expected to be analogous to CN and is likely to add to double and triple bonds without a barrier. On the other hand, HCCN is a carbene similar to CH_2 and is likely to react with closed-shell molecules *via* a distinct barrier. Hence, we would expect C_3N, but not HCCN, to behave similar to CN. For the second question, since the presence of N in the ring reduces its aromaticity and destabilizes the ring, it should reduce the probability of further growth of larger aromatic structures. However, if C_2H or CN add to the rings in an N-PAC not containing the nitrogen atom, the further growth likely would not be affected.

Professor Ashfold concluded: The overview of previous studies of CH_4 photochemistry in the introduction to paper 21[1] may leave the audience with the impression that our understanding of the primary photochemistry of this important species is even less secure than it actually is. As noted in the Introduction of paper 7,[2] however, we are still missing precise, wavelength dependent, product branching ratios. The early quantum yield estimates (listed under (1)–(4) in paper 21) were from analysis of a collection of indirect measurements. Subsequent direct experimental studies of CH_4 photolysis at 121.6 nm,[3-6] and, recently, in the wavelength range 128–133 nm,[7] aided by *ab initio* calculations of the potential energy surfaces (PESs) of the ground (S_0) and first excited (S_1) singlet states,[8,9] have advanced our knowledge considerably. C–H bond fission (yielding highly internally (rotationally) excited CH_3 fragments) is one major primary process following excitation to the S_1 state. H_2 elimination (yielding $CH_2(a^1A_1)$ co-fragments) is the other. Part of the difficulty in defining the product quantum yields is that some of the CH_3 fragments formed by photolysis at 121.6 nm have sufficient internal energy to decay further—by loss of an H atom, or by eliminating molecular H_2. The relative efficiencies of these two processes may well be a sensitive function of the detailed energy content in the primary CH_3 products and thus of the photo-excitation wavelength.

Kaiser *et al.*[1] also comment that C_2 radicals can be formed in Titan's atmosphere by the UV photolysis of $C_2H(X)$ radicals that are themselves formed by UV photolysis of C_2H_2 at wavelengths $\leq$217 nm. For completeness, it should be noted that the best estimate of the threshold energy for the latter process, $D_0(HCC-H) = 46074 \pm 8$ cm^{-1}, was provided by an early H (Rydberg) tagging study by the Bristol group[10] and that the dominant fragmentation pathway when C_2H_2 is photo-excited at shorter wavelengths yields electronically excited C_2H fragments in the low-lying $A^2\Pi$ state.[11,12]

1 R. I. Kaiser *et al.*, this volume.
2 K. L. Gannon *et al.*, this volume.
3 D. H. Mordaunt, I. R. Lambert, G. P. Morley, M. N. R. Ashfold, R. N. Dixon, L. Schnieder, K. H. Welge, *J. Chem. Phys.*, 1994, **93**, 2054.
4 J.-H. Wang and K. Liu, *J. Chem. Phys.*, 1998, **109**, 7105.
5 J.-H. Wang, K. P. Liu, Z. Y. Min, H. M. Su, R. Bersohn, J. Preses and J. Z. Larese, *J. Chem. Phys.*, 2000, **113**, 4146.
6 P. A. Cook, M. N. R. Ashfold, Y.-J. Lee, K.-H. Jung, S. Harich and X. Yang, *Phys. Chem. Chem. Phys.*, 2001, **3**, 1848.
7 Y. Zhang, K. Yuan, S. Yu and X. Yang, *J. Phys. Chem. Lett.*, 2010, **1**, 475.
8 A. M. Mebel, S.-H. Lin and C.-H. Chang, *J. Chem. Phys.*, 1997, **106**, 2612.
9 R. van Harrevelt, *J. Chem. Phys.*, 2006, **125**, 124302.
10 D. H. Mordaunt and M. N. R. Ashfold, *J. Chem. Phys.*, 1994, **101**, 2630.
11 P. Löffler, E. Wrede, L. Schnieder, J. B. Halpern, W. M. Jackson and K. H. Welge, *J. Chem. Phys.*, 1998, **109**, 5231.
12 M. N. R. Ashfold, G. A. King, D. Murdock, M. G. D. Nix, T. A. A. Oliver and A. G. Sage, *Phys. Chem. Chem. Phys.*, 2000, **12**, 1218.

Professor Gauyacq commented: As pointed out by Professor Ashfold, the branching ratios for the methane photolysis at Lyman-α 121.6 nm^{-1} given in eqn (1)–(4) of paper 21 should be corrected: the figures quoted in eqn (1)–(4) of this paper do not correspond to the results of the cited references of paper 21. An experimental effort has been devoted to this important issue as reviewed for instance by Romanzin *et al.* [*Adv. Space Res.*, 2005, **36**, 258] or by M. D. Lodriguito *et al.* [*J. Chem. Phys.*, 2009, **131**, 224320. The photolysis of methane has such an important impact on the gas phase chemistry of Titan atmosphere that the branching ratios should be considered with great care since the published data are still controversial.

In this context we would like to mention a related unpublished experimental work which we presented at the 4th NSF Workshop on Titan "Photolysis of methane at 121.6 nm and 118.2 nm in the laboratory: new input for Titan atmosphere photochemistry" by S. Boyé-Péronne and collaborators. In this work, we have performed a well-controlled experiment based on methane photolysis at 121.6 nm and at 118.2 nm (pure VUV beams) and on the direct measurement of the radical fragments (CH_3, triplet CH_2, singlet CH_2) by mass spectrometry following VUV photoionisation. This two-VUV colour experiment enabled us to discriminate among singlet and triplet CH_2 products by selective ionisation at both photolysis wavelengths. The most striking result is that the photolysis of methane at these two close wavelengths produces different 1,3-CH_2/CH_3 ratios. Therefore one should be very careful when

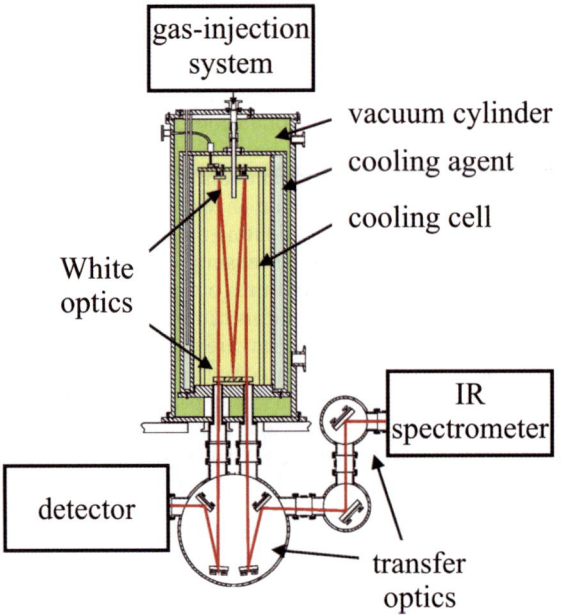

Fig. 1 Schematic illustration of the collisional cooling cell setup.[8]

comparing results from experiments performed at different photolysis wavelengths. This wavelength-dependent branching ratio should be taken into account in the modeling of the Titan atmosphere. A final remark following the comment by M. Ashfold is that the CH_3 radical should be formed with very high internal energy and could further dissociate into CH_2 + H or CH + H_2 (as measured by Mordaunt et al., J. Chem. Phys., 1993, **98**, 2054. Our experiment probes radical fragments in a typical timescale of 10–100 ns after photodissociation. We could rationalise our present results as follows: either the hot CH_3 dissociation kinetics is too slow so that the observed products are nascent products, or it is faster than 1 nanosecond so that the observed products are final products after subsequent dissociation. A paper on this work is currently under progress.

Dr Knox (on behalf of E. Kathrin Lang, Chia C. Wang and Ruth Signorell) opened the first general discussion of the papers: I would like to share with delegates some of the work that is underway in the group of Professor Ruth Signorell at the University of British Columbia. We are carrying out laboratory studies of aerosol particles relevant to the lower atmosphere of Titan. Professor Zwier highlighted

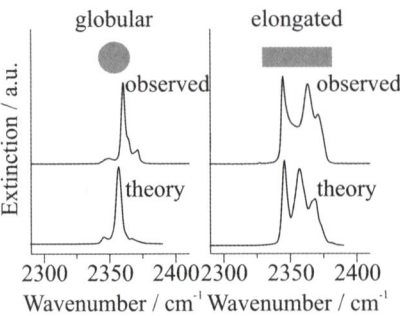

Fig. 2 The influence of particle shape on the ν_3 $^{12}CO_2$ band for carbon dioxide particles.[11]

the connection between this work and the spectroscopic studies being carried out by Professor Meuwly and co-workers during the discussion of papers 8–11. During the discussion of papers 18 and 19 Professor Lunine mentioned the phase transitions of ethane aerosols as they descend through the atmosphere of Titan. In our laboratory we are able to study the phase behavior of various hydrocarbon aerosols.[1–7]

We carry out measurements of aerosols in a collisional cooling cell, illustrated in Fig. 1, the temperature and pressure within which can be tuned to match a wide range of planetary atmospheres.[8] In the case of Titan we can simulate the conditions present in the lowest 60 km of the atmosphere, *i.e.* the troposphere and lower stratosphere. This corresponds to a temperature range of 71–94 K and a total pressure of 42–1466 mbar.[9,10] The cooling cell is connected to a gas-injection system which is used to introduce warm gas into the cold cell, leading to particle condensation. The gas-injection system is equipped with two valves which allows combinations of pure or mixed gases to be introduced into the cell simultaneously or sequentially. The cooling cell is coupled to a rapid-scan Fourier transform infrared spectrometer which is used to collect infrared absorption spectra from the suspended aerosols.

The Signorell group has previously demonstrated that infrared spectroscopy can be used to characterize a wide range of properties of aerosol particles by comparing the shape of the observed infrared absorption bands with those predicted from vibrational exciton modeling.[11] For example, the shape of carbon dioxide aerosol particles can be determined in this way by comparing the observed shape of the ν_3 band of carbon dioxide with the prediction of the shape from theory, as illustrated in Fig. 2.[11] The particle architecture can also be determined, for example in the case of mixed composition acetylene/carbon dioxide particles. Under certain conditions the shape of the observed band is consistent with an acetylene core and carbon dioxide shell, as shown in Fig. 3.[3] The crystal structure present in the particle can also be determined. It can be seen in Fig. 4 that under the prevailing conditions acetylene exists as a partially amorphous structure in the aerosol particles sampled, determined by examining the shape of the ν_5 band of acetylene.[6] Finally, the particle phase can be determined. This has been the subject of recent measurements, which have sought in particular to characterize the freezing kinetics of supercooled ethane droplets under a range of conditions, a topic of interest for understanding cloud processes and the formation of hydrocarbon lakes on Titan. When ethane is first

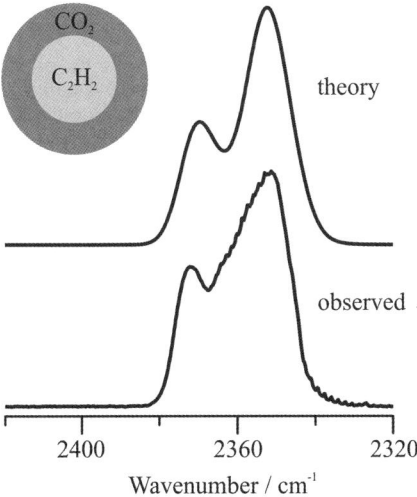

Fig. 3 The determination of particle structure from the shape of the ν_3 $^{12}CO_2$ band for mixed composition acetylene/carbon dioxide particles.[3]

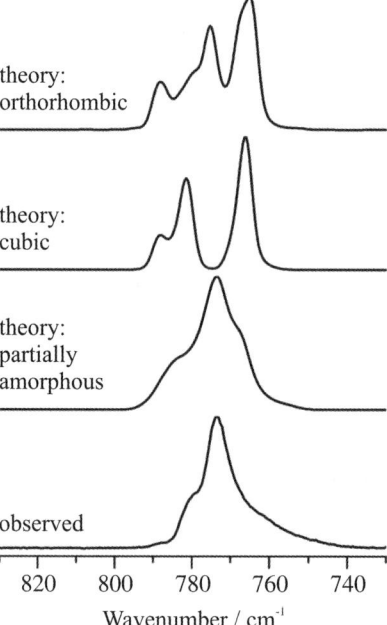

Fig. 4 The determination of particle crystal structure by comparing the observed shape of the ν_5 C_2H_2 band with predictions from vibrational exciton modeling.[6]

introduced to the cooling cell ($t = 0$ seconds) supercooled liquid droplets are formed, which give rise to broad, featureless absorption bands, as shown in Fig. 5 for the ethane ν_5 band. As ethane freezes, structure emerges in the shape of the absorption band; by following the time evolution of the shape of the absorption bands the progress of the phase transition of the ethane particles can be determined.[2]

I will describe one example of the type of measurement which can be performed using this approach. It has been determined that gas-phase methane can significantly stabilize supercooled ethane droplets.[5,7] Panel (a) of Fig. 6 shows the evolution of the ν_8 band of ethane as ethane droplets freeze in an atmosphere of nitrogen gas. It can be seen that freezing takes around 2000 seconds. As an increasing amount of gas-phase methane is introduced to the cell atmosphere freezing takes considerably longer. In panel (c) the partial pressure of methane in the cell is representative of that in the troposphere of Titan. It can be seen that crystallization is significantly delayed, and it is estimated that in this case it would take several days for freezing to be completed. This indicates that supercooled ethane droplets may exist for much longer periods in the lower atmosphere of Titan than previously thought. They may therefore play an important role in cloud processes and the formation of lakes.

I would like to acknowledge Professor Ruth Signorell, the principal investigator for these studies, Dr Chia Wang, who is now an assistant professor at National Sun-Yat Sen University in Taiwan, Kathrin Lang, a graduate student involved with the cooling cell measurements, and Thomas Preston, a graduate student and Dr. George Firanescu, a postdoctoral fellow, who are responsible for the computational work. Financial support from NSERC, the Canada Foundation for Innovation and the A. P. Sloan Foundation (Ruth Signorell) is gratefully acknowledged.

1 R. Signorell and M. Jetzki, *Phys. Rev. Lett.*, 2007, **98**, 013401.
2 Ó. F. Sigurbjörnsson and R. Signorell, *Phys. Chem. Chem. Phys.*, 2008, **10**, 6211.
3 C. C. Wang, P. Zielke, Ó. F. Sigurbjörnsson, C. R. Viteri and R. Signorell, *J. Phys. Chem. A*, 2009, **113**, 11129.
4 C. C. Wang, S. K. Atreya and R. Signorell, *Icarus*, 2010, **206**, 787.

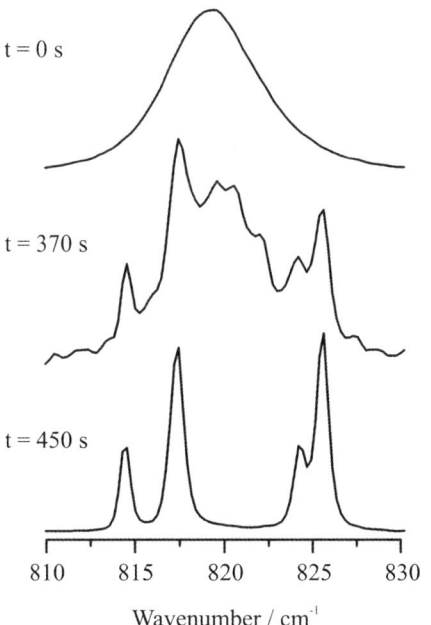

Fig. 5 The evolution of the infrared absorption spectrum of ethane during particle freezing.[2]

5 C. C. Wang, E. K. Lang and R. Signorell, *Astrophys. J. Lett.*, 2010, **712**, 40.
6 T. C. Preston, G. Firanescu and R. Signorell, *Phys. Chem. Chem. Phys.*, 2010, **12**, 7924.
7 E. K. Lang, K. J. Knox, C. C. Wang and R. Signorell, 2010, submitted.
8 G. Firanescu, D. Hermsdorf, R. Ueberschaer and R. Signorell, *Phys. Chem. Chem. Phys.*, 2006, **8**, 4149.
9 A. Coustenis, A. Salama, B. Schulz, S. Ott, E. Lellouch, T. Encrenaz, D. Gautier and H. Feuchtgruber, *Icarus*, 2003, **161**, 383.
10 R. de Kok, P. G. J. Irwin, N. A. Teanby, C. A. Nixon, D. E. Jennings, L. Fletcher, C. Howett, S. B. Calcutt, N. E. Bowles, F. M. Flasar and F. W. Taylor, *Icarus*, 2007, **191**, 223.
11 Ó. F. Sigurbjörnsson, G. Firanescu and R. Signorell, *Annu. Rev. Phys. Chem.*, 2009, **60**, 127.

Professor Lunine asked: Your experimental apparatus would seem suitable for the investigation of the hypothesis of D. M. Hunten (*Nature*, 2006, **443**, 669), that the bulk of the ethane produced photochemically over Titan's history is sequestered in sedimented aerosol particles on Titan's surface. The hypothesis, which I consider speculative, requires a large adsorption capacity (or "wettibility"?) of ethane on the photochemically produced aerosols. It is important to remember that the aerosols we are talking about are not the so-called "tholins"—the highest molecular weight products—but rather all of the products of Titan atmospheric chemistry that are solid under Titan surface conditions. Hence I am including here acetylene and its products such as benzene. A proper investigation to test Don's hypothesis would be of great value in understanding why there seems to be so little ethane on Titan's surface today.

Dr Knox replied: We have observed that ethane condenses from the gas-phase onto solid particles such as acetylene and carbon dioxide under the conditions relevant to the atmosphere of Titan.[1,2] Sequestration of ethane in sedimented aerosols, however, is something that we cannot observe with our current experimental setup. We also consider the hypothesis about sequestration in sedimented aerosols to be

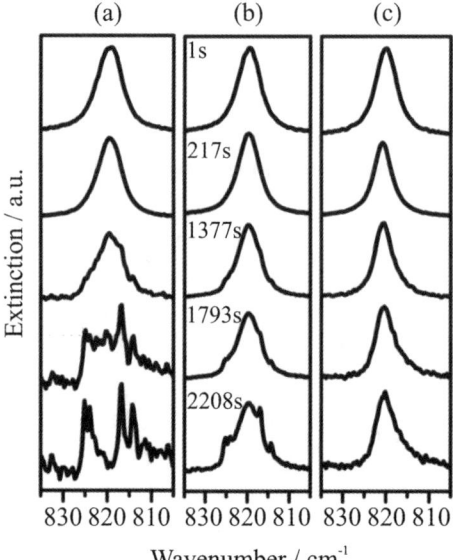

Fig. 6 Evolution of the ν_8 C_2H_6 band during the freezing of supercooled ethane droplets formed in (a) nitrogen gas, (b) a mixture of nitrogen/methane gas with a methane mole fraction of 0.007 and (c) a mixture of nitrogen/methane gas with a methane mole fraction of 0.011–0.026.[5,7]

speculative and it seems highly unlikely to us (see also Professor Atreya's comment, below).

1 C. C. Wang, E. K. Lang and R. Signorell, Astrophys. J. Lett., 2010, **712**, 40.
2 E. K. Lang, K. J. Knox, C. C. Wang and R. Signorell, 2010, submitted.

Professor Atreya commented: The Galileo probe mass spectrometer does not show evidence of SMUST aerosols in Jupiter. For example, the ethane mixing ratio does not increase with depth, whereas SMUST are supposed to vaporize and release it at increasing temperatures, hence depth.

Professor Mitchell asked: In the apparatus that you have shown, are the measurements of the droplets made in air or in vacuum ? Have you considered using small angle X-ray scattering (SAXS) to measure the size of the droplets. This would also give you the number density and the form of the droplets (smooth, aggregated *etc.*). The advantage of SAXS is that one does not need to know the index of refraction for the material as in laser scattering studies. Measurements can be done in air or in vacuum using thin mica windows to allow the X-rays to pass through.

Dr Knox responded: The pressure within our experimental chamber can be varied from around 10–1500 mbar, and the gas-phase composition can also be controlled, for example measurements can be carried out in nitrogen, helium or a mixture of these gases with hydrocarbons such as methane and ethane. Currently the cell is not configured to allow small angle X-ray scattering measurements, but this could form the basis of future work.

Professor Raulin asked: Can you look at the effect of small solid particles, other than frozen CO_2, on the processes of condensation and on the shape of the observed particles obtained after condensation?

Dr Knox answered: The process of aerosol particle formation *via* condensation from the gas-phase proceeds too rapidly to study using our method. However, the time evolution of the shape and phase of aerosol particles can readily be studied for a wide range of particles. For example, besides frozen carbon dioxide, aerosols of methane, acetylene, ethane, ethylene, propane, ammonia and sulfur dioxide, and various mixtures of the above, have been generated in the cell.[1-6]

1 R. Signorell and M. Jetzki, *Phys. Rev. Lett.*, 2007, **98**, 013401.
2 R. Signorell and M. Jetzki, *Faraday Discuss.*, 2008, **137**, 51.
3 C. C. Wang, P. Zielke, Ó. F. Sigurbjörnsson, C. R. Viteri and R. Signorell, *J. Phys. Chem. A*, 2009, **113**, 11129.
4 C. C. Wang, S. K. Atreya and R. Signorell, *Icarus*, 2010, **206**, 787.
5 T. C. Preston, G. Firanescu and R. Signorell, *Phys. Chem. Chem. Phys.*, 2010, **12**, 7924.
6 E. K. Lang, K. J. Knox, C. C. Wang and R. Signorell, 2010, submitted.

Professor Atreya asked: I suggest looking also into the thermodynamics of coexistence of methane ice particles with ethane droplets. As the solid ethane ice particles descend from their stratospheric region of origin, they change phase to liquid droplets in the middle troposphere, but can encounter substantial methane ice particles from approximately 25 km to 15 km or even lower. It is important to know whether the methane ice plays a role, in addition to methane gas that you are studying.

Dr Knox answered: Such measurements would be very interesting but somewhat challenging. In order to isolate the influence of solid methane particles from that of gas-phase methane, it is likely that it would be necessary to perform the measurements at temperatures lower than those found in the lower atmosphere of Titan; such measurements may still offer worthwhile insight however.

Mr Sebree asked: Are there cases in which the aerosols in your chamber show more homogeneous mixing rather than shell like structure? Are there time scales during which the different mixing structures interchange?

Dr Knox responded: Yes, both are the case, and both depend on the substances involved and the conditions. To give but one example: if carbon dioxide and acetylene gas are premixed and the mixture injected into the cell, homogeneously mixed particles condense and freeze with a metastable mixed crystal phase.[1] Over a timescale of tens of minutes the mixed phase decomposes into domains of the more stable pure crystal phases. If, however, carbon dioxide and acetylene gases are sequentially introduced to the cell, core–shell particles are formed, with the gas which is introduced first forming the core of the particle, and the gas which is introduced second coating the core and forming the shell.[1] Such particles are structurally stable on the timescale of the measurements.

1 C. C. Wang, P. Zielke, Ó. F. Sigurbjörnsson, C. R. Viteri and R. Signorell, *J. Phys. Chem. A*, 2009, **113**, 11129.

Professor Miller Dr Mousis: As I understand the scenario you proposed in Mousis *et al.* (*Astrophys. J.*, 2009, **691**, 1780–1786), you proposed that xenon and krypton would be preferentially complexed with H_3^+ in comparison with argon. This would mean that Xe and Kr would be depleted in the ices from which Titan formed and in which Ar would be clathrated, but still available for the gas mixture from which Saturn's atmosphere forms.

You used that scenario to explain the relative depletion of Xe and Kr with respect to Ar in Titan, yet being in jovian/solar proportions in Saturn. Two questions follow: (1) Could you clarify how the explanation in Mousis *et al.* (2009) fits in with the explanation contained in this paper? (2) Could you say something on the relative importance of thermodynamic stability as a driver, as you have used in both this paper and Mousis *et al.* (2009), and kinetic issues, which involve not how stable products might be but rather how quickly they might form?

Dr Mousis answered: The H_3^+ scenario requires that the abundance of this ion must be high enough at 10 AU to trap Xe and Kr in the solar nebula gas phase and impede these two noble gases to be incorporated in Titan's building blocks. The explanation contained in the paper works if the abundance of H_3^+ is not sufficiently high to allow a substantial trapping of Xe and Kr.

The results presented in the paper are based on equilibrium calculations and do not consider the kinetics of clathration. The reason is that, unfortunately, there is no laboratory experiment that investigates the kinetics of clathrate formation at thermodynamic conditions corresponding to those of the solar nebula.

Dr Bézard asked: Your calculations assume a 100% efficiency of the clathration process in the solar nebula, *i.e.* that all the cages of the clathrate hydrates are occupied by guest molecules. This looks to me as a very optimistic assumption. Under which conditions do you expect that this can happen?

Dr Mousis replied: A full efficiency for the clathration process in the solar nebula requires that guest molecules had the time to diffuse through porous water-ice planetesimals in the solar nebula. This remains plausible only if collisions between planetesimals have exposed essentially all the ice to the gas over timescales shorter or equal to the planetesimals' lifetimes in the nebula. However, we agree that the efficiency of collisions between planetesimals to expose all the "fresh" ice over such a timescale still remains unknown.

Dr Plattner said: For your simulations you have chosen to use the spherically averaged Kihara potential instead of an atomistic model and argue that atomistic models, similar to Kihara potentials, require the choice of parameters which are somewhat arbitrary. However, there is one important issue which cannot be addressed by Kihara potentials, namely the distortion of shape and size of the clathrate cages upon exchange of guest molecules. This issue could be captured at least qualitatively with an atomistic model, *e.g.* an atomistic force field. Do you have an estimate to what extent your results are affected by neglecting guest-dependent distortion of the clathrate cages?

Dr Mousis replied: Unfortunately, we have only investigated the influence of the clathrate cage distortion due to temperature variations and not due to the size of the guest molecule. However, we agree that the atom–atom description of the intermolecular guest–clathrate interactions is more accurate than the spherically averaged approach. We have already started to work on the atomistic approach and the influence of the guest-dependent distortion of the clathrate cages will be investigated in the near future.

Professor Suits opened the discussion of the paper by Dr Vuitton: I wonder if you can gain any additional information by using some of the approaches in proteomics, such as electron transfer dissociation or operation in negative ion mode?

Dr Vuitton answered: Electron-transfer dissociation, an effective technique for stimulating extensive fragmentation of multicharged ions, could be performed in our apparatus. However, we do not observe any multicharged ions in either the tholins' or HCN polymers' spectra. As a consequence, this technique is not appropriate to gain further insight on our samples.

We did perform some studies in the negative ion mode. They revealed the presence in the samples of new classes of molecules that are not present in the positive mode. Because of the lower ionization efficiency, the results were not as conclusive as in the positive ion mode but were consistent with the fact that tholins contain at best a small fraction of HCN polymers.

Dr Ellinger asked: Titan is often considered as an interesting laboratory for prebiotic chemistry. Did you find any trace of $(HNC)_5$, *i.e.* adenine ?

Dr Vuitton responded: Orbitrap measurements indicate that $C_5H_5N_5$ is present in both our Titan's tholins and our HCN polymers. Moreover, GC-MS shows unambiguously that the tholins do contain adenine. However, we have not determined yet if the $C_5H_5N_5$ detected in the HCN polymers is adenine or another isomer.

Professor Coates remarked: In the mass spectra you see, is there any tendency for the negative ions to be higher mass? I was wondering how relevant the results are to what we see in Titan's upper atmosphere.

Dr Vuitton responded: The ions generally extend to lower m/z in the negative ion mode than in the positive ion mode. However, this is related to the ionization efficiency of the electrospray and cannot be extrapolated to the detection of heavier negative than positive ions in Titan's upper atmosphere.

Dr Carrasco communicated: It was asked of Dr Vuitton how far the soluble part of the tholin extracted in methanol and analysed by ESI-MS was representative of the bulk tholin.
We studied this important validation issue in a previous paper [Carrasco *et al.*, Chemical characterization of tholins: solubility, morphology and molecular structure revisited, *J. Phys. Chem A*, 2009 **113**(42), 11195–11203. The solubility properties of the PAMPRE tholin in methanol were first quantified, showing that the analysed soluble part actually counted for ~30% of the bulk sample.
A comparison of the IR spectra of the primary tholin, its soluble part and its insoluble part confirmed that the IR chemical signature of the soluble part was in good agreement with IR spectrum of the the bulk tholin, despite a slight over-estimation of the 3 μm absorption band, typical of the CH_2/CH_3 functions.

Professor Raulin returned to the discussion of the paper by Dr Mousis: Can two species be trapped in the same water cage? If yes, is such trapping included in models of the formation of planetary bodies in the outer solar system?

Dr Mousis replied: Some small molecules such as H_2 can indeed complete the stabilization of the structure when they share the same cage with a larger molecule. To the best of our knowledge, this has not been yet included in formation models of planetary bodies of the solar system.

Professor Suits said: Concerning the missing argon: Can it simply remain trapped in the clathrates?

Dr Mousis answered: This possibility has been investigated by Thomas *et al.* (*Astron. Astrophys.*, 2007, **474**, L17–L20). It appears that argon is a poor clathrate former compared to xenon and krypton.

Dr Thissen asked: It was said in paper 17 by Professor Lunine that, when considering a permanent presence of methane in the atmosphere of Titan, the C_2H_6 is missing on Titan. Some hypothetical cataclysmic release of methane by cryovolcanism has been considered in ref. 1; this would lead to a description of Titan as a transient methane reactor, due to episodic clathrate destruction. My question to you is: "As you describe, in terms of rare gases, the average composition of the Titan clathrates, would the present composition of the atmosphere be in agreement with the release of material from a recent (100 000 to 10 000 000 year) cataclysmic event?"

1 M. Choukroun, O. Grasset, G. Tobie, C. Sotin, *Icarus*, 2010, **205**, 2581–2593.

Professor Lunine responded: I do not know the answer to the question (it would require more detailed modeling to assess this), but the amount of solid hydrocarbon accumulated in the form of dunes would suggest that methane photolysis has been going on for a significant part of geologic time and is not a novel, recently-started event.

Dr Mousis replied: These are several mechanisms that can explain the noble gas deficiency of Titan's atmosphere, irrespective of the release epoch of the methane (noble gas trapping into clathrates located on the surface of the satellite, trapping by H_3^+ ion in the nebula, partial devolatilization of planetesimals...). There is then apparently no conflict between the measured noble gas impoverishment in the atmosphere and a recent methane outgassing episode.

Professor Lunine asked: In considering whether noble gases are still sequestered today in clathrate hydrate in Titan's interior, it is important to remember that while the outer Ice I crust of Titan might contain large amounts of methane-bearing clathrate hydrate, this crust is almost certainly not primordial. Rather, it was probably formed early in Titan's history, as the initially warm and perhaps liquid water (or water–ammonia) outer layers solidified and accommodated methane from the interior. The original composition of the planetesimals which were incorporated in Titan, including any noble gas bearing clathrate hydrates, would have been significantly altered. A useful reference to this is Tobie *et al. Nature*, 2006, **440**, 61.

Dr Pernot opened the second general discussion: As a follow-up of the discussions about the chemical composition of tholins, I would like to present the main results from a paper published earlier this year about the systematic analysis of high-resolution mass spectra of the soluble fraction of tholins produced in the PAMPRE setup [Szopa *et al.*, *Planet. Space Sci.*, 2006, **54**, 394–404]. The paper is entitled "Tholinomics—chemical analysis of nitrogen-rich polymers" [P. Pernot *et al.*, *Anal. Chem.*, 2010, **82**, 1371–1380]. As it has been published in *Analytical Chemistry*, it might have escaped the attention of this community. The paper introduces an original development of the methods used in petroleomics, that has been applied to some of the same tholin samples as presented by V. Vuitton in paper 23.

Our aim was to search for a systematic representation of the chemical composition of the thousands of peaks in a high resolution mass spectrum of tholin. Using Kendrick mass defect analysis and van Krevelen (H/C, N/C) composition diagrams, we have shown that not only is a CH_2 motif involved (in agreement with earlier studies by Sarker *et al.* [*Astrobiology*, 2003, **3**, 719–726]), but also a N-bearing motif. However, the van Krevelen diagram is too congested to enable clear conclusions on the nature of the latter. The use of an augmented 3D van Krevelen diagram (N/C, H/C, log m/z) reveals a stratification of the sample into less than a dozen families of general formula $X–(CH_2)_m(HCN)_n$, where the balanced "copolymer" ($m = n$) plays a central role. Within each family, the polymer lengths n and m present Poisson-type distributions. The stoichiometry of the families differ by the carbon content of the root X. The "pure family" $(CH_2)_m(HCN)_n$ is the cornerstone of the samples. Other families can be formally derived from it by adding or subtracting sequentially one C atom. Please note that this formula provides a systematic, convenient and compact representation of most (non-oxygenated) species in the samples, but it should not be taken literally with regard to the copolymeric nature of tholins.

From the mass spectra, we finally identified the lightest species within each family. For the "carbon poor" families, these species present only a few plausible isomers and can be determined as linear and cyclic amino nitrile compounds of great astrobiological interest: biguanide, guanidin, acetamidine, aminoacetonitrile and methylimidazole.

I hope that this work will help to bridge the gap evoked in a previous discussion between reaction dynamics in crossed beams experiments and laboratory simulations of complex reactive media. At least, it provides the former with reasonably sized targets as fundamental species in molecular complexation. Of course, this stoichiometric study should not be directly extrapolated to mechanistic conclusions, but we hope that it might help us to get some clues about the formation and growth of tholins, whether in the lab or in Titan's atmosphere.

Professor Suits said: These are interesting trends you observe in the mass spectra. We must be careful, though, not to ascribe any direct mechanistic significance to these.

Dr Pernot replied: I agree. However, interpreting the regular statistical distributions of CH_2 and HCN "units" within each family might require mechanistic hypotheses.

Professor Lunine opened the discussion of the concluding paper by Professor Strobel: A key input to your conclusion that there is a downward flux of hydrogen in the lower atmosphere of Titan is the measurement of hydrogen in the upper atmosphere by the INMS instrument aboard Cassini. Could you elaborate on the accuracy of that determination and hence on the strength of your conclusion regarding a possible sink of hydrogen at Titan's surface?

Professor Strobel replied: With the lower boundary condition constrained by Cassini CIRS, Huygens GCMS, and Voyager IRIS mole fractions at 0.001 (with error bars of ±25%) and negligible H_2 flux at the surface and currently accepted photochemistry, one predicts the H_2 mole fraction profile to be about a factor of 2 lower than measured by INMS. To quote from one of my recent papers [D. F. Strobel, Molecular hydrogen in Titan's atmosphere: implications of the measured tropospheric and thermospheric mole fractions, *Icarus*, 2010, DOI:10.1016/j.icarus.2010.03.003, in press]: "While the absolute measured values of densities by INMS are apparently a factor of ~2.6 lower than densities measured by HASI and inferred from torques by the Cassini AACS, it is not expected that the ratios of densities, in particular, the H_2/N_2 ratio, is uncertain by more than approximately 20% and certainly not a factor of 2 [J. Cui, R. V. Yelle, V. Vuitton, J. H. Waite Jr. *et al.*, Analysis of Titan's neutral upper atmosphere from Cassini ion neutral mass spectrometer measurements, *Icarus*, 2009, DOI:10.1016/j.icarus.2008.12)." Without any adjustment in either or both the CIRS and INMS H_2 mole fractions, they jointly require a downward flux of H_2 into Titan's surface at a rate on the order of 10^{28} H_2 s^{-1}, comparable to the H_2 escape rate from the top of the atmosphere.

Dr Carrasco asked: Professor Strobel mentioned an issue with the calibration of the INMS instrument on-board with Cassini. It seems that there is an underestimation by a factor 2.5 to 3 of the absolute measurements given by this instrument for the neutral mass spectra.

My question is : does this factor also affect the absolute measurement of the INMS ion mass spectra ?

Professor Strobel replied: The INMS has not officially revised their absolute measurements by a factor of 2.5 to 3. On the Cassini orbiter there are a number of ways to measure low energy plasma densities: Langmuir Probe—electrons; INMS—positive ions; CAPS—negative ions in the ram direction; RWPS—electrons *via* the upper hybrid frequency. I have heard a talk by an INMS team member who argued that by increasing the positive ion densities by the above factor, one would obtain better agreement among all measurements with the constraint of charge neutrality. which should hold rigorously beyond the Debye length.

Dr Vuitton commented: The total positive ion density measured by INMS is in good agreement with the electron density measured by RPWS/LP (Wahlund *et al.*, 2009), indicating that the ion densities are robust. It is conceivable that even *if* the neutral densities measured by INMS are off, the ion densities are accurate depending on what the problem with the instrument (or data analysis) actually is.

Professor Strobel responded: Please note my reply immediately above; I would like to state that, at lower altitudes, negative ions also contribute to the total negative charge. A paper with an intercomparison of plasma densities measured by the various orbiter instruments on selected Titan flybys would be very helpful to address this question.

Closing remarks
Darrell F. Strobel

Received 11th August 2010, Accepted 11th August 2010
DOI: 10.1039/c005513c

1 Introduction

First, let me note that I observe two population distributions in the audience, namely, those who were born before the Voyager 1 flyby of Saturn and Titan in November 1980 and those were born afterwards. As one in the Maxwellian tail of the former distribution, it is gratifying to see a significant number in the latter distribution who will carry on the tremendous breath and quality of research presented at this *Faraday Discussion* on the Chemistry of the Planets.

Although this *Faraday Discussion* was proposed to cover all planets and satellites, participants clearly zeroed in on Titan, Saturn's largest moon and the second largest satellite in the solar system. Some 60% of the papers dealt with the chemistry of Titan, as noted by Prof. Atreya in his introductory remarks.[1] This focus on the chemistry of Titan's atmosphere is driven by data from the Cassini–Huygens mission, with an orbiting spacecraft that has flown past Titan some 70 times already and the remarkable results from the descent and successful landing of the Huygens probe on its surface. To date, European and the United States governments have invested more than $3/€2/£2 billion to study the Saturn system in depth. Titan's mostly nitrogen, mildly reducing, atmosphere with a few percent methane, a large suite of hydrocarbons, organic molecules, and nitriles is widely thought to be a present day natural laboratory for chemical synthesis and the evolution of the Earth's early atmosphere, analogous to the pioneering laboratory experiments on the origin of life that Professors Stanley Miller and Harold Urey carried out in the 1950s. As a consequence, Titan has been an object of intense interest to exobiologists and planetary scientists for more than three decades. The participants of this Faraday Discussion have confirmed that this interest has not diminished and that Titan still has more interesting chemistry to offer.

What is somewhat surprising was the paucity of papers on the chemistry of the Martian atmosphere. With recent planetary programs being driven by the strategic goal of the studying the origin of life by following 'the water', one would have thought that Mars would have attracted more attention at this Discussion meeting.

2 Historical comments

To understand the development of the intense interest in Titan a good starting point is the influential paper of John Lewis,[2] who pointed out that satellites in the outer solar system would be comprised principally of the ices of water, methane, and ammonia. In the case of Titan, he noted that ammonia could be photochemically converted to molecular nitrogen. In the summer of 1973, NASA sponsored a workshop on the atmosphere of Titan,[3] where Prof. Hunten proposed a possible Titan model with a 20 bar atmosphere of molecular nitrogen.[4] In contrast, Prof. Danielson argued for a thin stratosphere on Titan with a few mbar of methane and a thermal inversion[5] to explain the observations that methane and ethane are observed in emission in the thermal infrared.[6] As a compromise, Dr Pollack[7] constructed

Johns Hopkins University, Baltimore, MD, USA. E-mail: strobel@jhu.edu; Fax: +1 410 516 7933; Tel: +1 410 516 7829

a greenhouse model with a minimum methane surface pressure of 0.4 bar and a comparable abundance of molecular hydrogen. At the workshop I presented a preliminary photochemical model for Titan's atmosphere,[8] where the rapid escape of atomic and molecular hydrogen led to irreversible photochemical loss of methane and conversion to higher hydrocarbons with an efficiency of more than 95% and observable amounts of ethane and acetylene.

The Voyager 1 flyby in November 1980 led to a wealth of data which yielded definitive answers to the outstanding questions at the time about Titan's atmospheric composition and structure. The net result was a paradigm shift in our knowledge. I expect a similar result from the Cassini–Huygens mission. One immediate paradigm shift is that aerosol formation is initiated in Titan's thermosphere above $\sim$900 km, where complex ion/neutral chemistry occurs, rather than in the stratosphere (cf. Section 3).

When I wrote my first paper on hydrocarbon photochemistry in 1969, most rate coefficients were based on end product analysis, rather than direct step-by-step measurements, i.e., the intermediate steps and products were never measured. Theoretical calculations were avoided as totally unreliable, unless one was exceedingly desperate. Now, as the papers presented at this Discussion demonstrate, laboratory techniques are capable of measuring the disappearance of the reactants, identification of the products, and their relative yields, and for good measure backed up with theoretical calculations by readily available software packages to confirm the validity of experimental work. And in some cases, the theoretical results are superior to the laboratory findings. One jet-lagged participant 'sleepless' the first night, Dr Klippenstein, demonstrated the power of theoretical quantum chemistry by submitting a number of jobs during the early morning hours to contribute to the discussion the 'same' day.

The direct photochemistry of methane in outer solar system atmospheres is driven by solar Lyman-α radiation. In 1969 the dissociation yields from Lyman alpha photolysis of methane were not known and in spite of all the research performed over the past 40 years, we are still lacking the required details[9] (also see the comments in general discussions 1 and 4 by Prof. Michael Ashford).

3 Chemistry in Titan's upper atmosphere

As noted above, one of the real surprises so far, and completely unexpected by anyone, from the Cassini orbiter were the measurements of very complex positive and negative ions, especially the heavy (up to 13 800 amu per charge) negative ions fortuitously detected by the CAPS (Cassini Plasma Spectrometer) instrument.[10] This instrument was not designed to measure negative ions, but the orbiter flies through the atmosphere at speeds of $\sim$6 km s^{-1} producing, relative to the spacecraft, beams of cold negative ions in the ram direction. The heaviest negative ions are concentrated at the lowest altitudes probed by the Cassini spacecraft ($\sim$950–1050 km) and may in fact be more abundant at even lower altitudes. The INMS (Ion Neutral Mass Spectrometer) measured positive ions and neutrals throughout its mass range of 1–100 Daltons and CAPS detected even higher mass positive ions. Overall, the observational evidence to date points to this region of Titan's atmosphere as the place where aerosol formation is initiated.[11] Ultraviolet stellar occultation measurements by the Cassini UVIS (Ultraviolet Imaging Spectrograph) yield aerosol extinction profiles,[12] which has been interpreted[13] as implying an aerosol mass flux 2.7–4.6 $\times$ 10^{-14} g cm^{-2} s^{-1} from the thermosphere down to the stratosphere. For comparison if all the methane dissociated by solar Lyman-α led to aerosol formation, the production rate would be $\sim$8 $\times$ 10^{-14} g cm^{-2} s^{-1}.

As discussed by Prof. Yelle,[14] the complex ion/neutral chemistry leads also to the formation of observable amounts of ammonia. Their calculations suggest that ammonia should be most abundant in the high atmosphere and formed by the reaction $NH_2 + H_2CN \rightarrow NH_3 + HCN$, whose rate coefficient was theoretically calculated. For further insight into the complex interplay of ion and neutral chemistry, the reader may also consult.[15,16]

4 Modeling of atmospheric chemistry

A number of participants questioned why planetary scientists model preferentially the chemistry of planetary atmospheres with one-dimensional models rather than 2D or 3D (GCM – general circulation model) models. It is important to remember that the most significant spatial variation in a planetary atmosphere is the vertical stratification due to gravity and characterized by an e-folding pressure (and total density) macroscopic length scale known as the scale height (H), with typical magnitude ~10–100 km. Thus one writes the distribution of the three thermodynamic variables that describe planetary atmospheres: mass (ρ) or number (n) density, pressure (p), and temperature (T) as

$$\psi(\phi,\theta,z,t) = \langle\psi\rangle(z) + \bar{\psi}(\theta,z,t) + \psi'(\phi,\theta,z,t)$$

The first term is the vertical profile of thermodynamic variable ψ averaged over the globe on each height [or pressure surface if $dz = -H(d\ln p)$]. The second term is the departure from the first term averaged over longitude, ϕ, on a latitudinal circle at latitude θ and height z. The last term is the longitudinally varying departure from the first two terms and generally characterizes waves in the atmosphere. Observationally, we know that the first term dominates, with each succeeding term being of lesser importance than the previous term. Constituents with long chemical lifetimes are adequately described by only the first term. In many planetary atmospheres the third term is generally negligible. For example on Titan, I am aware of no detectable stratospheric variations in longitude. From the Cassini CIRS (Composite Infrared Spectrometer) instrument one infers an upper limit on the 3rd term T' of 1–2 K in comparison to a mean temperature (the 1st term) of ~170 K. For latitudinal variations of chemical species, Fig. 8 in the paper,[17] on Titan's winter north pole, one notes that species with shorter photochemical lifetimes are more enriched and species with exceedingly long photochemical lifetimes such as ethane and carbon dioxide are well-mixed throughout the atmosphere.

Rarely do we possess enough 3D knowledge of the distributions of atmospheric constituents to justify the computational costs (labor and computer time) of running GCM calculations. There are many instances where 2D calculations are recommended to model the polar enrichment displayed in Fig. 8 of Teanby *et al.*, provided the physical processes driving the fluid dynamics of the atmosphere are understood and there exists an adequate observational database on the atmospheres' dynamics. In the case of Titan, I believe that we are far from a physical understanding of the dynamics of its atmosphere. Also, the available wind data is totally inadequate. Returning to the original question of why only 1D models are used, given the computational speed of an 1D model one can test the sensitivity of chemical schemes to various input quantities and gain understanding of which quantities (rate coefficients, photo cross sections, *etc.*) are most critical. Also note that the calculation of the first term above $\langle\psi\rangle(z)$ is area weighted and hence proportional to the cosine of the latitude. Thus while the polar enrichment of molecule might be large, when weighted by the cosine of latitude the polar contribution might be minimal to the global average. This gives further support to the utility of 1D in the study of atmospheric chemistry. One can also consult an old paper of mine,[18] where I showed how the vertical eddy diffusion coefficient representing vertical transport can be related to the multi-dimensional dynamics of atmospheres.

5 The closure problem

A number of participants raised the issue of closure in chemical schemes designed to model chemistry in planetary atmospheres and why many chemical schemes were truncated in what appeared to be an arbitrary manner. The closure problem is well know in turbulence theory where one divides fluid dynamic fields into time averaged fields and perturbations from the mean and writes evolution equations for the mean fields in terms of correlations of the perturbation fields, which in turn require additional evolution equations for the correlated quantities. Since this procedure can be carried out *ad infinitum*, at some point a closure method must be introduced, for example, by writing correlated variables in terms of fluxes proportional to gradients of mean fields.

Using Titan chemistry as a concrete example to illustrate this chemical problem, we know that methane is photochemically converted irreversibly to more complex molecules at a global rate of $\sim 10^{28}$ s^{-1}, implying a "geologically" short chemical lifetime of ~ 15 MY and necessarily a continual resupply from Titan's interior. The photochemical destruction rate of methane is equivalent to 4×10^{-13} g cm^{-2} s^{-1}. I noted previously that Lavvas *et al.*[13] calculated a downward aerosol mass flux 2.7–4.6 $\times 10^{-14}$ g cm^{-2} s^{-1} at 500 km, which is approximately 10% of the methane mass loss rate. In addition most, but not all, more complex species condense out in the lower stratosphere and tropopause region (~ 45 km). Hence chemical schemes for gas phase composition must incorporate loss processes for the conversion of gas phase molecules into aerosol, cloud, and haze particles, which ultimately end up on the surface.

Generally, one attempts to model stable molecules that are capable of detection by current instrumentation on spacecraft and ground-based telescopes. According to Nixon *et al.*[19] we find that the best that we do currently is ~ 1 ppb for abundances. Clearly, modeling species which are present at the ppt level would be counterproductive at this time and chemical closure at the ppb abundance level would be advised with appropriate provision for mass conversion rates of molecules into particles.

As an aside, the paper presented by Prof. Casavecchia[20] for the reaction of N(^{2}D) with ethane, of minor importance in the chemistry of Titan's upper atmosphere, has 22 exit channels, all thermodynamically allowed. Fortunately, only two channels are important: yielding methanimine plus methyl and ethanimine plus H atoms. But this illustrates that laboratory experiments and theoretical calculations are supplying us a level of detail on chemical reactions that exceeds our ability to profitably incorporate them into atmospheric models.

In the case of gas giant atmospheres the escape of atomic and molecular hydrogen is negligible from a photochemical perspective and closure is reached by the downward transport of photochemical products into the thermochemical furnace of the deep atmosphere where complex molecules undergo pyrolysis and regenerate the parent saturated hydrides of reactive atoms (*e.g.*, methane, ammonia, phosphine) for transport upward to the photochemical region. If calculations are restricted to the photochemical region, closure requires not only a limitation on the number of species explicitly modeled, but also the application of "leaky" lower boundary conditions to represent exchange of mass between the thermochemical and photochemical regimes.

6 Chemistry in gas giant Jupiter atmospheres

In the gas giants in our own solar system and Jovian type exoplanets thermochemistry dominates in the deep atmosphere and photochemistry in the outer portions of their atmospheres. Moses *et al.* show both types of chemistry can be modeled and in the absence of impactor produced HCN, HCN would not be currently detectable on Jupiter.[21] While the number of constituents treated in this paper is large, we are in our infancy studying 'hot-Jupiters'. The list of known species is short: water, methane, carbon dioxide, and possibly carbon monoxide.[22] The study of hot-Jupiters is further complicated by the lack of knowledge concerning their thermal structure and the presence or absence of stratospheres. Exoplanet research is in part motivated by "are we alone" and are there terrestrial exoplanets that are potentially habitable. Prof. Lunine[23] argues that the search for these should be focused on planets orbiting M-dwarfs.

7 Uncertainty in chemical modeling

Peng et al. addressed the interesting hypothesis[24] if we knew input rate coefficients with a relative uncertainty of 10% can we accurately model atmospheric chemistry? As one 'anonymous' participant remarked "if we really need error bars of, at most, 10% on rate coefficients, we might as well go home and do something else". Currently such small error bars are exceedingly optimistic in the laboratory and unattainable by remote sensing measurements of constituents in planetary atmospheres. While progress is being made to measure rate coefficients at the low temperatures found in the outer solar system, many rate coefficients still need to be extrapolated from room temperature measurements. Furthermore, photochemistry is mostly driven by solar UV radiation which varies with the solar rotation period ($\sim$27 days, amplitude $\leq$ 20% at Lyman-α) and the solar activity cycle (11 yr, amplitude $\sim$33% at Lyman-α). Typical error bars are in excess 10%. In contrast, the solar constant is accurately known and is slightly time variable with an amplitude of $\sim$0.1%. In actual models, the closure problem discussed above and how it is treated will impact the model results. In addition, in photochemical models lower and upper boundary are not always selected at optimal locations for all species and compromises in boundary conditions must be made for some species.

8 Final comments

Kaiser et al.[9] investigated potential surface chemistry on Titan by irradiating ethane and propane ices and suggested that propane ice decomposition leads to polymer formation. With an estimated surface cosmic ray energy flux of $\sim$4.5 $\times$ 10^9 eV cm^{-2} s^{-1}, they predicted over a GY period about 1 m of polymer could be formed on Titan's surface. This amount should be compared with the total deposition of photochemical products from methane photolysis of $\sim$100 m in 1 GY.

Two recent papers published this year suggest that surface chemistry may be important on Titan. Clark et al.[25] reported no detection of acetylene, which was expected to be more abundant on Titan's surface, according to some models, than benzene which they did detect, and hence suggestive of active surface chemistry. In a separate study,[26] I found that various measurements of the molecular hydrogen mole fraction in Titan's troposphere and lower stratosphere (the lowest 70 km of its atmosphere) when coupled with INMS measurements of the molecular hydrogen density and mole fraction profiles above 950 km required a large downward flow of molecular hydrogen into Titan's surface ($\sim 10^{28}$ s^{-1} over the entire surface and comparable in magnitude to its escape rate from the top of the atmosphere). Taken together, these papers imply active surface chemistry, most probably abiotic in origin, involving metallic catalysts. But one cannot rule out the considerably more improbable exotic, methane-based life form scenario hypothesized by McKay and Smith,[27] where $3H_2 + C_2H_2 \rightarrow 2\ CH_4 + \Delta E$ is equivalent to $6\ O_2 + C_6H_{12}O_6 \rightarrow 6\ CO_2 + 6\ H_2O + \Delta E$ for water-based life. While conventional methane photochemistry can supply the molecular hydrogen escaping from Titan's atmosphere, it cannot supply an additional, equal amount to support the downward flux towards the surface. The integrated photochemical destruction rate of methane must be twice what conventional chemistry can supply.

References

1. S. K. Atreya, Introductory Lecture: The significance of trace constituents in the solar system, *Faraday Discuss.*, 2010, **147**, DOI: 10.1039/C005460G.
2. J. S. Lewis, Satellites of the outer planets: Their physical and chemical nature, *Icarus*, 1971, **15**, 174–185.
3. D. M. Hunten, *The Atmosphere of Titan*, Washington, NASA SP-340, 1974.
4. D. M. Hunten, A Titan atmosphere with a surface temperature of 200 K, in *The Saturn System*, ed. D. M. Hunten and D. Morrison, NASA Conf. Publ. 2068, Washington, DC, 1978, pp. 127–140.
5. R. E. Danielson, J. Caldwell and D. R. Larach, An inversion in the atmosphere of Titan, *Icarus*, 1973, **20**, 437–443.
6. F. C. Gillett, W. J. Forrest and K. M. Merrill, 8–13μ observations of Titan, *Astrophys. J. Lett.*, 1973, **184**, L93–95.
7. J. B. Pollack, Greenhouse models of the atmosphere of Titan, *Icarus*, 1973, **19**, 43–58.
8. D. F. Strobel, The photochemistry of hydrocarbons in the atmosphere of Titan, *Icarus*, 1974, **21**, 466–470.
9. R. I. Kaiser, P. Maksyutenko, C. Ennis, F. Zhang, X. Gu, S. P. Krishtal, A. M. Mebel, O. Kostkoc and M. Ahmed, Untangling the chemical evolution of Titan's atmosphere and surface—from homogeneous to heterogeneous chemistry, *Faraday Discuss*, 2010, **147**, DOI: 10.1039/C003599H.
10. A. J. Coates, A. Wellbrock, G. R. Lewis, G. H. Jones, D. T. Young, F. J. Crary, J. H. Waite, R. E. Johnson, T. W. Hill and E. C. Sittler Jr., Negative ions at Titan and Enceladus: recent results, *Faraday Discuss*, 2010, **147**, DOI: 10.1039/C004700G.
11. J. H. Waite, Jr., D. T. Young, T. E. Cravens, A. J. Coates, F. J. Crary, B. Magee and J. Westlake, The process of tholin formation in Titan's upper atmosphere, *Science*, 2007, **316**, 870.
12. M.-C. Liang, Y. L. Yung and D. E. Shemansky, Photolytically generated aerosols in the mesosphere and thermosphere of Titan, *Astrophys. J.*, 2007, **661**, L199–L202.
13. P. P. Lavvas, R. V. Yelle and V. Vuitton, The detached haze layer in Titan's mesosphere, *Icarus*, 2009, **201**, 626–633.
14. R. V. Yelle, V. Vuitton, P. Lavvas, S. J. Klippenstein, M. A. Smith, S. M. Hörst and J. Cui, Formation of NH_3 and CH_2NH in Titan's upper atmosphere, *Faraday Discuss*, 2010, **147**, DOI: 10.1039/C004787M.
15. V. Vuitton, R. V. Yelle and M. J. McEwan, Ion chemistry and N-containing molecules in Titan's upper atmosphere, *Icarus*, 2007, **191**, 722–742.
16. V. Vuitton, P. Lavvas, R. V. Yelle, M. Galand, A. Wellbrock, G. R. Lewis, A. J. Coates and J.-E. Wahlund, Negative ion chemistry in Titan's upper atmosphere, *Planet. Space Sci.*, 2009, **57**, 1558–1572.
17. N. A. Teanby, P. G. J. Irwin, R. de Kok and C. A. Nixon, Mapping Titan's HCN in the far infra-red: implications for photochemistry, *Faraday Discuss*, 2010, **147**, DOI: 10.1039/C001690J.
18. D. F. Strobel, Parameterization of linear wave chemical transport in planetary atmospheres by eddy diffusion, *J. Geophys. Res.*, 1981, **86**, 9806–9810.
19. C. A. Nixon, R. K. Achterberg, N. A. Teanby, P. G. J. Irwin, J.-M. Flaud, I. Kleiner, A. Dehayem-Kamadjeu, L. R. Brown, R. L. Sams, B. Bézard, A. Coustenis, T. M. Ansty, A. Mamoutkine, S. Vinatier, G. L. Bjoraker, D. E. Jennings, P. N. Romani and F. M. Flasar, Upper limits for undetected trace species in the stratosphere of Titan, *Faraday Discuss*, 2010, **147**, DOI: 10.1039/C003771K.
20. N. Balucani, F. Leonori, R. Petrucci, M. Stazi, D. Skouteris, M. Rosic and P. Casavecchia, Formation of nitriles and imines in the atmosphere of Titan: combined crossed-beam and theoretical studies on the reaction dynamics of excited nitrogen atoms $N(^2D)$ with ethane, *Faraday Discuss*, 2010, **147**, DOI: 10.1039/C004748A.
21. J. I. Moses, C. Visscher, T. C. Keane and A. Sperier, On the abundance of non-cometary HCN on Jupiter, *Faraday Discuss*, 2010, **147**, DOI: 10.1039/C003954C.
22. G. Tinetti, C. A. Griffith, M. R. Swain, P. Deroo, J. P. Beaulieu, G. Vasisht, D. Kipping, I. Waldmann, J. Tennyson, R. J. Barber, J. Bouwman, N. Allard and L. R. Brown, Exploring extrasolar worlds: from gas giants to terrestrial habitable planets, *Faraday Discuss*, 2010, **147**, DOI: 10.1039/C005126H.
23. J. I. Lunine, Titan and habitable planets around M-dwarfs, *Faraday Discuss*, 2010, **147**, DOI: 10.1039/C004788K.
24. Z. Peng, M. Dobrijevic, E. Hébrard, N. Carrasco and P. Pernot, Photochemical modeling of Titan atmosphere at the "10 percent uncertainty horizon", *Faraday Discuss*, 2010, **147**, DOI: 10.1039/C003366A.

25 R. N. Clark, J. M. Curchin, J. W. Barnes, R. Jaumann, L. Soderblom, D. P. Cruikshank, R. H. Brown, S. Rodriguez, J. Lunine, K. Stephan, T. M. Hoefen, S. Le Mouélic, C. Sotin, K. H. Baines, B. J. Buratti and P. D. Nicholson, Detection and Mapping of Hydrocarbon Deposits on Titan, *J. Geophys. Res*, 2010, DOI: 10.1029/2009JE003369.
26 D. F. Strobel, Molecular hydrogen in Titan's atmosphere: implications of the measured tropospheric and thermospheric mole fractions, *Icarus*, 2010, **208**, 878–886, DOI: 10.1016/j.icarus.2010.03.003.
27 C. P. McKay and H. D. Smith, Possibilities for methanogenic life in liquid methane on the surface of Titan, *Icarus*, 2005, **178**, 274–276, DOI: 10.1016/j.icarus.2005.05.018.

Poster titles

Addition of one and two units of C_2H to styrene: A threshold study of the $C_{10}H_9$ and $C_{12}H_9$ systems and implications towards Titan's atmopshere, **A. Landera, A. M. Mebel and R. Kaiser**, *Department of Chemistry and Biochemistry, Florida Internatinal University, USA*

An *ab initio*/RRKM study of the reaction mechanism and products of branching ratios of ethynyl radical (C_2H) with unsaturated hydrocarbons, **A. Jamal, A. M. Mebel**, *Department of Chemistry and Biochemistry, Florida Internatinal University, USA*

Phase behavior of supercooled ethane droplets in the troposphere of Titan, **K. J. Knox, E. K. Lang, C. C. Wang and R. Signorell**, *University of British Columbia, Canada*

Evidence for layered methane clouds in the troposphere of Titan, **C. C. Wang, K. J. Knox, S. K. Atreya and R. Signorell**, *University of British Columbia, Canada*

Role of dissociative recombination in Titan's ionosphere: a reappraisal, **S. Plessis, N. Carrasco, M. Dobrijevic and P. Pernot**, *CNRS/Université Paris Sud 11, France*

Hydrogen/deuterium exchanges in H_2O: XD ice mixtures, **A. Ratajczak, E. Quirico, A. Faure, B. Schmitt and C. Ceccarelli**, *Universite Joseph Fourier, France*

Correlating far-infrared spectra and structural features for pure and mixed ices, **M. W. Lee, N. Plattner and M. Meuwly**, *University of Basel, Switzerland*

Impact of the size distribution representation on transport in the Titan IPSL-GCM, **J. Burgalat, P. Rannou, S. Lebonnois and F. Montmessin**, *Université Reims Champagne-Ardenne, France*

Tropospheric clouds on Titan from far-infrared observations, **R. de Kok, P. J. G. Irwin and N. A. Teanby**, *SRON Netherlands Institute for Space Research, The Netherlands*

Titan's atmosphere: an optimal gas mixture for aerosol production?, **E. Sciamma O'Brien, N. Carrasco, C. Szopa, G. Cernogora and A. Buch**, *Université Versailles St-Quentin, France*

Absolute photoionization cross-section of the methyl, ethyl, propyl radicals, **J.-C. Loison**, *CNRS/Université Bordeaux 1, France*

Laboratory studies of cyanoacetylene and its photoreactivity in matrix environment, **R. Kolos, M. Turowski, M. Gronowski, S. Douin, S. Boyé-Péronne and C. Crépin**, *CNRS/Université Paris Sud 11, France*

Cavity ring down and thermal lens at low temperatures for laboratory studies of planetary atmospheres, **C. E. Manzanares, Y. Perez-Delgado and H. Diez-y-Riega**, *Baylor University, USA*

Oxygen complexes in the Earth's atmosphere: measurements during a lunar eclipse, **A. G. Muñoz and E. Pallé**, *Instituto de Astrofísica de Canarias, Tenerife, Spain*

Investigation of primary processes in Titan's upper atmosphere, **P. Lavvas, M. Galand, R. V. Yelle, A. N. Heays, B. R. Lewis, G. R. Lewis and A. J. Coates**, *University of Arizona, USA*

Primary photodissociation processes important in Titan's atmosphere, **W. Gichuhi, C. Huang, F. Zhang, R. I. Kaiser, V. V. Kislov, A. M. Mebel, R. Silva and A. G. Suits**, *Wayne State University, USA*

Thermo-photochemistry of exoplanet atmospheres, **O. Venot, M. Dobrijévic, F. Selsis, E. Hébrard, R. Bounaceur,Y. Bénilan, M. C. Gazeau and N. Champion**, *Université de Bordeaux 1, France*

Identification of complex organic molecules in PAMPRE tholins, **S. M. Hörst, R. V. Yelle, N. Carrasco, E. Sciamma-O'Brien, M. A. Smith, Á. Somogyi, C. Szopa, R. Thissen and V. Vuitton**, *University of Arizona, USA*

Isomer-specific spectroscopy of gas-phase 1-hydronaphthyl,2-hydronaphthyl, and 1,2,3-trihydronaphthyl radicals, **J. A. Sebree, T. S. Zwier, V. V. Kislov and A. M. Mebel**, *Purdue University, USA*

Aerosols in detached haze layer observed by ISS/Cassini, **T. Cours, P. Rannou, S. Rodriguez and A. Brahic**, *Université de Reims Champagne-Ardenne, France*

Titan north polar cloud studied with VIMS observations, **P. Rannou, T. Cours, J. Burgalat, S. Le Mouélic, S. Rodriguez, C. Sotin, P. Drussart and R. Brown**, *Université de Reims Champagne-Ardenne, France*

Low temperature study of Titan's atmosphere reactivity with a cold N_2–CH_4 plasma discharge, **N. Carrasco, C. Szopa, E. Sciamma O'Brien, J.-J. Correia, G. Cernogova, A. Jolly, E. T. Es-Sebbar, A. Buch and P. Pernot**, *LATMOS – Université de Versailles St Qunetin, France*

Titan's surface temperature from Cassini – CIRS observations, **V. Cottini, C. A. Nixon, D. E. Jennings, R. E. Samuelson, R. de Kok, P. G. J. Irwin and F. M. Flasar**, *Department of Astronomy, University of Maryland at College Park and Planetary Systems Laboratory, NASA Goddard Space Flight Centre, Maryland, USA*

Saturn's stratospheric hydrocarbons derived from Cassini/CIRS: constraints on photochemistry and dynamics, **S. Guerlet, T. Fouchet, J. Moses and B. Bézard**, *LEISA – Observatoire de Paris-Meudon, France*

The reactions of hydroxyl radicals with unsaturated hydrocarbons at low temperatures. Experiments (50–224 K) and theory, **J. Daranlot, K. M. Hickson, A. Bergeat, F. Caralp, W. Forst, R. Mereau and M. Costes**, *Université Bordeaux 1, France*

CH_4 and NH_3 measurements for exoplanet atmospheres, **R. Hargreaves, G. Li, M. Irfan and P. Bernath**, *University of York, UK*

Theoretical study of H_2O–H_2 for astrochemistry from *ab initio* intermolecular potentials, **Y. Scribano, D. Benoit, O. Akin-Ojo and A. Faure**, *Université de Bourgogne, France*

Nitrogen atom reaction kinetics to low temperatures (50–220 K), **J. Daranlot, K. M. Hickson, P. Caubet, M. Costes and A. Bergeat**, *Université Bordeaux 1, France*

Negative ions at Titan – density trends, **A. Wellbrock, A. J. Coates, G. H. Jones, G. Lewis, C. S. Arridge, B. Magee, F. J. Crary, E. C. Sittler Jr, D. T. Young and J. H. Waite**, *Mullard Space Science Laboratory, Department of Space and Climate Physics, University College London, UK*

Spatial distribution of molecular gas and aerosol mixing ratios in Titan's middle atmosphere derived from Cassini/CIRS spectra, **S. Vinatier, B. Bézard, P. Rannou, R. de Kok, C. M. Anderson, C. A. Nixon, R. E. Samuelson and the CIRS Investigation Team**, *NASA/GSFC, France*

The chemistry of exoplanetary atmospheres, **C. Walsh, T. J. Millar and J. Millar**, *Astrophysic Research Centre, School of Mathematics and Physics, Queen's University Belfast, UK*

Which new compounds could be detected in the atmosphere of Titan?, **M. Yáñez, Otilia Mó, A. Benidar, Y. Benilan, A. Jolly, N. Fray, M.-C. Gazeau, B. Khater, Y. Trolez and J.-C. Guillemin**, *Ecole Nationale Supérieure de Chimie de Rennes, France*

Collisional line profiles of sodium perturbed by H_2, **N. F. Allard, B. Pertev, G. Tinetti and J.-P. Beaulieu**, *GEPI, Observatoire de Paris and Institut d'Astrophysique de Paris, France*

Negative ion formation in a point-to-plane corona discharge fed by $N_2/CH_4/Ar$ mixture – a contribution to Titan's atmospheric chemistry, **G. Horvath, Y. A. Gonzalvo, N. J. Mason, M. Zahoran and S. Matejcik**, *The Open University, UK*

A laboratory simulation of the Martian atmosphere using electrical discharges, **J. Országh, N. J. Mason and Š. Matejčík**, *The Open University, UK*

Benzene in Titan's atmospheric envelope from Cassini/CIRS data, **G. Bampasidis, P. Lavvas, A. Coustenis, R. Carlson, C. Nixon, R. Achterberg, D. Jennings, S. Vinatier, F. Michael Flasar, X. Moussas, P. Preka-Papadema**, *University of Athens, Greece and LESIA, Observatoire de Paris, France*

Kinetic studies at room temperature on the cyanide anion CN^- with cyanoacetylene (HC_3N) reaction, **S. Carles, F. Adjali, C. Monnerie, J.-C. Guillemin and J.-L. Le Garrec**, *Universitéde Rennes 1, France*

A new pulsed CRESU apparatus for the study of gas phase collisional processes, **S. B. Morales, S. D. Le Picard, A. Canosa, B. R. Rowe, S. Cheikh Sid Ely and I. R. Sims**, *Université de Rennes 1, France*

Evaporation rates of methane and ethane on Titan, **A. Luspay-Kuti, V. F. Chevrier, F. Wasiak and D. G. Blackburn**, *Arkansas Center for Space and Planetary Sciences, USA*

High temperature kinetics of reactions involving the CN radical and small hydrocarbons: application to the atmosphere of hot Jupiter extrasolar planets, **A. Gardez, L. Biennier, G. Saidani, E. Hugo, R. Georges and B. R. Rowe**, *Institut de Physique de Rennes, Equipe Astrochimie Expérimentale, FRANCE*

List of participants

Dr Nigel Adams, *University of Georgia, U.S.A.*
Dr Christian Alcaraz, *LCP/CNRS-Univ.ParisSud11, France*
Dr Nicole Allard, *Observatoire de Paris, France*
Professor Michael Ashfold, *University of Bristol, United Kingdom*
Professor Sushil Atreya, *University of Michigan, U.S.A.*
Dr Bernabe Ballesteros, *University of Castilla-La Mancha, Spain*
Dr Jean-Philippe Beaulieu, *Institut d'Astrophysique de Paris, France*
Dr Abdessamad Benidar, *IPR université de Rennes 1, France*
Dr Astrid Bergeat, *Institut des Sciences Moléculaires, Université Bordeaux 1, France*
Dr Bruno Bézard, *LESIA, Observatoire de Paris, France*
Dr Ludovic Biennier, *Institut de Physique de Rennes UMR CNRS 6251, France*
Mr Jean-Yves Bonnet, *Laboratoire de Planétologie de Grenoble, France*
Dr Séverine Boyé-Péronne, *Institut des Sciences Moléculaires d'Orsay, France*
Mr Jeremie Burgalat, *Groupe de Spectrométrie Moléculaire et Atmosphérique - UMR 6089, France*
Dr André Canosa, *CNRS - Université de Rennes 1, France*
Dr Sophie Carles, *IPR - Rennes University, France*
Dr Nathalie Carrasco, *LATMOS, France*
Prof Piergiorgio Casavecchia, *Università degli Studi di Perugia, Italy*
Dr Sidaty Cheikh Sid Ely, *Institut de Physique de Rennes, France*
Professor Andrew Coates, *University College London, United Kingdom*
Dr Paul Cooper, *Royal Society of Chemistry, United Kingdom*
Dr Daniel Cordier, *Institut de Physique de Rennes, France*
Dr Valeria Cottini, *University of Maryland, U.S.A.*
Dr Remcode Kok, *SRON Netherlands Institute for Space Research, The Netherlands*
Dr Michel Dobrijevic, *Laboratoire d'Astrophysique de Bordeaux, France*
Dr Yves Ellinger, *UPMC, France*
Miss Berivan Esen, *London Metropolitan University, United Kingdom*
Dr Alexandre Faure, *Observatoire de Grenoble, France*
Mr Bérenger Gans, *Institut des Sciences Moléculaires d'Orsay, France*
Dr Antonio García Muñoz, *Instituto de Astrofísica de Canarias, Spain*
Miss Aline Gardez, *Institiut de Physique de Rennes - Equipe Astrochimie, France*
Professor Dolores, *Gauyacq, Institut des Sciences Moléculaires d'Orsay (ISMO), France*
Dr Robert Georges, *Universite de Rennes 1, France*
Mr Saidani Ghassen, *IPR, France*
Miss Sandrine Guerlet, *LESIA - Observatoire de Paris, France*
Dr Jean-Claude Guillemin, *Sciences Chimiques de Rennes, France*
Mr Robert Hargreaves, *University of York, United Kingdom*
Dr Eric Hébrard, *Laboratoire d'Astrophysique de Bordeaux, France*
Dr Kevin Hickson, *Institut des Sciences Moleculaires, Universite Bordeaux 1, France*
Ms Sarah Horst, *Lunar and Planetary Laboratory, The University of Arizona, U.S.A.*
Dr David Huestis, *SRI International, U.S.A.*
Mr Adeel Jamal Florida, *International University, U.S.A.*

Dr Diego Janches, *NorthWest Research Associates, U.S.A.*
Dr Elena Jiménez, *University of Castilla-La Mancha, Spain*
Professor Ralf Kaiser, *University of Hawaii at Manoa, U.S.A.*
Miss Annika Karlsson, *London Metropolitan University, United Kingdom*
Dr Stephen Klippenstein, *Argonne National Laboratory, U.S.A.*
Dr Kerry Knox, *University of British Columbia, Canada*
Mr Alexander Landera, *Florida International University, U.S.A.*
Dr Panayotis Lavvas, *University of Arizona, U.S.A.*
Dr Sébastien Le Picard, *University of Rennes 1, France*
Dr Myung Won Lee, *University of Basel, Switzerland*
Dr Jean-Christophe Loison, *CNRS - Université de Bordeaux, France*
Professor Jonathan Lunine, *University of Rome, Italy*
Ms Adrienn Luspay-Kuti, *Arkansas Center for Space and Planetary Sciences, U.S.A.*
Dr Carlos Manzanares, *Baylor University, U.S.A.*
Professor Nigel Mason, *The Open University, United Kingdom*
Professor Alexander Mebel, *Florida International University, U.S.A.*
Professor Markus Meuwly, *University of Basel, Switzerland*
Professor Steve Miller, *University College London, United Kingdom*
Prof James Brian Mitchell, *Université de Rennes I, France*
Dr Sébastien Morales, *Equipe Astochimie experimentales, France*
Dr Julianne Moses, *Space Science Institute, USA*
Dr Olivier Mousis, *Institut UTINAM, Observatoire de Besançon, France*
Dr Conor Nixon, *University of Maryland, U.S.A.*
Dr Françoise Pauzat, *CNRS, France*
Dr Pascal Pernot, *CNRS, France*
Dr Blaise Pertev, *Institut d'Astrophysique de Paris - UPMC Paris 6, France*
Professor John Plane, *University of Leeds, United Kingdom*
Dr Nuria Plattner, *Department of Chemistry, University of Basel, Switzerland*
Mr Sylvain Plessis, *Laboratoire de Chimie Physique (LCP), France*
Professor Pascal Rannou, *GSMA UMR 6089 University of Reims, France*
Mr Alexandre Ratajczak, *Laboratoire d'Astrophysique, France*
Professor Francois Raulin, *LISA - CNRS/UPEC & Univ. Paris-Diderot, France*
Dr Roland Thissen, *Laboratoire de Planétologie de Grenoble, France*
Dr Ella Sciamma-O'Brien, *LATMOS-CNRS, France*
Professor Daniel Schram, *Technische Universiteit Eindhoven, The Netherlands*
Dr Yohann Scribano, *CNRS ICB UMR 5209, France*
Professor Paul Seakins, *University of Leeds, United Kingdom*
Mr Joshua Sebree, *Purdue University, U.S.A.*
Professor Ian Sims, *IPR UMR 6251 du CNRS - Universite de Rennes 1, France*
Dr Edward Sittler, *NASA Goddard Space Flight Center, U.S.A.*
Dr Michael Spencelayh, *Royal Society of Chemistry, United Kingdom*
Professor Darrell Strobel, *Johns Hopkins University, U.S.A.*
Professor Arthur Suits, *Wayne State University, U.S.A.*
Dr Nicholas Teanby, *University of Oxford, United Kingdom*
Dr Giovanna Tinetti, *University College London, United Kingdom*
Miss Meryem Tizniti, *IPR UMR 6251, France*
Miss Olivia Venot, *Laboratoire d'Astrophysique de Bordeaux, France*
Dr Sandrine Vinatier, *NASA/GSFC, U.S.A.*
Dr Veronique Vuitton, *Laboratoire de Planétologie de Grenoble, France*
Miss Catherine Walsh, *Queen's University Belfast, United Kingdom*

Dr David Wassell, *Queen's University Belfast, United Kingdom*
Miss Anne Wellbrock, *University College London, United Kingdom*
Ms Charlotte Whalley, *University of Leeds, United Kingdom*
Professor Manuel Yáñez, *Universidad Autónoma de Madrid, Spain*
Dr Roger Yelle, *University of Arizona, U.S.A.*
Professor Timothy Zwier, *Purdue University, U.S.A.*

Index of contributors*

Achterberg, R. K., **65**
Adams, N. G., 83, 251, **323**, 379
Ahmed, M., **429**
Allard, N., **369**
Ansty, T. M., **65**
Ashfold, M., 83, 251, 379, 527
Atreya, S. K., **9**, 83, 251, 379, 527
Balucani, N., **189**
Barber, R. J., **369**
Beaulieu, J. P., **369**
Bézard, B., **65**, 251, 527
Biennier, L., 251, 379
Bjoraker, G. L., **65**
Blitz, M. A., **173**
Bonnet, J.-Y., **495**
Bouwman, J., **369**
Brassé, C., **419**
Brown, L. R., **65**, 369
Buch, A., **419**
Canosa, A., **155**
Carrasco, N., **137**, **495**, 527
Casavecchia, P., 83, **189**, 251, 527
Coates, A. J., **293**, 379, 527
Coll, P., **419**
Cordier, D., **509**
Cottin, H., **495**
Coustenis, A., **65**
Crary, F. J., **293**
Cui, J., **31**
de Kok, R., **51**
Dehayem-Kamadjeu, A., **65**
Deroo, P., **369**
Dobrijevic, M., **137**, 251
Dutuit, O., **337**, **495**
Ellinger, Y., 83, 251, 379, 527
Ennis, C., **429**
Faure, A., **337**, 251, 379
Flasar, F. M., **65**
Flaud, J.-M., **65**
Fox, J. L., **307**
Fray, N., **495**
Frisari, M., **495**
Gannon, K. L., **173**
Gauyacq, D., **527**
Glowacki, D. R., **173**
Griffith, C. A., **369**
Gu, X., **429**
Hargreaves, R., 527
Harvey, J. N., **173**
Hébrard, E., **137**

Hill, T. W., **293**
Hörst, S. M., **31**, 527
Huestis, D. L., **307**, 379
Irwin, P. G. J., **51**, 65
Jennings, D. E., **65**
Johnson, R. E., **293**
Jones, G. H., **293**
Kaiser, R. I., 83, 251, **429**, 527
Keane, T. C., **103**
Kipping, D., **369**
Kislov, V. V., **231**
Kleiner, I., **65**
Klippenstein, S. J., **31**, 83, 251, 527
Knox, K., 83, 527
Kostko, O., **429**
Krishtal, S. P., **429**
Landera, A., **479**
Lavvas, P., **31**, 527
Lee, M. W., **217**
Le Picard, S. D., **155**, 251
Le Roy, L., **495**
Leonori, F., **189**
Lewis, G. R., **293**
Liang, C.-H., **173**
Loison, J.-C., 251, 379
Lunine, J. I., **405**, **509**, 527
Maksyutenko, P., **429**
Mamoutkine, A., **65**
Manzanares, C., 83, 251, 527
Mason, N., 83, 251, 379, 527
Mathews, L. D., **323**
Mebel, A. M., **231**, 251, **429**, **479**, 527
Melin, H., **283**
Meuwly, M., **217**, 251
Miller, S., 251, **283**, 379, 527
Mitchell, J. B., 83, 251, 379, 527
Morales, S. B., **155**
Moses, J. I., 83, **103**, 251, 379, 527
Mousis, O., **509**, 527
Nixon, C. A., **51**, 65, 83
Osborne Jr. D., **323**
Peng, Z., **137**
Pernot, P., **137**, 251, 379, 527
Petrucci, R., **189**
Picaud, S., **509**
Pilling, M. J., **173**
Plane, J. M. C., 83, 251, **349**, 379, 527
Plattner, N., **217**, 527
Poch, O., **419**

Quirico, E., **495**
Ramírez, S. I., **419**
Raulin, F., **419**, 527
Romani, P. N., **65**
Rosi, M., **189**
Sams, R. L., **65**
Schmitt, B., **495**
Schram, D., 251, 379, 527
Sciamma-O'Brien, E., **495**
Seakins, P. W., **173**, 251
Sebree, J. A., **231**, 527
Sharpee, B. D., **307**
Sims, I. R., 83, **155**, 251, 379
Sittler Jr. E. C., **293**, 379, 527
Skouteris, D., **189**
Slanger, T. G., **307**
Smith, M. A., **31**
Sperier, A., **103**
Stallard, T., **283**
Stazi, M., **189**
Strobel, D. F., 83, 251, **553**, 527

Suits, A., 83, 251, 379, 527
Swain, M. R., **369**
Szopa, C., **495**
Teanby, N. A., **51**, **65**, 83
Tennyson, J., **283**, **369**
Thissen, R., 251, **337**, 379, **495**, 527
Tinetti, G., **369**, 527
Vasisht, G., **369**
Vinatier. S., **65**
Visscher, C., **103**
Vuitton, V., **31**, 83, 251, **337**, 379, **495**, 527
Waite, J. H., **293**
Waldmann, I., **369**
Wellbrock, A., **293**, 379
Whalley, C. L., **349**, 379
Wiesenfeld, L., **337**
Yelle, R. V., **31**, 83, 251, 379, 527
Young, D. T., **293**
Zhang, F., **429**
Zwier, T. S., 83, **231**, 251, 527

* The page numbers in **bold** type indicate papers submitted for discussions.